This symposium brought together specialists from a number of countries to discuss in Camerino, Italy, both fossil and living problematical animals. This book, however, has a much wider relevance to palaeontologists, zoologists and evolutionary biologists. Problematical taxa traditionally have been ignored by workers in these areas, but a recent work is showing that they hold a key position in understanding phylogenetic relationships, macroevolution, and the filling of ecological space. The unveiling of spectacular problematica from the Cambrian, especially the Burgess Shale, has tended to overshadow other instances. This book serves to redress the balance, by considering not only other examples from the Cambrian, but also other fossil deposits, as well as living forms that remain of problematical status. This book includes not only information on a wide variety of animal groups, reviews of special fossil preservation, major themes of phylogeny, and new data from molecular biology. The book is designed to provide not only an up-to-date review, but also act as a focus for further research.

The early evolution of Metazoa and the significance of problematic taxa

The early evolution of Metazoa and the significance of problematic taxa

Proceedings of an International Symposium
held at the University of Camerino
27 - 31 March 1989

Edited by

ALBERTO M. SIMONETTA
and
SIMON CONWAY MORRIS

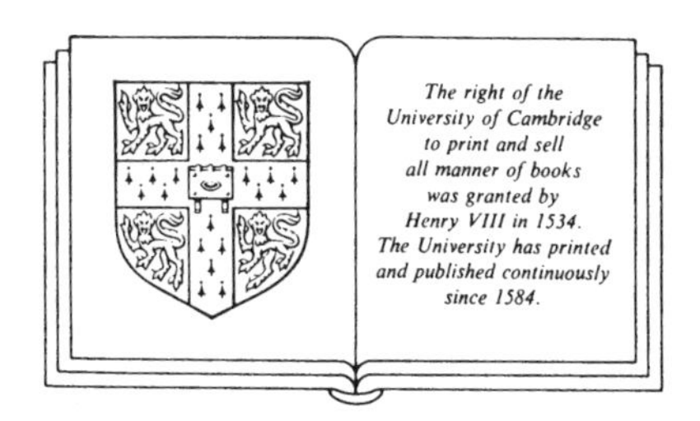

CAMBRIDGE UNIVERSITY PRESS
Cambridge
New York *Port Chester*
Melbourne *Sydney*

CAMBRIDGE UNIVERSITY PRESS
Cambridge, New York, Melbourne, Madrid, Cape Town, Singapore, São Paulo, Delhi

Cambridge University Press
The Edinburgh Building, Cambridge CB2 8RU, UK

Published in the United States of America by Cambridge University Press, New York

www.cambridge.org
Information on this title: www.cambridge.org/9780521111584

First published 1991
This digitally printed version 2009

A catalogue record for this publication is available from the British Library

ISBN 978-0-521-40242-2 hardback
ISBN 978-0-521-11158-4 paperback

Table of contents

PREFACE

Problematical taxa are one of the most intriguing, and most ignored, of the problems in biology. The tendency to relegate them to the sidelines of enquiry, and the dustbins of classification, is understandable, but such treatment threatens to remove an area of great interest to evolutionary biology. It was with these thoughts in mind that we decided that a symposium on problematical taxa would be timely, both in view of the existing expertise and the growing interest in this topic. The area is diffuse, and we were not seeking any clear-cut answers, rather, we hoped that a discussion of common problems would lend a better perspective to our individual efforts, and also foster research and collaboration. Both these aims seem to have been achieved. The Proceedings fall into three basic themes: those concerned with large scale problems, especially the origin and rise of metazoans; those interested in fossil problematica; and those who study living problematica. It is perhaps this last category which is most neglected, and it is sobering to realize that even with the armoury of investigative techniques a number of living species still present major taxonomic conundrums. And if this is true of recent forms, how much more so it is for the plethora of fossil taxa. Many of the best-known examples are from the Cambrian, but as this volume makes clear there are many other examples from younger sediments which await patient and imaginative analysis. Although the problem of problematica will remain with us for many years, our more sophisticated understanding of methods of study, phylogenetic analysis, and macroevolutionary patterns mean that even if the precise place of many of the problematica in metazoan phylogeny is still contentious, their actual significance and importance is now becoming more generally appreciated. Here we have new insights both into the scale of morphological diversification and the success of ecological strategies, as well as fresh data on the role of contingency and extinction in shaping the patterns of evolution.

The success of the Symposium was largely a result of the enthusiastic participation of speakers from Italy, the United Kingdom, Canada, United States, Sweden, U.S.S.R., China, Austria and Switzerland. It was also due to most generous financial support from the Italian National Research Council, the University of Camerino, the Department of Molecular, Cell and Animal Biology of the University of Camerino and the Regione Marche, and the willing help and assistance of Drs. Luisa Magnoni, Valeria Rivola, Anna Sagretti, Emilio Insom, Mario Marconi, and Mauro Marconi, who took the burden of most of the administrative and logistical troubles. Only five of the speakers have been unable to submit in time the finalized text of the papers they read: M. Fedonkin (Moskow), D. Briggs (Edinburgh), P. Omodeo (Rome), B. Runnegar (Los Angeles) and D. Collins (Ottawa).

Alberto M. Simonetta
Simon Conway Morris

Origin of metazoans
A phylogeny deduced from sequences of the 28S ribosomal RNA.

R. Christen[1], A. Ratto[1], A. Baroin[2], R. Perasso[2], K.G. Grell[3] & A. Adoutte[2]

Abstract

The 5'end part of the sequence of the 28S ribosomal RNA (rRNA) has a rate of evolution particularly well suited for use as a phylogenetic index. We have compared the rRNA sequences of a large number of protists, plants, triploblastic metazoans, sponges, cnidarians, ctenophores and a placozoan. After alignment, the most conserved parts of the sequences were selected in order to exclude as much as possible the possibility of multiple substitutions (which also includes convergent evolution). These sequences were then analysed by computer using either a distance matrix method or a maximum parsimony analysis.

Broad-scale phylogenies of eukaryotes showed early radiations for a number of flagellate groups which therefore cannot have been coincident with the origin of metazoans; these events were followed by an "explosive" period, in which an intense diversification of protists was observed, closely associated with the metazoans-fungi-metaphytes radiations. A more detailed study of metazoans suggested that triploblastic and diploblastic metazoans were the results of separate protozoan aggregations; however, a divergence from a common multicellular ancestor could not be totally excluded. In all cases our data suggested that the triploblastic-diploblastic radiation has preceded the diversification of diploblastic animals (i.e. the successive sponges, ctenophores, cnidarians radiations).

Introduction

Metazoans are usually viewed as having been derived from an ancestral aggregation of unicellular eukaryotes. They have been classified in about 35 phyla according to various biological criteria such as symmetry, body architecture and embryological development. Any single phylum contains living species which share a characteristic pattern of organization and which are the results of intra-phylum radiations that can often be observed in the fossil record. In contrast, most phyla were already present in the Precambrian or early Cambrian era, and fossils that could help to link these phyla are generally missing. As a result, general phylogenies that include the origins of all phyla have been mostly deduced from the observations of morphology and embryology of extant animals.

In recent years, an analysis based on the homologies of the sequences of molecules present in all species has become a preferred method to determine phylogenies (Zuckerkandl & Pauling, 1965; Wilson *et al.*, 1977, 1987; Woese, 1987; Dayhoff, 1983; Gray *et al.* 1984). Ribosomal RNA (rRNA), the function of which is conserved from bacteria to man, that contains both domains with high and low rates of divergence and is technically easy to sequence directly without cloning, has rapidly become the most widely used molecule to provide rapid and precise molecular phylogenies (Qu *et al.*, 1983; Hasegawa *et al.*, 1985; Gerbi, 1985; Pace *et al.*, 1986; Woese, 1987; Qu *et al.*, 1988).

In this study, we have used partial sequences from RNA of the large ribosomal subunit (28S-rRNA) to study the origin of metazoans. We have compared rRNA sequences from a large number of metazoans and protists, and analysed their phylogenetic relationships either by a least square distance method or by parsimony analysis, two methods generally used to reconstruct molecular phylogenies (Nei, 1987 for review). In distance matrix methods (DMM), the lengths of the branches are calculated for all pairs of species as the total number of nucleotide substitutions between each two sequences: a triangular matrix is thus obtained. A phylogenetic tree is then reconstructed by minimizing the differences between the lengths of the branches in the tree with respect to the distances provided by the matrix (we have used the computer version kindly proposed by J. Felsenstein in his Phylip package). It is also possible to derive a tree topology using the distance-matrix but a different methodology based on the coordinates of each species in an n-dimensions space (ATD computer program: Hénault & Delorme, 1988).

In maximum parsimony methods (MPM), a most parsimonious tree is chosen, the topology of which minimizes the total number of substitutions necessary to link all sequences. In this case, we have used the PAUP program of D. Swofford.

To analyze our data, we have always used simultaneously MPM and DMM; we found that the use of MPM was particularly difficult when some species were present as a unique representative of a deep monophyletic unit, since at many positions there were nucleotides typical of these species that became *de facto* non informative singularities; as a result, the total number of informative positions became low. In this case, MPM were either misleading or proposed a large number of alternative trees (among which the correct tree resided). For this reason, the choice of species was often particularly important for keeping as many informative positions as possible in MPM, while the use of the matrix distance method was straightforward. As a rule of thumb, the correct tree topology seemed to be obtained by MPM when there was no "isolated species": when two species belonging to the same radiation were present in an analysis, singularities were avoided. This result can be also stochastically obtained when a very large number of organisms are analyzed, but computer limitations seriously weaken this approach. When this particular approach was taken, similar phylogenies were obtained with all methods.

[1]UA 671 CNRS Station Zoologique, Villefranche sur mer; 06230 FRANCE.
[2]Laboratoire de Biologie Cellulaire 4 URA1134, Orsay Cedex; 91405 FRANCE.
[3]Institut für Biologie, Universität Tübingen; Morgenstelle 28. D7400 Tübingen GERMANY.

Results: General phylogeny of eukaryotic organisms

Before proceeding to a detailed analysis of the origin of metazoans, we have analyzed the general relationships between all eukaryotes; provided that the precautions mentioned above were taken, all methods showed essentially the same results (summarized in Figures 1 & 2) in agreement with previous studies (Perasso *et al.* 1989, Baroin *et al.* 1988; Sogin *et al.*, 1989). Several protists were seen as early radiations, among which were the flagellates *Trichomonas vaginalis* and *Crithidia fasciculata* as well as *Dictyostelium discoideum*. Late radiations comprised an amoeba, ciliates (as an individualized group), fungi, chromophytes and finally a cluster that contained all multicellular organisms, as well as several protists: *Chlorogonium elongatum* and *Pyramimonas parkeae* (two chlorophytes), *Cryptomonas ovata* and *Chilomonas paramecium* (two cryptophytes), *Acanthocystis longiseta* (an heliozoan) and *Porphyridium purpureum* (a rhodophyte).

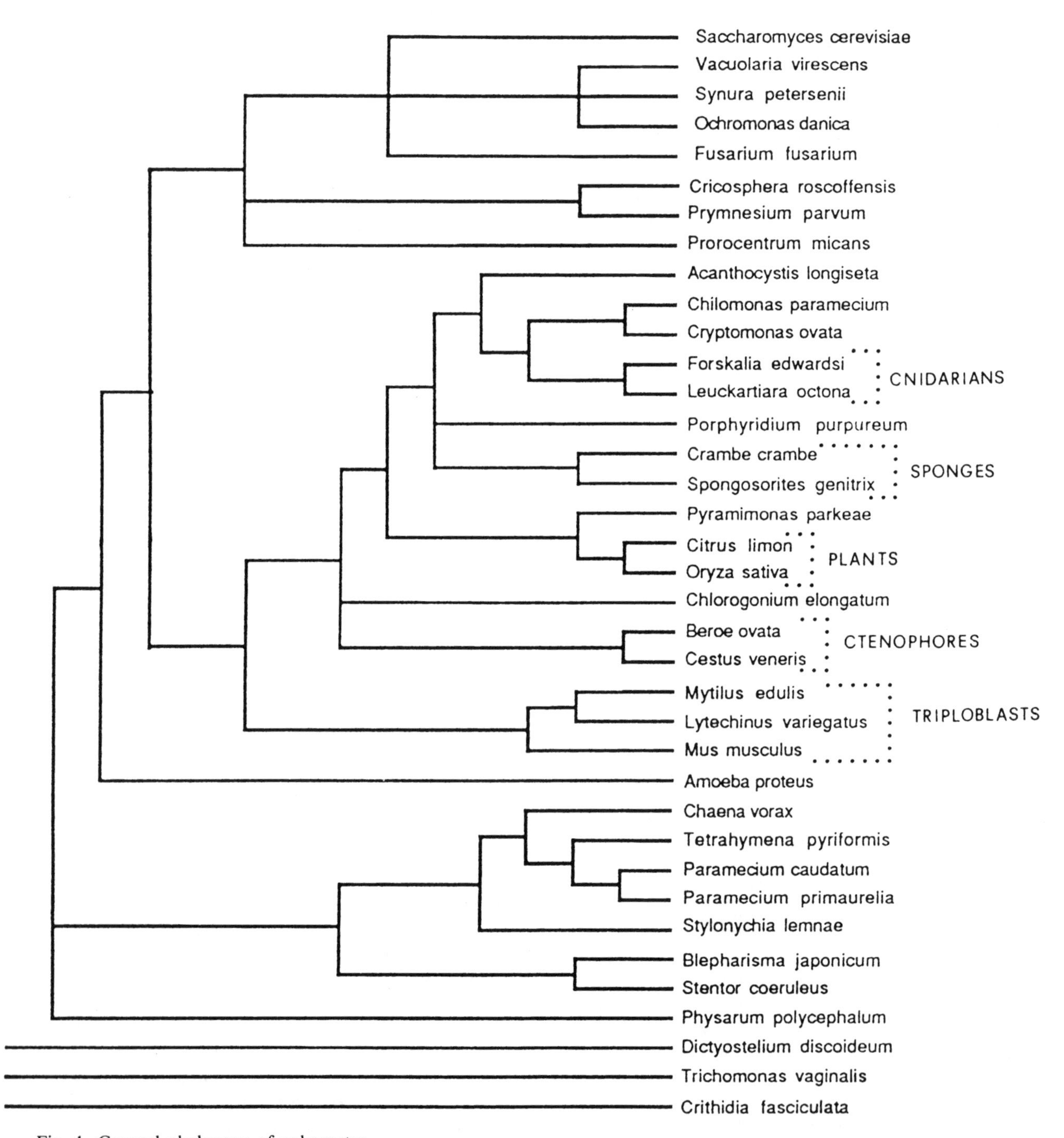

Fig. 1. General phylogeny of eukaryotes.

Sequences from 37 different eukaryotes were aligned and a matrix distance was calculated. This matrix was analyzed using the ATD computer program (see the text for details).

Three main monophyletic groups can be detected among which a group containing all multicellular organisms and several protists.

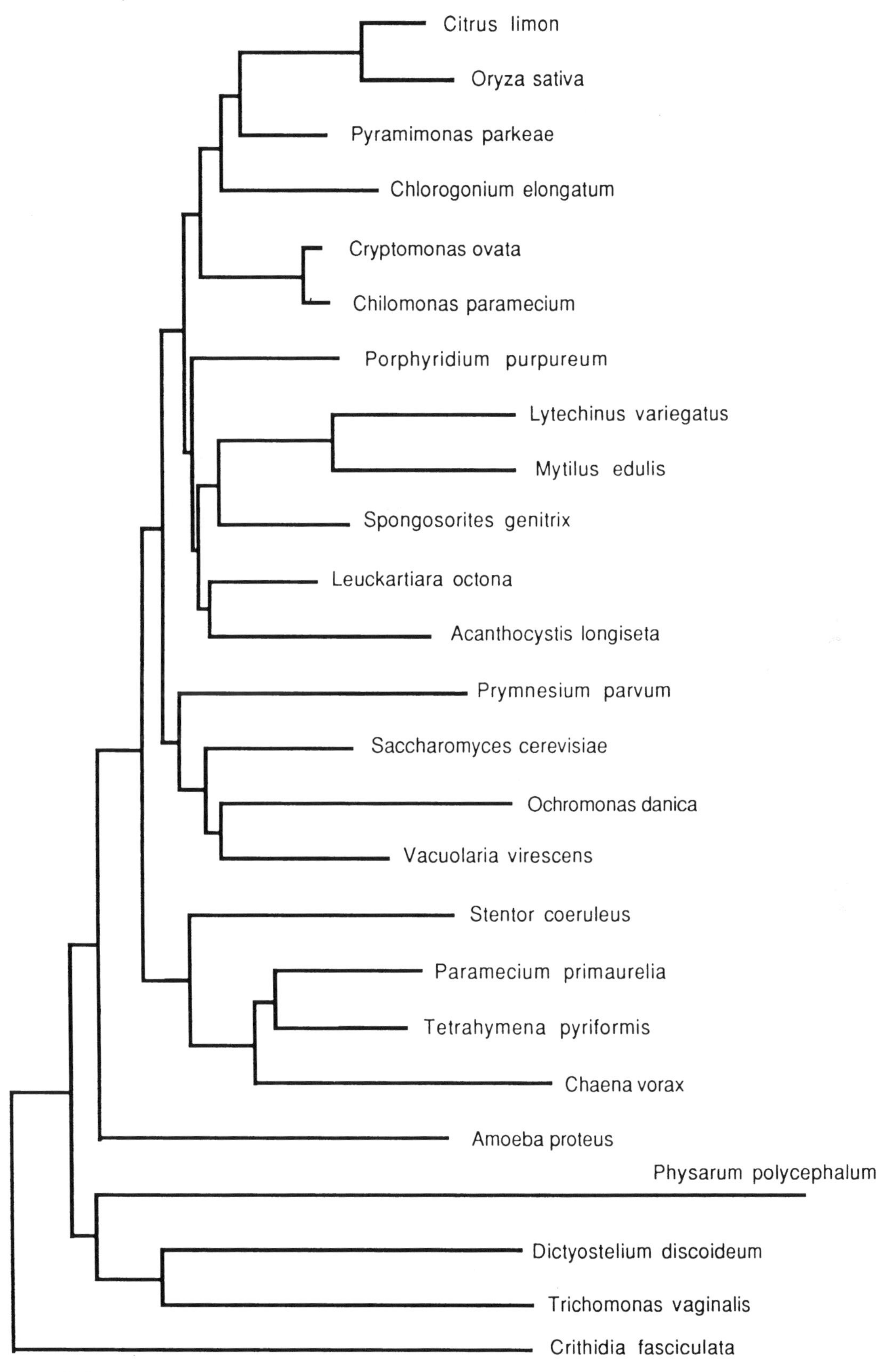

Fig. 2. General phylogeny of eukaryotes.

The same data were analyzed using the "FITCH" program contained in the Phylip package; the analysis was restricted to 25 species to maintain computer time within reasonable limits. Monophyletic groups identical to those shown in figure 1 are revealed by this analysis.

Similar trees were obtained when different species (chosen among the list of Fig. 1) were used.

Origin of multicellular eukaryotes

To examine in more details the origin of metazoans, the analysis was restricted to the protists closely related to multicellular organisms. This approach was necessary for two reasons: i) when only sequences from closely related organisms were examined, conserved domains could be more precisely defined, ii) multiple substitutions (possibly present in sequences from distant organisms) were avoided and the information/noise ratio was improved (in particular, parallel substitutions were largely avoided).

Independently of the method used, three main stems were detected, namely plants, triploblastic and

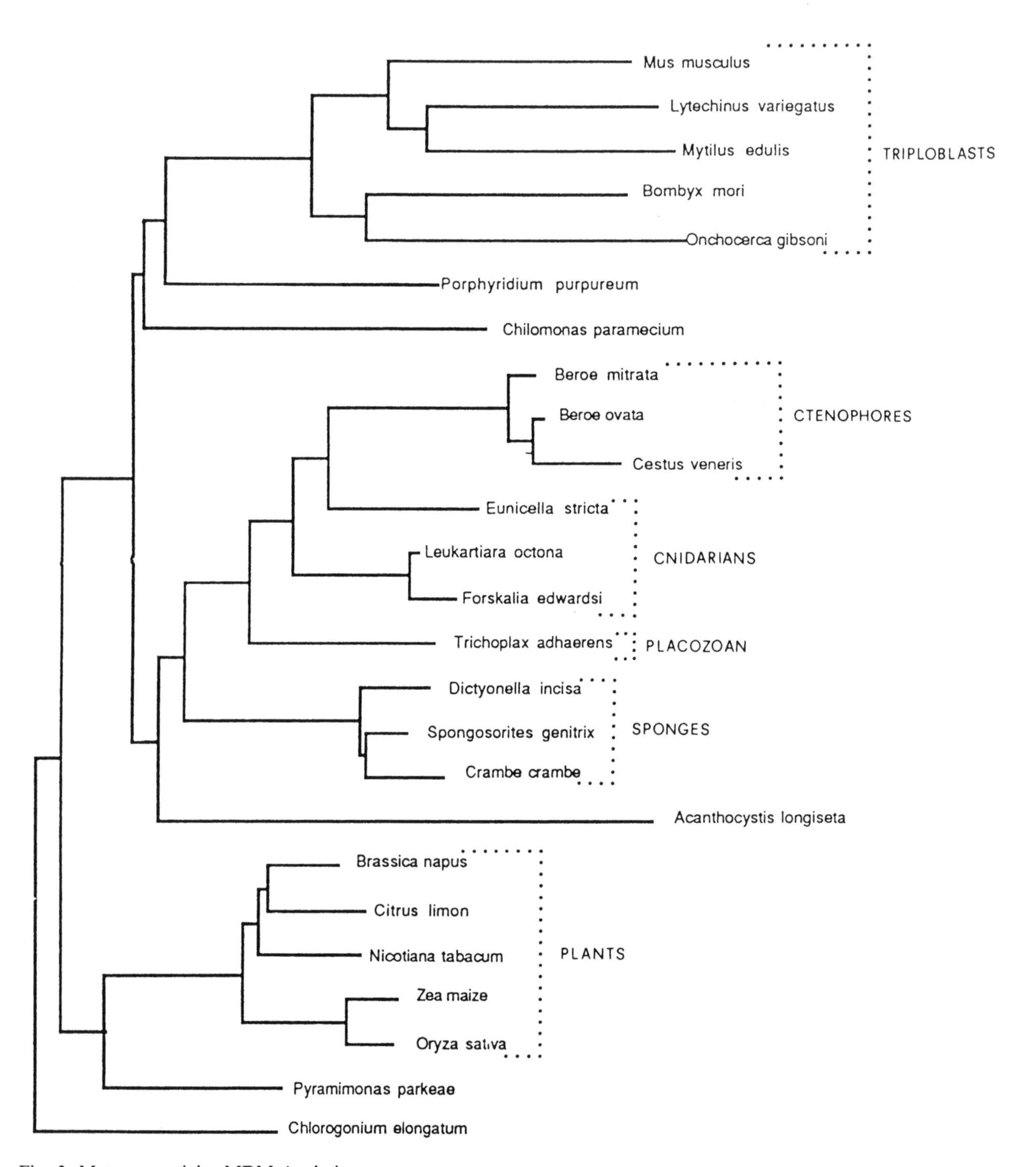

Fig. 3. Metazoan origin. MDM Analysis.

Sequences of eukaryotes belonging to the same monophyletic group (that comprises the metazoans) were analyzed using a distance matrix method (FITCH). Plants, triploblastic metazoans and diploblastic metazoans were seen as three different monophyletic divergences, separated by the radiations of a few protists: *Chilomonas paramecium, Porphyridium purpureum* and *Acantocystis longiseta.*

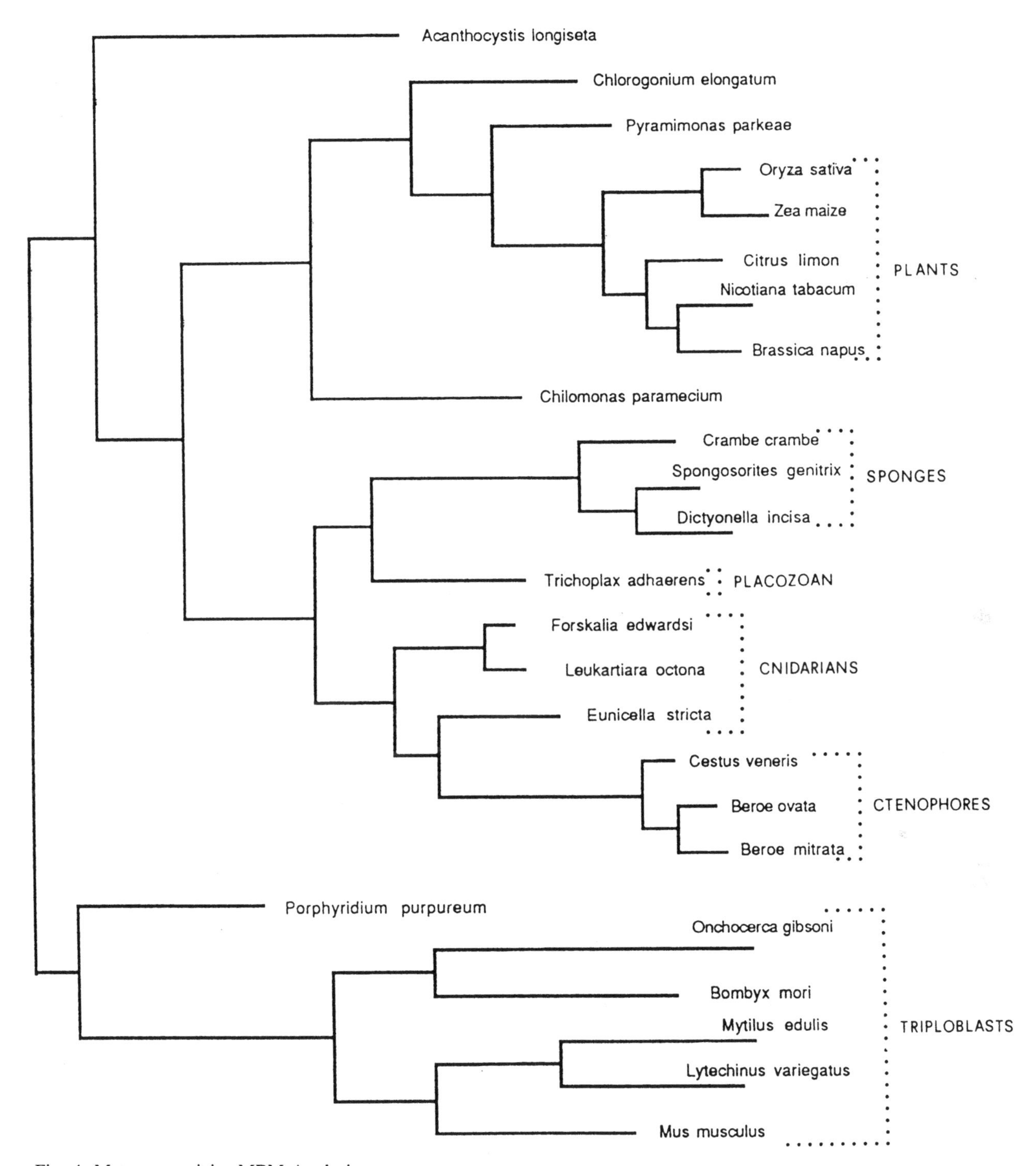

Fig. 4. Metazoan origin. MPM Analysis.
The same organisms as in Fig. 3 were analyzed using a maximum parsimony method (PAUP). The topology of the tree obtained was essentially the same as the one obtained by MDM in Fig. 3.

diploblastic metazoans. These main radiations were separated by the radiations of protists such as *Acanthocystis longiseta* and *Porphyridium purpureum* (see Figures 3 & 4). No protist could be clearly related with the metazoan radiations, in contrast to plants which were closely associated with the unicellular green algae, *Chlorogonium elongatum* and *Pyramimonas parkeae*. Among diploblasts, sponges were often seen as an early radiation, which could perhaps originate from a different multicellularization event than the rest of diploblastic organisms (see the discussion). *Trichoplax adhaerens* was clearly associated with the cluster of diploblastic organisms; it generally branched off quite early within this cluster and displayed a weak link either with sponges or with ctenophores.

Discussion

In all analyses of multicellular eukaryotes evolution, there is a consensus to separate plants from metazoa and it is usually thought (Ragan & Chapman, 1978) that plants share a common ancestor with unicellular green algae, an hypothesis supported by our data.

The presence of chloroplasts and the lack of a neuromuscular system as well as a completely different embryology are the major characteristics that differentiate plants from animals. In contrast, while the hypothesis of a protozoan origin of metazoans is now widely accepted (Hanson, 1977; Ohama *et al.*, 1984), the nature of this original ancestor and the mechanisms that led to cell aggregation are still obscure, as demonstrated by the diversity of hypotheses that have been proposed. Two major theories have been proposed to explain a common origin of the metazoans: the synctitial theory (from a multinucleated ciliate) and the colonial theory (from a colonial flagellate).

It is usually suggested that all metazoans derive monophyletically from a single common primitive multicellular ancestor, since sponges and other "lower" metazoans share fundamentally similar life cycles (Franzen, 1970; MacQuade, 1983). The universal presence of collagen in metazoans has also been used as an argument for the common origin of all metazoans (Runnegar, 1985). The prevailing view is that during the transition from unicellular eukaryotes to the highly complex living metazoans, life evolved from simple to more complex body plans, namely: from colony-like organisms, through a diploblastic body plan to a triploblastic body plan. In this scheme, sponges are seen as an example of the type of organisms that could possibly have arisen primitively.

Similarities in morphologies of choanoflagellates and choanocytes in sponges were used to support the hypothesis (now often rejected) that choanoflagellates could be close to "the" protozoan ancestor (Salvini-Plawen, 1982), although choanoflagellates do not reproduce sexually. The trochaea theory (Nielsen, 1985) also followed the idea of a monophyletic origin of metazoans and proposed a phylogeny based mainly on the structure of larval-ciliary bands and nervous system of living animals; ctenophores were described as a rather late radiation.

Placozoans and mesozoans, which have an extremely simple organization and a low number of differentiated tissues have also been taken as an example of the primitive type of cell aggregation that must have characterized the first metazoans (Grell, 1961; Whittaker and Margulis, 1978). Most studies concluded that the mesozoan morphological simplicity is in fact derived from a more complex organism as a result of parasitism. *Trichoplax adhaerens*, which is clearly not a plant, has been classified among metazoans, but a discrete phylum has been suggested (Placozoa, Grell 1961). *Trichoplax adhaerens* has probably one of the most simple diploblastic organisations among metazoans: It has two epithelia separated by a mesenchyme but no neuro-muscular system, no polarity (except a dorso-ventral axis), no bilateral symmetry, no organs, and only six types of cells. Eggs are nevertheless produced, followed by larval stages. Due to its peculiar body organization, *Trichoplax adhaerens* has been proposed as a living model of Bütschli's plakula theory for the origin of metazoans (Grell 1961). In conclusion, even if one accepts the idea that sponges or placozoans represent the kind of cell interactions that were present in the earliest forms of metazoans, the identity of the protistan organism that gave rise to this organism remains unknown.

Morphological and embryological data are extremely difficult to interpret. This is because an understanding of the mechanisms of morphological evolution (see Raff & Kaufman, 1983 for review) and an hypothesis concerning the morphology of ancestral organisms are a prerequisite to the proposal of any phylogeny which can then be seen as the most parsimonious explanation of the diversity of living species. In this respect, the new tools of molecular biology might provide an alternative approach: an independently derived phylogeny could help morphologists and embryologists to analyze the process of morphological evolution and to derive the basic laws of morphological evolution in metazoans. Phylogenies based on molecular sequences are in no way linked to hypotheses of morphological evolution, but only upon the assumption that sequences of different organisms have diverged since the time of separation of these species. Ideally, if we had a correct phylogeny and if we knew the morphology of the first metazoan, we might be able to hypothesize the underlying mechanisms that regulated (and still regulate) cell to cell interactions and differentiation.

Proteins such as cytochrome c (Schwartz & Dayhoff, 1978, 1979; Meatyard *et al.*, 1975; Hunt *et al.*, 1985) and histones (Elgin *et al.*, 1979; Glover & Gorovsky, 1979) were used to analyze the relationships between protists and multicellular organisms, but this approach was impeded by the difficulty of sequencing proteins. Because of its small size, the 5S rRNA has been one of the first nucleotide molecules to be extensively sequenced, with several hundreds of sequences now available, spanning the entire spectrum of living organisms (Hori & Osawa, 1979, 1986, 1987; Kuntzel *et al.*, 1981; Huysmans *et al.*, 1983; Wolters & Erdman, 1986; Erdman & Wolters, 1986). In their analysis of more than 350 5S rRNA sequences, Hori and Osawa (1987) concluded that ciliates, euglenophytes and metazoans diverged at approximatively the same date and that metazoans had a monophyletic origin because they formed a single cluster. However, the 5S rRNA (which has only a low content of information, about 120 nucleotides) has a relatively high rate of mutation and substitutions probably reach a saturation level (Qu *et al.*, 1988) when sequences of distantly related organisms are compared. It is noteworthy that notwithstanding a lack of resolution of the 5S rRNA, it is the absence of the protozoans the most closely related to metazoans (see below) which was the main reason for obtaining a monophyletic radiation of the metazoans.

In a previous analysis of protozoan phylogeny using 28S rRNA sequences (Baroin *et al.*, 1988) flagellates were seen as an early radiation, whereas ciliates as well as *Amoeba proteus* were placed relatively close to the metazoan-metaphyte radiations. When the three groups of algae and fungi were added, a broad cluster was obtained that also contained ciliates and metaphytes (Perasso *et al.*, 1989). In this study, we confirmed this result and with sequences from a large number of metazoans, we were able to examine more closely the relationships between metazoans and protozoans. Ciliates, fungi and chromophytes could clearly be assigned to monophyletic late radiations (Figure 1 & 2), but none of these groups belonged directly to the branch that gave rise to the multicellular organisms. The detailed analysis of metazoans origin showed that plants, diploblastic and triploblastic metazoans were probably the results of three different radiations that took place at approximatively the same time. We

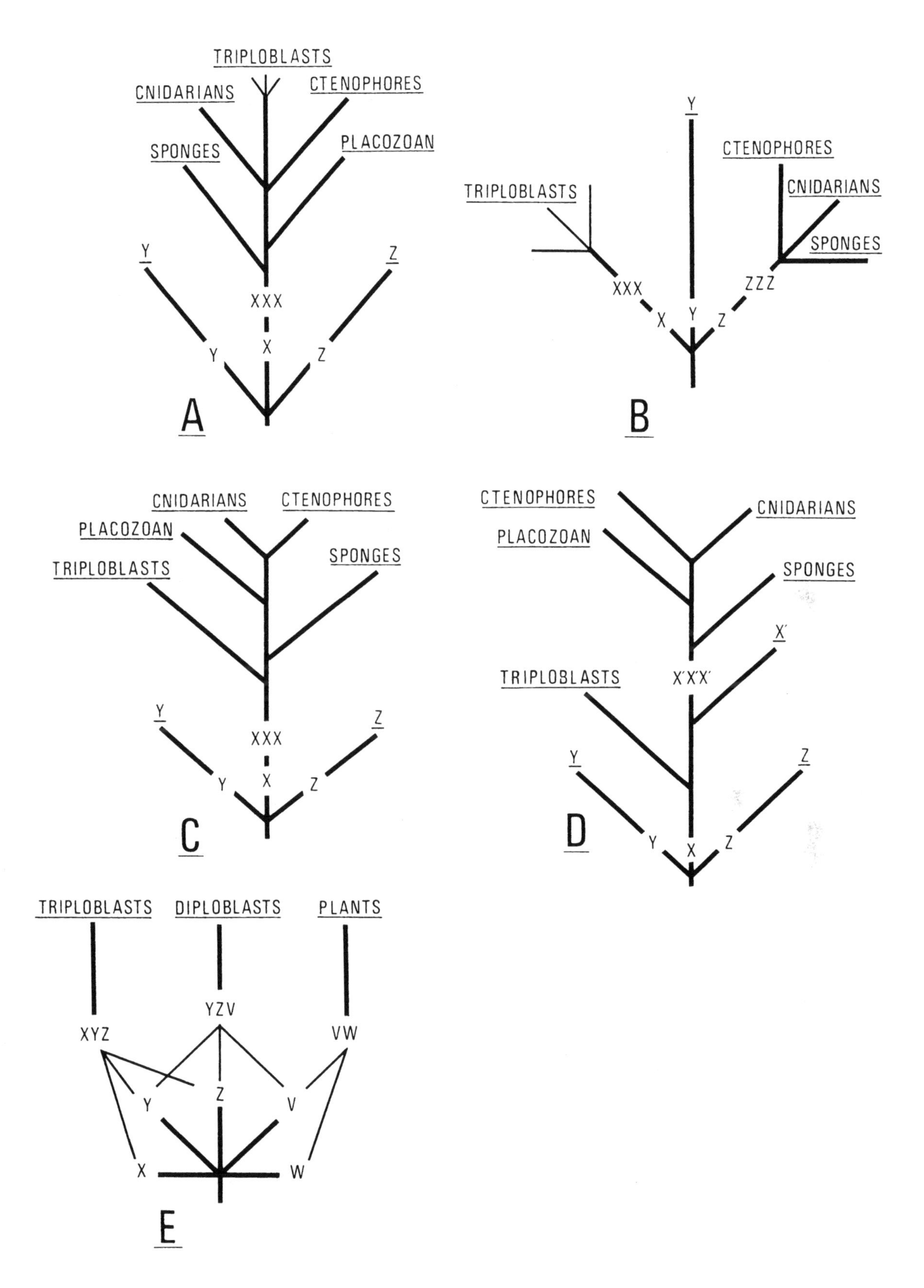

Fig. 5. Alternative phylogenies of metazoans.

Fig. 5A represents the scheme proposed by "classical" phylogenies.

Fig. 5B shows the alternative most favoured by our results: the independent aggregation of protists belonging to two different lineages gave rise to the living metazoans.

Fig. 5C: less likely but possible, a single event of aggregation was followed by an early divergence of the radiation that later produced the triploblastic metazoans.

Fig. 5D: metazoans were the results of two independent aggregation events, but the same protozoan lineage was involved with the origin of both groups (in the absence of a protist X', this phenomenon cannot be distinguished from the previous one).

Fig. 5E: multicellular organisms are the result of the "genomic fusion" of several symbiotic protozoans and not of the subsequent differentiation of a colony of protists.

should stress that metazoans would have probably be seen as grouped with ciliates and fungi if protists such as *Porphyridium purpureum, Acanthocystis longiseta* and the cryptomonadines had not been present in the analysis. This was also the case when more distant protists such as *Dictyostelium discoideum, Crithidia fasciculata* and *Trichomonas vaginalis* were included among the species that were treated. The dendrograms obtained in these circumstances showed a grouping of all triploblastics on the one hand and all diploblastics on the other hand, the latter falling within the broad cluster of protists described above. A study of 18S rRNA sequences also showed that metazoans could be divided into two groups (Field *et al.*, 1988): the two cnidarians that had been sequenced formed a monophyletic radiation (with a ciliate, a yeast and a metaphyte: corn) which was clearly distinct from the radiation that comprised all triploblastic animals.

Our results suggest a polyphyletic origin of the metazoans as the most likely hypothesis inferred from rRNA sequence analysis; however, considering the relatively low numbers of species which we have analyzed and the fact that errors are extremely difficult to evaluate, the hypothesis that all metazoans indeed descend from a common multicellular eukaryote cannot be definitively ruled out. If this is the case, the traditionnal tree as reproduced in Figure 6A is nevertheless definitively not sustained by our data. In Figure 6B, we present what seems to be the most likely sequence of events, with at least two independent multicellularization events, one leading to diploblastic metazoans and the other one leading to triploblastic metazoans, each event occurring as the result of the aggregation of two different protists lineages, and the two events being relatively close in time. Figure 6D shows an alternative possibility, in which the same protist lineage gave rise successively in time to two independent aggregations (a scheme not fundamentally different from the previous one). In Figure 6C, we show that if all metazoans derive from a single multicellular ancestor, then the radiation of the triploblastic metazoans would have been the first to take place, followed by the radiations of sponges, placozoan and last the divergence between cnidarians and ctenophores. This last scheme could be reconciled with the classical view that sponges have conserved the type of organisation present in the first primitive metazoan. Finally in Figure 6E, we present an unorthodox possibility that cannot be ruled out: metazoans (and possibly plants) were the results not of the aggregation of a single species of unicellular organisms, but the results of various symbiotic events between different types of protistan organisms. The presence of one (or several) protozoan in more than one symbiosis could explain some convergence between all multicellular organisms (presence of collagen for example). We should note that the mixing of genetic material between two different organisms was probably not as difficult more a billion years ago than it is now, and that transfection of genetic material still exists between present days symbionts such as a plant and *Agrobacterium*.

Acknowledgments

We thank our colleagues C. Carré (Marine Station of Villefranche sur mer) and J. Vacelet (Marine Station of Endoume) for their help in the collection and identification of the animals, and also H. Philippe for his computer programs. We are also grateful to the University of Camerino and the Agencies who sponsored this symposium for their hospitality.

REFERENCES

BAROIN, A., PERASSO, R., QU, L.H., BRUGEROLLE, G., BACHELLERIE, J.P. & ADOUTTE, A. 1988. Partial phylogeny of the unicellular eukaryotes based on rapid sequencing of a portion of 28S ribosomal RNA. *Proceedings of the National Academy of Sciences USA* **85**, 3474-3478.

DAYHOFF, M.O. 1983. Evolutionary connections of biological kingdoms based on protein and nucleic acid sequence evidence. *Precambrian Research* **20**, 299-318.

ELGIN, S.C.R., SCHILLING, J. & HOOD, L.E. 1979. Sequence of histone 2B of *Drosophila melanogaster. Biochemistry* **18**, 5679-5685.

ERDMAN, V.A. & WOLTERS, J. 1986. Collection of published 5S, 5.8S and 4.5S ribosomal RNA sequences. *Nucleic Acids Research* **14**, r1-r59.

FIELD, K.G., OLSEN, G.J., LANE, D.J., GIOVANNONI, S.J., GHISELIN, M.T., RAFF, E.C., PACEN,R. & RAFF, R.A. 1988. Molecular phylogeny of the animal kingdom. Science **239**, 748-753.

FRANZEN, A. 1970. Phylogenetic aspects of the morphology of spermatozoa and spermiogenesis. In *Comparative spermatology* (ed. B. Baccetti), pp. 29-46. New York: Academic Press.

GERBI, S.A. 1985. Evolution of ribosomal DNA. *Molecular evolutionary genetics.* (ed. R.J. MacIntyre). pp. 419-517. New York: Plenum.

GLOVER, C.V.C. & GOROVSKY, M.A. 1979. Amino acid sequence of *Tetrahymena* histone H4 differs from that of higher eukaryotes. *Proceedings of the National Academy of Sciences USA* **76**, 585-589.

GRAY, M.W., SANKOFF, D. & CEDERGREN, R.J. 1984. On the evolutionary descent of organisms and organelles: a global phylogeny based on a highly conserved structural core in small subunit ribosomal RNA. *Nucleic Acids Research* **12**, 5837-5852.

GRELL, K.G. 1981. *Trichoplax adhaerens* and the origin of the metazoa. In *Convegno internazionale: Origine dei grandi phyla dei metazoi. Atti dei Convegni Lincei* **49**, 107-122. Roma: Accademia Nazionale dei Lincei.

HANSON, E.D. 1977. *The origin and early evolution of animals.* Middletown CT: Wesleyan University Press.

HASEGAWA, M., IIDA, Y., YANO, Y., TAKAIWA, T. & IWABUCHI, M. 1985. Phylogenetic relationships among eukaryotic kingdoms inferred from ribosomal RNA sequences. *Journal Molecular Evolution* **22**, 32-38.

HENAUT, A. & DELORME, M.O. 1988. Distance matrix comparison and tree construction. *Pattern Recognition Letters* **7**, 207-213.

HORI, H. & OSAWA, S. 1979. Evolutionary change in 5S RNA secondary structure and a phylogenetic tree of 54 5S RNA species. *Proceedings of the National Academy of Sciences USA* **76**, 381-385.

HORI, H. & OSAWA, S. 1986. Evolutionary change in 5S RNA secondary structure and a phylogenetic tree of 352 5S RNA species. *BioSystems* **19**, 163-172.

HORI, H. & OSAWA, S. 1987. Origin and evolution of organisms as deduced from 5S ribosomal RNA sequences. *Molecular Biological Evolution* **4**, 445-472.

HUNT, L.T., GEORGE, D.G. & BARKER, W.C. 1985. The procaryotes-eucaryotes interface. *BioSystems* **18**, 223-240.

HUYSMANS, E.E., VANDERBERGH, A., & DE WACHTER, A. 1983. The nucleotide sequences of the 5S rRNAs of four mushrooms and their use in studying the phylogenic position of basidiomycetes among the eukaryotes. *Nucleic Acids Research* **11**, 2871-2880.

KUNTZEL, H., HEIDRICH, M. & PIECHULLA, B. 1981. Phylogenetic tree derived from bacterial, cytosol, and organelle 5S rRNA. *Nucleic Acids Research* **9**, 1451-1462.

MACQUADE, A.B. 1983. Origins of the nucleate organisms II. *BioSystems* **16**, 39-55.

MEATYARD, B.T, SCAWES, M.D., RAMSHAW, J.A.M. & BOULTER, D. 1975. Cytochromes cs from *Rhodymenia palmatum* and *Porphyra umbilicalis* and the amino acid sequences of their N-terminal sequences. *Phytochemistry* **14**, 1493-1498.

NEI, M. 1987. Phylogenetic trees. In *Molecular evolutionary genetics*, pp. 287-325, New York: Columbia University Press.

NIELSEN, C. 1985. Animal phylogeny in the light of trochaea theory. *Biological Journal of the Linnean Society* **25**, 243-299.

OHAMA, T., KUMASAKI, T., HORI, H. & OSAWA, S. 1986. Evolution of multicellular animals as deduced from 5S rRNA sequences: a possible early emergence of the Mesozoa. *Nucleic Acids Research* **12**, 5101-5108.

PACE, N.R., OLSEN, G.J. & WOESE, C.A. 1986. Ribosomal RNA phylogeny and the primary lines of evolutionary descent. *Cell* **45**, 325-326.

PERASSO, R., BAROIN, A., QU, L.H., BACHELLERIE, J-P. & ADOUTTE A. 1989. Origin of the algae. *Nature* **339**, 142-144.

QU, L.H., MICHOT, B. & BACHELLERIE, J.P. 1983. Improved methods for structure probing in large RNAs: a rapid "heterologous" sequencing approach is coupled to the direct mapping of nuclease accessible sites. Application to the 5' terminal domain of eukaryotic 28S rRNA. *Nucleic Acids Research* **11**, 5903-5920.

QU, L.H., NICOLOSO, M. & BACHELLERIE, J.P. 1988. Phylogenetic calibration of the 5' terminal domain of large rRNA achieved by determining twenty eukaryotic sequences. *Journal Molecular Evolution* **28**, 113-124.

RAFF, R.A. & KAUFMAN, T.C. 1983. *Embryos, genera and evolution*. New York: Macmillan publishing.

RAGAN, M.A. & CHAPMAN, D.J. 1978. *A biochemical phylogeny of the protists*. New York: Academic Press.

RUNNEGAR, B. 1985. Collagen gene construction and evolution. *Journal of Molecular Evolution* **22**, 141-149.

SALVINI-PLAWEN, L. VON 1982. A paedomorphic origin of the oligomerous animals? *Zoologica Scripta* **11**, 77-81.

SCHWARTZ, R.M. & DAYHOFF, M.O. 1978. Origins of prokaryotes, eukaryotes, mitochondria, and chloroplasts. *Science* **199**, 395-403.

SCHWARTZ, R.M. & DAYHOFF, M.O. 1979. Protein and nucleic acid sequence data and phylogeny. *Science* **250**, 1038-1039.

SOGIN, M.L., GUNDEERSON, J.H., ELWOOD, H.J., ALONSO, R.A. & PEATTIE, D.A. 1989. Phylogenetic meaning of the kingdom concept: an unusual RNA from *Giardia lamblia*. *Science* **243**, 75-77.

WHITTAKER, R.H. & MARGULIS, L. 1978. Protist classification and the kingdoms of organisms. *BioSystems* **10**, 3-18.

WILSON, A.C., CARLSON, S.S. & WHITE, T.J. 1977. Biochemical evolution. *Annual Review of Biochemistry* **46**, 573-639.

WILSON, A,C., OCHMAN, H. & PRAGER, E.M. 1987. Molecular time scale for evolution. *Trends in Genetics* **3**, 241-247.

WOESE, C.R. 1987. Bacterial evolution. *Microbiological Revews* **51**, 221-271.

ZUCKERKANDL, E. & PAULING, L. 1965. Molecules as documents of evolutionary history. *Journal of Theoretical Biology* **8**, 357-366.

Major factors in the rapidity and extent of the metazoan radiation during the Proterozoic-Phanerozoic transition

James W. Valentine[1]

Abstract

The early history of metazoans was characterized by evolution towards more complex body plans, implying growth in the numbers of cell types differentiated during ontogeny as the more complex forms appeared. Such broad anagenesis should be relatively slow. Radiations occurred during this trend, chiefly recorded by fossils during the Vendian and during the Proterozoic-Phanerozoic transition. The latter radiation occurred when body plan complexity had reached the level of the higher invertebrates (50 or so cell types) in a number of lineages, thus producing a host of novel body plans, recognized as phyla and classes today, all at about the same level of complexity. The rapidity of the radiation was promoted by a breadth of opportunity (relatively empty "adaptive space") in which those lineages with inherently fast-paced evolutionary dynamics (turnover rates) were favoured.

The Rise of Complexity

Advances in grades of construction of multicellular animals have involved the progressive differentiation of cell types. Perhaps the most useful single metric of complexity of animal bodies is cell-type number (Bonner, 1965). Counting cell types is an inexact process; one may be a splitter or a lumper and identify significantly different numbers of cell types from the same organism accordingly. Perhaps for this reason, there are few counts of cell types available; those few recorded by Bonner (1965, 1988) and used here are said to be based on a lumper's approach, and include chiefly the structural cells in metazoan bodies (and exclude such complications as presented by the immune system). In this tabulation, living multicellular organisms range from a few different cell types (sponges) to 120 or so (mammals).

Cell differentiation is mediated by information in the genome (in some cases the maternal genome) and by epigenetic signals or events. Kauffman (1969, 1974, 1987) has discussed a model for the regulation of differentiation with important evolutionary implications. Individual ontogenies begin with a zygote and then, through a series of differentiations, cell lineages radiate into a variety of cell types. One can imagine a tree-like figure that traces the sequence of differentiation from the "founding" cell through cleavage to, say, 120 branch tips representing the cell types of a mammal. As this is an ontogeny, not a phylogeny, the evolutionary steps that led from ancestral cell types to the many descendant ones are not necessarily recorded. One may, however, imagine a sort of family tree of all metazoan cell types, starting with the last common unicellular ancestor, and including all extant and extinct types. In such a tree the extant cell types form the branch tips, while many cell types ancestral to those living would be found embedded within the tree. Other extinct cell types would terminate extinct branches. This tree would trace the branching pattern of cell phylogeny.

During animal evolution the numbers of cell types in some lineages rose from few to many, and increasing amounts of information required to specify all the cell types had to be incorporated in the single zygote from which differentiation proceeded. Kauffman (1969, 1974, 1987) has discussed a model in which the amount of regulatory information required to specify new cell types must increase as about the square of the number of types. This information would presumably be chiefly encoded in the genome, although some could be expressed as a result of epigenetic interactions or encoded by maternal genomes. The growth in cell types should occur at a relatively slow pace, because of the novel gene regulatory patterns and epigenetic interactions that have to be worked out via selection. Thus the evolution of increasingly complex organisms should be a relatively slow and deliberate process. Rapid jumps or saltations would be the least likely pathway to expect in tracing the rise of complexity, and rapid radiations seem improbable if they involve significant increases in cell type numbers.

From the few available estimates of cell type numbers, Bonner (1988) indicates what amounts to a plateau in complexity among higher invertebrates. Although cnidarians and sponges may have perhaps 11 cell types (some sponges have fewer), the major invertebrate phyla — annelids, arthropods, mollusks and echinoderms — have something around 55. One might predict that lower bilaterians such as flatworms and nemertines have ranges of lumped cell type numbers intermediate between 11 and 55; there seems to be no data on these organisms. At any rate the important point here is that it requires about, say 50 times as many kinds of cells to make a higher invertebrate body of the sort that characterized the metazoan radiation during the Proterozoic - Phanerozoic transition, than to make a protozoan. If Kauffman's model is approximately correct, then such an increase in complexity implies an increase in regulatory capacity on the order of 2500 times. It is of course possible that this model overestimates the regulatory requirements; perhaps the number of additional instructions required falls off somewhat as cell type numbers increase. However, the principle is clear.

The Metazoan radiations

If this principle is generally correct, however imprecise the numbers may be, it predicts that the evolution of higher invertebrates would involve a significant time span after the appearance of metazoans. On

[1]Department of Geological Sciences, University of California, Santa Barbara, CA 93106 USA.

the other hand, once a complexity of 50 or so cell types had been achieved, the regulatory power inherent in the genomes (and with extensive epigenetic patterns in place) would be great enough to mediate the elaboration of a wide variety of body plans. The modifications required, while involving some repatterning of gene expression to be sure, would produce novel body plans of about the same complexities as the ancestral plans. Repatterning might occur via heterochronies or other switches in the pattern of gene expression that involve large sets of genes that already function harmoniously. Britten & Davidson (1971) have suggested a hierarchical, cascading architecture of gene regulation that might respond relatively rapidly to mutations that led toward novel morphologies at the same grade of complexity (applied to the metazoan radiation by Valentine & Campbell, 1975).

It is suggested, then, that between the (unknown) time of origin of the metazoans during the Proterozoic, and the relative explosive appearance of numerous new body plans during the Proterozoic-Phanerozoic transition, was a period characterized by the development of a series of ascending grades of anatomical complexity. Although there probably were radiations within grades — the radiation of what appear to be cnidarian-grade Vendian forms indicates as much — and these radiations may have been abrupt, the broad anagenetic trend towards increasing complexity was probably slow-paced, depending as it did upon the evolution of encoding instructions for new cell types. When cellular and genomic complexity reached the level we observe in higher invertebrates, body plans capable of a broad variety of ecologic functions became possible for the first time, and a relatively rapid radiation occurred that established the rich benthic ecosystems of the early Phanerozoic, with fairly elaborate communities in a variety of habitats. A number of lineages gave rise to novel body plans during the Proterozoic-Phanerozoic transition, producing a plethora of types among which living phyla were in the minority, and including such disparate architectures as those of brachiopods, arthropods and echinoderms. The rise to an anatomical complexity appropriate to these forms, with body cavities of coelom or haemocoel providing hydrostatic skeletons, need not have been precisely contemporaneous in the various lineages; their sequence of origin is not yet resolved in any case.

Additional factors

This appeal to a key level of complexity as underlying an abrupt metazoan radiation is not inconsistent with the fossil record but is not sufficient in itself to explain all of the evidence. For example, benthic foraminifera with agglutinated tests appear, suggesting a radiation at that time (Signor, 1988) that happened to include lineages with durable tests and thus that entered the fossil record. Among the lower metazoans, the archaeocyathans appear and radiate extraordinarily (Hill, 1972). We do not know whether archaeocyathans had existed without skeletons during the late Proterozoic, achieving an ecological breakthrough with the advent of skeletons, or whether the archaeocyathan body plan itself first arose in early Cambrian time. In either case, however, it seems unlikely that their radiation depended upon a genomic complexity comparable to that of the higher invertebrates. It is possible that all these events were tied together. One might hypothesize an expansion and stabilization in phytoplankton providing the base upon which benthic community elaboration could proceed, and that foraminifera and lower invertebrates may also have benefitted from the creation of new benthic ecosystem structures with more complicated food webs and detritus cycling. Equally plausible at present are models that interpret all of these features as a common response to physical environmental changes. Nevertheless, whatever hypothesis is invoked to explain the timing of these events, the fact remains that the creation of the many higher invertebrate body plans during this radiation coincides with the first known appearance of the higher invertebrate grade itself, suggesting that appropriately complex regulatory machinery had just evolved or was evolving even as the radiation proceeded in the various lineages.

Other factors that may have contributed to the character of the radiation include the relative openness of "adaptive space" at the time, providing a wide range of ecological opportunities awaiting the invention of appropriate body plans to exploit them. Some of this untenanted adaptive space had not previously been occupied; for example, penetrating vertical burrowing and deep bioturbation had not appeared previously. Much adaptive space may also have become available through the extinction of some of the late Proterozoic lineages (Seilacher, 1984). None of the structure and timing of those extinctions are now apparent, but at least the bulk of the Proterozoic fauna that we know disappeared, though there were clearly enough survivors to provide the stocks from which Phanerozoic faunas descended.

The fossil record indicates that many clades have a relatively narrow range of taxonomic turnover rates throughout their histories. These rates may be observed on the level of species (see Stanley, 1979), genera (Raup & Boyajian, 1988) and even families (see the figures in Sepkoski, 1981 and Sepkoski & Hulver, 1985). From a variety of theoretical models it has been predicted that clades with the highest turnover rates are most at risk of extinction, other things being equal (see Stanley, 1979; Raup, 1983; Walker, 1984; Van Valen, 1985; Valentine, 1989). These expectations appear to be met, in that clades with high turnover rates, which dominate early in metazoan history, are replaced in dominance by clades with low turnover rates (Van Valen, 1985; Valentine, 1989). The three faunas of Sepkoski (1981) are in fact composed chiefly of clades with similar turnover rates (Valentine, in press), with the highest rates among Cambrian dominants (trilobites), intermediate rates among post-Cambrian Palaeozoic dominants (brachiopods, crinoids) and lowest rates among Mesozoic-Cainozoic dominants (bivalves, gastropod). The most glaring exception to this trend occurs among cephalopods, a high-turnover group that dominates turnover statistics at the start of the Mesozoic, a time when as Van Valen (1984) has shown, the clade dynamics were re-set at a higher level following the extensive Permian-Triassic extinctions (see also Valentine, 1989, in press). It is quite plausible that in the relatively empty adaptive spaces of the early Cambrian and early Triassic, those clades with the highest speciation rates should come to dominate the fauna. They were, after all, the more likely to radiate first into the large numbers of available habitats. In this light, the high rates of diversification of body plans during the Proterozoic-Phanerozoic transition may simply reflect the success of clades with the least conservative evolutionary modes — those with the least restrictive genomes and the most flexible or pliable architectures — in dominating the diversifying clades; but it was also their fate to be among the early victims of extinction.

Another factor that may have been important during the radiation is that mutation rates among some, perhaps most clades may have been considerably higher than those observed today (see Britten, 1986; Buss, 1987). This source of genetic novelty may have permitted the rise and exploitation of complexity to occur at a more rapid pace, even given the adaptive advantages provided by the breadth of opportunity, than we would estimate for today's organisms.

Conclusions

Considering the propensities for self-organization and for increase in numbers both of individuals and of taxa that seem to be fundamental to life as we know it, it seems probable that evolution would have produced higher metazoans sooner or later on earth regardless of the details of their history. Presumably the metazoan radiation during the Proterozoic-Phanerozoic transition would not have been such an abrupt and extensive event, but for historical contingencies that produced a concomitance of circumstance: high mutation rates, broad environmental opportunities, and the rise of complexity to a point that body plans become capable of exploiting the situation with which they were faced. The result was the rapid creation of complicated benthic ecosystems and the establishment of the Phanerozoic Eon.

Acknowledgments

The work on which this paper is based was supported by Grants EAR87-21192 and EAR 84-17011 from the National Science Foundation, USA. The author is also grateful to the University of Camerino and to the sponsors of this Symposium for their hospitality.

REFERENCES

BONNER, J.T. 1965. *Size and Cycle.* Princeton, N.J.: Princeton University Press.

BONNER, J.T. 1988. *The Evolution of Complexity by means of Natural Section.* Princeton, N.J.: Princeton University Press.

BRITTEN, R.J. 1986. Rates of DNA sequence evolution differ between taxonomic groups. *Science* **231**, 1393-1398.

BRITTEN, R.J., & DAVIDSON, E. H. 1971. Repetetive and non- repetetive DNA sequences and a speculation on the origins of evolutionary novelty. *Quarterly Review of Biology* **46**, 111-133.

BUSS, L.W. 1987. *The Evolution of Individuality.* Princeton, N.J.: Princeton University Press.

HILL, D. 1972. Archaeocyatha. In *Treatise on Invertebrate Paleontology*, Part. E. vol. 1 (ed. C. Teichert), 2nd. ed., Boulder: Geol. Soc. Amer. and Univ. Kansas.

KAUFFMAN, S.A. 1969. Metabolic stability and epigenesis in randomly constructed genetic nets. *Journal of Theoretical Biology* **22**, 437-467.

KAUFFMAN, S.A. 1974. The large scale structure and dynamics of gene control circuits: An ensemble approach. *Journal of Theoretical Biology* **44**, 167-190.

KAUFFMAN, S.A. 1987. Self-organization, selective adaptation, and its limits: A pattern of inference in evolution and development. In *Neutral Models in Biology* (eds. M.H. Nitecki & A. Hoffman), pp.56-89. New York: Oxford University Press.

RAUP, D.M. 1983. On the early origins of major biologic groups. *Paleobiology* **9**, 107-115.

RAUP, D.M., & BOYAJIAN, G.E. 1988. Patterns of generic extinction in the fossil record. *Paleobiology* **14**, 109-125.

SEILACHER, A. 1984. Late Precambrian and Early Cambrian Metazoa: Preservational or real extinctions? In *Patterns of Change in Earth Evolution.* (eds. H.D. Holland & A.F. Trendall), pp. 159-168, Berlin: Springer-Verlag.

SEPKOSKI, J.J., Jr. 1981. A factor analytic description of the Phanerozoic marine fossil record. *Paleobiology* **7**, 36-53.

SEPKOSKI, J.J., JR., & HULVER, M.L. 1985. An atlas of Phanerozoic clade diversity diagrams. In *Phanerozoic Diversity Patterns: Profiles in Macroevolution.* (ed. J.W. Valentine), pp. 11-39, Princeton, N.J.: Princeton University Press.

SIGNOR, P. W. 1988. The Precambrian-Cambrian metazoan radiation: Significance of earliest Cambrian agglutinated skeletons. *Geological Society of America, Abstracts with Program* **20**, A 104.

STANLEY, S.M. 1979. *Macroevolution, Pattern and Process.* San Francisco: Freeman.

VALENTINE, J.W. 1989. Phanerozoic marine faunas and the stability of the earth system. *Palaeogeography, Paleoclimatology, Palaeoecology,* **75**, 137-155.

VALENTINE, J.W. In press. The macroevolution of phyla. In *Origins and Early Evolutionary History of the Metazoa* (eds. J.H. Lipps & P.W. Signor), New York: Plenum.

VALENTINE, J.W., & CAMPBELL, C.A. 1975. Genetic regulation and the fossil record. *American Scientist* **63**, 673-680.

VAN VALEN, L. 1984. A resetting of Phanerozoic community evolution. *Nature* **307**, 50-52.

VAN VALEN, L. 1985. A theory of origination and extinction. *Evolutionary Theory* **7**, 133-142.

WALKER, T.D. 1984. *The evolution of diversity in an adaptive mosaic.* Unpublished PhD. Thesis, Santa Barbara: University of California.

Divergence and persistence of highly ranked taxa

Richard R. Strathmann[1]

Abstract

The occurrence of problematical groups can be predicted from models that assume (1) random speciation and extinction and (2) an early divergence of phyla and classes. The models predict that many clades that diverged early would take a random walk to extinction and would be especially vulnerable to extinction when they contained few species. A substantial fraction of extinctions are predicted even among groups with 100 species when speciations and extinctions occur at rates of 0.2 or more per million years. Other extinctions might be associated with factors outside the models, such as relatively inferior traits or bottlenecks from mass extinctions. These models predict numerous extinct groups that are as distantly related to existing highly ranked taxa as these existing taxa are distantly related to each other. These models address relationship, but not morphological change. Sudden appearance of highly ranked taxa and their presently great morphological differences do not necessarily imply uniquely rapid morphological divergence at the time that phyla or classes diverged. Taxonomic ranks indicate a sequence of morphological divergences but are poor measures of the magnitude of morphological differences arising at each divergence. Great post-Palaeozoic changes in body plans are known and some appear to have been rapid.

Introduction

Most of the presently existing phyla of multicellular animals that are readily fossilized appeared early in the Cambrian during a short period (Cloud, 1968; Raup, 1972). Most of the existing classes appeared early in the Palaeozoic. This pattern was counter to the expectation of many evolutionary biologists, and the pattern of early and rapid appearance of higher taxa is generally seen as a problem that demands an explanation (Valentine & Erwin, 1987; Levinton, 1988). Problematical fossils are fossils whose placement within existing taxa of high rank is uncertain (Bengtson, 1986). Most problematical fossils also appear early in the fossil record for multicellular animals, but the clades have subsequently become extinct (Briggs & Conway Morris, 1986). They therefore fit the same general pattern of origination of highly ranked metazoan taxa and raise similar problems.

Part of the problem of early and rapid origin of highly ranked taxa is their persistence. Should we expect to find many extinct groups that are as unrelated to Recent phyla and classes as these highly ranked taxa are unrelated to each other? The occurrence and diversity of problematic extinct groups can be predicted from models with random speciation and extinction. Another part of the problem is the apparently rapid rates of morphological change during divergence of phyla and classes. However, taxonomic rank is a poor measure of morphological distance and its use as a measure has exaggerated the problem of divergence of phyla and classes. This argument is supported by examples of great changes in body plans within low taxonomic ranks.

Persistence of higher taxa

If all existing highly ranked taxa originated near the beginning of the Palaeozoic or earlier, then they represent either remarkably persistent clades, or they are the chance survivors of a large number of clades. Unless they are remarkably persistent, we should expect a large number of extinct clades that occur near the time of divergence of existing highly ranked taxa and that are distinct from any of the existing groups. If preserved, these would appear as problematical fossils. If we knew what influenced persistence, we could obtain a prediction about the number of early problematical groups.

One approach to this problem is to use stochastic models of speciation and extinction. Raup *et al.* (1973) and Raup (1978) offered stochastic models of speciation and extinction as minimally sufficient explanations of patterns of diversification and extinction of clades. Raup (1983) also used this approach to examine the early origin of major biologic groups. The appeal of these models is that they do not depend on peculiarities of different lineages. Strathmann & Slatkin (1983) extended probabilistic models to the persistence of higher taxa in three models with differing assumptions about speciation and extinction. These models addressed the 10 or more phyla estimated to have less than 1000 species today. These minor phylum models circumvented problems of estimating numbers of species per phylum or class over Phanerozoic time. They required only estimates of time since origin of the clade, rates of speciation and extinction, and numbers of species existing in the clade today. The results were not highly sensitive to errors in estimates for these parameters. This approach discarded the abundant information on trends in taxonomic diversity in fossil material (Valentine, 1985), but was free of biases in estimating species diversity from the fossil record. These minor phylum models did remain dependent on estimates of species durations that are based on change in form and are only indirectly related to speciation and extinction. What do these models tell us about persistence of clades, and what does persistence tell us about problematic fossils?

In these models the greatest hazard of extinction comes when clades have few species. This condition is encountered by all clades at their origin. When there are only a few species in a clade, the chances of a random walk to extinction are quite high. This is illustrated by the simplest minor phylum model, in which speciation rate (s) and extinction rate (e) remained constant

[1]Friday Harbor Laboratories and Department of Zoology, University of Washington, 620 University Road, Friday Harbor, WA 98250.

and equal throughout the history of a clade. If clades diverged 500 million years ago, what would the initial number of clades have been for there to be a 5% chance that 10 clades survived and have between 1 and 1000 species per clade at present? There would have been more than 162 initial clades if and s and e equalled 0.05 per million years and over 3577 initial clades if s and e had equalled 1.0 per million years. The range of speciation and extinction rates was taken from Stanley's (1979) estimates of species durations from the fossil record. Thus under these evolutionary processes, there would have been hundreds to thousands of initial clades required for 10 to reach the present with only 1 to 1000 species. Unequal but constant rates of speciation and extinction did not much change this result. Randomly varying rates of speciation and extinction within the estimated range of rates would increase the chances of extinction, and so would mass extinctions. Thus the simplest models require a large number of early Phanerozoic clades as unrelated to those now recognized as phyla as existing phyla are unrelated to each other. The same argument applies to early clades distinct from existing classes. Under this model, a very large number of problematic groups early in the history of the Metazoa is expected. Under this model one would expect many problematic groups to be known from a single species and time because the greatest risk of extinction is when a clade has few species.

The chances of persistence are greatly improved if there is an initial rapid increase in species per clade (Strathmann & Slatkin, 1983) We know that a rapid increase in species could not have occurred for much of the history of existing phyla and classes. The fossil record does not show a rapid continuing increase in diversity of species throughout the Phanerozoic (Sepkoski, 1979) Also many existing phyla and classes have few species today. An initially great excess of speciation over extinction might result from initially less severe predation or competition. With no extinctions, it would take less than seven doublings of species for a clade to expand to 100 species. If speciation rates greatly exceeded extinction rates early in the evolution of metazoans, then metazoan clades initiated early could have had a much higher chance of persistence than metazoan clades initiated from later divergence.

There is some support for the assumption that rates of speciation were high or rates of extinction low early in the Phanerozoic. For example, Sepkoski (1979) found that rates of origination and disappearance of early Phanerozoic families were consistent with diversity-dependence. Raup (1983) modelled times of divergence with species durations of 0.2 to 20 million years, a modern diversity of 150000 readily fossilizable species, and up to a 10 fold increase in species from Cambrian to Recent. Because the model placed the invertebrate radiation "deep in the Precambrian", Raup concluded that the early stages of the radiation were characterized by much higher branching rates.

Once a clade has even 100 species there is a good chance that it will persist 500 million years with equal speciation and extinction rates (Strathmann & Slatkin, 1983) With very low extinction rates (0.05 per million years) the survival probability is 0.98. Under this extreme assumption any clades that reached about 100 species would be unlikely to become extinct by chance, and problematical fossil groups with many species would not be expected. However, the model does not include later increases in rates of extinction, as in episodes of mass extinctions, and these could reduce numbers and extinguish major groups. With higher rates of speciation and extinction, the model predicts more problematical groups. With a rate of 0.2 speciations or extinctions per million years, a group with 100 species has a survival probability of 0.63 for 500 million years. With a rate of 1 speciation or extinction per million years, a group with 100 species has a survival probability of only 0.18 for 500 million years, and therefore a large number of problematical groups with numerous species is predicted even without later mass extinctions or inferior traits in the extinct groups.

In a third model, speciation and extinction rates depended on the number of species within the clade. With sufficiently strong density dependence, few clades would go extinct and there would be few extinct groups of problematical relationships. This assumption seems implausible because interactions between species are often no stronger within clades than between clades. Also, the assumption cannot be generally true because many clades do become extinct; for this model one must assume that some clades are special from their origin and others not.

The more plausible probabilistic models predict the existence of problematical fossil groups that appear early and are as unrelated to existing higher taxa as the existing higher taxa are unrelated to each other. In other words, the models predict extinct groups whose ancestors are not the latest common ancestor for members of any existing phylum; and similarly, the models predict extinct groups whose ancestors are not the latest common ancestor for any existing class. Distant relationship does not necessarily mean that one cannot infer relationships, but the difficulties certainly increase. The predicted extinct groups would be expected to lack traits that are now used to distinguish higher taxa.

The models do not reject the possibility that extinction or persistence of a clade depends on traits present in all members of the clade, but it is difficult to relate properties of organisms to survival of a clade. Valentine (1973) suggested that extinct phyla and classes were failed experiments, but one cannot infer inferiority in any particular respect from the mere fact of extinction. One needs to relate some observable trait to performance, and performance to extinction. Functional analysis is particularly uncertain for structures with no close modern analogues, and such structures are common in problematical fossils. Structures that appear to be subject to functional analysis may have an uncertain relation to performance, and performance may have an uncertain relation to survival. Unobserved features, such as chemical defenses or larval traits, could be more important than the preserved structures subject to functional analysis. Extinct clades may even have possessed many capabilities superior to any existing forms (Strathmann, 1978). Traits could be selected differently during bottlenecks of mass extinctions than at other times (Strathmann, 1978; Jablonski, 1986; Valentine 1986). Though escalation in criteria for performance during the Phanerozoic can be demonstrated for some traits, escalation is not equal everywhere and safe areas persist for organisms of lower capability (Vermeij, 1987). Because it is difficult to relate functional requirements to diversification and extinction of clades, simple models of chance are attractive.

In summary, simple models predict extinct groups

of problematical relationships, and the absence of problematical extinct groups would raise questions about the persistence of presently existing taxa of high rank.

Taxonomic rank and morphological divergence

What makes problematical fossils especially intriguing is the hint that there are many extinct body plans. Models of persistence and relationships between clades tell us that we should expect extinct phyla and classes, but they tell us little about the evolutionary change in development and form that would result in the sudden appearance of both the extinct and persisting body plans. The apparently sudden divergence of phyla and classes has prompted numerous hypotheses about late Precambrian and early Cambrian conditions. Valentine & Erwin (1987) review these hypotheses and add their own hypothesis that such rapid divergence required unusual evolutionary mechanisms, such as transfer of genetic material between distantly related lineages.

It is, of course, possible that the divergence was not as rapid as the fossil record suggests. Though the fossilizable parts may have diverged rapidly, there are many features of metazoan body plans that do not fossilize well and do not depend on hard parts or burrows. These include many features of the body cavities, body wall, circulatory system, nervous system, and excretory system. Thus much divergence could have been earlier than the Cambrian without much fossil record (Runnegar, 1982). Also much of the morphological isolation of some Recent higher taxa results from descent from a relatively late common ancestor whose modifications accumulated over a long period, as in the case of echinoids in contrast to other eleutherozoan echinoderm classes (Smith, 1984).

Still, even if one accepts a rapid divergence of highly ranked taxa, one need not accept taxonomic rank as a measure of morphological divergence. Taxonomic rank is usually established to indicate relationships, not morphological distance.

We often equate the characters that distinguish highly ranked taxa with the characters that make up a body plan. This gives a definition of a body plan that makes body plans seem conservative. It then surprises us that body plans might diverge rapidly. However, it is possible that these traits were not so conservative when the existing phyla were diverging (Riedl, 1978; Strathmann 1978), and it is certainly the case that many characters used to distinguish phyla and classes continue to be changed. Here are some examples of changes in body plans in lower ranked taxa. Some of the changes were post-Palaeozoic and rapid.

Discovery of some deep-sea echinoderms with a concentric water vascular system has prompted description of a new class of echinoderms (Baker *et al.*, 1986; Rowe, 1988). These echinoderms appear to be a sister group of *Caymanostella* within the Asteroidea (Smith, 1988) and would be placed in the asteroid family Caymanostellidae by some systematists. This group appears to have diverged from its sister asteroids since the Palaeozoic (Blake, 1987). A reason that higher taxa do not evolve late is that most biologists do not like paraphyletic taxa, but this does not mean that the traits that make up body plans are conservative.

Bivalved gastropods with equal valves (Kawaguti & Baba, 1959) are known from the Eocene and Recent (Kay, 1968; Runnegar, 1985). They are a family of sacoglossan opisthobranchs and derived from gastropod ancestors with helicoid shells. Here one of the traits that usually distinguishes molluscan classes has been changed. The helicoid larval shell remains on one valve of the adult as a vestige of gastropod development and ancestry, but nevertheless a character that diverged early, is preserved in fossils, and usually distinguishes molluscan classes has been changed.

An example of an apparently more fundamental change in traits is a deuterostome snail (*Paludina* = *Viviparus*) in which the mouth develops at a new site instead of at the blastopore (Verdonk & van den Biggelaar, 1983). However, processes and structures appearing earlier in development often change while later stages remain more conservative (Ballard, 1981; Sander, 1983; Alberch, 1985; Strathmann, 1988), in clear violation of von Baer's law. In this case there has been a change in development of a trait that is usually assumed to be extremely conservative and that is associated with an entire group of phyla, the protostomes, which includes both molluscs and annelids.

Rearrangements in the development of the water vascular system and other features of the body plan of clypeasteroid echinoids have opened new developmental and functional possibilities unique within echinoderms (Strathmann, 1975). Kier (1982) has presented evidence that the changes that derived sand dollars from a cassiduloid ancestor occurred in the Paleocene and Eocene within 20 million years. The dissociation of branching of the water vascular system and branching of arrays of skeletal elements in the clypeasteroid body plan is not just different from other echinoids; it is different from the architecture and development of all other echinoderms. If this trait had evolved early, one might use it in distinguishing groups of echinoderm classes. Because it evolved late, it helps to distinguish an order.

It is premature to invoke unusual evolutionary processes to account for a rapid morphological divergence of the metazoan phyla and classes. Taxonomic ranks are an approximate indication of a sequence of morphological divergences. Earlier divergence in structure is not necessarily a greater divergence. Divergence in a character used to distinguish higher ranks is not necessarily greater than divergence in characters distinguishing lower ranks.

For an unbiased comparison of rates of morphological divergence, one could first list a set of characters that does not depend on knowledge of distinguishing traits of existing higher taxa. These could be selected on the basis of structural or functional principles. One could then compare rates of divergence in these traits, in so far as they can be inferred from the fossil record.

Acknowledgements

This study was supported by NSF grant OCE-8606850, the Friday Harbor Laboratories, and the Zoology Department of the University of Washington. The models were from M. Slatkin. Comments from B.S. Beall, H. Van Iten, A.J. Kohn, R.M. Rieger, B. Runnegar, and M. Strathmann were also of great help. I am also grateful to the University of

Camerino and to the other agencies sponsoring this Symposium for their hospitality.

REFERENCES

ALBERCH, P. 1985. Problems with the interpretation of developmental sequences. *Systematic Zoology* **34**, 46-58.

BAKER, A.N., F.W.E. ROWE, & H.E.S. CLARK. 1986. A new class of Echinodermata from New Zealand. *Nature* **321**, 862-864.

BALLARD, W.W. 1981. Morphogenetic movements and fate maps of vertebrates. *American Zoologist* **21**, 391-399.

BENGTSON, S. 1986. The problem of the problematica. In *Problematic Fossil Taxa* (ed. A. Hoffman and M. H. Nitecki), pp. 3-11. Oxford: Clarendon Press.

BLAKE, D.B. 1987. A classification and phylogeny of post- Palaeozoic sea stars (Asteroidea: Echinodermata). *Journal of Natural History* **21**, 481-528.

BRIGGS, D.E.G., & S. CONWAY MORRIS. 1986. Problematica from the Middle Cambrian Burgess Shale of British Columbia. In *Problematic Fossil Taxa* (ed. A. Hoffman and M. H. Nitecki), pp. 167-183. Oxford: Clarendon Press.

CLOUD, P.E., JR. 1968. Pre-metazoan evolution and the origins of metazoa. In *Evolution and Environment* (ed. E. T. Drake), pp. 1-72. New Haven: Yale University Press.

JABLONSKI, D. 1986. Larval ecology and macroevolution in marine invertebrates. *Bulletin of Marine Science* **39**, 565-587.

KAWAGUTI, S., & K. BABA. 1959. A preliminary note on two-valved sacoglossan gastropod, *Tamanovalva limax*, n. gen., n. sp., from Tamano, Japan. *Biological Journal of Okayama University* **5**, 177-184.

KAY, E.A. 1968. A review of the bivalved gastropods and a discussion of evolution within the Sacoglossa. *Symposia of the Zoological Society of London* **22**, 109-134.

KIER, P.M. 1982. Rapid evolution in echinoids. *Paleontology* **25**, 1-9.

LEVINTON, J. 1988. *Genetics, Palaeontology, and Macroevolution*. Cambridge: Cambridge University Press, 637 pp.

RAUP, D.M. 1972. Taxonomic diversity during the Phanerozoic. *Science* **177**, 1065-1071.

RAUP, D.M. 1978. Cohort analysis and survivorship. *Paleobiology* 4, 1-15.

RAUP, D.M. 1983. On the early origins of major biologic groups. *Paleobiology* **9**, 107-115.

RAUP, D.M., S. J. GOULD, T. M. SCHOPF, & D. S. SIMBERLOFF. 1973. Stochastic models of phylogeny and the evolution of diversity. *Journal of Geology* **81**, 525-542.

RIEDL, R. 1978. *Order in Living Organisms*. (trans. R. P. S. Jefferies), Chichester: John Wiley, 313 pp.

ROWE, F.W.E. 1988. Review of the extant class Concentricycloidea and reinterpretation of the fossil class Cyclocystoidea. In *Echinoderm Biology* (ed. R.D. Burke, P.V. Mladenov, P. Lambert & R.L. Parsley), pp. 3-15. Rotterdam: A.A. Balkema.

RUNNEGAR, B. 1982. A molecular clock date for the origin of the animal phyla. *Lethaia* **15**, 199-205.

RUNNEGAR, B. 1985. Origin and diversification of the Mollusca. In *The Mollusca* (ed. E. R. Trueman and M. R. Clark) pp. 1-57. Orlando: Academic Press.

SANDER, K. 1983. The evolution of patterning mechanisms: gleanings from insect embryogenesis and spermatogenesis. In *Development and Evolution* (ed. B.C. Goodwin, N. Holder, and C.C. Wylie), pp. 137-159. Cambridge: Cambridge University Press.

SEPKOSKI, J.J., JR. 1979. A kinetic model of Phanerozoic taxonomic diversity. II. Early Phanerozoic families and multiple equilibria. *Paleobiology* **5**, 222-251.

SMITH, A.B. 1984. *Echinoid Palaeobiology*. London: George Allen & Unwin, 190 pp.

SMITH, A.B. 1988. To group or not to group: the taxonomic position of *Xyloplax*. In *Echinoderm Biology*, (ed. R.D. Burke, P.V. Mladenov, P. Lambert, and R.L. Parsley), pp. 17-23. Rotterdam: A.A. Balkema.

STANLEY, S.M. 1979. *Macroevolution*. San Francisco: Freeman.

STRATHMANN, R.R. 1975. Limitations on diversity of forms: Branching of ambulacral systems of echinoderms. *American Naturalist* **109**, 177-190.

STRATHMANN, R.R. 1978. Progressive vacating of adaptive types during the Phanerozoic. *Evolution* **32**, 907-914.

STRATHMANN, R.R. 1988. Larvae, phylogeny, and von Baer's law. In *Echinoderm Phylogeny and Evolutionary Biology* (ed. C.R.C. Paul and A.B. Smith), pp. 53-68. Oxford: Clarendon.

STRATHMANN, R.R., & M. SLATKIN. 1983. The improbability of animal phyla with few species. *Paleobiology* **9**, 97-106.

VALENTINE, J.W. 1973. *Evolutionary Palaeoecology of the Marine Biosphere*. Englewood Cliffs, New Jersey: Prentice-Hall.

VALENTINE, J.W., ed. 1985. *Phanerozoic Diversity Patterns*. Princeton: Princeton Univ. Press.

VALENTINE, J.W. 1986. The Permian-Triassic extinction event and invertebrate developmental modes. *Bulletin of Marine Science* 39, 607-615.

VALENTINE, J.W., & D.H. ERWIN. 1987. Interpreting great developmental experiments: the fossil record. In *Development as an Evolutionary Process*, (ed. R.A. Raff and E.C. Raff), pp. 71-107. New York: Liss.

VERMEIJ, G.J. 1987. *Evolution and Escalation*. Princeton: Princeton University Press.

VERDONK, N.H., & J.A.M. VAN DEN BIGGELAAR. 1983. Early development and the formation of the germ layers. *The Mollusca* **3**, 91-122.

Problematic taxa: a problem for biology or biologists?

S. Conway Morris[1]

Abstract

Problematic metazoan taxa are receiving renewed attention. Here I draw attention to their wider role in some aspects of evolutionary biology, especially their importance in delimiting the range of morphological design and their possible use in establishing whether functional superiority is an important factor in survival during geological time.

Introduction

The twin observations, that metazoans must share a common ancestor (e.g. Salvini-Plawen, 1978; Runnegar, 1985) and that descent of component species was by an orderly genealogy (Darwin, 1859), implies in one sense that no metazoan taxon can ever be regarded as strictly problematic. This judgement, however, is relative given that the taxonomic hierarchy conceals a series of levels of ignorance, albeit codified from those of minor uncertainty (genera) to almost complete indecision (phyla). That the hierarchy is an artificial and imperfect device for expressing taxonomic relationships is apparent on at least two counts. First, consider the debates of how best to allocate a group to the most appropriate taxonomic level. For example, are the Brachiopoda, Ectoprocta and Phoronida (and according to some workers also Entoprocta, see Nielsen, 1977, 1985) separate phyla, or classes within the phylum Lophophorata (Emig, 1977, 1985), or is it even worth reducing this group to class status by incorporating other major taxa? The same conceptual problems may be identified within the Arthropoda, (e.g. Whittington, 1979; Patterson, 1978), and here too the kernel of the problem revolves around whether the few shared characteristics arose by convergence or from a common ancestor. If the latter then the phylum Arthropoda may be treated as the largest natural group capable of sensible recognition, but it is equally the most profound statement of taxonomic ignorance because at this point the hierarchy is unable to indicate clearly yet wider relationships, be they to annelids, molluscs, even kinorhynchs (Moritz & Storch, 1972), let alone priapulid worms (Dzik & Krumbiegel, 1989).

Second, witness the turmoil caused to orthodox taxonomy when an intermediate form, say a taxon that clearly shares the characters of two major taxa, is discovered (e.g. Parsley & Mintz, 1975: but see Paul & Smith, 1984, p.468). The continuous process of taxonomic adjustment, a concept rarely grasped by nonbiologists who often appear to regard the hierarchy as immutable, may and should allow incorporation of such discoveries. Such problems, however, become more evident from the study of major groups undergoing substantial adaptive radiations. In many such cases those taxa that are more or less convincingly attributable to extant groups, thereby earning the sobriquet of ancestors, receive disproportionate attention in comparison with the welter of "difficult" or "problematic" taxa. These latter taxa, to a later observer, typically are seen to possess a puzzling mosaic of characters that defy straightforward classification. This problem only arises because the taxonomic hierarchy is largely a product of hindsight. To the observer of a group in the throes of an adaptive radiation, however, no such luxury of allocation within a preexisting hierarchy is either possible or, more importantly, desirable. It is here that the central importance of so-called problematic taxa lies: by their proper documentation we can assess more reliably both the range of possible morphologies and ecologies available to a clade, and also seek to answer the central problem of efficiency of biological design and its susceptibility to extinction.

Cladistic methodology offers one escape from this system in as much as character acquisition in the first instance is assumed to be orderly and serial, although this is not to belittle the problems of convergence and secondary loss of characters. Perhaps in the initial stages of a radiation, when the number of species is small, the opportunities for these complexities are correspondingly diminished. A cladogram also offers via its nested set of branches the option of superimposing a taxonomic hierarchy that can be compared with that obtained by other methods. It is widely appreciated, however, that hierarchies so generated by cladistic analysis tend to be exceptionally cumbersome because of the large number of levels normally generated. Nevertheless, cladistic methods offer a more objective approach to understanding problematic taxa, although it will be obvious that this methodology can offer no superiority when the taxon in question has no significant features in common with any other taxon. But this must be a question of relative ignorance: in fossils most often this is a reflection of incomplete preservation, in Recent animals it is due to insufficient knowledge of key areas such as early ontogeny or molecular sequences. Given the latter features are either difficult or more often impossible to identify in fossils, so we must expect palaeontologists to bear the brunt of the study of the Problematica.

(a) Fossil problematica

(i) Introduction

The history of early metazoan evolution has been reviewed elsewhere (e.g. Conway Morris, 1989a,b, 1990a), where emphasis has been laid on the problematic status of many of the Ediacaran (ca. 560-620 Ma) and Cambrian (ca. 550-500 Ma) fossils. Ediacaran taxa

[1]Department of Earth Sciences, University of Cambridge, Downing Street, Cambridge CB2 3EQ, United Kingdom.

are almost entirely soft-bodied, whereas the Cambrian biotas are best known from skeletal remains following the onset of skeletogenesis close to the Precambrian-Cambrian boundary (Conway Morris, 1987). Many of the early skeletal assemblages are noteworthy for the abundance and diversity of problematic fossils. However, the geographically widespread and stratigraphically long-ranging soft-bodied faunas of the Burgess Shale type (Conway Morris, 1989b,c) have contributed substantially to this area, as well as altering radically our broader notions of Cambrian evolution and palaeoecology. While Ediacaran and Cambrian faunas have played an important role in recent discussions of problematica, the early Phanerozoic does not hold a complete monopoly of this area and a number of younger fossils continue to present major taxonomic conundrums.

(ii) Ediacaran fossils

Concerning Ediacaran fossils there lies a spectrum of opinion, from those who would accommodate most, if not all, taxa in known groups (e.g. Glaessner & Wade, 1966; Glaessner, 1984), to workers who regard only some as so attributable and others as only comparable in terms of broad organizational grade (e.g. Fedonkin, 1981, 1983, 1986, 1987), to ultimately palaeontologists who argue that even a position in the Metazoa is not secure (Seilacher, 1984, 1989). This is not the place to review the merits of the competing schemes in any detail. The consensus of opinion regards Ediacaran faunas as dominated by cnidarians, but there seems to be a gradation from apparently bona-fide examples to ones that still dwell on the edges of, if not within, taxonomic limbo. For example, a number of the medusiforms are placed either in the Scyphozoa or their purported near relatives, the Conulata (see also Babcock and van Iten in this volume). The characteristic preservation of Ediacaran fossils as casts and moulds in arenaceous lithologies does not lend itself readily to the preservation of diagnostic organs (if ever present), but it must be admitted also that some efforts to reconcile the preserved features of the fossils with Recent scyphozoan anatomy seem forced. Even granted that some Ediacaran medusiforms are genuine scyphozoans, a number of others lack sufficient characters for anything more than tenuous assignment. Other medusiforms, however, seem to be emerging as major monophyletic groups. One of these might be the triradially symmetrical trilobozoans (Fedonkin, 1981, 1983, 1986, 1987) in which the disc-like body houses a threefold arrangement of pouches or cavities, apparently sometimes connected to a system of internal canals.

Similar problems attend the pennatulacean organisms. Forms like *Charniodiscus*, with bulbous holdfast and elongate stalk bearing foliate extensions with lateral divisions approach Recent pennatulaceans fairly closely (Jenkins & Gehling, 1978). Even here, however, there is not exact correspondence, most particularly in the lack of separation between the lateral branches in the Ediacaran example. Other forms, such as *Charnia*, depart quite substantially from any pennatulacean, but still possess the attachment stalk and holdfast. Yet more enigmatic are bag-like foliate organisms such as *Pteridinium, Inkrylovia* and *Rangea*. Whether these represent a single group, or a polyphyletic assemblage is still uncertain, as is their nearness to either pennatulaceans or any other type of cnidarian.

The putative Ediacaran triploblasts are also a conundrum in terms of convincing comparisons to extant groups. Fedonkin (1981, 1983, 1986, 1987) has proposed that a number of transversely segmented worms should be regarded as platyhelminthes, but this remains conjectural. Indeed apart from some possible trace fossils (Alessandrello *et al.*, 1988) and sub-fossil occurrences (Conway Morris, 1985) the fossil record of platyhelminthes is non-existent. Many other segmented taxa are placed in the Arthropoda. Perhaps the most convincing example is *Spriggina* (and perhaps *Marywadea*), although an annelid affinity still finds favour in some quarters (Glaessner, 1984). In support of the formal proposal is the headshield-like structure and the complex segmented body. Other putative arthropods, such as *Vendia, Praecambridium* and *Parvancorina*, consist of relatively large shields with either segmental units (the first two genera) or supposed appendages (*Parvancorina*), although their arrangement and angular relationship in this last taxon make it difficult to reconcile them easily with an arthropodan arrangement. In any event, even if some or all of these taxa are arthropods, it is not easy to place them in a phylogenetic context that would lead to Cambrian representatives, other than trying to regard them in some sense generalized ancestors.

(iii) Cambrian fossils

Although the early skeletal faunas contain representatives of several well-known groups, e.g. brachiopods, echinoderms, molluscs, sponges, and probably chaetognaths (as protoconodonts), there is also a wide variety of animals with a more enigmatic status. Recently, there has been extensive interest in an informal category, referred to as small skeletal fossils, most of which are millimetric in size and typically recoverable by acid digestion of carbonates on account of being either composed of primary phosphate or showing secondary diagenetic replacement of original calcite and aragonite. Continuing research, especially on faunas from China, South Australia and the Soviet Union, is demonstrating several major groups (Qian & Bengtson 1989; Bengtson *et al.*, 1990; Conway Morris, 1989d). These include the coeloscleritophorans, cambroclaves, tommotiids and anabaritids, all of whose wider relationships remain problematic unless the triradial symmetry of anabaritid tubes indicates a relationship with the Ediacaran trilobozoans. Other small skeletal fossils, such as *Fomitchella* (Bengtson, 1983), *Hadimopanella* (Peel & Larsen, 1984, but see Hinz *et al.*, 1990) and *Tumulduria* (Bengtson *et al.*, 1987) remain enigmatic. A recurrent problem with interpretation of many of these fossils is that they represent dispersed scales, spicules or sclerites derived from the post-mortem disintegration of complex aggregates or scleritomes. Fused assemblages and recognition of articulatory facets of points of juxtaposition help to constrain former arrangements, but given the skeletal distinctiveness of these forms, the key lies in the recognition of intact assemblages in soft-bodied faunas. Unfortunately, the taphonomic circumstances that favour the latter are typically in deeper waters than those where the majority of small skeletal taxa flourished. However, recognition of articulated specimens of various coeloscleritophorans (halkieriids, wiwaxiids, chancelloriids) and *Microdictyon* in association with soft-tissue, representing either an entire animal (Chen *et al.*, 1989) or possibly an appendage, are of considerable value. In the case of *Wiwaxia*, for example, the sclerite arrangement allows assignment of isolated sclerites to particular areas of the body, while the recognition of soft-body organs such as a possible radula suggests a possible affinity with the Mollusca (Bengtson & Conway Morris, 1984). However, if the Coeloscleritophora is a monophyletic clade, as is deduced from similarities in sclerite growth and composition, then it is clear that its morphological range was extensive. Thus, articulated series of *Chancelloria* show the rosettes to have been studded over the body-wall of a sessile, vase-like organism (Walcott, 1920; S.B. Bengtson, unpublished observations) strikingly different from the vagrant wiwaxiids and halkieriids.

Other Cambrian skeletal groups whose affinities remain problematic include the coniform Agmata (Yochelson, 1977; Fritz & Yochelson, 1988) and the archaeocyathids. The former have a characteristic infill of apically directed laminae about a central lumen, and may have an outer calcareous wall (*Salterella*). Attempts at reconstructing soft-tissue distributions remain unconvincing (e.g. Glaessner, 1976, fig. 1A), and it has not been established beyond doubt as to whether each cone represents a single organism. The calcareous archaeocyathids have had a chequered phylogenetic career, with suggestions that they were sponges (Debrenne & Vacelet, 1984), algae (Sepkoski, 1979) or members of an extinct phylum (Zhuravleva & Myagkova, 1972) all being proposed. Given that this assemblage is probably polyphyletic, it would be unwise to suggest a single assignment. Some forms, especially amongst the so-called irregular archaeocyathids, are assignable to sponges (R.A. Wood, pers. comm.), while some of the radiocyathids (Nitecki & Debrenne, 1979; Zhuravlev & Nitecki, 1985; see also Zhuravlev, 1986) could be interpreted as algae comparable to certain dasycladaceans. However, at the moment there remains a large residue of archaeocyathids whose affinities remain quite problematical.

The role of problematical taxa in Cambrian faunas has come into sharpest focus, however, amongst the soft-bodied faunas of the Burgess Shale-type, of which the most prolific source and original locality is the Phyllopod bed of the Burgess Shale, an informal stratigraphic unit in the "thick" Stephen Formation of Middle Cambrian age, British Columbia. The twenty or so taxa which defy assignment to known groups have been reviewed elsewhere (Conway Morris, 1989) and require no reiteration here. It is worth emphasizing, however, that recent discoveries at more than 30 localities ranging through much of the Lower and Middle Cambrian show that the Burgess Shale-type fauna has a recognisable integrity (Conway Morris 1989c), and that problematic taxa are an integral part of its make-up. Some examples of the latter include *Anomalocaris* from the Zawisyn Formation (Lower Cambrian) of Poland and slightly younger Qiongzhusi Formation of China (Hou &Bergström, this volume) and 10 other localities extending into the *Ptychagnostus punctuosus* Interval Range Zone of the Middle Cambrian, *Microdictyon* and *Dinomischus* from the Qiongzhusi Formation (Lower Cambrian) of South China (Chen *et al.*, 1989a,b), *Eldonia* from the Spence Shale and Marjum Formation (Middle Cambrian) of Utah, and *Wiwaxia*, also from the Spence Shale (Conway Morris & Robison, 1988).

These Burgess Shale-type problematica are remarkable for both the range of morphologies, and in general their distinctiveness from known groups. Segmented taxa such as *Anomalocaris* and *Opabinia* recall arthropodan arrangements, but notwithstanding a proposal by Bergström (1986) that they be included in the Arthropoda, other workers remain more impressed by the differences (Whittington & Briggs, 1985; Briggs & Whittington, 1987). *Dinomischus* has a distant similarity to the Entoprocta, but the similar anatomy of elongate stalk and calyx housing a recurved gut is probably a result of convergence. Amongst the forms *Banffia, Amiskwia* and *Odontogriphus* all are sufficiently strange to make assignment to any known group highly problematic, while *Hallucigenia* is so peculiar that even if it represents only a fragment of a larger organism, it is difficult to see any significant points of similarity to other taxa.

(iv) Post-Cambrian fossils

Amongst younger soft-bodied Palaeozoic Lagerstätten there are a number of reports of problematic forms, although few, if any, seem to be directly comparable to those of the Burgess Shale type. The three principal deposits of interest are the Hunsrückschiefer (Emsian, Devonian) of West Germany, the Bear Gulch Limestone (Chesterian, Mississippian) of Montana, U.S.A., and the Mazon Creek siderite nodules (Westphalian, Pennsylvanian) of Illinois, U.S.A. The pyritized faunas of the Hunsrückschiefer appear to contain several problematical forms (pers. comm. W. Stürmer), although none have been described.

The Bear Gulch Limestone has yielded one of the most extraordinary of the problematica in the form of *Typhloesus wellsi* (Conway Morris, 1990b), formerly referred to as "conodontochordates" on the mistaken twin suppositions that the animal was a chordate that bore conodonts as an in situ feeding apparatus. A detailed re-assessment of its anatomy, which includes a voluminous but blind gut, a remarkable posterior fin with a unique type of fin ray bracing, and an enigmatic paired discoidal organ (the so-called ferrodiscus), all support the extraordinary status of this pelagic animal that hunted or scavenged conodonts, worms and fish (Conway Morris, 1990b). The anatomy is so peculiar that its nearest relatives could be sought amongst a number of phyla, from which *Typhloesus* could be derived by the same number of arbitrary steps. No doubt if more data on its anatomy or other aspects, e.g. biochemistry, were available then speculation on its affinities could be better constrained. At the moment, however, its remarkable anatomy renders it as one of the principal Palaeozoic problematica. Much of the remainder of the Bear Gulch Limestone fauna is readily assignable to known groups, but amongst the worms there is at least one problematic form. This species (*Soris labiosus*) is not well known on account of relatively indifferent preservation, but the smooth vermiform body appears to lack chaetae and at its anterior end bears a paired jaw (Schram, 1979).

The final major repository of Palaeozoic problematica, the Mazon Creek deposit, has yielded several distinctive forms. *Escumasia roryi* has a saccate body bearing two tentacles and supported on a peduncle (Nitecki & Solem, 1973). Little is known of its internal anatomy, although the mouth was presumably located between the tentacles; possibly an anus opened on the side of the body. As with *Dinomischus* it vaguely approaches a number of other phyla, but in the absence of critical anatomical detail further speculation is difficult. *Etacystis communis* has an even more remarkable morphology, with a bladder-like unit and papillate extension arising from a stolon (Nitecki & Schram, 1976). Proposed comparisons with hemichordates are tenuous, but the possibility that *E. communis* is part of a larger ?colonial organism may constrain discussion. The most celebrated of the problematica, however, is *Tullimonstrum*, which with *Typhloesus* and *Hallucigenia*, is widely regarded as typifying Palaeozoic problematica. In their original description Johnson & Richardson (1969) discussed most of the salient features of this taxon, and the difficulty in finding a wider taxonomic home. Since the proposals that *Tullimonstrum* be assigned to the heteropod gastropods (Foster, 1979) or compared more closely with the conodonts or molluscs (Beall, this volume) have been advanced, but in my opinion our current knowledge of its anatomy make these proposals difficult to substantiate.

Apart from these soft-bodied groups from Palaeozoic Lagerstätten, there are a number of other groups whose affinities remain undetermined. These include a number of isolated occurrences, such as the sclerite-bearing *Dimorphoconus* from the Lower Ordovician of Shropshire (Donovan & Paul, 1985), to which may well be added a number of other vermiform taxa which require detailed reassessment. In addition, tubicolous groups such as *Sphenothallus* (Mason & Yochelson, 1985; Fauchald *et al.*, 1986; Bodenbender *et al.*, 1989) and the conulariids (Babcock and van Iten, this volume) are often regarded as being of problematical status. Finally, there are a number of calcareous organisms, often in the form of sheets and crusts (Babcock, 1986), whose affinities remain uncertain, although comparisons are most often drawn with algae or sponges.

(v) Recent problematica

The above discussion has stressed that the problematical status of many of the Ediacaran and Palaeozoic fossils considered here results from a lack of information. This is not to belittle the morphological distinctiveness of many taxa, but as typically spot occurrences without either obvious ancestors or descendants, so it is to be expected that their phyletic affinities remain uncertain. However, such reasoning is hardly applicable to living taxa, where one might expect that the full arsenal of investigative techniques (e.g. anatomy, biochemistry, genetics) would indicate nearest ancestry. At the moment, however, such is not the case for at least two groups, the largely pelagic Chaetognatha and to a somewhat lesser extent in the meiofaunal Lobocerebratidae. The former has a very sparse fossil record that may include the Cambrian-Ordovician protoconodonts (Szaniawski, 1982) and a possible soft-bodied occurrence in the Carboniferous Mazon Creek nodules (Schram, 1973). Recently Kraft & Mergl (1989) have also described a possible example from the Lower Ordovician of Czechoslovakia, but in my view this assignment is tentative. Concerning Recent chaetognaths, a broad consensus regards them as deuterostomous, but they provide a morphological analogy with the Bear Gulch Limestone *Typhloesus* where the distinctiveness of the chaetognath body-plan, which being effectively invariant isolates them from all other phyla. Much the same comments are applicable to the Lobocerebratidae, and Rieger *et al.* (this volume) review the possible annelidan affinities of this group but also speculate whether inclusion in a new phylum is a more sensible alternative. Yet other Recent problematica are addressed by Rieger *et al.* (this volume). Attention should be drawn also to work by Riser (1988, 1989) on nemerteans that appear to lack the proboscis, and might represent a primitive condition that requires a rethinking of the concept of this phylum. That other problematical taxa await to engage the attention of zoologists will also be apparent from continuing study of deep-sea faunas (Thiel & Schriever 1989).

Discussion

The relative abundance of problematic taxa in the early Phanerozoic, especially during the Ediacaran and Cambrian intervals, is of considerable significance in understanding the style and tempo of early metazoan evolution. Anthromorphologically it is customary to designate these bizarre animals as "experiments", that arose because of the ecological opportunities. Such a notion would seem to be readily applicable to the earliest stages of the filling of the "ecological barrel", especially if the Cambrian radiations were in part a response to ecological vacancies occasioned by a mass extinction of the pre-existing Ediacaran faunas (Conway Morris, 1989a). Even the Carboniferous problematica might represent the adaptive opportunities grasped by organisms able to survive in the eurytopic environments of the Bear Gulch Limestone and Mazon Creek deposits, although in both cases taphonomic bias in favour of soft-part preservation in such environments (lithographic limestone and very early siderite nodule growth) may be of greater significance. In contrast to arguments based on ecological opportunities, an alternative school has proposed that morphological novelty represents either fluidity in the developmental programs of the genome or even lateral transfer of genetic material.

A further point of discussion is whether the rise of problematical groups should be taken as one of the key examples of macroevolution whereby morphological distinctiveness is achieved in only a few steps from ancestral species, or whether their origin fits into microevolutionary hypotheses of speciation. In the context of the Cambrian faunas Valentine (1981) has expressed the belief that given the limited number of species combined with the wide occupation of morphospace (admittedly not precisely defined), saltatory jumps are inevitable to explain the extent of morphological diversification. Much revolves, however, on species recognition and whether the use of the taxonomic hierarchy in hindsight (see above) is appropriate during this initial diversification. It is becoming increasingly clear that notwithstanding the morphological range of Cambrian taxa, as studies progress so evidence for morphological continuity across this spectrum becomes evident. Good examples of this are apparent in the Burgess Shale-type arthropods, and elsewhere amongst Cambrian faunas in the brachiopods and apparently related taxa and the tommotiids.

The last example seems to be particularly instructive in that the end-points of the morphological range of the phosphatic sclerites are strikingly distinct, but as studies of the small skeletal fossil assemblages have progressed so intermediate forms have become evident (Bengtson *et al.*, 1990). It seems, therefore, as the origin of taxa, at this juncture of time potentially classes or even phyla, is no different to the intrapopulational variation from which derive new species. I conclude that the origin of any taxon, even if of problematical status, is by microevolutionary processes, even if it leads to macroevolutionary consequences. At the initial stages of diversification the winnowing out of say 90 per cent of the first wave of species allows the subsequent re-radiation of the survivors, who may owe their persistence to good luck or perhaps competitive superiority.

As this process is repeated, so the chances for recognition of ancestral lineages improves, both because of the increasing sample available as time progresses and the number of steps (speciations) made. As each of the surviving lineages carries with it its ontogenetic "burden" of characters, so assignation to given groups becomes increasingly rare. Thus, it is that problematical forms fade from the fossil record: by the Mesozoic practically every fossil has an ancestor identifiable to usually family or order level.

Problematic taxa, therefore, are regarded here as a problem for biologists, not biology. They are an inevitable consequence of the dynamics of adaptive radiations combined with a poor fossil record. This does not mean, however, that they lack significance in our deliberations. To the contrary, I would suggest that they will be of particular value in addressing the following major questions:

(i) Are there limits to morphospace, and in the initial radiations is its occupation clustered or more or less randomly dispersed?

(ii) Linked to the first point, how important is convergence of form and is it realistic to speak of a finite number of body-plans?

(iii) Is there any directionality in the evolutionary process: can we regard some problematica as adaptations to extreme environments (however visualized), analagous to the bizarre constructional morphologies that typify many deep-sea faunas (Hickman, 1981)?

(iv) Notwithstanding reference to problematical fossils as "experiments", is this anthropomorphic judgement an indication of competitive inferiority, or are the number of surviving taxa so small in comparison with those that go extinct that chance is the controlling factor?

If the last question is answered in favour of chance, then in the furnaces of metazoan creation the forger of the grand designs that have come to dominate our planet may be free to draw almost at random on a near limitless supply of morphological diversity. What seems to us an almost unending cavalcade of organisms, as is evident from both the fossil record and teeming biosphere of today, is probably only a minute fraction of all the possibilities that were filtered through the iron gates of extinction and the subsequent opportunities for diversification.

REFERENCES

ALESSANDRELLO, A., PINNA, G. & TERUZZI, G. 1988. Land planarian locomotion trail from the Lower Permian of Lombardian pre-Alps. *Atti della Società Italiana di Scienze Naturali e del Museo Civico di Storia Naturale, Milano* **129**, 139-45.

BABCOCK, J.A. 1986. The puzzle of alga-like problematica, or rummaging around in the algal wastebasket. In *Problematic Fossil Taxa* (ed. A. Hoffman & M.H. Nitecki), pp. 12-26. New York: Oxford University Press.

BENGTSON, S. 1983. The early history of the Conodonta. *Fossils and Strata* **15**, 5-19.

BENGTSON, S., FEDOROV, A.S., MISSARZHEVSKY, V.V., ROZANOV, A.YU, ZHEGALLO, E.A. & ZHURAVLEV, A.YU. 1987. *Tumulduria incomperta* and the case for Tommotian trilobites. *Lethaia* 20, 361-70.

BENGTSON, S., CONWAY MORRIS, S., COOPER, B.J., JELL, P. & RUNNEGAR, B. 1990. Early Cambrian skeletal fossils from South Australia. *Memoir of the Association of Australasian Palaeontologists* **9**, 1-364.

BERGSTRÖM, J. 1986. *Opabinia* and *Anomalocaris,* unique Cambrian "arthropods". *Lethaia* **19**, 241-46.

BODENBERDER, B.E., WILSON, M.A. & PALMER, T.J. 1989. Paleoecology of *Sphenothallus* on an Upper Ordovician hardground. *Lethaia* **22**, 217-25.

BRIGGS, D.E.G. & WHITTINGTON, H.B. 1987. The affinities of the Cambrian animals *Anomalocaris* and *Opabinia. Lethaia* **20**, 185- 186.

CHEN JUN-YUAN, HOU XIAN-GUANG, & LU HAO-ZHI. 1989a. Early Cambrian netted scale-bearing worm-like sea animal. *Acta Palaeontologica Sinica* **28**, 1-16.

CHEN JUN-YUAN, HOU XIAN-GUANG, & LU HAO-ZHI. 1986b. Early Cambrian hock glass-like rare sea animal *Dinomischus* (Entoprocta) and its ecological features. *Acta Palaeontologica Sinica* **28**, 58-71.

CONWAY MORRIS, S. 1985. Non-skeletalized lower invertebrates: a review. In *The Origins and Relationships of Lower Invertebrates* (ed. S. Conway Morris, J.D. George, R. Gibson & H.M. Platt), pp. 343-59. Oxford: Clarendon.

CONWAY MORRIS, S. 1987. The search for the Precambrian-Cambrian boundary. *American Scientist* **75**, 156-67.

CONWAY MORRIS, S. 1989a. Early metazoans. *Science Progress, Oxford* **73**, 81-99.

CONWAY MORRIS, S. 1989b. Burgess Shale faunas and the Cambrian explosion. *Science, Washington* **246**, 339-346.

CONWAY MORRIS, S. 1989c. The persistence of Burgess Shale-type faunas: implications for the evolution of deeper-water faunas. *Transactions of the Royal Society of Edinburgh: Earth Sciences* **80**, 271-283.

CONWAY MORRIS, S. 1989d. Lower Cambrian anabaritids from South China. *Geological Magazine* **126**, 615-632.

CONWAY MORRIS, S. 1990a. Late Precambrian and Cambrian soft-bodied faunas. *Annual Review of Earth and Planetary Sciences* **18**, 101-122.

CONWAY MORRIS, S. 1990b. *Typhloesus wellsi* (Melton & Scott 1973), a bizarre metazoan from the Carboniferous of Montana, USA. *Philosphical Transactions of the Royal Society of London,* **B 327**, 595-624.

CONWAY MORRIS, S. & ROBISON, R.A. 1988. More soft-bodied animals and algae from the Middle Cambrian of Utah and British Columbia. *The University of Kansas Paleontological Contributions, Paper* **122**, 1-48.

DARWIN, C. 1859. *On the Origin of Species by means of Natural Selection or the Preservation of Favoured Races in the Struggle for Life.* London: Murray.

DEBRENNE, F. & VACELET, J. 1984. Archaeocyatha: is the sponge model consistent with their structural organization? *Palaeontographica Americana* **54**, 358-69.

DONOVAN, S. & PAUL, C.R.C. 1985. A new possible armoured worm from the Tremadoc of Sheinton, Shropshire. *Proceedings of the Geologists' Association* **96**, 87-91.

DZIK, J. & KRUMBIEGEL, G. 1989. The oldest "onychophoran" *Xenusion*: a link connecting phyla? *Lethaia* **22**, 169-82.

EMIG, C.C. 1977. Un nouvel embranchement: les lophophorates. *Bulletin de la Societé Zoologique de France* **102**, 341-44.

FAUCHALD, K., STÜRMER, W. & YOCHELSON, E.L. 1986. *Sphenothallus* "vermes" in the early Devonian Hunsrück Slate, West Germany. *Paläontologische Zeitschrift* **60**, 57-64.

FAUCHALD, K., STÜRMER, W. & YOCHELSON, E.L. 1988. Two worm-like organisms from the Hunsrück Slate (Lower Devonian), southern Germany. *Paläontologische Zeitschrift* **62**, 205-12.

FEDONKIN, M.A. 1981. White Sea biota of Vendian (Precambrian non skeletal fauna of the Russian Platform North). *Trudy Akademii Nauk* **342**, 1-100.

FEDONKIN, M.A. 1983. Organic world of the Vendian. Stratigraphy, palaeontology. *Itogi Nauki i Tekhniki* (VINITI) **12**, 1-128.

FEDONKIN, M.A. 1986. Precambrian problematic animals: their body plan and phylogeny. In *Problematic Fossil Taxa* (ed. A. Hoffman & M.H. Nitecki), pp.59-67. New York: Oxford University Press.

FEDONKIN, M.A. 1987. Soft-bodied Vendian fauna and their position in metazoan evolution. *Trudy Paleontologischii Institut Akademii Nauk* **226**, 1-176.

FOSTER, M.W. 1979. A reappraisal of *Tullimonstrum gregarium*. In *Mazon Creek Fossils* (ed. M.H. Nitecki), pp. 269-301. New York: Academic Press.

FRITZ, W.H. & YOCHELSON, E.L. 1988. The status of *Salterella* as a Lower Cambrian index fossil. *Canadian Journal of Earth Sciences* **25**, 403-16.

GLAESSNER, M.F. 1976. Early Phanerozoic worms and their geological and biological significance. *Journal of the Geological Society, London* **132**, 259-75.

GLAESSNER, M.F. 1984. *The Dawn of Animal Life. A Biohistorical Study*. Cambridge: University Press.

GLAESSNER, M.F. & WADE, M. 1966. The late Precambrian fossils from Ediacara, South Australia. *Palaeontology* **9**, 599-628.

HICKMAN, C.S. 1981. Giants, dwarfs, and bizarre constructional patterns in the marine fossil record. *Geological Society of America, Abstract with Programs* **13**, 473.

HINZ, I., KRAFT, P., MERGL, M. & MÜLLER, K.J. 1990. The problematic *Hadimopanella, Kaimenella, Milaculum* and *Uthaphospha* identified as sclerites of Palaeoscolecida. *Lethaia,* **23**, 217-221.

JENKINS, R.J.F. & GEHLING, J.G. 1978. A review of the frond-like fossils of the Ediacara assemblage. *Records of the South Australian Museum* **17**, 347-59.

JOHNSON, R.G. & RICHARDSON, E.S. 1969. Pennsylvanian invertebrates of the Mazon Creek area, Illinois: the morphology and affinities of *Tullimonstrum. Fieldiana, Geology* **12**, 119-49.

KRAFT, P. & MERGL, M. 1989. Worm-like fossils (Palaeoscolecida: ?Chaetognatha) from the Lower Ordovician of Bohemia. *Sbornik Geologickych ved Paleontologie* **30**, 9-36.

MASON, C. & YOCHELSON, E.L. 1985. Some tubular fossils (*Sphenothallus*: "Vermes") from the Middle and Late Paleozoic of the United States. *Journal of Paleontology* **59**, 85-95.

MORITZ, K. & STORCH, R. 1972. Zur Feinstruktur der integumentes von *Trachydemus gigantus* Zelinka. *Zeitschrift für Morphologie (und Okologie) der Tiere* **71**, 189-202.

NIELSEN, C. 1977. Phylogenetic considerations: the protostomian relationships. In *Biology of Bryozoans* (eds. R.M. Woollacott & R.L. Zimmer), pp. 519-34. London: Academic Press.

NIELSEN, C. 1985. Animal phylogeny in the light of trochaea theory. *Biological Journal of the Linnean Society* **25**, 243-99.

NITECKI, M.H. & DEBRENNE, F. 1979. The nature of radiocyathids and their relationship to receptaculids and archaeocyathids. *Geobios* **12**, 5-27.

NITECKI, M.H. & SCHRAM, F.R. 1976. *Etacystis communis*, a fossil of uncertain affinities from the Mazon Creek fauna (Pennsylvanian of Illinois). *Journal of Paleontology* **50**, 1157-61.

NITECKI, M.H. & SOLEM, A. 1973. A problematic organism from the Mazon Creek (Pennsylvanian) of Illinois. *Journal of Paleontology* **47**, 903-7.

PARSLEY, R.L. & MINTZ, L.W. 1975. North American Paracrinoidea: (Ordovician: Paracrinozoa, new, Echinodermata). *Bulletins of American Paleontology* **68**, 1-115.

PAUL, C.R.C. & SMITH, A.B. 1984. The early radiation and phylogeny of echinoderms. *Biological Reviews* **59**, 443-81.

PATTERSON, C. 1978. Arthropods and ancestors. *Antenna; London* **2**, 99-103.

PEEL, J.S. & LARSEN, N.H. 1984. *Hadimopanella apicata* from the Lower Cambrian of western North Greenland. *Rapport Grønlands Geologiske Undersøgelse* **121**, 89-96.

QIAN YI & BENGTSON, S. 1989. Palaeontology and biostratigraphy of the early Cambrian Meishucunian Stage in Yunnan Province, South China. *Fossils and Strata* **24**, 1-156.

RISER, N.W. 1988. *Arhynchonemertes axi* gen. n., sp.n. (Nemertinea) -an insight into basic acoelomate bilaterian organology. In *Free-living and Symbiotic Plathelminthes* (ed. P. Ax, U. Ehlers & B. Sopott-Ehlers), pp. 367-373. Stuttgart: Gustav Fischer.

RISER, N.W. 1989. Speciation and time-relationships of the nemertines to the acoelomate metazoan Bilateria. *Bulletin of Marine Science* **45**, 531-38.

RUNNEGAR, B. 1985. Collagen gene construction and evolution. *Journal of Molecular Evolution* **22**, 141-49.

SALVINI-PLAWEN, L. VON. 1978. On the origin and evolution of the lower Metazoa. *Zeitschrift fur Zoologische Systematik und Evolutionsforschung* **16**, 40-88.

SCHRAM, F.R. 1973. Pseudocoelomates and a nemertine from the Illinois Pennsylvanian. *Journal of Paleontology* **47**, 985-89.

SCHRAM, F.R. 1979. Worms of the Mississippian Bear Gulch Limestone of central Montana, USA. *Transactions of the San Diego Society of Natural History* **19**, 107-120.

SEILACHER, A. 1984. Late Precambrian and early Cambrian Metazoa: Preservational or real extinctions? In *Patterns of Change in Earth Evolution* (eds. H.O.Holland & A.F. Trendall), pp. 159-68. Berlin: Springer.

SEILACHER, A. 1989. Vendozoa: Organismic construction in the Proterozoic biosphere. *Lethaia* **22**, 229-39.

SEPKOSKI, J.J. 1979. A kinetic model of Phanerozoic taxonomic diversity II. Early Phanerozoic families and multiple equilibria. *Paleobiology* **5**, 222-251.

SZANIAWSKI, H. 1982. Chaetognath grasping spines recognized among Cambrian protoconodonts. *Journal of Paleontology* **56**, 806-810.

THIEL, H. & SCHRIEVER, G. 1989. The DISCOL enigmatic species: a deep-sea pedipalp? *Senckenbergiana Maritima* **20**, 171-75.

VALENTINE, J.W. 1981. Emergence and radiation of multicellular organism. In *Life in the Universe* (ed. J. Billingham), pp. 229-57. Cambridge, Mass.: MIT Press.

WALCOTT, C.D. 1920. Middle Cambrian Spongiae. *Smithsonian Miscellaneous Collections* **67**, 261-364.

WHITTINGTON, H.B. 1979. Early arthropods, their appendages and relationships. In *The Origin of Major Invertebrate Groups* (ed. M.R. House), pp. 253-68. London: Academic Press.

WHITTINGTON, H.B. & BRIGGS, D.E.G. 1985. The largest Cambrian animal, *Anomalocaris*, Burgess Shale, British Columbia. *Philosophical Transactions of the Royal Society of London* **B309**, 569-609.

YOCHELSON, E.L. 1977. Agmata, a proposed extinct phylum of early Cambrian age. *Journal of Paleontology* **51**, 437-454.

ZHURAVLEV, A. YU. 1986. Radiocyathids. In *Problematic Fossil Taxa* (eds. A. Hoffman & M.H. Nitecki), pp. 35-44. New York: Oxford University Press.

ZHURAVLEV, A. YU. & NITECKI, M.H. 1985. On the comparative morphology of the archaeocyathids and receptaculitids. *Paleontological Journal* **1985(4)**, 134-36.

ZHURAVLEVA, I.T. & MYAGKOVA, YE.I. 1972. Archaeata: a new group of Paleozoic organisms. *Proceedings of the International Geological Congress* **24**, 7-14.

Metazoan evolution around the Precambrian-Cambrian transition

Jan Bergström[1]

Abstract

It has been said that metazoan phyla were initially so close to each other that recognition of them is largely a result of hindsight. The opposite viewpoint is defended here: coelomate phyla in particular originated through macroevolutionary events and received their typical features immediately. Virtually all of them seem to be derived directly from a basal stock of procoelomate sole-creepers. This helps explain 1) the virtual absence of Vendian body fossils representing the Bilateria, but the presence of Vendian trace fossils of apparent bilaterian origin, 2) the 'Cambrian fossil explosion' 3) the fact that the gaps between phyla were virtually as wide in the Cambrian as they are today, 4) the mixture of protostomian and deuterostomian characters in many phyla, and the difficulty in sorting out the evolutionary relationships between phyla by methods of comparative anatomy. The Bilateria, including acoelomates, aschelminths and coelomates, seem not to share a common origin with the Radiata (cnidarians) or Parazoa (sponges), and there is no place for the Ediacaran quilted organisms in the model. The results indicate that even the oldest problematic animal fossils belonged to phyla which should be distinctly definable. Apart from phyla present today, we may think of a procoelomate phylum representing the stock of slug-like bilaterian animals from which the coelomates ultimately evolved. It is most likely that representatives of this stock still lived in the Cambrian, and that some of these developed epidermal sclerites for protection. Such animals would be very similar to ancestral molluscs and may include certain problematic Palaeozoic sclerite-bearing forms. In summary, the Vendian bilaterians thus appear to have been represented by soft-bodied acoelomates, pseudocoelomates and procoelomates. Roughly at the Vendian-Cambrian transition, the latter gave rise to the coelomates through a rapid adaptive radiation, often including the development of skeletons; these coelomates dominate Phanerozoic faunas.

Introduction

The 'Cambrian explosion' is something we definitely see in the fossil record, but is it a multiplication of organisms or only of fossils? This question has been under debate for a long time. I will here restrict myself to multicellular animals, but protists exhibit similar patterns. The evidence appears conflicting. On the one hand, acritarchs, macrofossils and trace fossils all indicate a massive expansion and modernization during the Precambrian-Cambrian transition interval (e.g. Seilacher, 1956; Conway Morris, 1985; Landing, 1988; Bergström, 1989). On the other hand, metazoan phyla were as distinct from each other in the early Cambrian as they are today, and there are many types of trace fossils in the Vendian. The latter facts have often been interpreted as indicating an extended history in the Proterozoic of soft-bodied representatives of extant phyla. This interpretation has recently received some questionable support from studies on biochemical evolution (Runnegar, 1982a, b). There are two important questions in this connection: were biological and biochemical evolutionary speeds uniform throughout the relevant time interval, and did splits in biochemical evolutionary trees necessarily coincide with separations between anatomically distinguishable phyla?

Two other questions relate to the characters of the body fossils found in the Vendian and early Cambrian. One refers to the large Vendian organisms. How many of them, if any, are true radiates (Cnidaria)? Are the rest a single group of quilted organisms ('vendozoans') unrelated to metazoans, as suggested by Seilacher (1983, 1984, 1985, 1989), or are some of them 'petalonamids', others cnidarians, flatworms, segmented worms, (pro)arthropods and echinoderms, as proposed by other students of these forms? The other question is related to more conventional problematic fossils. Do they belong to clearcut phyla, and, if so, where do they belong? Or are they intermediate forms in a plexus of slowly emerging phyla?

I think that we can begin to discern a pattern in this complex of problems. This is far from saying that we have arrived at a definite solution.

The lack of a defined Vendian-Cambrian boundary and the great difficulties in correlating around this level (e.g. Moczydlowska & Vidal, 1988; Landing, 1988) are a problem but do not affect the observation that the immense faunal changeover was very rapid when seen in a wider perspective.

What is a coelomate phylum?

It is a well known fact that 'coelomate' phyla are so sharply separated from each other, and in several cases have such a mixture of characters, that it appears hopeless to reconstruct an evolutionary tree which can be agreed upon from the characters alone. It has often been thought that the differences have grown with geological time. Levinton (1988) forcefully claims that macroevolution and high rank taxa are terms only to indicate that long time has passed since the rise of a group. Conway Morris (e.g. 1988, p. 79; cf. 1989, p. 345) has phrased this view elegantly: "... the established major clades are graced with defined positions in the taxonomic hierarchy as phyla or classes. Their recognition is largely dependent on hindsight."

Although this seems to be a generally held opinion backed up by certain logical reasoning, there is reason to doubt its correctness. Phyla were as separate in the Cambrian as they are today. Groups such as Cambrian crustaceans, acrotretid brachiopods and monoplacophoran molluscs were in no way closer to each other than are modern representatives of the phyla. Problematic groups are difficult to place, but this is because they are poorly understood and not necessarily because they fill in the gaps, as maintained by Conway Morris (1988, 1989). His belief is not backed

[1]Department of Palaeozoology, Swedish Museum of Natural History Box 50007, S-104 05 Stockholm, Sweden.

up by any arguments. If phyla have not drifted apart since the beginning of the Cambrian, how can we possibly derive them from a common Proterozoic origin by extrapolating evolutionary trends?

I frankly believe that 'coelomate' phyla are something else than the mere taxonomic appreciation of a long existence of a clade. During an attempt to combine results from the fields of anatomy and biochemical sequence analyses, I found that evolutionary trees from cytochrome c and two other compounds gave most surprising results. Nothing could be seen of the supposed protostomian-deuterostomian dichotomy (Bergström, 1986). This conclusion is verified by studies on additional globins (Goodman *et al.*, 1988), 18S rRNA (Fig. 1; Field *et al.*, 1988), and 28S rRNA (Christen *et al.* this volume). By using the cytochrome c tree as a 'Christmas tree' and metazoan characters as 'decoration', I arrived at a new model of metazoan evolution. In this model, protostomian characters were primitive, deuterostomian ones advanced and derived through the loss of rigid protostomian steering of egg cleavage and ontogeny. Like Jägersten (1968), but along a different path, I arrived at the conclusion that creeping on a ciliated ventral sole is the primitive mode of locomotion in bilaterian metazoans. Vendian ichnofossils actually are dominated by surface trails (Runnegar, 1982c; Bergström, 1990), and burrows are common only close to the Cambrian.

This should not be taken as a refutation of Clark's (1964, 1979) ideas on coelomate evolution. Clark rightly pointed out the importance of coelom for burrowing, and he was careful not to state that all coeloms formed in the same way or for the same reason. Still, there seem to be complementary alternatives to Clark's, as non-segmented coelomate worms have now been shown to burrow effectively through direct peristalsis (Elder, 1980; Mettam, 1985).

The new model has considerable bearings on the rise of coelomate phyla and on the 'Cambrian explosion' (Bergström, 1989). The mixture of developmental and anatomical characters in various phyla is most parsimoniously explained if virtually all 'coelomate' phyla are derived directly from a shared ancestral stock with locomotion on a ciliated sole, with spiral egg cleavage, and with a pelagic Trochophoralike larva. This explains the wide distribution of these characters, fully developed or as rudimentary remnants. A coelom developed where needed, as did segmentation (Clark, 1964, 1979). The protostomian characters are plesiomorphic; the loss of any of them automatically meant the introduction of a different, i.e. a deuterostomian character.

The direct derivation from a stock of sole-creepers meant that the origination of a new phylum was associated with a major shift either in the mode of locomotion or in the mode of feeding, or in both. This is appropriately considered as macroevolution because such a shift must have been associated by a major reorganization of the basic anatomy and morphology. New phyla were born in a virtually revolutionary manner, and the new modes of life that were introduced opened completely new niche fields. If the Cambrian explosion was triggered by environmental factors such as for instance a rise of the oxygen content of the atmosphere, it is easy to see how the coelomate metazoans could meet the challenge on broad front and in short time by producing a number of "experimental" phyla from a basal procoelomate stock. It is much more difficult to understand both the strength of the Cambrian explosion and the mixture of characters in coelomate phyla if metazoans evolved successively through phyla (e.g. arthropods from segmented worms), adding characters slowly, as is generally thought to have been the case.

The mode of origination logically leads to a situation in which similarities between coelomate phyla are due to retention of plesiomorphic characters or to attainment of similar adaptations, while dissimilarities are caused by different adaptations. Phyletic relationships between them cannot be recognised from comparisons of anatomic features. Hence Bengtson's characterisation was very appropriate before molecular sequencing led to the present possibilities: 'A phylum is a group of organisms of uncertain taxonomic affinities, that is, a problematic taxon' (Bengtson, 1986, p. 3).

Some general patterns can be identified in the births of phyla; in the past such patterns have often been considered to indicate relationships. One pattern is found in the tentaculate phyla (brachiopods, bryozoans, phoronids), endoprocts, and hemichordates. In all of these, the adults utilise more or less the same mode of feeding as the trochophore larva, and particularly the endoprocts are so simply constructed that they have been likened to trochophoras on a stalk (Meglitsch, 1972, p. 251). Most likely, the birth of each of these phyla was a paedomorphic process, in which reproduction occurred before larval features were abandoned. Functional needs led to the development of oligomery. A second general pattern is found in worms (annelids, echiurids, pogonophorans, sipunculids) which adopted a burrowing mode of life perhaps to avoid carnivores on the surface. The coelom is well developed as an adaptation to burrowing. A third pattern led to the arthropods, segmented animals moving with the aid of appendages. Only the molluscs do not fit into such a pattern: the basic procoelomate mode of life is more or less retained in the classes Polyplacophora, Monoplacophora and Gastropoda. Molluscs provide the exception that proves the rule: in a way it can be said that they are surviving procoelomates, which have not undergone any revolutionary shift in the mode of locomotion or feeding affecting the group as a whole. Still I think that it is practical to retain the Mollusca as a distinct phylum. The retention of plesiomorphic characters explains why molluscs are often thought to be related to flatworms.

There are two things I want to stress in connection with this model. First, I suggest that the radiation was not through macromutations (although macroevolutionary). The paedomorphic shift to ciliary feeding mentioned above, for instance, could have been brought about through simple changes in ontogenetic growth ratios. Second, biochemical evolutionary trees can possibly pinpoint phyletic forkings in the tree, but they cannot mark the position where the mode of life shifted from ciliary creeping to something else. The birth of a phylum must be where we can first recognize the characteristics of the phylum, not where lineages of slug-like procoelomates diverged from one another.

Ultimately I want to define coelomate phyla as the procoelomates and their descendants. Some groups, such as arthropods, molluscs, leeches and endoprocts, have only a rudimentary coelom or lack it altogether.

Other animals

There are two other main groups of bilaterian animals, the acoelomates and the pseudocoelomates. Biochemical evidence shows that these groups split off from the shared origin before the coelomate radiation (Fig. 1). They have a poor fossil record, but members are preserved for instance in the Burgess Shale (Conway Morris, 1977b). Flatworms, nemertines and, probably, gnathostomulids are acoelomates. Pseudocoelomates are the rotifers, gastrotrichs, nematodes, nematomorphs, kinorhynchs, loriciferans, priapulids, and acanthocephalans, in addition possibly also the tardigrades and chaetognaths. The pseudocoelomate groups are commonly regarded as phyla, sometimes united as the superphylum Aschelminthes. It is quite clear, however, that they are not as distinct from each other as are coelomate phyla. Loriciferans, for instance, are obviously closely related to kinorhynchs and priapulids, and acanthocephalans apparently evolved from rotifers (Clement, 1985; Lorenzen, 1985). Such relationships cannot be recognized between coelomate phyla. I therefore regard it as entirely possible that they may be the result of a gradual separation of a basal stock, and it can be practical to distinguish the groups as members of a single phylum or superphylum Aschelminthes.

Bilaterian animals are commonly united with the Radiata to form the Eumetazoa. The two groups are strikingly dissimilar: the Bilateria have a bilateral symmetry, have three cell layers, and are primarily benthic; the Radiata are either radially symmetrical or disymmetrical, have only two cell layers, and appear to be primarily pelagic. Cnidarians are commonly reported from Vendian strata. However, 18S ribosomal RNA sequence appearances (Ghiselin, 1988, table 2) indicate that Bilateria and Radiata are unrelated (Fig. 1). The third main group of multicellular animals is the Parazoa, the sponges. They have unique characteristics, and it is commonly claimed that they are not related to other multicellular animals.

It is interesting that 28S rRNA sequencing has now led to the recognition that ctenophores, cnidarians, placozoans and possibly also sponges belong together in a natural group which appears not to share an origin with bilaterian animals (Christen *et al.*, this volume). This means that there is no longer any basis for speculations on an origin of bilaterians among cnidarians or any other sessile animals (cf. Fedonkin, 1985, 1987).

Vendian quilted organisms - are they metazoans?

For some decades it has been generally accepted that some of the Ediacara type organisms are metazoans (e.g. Fedonkin, 1981, 1985, 1987; Runnegar, 1982c; Palij *et al.*, 1983; Glaessner, 1983, 1984; Jenkins, 1984), although there has been no agreement on the classification that should be adopted. Some forms have been referred to segmented worms or arthropods, others occasionally to echinoderms and flatworms.

However, Seilacher (1983, 1984, 1985, 1989) recently made the revolutionary statement that he regarded both petalonamids and supposed worms and arthropods of the Ediacaran faunas to be immobile, tough-skinned (cf. Norris, 1989) and 'quilted' organisms of a completely unique 'vendozoan' type, unrelated to metazoans. He based this conclusion on the evidence of the type of preservation characteristics of Ediacara animals. There is a contrast between soft-bodied metazoans leaving their surface locomotory trails but no casts of their bodies, and the 'vendozoans' leaving imprints of their leaf-like bodies but no surface locomotory trails (although in fact their surface shrinkage trails were described by Runnegar, 1982c). The preservation of 'vendozoans' is also different from that of at least some Vendian supposed cnidarians; Wade (1968) distinguished between 'resistant' and 'non-resistant' animals.

Seilacher's suggestion makes it important to look more closely at other aspects of the 'vendozoans'. As far back as 1969, in the description of *Vendia sokolovi*, Keller (see Keller in Rozanov *et al.*, 1969) pointed out that the organism was markedly asymmetric. Glaessner & Wade (1971) explained this asymmetry as due to deformation, but Bergström (1981, p. 31) noted that the alternation of 'segments' of the two sides cannot have so arisen. Fedonkin (1985, p. 36) also assures us that 'segments alternate rather than being opposed' in Vendian 'Bilateria' generally. The thread was taken up again recently (Bergström, 1989, 1990). I agree with Fedonkin in his conclusion that not only *Vendia* but also other forms of the supposed arthropods and annelids (*Parvancorina, Praecambridium, Spriggina, Vendomia*) are genuinely asymmetric (Fig.2). In addition, the integument appears not to be separated into segmental sclerites. The probable immobility throws doubt on Runnegar's (1982c) interpretation of the 'worm' *Dickinsonia* as a sediment-eater. It is worth noting that Ford (1958) suggested the petalonamid *Charniodiscus* to be an alga, and that Pflug (1974a) described algal type cell structures from a petalonamid.

The combined strength of the sedimentological, ichnological, histological and morphological arguments, and (although more circumstantial) the poor fit of the 'vendozoans' in the metazoan evolutionary tree, makes Seilacher's interpretation most interesting. The 'vendozoans' are no metazoans, at least not bilaterian animals, but a group of highly problematical organisms, thin and flexible but tough-skinned, often quilted in a segment-like way, benthic, immobile, in some cases stalked. It seems that Pflug (e.g. 1974b) may have been right in suggesting a relationship between petalonamids, as he defined them, and various Vendian organisms which he thought were metazoans.

Supposed cnidarians in the Vendian

Among the Vendian fossils generally regarded as body fossils, supposed cnidarians dominate (e.g. Fedonkin, 1983; Jenkins, 1984). They have been referred to a number of different classes: Hydrozoa, Scyphozoa, Conulata, Trilobozoa, Cyclozoa, Inordozoa and Anthozoa. Seilacher (1984) gave short shrift to these supposed 'medusoids': he thought that some of them are body fossils of unknown affinity, others burrows, some of which were perhaps made by actinians, i.e. cnidarians.

There is probably much sense in Seilacher's suggestion. It may go too far to include all supposed cnidarians in the same pigeon-hole. Seilacher did not do so. The trilobozoans are round impressions with a

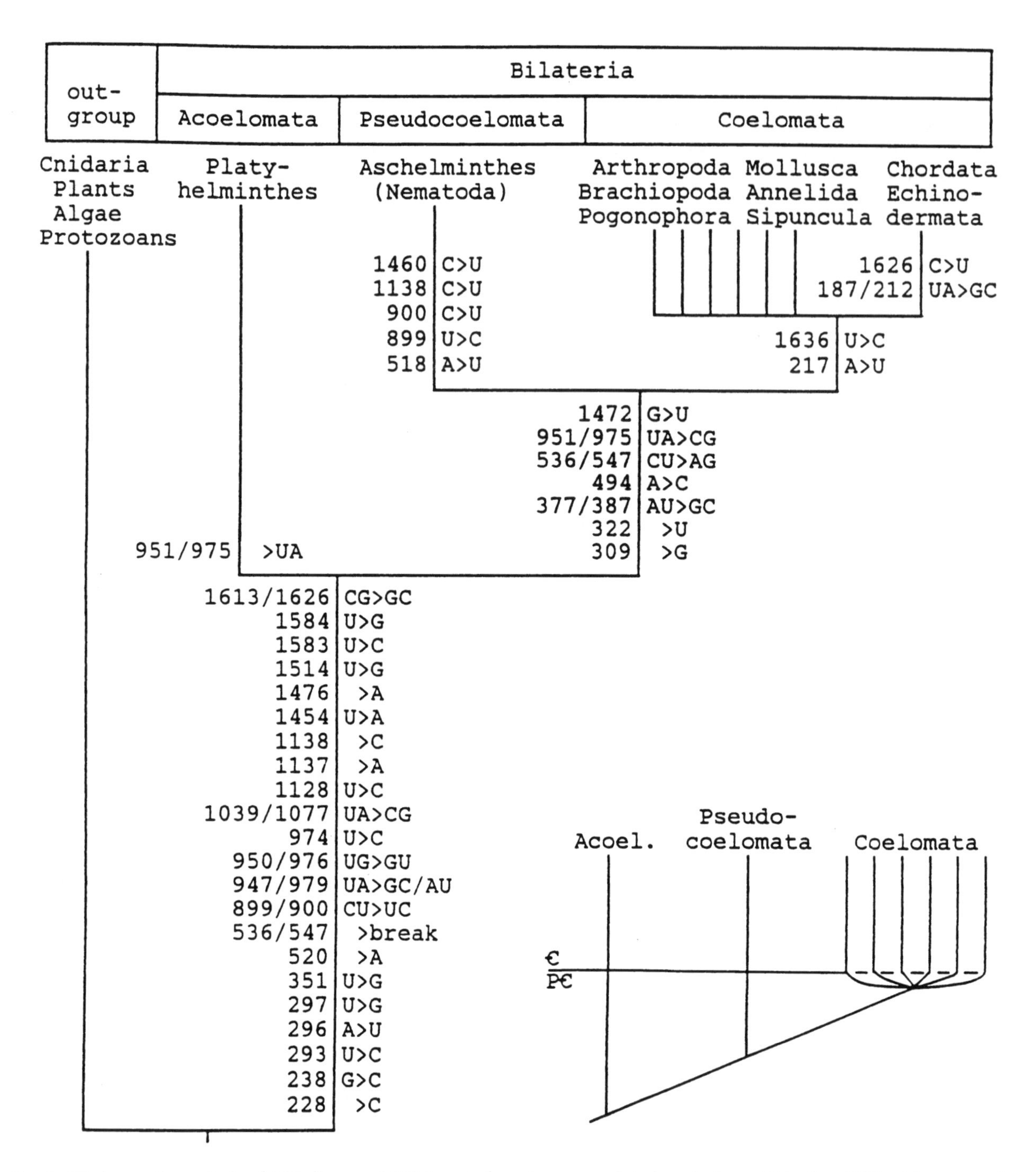

Fig. 1. Phylogenetic analysis of 18S ribosomal RNA mutations (sequence data largely from Ghiselin, 1988). A large number of synapomorphic characters separate bilaterian animals from radiates (cnidarians; the single line leading to outgroups represents an unknown number of lineages). It is notable that the resolution is not high enough to give a branching order among the coelomate phyla (see Field *et al.* 1988, figs 2-5), which indicates that the radiation was of short duration. Only the deuterostomes (represented by chordates and echinoderms) stand out as a distinct group, but the point of departure from protostome coelomates can not be discerned. The figure in the lower right corner shows how an adaptive radiation of the coelomates is interpreted to correspond to the Cambrian "fossil explosion" event; acoelomates, pseudocoelomates and procoelomates represented the Bilateria in the Vendian and left simple types of trace fossils. Code letters for nucleotides: A adenine, C cytosine, G guanine, and U uracil.

triradiate symmetry. They differ from imprints of medusae not only in their symmetry, but also in their distinct high relief. This indicates that the trilobozoans differed considerably from cnidarians in the composition of their tissues, so much so that they may 'be considered as vendozoans' or quilted organisms. Some other medusoids, such as *Armillifera*, may be interpreted in the same way. Hofmann (1988) recently suggested that the supposed chondrophore *Chondroplon* is a deformed *Dickinsonia*. The basal disc-shaped 'anchor' found in the 'vendozoan' *Charnia* has a scar of a shaft on one side, a pattern of radial striae, and the surface can show concentric wrinkles due to compaction (Fig. 2). At least two of these features are seen in medusoids such as *Charniodiscus, Cyclomedusa, Hiemalora, Medusinites, Paliella, Pinegia* and *Tirasiana*. Are these later forms also basal discs? The radiating striae have no counterpart in younger fossil medusae, but may in-

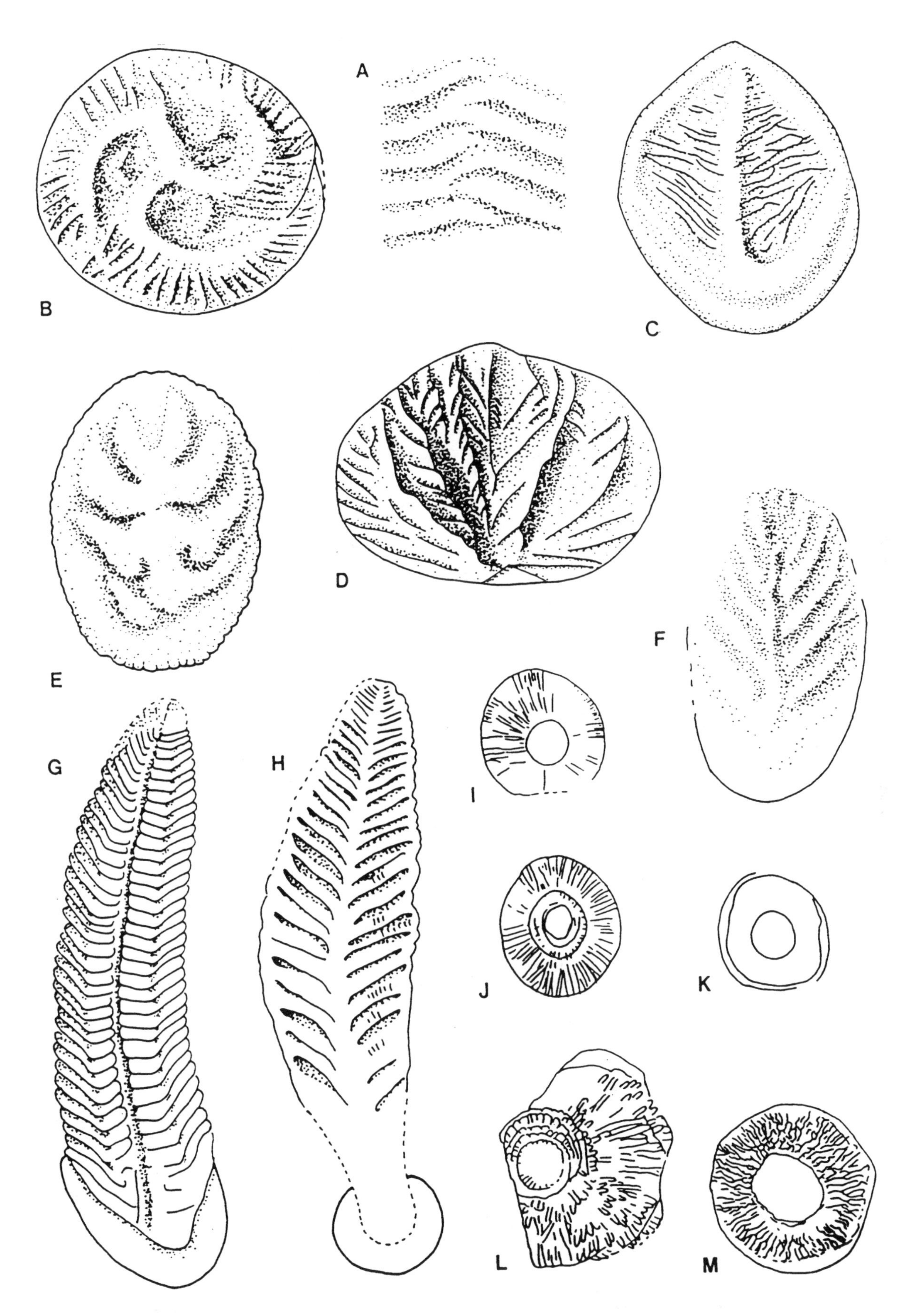

Fig. 2. Vendian petalonamids s.l. or "vendozoans": a central portion of *Dickinsonia* to show alternation of lateral "segments"; B *Tribrachidium*; C *Parvancorina*; D "bush-like form" from Newfoundland; E *Praecambridium*; F *Vendia*; G *Spriggina*; H *Charnia*; I *Charniodiscus*, medusa-like holdfast; J *Cyclomedusa*, K *Medusinites*; L *Eoporpita*; M for comparison the Lower Cambrian bag-shaped holdfast of *Protolyella princeps* with scar after broken shaft. Drawings based on: A. Fedonkin (1987, Pl. 9), B a combination of differently illuminated photographs, C. McAlester (1968, fig. 1-13D), D. Anderson & Conway Morris 1982; E. Stanley 1976, F-I, K-L Glaessner (1979); J. Cloud & Glaessner (1982); M. Nathorst (1881).

dicate a similarity with the fibrous periderm of 'petaloorganisms' (Pflug, 1984b). The supposed chondrophorine hydroid *Eoporpita* shows similar features (Fig. 2). The radiating rays are usually interpreted as tentacles, which appears implausible with respect to their apparently cylindrical cross section. It is possible that a number of medusoids are vendozoan holdfasts in various states of decay. *Protolyella* (Fig. 2) appears to be a similar type of holdfast from the Cambrian.

Others share characters with *Chuaria* and may have been individual algae: *Beltanelliformes, Bronicella, Hagenetta, Nemiana* and *Sewkia. Nimbia* looks more like an algal colony (cf. Cloud, 1968; Germs, 1972). I have to agree with Seilacher (and with Norris 1989): none of the medusoids in the Ediacaran fauna can be interpreted with certainty as a jellyfish. The question thus stands: is there a single species that can be positively identified as a cnidarian? *Conomedusites* has a four-sided radial symmetry which makes it possibly acceptable as a member of the Conulata and of Cnidaria, but symmetry is of course no definite proof for the assignation. Anyway, this seems to be the closest we can come to a proof for the moment.

Cnidarians are carnivores. A rich cnidarian fauna implies, therefore, the existence of a coexisting fauna upon which to prey. However, other faunal elements became really important only in the Phanerozoic, as far as we know. It is therefore difficult to understand why cnidarians should have been so much more successful in the Vendian than in the Phanerozoic. The ultimate solution may be that the organisms were no cnidarians, but that they had a similar radial symmetry.

We can now try to list the main categories of the Vendian 'quilted organisms':

1) Leaf-shaped, almost bilateral petalonamids with shaft and holdfast (anchor) (Pteridiniidae, Rangeidae, Charniidae); several 'medusoids' and the supposed hydrozoans *Eoporpita* and *Protolyella* may belong in this category as holdfasts.
2) Petalonamids with 'leaf' folded to vase-shaped structure, with shaft (Erniettidae).
3) Flat, disymmetric forms without shaft (class Dipleurozoa Harrington & Moore, 1956: Dickinsoniidae including the deformed specimen described as the hydrozoan *Chondroplon*, and 'spindle-shaped form' from Newfoundland)
4) Small, almost bilaterally symmetrical forms with head-like end and segment-like lateral folds (Spriggínidae, Vendomiidae).
5) Small, almost bilaterally symmetrical forms with sharp-ridged topography similar to that of trilobozoans (Parvancorinidae, 'bush-like form' from Newfoundland).
6) Small triradiate forms (class Trilobozoa Fedonkin 1983: *Tribrachidium*, Albumaresidae).

Sclerites and sclerite-carrying animals

A number of the problematic fossils found in the Lower Cambrian consist of sclerites (e. g. Bengtson & Missarzhevsky, 1981; Rozanov, 1986). Based on the morphology of the Middle Cambrian *Wiwaxia*, Bengtson & Conway Morris (1984) reconstructed this animal and *Halkieria* as slug-like animals covered with sclerites except on the locomotory underside. This reconstruction was accepted by Dzik (1986). What is particularly interesting is that this morphology, except for the sclerites, is just what can be expected not only from

	amery and oligomery	pseudomery	metamery (apomorphy)
COELOMATA	Deuterostomia ←		(Chordata)
	Brachiopoda ←		
	Phoronida ←		
	Bryozoa ←		→ Arthropoda
	Echiurida ←		
	Mollusca ←	Mollusca ←	→ Annelida
	Sipunculida ←	(procoelomate stem)	→ Pogonophora
	Endoprocta ←		
PSEUDO-COELOMATA		Nematomorpha	
		Nematoda	→ Tardigrada
		Gastrotricha	
		Loriciferida	
		Priapulida	→ Kinorhyncha
		Acanthocephala	
	Chaetognatha ←	Rotifera	
ACOELO-MATA		Nemertina	→ Annulo-nemertes
		Platyhelminthes (bilaterians)	

Fig. 3. Acoelomates and pseudocoelomates are predominantly pseudosegmented; pseudomery apparently is the primitive bilaterian condition. Coelomate evolution is characterized by macroevolutionary steps away from pseudomery. Either organ repetition became highly integrated to form the true segmentation (metamery), or repetition was lost more or less completely (oligomery, amery). Pseudomery is preserved only in some molluscs in combination with other plesiomorphic features elsewhere found in the Platyhelminthes.

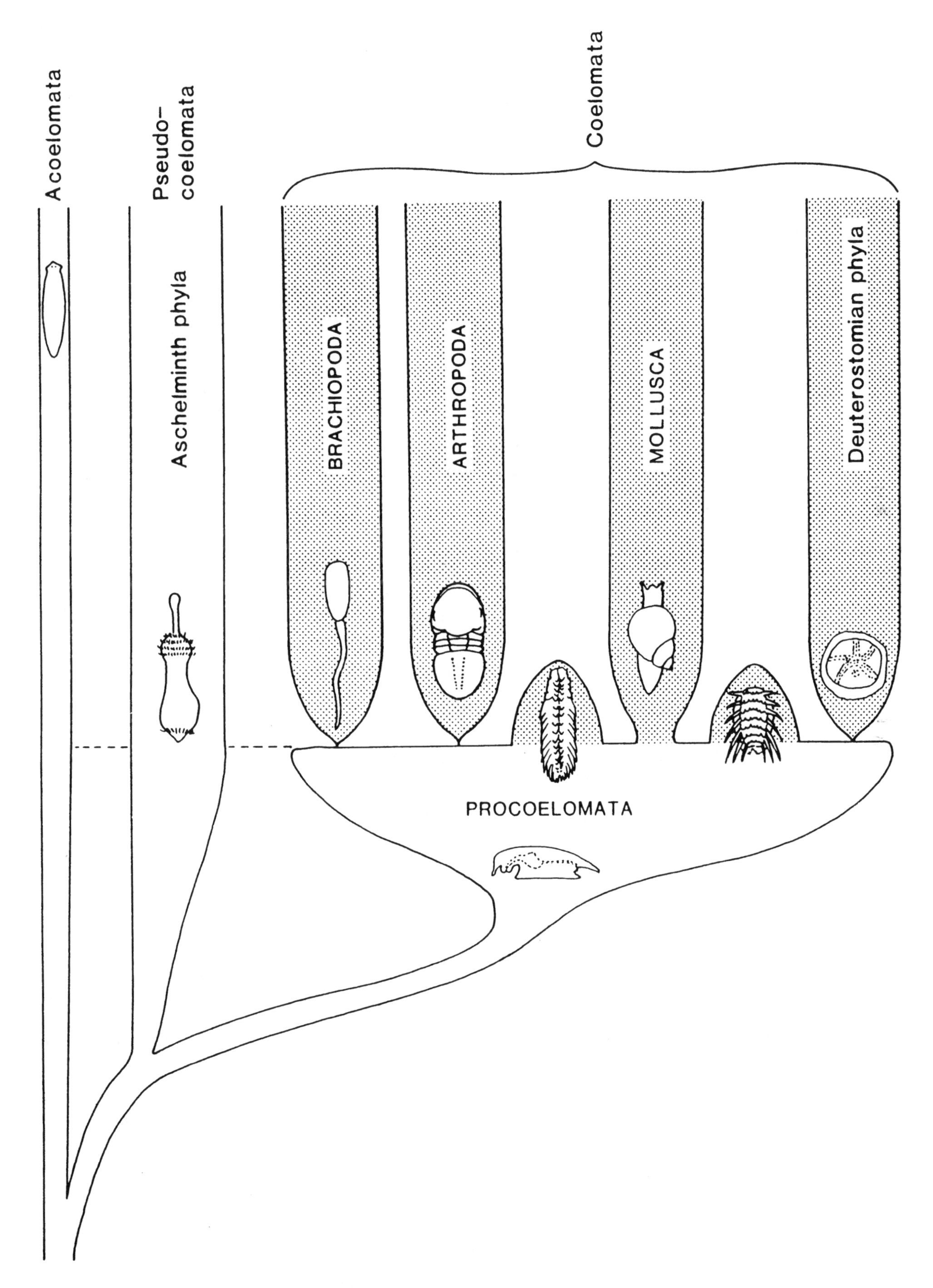

Fig. 4. Sketch of bilaterian evolutionary tree showing major radiation event at the Precambrian-Cambrian transition. Most of the phyla arose through macroevolutionary events (narrow base), many of which included the acquisition of fossilizeable skeletons (shaded). Only a few of the resulting phyla are indicated.

molluscan ancestors (e.g. Stasek, 1972; Pojeta in Boardman *et al.*, 1987, pp. 284-286) but also from the ancestors of almost all bilaterian phyla (cf. above; Bergström, 1986, 1989). The repetitive arrangement of the sclerites indicates some degree of pseudosegmentation, as would be anticipated. Such a repetition is seen also in polyplacophoran molluscs and even in some aplacophoran larvae (cf. Wingstrand, 1985). Indeed, the arrangement of the sclerites can easily be compared with that of the calcareous spicules found in the mantle of some aplacophorans (Pojeta in Boardman *et al.*, 1987, Figs. 14.2A, 14.7, 14.8). As Bengtson & Conway Morris (1984) found a radula-like mouth apparatus in *Wiwaxia*, it may even be pertinent to ask if this animal may be a mollusc. However, I find the alternative question equally sensible: are aplacophorans molluscs, or are they an independent branch or independent branches stemming from a procoelomate stock? The characters they share with more ordinary molluscs are mostly, if not entirely, plesiomorphic.

To my mind, it seems possible that at least many of the sclerite-covered animals designated as coeloscleritophorans by Bengtson & Missarzhevsky (1981) and as machaeridians by Dzik (in Hoffman & Nitecki, 1986) were in fact members of the ancestral bilaterian slug-like procoelomates which had survived into the Phanerozoic and had developed hard parts during the 'Cambrian explosion' (Fig. 4; Bergström, 1989).

Some Burgess Shale problematica

The Burgess Shale provides us with a well-known 'window' to the Middle Cambrian fauna. A similar occurrence in Yunnan, China may provide us with a still older, Early Cambrian window, but it is still under description (Chen Junyuan & Erdtmann, this volume; Hou Xian-guang & Bergström, this volume). The Burgess fauna contains several elements which are difficult to place systematically (Briggs & Conway Morris, 1986). Some of them are more or less safely placed in a still existing phylum and are not considered here. Others are much more difficult. *Wiwaxia* is only one of them. Referring to my model of bilaterian evolution (above; Bergström, 1986, 1990), it is most unlikely that they are in any way intermediates between phyla, except possibly between procoelomates and the basically similar molluscs. Two alternatives are left: they represent phyla of their own, or they belong to phyla already defined. The first alternative is probably exceptional and should not be chosen unless we are very definite about the characters of a fossil.

Amiskwia, originally thought to be a chaetognath, was considered by Owre & Bayer (1962) to be a pelagic nemertean, but Conway Morris (1977a) and Briggs & Conway Morris (1986) consider it to be an offshoot from turbellarians forming a phylum of its own. As nemerteans are probably closely related to turbellarians, the difference between the two latter opinions is small. I find it practical to include it provisionally with nemerteans or flatworms; the known characters do not suffice for the distinction of phylum features.

Anomalocaris, competing in size with the largest Middle Cambrian paradoxidid trilobites, and *Opabinia* show certain similarities with arthropods, notably in their segmentation, sclerite cover and frontal limbs of *Anomalocaris*. However, other features appear alien, and the listed characters do not suffice for a definite placement. Thus pseudosegmentation is widely distributed among bilaterian animals, and more advanced segmentation is found not only in arthropods, but also in vertebrates, annelids, tardigrades, kinorhynchs, and nemerteans (Fig. 3; Berg, 1985 for nemerteans). Thus, advanced segmentation is found among acoelomates (nemerteans), pseudocoelomates (kinorhynchs and presumably tardigrades) and coelomates (arthropods, vertebrates, annelids). Tardigrades have walking limbs like arthropods. Sclerites arranged as if they were covering segments and therefore comparable with those of *Anomalocaris* and *Opabinia* are found in some aschelminths: the Rotatoria and Kinorhyncha. Also the Loriciferida and Priapulida have sclerites. The disymmetry in the mouth of *Anomalocaris* has a counterpart in the disymmetry of body organisation (muscles, nerves) in many aschelminths. Briggs & Conway Morris (1986) believe that the two genera represent two new phyla; I agree with Collins (personal communication) that they are probably related to each other, and regard them as possible aschelminths. Reinhard Möbjerg Kristensen (personal communication 1988), a specialist on extant aschelminths, has independently reached the same conclusion.

Briggs & Conway Morris (1986) further suggest that *Banffia, Dinomischus* and *Hallucigenia* also represent new phyla, while *Nectocaris, Odontogriphus* and two unrelated specimens referred to *Redoubtia* may represent major extinct groups. Referring to what was said above about the origin of phyla, and not going into detail, I am reluctant to accept the idea (Briggs & Conway Morris, 1986; Conway Morris, 1989) that a fossil belongs to a new phylum (or major new group) just because it is poorly understood.

Conclusion

It is concluded that many of the Vendian organisms usually considered as metazoans more probably belong to a group of 'quilted organisms' which may not be animals and exhibit no close similarity to bilaterians. The 'quilted organisms' may include the vendozoans in the sense of Seilacher (including petalonamids) and many 'medusoids'. It is not clear if these organisms are to be considered as animals. It is unlikely that they have anything to do with the origin of bilaterian animals. Some 'quilted organisms' appear to have lingered on into the Cambrian (e.g. Fig. 2M; Glaessner, 1984; Conway Morris, 1989).

Definite metazoans are represented by very few Vendian body fossils. The bilaterians were probably represented in the Vendian by three main groups, the acoelomates, the pseudocoelomates, and the procoelomates. All of these were soft-bodied animals with a low fossilization potential. The primitive mode of locomotion should have been ciliary creeping, but particularly among the pseudocoelomates there may have been burrowing forms. This pattern is well mirrored by trace fossils.

The transition to the Phanerozoic included a major shift in faunal composition, probably induced by environmental factors. What can be seen is a major adaptive radiation of coelomate phyla (Fig. 4). The radiation was probably similar to radiations in the Phanerozoic: a quite rapid initial radiation from a shared origin. What made this radiation unique is that it led to the population of a number of basic niche groups which had never been exploited before and were never emptied afterwards. The associated anatomical rearrangements created a number of new phyla. Many of these had hard skeletal parts, which made fossilization much easier than before. It is possible that there was a contemporary radiation of acoelomates and pseudocoelomates, but there is no fossil evidence for it.

Conway Morris (e.g. 1989) maintains that the radiation was fully unpredictable and that the result was a plexus of animal groups without sharply defined phyla. My view is fundamentally different (cf. Bergström, 1989). The radiation of coelomates from a pseudosegmented origin fell into two distinct morphological fields: the repetitive pattern was either lost (in filter-feeding sessile groups) or refined to form true segmentation (agile groups) These two patterns arose repeatedly. The existence of several groups (distinct phyla) is in no way proof of a morphological continuum; each group (phylum) had its closest relatives in slug-like animals, not in other groups sharing the same pattern. This pattern helps explain the abruptness of the Cambrian fossil explosion and is consistent with biochemical phylogeny and the distribution among phyla of primitive and advanced characters.

With Runnegar (1982b), I can conclude that the Cambrian explosion was one of both fossils and of animals of new designs, and that there must have been an external triggering agent to cause the timing.

REFERENCES

Anderson, M.M. & Conway Morris, S. 1982. A review, with description of four unusual forms, of the soft-bodied fauna of the Conception and St. John groups (Late-Precambrian), Avalon Peninsula, Newfoundland. *Third North American Paleontological Convention (Montreal), Proceedings* **1**, 1-8.

Bengtson, S. 1986. Introduction: The problem of the problematica. In *Problematic fossil taxa* 6eds. A. Hoffman and M.H. Nitecki), pp. 3-11. New York: Oxford University Press.

Bengtson, S. & Conway Morris S. 1984. A comparative study of Lower Cambrian *Halkieria* and Middle Cambrian *Wiwaxia*. *Lethaia* **17**, 307-329.

Bengtson, S. & Missarzhevsky, V.V. 1981. Coeloscleritophora, a major group of enigmatic Cambrian metazoans. *U.S. Geological Survey Open-File Report* **81-743**, 19-21.

Berg, G. 1985. *Annulonemertes* gen. nov., a new segmented hoplonemertean. In *The Origins and Relationships of Lower Invertebrates.* (eds. S. Conway Morris, J.D. George, R.Gibson & H.M. Platt) *The Systematics Association Special Volume* **28**, 200-209.

Bergström, J. 1981. Morphology and systematics of early arthropods. *Abhandlungen des naturwissenschaftlichen Vereins Hamburg, NF* **23**, 7-42.

Bergström, J. 1986. Metazoan evolution - a new model. *Zoologica Scripta* **15**, 189-200.

Bergström, J. 1989. The origin of animal phyla and the new phylum Procoelomata. *Lethaia* **22**, 259-269.

Bergström, J. 1990. Precambrian trace fossils and the rise of bilaterian animals. *Ichnos* **1** (in press).

Boardman, R.S., Cheetham, A.L. & Rowell, A.J. (eds) 1987. *Fossil invertebrates.* Palo Alto: Blackwell Scientific Publications.

Briggs, D.E.G. & Conway Morris, S. 1986. Problematica from the Middle Cambrian Burgess Shale of British Columbia. In *Problematic Fossil Taxa* (eds. A. Hoffman & M.H. Nitecki), pp.167-183. New York: Oxford University Press.

Clark, R.B. 1964. *Dynamics in metazoan evolution.* Oxford: Clarendon Press.

Clark, R.B. 1979. Radiation of the Metazoa. In *The origin of major invertebrate groups.* (ed. M.R. House). *The Systematic Association Special Volume* **12**, 55-102.

Clément, P. 1985. The relationships of rotifers. In *The origins and Relationships of Lower Invertebrates* (eds. S. Conway Morris, J.D. George, R. Gibson & H.M. Platt). *The Systematics Association Special Volume* **28**, 224-247.

Cloud, P. 1968. Pre-metazoan evolution and the origins of the Metazoa. in *Evolution and environment* (ed. E.T. Drake), pp.1-72. New Haven: Yale University Press.

Cloud, P. & Glaessner, M.F. 1982. The Ediacarian Period and System: Metazoa inherit the Earth. *Science* **217**, 783-792.

Conway Morris, S. 1977a. A redescription of the Middle Cambrian worm *Amiskwia sagittiformis* Walcott from the Burgess Shale of British Columbia. *Paläontologische Zeitschrift* **51**, 271-287.

Conway Morris, S. 1977b. Fossil priapulid worms. *Special papers in Palaeontology* **20**, 1-95.

Conway Morris, S. 1985. The Edicarian biota and early metazoan evolution. *Geological Magazine* **122**, 77-81.

Conway Morris, S. 1989. The Burgess Shale faunas and the Cambrian explosion. *Science,* **246**, 339-346.

Dzik, J. 1986. Turrilepadida and other Machaeridia. in *Fossil Problematic Taxa* (ed. A. Hoffman and M.H. Nitecki), pp. 116-134. New York: Oxford University Press.

Elder, H.Y. 1980. Peristaltic mechanisms. In *Aspects of Animal Movements*, vol. 5 (eds. H.Y. Elder and E.R. Trueman), pp. 71-92. Cambridge: University Press.

Fedonkin, M.A. 1981. Byelomorskaja biota Vend [White Sea Vendian biota]. *Akademia Nauk SSSR* **342**. [In Russian].

Fedonkin, M.A. 1983. Organicheskij mir Venda [Organic world of the Vendian]. *Itogi Nauki Tekhniki. VINITI AN SSSR* **12**, 1-127. [In Russian].

Fedonkin, M.A. 1985. Precambrian metazoans: the problems of preservation, systematics and evolution. *Philosophical Transactions of the Royal Society of London* **B311**, 27-45.

Fedonkin, M.A. 1987. Besskeletnaja fauna venda i ee mesto v evoliotsii Metazoa. [The Vendian non-skeletonized fauna and its place in the evolution of the Metazoa]. *Akademia Nauk SSSR, Trudy Paleontologicheskogo Instituta* **226**, 174 pp. [In Russian].

Field, K.G., Olsen, G.J., Giovannoni, S.J., Ghiselin, M.T., Raff, E.C., Pace, N.R. & Raff, R.A. 1988. Molecular phylogeny of the animal kingdom. *Science* **239**, 748-753.

Ford, T.D. 1958. Pre-Cambrian fossils from Charnwood Forest. *Proceedings of the Yorkshire Geological Society* **31**, 211-217.

Germs, G.J.B. 1972. Thin concentric structures of biologic origin from the Nama System, South West Africa. *Geological Society of America Bulletin* **83**, 463-466.

Ghiselin, M.T. 1988. The origin of molluscs in the light of molecular evidence. *Oxford Surveys in Evolutionary Biology* **5**, 66-95.

Glaessner, M.F. 1979. Precambrian. In *Treatise on Invertebrate Paleontology* Part A (Introduction) (eds. R.A. Robison and C. Teichert), pp. A79-A118. Boulder: Geological Society of America and Lawrence: University of Kansas Press.

Glaessner, M.F. 1983. The emergence of Metazoa in the early history of life. *Precambrian Research* **20**, 427-441.

Glaessner, M.F. 1984. *The dawn of animal life. A Biohistorical study.* Cambridge: University Press.

Glaessner, M.F. & Wade, M 1971. *Precambridium* - a primitive arthropod. *Lethaia* **4**, 71-77.

Goodman, M., Pedwaydon, J., Czelusniak, J., Suzuki, T., Gotoh, T., Moens, L., Shishikura, F., Walz, D. & Vinogradov, S,. 1988. An evolutionary tree for invertebrate globin sequences. *Journal of Molecular Evolution* **27**, 236-249.

Harrington, H.J. & Moore, R.C. 1956. Dipleurozoa. In *Treatise on Invertebrate Paleontology,* Part F (Coelenterata) (ed. R.C. Moore), Boulder: Geological Society of America and Lawrence: University of Kansas Press.

Hoffman, A. & Nitecki, M.H. [eds.] 1986. *Problematic fossil taxa.* New York: Oxford University Press.

Hofmann, H.J. 1988. An alternative interpretation of the Ediacaran (Precambrian) chondrophore *Chondroplon* Wade. *Alcheringa* **12**, 315-318.

Jägersten, G. 1968. *Livscykelns Evolution hos Metazoa. En generell Teori.* Lund: Scandinavian University Books. 295 pp. [translated into English in 1972: *Evolution of Metazoan Life Cycle, a Comprehensive Theory.* New York: Academic Press].

Jenkins, R.J.F. 1984. Interpreting the oldest fossil cnidarians. *Palaeontographica Americana* **54**, 95-104.

Landing, E. 1988. *Genetics, Paleontology and Macroevolution.* Cambridge: University Press.

LORENZEN, S. 1985. Phylogenetic aspects of pseudocoelomate evolution. In *The Origins and Relationships of Lower Invertebrates* (ed. S. Conway Morris, J.D. George, R. Gibson, H.M. Platt). *The Systematics Association, Special Volume* **28**, 224-247.
MCALESTER, A.L. 1968. *The history of life*, 2nd edn. Prentice- Hall.
MEGLITSCH, P.A. 1972. *Invertebrate Zoology*, 2nd edn. Oxford: Clarendon Press.
METTAM, C. 1985. Functional constraints in the evolution of the Annelida. In *The Origins and Relationships of Lower Invertebrates* (ed. S. Conway Morris, J.D. George, R. Gibson, H.M. Platt). *The Systematics Association, Special Volume* **28**, 297-309.
MOCZYDLOWSKA, M. & VIDAL, G. 1988. How old is the Tommotian? *Geology* **16**, 166-168.
NATHORST, A.G. 1881. Om aftryck af medusor i Sveriges kambriska lager. *Kongliga Vetenskaps-Academiens Handlingar* **19** (1).
NORRIS, R.P. 1989. Cnidarian taphonomy and affinities of the Edicara biota. *Lethaia* **22**, 381-393.
OWRE, H.B. & BAYER, F.M. 1962. The systematic position of the Middle Cambrian fossil *Amiskwia* Walcott. *Journal of Paleontology* **36**,1361-1363.
PALIJ, V.M., POSTI, E. & FEDONKIN, M.A. 1983. Soft-bodied Metazoa and animal trace fossils in the Vendian and early Cambrian. In *Upper Precambrian and Cambrian palaeontology of the East-European Platform* (eds. A. Urbanek and A.Yu. Rozanov). Warsaw [translated from Russian edition published in 1979].
PFLUG, H.D. 1974a. Feinstruktur und Ontogenie der jungpräkambrischen Petalo-organismen. *Paläontologische Zeitschrift* **48**, 77-109.
PFLUG, H.D. 1974b. Vor- und Frühgeschichte der Metazoen. *Neues Jahrbuch für Geologie und Paläontologie*, Abhandlungen **145**, 328-374.
ROZANOV, A.YU., MISSARZHEVSKY, V.V., VOLKOVA, N.A., VORONOVA, L.G., KRYLOV, I.N., KELLER, B.M., KOROLYUK, I.K., LENDZION, K., MICHNIAK, R., PUKHOVA, N.G. & SIDOROV, A.D. 1969. *Tommotskij yarus i problema nizhnej granitsy kembriya [The Tommotian Stage and the Cambrian lower boundary problem.] Trudy Geol. Inst. An SSSR* **206**, 380 pp. [Translated into English in 1981, New Delhi: Amerind Publ. Corp.].
ROZANOV, A.YU. 1986. Problematica of the Early Cambrian. In *Problematic Fossil* (ed. A. Hoffman and M.H. Nitecki), pp. 87-96. New York: Oxford University Press.
RUNNEGAR, B. 1982a. A molecular-clock date for the origin of the animal phyla. *Lethaia* **15**, 199-205.
RUNNEGAR, B. 1982b. The Cambrian explosion: animals or fossils? *Journal of the Geological Society of Australia* **29**, 395-411.
RUNNEGAR, B. 1982c. Oxygen requirements, biology and phylogenetic significance of the Late Precambrian worm *Dickinsonia*, and the evolution of the burrowing habit. *Alcheringa* **6**, 223-239.
SEILACHER, A. 1956. Der Beginn des Kambriums als biologische Wende. *Neues Jahrbuch für Geologie und Paläontologie, Abhandlungen* **103**,155-180.
SEILACHER, A. 1983. Precambrian metazoan extinctions. *Geological Society of America, Abstracts with Program* **15**, 683.
SEILACHER, A. 1984. Late Precambrian and early Cambrian Metazoa: Preservational or real extinctions. In *Patterns of Change in Earth Evolution* (eds. H.D. Holland & A.F. Trendall), pp.159-168. Dahlem Konferenzen 1984. Berlin: Springer-Verlag.
SEILACHER, A. 1985. Discussion of Precambrian Metazoans. *Philosophical Transactions of the Royal Society of London* **B311**, 47-48.
SEILACHER, A. 1989. Vendozoa: Organismic construction in the Proterozoic biosphere. *Lethaia* **22**, 229-239.
STANLEY, S.M. 1976. Fossil data and the Precambrian-Cambrian evolutionary transition. *American Journal of Science* **276**, 56-76.
STASEK, C.R. 1972. The molluscan framework. *Chemical Zoology* **7**: 1-44.
WADE, M. 1968. Preservation of soft-bodied animals in Precambrian sandstones at Ediacara, South Australia. *Lethaia* **1**, 238-267.
WINGSTRAND, K.G. 1985. On the anatomy and relationships of recent Monoplacophora. *Galathea Report* **16**, 94 pp.

Cladistic analysis of metazoan phyla and the placement of fossil problematica

Frederick R. Schram[1]

Abstract

There have been numerous explanations of the phylogenetic relationships of problematic fossils as well as invertebrate phyla advanced in the past. All of these have been framed essentially in a context of evolutionary systematics, with each authority choosing to weight what few characters that they deemed appropriate. As a consequence, none of these theories agrees with each other. No analysis of metazoan relationships has attempted until now to assess and polarize as many characters as possible for all phyla. With such analysis, we can hope now to place problematic organisms.

Introduction

Probably no subject has ever been marked by so much subjective speculation as that concerning relationships of invertebrate phyla. Hardly any two authorities agree. Furthermore, the abundance of rival interpretations of individual aspects of invertebrate anatomy and the confusing array of names applied to all manner of "hypothetical ancestors" or paper animals is intimidating.

Invertebrate phylogeneticists often have been concerned with identifying archetypes, or hypothetical forms, that were supposed to embody the essence of a particular animal kind. These "Platonic" archetypes eventually came to be viewed by some as actual ancestors. As an example, one of the most famous archetypes is HAM, the "hypothetical ancestral mollusk," the existence of which some malacologists (see e.g., Runnegar and Pojeta, 1985) firmly believe served to suppress consideration of important fossil forms in developing an adequate phylogeny of the Mollusca.

Perhaps the most productive of the early phylogeneticists was Ernst Haeckel (1874). He originated several ideas, such as the blastea and gastrea theories, that became a central focus for phylogenetic speculations about invertebrates. One view that focused on the blastea was the idea that all metazoans evolved from hollow, ball-like colonies of flagellates. The planula theory (Metschnikoff, 1886) derived higher invertebrates from an epibenthic planula larva-like organism, i.e., an animal in which the hollow ball of cells was in turn filled with cells delaminated from the surface layer. This planula organism was in turn the archetype for the development of cnidarians (Fig. 1A). Another approach assumed that a ciliate rather than a flagellate stock (Hanson, 1958; Hadzi, 1963) gave rise to a syncitial, acoeloid stem. A consequence of the ciliate theory was to view cnidarians as a kind of degenerate platyhelminth (Fig. 1A and 1C).

An alternative approach to those above was the gastrea theory (favoured by Haeckel himself) that envisioned a spherical blastea giving rise to a cuplike, two-layered gastrea. The conceptualization of this animal, which looked much like a cnidarian, took its inspiration from the common gastrula stage shared among higher metazoans. This theory, in turn, gave rise to some sub-theories, among which (Fig. 1B) some workers speculated that the partitions of the gastrocoel seen in anthozoans prefigured the formation of segmented coelomic septa. This idea was most recently elaborated by Jägersten (1955), Marcus (1958), Remane (1963), and Sharov (1966).

With regard to the coelom itself, all sorts of contending theories developed that focused on how that cavity might have formed, whether they be gonocoels, nephrocoels, or gastrocoels. As a result, many "family trees" of the animal kingdom have been produced since the days of Haeckel; and the well-known synthesis of Hyman (1940) (Fig. 2) is but one example of what these look like.

All these theories, regardless of their details, had certain things in common. First, all phyla were typically dealt with, but only a few characters were used to arrange them. Second, none of these theories really attempted to make testable statements about animal relationships, but rather prefered to focus on producing "definitive stories" about animal history.

We can now perceive the error of these early naturalists. Several authorities have pointed out that three distinct phases should occur in the course of a phylogenetic analysis (e.g., Tattersall & Eldredge, 1977). The first step is to produce a cladogram. This entails the examination of as many characters as possible, the polarizing of these features by one or more methods of analysis, the assembly of a character matrix, and the analysis of this matrix in a manner designed to produce the shortest tree with the least amount of convergence or character reversal and the highest level of congruence of characters on its branches.

The second step is to produce a phylogram, or evolutionary tree. Some authorities would argue that there is little or no difference between a cladogram and an evolutionary tree, and others would say that there is no need to use anything other than a cladogram. A cladogram, strictly speaking, is a graphic arrangement of taxa that displays their genealogical arrangement. An evolutionary tree is an attempt to qualify the information in the cladogram and interpret its meaning. For example, cladograms can be unrooted, that is, have no implicit statement about the direction of evolution within the group in question. A phylogenetic tree is always rooted, and implies a definite sequence of events.

The third phase of a phylogenetic analysis is to develop a scenario, that is, produce a "story" or evolutionary narrative based on the tree.

[1]Scripps Institution of Oceanography A-002 c/o R.R. Hessler La Jolla - California 92095 - USA.

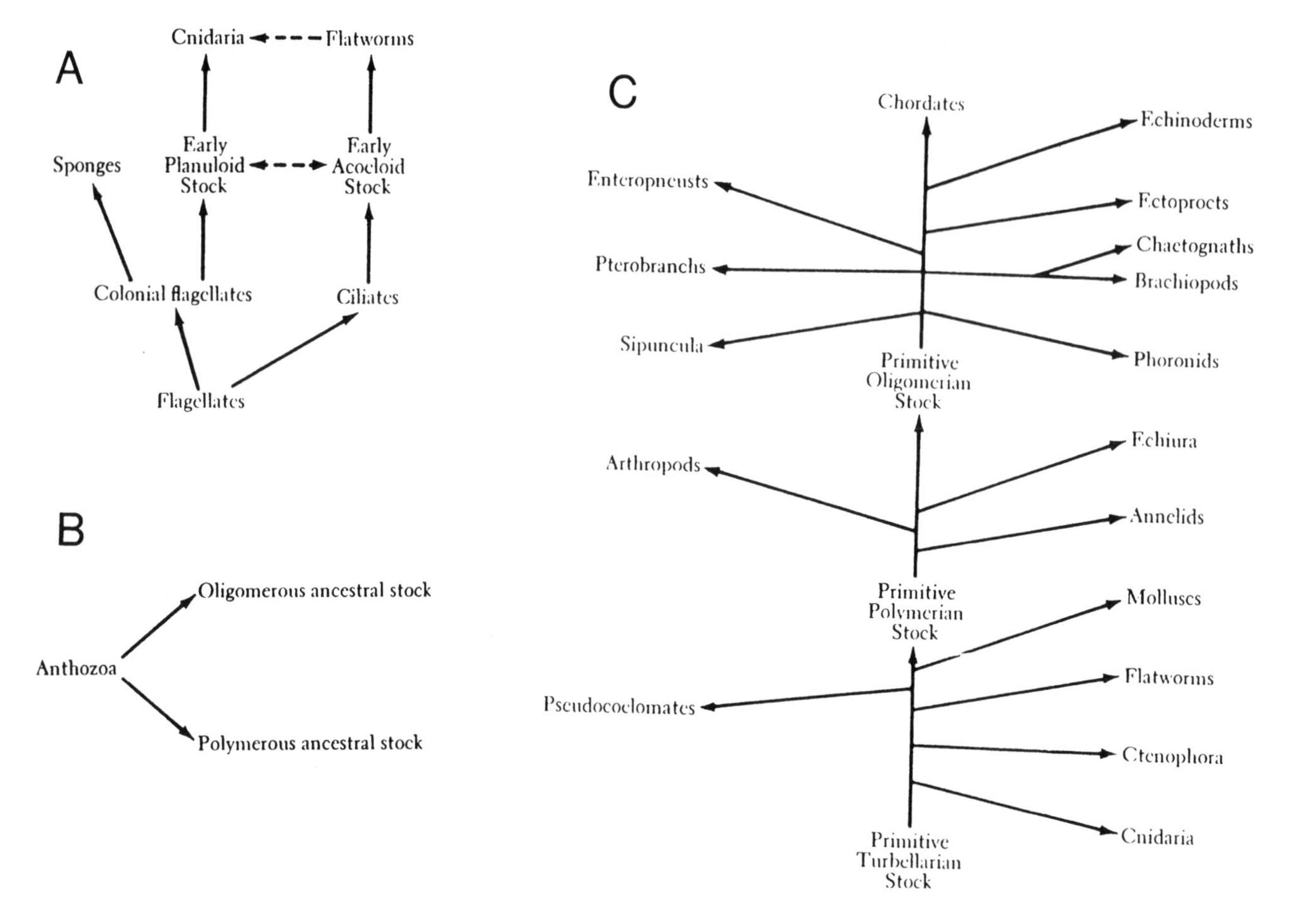

Fig. 1. Various phylogenetic schemes proposed for parts of the animal kingdom. A. Contrasting interpretations of the colonial flagellate or syncytial cilate theories, with possible alternative pathways indicated. B. Interpretation of the origin of segmentation from endocoelic cyclomerism in cnidarians. C. Principle phylogenetic relationships based on Hadzi (1963).

Each of these steps requires an increasing level of subjectivity. The first step, production of the matrix and cladogram, is least subjective in that its assumptions about the direction of character changes and its method of analysis are subject to testing. However, this method only minimizes subjectivity and does not totally eliminate it. The second step, producing an evolutionary tree, introduces more subjectivity into the analysis in that in this step one makes assumptions about the evolutionary events. The third step, the outline of a scenario, is the most subjective of the three, and in fact some authorities would not even allow it to pose as science since it is marked by a lack of clear, falsifiable hypotheses. Any number of scenarios, or stories, could be developed from the consideration of a single evolutionary tree.

The classic, traditional speculations on animal evolution started out with step three, the scenario, the most subjective element in the process, and then tried to create an evolutionary tree and data set to support that scenario.

Current theories

More recent discussions of invertebrate phylogeny seem to separate into two distinctly different approaches. One group contains workers who are marked by a kind of intellectual despair. These workers focus on the differences that separate animal groups and, in so doing, conclude that little or nothing can be said about their relationships. The extreme of this school is exemplified by Nursall (1962), who concluded that no tree of metazoan relationships can be drawn, that all phyla arose independently from various protistans, that in some cases the divisions between them extend back to the level of the monerans. A scheme of relationships in this context would resemble a sort of phylogenetic grass rather than a tree (Fig. 3).

One advantage of these approaches is that this school is adamantly opposed to the production of hypothetical archetypes. An example of this is Anderson (1983), who felt that the consideration of archetypes all too often ignores the need to provide functional intermediates between phyla. Though Anderson grouped phyla which he felt are related; he did not hesitate, based mostly on differences in developmental patterns, to suggest that metazoans may be polyphyletic. He felt that sponges, cnidarians, and ctenophores may have been independently derived from choanoflagellates and/or zooflagellates; that the phyla that exhibit true or some modified form of spiral cleavage may have come from ciliates; and that deuterostomes and other minor groups may have had yet other independent ancestries from within protistans (Fig. 4).

Inglis (1985), while he did not agree with Anderson's line of reasoning, nevertheless came to similar conclusions. Inglis advanced a concept he termed "evolutionary waves." He postulated that groups of phyla arose with multiple, parallel evolutions of major technical improvements. He recognized five such

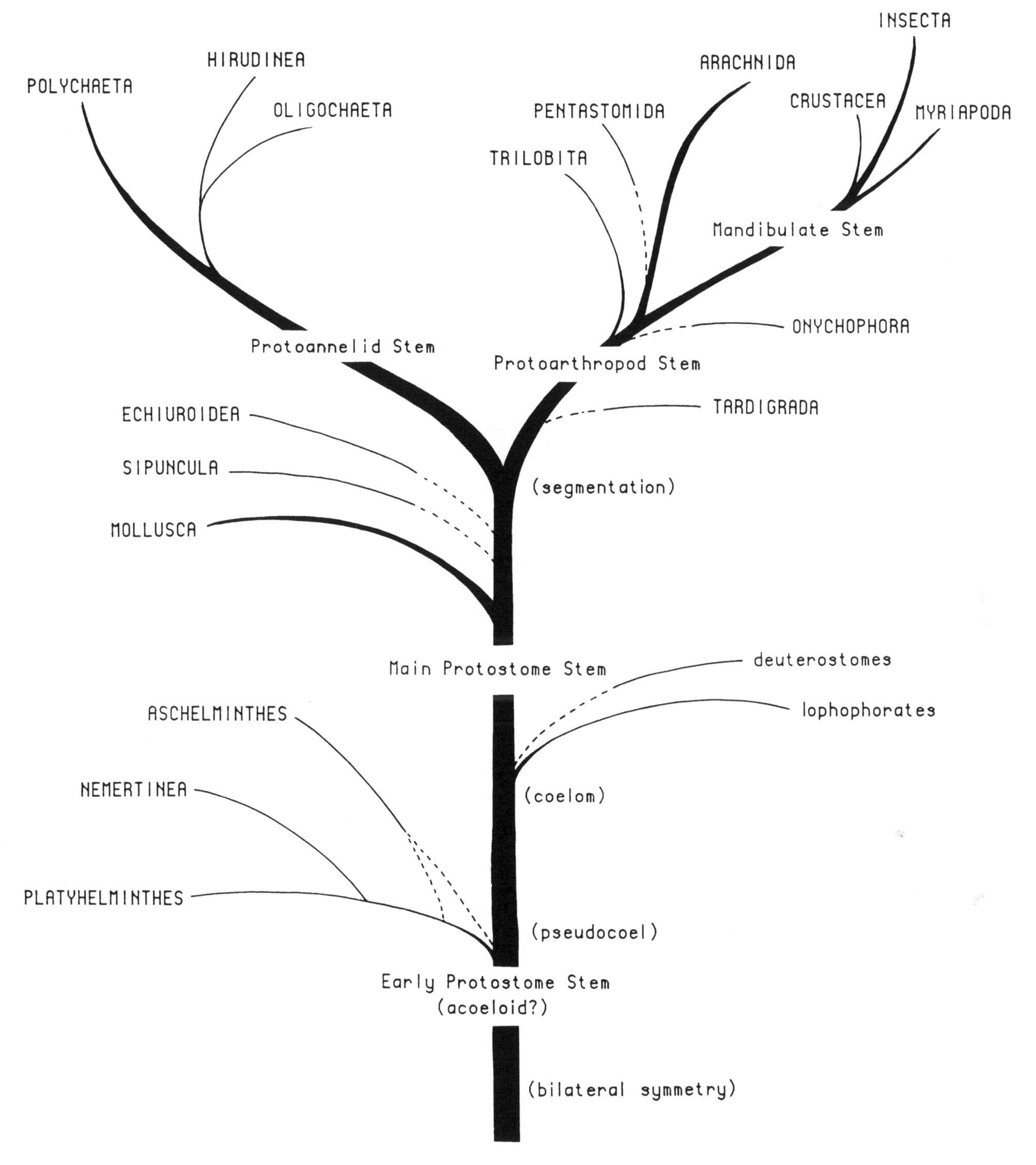

Fig. 2. Phylogenetic tree of animal kingdom derived from the views of Hyman (1940).

events: sponges and placozoans, radiates, deuterostomes, protostomes, and what he calls paracoelomates (i.e., mesozoans, pseudocoelomates, and acoelomates).

Investigators like Nursall, Anderson, and Inglis often leave us frustrated in our attempts to understand the genealogy of invertebrates in that they essentially tell us that, because of the great differences between phyla, we can never know the relationships of animal groups. Yet, as students of evolution we feel that we ought to be able to propose hypotheses about invertebrate relationships based on shared similarities. Schram (1983) termed this the Phylogenetic Uncertainty Principle, where, in delineating lines of evolution, one can either group them and be uncertain about their proximity, or one can hypothesize archetypes and be uncertain that such creatures ever existed or in fact connected the lines one seeks to unite.

The opposing school to the above approach is dedicated to the idea of producing evolutionary trees and scenarios in which the node points are various kinds of hypothetical ancestors. Many of these ideas were proposed in book-length or near book-length treatments (e.g., Remane, 1967; Siewing,1969; Jägersten, 1972). Often these workers called attention to the importance of having fully functional intermediates between phyla. Nevertheless, the complexity of their phylogenetic schemes with serial, hypothetical ancestors (e.g., Salvini-Plawen, 1980) are often marked with strange terminology [(e.g., why call Ctenophora "Collaria" as in Salvini-Plawen, 1978)?].

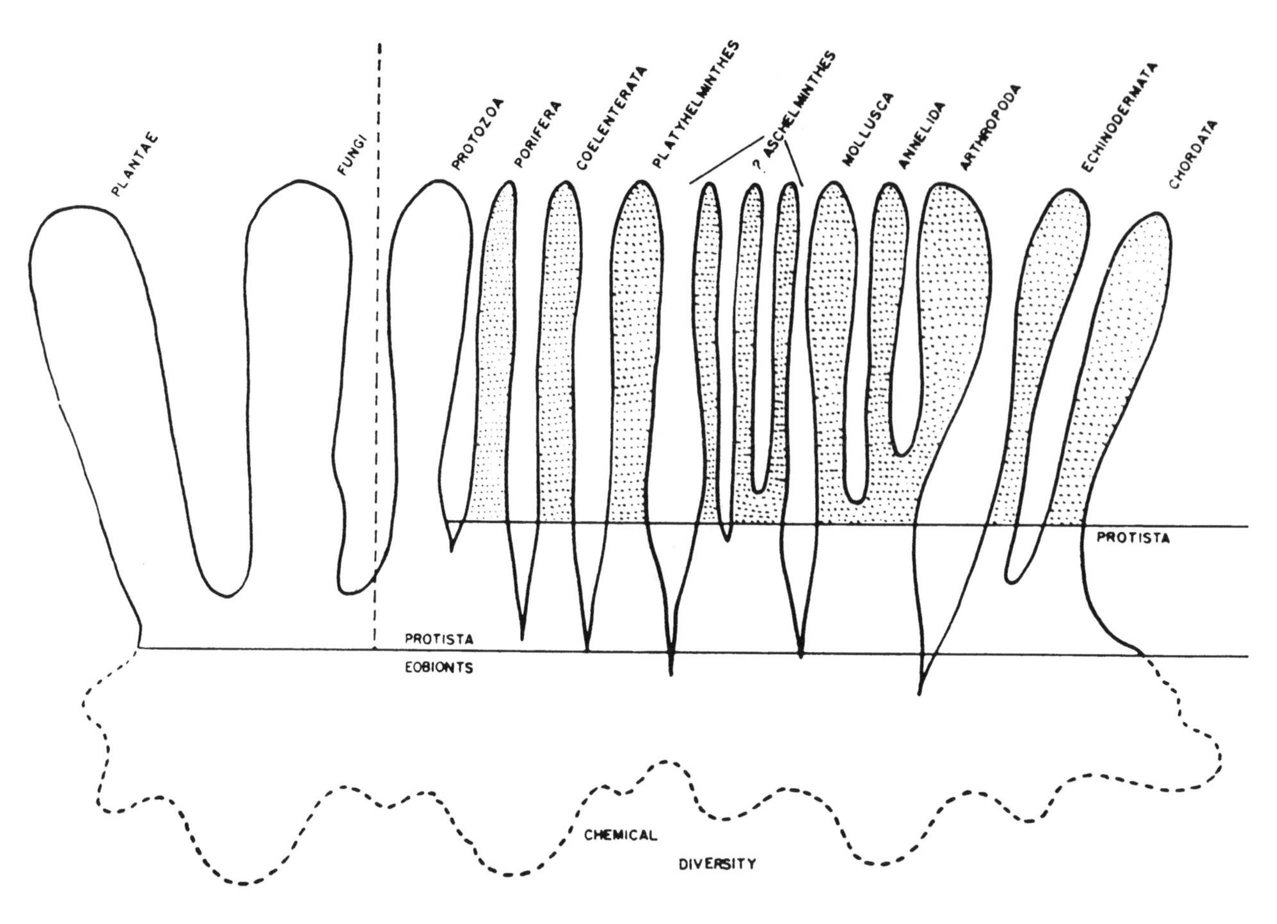

Fig. 3. Phylogenetic relationships of invertebrates proposed by Nursall (1962).

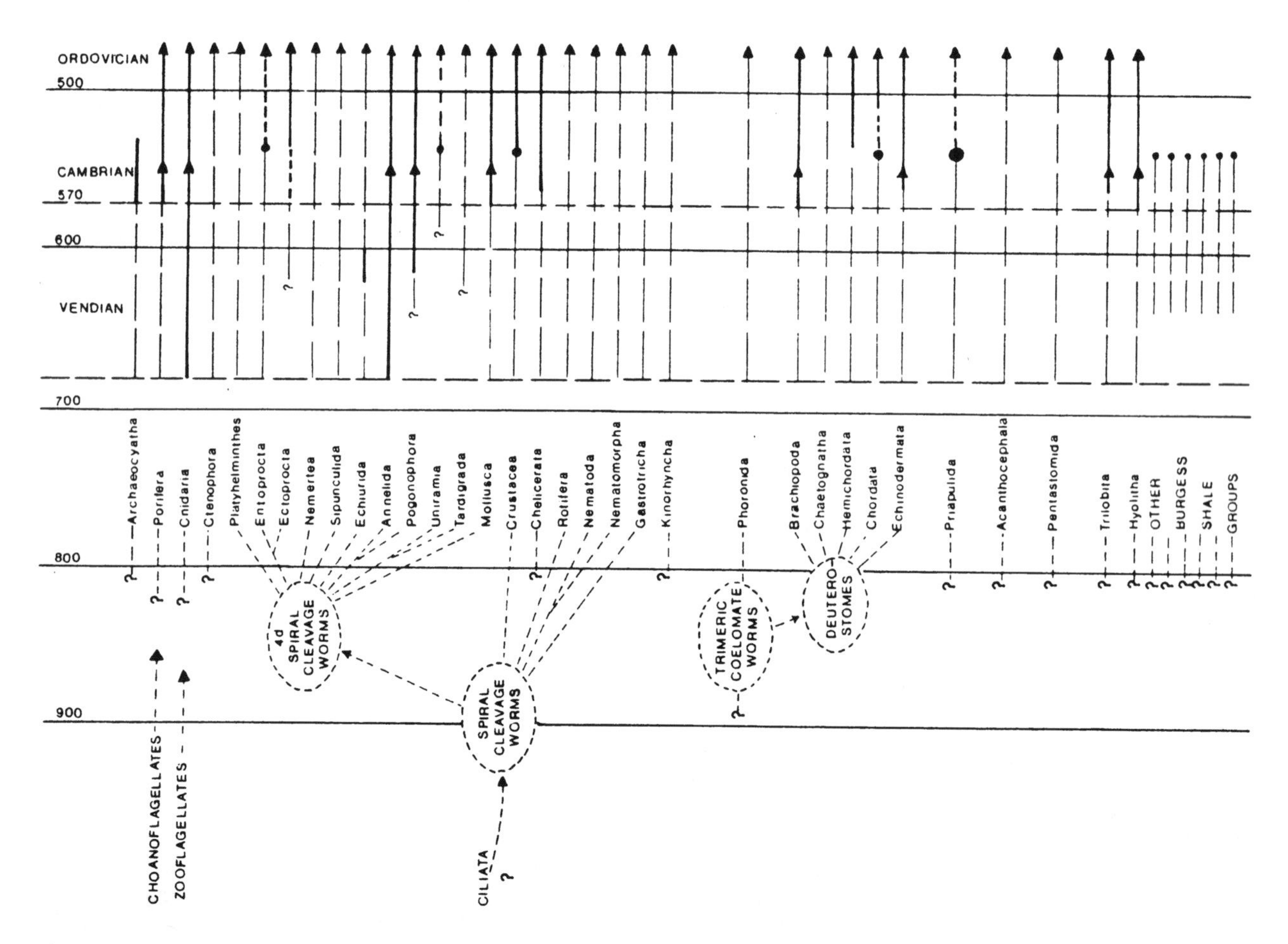

Fig. 4. Phylogenetic relationships of invertebrates proposed by Anderson (1983).

However, the work of this school is not totally devoid of useful observations. Many interesting conclusions deserve consideration: for example, the equating of the coelom of protostomes with the metacoel of the lophophorates (Salvini-Plawen, 1982), the recognition of mesoderm in ctenophores (Siewing, 1977), or the concept of an archimeric coelom (Siewing, 1980). Sometimes, the conclusions based on hypothetical archetypes are strange and quickly brought into question, e.g., the contention by Siewing (1969) that the basic spiralian was a polychaete (refuted by Ivanova-Kazas, 1981).

However, one of the most stimulating papers of this genre was Nielsen (1985). Nielsen took an essentially historical, scenario approach to animal evolution and proposed what he called the trochaea theory. This idea was one of the first effective attempts to integrate larval and adult features into a coherent system. Several archetypes were proposed in the trochaea scheme, but a large number of characters were utilized in the analysis and an effort was made to define derived character states. The result was something very near to a cladogram, and indeed has been a major inspiration for the following analysis.

Cladistic analysis

In the light of the difficulties outlined above, it might appear that analysis of invertebrate phylogeny in general, let alone the placement of problematica, is an exercise in futility. Yet, if the study of invertebrate phyla, both of living organisms as well as problematic fossil forms, is indeed science, and Evolutionary Theory is the grand paradigm that holds all of neo- and palaeobiology together, then there is no reason why the analysis of animal interrelationships cannot also be done in a more scientific manner.

Towards this end a data set was developed for cladistic analysis (Table 1) that focuses largely on anatomical and larval features that have figured in the literature of invertebrate relationships. Initially over 90 characters were scored for all metazoan phyla. This included a large amount of recently acquired data that has come to light as a result of studies that employ scanning and transmission electron microscopy, e.g., mono- versus multiciliate epidermal cells. However, these characters often were found to be highly homoplastic in the initial phases of this analysis. The problem is that to use such characters consistently across all phyla, every one must be equally examined for all of these features; this is often not yet possible at this time. Furthermore, in this data set no biochemical data were utilized. Studies that compare cytochrome c, haemoglobins, ribosomal RNA, and DNA hybrids have been yielding some interesting results (see, e.g., Bergström, 1986). However, this kind of data cannot be applied in the kind of analysis discussed here because, again, too few groups have been studied with these methods. Information of this kind obtained to date is often contradictory (compare, e.g., the results of Bergström, 1986, to those of Field *et al.*, 1988). The use of these methodologies is just beginning and it is likely that they will produce data in a few years that can be combined with traditional lines of evidence to yield a more comprehensive picture of invertebrate evolution.

In developing the final data matrix (Table 1), an effort was made to polarize these characters (Table 2) using the methods of multiple outgroup and ingroup analysis. The resulting data matrix was analyzed using PHYSYS and rooted to an ancestor scored plesiomorphic for all characters. Subsequently, the data was reanalyzed using PAUP, in that case the data being treated as unpolarized and a consensus tree developed for 100 runs. It was found that PAUP and the PHYSYS WAGNER.S option gave comparable results. The analysis of WAGNER.S is used here. WAGNER.S uses global branch-swapping to select the most parsimonious tree.

Cladogram of invertebrates

The computer analysis of these characters produced the cladogram seen in Fig. 5. Some surprises occur with respect to where particular phyla ordinarily occur in light of the traditional scenarios of invertebrate relationships.

The first point to note is that there is but a single stem to this cladogram. The idea that eumetazoans and poriferans may have arisen independently from protistans is not supported by this data matrix. I would agree with Nielsen (1985) that the presence of collagen, acetylcholine/cholinesterase systems, 9 + 2 flagellated or ciliated sperm with condensed chromatin and mitochondria, and location of reproductive cells and tissues internally serve to unite all metazoans into a single clade. The resolution of the relationships of the placozoans and the mesozoans to each other and the main stem must await more information on structure and modes of reproduction and development, especially for placozoans.

Ctenophores and cnidarians are not sister groups within some larger group often referred to as radiates. Rather, the ctenophores are a sister group to all the more derived invertebrates. The recognition of a mesodermal layer and determinate cleavage, along with the beginnings of a subepidermal musculature in ctenophores allies them with advanced invertebrates. The development of radial and biradial symmetry in the Ctenophora is merely a convergence to that seen in the Cnidaria, as well as Echinodermata.

Another surprise is noted in the relationship of the traditional acoelomate, pseudocoelomate, and eucoelomate clades to each other. Traditional views of animal history hold that acoelomates are a grade of organization ancestral to all others. However, the cladogram of Fig. 5 suggests that after the essential bilateral symmetry and the development of an anus was established, one line was structured on a highly organized mode of development, with spiral quartet cleavage and 4d mesoderm as a foundation. This includes the acoelomates and eucoelomates. Another line exhibits a variety of divergent monet and duet types of spiral cleavage (wherein cell determination is set at some stage before the first quartet stage), as well as the primitive radial cleavage. These latter phyla, with an orthogonal nervous system, a pseudocoel, and general lack of primary larvae, are the pseudocoelomates.

The acoelomates (platyhelminths and gnathostomulids) are postulated to form a sister group to the main "spiralian" line. They are characterized by a reversion to mainly ciliary locomotion, the loss

Table 1. Data matrix of characters used in analyses to produce cladograms in Figs. 5, 6, and 7. Characters defined in Table 2; 1 - derived, 0 - primitive, 9 - unknown.

Taxa	Characters
Ancestor	00000000000000000000000000000000000000000000000000 00000000000000000000000000
Mesozoa	10111000000000000000000000000000000000000000000000 00000000000000000000000000
Placozoa	10900000000000000000090000000000000000000000000000 00000000000000000000000000
Porifera	11000000110000000000000000000000000000000000000000 00000000000000000000000000
Cnidaria	11001111000001100000000000000000000000000000000000 00000000000000000000000000
Ctenophora	11000111001111001000000000000000000000000000000000 00000000000000000000000000
Gnathostomulida	11010011001910000190010000000010000000000000000000 00000000000000000000000000
Platyhelminthes	11010011001110000110000000000010000000000000000000 00000000000000000000000000
Gastrotricha	11010011001110010001010100010011001000000000000000 00000000000000000000000000
Rotifera	11010111001110010001110101110011110001000000000000 00000000000000000000000000
Acanthocephala	10010111001110010001110101010011110110000000000000 00000000000000000000000000
Loricifera	11010111001110010909910910090099000000111000000000 00000000000000000000000000
Kinorhyncha	11010111001110010901110910010011000000100000000000 00000000000000000000000000
Priapulida	11010111001110010001110010010011000000111000000000 00000000000000000000000000
Nematomorpha	11010111001110010001111110090000000000000000000000 00000000000000000000000000
Nematoda	11010111001110010001111110011100000000000000000000 00000000000000000000000000
Chaetognatha	11010111001010010001111001100000000000000000000000 00000000000000000000000000
Mollusca	11010111001110010110000000000000000000000110000010 00100000000101010000000000
Nemertinea	11010111001110010110000000000001001000000110000000 00000000000001110000000000
Sipuncula	11010111001110010110000000000000000010000101000010 00100000000100000000000000
Echiura	11010111001110010110000000000000010000000100000010 00100000000100000110000000
Annelida	11010111001110010110000000000000000000000110000010 00100000000110000000000000
Pogonophora	10010111001110000190000000000010000000000110000010 00000000001100001000000000
Pentastomida	11010111001110010190000000000000000000000110000000 00000000001100000000000000
Tardigrada	11010111001110010190000000010100000000000110000000 01000000001100010001000000
Onychophora	11010111001110010190000000000100000000000110000000 00000000001100010101000101
Uniramia	11010111001110010190000000000100000000000110000000 00000000001100010001100111
Cheliceriformes	11010111001110010190000000000100000000000110000000 00000000001100010009019110
Crustacea	11010111001110010100000000000100000000000110000000 00000000001100010000111000
Phoronida	11010111001110010000000000000000000000000101111100 10100000000000000000000000
Ectoprocta	11010111001110010000000000000000000000000101111001 90000000000000000000000000
Entoprocta	11010111001110010110100000000010000000000101100101 00000000000000000000000000
Brachiopoda	11010111001010010000000000000000000000000101111100 11100000000000000000000000
Echinodermata	11000111001011010000000000000000000000000100001100 11010000110000000000000000
Enteropneusta	11010111001010010000001000000000010000000100001100 11011111000000000000000000
Pterobranchia	11010111001010010000000000000000000000000101119100 19011100000000000000000000
Urochordata	11011111001010010000000000000000000000000001000100 00001011001000000000000000
Cephalochordata	11010111001010010000000000000000000000000110000100 01001011001000000000000000

Table 2. Characters used in the analysis of living invertebrate phyla. Expression of these characters for each taxon is provided in Table 1, and used to generate cladograms in Figs. 5,6, and 7.

Primitive State	Derived State
1. no collagen, no acetylcholine/transmission, reproductive cells (if separate) on surface or exterior	1. collagen, acetylcholine/cholinesterase system, flagellated or ciliated sperm with condensed chromatin & mitochondria, reproductive cells or tissues internal
2. no gut	2. gut or special cells or areas for digestion
3. generalized free-swimming larva	3. nematogen/infusorigens
4. no symmetry	4. bilateral symmetry
5. no alteration of generations	5. metagenesis of some type
6. locomotion by action of cilia or flagella	6. locomotion by muscle action
7. no basal membranes, nerve cells, or gap junctions	7. epithelia with basal nerve synapses, gap junctions between cells
8. no endoderm	8. endoderm in embryo
9. no special water channels or choanocytes	9. pores & canals for water circulation facilitated by choanocytes
10. embryonic layers do not invert	10. embryonic inversion
11. no mesoderm	11. mesoderm in embryo
12. cleavage indeterminate	12. cleavage determinate
13. muscles ectodermal in origin	13. muscles subepidermal
14. no symmetry	14. radial or biradial
15. no nematocysts	15. nematocysts
16. no anus	16. anus
17. no colloblasts	17. colloblasts
18. radial cleavage	18. spiral quartet cleavage
19. ectomesoderm	19. 4d mesoderm
20. nervous system as poorly polarized nerve net	20. orthogonal nervous system with anterior nerve ring and several longitudinal cords
21. no body cavity	21. pseudocoel
22. generalized free-swimming larvae	22. no primary larva
23. body wall muscles both longitudinal and circular, when present	23. longitudinal body wall only
24. radial cleavage	24. aberrant spiral cleavage (monets and duets)
25. buccal chamber not eversible	25. buccal introvert
26. cuticle solid, if present	26. cuticularized epidermis with tubules
27. no retrocerebral organ	27. retrocerebral organ
28. no eutely	28. eutely
29. no renettes, amphids, or phasmids	29. renette cells, amphids, phasmids and phasmids
30. no molting	30. ecdysis with ecdysone
31. no special excretory organ	31. protonephridia
32. flame cells (ciliate)	32. flame bulbs (flagellate)
33. no lemnisci	33. lemnisci
34. no proboscis	34. proboscis
35. free cilia on epidermis	35. cilia ensheathed with cuticle
36. no uterus bell	36. uterus bell
37. epidermis solid	37. epidermal lacunae
38. cuticle uniform	38. cuticle with several unit membranes
39. no scalids	39. scalids on introvert
40. no lorica	40. lorica at some stage
41. no caudal appendages	41. caudal appendages at some stage
42. no body cavity	42. coelom (metacoel)
43. mesoderm not segmented	43. segmented or serial structures derived from mesoderm
44. gut straight	44. gut coiled or looped, anus anterior
45. no lophophore	45. lophophore
46. downstream particle	46. upstream particle capture in adults
47. downstream particle capture	47. upstream particle capture in larvae
48. larval apical organ does not degenerate	48. larval apical organ degenerates
49. generalized free-swimming larva	49. larvae as trochophore or trochophore-like
50. no special photoreceptor	50. photoreceptor cell with tuft of tightly packed cilia
51. coelom undivided	51. archimeric coelom
52. coelom as a schizocoel, if present	52. enterocoel if present
53. no special excretory organ	53. metanephridia
54. generalized free-swimming larva	54. tornaria/bipinaria larva
55. no pharygeal slits	55. pharyngeal slits
56. no buccal diverticulum	56. buccal diverticulum
57. no single, dorsal nerve cord	57. dorsal nerve cord from tube-like infolding of ectoderm
58. pharyngeal slits simple	58. pharyngeal slits divided

59. no calcareous endoskeleton
60. no water-vascular system
61. no notochord
62. brain not derived from any part of larval apical organ
63. no particular cell source for mesoderm and coelom
64. coelom large and pervasive
65. no rhynchocoel
66. no larval prototrochal lobes
67. no hemocoel
68. body grows uniformly
69. only circular and longitudinal body muscles
70. no anal vesicles
71. no appendages
72. no antennae
73. no special excretory organ
74. no appendages
75. no tracheae
76. no special excretory organ
77. no special jaws

59. calcareous endoskeleton
60. water-vascular system
61. notochord
62. brain in part derived from larval apical organ; main nerve cord ventral
63. teloblasts give rise to for mesoderm and coelom mesoderm and coelom
64. coelom restricted to around circulatory system
65. rhynchocoel
66. larva with prototroch developed as ciliated lobes (pilum or velum)
67. hemocoel (unlined cavity in mesoderm)
68. apical growth of larva
69. oblique body wall muscles
70. anal vesicles
71. lobopods or uniramous limbs
72. deutocerbral antennae
73. segmental, excretory glands
74. biramous limbs
75. tendency to develop tracheae
76. tendency to develop Malpighian tubules
77. whole limb jaw

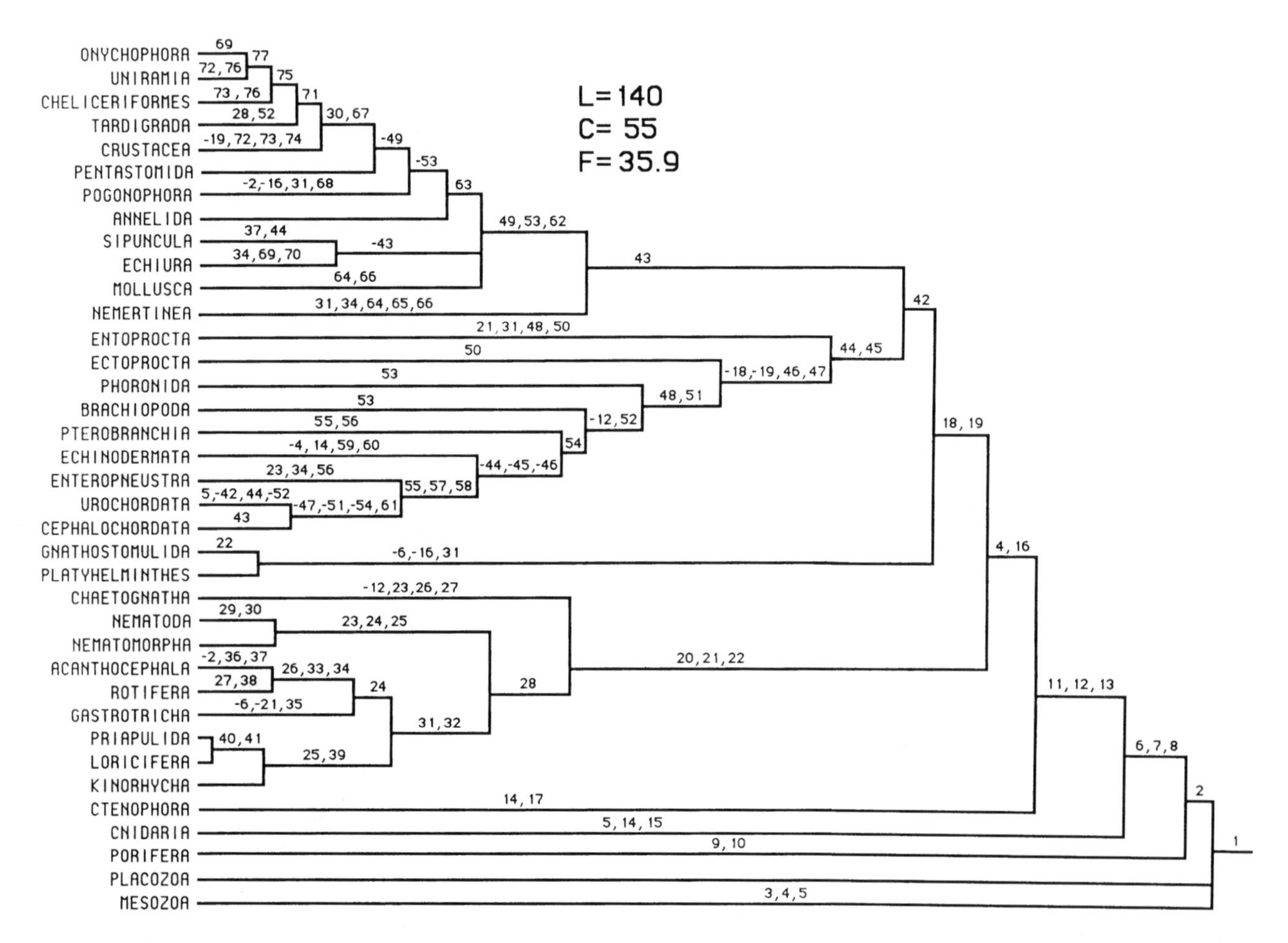

Fig. 5. Cladogram of animal kingdom based on analysis of data matrix of Table 1 with PHYSYS (WAGNER.S) to produce shortest, most parsimonious tree. Characters are those listed in Table 2. Arrangement shown for arthropods only one of several equally parsimonious possibilities.

of an anus, and utilization of protonephridia for excretion and osmoregulation.

The eucoelomates have two branches. One of these lines include taxa that have traditionally been included within the deuterostomes and the lophophorates. This is a lineage founded on the development of a sessile life style, lophophore, and looped gut with anterior anus. A transition series is seen in this clade whereby the lophophorate characters are lost and a mobile life-style is regained. In this analysis, the lophophorates and the phylum Hemichordata (pterobranchs and enteropneusts) are sorted as paraphyletic groups.

The sister group to the deuterostome/lophophorate clade consists of phyla that tend to develop segmental or serial structures in tissues derived from mesoderm. This feature became increasingly important in the evolution of this clade. This lineage includes the trochophorate/annulate phyla as well as the nemertines.

The reader should remember that the above-postulated relationships are dependent on the particular data matrix used here. New knowledge of the structure and development of all the phyla will produce factual information that will undoubtedly alter the matrix and thus this scheme.

What to do with Problematica

Can the above method be applied towards elucidating the systematic affinities of fossil problematica? One of the principle difficulties in dealing with problematica is that so little is known of their anatomy. In studying such organisms, one deals with this lack of information. The literature on these creatures often abounds with subjective speculations about their supposed affinities.

However, computer programs are structured to handle taxa with unknown (= missing) characters. In typical cladistic analyses of living groups, as that detailed above, some taxa have a few characters for which nothing is known (see e.g., the scoring of the Loricifera in the matrix of Table 1). Fossil problematica, however, might have many, if not most, character states unknown. Nevertheless, if problematica are handled in a manner akin to that used for living taxa, a great advantage could be gained in trying to understand their affinities.

To test this, three problematic fossil taxa, whose phylum affinities have been the subject of speculation, were arbitrarily selected for treatment in a cladistic analysis. These were *Anomalocaris canadensis* (see Whittington & Briggs, 1985) from the Cambrian Burgess Shale fauna of British Columbia, and *Tullimonstrum gregarium* (see Johnson & Richardson, 1969) and *Etacystis communis* (see Nitecki & Schram, 1976) of the Pennsylvanian Mazon Creek faunas of Illinois. Two data sets were developed (Table 3), one set in which any unknown character was scored as a number 9, and another in which reasonable assumptions were made about what character states in the fossils may have been. In the former case, most character states were designated 9, whereas in the latter, only those characters that dealt with larval anatomy were scored 9. Each of these data sets were then appended to the basic matrix for all invertebrate phyla and subjected to a WAGNER.S analysis.

Table 3. Data matrices of characters combined with the matrix of Table 1 and used to analyze the fossil problematica and produce the cladograms in Figures 6 and 7. Matrix A, which factors in Figure 6, has all unknown characters designated with a 9, while matrix B, which factors in Figure 7, contains the best approximations for as many of the adult characters as possible. Characters are those listed in Table 2.

Taxa	Characters
A	
Anomalocaris	11919191091910019990090909999199909999000110099990 09990900000999099009099099 0
Tullimonstrum	11919191091910919990999909999999919999000110099990 09990990009990999009009000 0
Etacystis	11919191099990999999999990999999999099990009091999 90 19999999009999099099000000 0
B	
Anomalocaris	11010111001110010190090000090199000000000110009990 09900000000999099090019099 0
Tullimonstrum	11010111001910010990000900090099901000000110009990 00900090000990999099009099 0
Etacystis	11019111001910010990090900090099000000000101199990 19990990000990090000000000 0

In Fig. 6, based on a matrix with unknown character states scored as missing data, the three taxa are placed in entirely reasonable positions. *Anomalocaris* is in an unresolved trichotomy with Crustacea and other arthropods, not too surprising given its worm-like body and arthropod-like cephalic limbs. *Tullimonstrum* is in another trichotomy with Annelida and higher annulates, an affinity, to my knowledge, never suggested before (see also Beall, this volume). *Etacystis* is part of an unresolved quadrichotomy with Ectoprocta, Phoronida, and the higher deuterostome/lophophorate line, not too different from the original suggestion of Nitecki & Schram (1976) that *Etacystis* might be allied with hemichordates.

The alternative matrix, that for which reasonable assumptions were made for many of the adult character states of the problematica, produced the cladogram of Fig. 7, one not unlike that of Fig. 6. In this tree, *Anomalocaris* is placed as a sister taxon to Crustacea, *Tullimonstrum* has shifted to a polychotomy in proximity to the Nemertinea (mirroring some original suggestions of Johnson & Richardson, 1969), and the position of *Etacystis* is unchanged over that seen in Fig. 6.

The treatment of fossil problematica with cladistic methods of analysis presents some distinct advantages over previous approaches. First, as with cladistics in general, the assumptions about character states are clearly obvious and thus more easily subjected to scrutiny than previous approaches that speculate only in a narrative fashion about taxonomic affinities of problematica. Second, because it forces one to deal with multiple characters, this method avoids arguments of affinity based on only one or two characters. Third, by dealing with large numbers of characters, we can

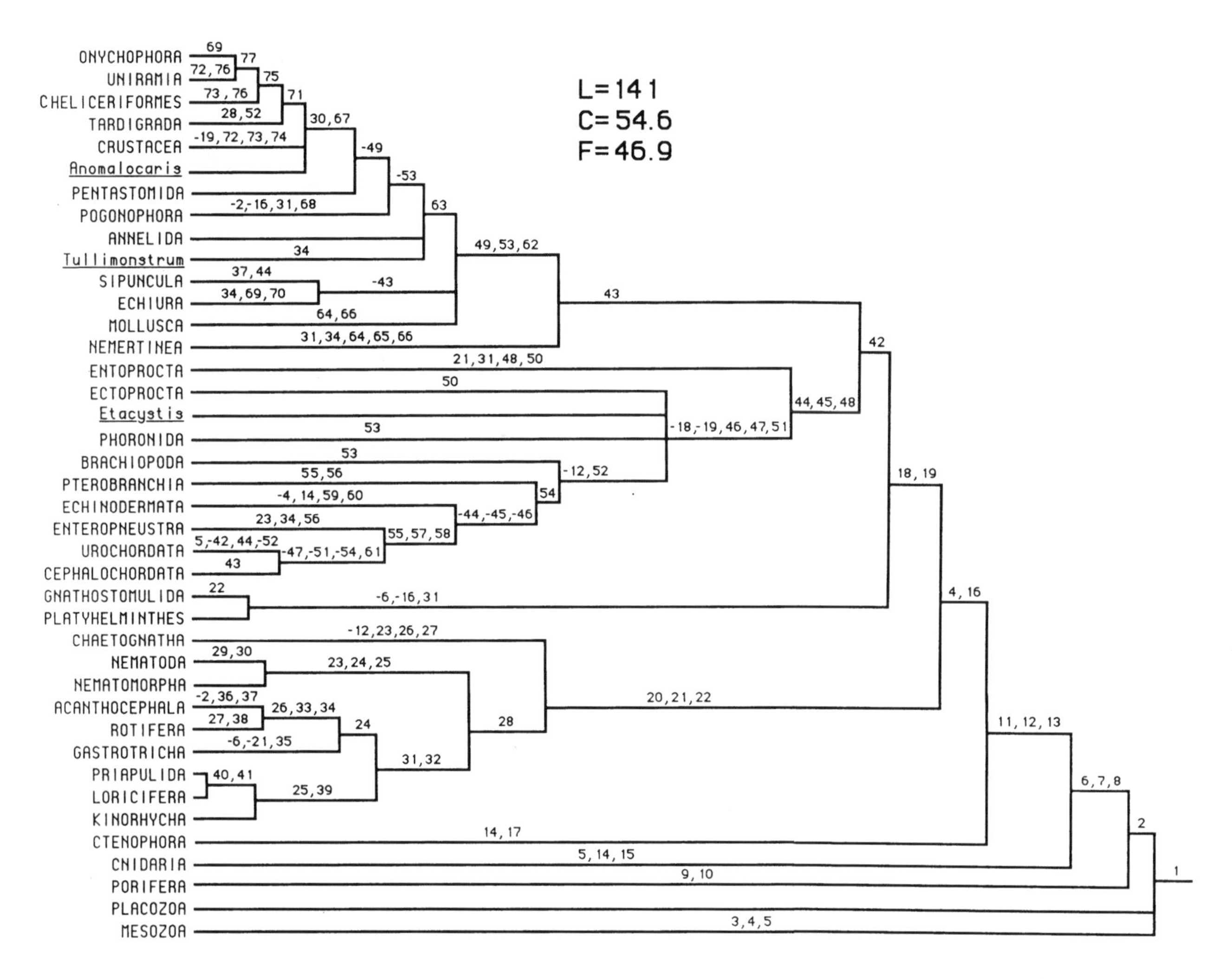

Fig. 6. Cladogram of animal kingdom including three problematica based on analysis of data matrices of Tables 1 and 3A with PHYSYS (WAGNER.S). Characters are those listed in Tables 2. No assumptions made as to the probable state of any unknown characters among the problematica.

clearly see what information is missing from our knowledge about the anatomy of problematica, and this provides us with an opportunity to formulate strategies in which we might be able to obtain that information. It is more meaningful to study a problematic taxon if you have some idea of what you should be looking for, given what you may already know about its anatomy, than if you approach each problematicum as a kind of tabula rasa.

Scenario of invertebrate evolution

If we accept the above cladistic analyses as hypotheses that can and should be tested with further observations and data, then we can now develop a corresponding scenario of events that may have occurred in the course of invertebrate evolution. Here is one possible story.

Early stages in the evolution of the animal kingdom centred on increasing the complexity of structure and biochemistry that facilitated the interaction of cells. This sequence of events began at a protistan grade, wherein individual cells, even those within colonies, were independent entities. An initial step in the evolution of metazoans involved the development of organization and partial specialization of cell types (at least as far as reproduction is concerned), such as that seen among the mesozoans and placozoans.

Next came a stage whereby specific body functions were restricted to specific cells. Sponges (see also Wood, this volume) reflect this grade, though they lack true tissues since poriferan cells often are capable of performing various functions at different times. The next step in metazoan evolution involved advanced cell specialization to produce true tissues. Cnidarians and ctenophores mark different phases in this sequence of events that involved not only development of tissues (i.e., the specialization of cells) but also development of embryonic germ layers (i.e., the specialization of a process that could produce tissues).

Animal evolution up to that point appears to have focused on internal factors, i.e., facilitation of cell interactions, specialization of cell function, and perfection of a process for differentiation of distinct cell and tissue types. A stage was achieved whereby moderately complex body forms had evolved that could handle a diversity of relatively sophisticated functions related to survival and reproduction.

At that point, animal history underwent a shift in focus away from internal factors related to basic body organization, toward evolving diverse body plans that

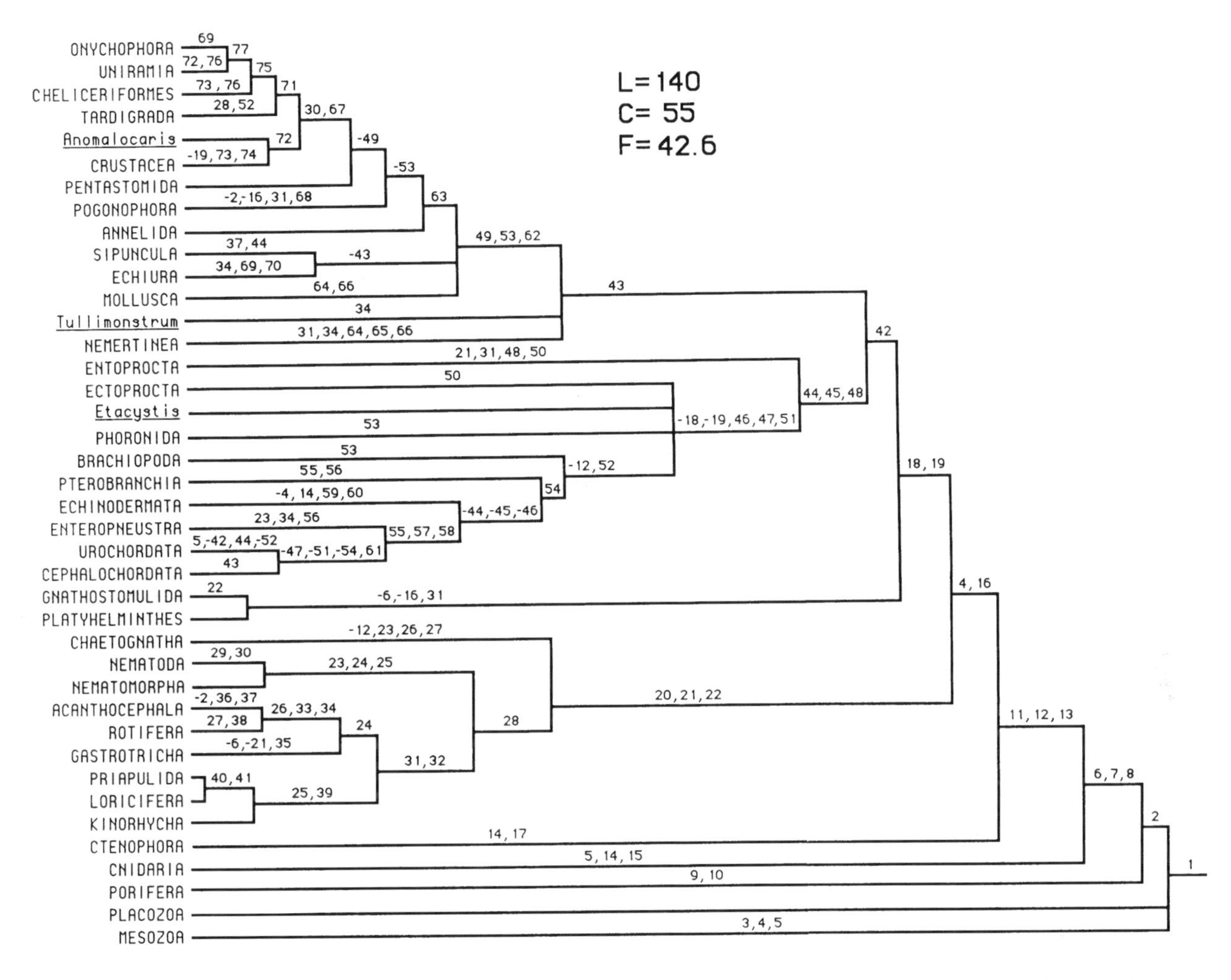

Fig. 7. Cladogram of animal kingdom including three problematica based on analysis of data matrices of Tables 1 and 3B with PHYSYS (WAGNER.S). Characters are those listed in Tables 2. Characters for problematica scored with best estimates as to their adult state if not directly known.

could more effectively interact with the external environment. The appearance of bilateral symmetry marks this transition. Two different clades then evolved, which reflect different solutions to the ontogenetic problems engendered by this shift.

One clade, represented by the pseudocoelomates, is often characterized by strange monet and duet patterns of early cell cleavage, and perhaps, as a result, lacks primary larval phases in their life cycles. Such "larvae" as they possess are rather specialized juveniles. The pseudocoelomates function with nervous systems (termed orthogonal) that are essentially modified from a radial form. Several nerve cords typically occur, none of them really predominating over any of the others, and all of the cords are loosely controlled by a nerve ring that is poorly or weakly developed as a "brain". The fact that the processes of embryonic differentiation seem poorly developed among pseudocoelomates may be responsible for the structural peculiarities we see among them. Their body plans have many parallels to those seen among the more advanced coelomates. There are specialized muscle layers for locomotion, but these are often incomplete (such as lack of circular muscle layers in the nematodes, nematomorphs, and chaetognaths). There is a body cavity, but there is a lack of the elaborate specialization of muscles that coelomates have, both in the body wall and around the gut tissues. Pseudocoelomates develop some very special and peculiar sense organs to deal with specific sensory needs, such as amphids, phasmids, retrocerebral organs, caudal appendages (?), and sensory spines, but they never evolved any effective degree of cephalization.

The other clade, which includes acoelomates and coelomates, differs from the pseudocoelomate line in having a highly regulated and determinate mode of ontogeny based on spiral quartet cleavage and 4d mesoderm. These distinctive features of development were often lost or greatly modified in the course of the evolution of this clade. Nevertheless, the hallmark of this clade as a whole, undoubtedly related to the specialized mode of development they passed through, is highly complex tissue types. The phyla in this clade are typically characterized by the possession of some kind of distinct larval stage.

At least three clades resulted from the specialization of cleavage pattern and specification of mesoderm formation. One group, the acoelomates, never developed particularly complex body plans, but did perfect increasingly complex life cycles that focused on meiofaunal or parasitic life-styles. Perhaps this emphasis was related to a lack of body cavity.

The other two lineages were based on a coelomic body plan. One of these, the deuterostome/lophophorate line, focused initially on the evolution of a primarily sessile life-style. Many aspects of the original spiral cleavage and 4d mesoderm formation were lost. This loss perhaps reflected an underlying process operating in the genome of these animals, since the more derived phyla of this clade exhibit paedomorphic processes in the course of their evolution. That is, either juvenile features persisted into the adult phase, or gonad function was shifted into a juvenile phase in order to achieve major morphological changes in body plan. Sustained ontogenetic paedomorphosis, however, brought a return to a mobile life-style among the advanced "deuterostome" phyla of this lineage.

The evolution of the rest of the coelomate phyla was apparently founded on mobile life-styles. In addition, these phyla seem to possess a genome that is organized in a way that "pre-disposes" these groups to differentiate at least serial, if not truly segmental, body structures. This capacity, when linked in some of the phyla with paedomorphic processes in ontogeny, resulted in an arthropod clade noted for great diversity in body forms as well as an overwhelming number of species. The scheme shown here (only one of several equally parsimonious possibilities) could imply that the phylum Arthropoda should be redefined in order to include groups that are traditionally excluded, such as Tardigrada and Onychophora, and thus recast the ideas of Manton (1977) and Anderson (1973). The trochophorate phyla de-emphasize the serial potential of their mesoderm, viz. the mollusks and echiuran/sipunculans.

Whether or not this scenario is a true and accurate description of what really happened in the course of animal history is almost irrelevant. What is important is that we subject the analysis of animal relationships to scientific methods, and that stories based on such analyses need not be pure whimsy.

REFERENCES

ANDERSON, D. T. 1973. *Embryology and Phylogeny on Annelids and Arthropods* Oxford: Pergamon Press.

ANDERSON, D. T. 1983. Origins and relationships among the animal phyla. *Proceedings of the Linnean Society of New South Wales* **106**, 151-66.

BERGSTRÖM, J. 1986. Metazoan evolution--a new model. *Zoologica Scripta* **15**, 189-200.

FIELD, K. G., OLSEN, G. J., LANE, D. L., GIOVANNONI, S. J., GHISELIN, M. T., RAFF, E. C., PACE, N. R., RAFE, R. A. 1988. Molecular phylogeny of the animal kingdom. *Science* **239**, 748-753.

HADZI, J. 1963. *The Evolution of the Metazoa*. New York: Macmillan.

HAECKEL, E. 1874. Die Gastrea Theorie, die phylogenetische Classification des Thierreiches und die Homologie der Keimblatter. *Jenaische Zeitschrift für Naturwissenschaft* **8**, 1- 55.

HANSON, E. D. 1958. On the origin of the Eumetazoa. *Systematic Zoology* **7**, 16-47.

HYMAN, L. H. 1940. *The Invertebrates: Protozoa Through Ctenophora*. New York: McGraw-Hill.

INGLIS, W. G. 1985. Evolutionary waves: patterns in the origins of animal phyla. *Australian Journal of Zoology* **33**, 153-78.

IVANOVA-KAZAS, O. M. 1981. Filogeneticheskoye znachyeniye cliralnogo drolyeniya. [Phylogenetic significance of spiral cleavage.] *Biologiya Morya* **1951(5)**, 3-14. [in Russian].

JÄGERSTEN, G. 1955. On the early phylogeny of the Metazoa. The bilaterogastrea theory. *Zoologiska Bidrag fran Uppsala*. **30**, 321-354.

JÄGERSTEN, G. 1972. *Evolution of the Metazoan Life Cycle*. London: Academic Press.

JOHNSON, R. G., & RICHARDSON, E. S. 1969. Pennsylvanian invertebrates of the Mazon Creek area, Illinois: The morphology and affinities of *Tullimonstrum*. *Fieldiana: Geology* **12**, 119-149.

MARCUS, E. 1958. On the evolution of the animal phyla. *Quarterly Review of Biology* **33**, 24-58.

MANTON, S. M. 1977. *The Arthropoda*. Oxford: Clarendon Press.

METSCHNIKOFF, E. 1886. *Embryologische Studien an Medusen*. Holder, Vienna.

NIELSEN, C. 1985. Animal phylogeny in light of the trochaea theory. *Biological Journal of the Linnean Society, London* **25**, 243-99.

NITECKI, M. H., & SCHRAM, F. R. 1976. *Etacystis communis*, a fossil of uncertain affinities from the Mazon Creek fauna. *Journal of Paleontology* **50**, 1157-1161.

NURSALL, J. R. 1962. On the origin of the major groups of animals. *Evolution* **16**, 118-23.

REMANE, A. 1963. The evolution of the Metazoa from colonial flagellates vs. plasmodial ciliates. In *The Lower Metazoa*. (ed. E. C. Dougherty), pp. 23-32. Berkeley: University of California.

REMANE, A. 1967. Die Geschichte der Tiere. In *Die Evolution der Organismen*, Vol. 1 (ed. G. Heberer), pp. 589-677. Stuttgart: Fischer.

RUNNEGAR, B., & POJETA, J. 1985. Origin and diversification of the Mollusca. In *The Mollusca*, Vol. 10 (eds. E. R. Trueman & M.R. Clark) pp. 1-57. New York: Academic Press.

SALVINI-PLAWEN, L. VON, 1978. On the origin and evolution of the lower Metazoa. *Zeitschrift für zoolologische Systematik und Evolutionsforschung* **16**, 40-88.

SALVINI-PLAWEN, L. VON, 1980. Was ist eine Trochophora? Eine Analyse der Larventypen mariner Protostomier. *Zoologische Jahrbuch Abteilung Anatomie* **103**, 389-423.

SALVINI-PLAWEN, L. VON, 1982. A paedomorphic origin of the oligomerous animals? *Zoologica Scripta* **11**, 77-81.

SCHRAM, F. R. 1983. Method and madness in phylogeny. *Crustacean Issues* **1**, 331-350.

SHAROV, A. G. 1966. *Basic Arthropodan Stock with Special Reference to Insects*. New York: Pergamon.

SIEWING, R. 1969. *Lehrbuch der Vergleichenden Entwicklungsgeschichte der Tiere*. Hamburg: Paul Parey.

SIEWING, R. 1977. Mesoderm in Ctenophores. *Zeitschrift für zoologische Systematik Evolutionsforschung* **15**, 1-8.

SIEWING, R. 1980. Das Archicoelomatenkonzept. *Zoologishe Jahrbuch Abteilung Anatomie* **103**, 439-82.

TATTERSALL, I., & ELDREDGE, N. 1977. Fact, theory, and fantasy in human paleontology. *American Scientist* **65**, 204-211.

WHITTINGTON, H. B., & BRIGGS, D. E. G. 1985. The largest Cambrian animal, *Anomalocaris*, Burgess Shale, British Columbia. *Philosophical Transactions of the Royal Society of London* **B 309**, 569-609.

Is fossil evidence consistent with traditional views of the early metazoan phylogeny?

Jerzy Dzik[1]

Abstract

The best known of the alleged Cambrian medusoids, *Velumbrella*, is a skeletal fossil. It is proposed to be a relative of *Eldonia* and *Dinomischus*, and is placed together with them in a separate new class Eldonioidea of the lophophorates. The supposed Cambrian coral *Tabulaconus* is probably a successor of the Tommotian *Cysticyathus*, which does not show features that could substantiate such a taxonomic placement. Ordovician *Conchopeltis* is closely related to scenellids and, having a bilaterally symmetrical, probably aragonitic shell with radially arranged muscle scars cannot be interpreted as a chondrophoran pneumatophore. The small size of ancestral conchiferan molluscs, assumed on fossil evidence, is a preservational artifact that resulted from very slow rate of sedimentation connected with phosphatization in the most fossiliferous strata of the earliest Cambrian. Adult specimens occurring rarely in the same strata do not differ in size from later molluscs. Relatively large size is also characteristic of the most primitive articulate, *Xenusion* from the basal Cambrian of the Baltic region. It is suggested that major evolutionary transformations between phyla operated between organisms of centimetre sizes, so their record is potentially recognizable in the fossil evidence.

Introduction

To derive a phylogenetic tree from morphologic data an assumption is necessary: that there is some correspondence between time that has passed since separation of lineages under consideration and the present morphologic distance between them. If such a correspondence really exists, one may expect that going back in time the morphologic differences within any monophyletic unit should generally decrease. It follows also that, however incomplete is the fossil record of evolution, the morphologic differences between oldest known representatives of any taxon and its ancestor should be smaller than between any Recent forms and the ancestor, simply because the time interval is shorter.

When this way of reasoning is accepted, it becomes somewhat surprising that the tremendous increase in knowledge of Vendian and Early Palaeozoic fossils in the last few decades has influenced so little our understanding of the early evolution of the metazoans. Usually we tend to blame incompleteness of the fossil evidence, but another possibility has also to be considered, namely that something is wrong with basic hypotheses that we are attempting to test. Virtually all interpretations of the early phylogeny of the Metazoa, which are based on analyses of the anatomy and biochemistry of Recent organisms, place the coelenterates and flatworms at the base of the tree (see Bergström 1986) and/or declare that the earliest representatives of the major groups were microscopic in size (see Nielsen 1985). In this review I intend to examine, on the basis of a few examples, whether these two features of the metazoan phylogeny necessarily result from the available evidence or perhaps they are unnecessary assumptions that we tend to fit with an obviously incomplete fossil data.

How ancient are coelenterates?

A remarkable feature of the Ediacarian assemblages of supposed jelly-fishes is an apparent lack of tetraradially organized forms. This does not allow us to consider them reasonable ancestors of later scyphozoans. The only fossil declared to show this kind of radial symmetry in the Vendian is *Conomedusites* from Ediacara (Glaessner, 1971) but its morphology is far for being convincingly scyphozoan. Its conical body seems to be split into four lobes which were movable in relation to each other allowing various arrangements (see Glaessner 1971; Pl. 1 9-10). It has been compared with allegedly the most primitive, widely conical "conulariid" *Conchopeltis* from the Trenton (Late Caradoc) of New York. However, it has been already shown by Oliver (1984) that *Conchopeltis* is not a radial organism but shows clear bilateral symmetry. It has little to do with conulariids, as already pointed out by Kozlowski (1960), and supposed tentacles represent rather a cuticular fringe along the shell margin.

This interpretation may find support in a new finding of a *Conchopeltis*-like fossil in the Baltic Caradoc (Fig. 1). Morphologically it is transitional between Cambrian *Scenella* and *Conchopeltis*. It is of interest to find radially arranged riblets on its interior that are interpreted as bordering muscle scars resembling those in *Scenella* (Rasetti, 1954). The shell was probably aragonitic as it is preserved in the same manner as associated snails, in contrast to trilobites and brachiopods cooccurring in the same block. It lacks any remnants of organic matter, despite good preservation of graptoliths and other organic fossils in the rock. To establish definitely the nature of the fossil it would be necessary to know either the shell microstructure or its early ontogeny. That is not possible in the case of the specimen under consideration, but in another boulder of similar age a minute plate has been found which somewhat resembles it in outline and presence of inconspicuous radial ribs on its interior (Fig. 2). Like in associated snail conchs an original shell matrix is replaced with an iron mineral, which is not the case regarding trilobites and brachiopods in this kind of rock. Its external side is tubercular and the shell is bent transversely in a way resembling polyplacophorans, so it may actually be an anal plate of a chiton. The apex is not preserved well enough to prove conchiferan or polyplacophoran affinities but definitely shows that it is not a coelenterate.

Because of inferred aragonitic wall composition,

[1]Zakład Paleobiologii PAN, Aleja Żwirki i Wigury 93, 02-009 Warszawa, Poland.

Fig. 1. *Conchopeltis*-like fossil from Chasmopskalk erratic boulder (late Caradoc) of Baltic origin, Jóźwin near Konin, Poland. A. Part. B. Counterpart, Both x 1.

apparent growth lines and structures interpreted as muscle attachments in both these fossils any possibility that they are chondrophorans can be rejected and I extend this also to morphologically close *Conchopeltis* and *Scenella*-like fossils. In some of them original mineralization is apparent (see Yochelson & Stanley, 1981; Fig. 1E for imprints of calcite prisms of external shell layer) which is hardly compatible with the notion of the chondrophoran pneumatophore. Shells of *Plectodiscus* are known to be overgrown with cementing organisms (Yochelson *et al.*, 1983), which require they were stiff and resistant for decay, features unlikely to develop if they were not mineralized. Horný (1985) found clearly bellerophontid larval shell in a Silurian fossil morphologically undistinguishable from *Palaeolophacmaea*.

Among the most controversial Early Palaeozoic fossils assigned to the Chondrophora by Stanley (1986) is the Middle Cambrian *Velumbrella*, originally described by Stasińska (1960) as a jelly-fish (see also Bednarczyk, 1970). It has been already pointed out by Fedonkin (1987) that actually *Velumbrella* is a skeletal fossil. It is known from numerous specimens. (Stasińska, 1960) lists 110 imprints (and more can be traced in private collections) which are not preserved on the bedding plane, as is usual for fossil jelly-fishes, but well within an unbedded coarse somewhat conglomeratic sandstone with small quartz pebbles. The *Velumbrella* discs are variably oriented, sometimes bent, and many small fragments of crushed specimens cooccur (see Fig. 3). This indicates high energy and shallow water conditions of sedimentation. It was

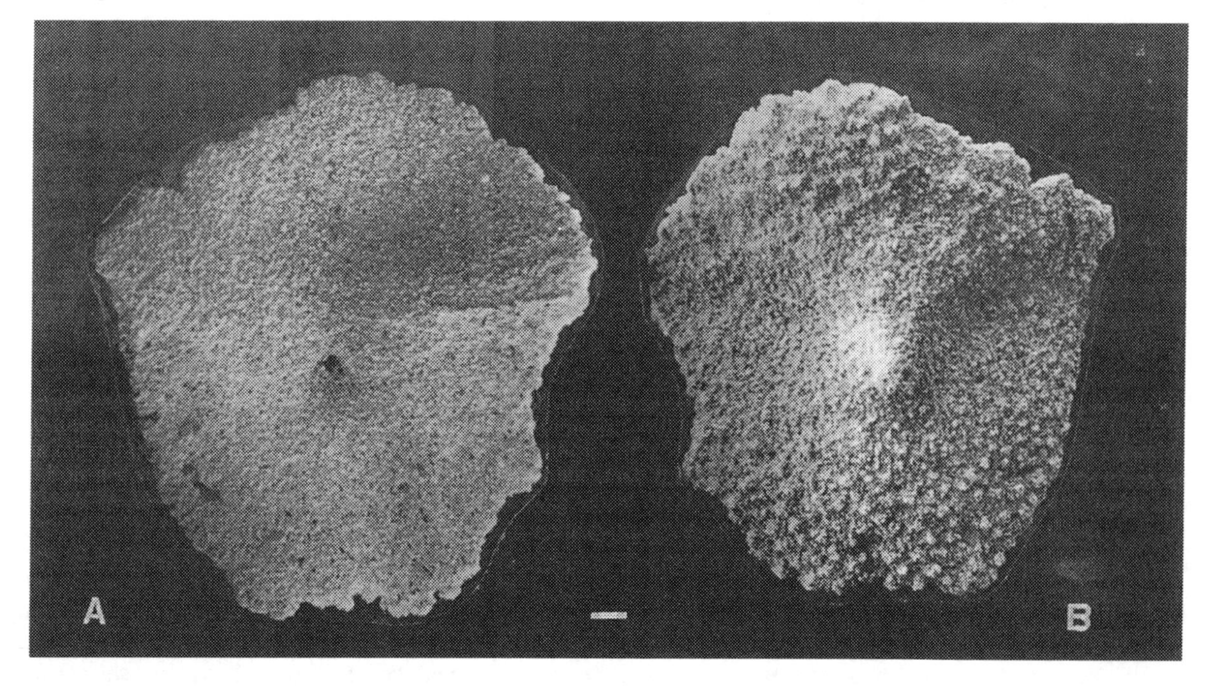

Fig. 2. Juvenile specimen of a species possibly related to that in Fig. 1 with original shell matrix replaced with an iron mineral; erratic boulder of pink micritic limestone (probably Vasalemma Baltic stage; late Caradoc), Orłowo near Gdynia, Poland. A. Inner side, note radial riblets. B. External view. Both x 32.

definitely not an environment appriopriate for preservation of jelly-fishes or organic chondrophoran pneumatophores. Discs of *Velumbrella* were clearly stiff, possibly mineralized, although still somewhat flexible, so indicating some organic matter.

Together with *Velumbrella czarnockii* Stasińska, 1960 another, probably related, species occurs, represented in the collection by two fragmentary specimens. These are labelled *Brzechowia* sp. by Jan Czarnocki, who assembled the whole collection. Specimens of "Brzechowia" are apparently two-walled. A space between the walls is filled with sand which requires, as long as the margins of both valves are not displaced despite their breakage prior to burial, a firm connection between them, presumably in the centre of valves. Another possibility, that the discs had been bent in half prior to their burial is less likely; at least it is not supported by any signs of bending in proximity of the breakage margin. It seems likely that *Velumbrella* was also bivalved, although all specimens are disarticulated and only in two cases (Stasińska, 1960: PI. II; here Fig. 3A) paired valves occur in partial overlap.

In both *Velumbrella* and "Brzechowia" the valves grew by marginal accretion. In the *Velumbrella* discs the central area (8 mm in diameter) does not show radial ornamentation and may represent an attachment area for both valves. "Brzechowia" differs from it in lacking a concentric depression in about half of the disc radius as well as radial ribs (both features occurring also in the Cambrian parapsonematid described by Popov 1967), although radial organization is clearly noticeable close to the margin. Concentration of growth lines close to the margin in "Brzechowia" and almost uniform size of all well preserved specimens of *Velumbrella* indicate that their mature size was determined ontogenetically.

Fig. 3. *Velumbrella czarnockii* Stasińska, 1960, earliest Middle Cambrian (Bednarczyk 1970) of Brzechów, Holy Cross Mts, Poland: two associated discs possibly belonging to a single specimen. B. "Brzechowia" sp. from the same locality with both valves still in articulation, note detritus of *Velumbrella* disks below. Both x 1.

The morphologic data listed above do not allow by themselves to establish the systematic position of these fossils. They show, however, that they definitely were not scyphozoans and give little support for their interpretation as chondrophorans. Supposedly near-shore, shallow water environment of fossilization, two-walled organization with walls connected (and attached to the substrate?) in the centre, scleritized (mineralized?) externally except for growing margin, suggest rather a benthonic mode of life of these organisms.

The Cambrian skeletal fossil morphologically closest to *Velumbrella* is *Yunnanomedusa* from the Chiungchussu fauna of Yunnan, China (Sun & Hou, 1987). The degree of scleritization of its disc was somewhat lower than in *Velumbrella* and radial ribs, being more numerous (about 44 instead of 20) are also less distinct. Even less scleritized is *Stellostomites* from the same strata, with about 60 radial striae. Conway Morris & Robison (1988) consider these Chinese genera synonymous with *Eldonia*. At the end of a morphocline of these Chinese discoidal fossils can be placed *Rotadiscus*, unless soft parts studied recently by Conway Morris (personal communication) will counterevidence this. It has a strongly scleritized disc without any prominent radial ornamentation but, instead, with distinct concentric growth lines and rugae. In all these fossils can be identified a central attachment area (interpreted as a mouth by Sun & Hou, 1987). The opposite end of the morphocline can be supplemented by the disc of "Brzechowia", *Eldonia* from the Middle Cambrian Burgess Shale of British Columbia (Durham, 1974) and the Spence Shale and Marjum Formation of Utah (Conway Morris & Robison, 1988) as well as a Siberian Late Cambrian fossil (perhaps assignable to *Parapsonema* but erroneously described under the name of the echinoderm genus *Camptostroma* by Popov, 1967; see Conway Morris & Robison, 1982) which have the margin of discs lobate. Within the central area of discs of *Eldonia* (Durham, 1974) and *Stellostomites* (Sun & Hou, 1987; p. 265) is preserved a helically coiled structure, interpreted as a gut by Durham (1974), that indicate that the morphocline represents a monophyletic group (Conway Morris & Robison, 1988). Assuming that the Durham's reconstruction of *Eldonia* is correct, in searching for affinities for this group one has to look for organisms with an U-shaped intestine and a conical body with radially lobate margins.

Transversely striated scleritized lobes around the margin of a conical cup and a U-shaped intestinum located in the centre of the cup are shared with the velumbrellids by another problematic fossil of the Cambrian, *Dinomischus*. The cup of *Dinomischus*, having a radius of about 2 cm, is thus much smaller than most of the velumbrellids. It bears approximately 10 scleritized "bracts" (Chen *et al.*, 1989: p. 69) and is attached to a long stalk. The structure interpreted by Chen *et al.* (1989) as an anal tube was actually the basal part of the stalk, strongly bent and partially hidden under the cup (Conway Morris, 1989: p. 270). I propose to homologize the stalk of *Dinomischus* with the attachment area in discs of the velumbrellids and the radial lobes of *Eldonia* with "bracts" of *Dinomischus*.

I conclude thus that the velumbrellids and *Dinomischus* form a monophyletic group. Unlike other problematic Vendian and Cambrian fossils enough anatomical data are available to characterize it in zoological terms and to propose its placement, at least provisionally, in the classification scheme of the Metazoa. The U-shaped intestine, mouth armed with two branched tentacles (known in *Eldonia*) strongly suggest

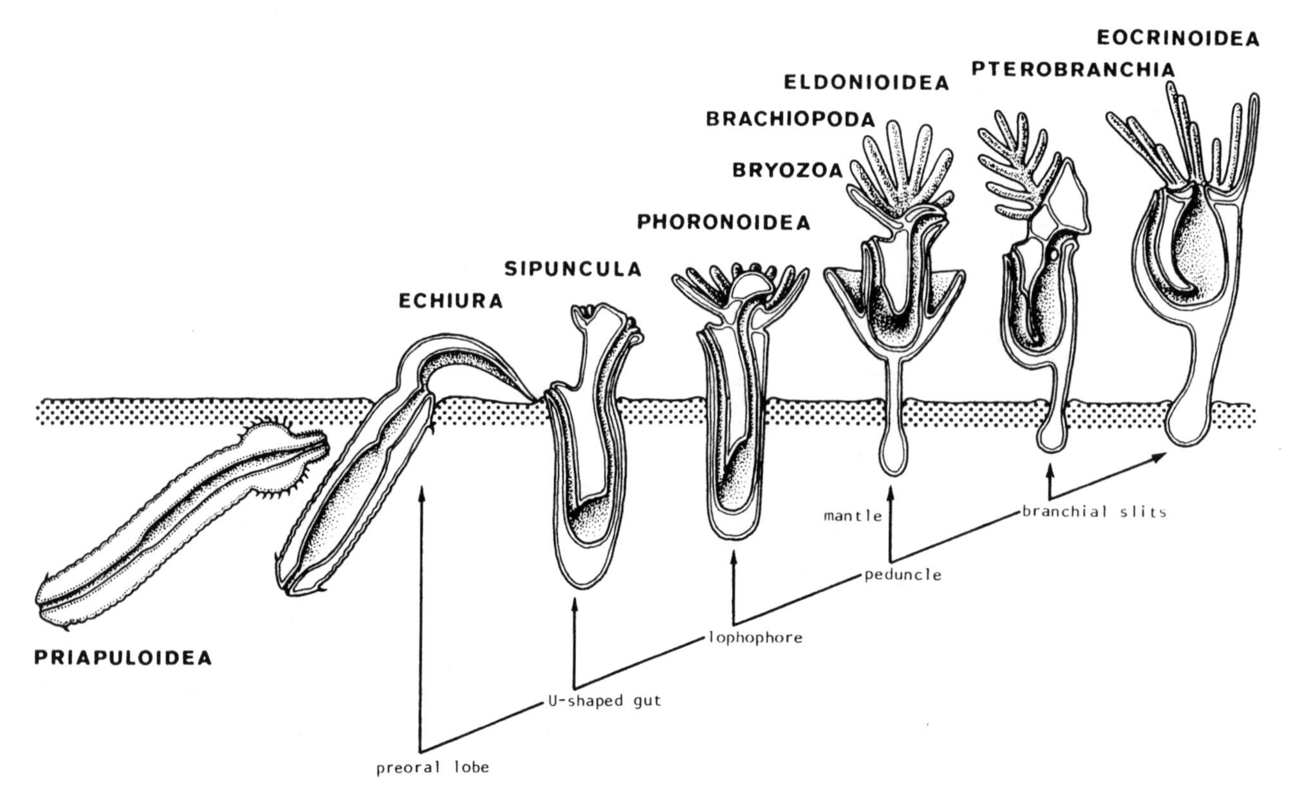

Fig. 4. Interpretation of relationships of sedentary coelomates derived from an assumption that the priapulids represent their basal stock. According to the proposed scenario there was a shift from active predatory life, with hydraulic locomotion, through detritophagy of the echiurids and sipunculids, to filtratory feeding of the lophophorates (from Dzik, in press b).

their placement among the lophophorates (phylum Tentaculata) (Fig. 4). Homology of the scleritized disc, surrounding both the anus and mouth, to organs of other lophophorates is the crucial point in the interpretation of the anatomy and relationships of the group. Conway Morris (1977b) discussed and rejected possible entoproct relationships of *Dinomischus*. It seems thus unlikely that these are giant solitary entoprocts. Instead, I propose to homologize the disc secreting organ with the mantle of the brachiopods and the bryozoans. The presence of a bryozoan-type protoecium in Ordovician *Cornulites*-like fossils suggests that at least in the evolution of the Bryozoa development of a skeleton, secreted by the mantle had preceded the origin of coloniality (Dzik, in press a). I propose, therefore, to classify the velumbrellids and *Dinomischus* in a new class, Eldonioidea, characterized by a conical, lobate mantle instead of cylindrical (as in the Bryozoa) or bivalved (as in the Brachiopoda) ones. This implies that the mantle is a shared derived character of the advanced Tentaculata (lacking in, although possibly secondarily lost by, the Phoronoidea). Two distinct groups of ordinal rank can be distinguished within the class: the Dinomischida, new order, characterized by a long pedicle and deeply lobate mantle, and the Velumbrellida, new order, with a circular, disc shaped body with reduced peduncle. Within the latter group at least two branches, both showing diverse modes of scleritization of the disc, can be discerned: the Eldoniidae Walcott, 1911, with radial ribs [including *Eldonia* Walcott, 1911, *Velumbrella* Stasińska, 1960, and *Yunnanomedusa* Sun & Hou, 1987 (= *Stellostomites* Sun & Hou, 1987)] and the Rotadiscidae, new family, with almost smooth discs ornamented only by growth lines (*Rotadiscus* Sun & Hou, 1987).

With *Velumbrella* and other alleged chondrophorans removed from considerations, little remains in the whole Early Palaeozoic, which can be reasonably compared with planktonic coelenterates. The oldest known undoubted coelenterates are thus sedentary polyps, well recognized from the beginning of the Early Ordovician when both the conulariids (see also Babcock, van Iten, this volume) and corals appear. Supposed Cambrian conulariids await thorough redescription and their presence among the oldest known, Vendian and Early Cambrian, metazoans is rather doubtful, although some tubular middle Early Cambrian fossils seem to show a conulariid-like tetraradial symmetry (Qian & Bengtson, 1989). The coelenterate nature of alleged Vendian pennatulaceans, Petalonamae, has been already soundly questioned by Seilacher (1984) and the new reconstructions proposed by Jenkins (1985) present creatures with petaloids attached to imperforate membranes, which are more likely to be photosynthesizers rather than sessile predators or even filter-feeders.

There are some fossils in the Early Cambrian that resemble corals in having a conical calcareous exoskeleton filled with tabulae. *Tabulaconus* from the late Early Cambrian of Alaska is the best known among them (Debrenne *et al.*, 1987), but the oldest one remains *Cysticyathus*. This genus frequently occurs in archaeocyathid buildups in the Tommotian of the Lena River, Yakutia. The wall microstructure of the fossil is definitely not of the archaeocyathan type, lamellar with tabulae of meniscal shape continuously passing into the wall layers (Fig. 5). Although unlike associated archaeocyathids, *Cysticyathus*, as well as its probable successors *Tabulaconus* and *Bacatocyathus*, do not necessarily need to be interpreted as corals. The calyx is somewhat too irregular in shape and its wall, lacking any radial septa or septal spines, shows at least in distal parts structures suggestive of porosity secondarily covered from inside with laminar calcitic layers (Fig. 5B) This makes its attribution to corals and coelenterates most unlikely. It could well be of sponge origin.

I do not pretend to state that coelenterates are completely lacking in the Vendian and Early Cambrian. After closer examination some problematic fossils may prove to belong to the phylum. Especially worthy reconsideration in this respect is *Parapsonema*, ranging from the Late Cambrian (see Popov, 1967) until the Late Devonian, and the Middle Cambrian *Fasciculus* (Simonetta & Delle Cave, 1978, Collins *et al.*, 1983). In any case it remains clear that at the beginning of the Phanerozoic coelenterates were much less important than one would expect, keeping in mind their role in most interpretations of the early phylogeny of the Metazoa. This is not unreasonable, however, from a purely ecological point of view. Until pelagic environments became really rich in nektonic and planktonic metazoans, the organization of benthic or planktonic predators, more or less passively waiting for freely living prey, did not make much sense.

Body size of the oldest metazoans

In Recent organisms anatomical simplicity is usually connected with small size. It is understandable thus that in most of neontologically biased interpretations of the phylogeny the smallest of Recent organisms or early ontogenetic stages of others are the main source of information on the anatomy of hypothetical ancestral forms. The resulting expectation that the oldest known organisms should also be small in size, however, is not met by fossil evidence. Size distribution of Middle Cambrian Burgess Shale fossils shows clear predominance of macroscopic sizes (Briggs & Whittington, 1985), although a taphonomic bias cannot be excluded (Conway Morris, personal communication). There are also arguments to the contrary from neontology itself. Cladistic analysis of Recent microscopic worms shows that "pseudocoelomates have evolved from relatively large ancestors with body sizes measured in centimeters rather than millimeters" (Lorenzen, 1985: p. 210). Locomotory mechanisms predominating among the metazoans, especially the development of gait, require also macroscopic sizes of ancestral forms developing particular method of locomotion (see Clark, 1979; Elder, 1980).

The most elaborated attempt to prove, on the basis of fossil data, that the oldest members of a large branch of evolutionary tree, which is now represented by organisms of wide range of sizes, were initially of millimetre size refers to the earliest molluscs. It is generally assumed as proven that until the Ordovician they were of microscopic sizes. Runnegar & Jell (1976; Fig. 5; also Runnegar, 1983: Fig. 30) computed then available data on shell size of Early Palaeozoic molluscs, which seemingly showed almost exponential increase in size starting from not more than a couple of millimetres in the earliest Cambrian.

Considering this particular problem it is necessary to refer to the geological background of data used in

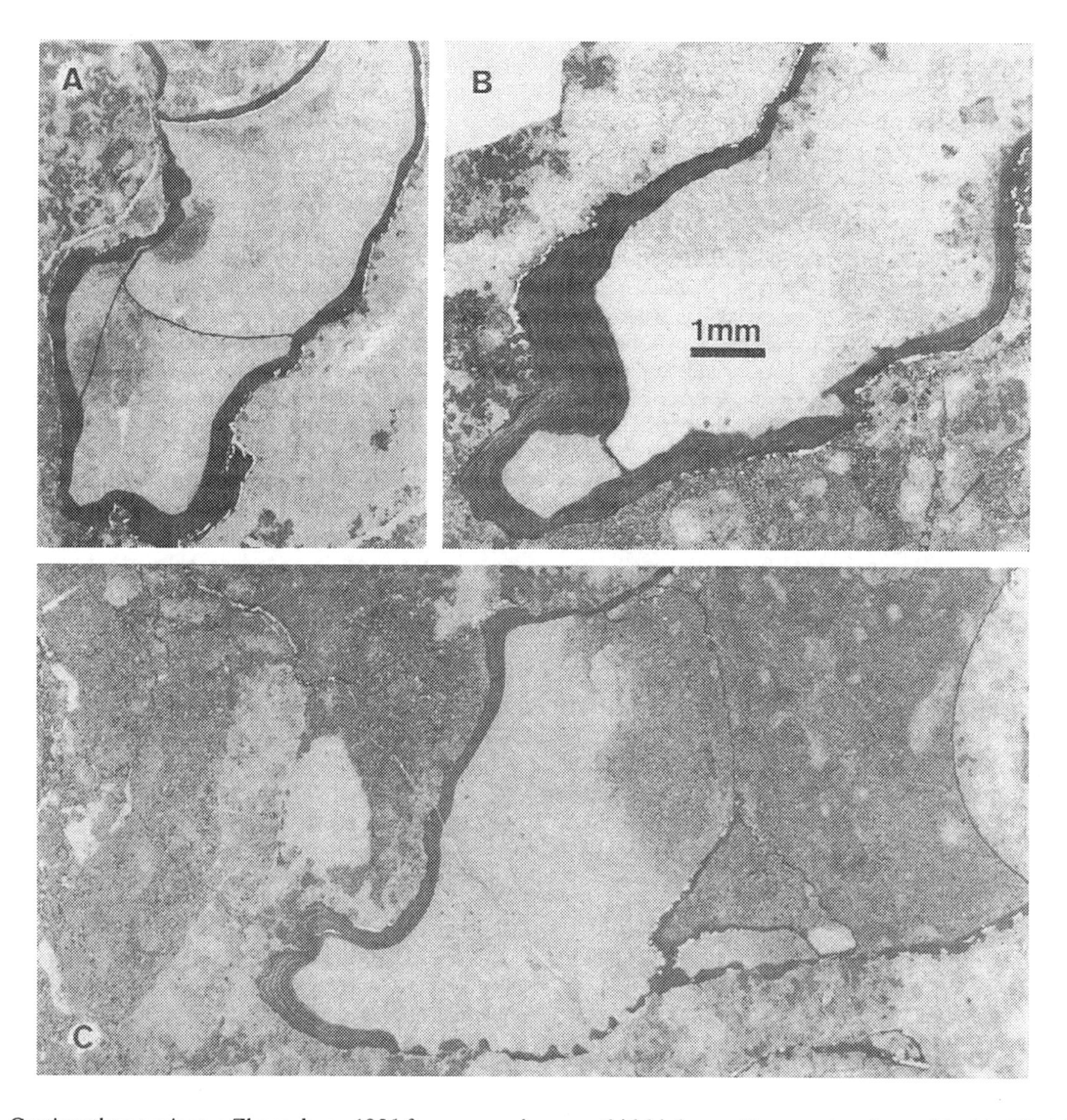

Fig. 5. *Cysticyathus tunicatus* Zhuravleva, 1956 from an archaeocyathid bioherm, Tommotian (loose block), Tiktirikteech on the Lena River, Yakutia. All x 10. A. Specimen with irregularly distributed tabulae. B. Another specimen with well visible laminated wall structure and continuity between wall and tabula. C. Specimen with wall structure suggestive of original porosity subsequently covered from inside by lamellar calcitic tissue.

the computation. Their basic source is the Siberian sections of the Tommotian and Atdabanian. Virtually all published materials concerning Tommotian molluscs were derived from rocks disaggregated by chemical means. They are usually phosphatic steinkerns of conch apices, juveniles or larval shells. The most fossiliferous Tommotian limestone sections are more or less stratigraphically condensed, abound in sedimentary discontinuities, with numerous hard-grounds and pockets filled with glauconitic limestones enriched in phosphatic debris (Fig. 6A, C). In such a conditions of reduced rates of sedimentation, with winnowing and reworking of the sediment, it would be difficult for any macroscopic mollusc shell to be preserved completely. As a result only microscopic shells can be found in residues, exactly as occurs in similar lithologically sections of the Baltic Ordovician where thousands of juvenile and larval shells can be found in a single sample while adult specimens are found in other beds.

In the Tommotian sections at the Lena River abundant assemblages of adult mollusc occur in the surroundings of archaeocyathid buildups, where locally rate of sedimentation was higher (Fig. 6B). At least in two such buildups, in localities Tiktirikteech and Bydjangaia, I was able to recognize such assemblages represented by the same species as are present in acid-resistant residues. Undoubtedly adult, with concentrations of growth lines at their apertures, specimens of *Bemella* (Fig. 7A-D) reach 25 mm in length. This is hardly different from sizes typical for Ordovician molluscs. The supposed small size of the earliest molluscs is thus a preservational artefact.

Another even more persuasive example of relatively large size of ancestral forms is provided by the oldest known lobopodian *Xenusion* (Fig. 8). The second specimen of the Geiseltalmuseum, Halle supplemented earlier interpretations with some new data enabling a tentative reconstruction of the entire body (Dzik & Krumbiegel, 1989). There are two remarkable points in its morphology: (1) An unbelievably simple organization, with a gradient in development of segmentation but without any signs of specialization in homonomous appendages arming the tubular body with terminal mouth. (2) A large size of the animal, counting at least twenty centimetres in length. The organization of the body of *Xenusion*, with pairs of dorsally located

Fig. 6. Source strata for earliest Tommotian molluscs in Yakutia. A. Sedimentary discontinuity covered by a layer enriched in small shelly fossils which fill also burrows below (note laminated limestone above, which is almost barren palaeontologically), Tiktirikteech, shore of the river. B. Archaeocyathid bioherm (at right) flanked by limestone beds with large mollusk shells (marked with asterisk), Tiktirikteech, mouth of the creek. C. Polished section across a discontinuity surface with burrow filled with sediment enriched in phosphatic fossils, Isyt'; x 2.

Fig. 7. Mollusc shells from a glauconitic limestone bordering an archaeocyathid buildup at the locality Bydjangaia (near Isyt'), Yakutia. A-D. Two subadult specimens of *Bemella jacutica* (Missarzhevsky, 1966); adults are 25 mm in diameter; x 2. E-F. Probably adult *Latouchella korobkovi* (Vostokova, 1962); x 3.

spiniferous "humps" and ventrolateral appendages in each segment, is suggestive of being derived from an original tetraradial arrangement. If taken together with the cylindrical shape, annulation of the cuticle, and terminal location of the mouth it reminds the organization of the nemathelminthan worms (Dzik & Krumbiegel 1989). This seems noteworthy as it has been already shown by Conway Morris (1977a) that in the Cambrian the priapulids were much more diverse than today and from their morphological organization all other groups of Nemathelminthes can be derived. It is now clear that anatomical (not necessarily physiological) simplification and reduction in size were the predominant feature in the phylogeny of the phylum (Lorenzen, 1985). The Priapuloidea may thus appear to be close to the roots of the metazoan phylogenetic tree. This implies that transitions between phyla took place at the level of relatively large, hydraulically propelled animals. Such a conclusion is optimistic for palaeontologists, giving fossils a chance not only to contribute to our understanding of details of phylogeny in some groups but also to help us in solving the basic problems of the phylogeny.

Conclusions

Complete lack of fossil flatworms (Conway Morris, 1985) and at best a low diversity of Cambrian coelenterates suggest that traditional phylogenetic trees placing the coelenterates and flatworms at the base should be critically reevaluated. At present, it is only possible to offer little more than a very provisional review on the ancestral anatomical organization for major groups of the Metazoa known from the Early Palaeozoic. It is clear that already by the Vendian two basic branches of free-living metazoans were established. The first one was represented by animals crawling over the sediment surface, ancestors of the molluscs and nemertineans (perhaps also flatworms). The second was represented by burrowers in the sediment, ancestors of the priapulids. Members of the first branch were propelled either directly by direct waves of muscular contraction in the foot or used hydraulic mechanisms (see Clark, 1979; Elder, 1980) with retrograde waves passing along vacuolized (filled with haemolymph in lacunae) foot, while in the second branch direct waves of muscle contraction pumped fluids through an unsegmented body cavity (Hunter *et al.*, 1983). This was used to propel the body in, as well as above, the sediment. Probably by secondary adaptation of the latter mechanism to active life on the surface of the sediment, gait of the onychophoran type developed. This resulted in development of appendages and increasingly deep segmentation of internal organs (Dzik & Krumbiegel, 1989). Sedentary life with adaptations to collect detritus from the sediment surface resulted perhaps in development of a preoral lobe of the echiurid type, and

Fig. 8. *Xenusion auerswaldae* Pompeckj, 1927 from erratic boulders of the Kalmarsund Sandstone, basal Cambrian of the Baltic region; latex casts, both x 1. A. Specimen housed at Geiseltalmuseum, Halle (see Dzik & Krumbiegel 1989), three-dimensionally preserved exuvium probably representing anterior part of the body. B. Holotype housed at the Museum für Naturkunde, Berlin; probably posterior end of the body. C-D. Reconstruction of the body in dorsal and lateral views.

U-shaped digestive tract and tentacles of the sipunculid type, which further might have allowed development of a lophophore leading, finally, to the echinoderm organization level (Nichols, 1967). In this highly speculative scenario (Fig. 4) the flatworms and coelenterates are not included and their position in the phylogenetic tree remains mysterious. A possibility that they represent secondarily simplified successors of more anatomically complicated forms should not be excluded.

Acknowledgements

Significant part of this paper is based on materials collected during an expedition to Yakutia organized in 1987 by Professor A.J. Rozanov (Paleontological Institute of the Academy of Sciences of the USSR). I deeply appreciate the opportunity to participate in this scientific adventure.

The specimens of *Velumbrella* have been kindly lent to me by Mrs. Jadwiga Zwierz, M.Sc., the curator of palaeontology at the Geological Institute, Warsaw. I am also grateful to the University of Camerino for its hospitality during the present symposium.

REFERENCES

BEDNARCZYK, 1970. Trilobites fauna of the lower *Paradoxides oelandicus* Stage from the Brzechòw area in the western part of the Święstokrzyskie Mts. *Bulletin de l'Academie polonaise des Sciences, Série des Sciences géologiques et géographiques* **18**, 29-35.

BERGSTRÖM, J. 1986. Metazoan evolution - a new model. *Zoologica Scripta* **15**, 189-200.

BRIGGS, D.E.G. & WHITTINGTON, H. B. 1985. Modes of life of arthropods from the Burgess Shale, British Columbia. *Transactions of the Royal Society of Edinburgh* **76**, 149-160.

CHEN JUN-YUAN, HOU XIAN-GUANG & LU, HAO-ZHI. 1989. Early Cambrian hock glass-like rare sea animal *Dinomischus* (Entoprocta) and its ecological features. *Acta Palaeontologica Sinica* **28**, 59-71.

CLARK, R.B. 1979. Radiation of the Metazoa. In *The Origin of Major Invertebrate Groups*. (ed. M.R.House). London: Academic Press.

COLLINS, D., BRIGGS, D. & CONWAY MORRIS, S. 1983. New Burgess Shale fossil sites reveal Middle Cambrian faunal complex. *Science* **222**, 163-167.

CONWAY MORRIS, S. 1977a. Fossil priapulid worms. *Special Papers in Palaeontology* **20**, 1-95.

CONWAY MORRIS, S. 1977b. A new entoproct-like organism from the Burgess Shale of British Columbia. *Palaeontology* **20**, 833-845.

CONWAY MORRIS, S. 1985. Non-skeletalized lower invertebrate fossils: a review. *Systematic Association Special Volume*, **28**, 210-223.

CONWAY MORRIS, S. 1989. The persistence of Burgess Shale-type faunas; implications for the evolution of deeperòwater fauñas. Transactions of the Royal Society Edinburgh, Earth Sciences **80**, 271-283.

CONWAY MORRIS, S. & ROBISON, R.A. 1988. More soft-bodied animals and algae from the Middle Cambrian of Utah and British Columbia. *The University of Kansas Paleontological Contributions, Paper* **122**, 23-48.

DEBRENNE, F.M., GANGLOFF, R.A. & LAFUSTE, J.G. 1987. *Tabulaconus* Handfield: microstructure and its implication in the taxonomy of primitive corals. *Journal of Paleontology* **61**, 1-9.

DURHAM, J.W. 1974. Systematic position of *Eldonia ludwigi* Walcott. *Journal of Paleontology*, 48,750-755.

DZIK, J. (in press a). Possible solitary bryozoans from the early Paleozoic. *Société des Sciences Naturelles de l'Ouest de la France. Memorie hors serie.*

DZIK, J. (in press b). *Dzieje Zycia na Ziemi. Wprowadzenie do Paleobiologii. (History of Life on the Earth: An Introduction to Paleobiology)* Warszawa: Panstwowe Wydawnictwo Naukowe.

DZIK, J. & KRUMBIEGEL, G. 1989. The oldest "onychophoran" *Xenusion*: a link connecting phyla? *Lethaia* **22**, 169-181.

ELDER, H.Y. 1980. Peristaltic mechanisms. In *Aspects of Animal Movement*. (eds. H.Y. Elder & E.R. Trueman) Cambridge: Cambridge University Press.

FEDONKIN, M.A. 1987. Besskeletnaia fauna venda i jeio miesto v evolucji Metazoa (Non-skeletal fauna of the Vandian and its place in the evolution of Metazoa). *Trudy Paleontologicesko-vo Instituta AN SSSR* **226**, 1-178.

HORNÝ, R. 1985. *Bellerophacmaea* gen. n. (Mollusca) and *Palaeolophacmaea* Donaldson, 1962 (Coelenterata), an example of homeomorphy in invertebrate paleozoology. Časopis Národniho Muzea v Praze, otdíl přiod. **154**, 119-124.

GLAESSNER, M.F. 1971 The genus *Conomedusites* Glaessner Wade and the diversification of the Cnidaria. *Paläontologische Zeitschrift* **45**, 7-17.

HUNTER, R.D., MOSS, V.A. & ELDER, H.Y. 1983. Image analysis of the burrowing mechanisms of *Polyphysia crassa* (Annelida, Polychaeta) and *Priapulus caudatus* (Priapulida). *Journal of Zoology, London* **199**, 305-323.

JENKINS, R.J.F. 1985. The enigmatic Ediacaran (late Precambrian genus *Rangea* and related forms. *Paleobiology* **11**, 336-355.

KOZŁOWSKI, R. 1968. Nouvelles observations sur les conulaires. *Acta Palaeontologica Polonica* **8**, 497-529.

LORENZEN, S. 1985. Phylogenetic aspects of pseudocoelomate evolution. *Systematics Association Special Volume* **28**, 210-223.

NICHOLS, D. 1967 The origin of echinoderms. *Symposia of the Zoological Society London* **20**, 209-229.

NIELSEN, C. 1985. Animal phylogeny in the light of the Trochea theory. *Biological Journal of the Linnean Society* **25**, 243-299.

OLIVER, W.A. JR. 1984. *Conchopeltis*: its affinities and significance. *Palaeontographica Americana* **54**, 141-147.

POPOV, I.N. 1967. Novaja kiembrijskaja scifomeduza. (A new Cambrian scyphomedusa). *Paleontologiceskij Zhurnal* **1967**, 122-123.

QIAN YI & BENGTSON, S. 1989. Palaeontology and biostratigraphy of the Early Cambrian Meishucunian Stage in Yunnan Province. *Fossils and Strata* **24**, 1-156.

RASETTI, F. 1954. Internal shell structures in the Middle Cambrian gastropod *Scenella* and the problematic genus *Stenothecoides*. *Journal of Paleontology* **28**, 59-66.

RUNNEGAR, B. 1983. Molluscan phylogeny revisited. *Memoirs of the Association of Australasian Paleontologists* **1**, 121-144.

RUNNEGAR, B. & JELL, P.A. 1976. Australian Middle Cambrian molluscs and their bearing on early molluscan evolution. *Alcheringa* **1**, 109-138.

SEILACHER, A. 1984. Late Precambrian and Early Cambrian Metazoa: Preservational or real extinctions. In *Patterns of Change in Earth Evolution* (eds. H.D. Holland and A.F. Trendall). pp. 159-169. Berlin: Dahlem Konferenzen, Springer-Verlag.

SIMONETTA, A. & DELLE CAVE, L. 1978. Notes on new and strange Burgess shale fossils (Middle Cambrian of British Columbia). *Atti della Società Toscana di Scienze Naturali, Memorie* **85A**, 45-49.

STANLEY, G.D. JR. 1986. Chondrophorine hydrozoans as problematic fossils. In *Problematic Fossil Taxa* (eds. A. Hoffman and M.H. Nitecki). pp.68-86. New York: Oxford University Press.

STASIŃSKA, A. 1960. *Velumbrella czarnockii* n. gen. n. sp. - Méduse du Cambrien inférieur des Monts de Sainte-Croix. *Acta Palaeontologica Polonica* **5**, 337-396.

SUN WEI-GUO & HUO XIAN-GUANG 1987. Early Cambrian medusae from Chengjiang, Yunnan, China. *Acta Palaeontologica Sinica* **26**, 257-270.

YOCHELSON, E.L. & STANLEY, G.D.JR. 1981 An Early Ordovician patelliform gastropod, *Palaeolophacmaea*, reinterpreted as a coelenterate. *Lethaia* **15**, 323-330.

YOCHELSON, E.L., STÜRMER, W. & STANLEY, G.D.JR. 1983. *Plectodiscus discoideus* (Rauff): a redescription of a chondrophorine from the Early Devonian Hunsrück slate, West Germany. *Paläontologische Zeitschrift* **57**, 39-68.

Lower Cambrian fossil Lagerstätte from Chengjiang, Yunnan, China: Insights for reconstructing early metazoan life

Chen Jun-yuan[1] & Bernd-D. Erdtmann[2]

Abstract

An Early Cambrian mixed shelly and soft-bodied fauna was discovered in 1984 in a dark grey (yellow weathering) silty mudstone sequence of the Qiongzhusi Formation at several localities near Chengjiang, ca. 50 km SE of the Yunnan capital of Kunming in SW China. Since then it has been partly described in a series of papers (mostly having appeared in recent issues of *Acta Palaeontologica Sinica*). Here it will be synoptically evaluated with emphasis on the evolutionary and palaeoenvironmental as well as bio- and event-stratigraphical significance. In this context also attempts are being put forth to elucidate the systematic and constructional morphological positions of *Microdictyon, Dinomischus*, and other taxa. The former is referable to a bilateral multi-legged worm-like animal whose exact sytematic position, however, cannot as yet be satisfactorily determined. The biostratigraphic reference of this fauna is given by a slightly earlier occurrence of the Earliest Cambrian trilobites *Parabadiella* and *Mianxiandiscus*, and the encompassing range of *Eoredlichia* with this regionally widespread soft-bodied fauna. Direct correlation with distinct Siberian platformal Tommotian or Atdabanian Stages is not satisfactory, and references to the initial occurrences of archaeocyathids in the superjacent Canglangpu Formation are too tenuous as evidence for a Tommotian age. The taphonomic overprint is low, i.e. post-mortem factors have not distorted the original living communities. Death was induced in the recurrent communities by probably polycyclic anoxia having invaded the biotopes. Bacterial decay of carcasses was retarded by rapid distal tubiditic or tempestitic burial and subsequent substrate sealing by microbial mats or by long-term residence of anoxic watermasses. Asphyxia of biota is documented by observed signals of death struggle (everted proboscis of worms, extended mantle setae of brachiopods, etc.). Even during substrate inhabitation periods oxygen availability may have been tenuous because almost no bioturbation is evident and attached benthos appeared to have developed multi-tiered feeding levels, as documented by various types of holdfast structures. In an evolutionary assessment the Chengjiang biota demonstrate a surprising conservative stock as only few of the younger Burgess Shale organisms show indications for an advancement of morphological traits already introduced with the Chengjiang fauna. The potential interrelations between Late Proterozoic and Early Cambrian rise of metazoans, their ecosystems and surmised eustatic cycles, stable isotope excursions, phosphogenesis and biomineralization events are discussed.

Introduction

The term "Fossil Lagerstätte" was first introduced into the scientific community by Seilacher (1970), and the term has since then been broadly accepted in an extended sense to represent rock bodies or stratal surfaces which contain special information on historical biology and palaeontology.

Among the select number of world-renowned "Fossil Lagerstätten", the occurrence of the Chengjiang soft-bodied biota stands out for the reason of representing one of the most spectacular evolutionary episodes of metazoan life, at a time when the complex physiological evolution of metazoan life was accompanied by an apparently "spontaneous" invention of skeletal hard parts more or less simultaneously at the beginning of the Cambrian Period. This as yet poorly understood episode has been referred to as the "biomineral event" by Brasier (1986a, b). This event is expressed by two distinct episodes of skeletal development spanning an interval of at least 20 to 30 million years. The first episode is documented by the rise of the so-called "small skeletal fauna" during the Meishucun Stage of the earliest Cambrian. The second episode is marked by the debut of skeletonized trilobites during the Qiongzhusi (Chiungchussu) Stage and by archaeocyathids during the subsequent Canglangpu Stage. The Chengjiang Lagerstätte was deposited just at the beginning of the second biomineral event. Its study, therefore, may be expected to yield new insights into marine life and palaeoecology during the Early Cambrian.

Palaeoenvironment and taphonomy

The Chengjiang Lagerstätte is basically shale-hosted, and originally was known from one locality, the Maotian Hill ("Maotianshan") of Chengjiang County. Here soft-part preservation was discovered in several levels of a unit, spanning as much as 50 m of vertical section, within the lower part of the *Eoredlichia* Zone of the Lower Cambrian Qiongzhusi Formation (Figs. 1, 2). This kind of soft-part preservation extends laterally for as much as 100 km, and recently fossils of similar type and age were also discovered at other locations in Jinning (Hou & Sun, 1988) and Wuting counties, Eastern Yunnan. Here the Lagerstätte, however, seem to possess less significance both quantitatively and qualitatively.

The Chengjiang Lagerstätte consists of finely laminated mudstones, which may represent part of an outer shelf detrital belt deposited in a quiet water environment. The predominant occurrence of macrobenthos indicates that oxygen was generally abundantly available, although evidence of bioturbation is weak and might indicate that only the uppermost layer of sediment was oxygenated. The preservation of non-mineralized biota and of articulated material is believed to have resulted from repeated and rapid burial events. The rapid burial under sheets of sediment may have prevented the destruction of the soft-body fossil carcasses by bioturbation, current activity, scavengers, and carnivores, and provided a chance for the conservation of the non-mineralized and articulated organisms. Furthermore, sulphate reduction is known to consume organic matter (Fisher & Hudson, 1985), and it is an ef-

[1]Nanjing Institute of Geology and Palaeontology, Academia Sinica, Nanjing, P.R. China
[2]Institut für Geologie und Paläontologie, Technische Universität Berlin, EB 10, Ernst-Reuter-Platz 1, D-1000 Berlin 10, F.R.G.

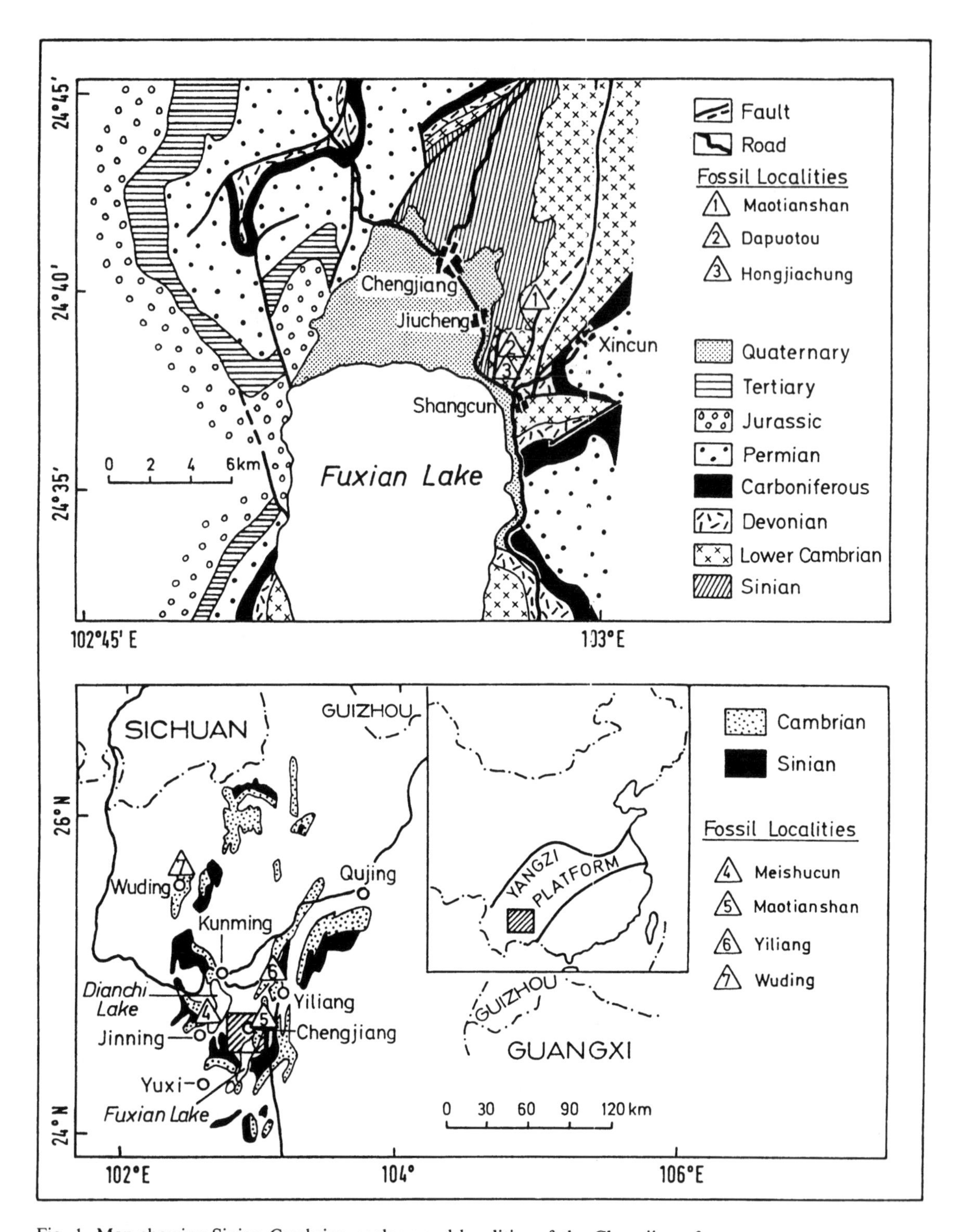

Fig. 1. Map showing Sinian-Cambrian geology and localities of the Chengjiang fauna.

fective factor of early diagenesis, particularly in organic-rich muds. That is the reason why soft bodied preservation is rare in mud-supported sediments. Chemical analysis indicates extremely low contents of pyrite in the Chengjiang Lagerstätte beds, suggesting that sulphate reduction may have been weak or absent. Therefore, it is believed that this Lagerstätte formed at a level below the sulphate-reduction zone.

During the Qiongzhusi Stage well developed anoxic conditions are reflected in the widespread occurrence of carbon-rich deposits at Maotianshan. An accompanying transgressive event may have periodically enhanced the upwelling potential of the oxygen-deficient waters over the outer part of the central Yunnan continental shelf. Therefore, frequent mass mortalities, probably by anoxic poisoning, are expected to have played a significant role in the soft-part preservation of the Chengjiang Lagerstätte. Generally chances for burial by sediment are greater for sessile benthos, particularly for sponges than for vagile forms. Mobile animals, however, usually are capable of escaping from suffocation and burial by sudden sedimentation. Therefore, anoxic events may have been an important factor for soft-part preservation because such sudden "poisoned clouds" may have left the animals stunned or suffocated shortly before or after actual burial. Presumably this provides an explanation for the reason why no escape burrows have been observed in association with the Chengjiang Lagerstätte. Coiled preservation in many of the soft-bodied worms and in some of the arthropods suggests anaerobic suffering and asphyxia before death. Numerous specimens show a raised relief and an extruded oesophagus; this may indicate that the animals were in metabolic stasis before their death. Features such as coiling and metabolic stasis have been observed by Dean *et al.* (1964) in modern annelids and arthropods which combat anoxic suffocation. It is proposed that many of the animals in the Chengjiang Lagerstätte may have been literally buried alive, although they were stunned by asphyxia prior to death. The extended mantle setae protruding from the valves of many inarticulate brachiopods are also interpreted as evidence of these animals having been alive at the moment of their burial. Furthermore, the soft-bodied pelagic medusiform animals were commonly buried in groups on certain bedding planes, this is also interpreted as evidence of these gregarious animals having been killed or stunned during the recurrent anoxic poisoning events immediately prior to their burial.

The Chengjiang Lagerstätte fossils are mostly preserved as aluminosilicate rather than carbonaceous films (according to SEM-EDAX analysis). A blueish stain is quite characterstic for the surfaces of most of the segmented worm-like animals, brachiopod pedicles and sea anemones from the Maotianshan localities. It is presumed that this bluish coating is the product of organic bacterial sheaths having colonized these metazoan carcasses immediately after they were covered by a thin blanket of sediment. This seems analogous to the Late Cambrian "orsten" soft part mummifications from Sweden reported by Müller (1979). All of the shelly skeletons are preserved in a decalcified state, with a broadly similar structure and morphology to that of the soft-bodied preservation. Many of the fossils are also severely weathered into reddish, ferric oxide-stained films, which appear in striking contrast to the yellowish weathered matrix. The ferrous oxides are likely to derive from oxidation of finely dispersed pyrite on bedding and fossil surfaces, and the origin of the pyrite is probably produced by the activities of anaerobic bacteria during the early diagenesis which generated a framboidal coating on the surfaces of these fossils.

Geological setting and stratigraphy

The Chengjiang area of east-central Yunnan lies in the southwestern interior of the Yangtze Platform, from where numerous fossiliferous sections across the Sinian to Early Cambrian (Fig. 2) are recorded. Here the sedimentary development and depositional environments may be traced for a stratigraphic interval beginning with the Sinian Nantuo glaciation, and continuing into the well defined early Middle Cambrian trilobite-bearing sequence. This time interval is the most critical period for the development of metazoan life. Moreover, global eustatic changes are signaled by multiple lines of evidence. The small shelly fossils, predating the trilobites, are delimited to a sequence of dolostones, phosphatic limestones and siltstones ranging through 35 m of section; their radiation represented by numerous taxa, however, broadly coincided with a period of phosphogenesis in the uppermost 30 cm of the Zhongyicun Member (Luo *et al.*, 1982; 1984; Yu, 1987). A significant drop of the sea level is indicated by a hiatus which documents subaerial exposure at the top of the Meishucun Formation (Fig. 2). During the subsequent rise of sea level, silts and muds, rich in organic matter were deposited. The extent of the diachroneity of the post-Meishucun transgression suggests the possible existence of a fairly high-relief land surface on the Yangtze Platform. Eastern Yunnan, probably having been a low- relief land area, was submerged earlier, with siliciclastic sedimentation resuming at the base to the Qiongzhusi Formation. Its lower part is represented by the fine grained clay-silt known as the Badaowan Member, which contains a rich trace fossil assemblage and a moderately abundant small shelly fauna, but it lacks trilobites. The Yuanshan Member (upper Qiongzhusi Formation and just below the first appearance of the soft-bodied biota) contains the first trilobites, which diversified rapidly during subsequent times so as to dominate the great majority of Cambrian shelly assemblages. A zonal scheme for the Lower Cambrian in the Meishucun-Chengjiang region was established by Zhou & Yuan (1981).

Detailed resolution of stratigraphic correlation is critical to a clearer understanding of the Early Cambrian metazoan radiation. However, as revealed by various authors, biofacies and biogeographical differentiation were already quite highly developed during this period of the Meishucun Stage and the Qiongzhusi Stage; this may diminish the significance and applicability of the small shelly fauna and trilobites for worldwide correlation. A debatable correlation scheme has been promoted by some authors (Conway Morris, 1987; Brasier, oral communication to B.-D.E., 1989) suggesting that the "small skeletal biota" of the upper Meishucun Stage of China could be late Atdabanian in age. This correlation is unlikely to be acceptable because the evidence of a first archaeocyathid record from the uppermost Qiongzhusi Formation (see fig. 2) indicates that the Meishucun Stage in China may even be older than the Tommotian of Siberia. As recognized by many authors, the archaeocyathids seem to be useful for global correlation due to their biological evolution and morphological differentiation. The fact that archaeocyathids appear at the base of the Tommotian in Siberia is secure, and with this being considered, that there are close facies ties between the Tommotian of Siberia and the Canglangpu of SW China (Yuan, pers. communication), a correlation between the Tommotian and the Canglangpu ages could be envisioned. Archaeocyathids first occur in the fossil record of China in strata correlative with the uppermost Qiongzhusi beds, then rapidly diversify during the early Canglangpu Stage. Four archaeocyathid assemblages were proposed by Yuan & Zhang (1981) in ascending order as follows: Chiungchussu (Qiongzhusi), Liangshuijing, Yingzuiyan, and Tieheban. Archaeocyathid genera appearing at the base of the Tommotian of Siberia could be correlated with those of Canglangpu age in China. Despite the presumed similarity of sedimentary environment, however, no archaeocyathids occur in the Meishucun Formation in China, for the reason that the Meishucun Stage was probably of pre-archaeocyathid age (although unsuitable environments may also be a reason for the lack of archaeocyathids in the Meishucun strata, B.-D.E.). If this correlation is correct, the new fauna is of paramount importance because of its very close stratigraphic position near the base of the

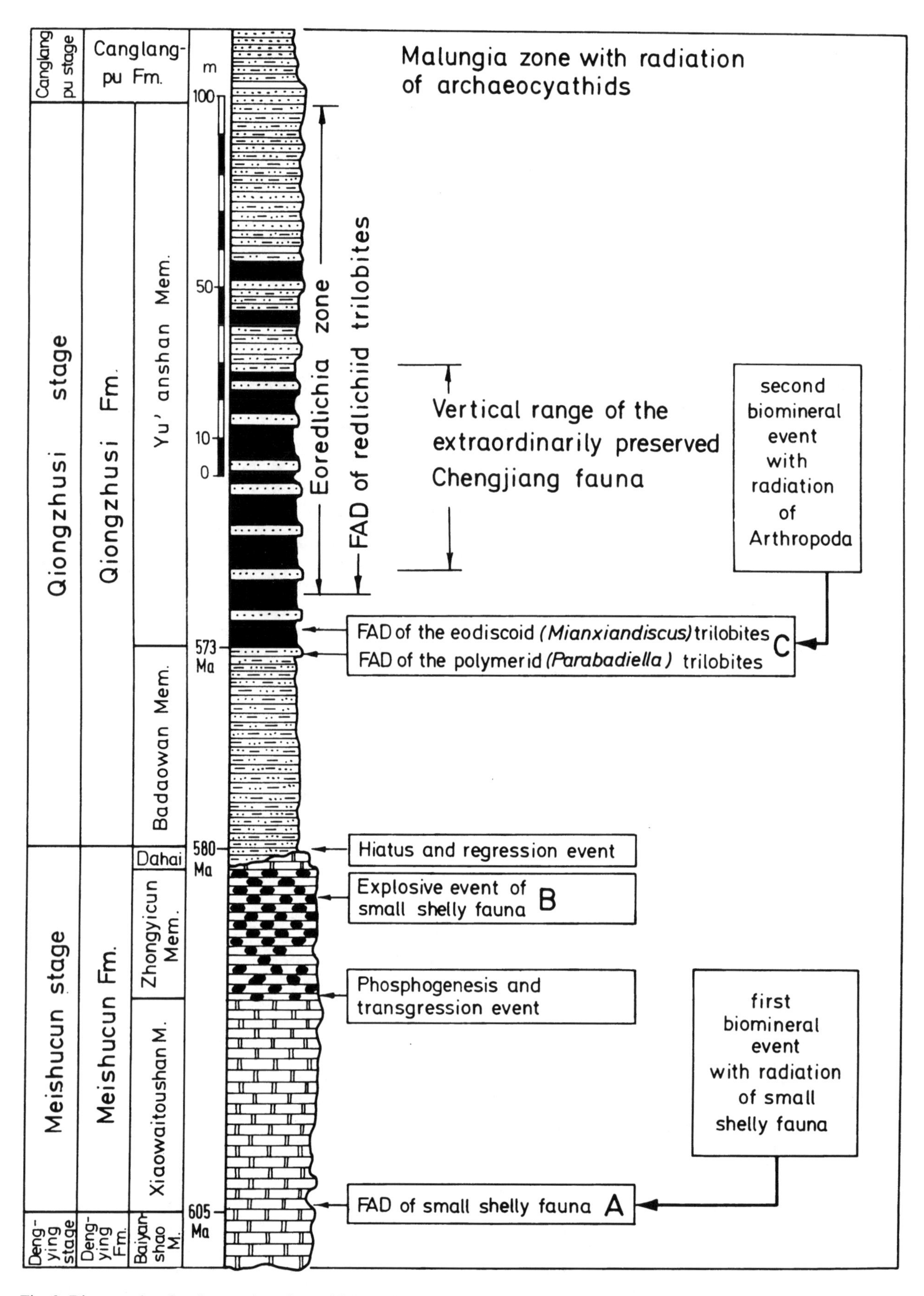

Fig. 2. Diagram showing the stratigraphy and lithogenic and biological events of Earliest Cambrian in the central Yunnan area.

Cambrian System, as documented by the occurrence of the earliest trilobites such as the eodiscid *Mianxiandiscus* the redlichiid *Eoredlichia* and the polymeriid *Parabadiella* (Fig. 2). The soft-bodied fauna described herein occurs at several levels within a unit, as much as 50 m thick, that represents the lower part of the *Eoredlichia* Zone of the Qiongzhusi Formation (see Fig. 2). These rocks are predominantly finely laminated mudstones of an outer shelf environment. Intercalated lenticular beds of 1-20 cm thick fine-grained sandstones are interpreted as distal storm deposits. Similar soft-bodied and coeval fossils were recently also discovered at neighbouring locations near Meishucun (Hou & Sun, 1988) and in Wuding county (Fig. 1).

Composition of biotic assemblages

The Chengjiang soft-bodied biota occurs only 25 m above the first appearence of earliest Cambrian trilobites (Fig. 2), and therefore, at least for the time being, this series of fossil Lagerstätten represents the oldest record of soft-part preservation in the Phanerozoic (not counting the Ediacaran and correlative Lagerstätten). The fossil assemblage consists of sponges (Chen *et al.*, 1989b), chancellorids, medusiform metazoans (Sun & Hou, 1987a), sea anemones (this paper), chondrophorids, priapulid worms (Sun & Hou, 1987b), netted scale-bearing worm-like animals (Chen *et al.*, 1989a), hyolithids, tentacled worm-like animals (Hou & Chen, 1989a), possible entoprocts (Chen *et al.*, 1989c), inarticulate brachiopods (Chen, Hou & Erdtmann, 1989), annelid-arthropod intermediate animals (Hou & Chen, 1989b) and non-trilobite arthropods (Hou, 1987a, b, c; Hou *et al.*, 1989), and associated rare skeletonized arthropods, such as eodiscid and redlichiid trilobites (Zhang, 1987), a puzzling group of metameric animals (*Anomalocaris*), and a diverse group of algae (this paper).

Algae

Algae are predominant among the Chengjiang biota. Among the most common are thalli, consisting of few flexuous stipes with an affinity to *Yuknessia* (Pl. 2, fig. 1). Stipes are long, slender, without evidence of jointing or branching, with a width between 0.5 and 0.7 mm. *Yuknessia* was described from the Middle Cambrian Burgess Shale of British Columbia and from coeval beds in Utah (Walcott, 1919, Johnson, 1966; Conway Morris & Robison, 1988). Walcott (1919) regarded this form as a green alga and this assignment was also accepted by Johnson (1966) and Conway Morris & Robison (1988).

The delicate, new alga *Sinocylindraena yunnanensis* gen. et sp. nov. occurs as long, simple threads of 0.3 mm uniform width (Pl. 2, fig. 5). It is preserved as irregular coils indicating that this alga may have been a highly flexible and effective floater. The surface of the thread is smooth, showing no evidence of microstructure; thus, the exact taxonomic affinity of this alga must at present remain uncertain.

One of the most distinctive forms among the algae is the new taxon *Megaspirellus houi* (Pl. 1, Figs. 3, 4; Pl. 2, Figs. 3, 4), represented by a helically coiled thread. The helix is parallel-sided, having a diameter of 3 mm; the diameter of the thread is ca. 0.5 mm; the whorls are open-coiled. No evidence of cell structure is visible on the thread; thus, a relatively substantial sheath probably covered the presumed trichome. Although a large number of spirally coiled tube-like fossils have been described from various Upper Precambrian and Lower Cambrian localities (Patiletov *et al.*, 1981; Song, 1984; Luo *et al.*, 1982; Peel, 1988), none of these structures is larger than 0.6 mm. In spite of a general similarity between the present taxon and the other spirally coiled forms, the size separation between them is so great that they may warrant separation at a higher taxonomic level. The biological affinity of the helically coiled fossils remains obscure so far, but on the basis of a general morphological similarity with the modern *Spirulina*, an affinity of this form with Cyanobacteria is suggestive.

Sponges

The sponges are among the most spectacularly preserved fossils, with at least a dozen genera and more than two dozen species having been found so far. They form the second most diversified metazoan group in the Chengjiang beds. The majority belongs to the Demospongea, with a fair spectrum of intraspecific morphological and size variation (Chen *et al.*, 1989b). Among the Demospongea, species belonging to *Leptomitus, Leptomitella* and *Paraleptomitella* (Leptomitidae) are dominant (Chen *et al.*, 1989b). Their thin-walled skeletons are tubular to balloon- or fan-shaped. The outer layer of the skeleton, both in *Leptomitella* and *Leptomitus*, consists of a vertical thatch-like pattern of delicate monaxons and inserted vertical rods. The rods are composed of rigid straight axes which are interlaced with one another at their ends. The rigid axes of *Paraleptomitella* are slightly curved, and interlock with one another to form an elongated net-like fabric. The inner layer consists of bundled (*Leptomitella* and *Paraleptomitella*) or unbundled monaxons (*Leptomitus*). The remaining demosponges include at least 15 taxa, ranging from globose, bladder-like to frondescent varieties close to *Hamptonia*, to large tabular sponges with monaxons that form a four-layered and diagonally arranged double-netted fabric (Chen *et al.*, in press). Additional forms are breviconic fronds of *Halichondrites* affinity, those with a fused, spiculate network comparable to *Vauxia*-related species, and tubercle-bearing species of *Sentinella*. The spherical sponges are a dominant component among the non-leptomitid Demospongea. They are small, lacking a spongocoel, with radiating, unbundled, delicate monaxons of *Sphaeriella* affinity. The genus has been described so far only from the Mississipian of Alabama, USA (Rigby & Bryant, 1979). Among the non-leptomitid sponges has been found the characteristic *Choia*, possessing a short conical body shape. Its concave surface was originally interpreted as a spongocoele or atrium surface, until Rigby (1986) showed that the convex side was orientated upwards in life. Hexactinellid sponges are quite rare. Several articulated mesh fragments of *Protospongia* display regular stauract spicular meshes whose individual spicules are arranged in rectangular patterns, consisting of at least five orders of squares along axial rows.

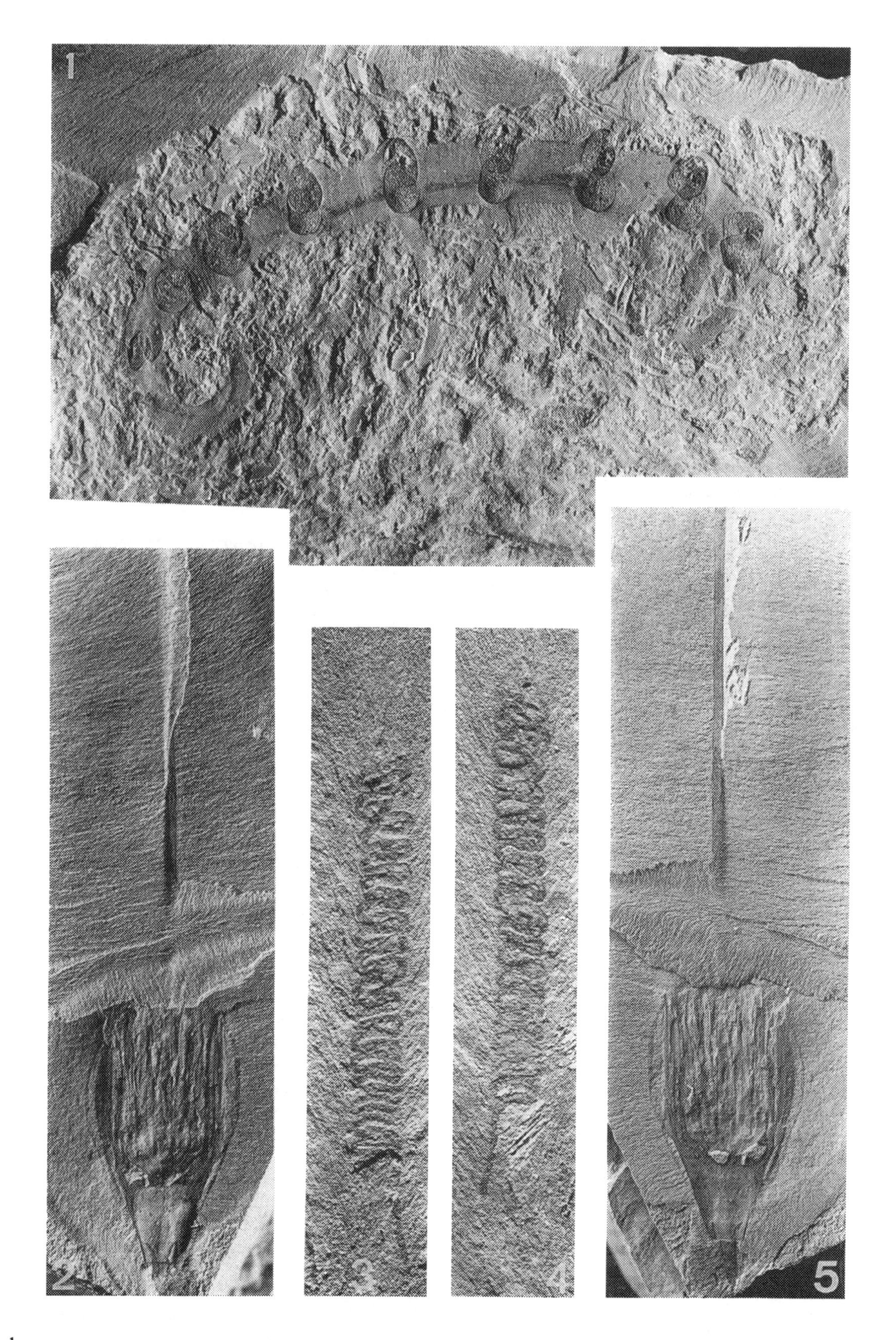

Plate 1

Fig. 1. *Microdictyon sinicum* Chen, Hou & Lu, 1989. Holotype. Cat. No. 108286, Coll. No. M2, x 3.

Fig. 2,5. *Dinomischus venustus* Chen, Hou & Lu, 1989. Holotype, Cat. No. 108478, Coll. No. M2, x 2.5. 2, part; x 5.

Fig. 3,4. *Megaspirellus houi* gen. et sp. nov. Holotype, Cat. No. 108501, Coll. No. M2, x 4. 3, Counterpart; 4, part of the holotype.

Plate 2

Fig. 1. *Yuknessia* sp., Cat. No. 108562, Coll. No. M2, x 3.

Fig. 2. *Stellasomites eumorphus* Sun & Hou, 1987 Cat. No. 108503, Coll. No. M3, x 1.

Fig. 3,4. *Megaspirellus houi* gen. et sp. nov. Same specimen of Pl. 1, figs 3 and 4, wetted with glycerine.

Fig. 5. *Sinocylindra yunnanensis* gen. et sp. nov. Holotype, Cat. No. 108504, Coll. No. cf. 3, x 3.

Plate 3

Fig. 1,2. *Xianguangia sinica* gen. et sp. nov. Holotype, Cat. No. 108506, Coll. No. M2, x 2. 1, part, and 2. counterpart of the holotype.

Fig. 3,4. Xianguangia sinica gen. et sp. nov. Paratype, Cat. No. 108505, Coll. No. M2, x 2. 3, part, and 4. counterpart of the paratype.

Medusiform metazoans

Among the soft-bodied radial-symmetric components, dominate supposed medusoids (Sun & Hou, 1987a). They possess a relatively large medusiform body which is radially divided into numerous partitions. The structures were previously regarded as radial canals of medusoid affinity (Sun & Hou, 1987a). The radial structures are here interpreted as mesenteries suspending the gut within a coelomic cavity. Evidence observed indicates that the radial structures terminate at the wall of the gut in "suspension points". The structure regarded as gut by Walcott (1911) and Clark (1912) or "intestine" by Conway Morris & Robison (1988) in *Eldonia* are also documented abundantly among the Chengjiang biota. The gut appears to be coiled in a characteristic manner, always lying within the outline of the coelomic cavity, being commonly preserved as a black organic film. The appearance of a curved gut within the coelomic cavity in the Chengjiang specimens recalls *Eldonia* (Conway Morris & Robison, 1988; Chen, Hou & Erdtmann, 1989). Although their medusiform body has been interpreted as indicating a medusoid affinity by some authors (Sun & Hou, 1987a, Chen, Hou & Erdtmann, 1989), the spiral gut seems to be an internal structure within the coelomic cavity; thus the creatures could not possibly have been coelenterates. *Eldonia* has long been considered to be a holothurian (since Walcott, 1911) by various authors (Durham, 1974). However, Paul & Smith (1984) doubt this assignment, and in their opinion *Eldonia* lacks even the most basic echinoderm characteristics. The biological affinity of *Eldonia* is best regarded as problematic, because it does not fit into any recognized phyla.

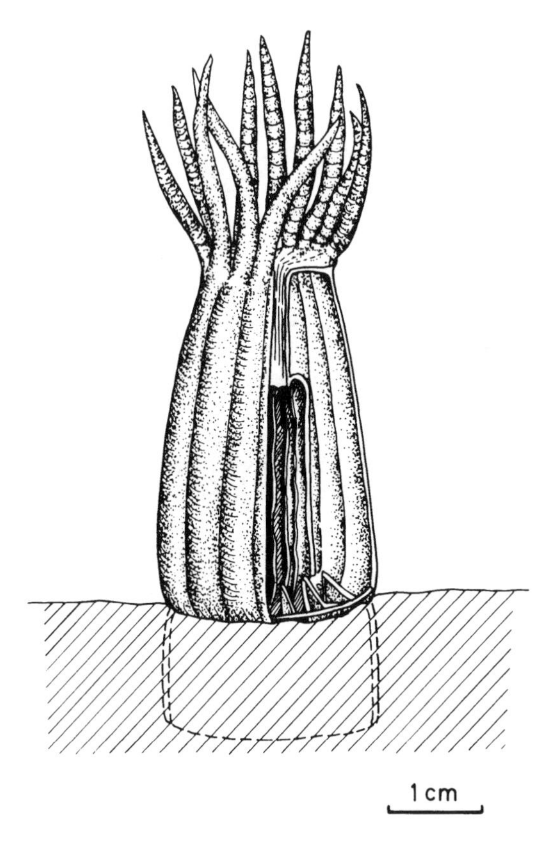

Fig. 3. Reconstruction of sea anemone *Xianguangia*.

These planktonic medusiform organisms were entombed in groups on certain bedding planes. They are commonly flattened, with their oral surfaces placed parallel to the bedding planes and their cone-shaped, aboral surface commonly rising into the overlying stratum. The high cone-shaped form of the aboral surface is demonstrable in rare specimens which are obliquely compacted (Pl. 2, Fig. 2). The type of preservation seems to indicate that the majority of specimens may have sunk toward the seafloor, without visible disturbances by currents before burial. The organisms were apparently gregarious, evidently having been killed off in large numbers under unknown circumstances, possibly by upwelling anoxic waters.

Sea anemones

There are only two specimens of potential Actiniaria (sea anemones), which are referred to a single new species *Xianguangia sinica* (pl. 3, Figs. 1-4). This is a relatively large conical form, about 53 mm long and 23 mm wide, with a non-skeletonized body consisting of a pedal disc, a column, and about 16 tentacles surrounding the mouth (Fig. 3). The morphologic features of this creature seem to be quite reminiscent of modern gonactinitids, which are considered to be the most primitive sea anemones possessing eight complete septa and a circlet of eight tentacles.

Chondrophorids

Today the order Chondrophorida consists of two families, each containing a single genus, *Velella* and *Porpita*, respectively. The genus *Velella* is asymmetrical, having a flattened disk with concentric air chambers and an erect chitinous "sail". The animals are functionally adapted to make use of prevailing winds for unidirectional movements, while the symmetric and sail-less *Porpita* probably is favoured by multidirectional currents, making use of its tentacles for adaptation to both currents and waves for propulsive movements. Chondrophorids form only a minor constituent of the Chengjiang communities. They are all of a *Porpita*-type, represented by two species. One of these, known as *Rotadiscus grandis* bears a large flattened disk with concentric air chambers which are radially divided into numerous partitions (Sun & Hou, 1987a), the maximum size of its disk reaching 12 cm. The other yet undescribed form bears a strong resemblance to a modern *Porpita* (Chen, Hou & Erdtmann, 1989).

Priapulid worms

Priapulid worms, a minor protostome phylum, are one of the most predominant groups at Chengjiang (Sun & Hou, 1987b). They possess a retractable proboscis (prosoma) studded with spinous papillae on the epidermis in most specimens. An intestine may be observed in several specimens that show a raised positive relief. A remarkably strong positive relief, supported by carbon and fine sediment composition, frequently appears in the anterior portion of the intestinal tract, strongly suggesting that swallowed prey is still contained within it.

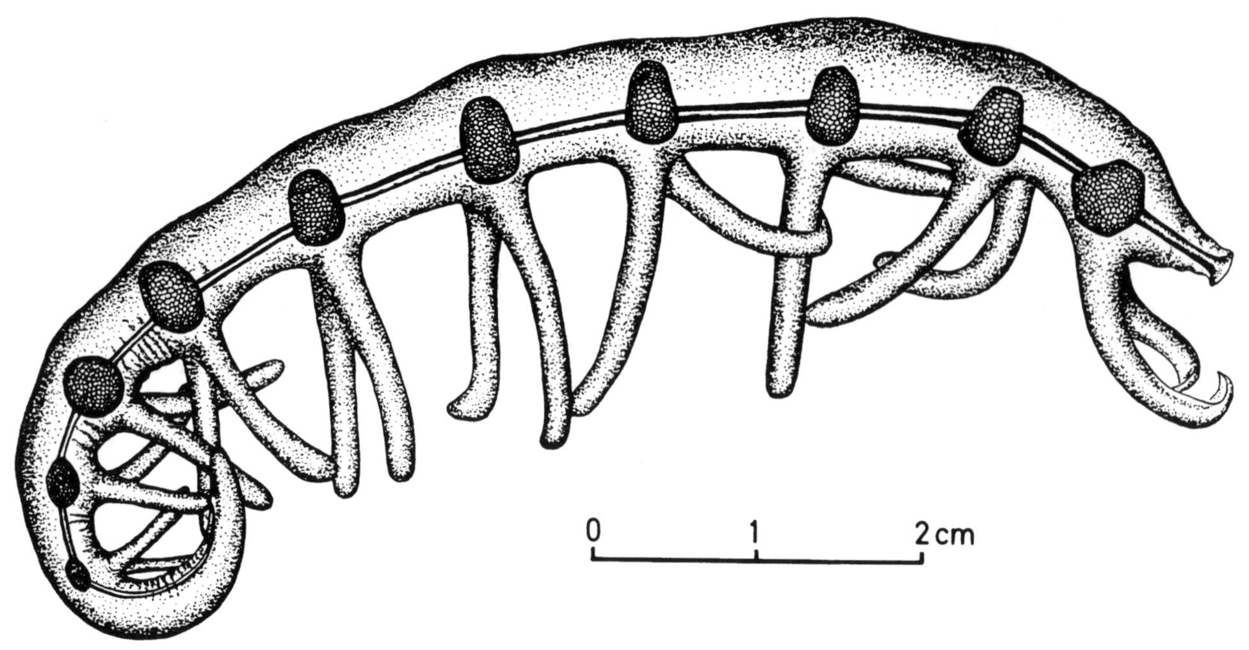

Fig. 4. Reconstruction of netted-scale worm-like animal *Microdictyon*; bar is 10 mm.

Microdictyon

The discovery of a worm-like animal closely related to *Microdictyon* (Chen *et al.*, 1989a; Pl. 1, Fig. 1), in enrolled death position, may provide us with new information on the nature and derivation of certain "small shelly fossils" (fig. 4). The function of the round to oval polygonal plates, however, remains obscure. The plates may have provided or protected a space for muscle attachment, linked to leg motion. The elongated soft legs of *Microdictyon* are thought to have served for attachment, or for a sliding or swimming locomotion (Chen *et al.*, 1989a). *Microdictyon* may have symbiotically associated with pelagic medusiform organisms, because some specimens are found directly attached to medusoids. It is possible, therefore, that the soft legs may have served as attachment organs to the medusoid hosts. Thus the animal may be considered to belong to the passive epiplankton, allowing a broader geographical range and occupying a broader environmental spectrum, a point consistent with the distribution of the fossil skeletal parts (Bengtson *et al.*, 1986; Tong, 1989). Hinz (1987), who investigated isolated material by SEM techniques, observed hollow spines fringing the round to oval holes in *Microdictyon*. These conical spines apparently showed signs of abrasion which probably should be referred to post mortem transport rather than to functional wear. A protective function of these studded plates may appear to be the most reasonable intepretation. *Microdictyon* was referred to the Problematica by Bengtson *et al.* (1986). However, Chen *et al.* (1989a), ascribed this genus to an extinct group of worm-like higher invertebrates. Its ten pairs of scales and unsegmented soft legs are regarded as evidence for a probable internal segmentation.

Vermiform animals with tentacles

One of the most distinctive groups among the worm-like animals is *Facivermis yunnanicus* (Hou & Chen, 1989a). A possible annelid, it possesses 5 pairs of segmented tentacles arising dorsally from the anterior portion of the trunk. The body is clearly segmented into rings, with a spinous, papillae-bearing proboscis. *Facivermis* shows affinities to living nereid polychaetes. Both possess an eversible proboscis and 5 pairs of segments, but they differ in the dispositions of the tentacles. All 5 pairs of tentacles in the living nereids are on the head, while the tentacles of *Facivermis* are disposed dorsally on the anterior portion of the trunk. The different positions of the tentacles in the two groups may indicate an evolutionary development of a fused head from the anterior segmented trunk (Hou & Chen, 1989a).

Dinomischus

There is a group of soft-bodied animals, the entoprocts, of which *Dinomischus* (fig. 5) may be the only known fossil genus, represented by three specimens from the Middle Cambrian Burgess Shale (Conway Morris, 1977). Two additional specimens of *Dinomischus* have now been collected from Chengjiang. They possess a cup- shaped body, which was anchored by a stem (Chen *et al.*, 1989c; Pl. 1, Figs. 2, 5). The stalk is considered to have enabled this creature to occupy an elevated epifaunal tier. *Dinomischus* appears to have been a suspension feeder that depended on horizontal, unidirectional currents. The laterally compressed calyx, as a feeding device, can be inferred to have been orientated with its anterior side against the currents, where the bracts diverged from one another, thus allowing the currents to enter. The formation of eddies within the bract circlet might be expected, as the incoming currents were held back by the closed lateral and posterior sides of the bract circlet. A tall and slender tube extending beyond the bract circlet may either be interpreted as the proximal portion of the stalk which could have been twisted below the calyx, or, alternatively, as an anal tube. The latter interpretation is advanced because it would allow for an efficient dispersion of faeces, although the first-

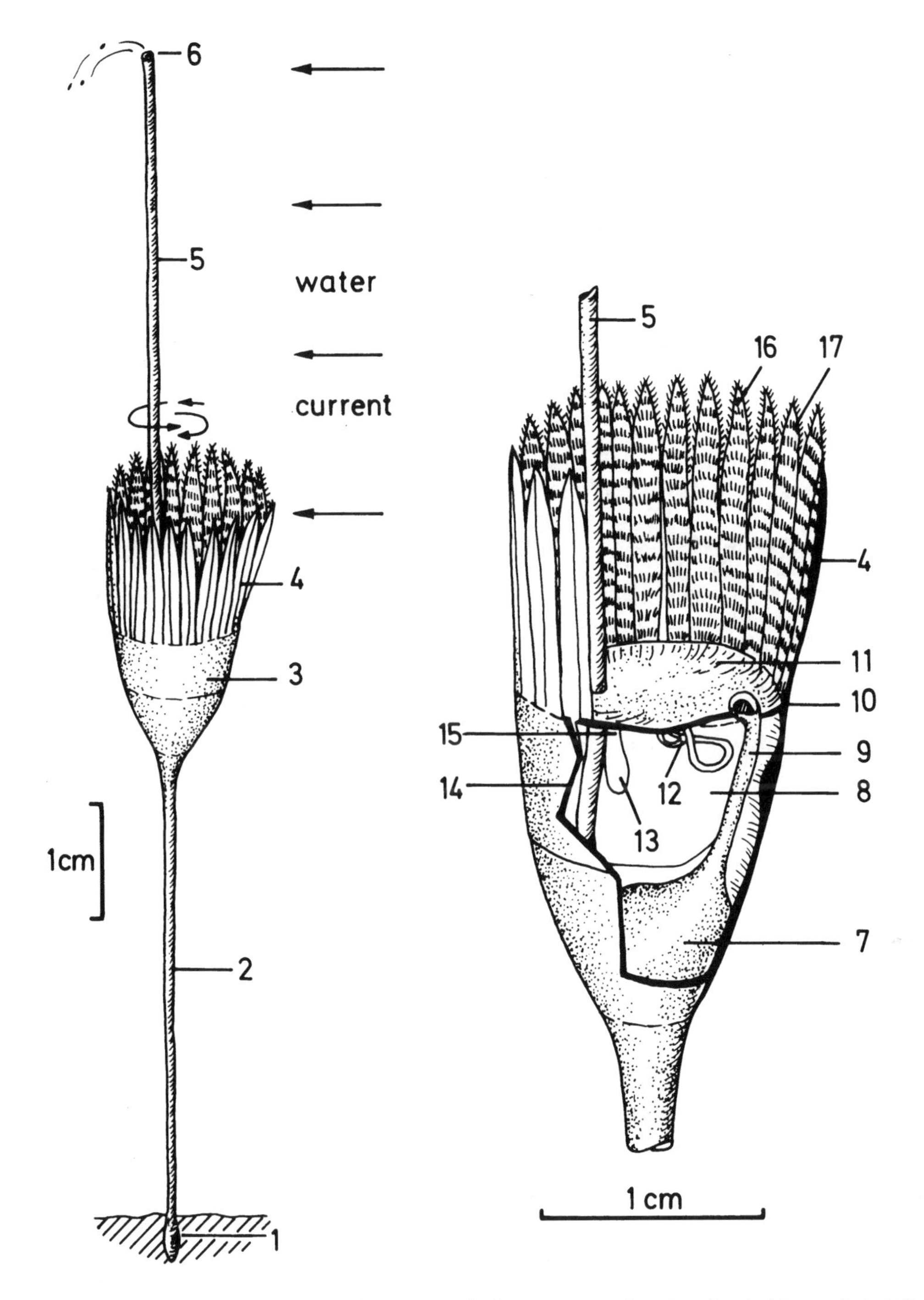

Fig. 5. Reconstruction of the entoproct *Dinomischus*; arrows indicate current direction; bar is 10 mm. 1. holdfast; 2. stalk; 3. calyx; 4. bracts; 5. excretory organ; 6. anus; 7. stomach sac; 8. visceral mass; 9. oesophagus; 10. mouth; 11. vestibule; 12. protonephridium; 13. gonad; 14. intestine; 15. ganglion; 16. cilia; 17. wrinkle.

mentioned interpretation would be congruent with previous reconstructions of *Dinomischus* (Conway Morris, 1977). However, this is not supported by the observation from the Chengjiang specimens made by the senior author showing that the tube lies precisely in a parallel manner with the bedding plane at a line related to the posterior-anterior plane of the calyx body, and it also displays a straight extension parallel to the axial elongation of the animal. As shown by the evidence of a laterally compacted specimen (Chen *et al.*, 1989c, Pl. 1, Fig. 5), the tube is clearly occurring at a line between the central and posterior position of the calyx. Alternatively, it may be observed that a sheet of sediment intervenes between the calyx and a stem part in the holotype specimen. A distinct rock cleavage plane is in evidence between the bedding surfaces including the calyx and the one containing the disjunct stem portion (see Fig. 5 or in Chen *et al.*, 1989c, Pl. 1, Figs. 2, 5) separating these two body parts. If the twisted stem interpretation is correct, however, the total length of the stalk would be rather long, at least four times as long as the stem of the Burgess Shale *Dinomischus isolatus*. An anal extension, on the other hand, occurs both in the holotype and paratype, both showing no supporting evidence (apart from separation by a layer of sediment) of this structure to represent a twisted portion of the stalk. On the contrary, the evidence from the paratype (a laterally compacted specimen) convinces one of the authors (CJY) that the anal tube arises in such a position that if the stalk had been recurved then it would not be aligned along its projected extension. This suggests that stalk and tube are unrelated. The junior author (B.D.E.), however, observes that the microstructure of the tube and stem are identical. Unfortunately, the slabs containing both specimens are broken below the short proximal stalk in the holotype and across the base of the calyx in the paratype. An X-ray analysis may provide a definitive solution at a future date.

Brachiopods

All the brachiopods belong to the inarticulate obolellid and linguloid forms. The obolellids, previously erroneously referred to the hydrozoan *Heliomedusa orienta* (Sun & Hou, 1987a, Pls. 1,2), are exceptionally well-preserved with mantle setae and mantle canals (Chen, Hou & Erdtmann, 1989). These were mistaken (see also Conway Morris & Robison, 1988) previously as marginal tentacles and radial canals of medusoids by Sun & Hou (1987). The linguloid forms show a long, annulated peduncle (see Fig. 7).

Annelid-arthropod intermediate animals

Onychopora and Tardigrada are the living representatives of presumed side branches, somewhat intermediate between annelids and arthropods. The fossil representatives of the intermediate forms are extremely rare, but interesting for many reasons. Among all the fossils discovered in the famous Burgess Shale, none has aroused more interest than *Aysheaia*, owing to its intermediate characters between those of arthropods and annelids. However, opinions on its phylogenetic affinities are quite diverse, ranging from an onychophoran (Robison, 1985) to a tardigrade affinity (Tiegs & Manton, 1958; Bergström, 1979). A new specimen (see Fig. 8), known as *Luolishania longicruris*, is broadly similar to *Aysheaia* (Hou & Chen, 1989b). However, the differences between the two taxa are so great that they warrant separation at a taxonomic level above the genus. *Luolishania* is differentiated from *Aysheaia* by having (i) a greater number of uniramous limbs (15 pairs); and (ii) a subcircular projection behind the last pair of limbs. The trunk somites are spaced evenly on its central portion, the limbs lack spines and are elongate, with their length being triple the width of the adjacent trunk, while in contrast the length of the limbs in *Aysheaia* is smaller than their width. In spite of a general similarity in appearance and size, which suggests an onychophoran affinity, *Aysheaia* also resembles the Tardigrada in several features: the terminal mouth and the terminal last pair of legs. On the other hand, it is obvious that *Luolishania* bears a strong overall similarity to the Onychophora. The last pair of legs, both in *Luolishania* and in Onychophora, never extend into the posterior end of the body. Therefore, we believe that *Luolishania* may be an archaic predecessor, from which Onychophora, Hexapoda and Myriapoda could have evolved. *Luolishania* possesses long and slender clawed lobopods, which suggests a non-burrowing mode of life. We believe that the soft substrate environment would have been troublesome for the locomotion of this animal. The general morphology of *Luolishania* with long clawed legs and an associated variety of sponges suggests that the animal might have preyed on sponges (Whittington, 1978).

A close biological affinity between *Luolishania* and *Xenusion* from Baltic Lower Cambrian erratic boulders seems convincing because both show an apparently heteronomous segmentation, and both possess numerous paired limbs. *Xenusion* is known only from two specimens, which are intepreted as opposite ends of the same animal, *X. auerswaldae* (Dzik & Krumbiegel, 1989). The holotype, described by Pompeckj (1927), represents a major portion of the posterior, showing a possible posterior projections behind the last pair of limbs. These projections are a feature that separates *Luolishania* and *Xenusion* (as well as *Aysheaia*) at a suprageneric taxonomic level. *Luolishania* also differs from *Xenusion* in its smaller size, in the presence of fewer and more slender pairs of limbs (15 pairs), and in the presence of more tubercles in each segment (three in each). The differences are considered to be less significant than those between *Luolishania* and *Aysheaia*, so allowing inclusion of *Xenusion* in the same class or order with *Luolishania*. Dzik & Krumbiegel (1989) suggested the presence of spines on the tips of these tubercles and interpreted the hypothetical spines to possess a defensive function. The explanation seems difficult to accept because (i) the actual presence of spines is doubtful; (ii) for the slender bodied creature it would have been difficult to enroll its body for protection of its ventral body side in the case of attack by predators.

Arthropods

The arthropods are the predominant group in the Chengjiang fauna in terms of both number of specimens and species. They exhibit a fair spectrum of morphological diversification and size variation ranging from less 1 mm up to half a meter. Skeletonized arthropods such as trilobites were only a minor component in terms of diversity and abundance among the Chengjiang arthropods, with only three species belonging to *Eoredlichia, Yunnanocephalus* and *Kuanyangia* (Zhang, 1987). Complete carapaces of the trilobites are common, but no evidence of soft part preservation is available. The bivalved bradoriids are microscopic, displaying maximum lengths of only 4 to 5 mm, and the most numerous among the arthropods, being represented by a great number of complete carapaces which, however, show no evidence of soft part preservation. The problem of why the soft parts, even of the bradoriids, are not preserved is difficult to explain. It might be possible that the appendages and antennae of the bradoriids were more sensitive to degradation than those in the other arthropods, or that the fine coating of these carcasses by phosphate or other inorganic substances was too slow, or did not form on the ventral portion prior to final burial.

Naraoia predominates among the soft-bodied arthropods, (Zhang & Hou, 1985) being represented by a great number of carapaces and by several hundred specimens which show appendages, guts and antennae. The taxonomic position of *Naraoia* has not yet been determined satisfactorily. *Naraoia* is considered to be a trilobite (Whittington, 1977) or to belong to a separate group of Trilobitomorpha (Bergström, 1979). *Alalcomenaeus? illecebrosus* is one of the relatively abundant arthropods, possessing a pair of large appendages on its head, each with three flagellae (Hou, 1987a). A number of external features suggests *A.? illecebrosus* to have been a nektobenthic form. There are (i) the elongated body with a thick and laterally compressed cross section; (ii) the presence of a lobate branch in each of the trunk somites; (iii) the paddle-shaped telson. *Jiangfengia multisegmentalis* belongs to these rare arthropods, with a streamlined body considered to be suitable for swimming. This form bears an elongate and relatively narrow exoskeleton consisting of a cephalic shield, 22 trunk somites, and a telson (Hou, 1987a). The remaining soft-bodied arthropod fauna consists of at least 17 additional taxa, including megathoracic forms such as *Fuxianhuia, Kuamaia, Rhombicalvaria*, and bivalved forms such as *Isoxys, Combinivalvula*, and *Vetulicola*, all being relatively common (Hou, 1987b, c). The three megathoracic forms mentioned above bear a broad exoskeleton, with a maximum width of 13 mm in *Fuxianhuia*, and 40 mm both in *Kuamaia* and in *Rhombicalvaria*. The external features of their exoskeletons suggests them to have been vagile benthos. The genera *Retifacies*, *Acanthomeridium* and *Urokodia* (Hou *et al.*, 1989b) are less common and are distinguished from one another by the broadly distinctive morphologies of their carapaces (Hou *et al.*, 1989b) and probably by their differing modes of life. *Retifacies* is interpreted to have been vagile benthos, possessing a rather broad exoskeleton, up to 75 mm in width.

Anomalocaris

Anomalocaris was a metameric animal, and bore an anterior pair of large jointed limbs, a unique jaw, and a series of lobes along each side of the trunk. Its morphology is regarded as a strange combination of characters which precludes fitting *Anomalocaris* into any recognized phylum (Whittington & Briggs, 1985). To date, it is the largest known Cambrian animal; the dimensions of the limbs and of the jaw suggest a length of perhaps half a metre. The creature would have been a formidable beast, swimming just above the sea bottom with the anterior limbs seeking out its prey of soft bodied benthos.

Although fragments of *Anomalocaris* have been reported from several localities in Lower and Middle Cambrian strata (Whittington & Briggs, 1985), until now no evidence was available of its occurrence outside the North American continent. The discovery of limb fragments of *Anomalocaris* at Chengjiang confirms a broader geographic extension for this macrophagous predator. This provides further evidence (see Dzik & Lendzion, 1988) that macrophagous predation may have appeared at an earlier geological time: more or less coincident with the first appearance of robust skeletons in the arthropods, *i.e.* trilobites.

Palaeoecological setting

The Chengjiang biota bears striking similarities with a number of Lower and Middle Cambrian soft-bodied assemblages (Whittington, 1985; Conway Morris, 1989), especially with the well known Middle Cambrian Burgess Shale biota. The Burgess Lagerstätte is considered to be 530 Ma old, represented by five different fossil assemblages from four different stratigraphic levels at more than a dozen localities in an area extending for 20 kilometres in the relatively deep, and open shelf environment immediately adjacent to the submarine Cathedral Escarpment (Collins, Briggs & Conway Morris, 1983). Regardless of their differences in age (perhaps as much as 40 Ma), in sedimentary and biological facies, and in their geographical separation, the similarities in taxonomic composition of the Burgess and Chengjiang Lagerstätten are significant. At least 16 genera are in common, and 10 more are closely related. The similarities suggest a rather conservative and slow evolutionary development for this intervening period since the first appearence of the mineralized trilobites. Based on the few points of Early to Middle Cambrian soft bodied fossil documentation, it is reasonable to hypothesize, as we believe, that the apparent conservatism may be explained by the common shallow water derivation (see also Conway Morris 1989) of all these localities (Chengjiang, North Greenland, Burgess; see also Conway Morris 1984; Conway Morris *et al.* 1987). However, it remains doubtful whether deeper water benthic or open oceanic habitats were occupied by metazoan communities at that time. For the period of ca. 40 Ma the metazoans may have kept an evolutionary pace which presents a striking contrast to the period approximately of 30 Ma predating the biomineralization episode of trilobites when the complex, yet physiological explosive evolution of the metazoan life was accompanied by a seemingly "spontaneous" invention of skeletal hard parts of several of the primordial "phyla". Only future discoveries may corroborate or disprove this observation.

The biota documented by the Lower and Middle Cambrian Lagerstätten tend to be similar over a broader geographic area and perhaps a broader environmental spectrum as well, suggesting that the Cambrian biota actually were or appear to have been ecologically "generalized", with broader niches and relatively simple trophic requirements and structures as compared to the ecological specializations of the younger Cambrian and Ordovician biota.

The fossil assemblages of the Chengjiang Lagerstätte can be regarded as autochthonous thanatocoenoses, because they appear to have been mostly buried without obvious transportation. The fossil assemblage combines benthic and pelagic communities. The fossil evidence suggests that the trophic structure of the benthic community was dominated by suspension and deposit feeders, forming a food chain of primary consumers. The major food source for the primary consumers was likely the detritus derived from phyto- and zooplankton. The planktonic detritus may have provided a rain of dead or fragmented zooplankton and faeces. The presence of primary producers is indicated by abundant probable planktonic fossil algae and acritarchs. Sponges, brachiopods, sea anemones and stalked *Dinomischus* were representatives of suspension feeders, which appear to have been comparatively specialized. The sponges were the largest group, and competitive interaction may have been intense because they were dependent on the same or similar nutrient resources. The competition may have led this community to develop a vertically stratified structure, with species of different body lengths exploiting different levels of food sources in the water column (Conway Morris, 1979; Chen, Hou & Erdtmann, 1989), leading to three tiering levels, with the highest tier of +30 cm (Chen *et al.*, 1989b).

The deposit feeders were most abundant trophic group within the Chengjiang communities, and are represented by bradoriids, trilobites and a few worm-like animals. Among them the bradoriids predominate in terms of number of individuals and biomass; they were probably a major food source of most carnivores.

Although some authors believe that macrophagous predators were absent or rare in those Cambrian communities (Glaessner, 1972; Valentine, 1973), increasing information suggests a fair number of macrophagous predators during the Early Cambrian (Conway Morris & Jenkins, 1985; Conway Morris, 1986). The Chengjiang Lagerstätte has yielded rare evidence for predators. This demands caution, however, since our knowledge is as yet speculative. Among the Chengjiang biota a number of species of non-trilobite arthropods, *Luolishania* (Hou & Chen, 1989b), priapulid worms, and *Anomalocaris* are all possible candidates for such carnivores. The jaws of this big creature could have inflicted serious wounds upon the mineralized plates of trilobites (cf. Whittington & Briggs, 1985).

The arthropods generally possessed the largest sizes among the Chengjiang biota. The phylum possesses a unique exoskeleton which is ideally suited for segmented organisms because it can be moulted and discarded (Valentine, 1973). This may be the reason why this group has evolved quite successfully for its entire evolutionary history. Periodic moulting provides arthropods with a high potential for developing multiple functions for their appendages i.e. for walking, burrowing, swimming, sensory, mastication etc.. However, it would be disadvantageous, if the skeletons were to have become thicker or highly mineralized because of higher cost of energy expenditure. The trilobites were the only major group of Cambrian arthropods to adopt a highly mineralized skeleton. We believe that trilobites could well have evolved much earlier but simply did not become sufficiently skeletonized until the Qiongzhusi Stage. The increasing intensity and sophistication of predation may have been a significant factor in leading to the development of protective exoskeletons in trilobites. Although evidence of wounds in Cambrian trilobites suggests that they suffered repeated attacks by effective predators, such attacks may not always have resulted in the death of the wounded animals (Conway Morris, 1986).

A SCENARIO FOR EARLY EVOLUTION

The Chengjiang Lagerstätte provides us with a window into the life of the Early Cambrian. As indicated (Fig. 2) by the contact between the Meishucun Formation and the Qiongzhusi Formation, also a clear biological discontinuity with the underlying Meishucun Stage occurs at this level. The Meishucun Formation. harbours the "small skeletal fauna" and, after a short unfossiliferous interval, the earliest trilobites appear in the Badaowan Member. This appearance precedes the rise of the Chengjiang biota by only a short interval. Such observation seems to fit well with the concept of punctuational change. Eldredge & Gould (1972; see also Gould, 1984) apply this concept as punctuational change to emphasize both the stability of systems and the concentration of changes within short episodes that break old equilibria and quickly reestablish new ones. A pattern of punctuational changes may have characterized the Precambrian-Cambrian transition, when the metazoans proceeded in a sequence of four equilibrium phases represented by the Huainanian, Ediacaran, Meishucunian and Chengjiangian phases, with rapid turn-over intervals between them (see Fig. 6).

HUAINANIAN PHASE. The Huainanian bio-evolutionary phase is represented by the pre-Ediacaran (in reference to a post-Nantuo glacial age of Ediacara and a secured pre-Nantuo tillite age of the Huainan biota) worm-like fossils which were found in the Huainan area, Anhui Province, in the strata lying about 300 m below the Latest Proterozoic Fengtai tillites (Chen, 1988; Fu, 1989). Until recently, it had been widely accepted that the Ediacaran contained the earliest records of metazoan life. Now, after the discovery of the Huainanian biota the presence of an extended pre-Ediacaran history of metazoans becomes apparent. The Huainanian phase is represented by an extremely low diversity of worm-like creatures, containing only two morphological types known as *Pararenicola huaiyuanensis* and *Protoarenicola baiguanshanensis*. Both of them possess a radial symmetry and are evidently associated with shallow subcircular burrow holes, strongly suggesting that the organisms were endichnial deposit feeders. The adaptation to an infaunal life is interpreted as a protective response to probable high flux of UV-rays. Field investigations indicate that the density of these fossils tends to be greatest adjacent to stromatolitic mounds. The decreasing density away from stromatolitic mounds is interpreted as resulting from decreasing availability of oxygen and possibly also of nutrients.

EDIACARAN PHASE. The widespread Ediacaran biota is unique, and they show no obvious connections with the preceding Huainanian biota. The Ediacaran fauna is dominated by medusoid cnidarians, sessile colonial coelenterates and worm-like organisms (Glaessner, 1984), although serious doubts have been put forth regarding the existence of any Phanerozoic analogues for the so called Vendozoa (Seilacher, 1989) of Ediacaran type. Post-glacial transgression in the Early Vendian might have considerably affected metazoan evolution, accelerating the areal extension of the epeiric seas and raising the temperature, allowing territorial expansion of oxygen-releasing photosynthesis. The widespread occurrence of phosphate deposits in post-Nantuo glacial sequences suggests intensive upwelling of nutrient-rich watermasses, which then released substantial quantities of phosphorus into the shallow water realms. The apparent rise of available phosphorus may have been an important stimulus for the Ediacaran radiation.

MEISHUCUNIAN PHASE. The innovative evolutionary episode within the Meishucun Stage is expressed by the development of mineralized skeletal parts in diverse groups. The widespread "small shelly biota", characterizing the Meishucun Stage, may represent a "meio-fauna", although a fair proportion of the small skeletal plates are constituents of potentially large poly-

Series	Stages	Eustatic Events	Biological Events
550 Ma	Canglangpu	maximum transgression	Radiation of archaeocyathids
	Qiongzhusi	Transgression and anoxic event	Second biomineral event with radiation of Chengjiangian biota
Early Cambrian		**Regression event**	
570 Ma	Meishucun	Phosphogenesis and transgression event	First biomineral event with radiation of "small shelly" Meishucunian biota
	Tonying	**Regression event**	
	Toushantuo	Phosphogenesis: anoxic and transgression event	Radiation of Ediacaran biota
Late Proterozoic	Nantuo	**Glaciation and regression event**	
(Sinian)	Datangpo	Interglaciation: anoxic and transgression event	Radiation of Huainanian biota
650 Ma	Gucheng	**Glaciation and regression event**	

Fig. 6. Diagram showing eustatic and biological events in the Precambrian-Cambrian transition interval.

plate exoskeletons. The equivalent fauna of the Meishucun type has previously been referred to as "Tommotian", although the stratigraphic correlations of these "early skeletal fossils" are controversial. There appears to be considerable doubt about the isochronicity of the initial occurrence of the "small skeletal fossil fauna" in the different parts of the world, and first appearances may simply reflect biofacies controls. Because of the advanced development and diversification of biofacies and biogeographical differentiation, the usefulness of the small shelly fauna for worldwide correlation is questionable.

The search for the causes of the Early Cambrian radiation has evoked an intense discussion. Holser (1977) documented a $\delta^{34}S$ positive shift, known as the Yudomski event, in the latest Precambrian. This event is regarded as having important implications for the explanation of biological innovation near the Precambrian-Cambrian boundary (Conway Morris, 1987). However, we consider the significance of a direct relation between the Yudomski event and biological innovation to be exaggerated by a misinterpretation of the duration of the Yudomski event. Examination of $\delta^{34}S$ values in sedimentary barite from various stratigraphic levels in the Late Precambrian Toushantou, Tongying, and Lower and Middle Cambrian strata in western Hunan, Anhui and Zhejiang provinces, China, demonstrates that the onset of the $\delta^{34}S$ positive anomaly occurred at the Lower Vendian (or even earlier). Furthermore, the anomaly may have persisted into the Middle Cambrian (Hu *et al.*, 1986; Chen & Zhang, 1988) (Fig. 1).

Extensive formation of Late Precambrian dolomites would have led to a decrease of the Mg/Ca ratio in the sea water; the changes in this ratio could have influenced the evolving biota and accompanying biomineral events of the Early Cambrian. The major shift in precipitation of carbonates across the Precambrian-Cambrian interval from dolomite to limestone as recorded in the Yangtze Platform sequence occurs in the Quiongzhusi Stage. Thus, it apparently postdates the onset of the Early Cambrian biomineral event by a few million years. Hence the evidence for a relation between changes in the Mg/Ca ratio and the onset of biomineralization is still rather tenuous.

Brasier (1982) proposed a possible connection between the Cambrian evolutionary explosion and global sea-level changes. A rapid global fluctuation in sea level has been documented for the terminal Precambrian, and is represented by a conspicuous hiatus between the Tongying and overlying Meishucun Formations over much of the Yangtze Platform. This Meishucunian transgression could have influenced the evolving biota. The apparent demise of the Ediacaran fauna could have been linked to a rapid drop in sea level. A transgressive event at the beginning of the Meishucun Stage may have considerably affected the Cambrian explosion through accelerating the areal extent of epeiric seas and upwelling of nutrient-rich water masses which released substantial quantities of phosphorus into the shallow marine realm (Chen & Zhang, 1988). The rising levels of available phosphorus could have led to a substantial increase in biomass via an explosive acceleration of primary productivity (Cook & Shergold, 1984).

CHENGJIANGIAN PHASE. The Chengjiang Lager-

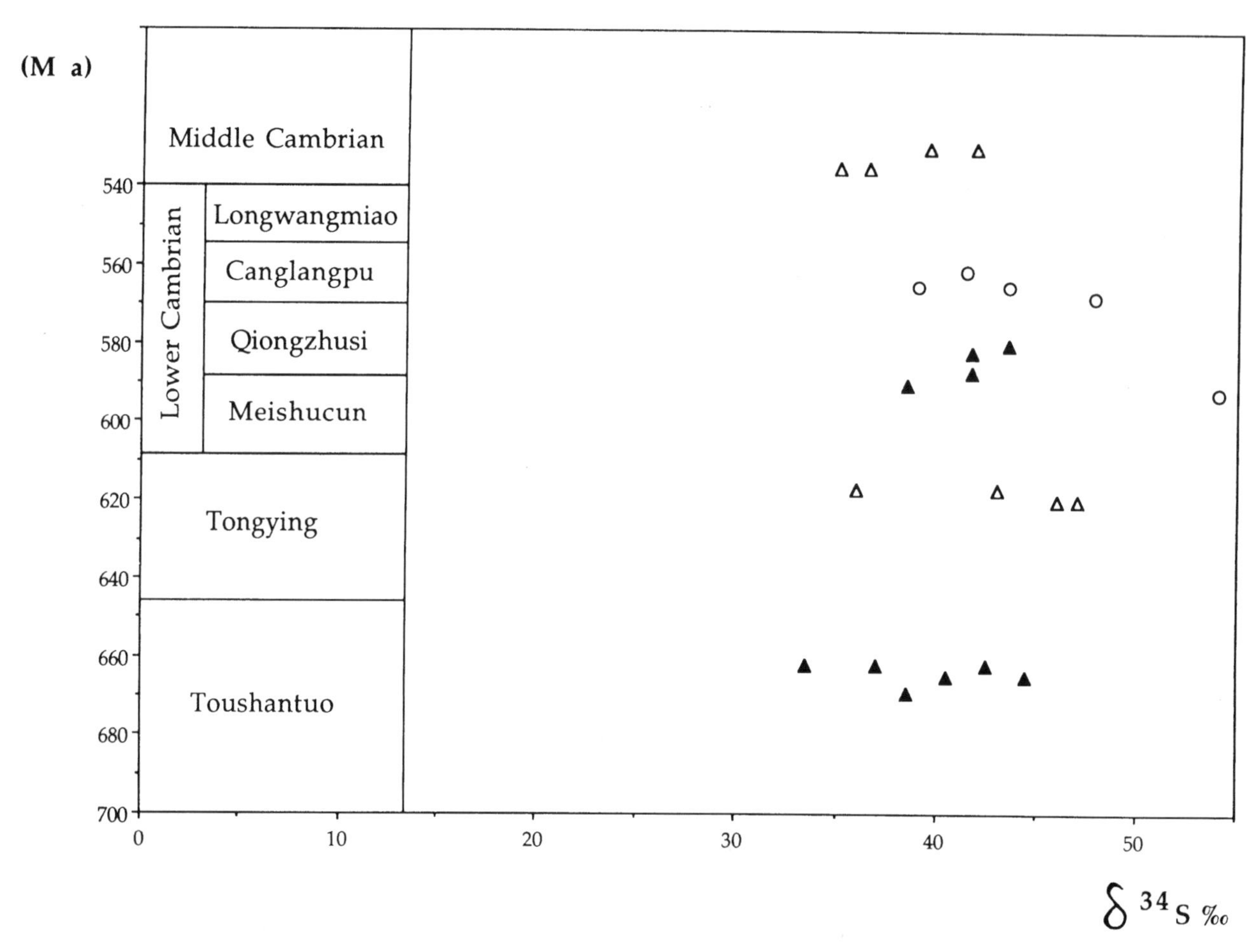

Fig. 7. Diagram showing distribution of $\delta^{34}S$ values of the barite deposits from Precambrian and Cambrian strata, in Southern China. Sample locality: open triangle-from Fuyang and Linyan Counties, Zhejiang Province, Solid triangle-from Xinguan, Hunan Province; open circle, from Dongzhi, Anhui Province.

stätte provides us with a window into the second episode of biomineralization, marked by the debut of skeletonized trilobites during the Qiongzhusi Stage. The Chengjiang biota inhabited soft muddy bottoms in a relatively deep and tranquil subtidal environment. The nonskeletonized biota could have existed for a considerable period before the onset of trilobites. However, strata below the Qiongzhusi beds are devoid of macrofossil evidence, because of their minimal chance of being preserved as body fossils. The connection of the appearance of trilobites shortly after or with the rise of effective macrophagous predators is considered to be a rational explanation. The onset of skeletonization is expressed by the approximately synchronous appearance of at least three probably unrelated groups of trilobites: the eodiscoids, olenellids, redlichiids and remaining polymerids, which evolved independently from unrelated forerunners of the arthropods under the stringent conditions of effective predation.

In south-west China a noticeable environmental shift, represented by sea level fluctuations and by changes in the characters of sedimentation occurs near the boundary between the Qiongzhusi Stage and the underlying Meishucun Stage. The widespread occurrence of carbon-bearing deposits during the Qiongzhusi Stage probably reflects a transgressive event with a shelfward movement of the oceanic oxygen minimum zone. This may have had an indirect bearing on biotic changes. The anoxic event could have led to a substantial increase both in biomass via phytoplankton and in the amount of dead organic matter raining down from surface waters; so it may be regarded as a trigger mechanism for the adaptive radiations and for the establishment of a relatively stable ecosystem, which is so characteristic for much of the remaining Cambrian Period. Hsü *et al.* (1985) document a negative carbon-isotopic event near the Qiongzhusi-Meishucun boundary, and speculate that the event could signify a mass extinction and associated sharp drop of oceanic productivity, followed by a subsequent biological radiation as a response to new ecological opportunities. We are sceptical about their proposals for the following reasons: Negative $\delta^{13}C$ values occur in samples collected close to the hiatus both in the Yunnan and Yangtze Gorges areas. It is likely that the negative isotopic $\delta^{13}C$ anomaly implies diagenetic reactions, probably involving either early diagenetic bacterial fermentation or late diagenetic alteration probably due to vadose groundwater during the hiatus. Furthermore, the carbon isotope anomaly definitely occurs at different stratigraphical levels in either sections. The small shelly fauna of the Meishucun analogues is known to have persisted during much of the Early Cambrian, so that the presence of a hypothetical mass extinction near the Meishucun-Qiongzhusi boundary seems to be unlikely. The biological explosion at the Qiongzhusi Stage is believed to have been initiated by a territorial invasion of metazoans into a new environment, such as into soft muddy substrates and into relatively deeper tranquil environments. However, these biota may have evolved slightly earlier than their first appearance in the fossil record, with their actual fossilisation requiring a favourable coincidence of preservational and taphonomic environments. The rise of effective predation activities were of fundamental significance as a trigger for advancing the second biomineral event and a further differentiation of the contemporary ecosystem. In the meantime the small shelly faunas of the Meishucun type survived into the Qiongzhusi Stage (and beyond), perhaps because they lived in an unchanging environment. None of the microscopic molluscs occurs in association with the Chengjiangian biota; thus it is believed that either the ecospace or the geochemical conditions of the taphotope between the Meishucun and the Chengjiang sections was essentially different. The small shelly fauna is delimited to a sequence conspicuously consisting of dolostones and phosphatic limestones - not of fine - grained clastics, which was the apparent habitat and taphotope of the Chengjiang biota.

SYSTEMATIC PALAEONTOLOGY

ALGAE
Genus *Megaspirellus* gen. nov.
Type species: *Megaspirellus houi* gen. et sp. nov.
Diagnosis: Macroscopic unmineralized algae, thread-like form. Much of the thread is open coiled, in a parallel-sided helix but at least one of its terminations is straight. Helix has a diameter of 2.5 mm; the thread is about 0.3 mm in diameter, smooth and without annulation.
Remarks: The present genus *Megaspirellus* can be closely compared to *Spirellus* Jiang in Luo *et al.*, 1982, and *Obruchevella* Reitlinger, 1948, in terms of coiling style and smooth thread. However, it can readily be differentiated from them by its greater dimensions. In *Spirellus* and *Obruchevella* the helix is seldom wider than 15 μm, and the thread less than 50 μm. The new genus differs from *Jiangispirellus* Peel, 1988 by the larger dimension of the smooth thread. The thread of *Jiangispirellus*, however, is annulated and the annulated thread is intepreted as a naked trichome without an enclosing sheath.

Megaspirellus houi gen. et sp. nov.
(Pl. 1, Figs. 3, 4; Pl. 2, Figs. 3, 4)
Holotype: Nanjing Institute of Geology and Palaeontology 108501.
Description: Only a single specimen is available for the present study. The fossil is represented by part and counterpart, lying parallel to the bedding plane and is tentatively interpreted as a filament. The preserved length of the filament is 24 mm, of which 19 mm is open coiled into a parallel-sided helix. One of the filamentous terminations, about 6 mm long, displays a straight extension parallel to the elongation of the helix. The helix has a maximum flattened width of 2.5 mm, consisting of 19 individual whorls, the individual whorls were not in contact with each other originally. They possessed a uniformly circular cross section. The surface of the filament is smooth, with a flattened width of 0.3 mm.
Remarks: The present species differs from the closely related species of *Spirellus, Obruchevella* and *Jiangispirellus* by its substantially larger dimension of the thread.

Yuknessia Walcott, 1919
Type species: *Yuknessia simplex* Walcott, 1919.
Remarks: The biological affinities of *Yuknessia* remains obscure; macroscopic features show that it resembles some modern tubular green algae.

Yuknessia sp.
(Pl. 2, Fig. 1)
Description: *Yuknessia* is dominant among the Chengjiang algae. Thallus small, consisting of only few flexuous stipes; height between 1 cm and 2 cm. Stipes slender, without evidence of jointing or branching; width expanding gradually from about 0.5 mm to 1 mm.
Remarks: Our specimens most closely resemble *Yuknessia simplex* Walcott, 1919 in dimensions of thallus and stipes. However, they differ from type specimen of *Yuknessia simplex* in possessing many fewer stipes in the thallus.

Sinocylindra gen. nov.
Type species: *Sinocylindra yunnanensis* gen. et sp. nov.
Diagnosis: Unmineralized thread cylindrical and unbranched, with uniform diameter of up to 0.3 mm. Surface smooth, lacking microscopic structure.

Sinocylindra yunnanensis
(Pl. 2, fig. 5)
Holotype: Nanjing Institute of Geology and Palaeontology 108504
Description: The holotype is preserved as an irregularly winding coil on an even bedding plane indicating that the thread was extremely flexible. The thread is cylindrical and unbranched, with a uniform flattened width of 0.3 mm, and a preserved length measuring 20 mm.
Remarks: The macroscopic features of the present species resemble some members of the Family Oscillatoriaceae. However, the dimension of the thread in *Sinocylindra yunnanensis* is considerably larger, so the placement of the present species in this family is as best regarded as uncertain.

METAZOA
Phylum Coelenterata
Class Anthozoa
Order Actiniaria
Genus *Xianguangia* gen. nov.
Type species: *Xianguangia sinica* gen. et sp. nov.
Diagnosis: Body moderately large, conical and unmineralized, consisting of a pedal disc, column and a row of about 16 tentacles surrounding mouth.

Xianguangia sinica gen. et sp. nov.
(Pl. 3, Fig. 1-4)
Material: Two flattened individuals.
Etymology: Generic name referring to Hou Xianguang, discoverer of the Chengjiang Lagerstätte.
Holotype: Nanjing Institute of Geology and Palaeontology 108506. Paratype Nanjing Institute of Geology and Palaeontology 108505
Description: The holotype is represented by part and counterpart, lying parallel to the bedding plane, displaying a low relief, positive in part and negative in counterpart. It is conical, 52 mm high and 23 mm in maximum width as flattened. The pedal disc is smooth, separated from the column by a conspicuous constriction (2 mm wide), known as limbus. The disc meas-

ures 9 mm in height and is 21 mm wide. The column gradually tapers distally from 23 mm to 15 mm over a length of 24 mm. The column wall may have been thin, so permitting the internal partitions (septa) within the body cavity to be discernible. These partitions appear as folded and longitudinal projections totalling 16 septa. The tentacles are flexible and all of about the same size, forming a single ring. They appear as simple slender conical lappets which taper distally to a point. The number of the tentacles seems to relate to that of the internal compartments. The surface of the tentacles is smooth on the external side and wrinkled laterally on the inside. The bluish coating on the preserved surface of the specimens is presumably attributed to post mortem phosphatization.

The paratype specimen closely resembles the holotype. It measures 50.5 mm in height and 20 mm in maximum width. The pedal disc is about 10.5 mm high and 20 mm wide, and is separated from the column by a distinct limbus of 2 mm width. The column measures 19 mm in height and 20.5 mm in maximum width. Its surface is folded into 7-8 longitudinal segments.
Remarks: The external morphology of *Xianguangia sinica* strongly supports its affinity to Recent actiniarian sea anemones. The surface of the column appears to having possessed longitudinal projections, which are interpreted as part of the internal anatomy. The number of projections appears to correspond to that of the internal spaces within the coelenteron of typical sea anemones. These compartments are considered to be separated from each other by septa or mesenterial structures. Thus, the present species appears to have had 16 compartments, and a corresponding number of tentacles. It demonstrates a great resemblance to modern gonactinitid sea anemones. The family Gonactinitidae are regarded as the most primitive group among the modern sea anemones, characterized by 8 complete septa and a few rows of 8 tentacles each (Friese, 1972).

Acknowledgements

We thank Drs. C. Teichert, J. Bergström, S. Conway Morris and D. Collins for reading an earlier draft of this paper; Dr. J.K. Rigby for advice and suggestions concerning the fossil sponges. The following aid is greatly appreciated; Miss Ma Yuying (Nanjing) and Miss S. Hinderland (TU Berlin) for typing, Mao Jilian and Deng Dongxing (Nanjing) and B. Kleeberg (TU Berlin) for photography, Li Gejun (Nanjing) and Mrs. B. Dunker (TU Berlin) for drafting. The costs of field studies and laboratory investigation were jointly covered by the Presidential Scientific Foundation of Academia Sinica and NSF (China) Grant no. 4870119. C.J.Y. acknowledges gratefully generous financial support from the University of Camerino and by the Italian National Research Council to attend the International Conference on Problematic Fossil Taxa and Early Metazoan Evolution at Camerino, Italy in April 1989 and B.-D.E. acknowledges advice and suggestions given by M.D. Brasier (University of Oxford) and by S. Conway Morris (Cambridge).

All catalogue numbers refer to the Nanjing Institute of Geology and Palaeontology, Academia Sinica. Collection numbers with prefixed "M" are from the quarries along the western slope, and "Cf" from the quarries along the northwestern slope of Miaotian Hill, Chengjiang County. All the fossils were embedded in the mudstone layers belonging to the lower part of the Yuanshan Member dated as the lower *Eoredlichia* Zone in the trilobite zonal sequence. The illustrated fossils are not retouched.

REFERENCES

Bengtson, S., Matthews, S.C. & Missarzhevski, V.V. 1986. The Cambrian netlike fossil *Microdictyon*. In *Problematic Fossil Taxa* (eds. A. Hoffman & M.H. Nitecki), pp. 97-115. *Oxford Monographs on Geology and Geophysics* **5**. New York: Oxford University Press.

Bergström, J. 1979. Morphology of fossil arthropods as a guide to phylogenetic relationships. In *Arthropod Phylogeny* (ed. A.P. Gupta), pp. 3-36. New York: Van Nostrand Reinhold.

Brasier, M.. 1982. Sea-level changes, facie changes and the Late Precambrian-Early Cambrian evolutionary explosion. *Precambrian Research* **17**, 105-123.

Brasier, M.D. 1986a. Why do lower plants and animals biomineralize? *Paleobiology* **12 (3)**, 241-250.

Brasier, M.D. 1986b. Precambrian-Cambrian boundary biotas and events. In *Global Bio-events* (ed. O.H. Walliser), *Lecture Notes in Earth Sciences* **8**, 109-117. Berlin: Springer Verlag.

Chen Jun-yuan, 1988. Precambrian metazoans of the Huai River drainage area (Anhui, E. China): their taphonomic and ecological evidence. *Senckenbergiana lethea* **69**, 189-215.

Chen Jun-yuan, Hou Xian-guang & Erdtmann, B.-D. 1989. New soft-bodied fossil fauna near the base of the Cambrian System at Chengjiang, eastern Yunnan, China. Chinese Academy of Sciences "Developments in Geoscience", Contribution to 28th International Geological Congress, 1989, Washington, D.C., U.S.A., pp. 265-277. Bejing: Science Press.

Chen Jun-yuan, Hou Xian-guang, Li Guo-xiang, (in press) New Lower Cambrian Demosponges *Quadrolaminiella* (gen. nov.) from Chengjiang, Yunnan. *Acta Palaeontologica Sinica.*

Chen Jun-yuan, Hou Xian-guang & Lu Hao-zhi, 1989a. Early Cambrian netted scale-bearing worm-like sea animal. *Acta Palaeontologica Sinica* **28**, 1-12. [in Chinese with English summary].

Chen Jun-yuan, Hou Xian-guang & Lu Hao-zhi, 1989b. Lower Cambrian Leptomitids (Demospongea) Chengjiang, Yunnan. *Acta Palaeontologica Sinica* **28**, 58-71. [in Chinese with English summary].

Chen Jun-yuan, Hou Xian-guang & Lu Hao-zhi, 1989c. Early Cambrian hock glass-like rare animal *Dinomischus* and its ecological features. *Acta Palaeontologica Sinica* **28**. [in Chinese with English summary].

Chen Jun-yuan & Zhang Jun-ming, 1988. Tectonic-sedimentary evolution of the Jiangnan marginal basin of the Late-Proterozoic and Early Palaeozoic time. In *Chinese-Soviet Symposium: Geology, Geophysics and Metalogeny of the Transition Zone from the Asiatic Continent to the Pacific Ocean*, pp. 20-21.

Clark, H.L., 1912. Fossil holothurians. *Science* **35**, 274-278.

Collins, D., Briggs, D. & Conway Morris, S. 1977. New Burgess Shale fossil sites reveal Middle Cambrian faunal complex. *Science* **222**, 163-167.

Conway Morris, S. 1977. A new entoproct-like organism from the Burgess Shale of British Columbia. *Palaeontology* **20**, 833-845.

Conway Morris, S. 1979. The Burgess Shale fauna (Middle Cambrian). *Annual Review of Ecology and Systematics* **10**, 423-467.

Conway Morris, S. 1986. The community structure of the Middle Cambrian phyllopod bed (Burgess Shale). *Palaeontology* **29**, 423-467.

Conway Morris, S. 1987. The search for the Precambrian-Cambrian Boundary. *American Scientist* **75**, 157-167.

Conway Morris, S. 1989. The persistence of Burgess Shale-type faunas: implications for the evolution of deeper-water faunas. *Transactions of the Royal Society of Edinburgh: Earth Sciences* **80**, 271-283.

Conway Morris, S. & Jenkins, R.J.F. 1985. Healed injuries in Early Cambrian trilobites from South Australia. *Alcheringa*, 9 (3-4), 169-177.

Conway Morris, S., Peel, S., Higgins, H.K., Soper, N.J. & Davis, N.C., 1987. A Burgess Shale-like fauna from the Lower Cambrian of North Greenland. *Nature* **326**, 181-183.

CONWAY MORRIS, S. & ROBISON, R.A. 1988. More soft-bodied animals and algae from the Middle Cambrian of Utah and British Columbia. *The University of Kansas Paleontological Contributions, Paper* **122**, 1-48.

COOK, P.J. & SHERGOLD, J.H. 1984. Phosphorous, phosphorites and skeletal evolution at the Precambrian-Cambrian Boundary. *Nature* **308**, 231-236.

DEAN,D., RANKIN, J.S. & HOFFMANN, E. 1964, A note on the survival of polychaetes and amphipods in stored jars of sediments. *Journal of Paleontology* **38**, 608-609.

DURHAM, W.J. 1974. Systematic position of *Eldonia ludwigi* Walcott. *Journal of Paleontology* **48**, 750-755.

DZIK, J. & KRUMBIEGEL, G. 1989. The oldest "onychophoran" *Xenusion*: a link connecting phyla? *Lethaia* **22**, 169-181.

DZIK, J. & LENDZION, K. 1988. The oldest arthropods of the East European platform. *Lethaia* **21**, 29-38.

ELDREDGE, N. & GOULD, S.J. 1972. Punctuated equilibria: An alternative to phyletic gradualism. In *Models in Palaeobiology* (ed. T.J.M. Schopf), pp. 82-115. San Francisco: Freeman.

FRIESE, U.E. 1972. *Sea anemones* : T.E.H. Publications.

FISHER, I.S.J. & HUDSON, J.D. 1985. Pyrite geochemistry and fossil preservation in shales. *Philosophical Transactions of the Royal Society of London* **B 311**, 167-169.

FU JUN-HUI, 1989. New materials of Late Precambrian Huainan biota fossil in Shouxian, Anhui. *Acta Palaeontologica Sinica* **28**, 72-78 [in Chinese with English abstract].

GLAESSNER, M.E. 1972. Precambrian palaeozoology. In *Stratigraphic problems of the Late Precambrian and Early Cambrian* (eds. J.B.Jones & B. McGowran), *Special Paper of the Centre for Precambrian Research, University of Adelaide* **1**, 43-52.

GLAESSNER, M.E. 1984. *The dawn of animal life. A biohistorical study*. Cambridge: Cambridge University Press.

GOULD, S.J. 1984 Toward the vindication of punctuational change, In *Catastrophes and Earth History* (eds. W.A. Berggren & J.A. van Couvering), pp. 9-39. Princeton, N.J.: Princeton University Press.

HINZ, I. 1987. The Lower Cambrian microfauna of Comley and Rushton, Shropshire, England. *Palaeontographica*, Abt. A 198, 41-100.

HOLSER, W.T. 1977. Catastrophic chemical events in the history of the Ocean. *Nature* **267**, 403-408.

HOU XIAN-GUANG, 1987a. Two new arthropods from Lower Cambrian, Chengjiang, eastern Yunnan. *Acta Palaeontologica Sinica*, **26**, 236-256.

HOU XIAN-GUANG, 1987b. Three new large arthropods from Lower Cambrian Chengjiang, Eastern Yunnan. *Acta Palaeontologica Sinica*, **26**, 272-285.

HOU XIAN-GUANG, 1987c. Early Cambrian large bivalved arthropods from Chengjiang, Eastern Yunnan. *Acta Palaeontologica Sinica* **26**, 286-298.

HOU XIAN-GUANG & CHEN JUN-YUAN 1989a. Early Cambrian tentacled worm-like animal from Chengjiang, eastern Yunnan. *Acta Palaeontologica Sinica* **28**, 32-41.

HUO XIAN-GUANG & CHEN JUN-YUAN, 1989b. Early Cambrian arthropod- annelid intermediate *Luolishania longicruris* (gen. et sp. nov.) from Chengjiang, eastern Yunnan. *Acta Palaeontologica Sinica* **28**, 207-213.

HOU XIAN-GUANG, CHEN JUN-YUAN & LUO HAO-ZHI, 1989. Early Cambrian new arthropods from Chengjiang, eastern Yunnan. *Acta Palaeontologica Sinica* **28**, 42-57.

HOU XIAN-GUANG & SUN WEI-GUO, 1988. Discovery of Chengjiang fauna at Meishucun, Jinning, Yunnan. *Acta Palaeontologica Sinica* **27**, 1-12.

HSÜ, K.J., OBERHÄNSLI, J.Y., GAO SUN-SHU, CHEN HAI-HONG & KRÄBENBÜBL, U. 1985. "Strangelove Ocean" before the Cambrian explosion. *Nature* **316**, 809-811.

HU YONG-JIA, LU YUAN-QUING, LIAONG XUE-DONG & XIANG YANG, 1986. Sulphur isotope composition of the marine sedimentary barite of Cambrian-Sinian system in South China and its significance in geology. *Hunan Geology* **5 (1)**, 51-56.

JOHNSON, J.H. 1966. A review of Cambrian algae. *Colorado School of Mines, Quarterly* **61**, 1-162.

LUO HUI-LIN, JIANG ZHI-WEN, WU XI-CHE, SONG XUE-LIANG, OUYANG LIN *et al*, 1982. *The Sinian-Cambrian Boundary in eastern Yunnan, China*. Yunnan Peoples Publishing House.

LUO HUI-LIN, JIANG ZHI-WEN, WU XI-CHE, SONG XUE-LIANG, OUYANG LIN, XING YU-SHENG, LIU GUI-ZHI, ZHANG SHI-SHAN & TWO YONG-HE, 1984. *Sinian-Cambrian boundary stratotype section at Meishucun, Jinning, Yunnan, China*. Yunnan, Beijing: Peoples Publishing House. 154 pp.

MÜLLER, K.J., 1979 Phosphatocopine ostracodes with preserved appendages from the Upper Cambrian of Sweden. *Lethaia* **12**, 1-27.

PATILETOV, V.G., LUCHININA, V.A., SHENFIL, V. YU & YAKSHIN, M.S. 1981. New data on Precambrian fossil algae of Siberia. *Doklady Akademia Nauk SSSR* **261**, 982-984.

PAUL, C.R.C. & SMITH, A.B. 1984. The early radiation and phylogeny of Echinoderms. *Biological Reviews*, 59,443-481.

PEEL, J.S. 1988. *Spirellus* and related helically coiled microfossils (Cyanobacteria) from the Lower Cambrian of North Greenland. *Rapport Gronland Geologiske Undersogelse* **137**, 5-32.

POMPECKJ, J.F. 1927. Ein neues Zeugnis uralten Lebens. *Paläontologische Zeitschrift* **9**, 289-313.

REITLINGER, E.A., 1948. Kembriijskie foraminifery Yakutii [Cambrian foraminifers of Yakutia]. *Byulleten' Moskovskogo Obshchestva Ispytatelej Prirody, Otdel Geologii* **23**, 77-81.

RIGBY, J.K. 1986. Sponges of the Burgess Shale (Middle Cambrian), British Columbia. *Palaeontographica Canadiana* **2**, 1-105.

RIGBY, J.K. & BRYANT, T.L.P. 1979. Fossil sponges of the Missisipian Fort Wayne Chert in northeastern Alabama. *Journal of Paleontology* **53 (4)**, 1005-1012.

ROBISON, R.A. 1985. Affinities of *Aysheaia* (Onychophora), with description of a new Cambrian species. *Journal of Paleontology* **59 (1)**, 226-235.

SEILACHER, A. 1970. Begriff und Bedeutung der Fossil-Lagerstätten. *Neues Jahrbuch für geologie und Paläontologische Monatshefte* **1970**, 34-39.

SEILACHER, A. 1989. Vendozoa: Organismic construction in the Proterozoic biosphere. *Lethaia* **22**, 229-239.

SONG XUE-LIANG, 1984. *Obruchevella* from the Early Cambrian Meishucun Stage of the Meishucun section, Jinning, Yunnan, China. *Geological Magazine* **121**, 179-183.

SUN WEI-GUO & HOU XIAN-GUANG, 1987a Early Cambrian medusae from Chenjiang, Yunnan China. *Acta Palaeontologica Sinica* **26 (3)**, 299-305.

SUN WEI-GUO & HOU XIAN-GUANG, 1987b. Early Cambrian worms from Chengjiang, Yunnan, China: *Maotianshania* gen. nov. *Acta Palaeontologica Sinica* **23 (3)**, 257-271.

TIEGS, O.W. & MANTON, S.M. 1958. The evolution of the Arthropoda. *Biological Reviews* **33**, 255-337.

TONG HAO-WEN, 1989. A preliminary study on the *Microdictyon* from the Lower Cambrian of Zhemba, South Shaanxi. *Acta Micropalaeontologica Sinica* **6 (1)**, 97-101.

VALENTINE, J.W. 1973. *Evolutionary palaecology of the marine biosphere*. Englewood Cliffs: Prentice-Hall.

WALCOTT, C.D. 1911. Cambrian geology and paleontology II, No.5 - Middle Cambrian holothurians and medusae. *Smithsonian Miscellaneous Collections* **57 (3)**, 41-68.

WALCOTT, C.D. 1919. Cambrian geology and palaeontology IV, No. 5- Middle Cambrian algae. *Smithsonian Miscellaneous Collections* **67 (5)**, 217-260.

WHITTINGTON, H.B. 1977: Cambrian trilobite *Naraoia*, Burgess Shale, British Columbia. *Philosophical Transactions of the Royal Society, London* **B 280**, 409-443.

WHITTINGTON, H.B. 1978. The lobopod animal *Aysheaia pedunculata* Walcott, Middle Cambrian, Burgess Shale, British Columbia. *Philosophical Transactions of the Royal Society, London* **B 284**, 165-195.

WHITTINGTON, H.B. 1985a. *The Burgess Shale* New Haven: Yale University Press.

WHITTINGTON, H.B. & BRIGGS, D.E.G. 1985b. The largest Cambrian animal *Anomalocaris*, Burgess Shale, British Columbia. *Philosophical Transactions of the Royal Society, London* **B 309**, 569-609.

YU WEN, 1987. Yangtze macromolluscan fauna in Yangtze Region of China with notes on Precambrian-Cambrian boundary. In *Stratigraphy and Paleontology of Systematic Boundaries in China. Precambrian-Cambrian Boundary*, Vol.1, pp.19-275. Anhui Science and Technology Publishing House.

YUAN KE-XING & ZHANG SEN-GUI, 1981. Discovery of Tommotian fauna in southwest China and its bearing on the problem of the Precambrian-Cambrian boundary. In *Short Papers for the Second International Symposium of the Cambrian System* (ed. M.E. Taylor). *U.S. Geological Survey Open-File Report* 81-743, 249.

ZHANG WEN-TANG & HOU XIANG-GUANG, 1985. Preliminary notes on the occurrence of the unusual trilobite *Naraoia* in Asia. *Acta Palaeontologica Sinica* **24 (6)**, 591-595.

ZHOU ZHI-YI & YUAN JIN-LIANG, 1981. The biostratigraphic distribution of Lower Cambrian trilobites in Southwest China. In *Short Papers for the Second International Symposium of the Cambrian System* (ed. M.E. Taylor) *U.S. Geological Survey Open-File Report* 81-743, 250-251.

Middle Cambrian biotic diversity: examples from four Utah Lagerstätten

Richard A. Robison[1]

Abstract

Exceptionally diverse biotas are preserved in four Lagerstätten of the Spence, Wheeler, and Narjum formations in Utah. These deposits accumulated in marine, muddy shelf environments on a rapidly subsiding, passive margin of equatorial Laurentia during the Middle Cambrian. One Wheeler Lagerstätte accumulated in a protected, shallow-subtidal environment near the lee side of peritidal shoals on the seaward edge of a carbonate platform. A second Wheeler Lagerstätte and those of the Spence and Marjum formations all accumulated in open-shelf, ramp environments below wave base. Many of the fossils, especially those of soft-bodied or lightly skeletized organisms, were preserved by rapid burial in fine, smothering sediment, probably during storm events.

All of the Utah biotas are broadly similar to that of the celebrated Burgess shale in British Columbia and provide important evidence that many taxa in the Burgess Lagerstätte were widespread in shelf environments of Laurentia. The morphology and stratigraphic distribution of fossils in the Utah Lagerstätten, ranging from lower to upper Middle Cambrian, provide documentation of evolution within major biotic groups during an interval of many millions of years. In general, soft-bodied and lightly skeletized groups show more evolutionary conservatism than do such shelly groups as trilobites, brachiopods, and echinoderms. The Utah fossils also provide further information about trophic structure of the Cambrian ecosystem.

1. Introduction

The Cambrian period is important in the history of life because of rapid evolution and diversification of the marine biota, especially the Metazoa. Most studies of this biota have been concerned only with its easily preserved, shelly component, which is dominated by trilobites and brachiopods. Several Cambrian Lagerstätten[2] with soft-bodied or lightly skeletized organisms are now known (Conway Morris, 1985b, 1989; Conway Morris *et al.*, 1987; Hou & Sun, 1988; Hou, 1989), and they are providing much additional information about biotic diversity, ecology, and evolution. Foremost among these is the celebrated Burgess shale in the Stephen Formation of British Columbia (Whittington, 1985, references therein; Conway Morris, 1986; Gould, 1989). Although less acclaimed, other Cambrian lagerstätten of Laurentia have biotas that are broadly similar to that of the Burgess shale. These are important in demonstrating that the Burgess shale biota is unusual only in preservation and not in taxonomic composition.

Among what Conway Morris (1985b) listed as minor Middle Cambrian Lagerstätten are those of the Spence, Wheeler, and Marjum formations of Utah. Recent collecting in these formations has produced remarkably diverse assemblages of both shelly and soft-bodied fossils, which provide further strong support for the notion that the biota of the Burgess shale is representative of marine-shelf environments of Laurentia (Conway Morris & Robison, 1982). They also show that many soft-bodied genera and some species are widespread and have long stratigraphic ranges. Collecting from these formations is continuing and, therefore, this paper is a progress report on the observed diversity and significance of their biotas. Species from four Lagerstätten, two in the Wheeler and one each in the Spence and Marjum formations, are listed in the Appendix, together with selected references to recent descriptions and discussions.

Illustrated specimens are deposited with collections of the University of Kansas Museum of Invertebrate Paleontology (KUMIP), U.S. National Museum of Natural History (USNM), and Department of Geology, Brigham Young University (BYU).

2. Geological setting

According to most recent Cambrian palaeogeographic reconstructions (e.g., Scotese, 1986), the former continent of Laurentia was isolated over the equator (Fig. 1.1). Middle Cambrian lithofacies of present-day western North America indicate the presence of a broad, rapidly subsiding, continental shelf in that sector of Laurentia. Thick deposits of shallow-water, carbonate sediment produced an extensive platform. Along the outer platform margin, carbonate sand-shoal and microbial-reef complexes were flanked by shallow, commonly restricted shelves on their shoreward side and deeper open ramps on their oceanward side (Robison, 1960, 1976; Aitken, 1978). During much of the Middle Cambrian two almost mutually exclusive shelf biotopes were probably separated by salinity and temperature barriers across the carbonate shelves. An inner restricted-shelf biofacies is characterized by sparse, low-diversity, polymeroid trilobites. Other biota is mostly limited to brachiopods, hyoliths, and calcareous algae. An outer open-shelf biofacies is characterized by common to abundant fossils in high-diversity faunas dominated by trilobites. Fossils are usually most abundant near the seaward margin of the carbonate platform facies.

The Spence Shale, with its type locality in southeastern Idaho, was originally described as a member of the Ute Limestone by Walcott (1908). It has an abundant biota from which many taxa have been described (e.g., Resser, 1939; Appendix, references therein). Its regional stratigraphic relationships, however, are not well understood, partly because of poor exposure throughout much of southern Idaho and

[1]Department of Geology, University of Kansas, Lawrence, Kansas 66045, U.S.A.

[2]Lagerstätten (singular, Lagerstätte) has been defined (Seilacher *et al.*, 1985, p. 5) as "rock bodies unusually rich in palaeontological information".

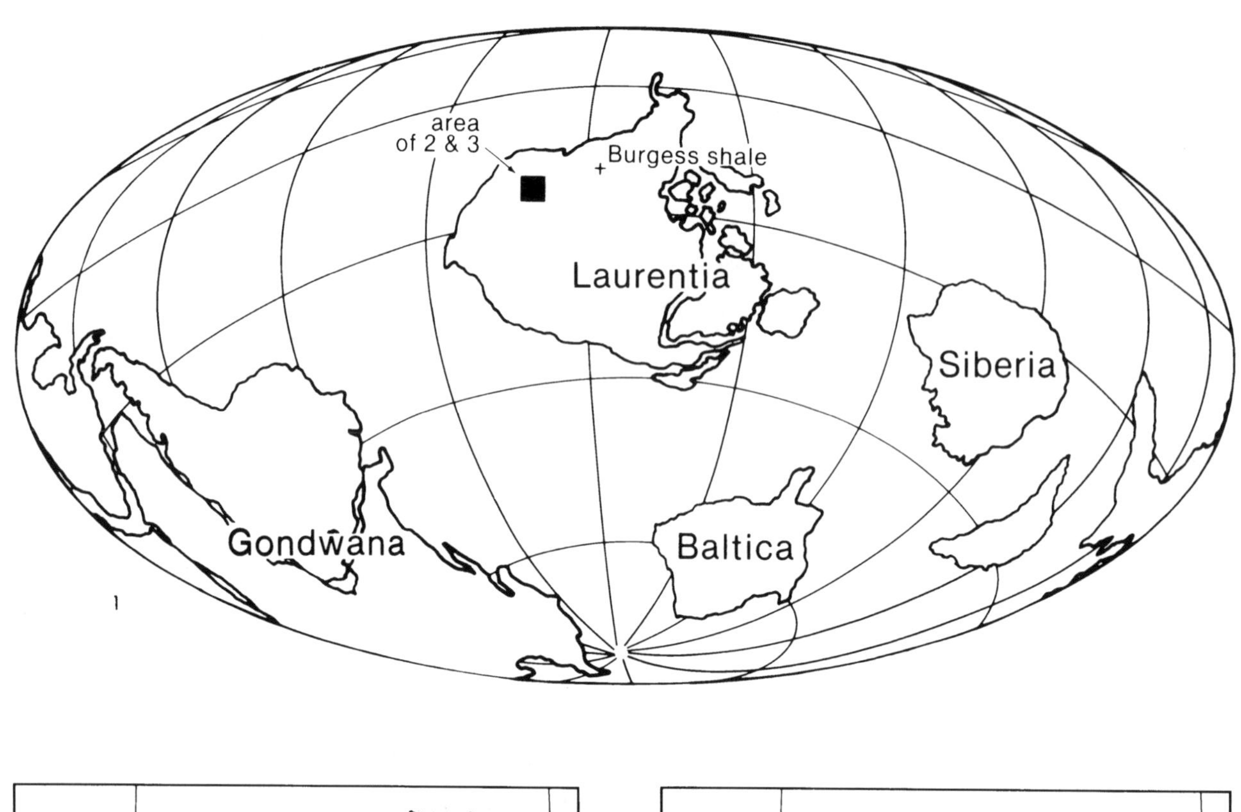

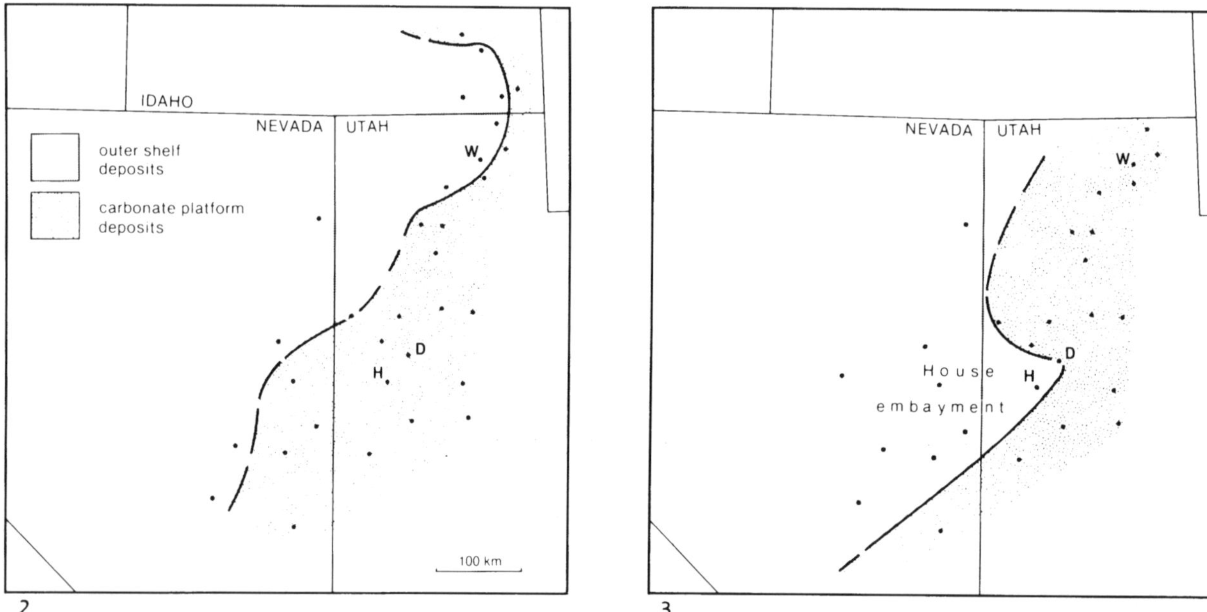

Fig. 1. Geologic setting. 1, Middle Cambrian paleogeography (after Scotese, 1986). 2, Lithofacies map of part of Utah and adjacent areas during the early Middle Cambrian (*Peronopsis bonnerensis* Biochron); outer shelf deposits of southeastern Idaho and northern Utah are assigned to the Spence Shale; black dots are control from stratigraphic sections (D = Drum Mountains, H = House Range, W = Wellsville Mountains). 3, Lithofacies map of eastern Nevada and western Utah during the middle Middle Cambrian (*Ptychagnostus atavus* Biochron); the outer shelf deposits of west-central Utah are assigned to the upper Wheeler Formation.

northern Utah where the name has been used. Although the unit was reassigned as the Spence Tongue of the Lead Bell Shale by Oriel & Armstrong (1971), I elevate the unit to formational rank in northern Utah pending further clarification of regional stratigraphic relationships.

The biota of the Spence Shale contains diverse early Middle Cambrian trilobites of the *Peronopsis bonnerensis* Zone (Robison, 1976). Its age is close to that of the Burgess shale, but the Burgess contains trilobites of the overlying *Ptychagnostus praecurrens* Zone (Robison, 1984a). The Spence Lagerstätte discussed here is primarily exposed along the west face of the Wellsville Mountains of northern Utah (Fig. 1.2), from about 2 to 9 kilometers north of Brigham City. In that area, the Spence Shale averages about 70 meters in thickness. Most of the soft-bodied fossils have been recovered from noncalcareous, laminated mudstone in the middle and upper parts of the formation.

Distribution of lithofacies indicate that predominantly noncalcareous marine mud accumulated over much of the area of northwestern Utah, includ-

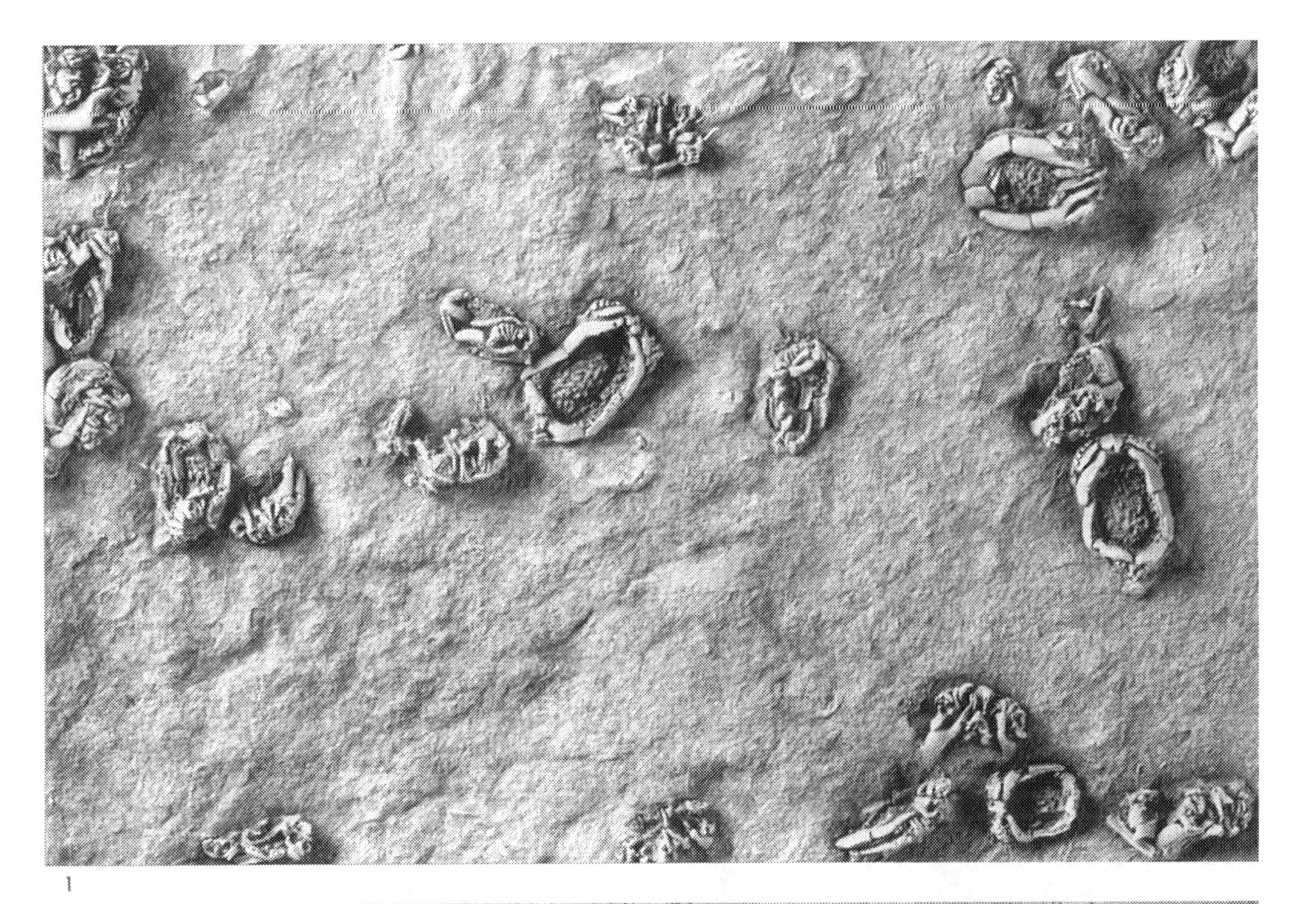

1

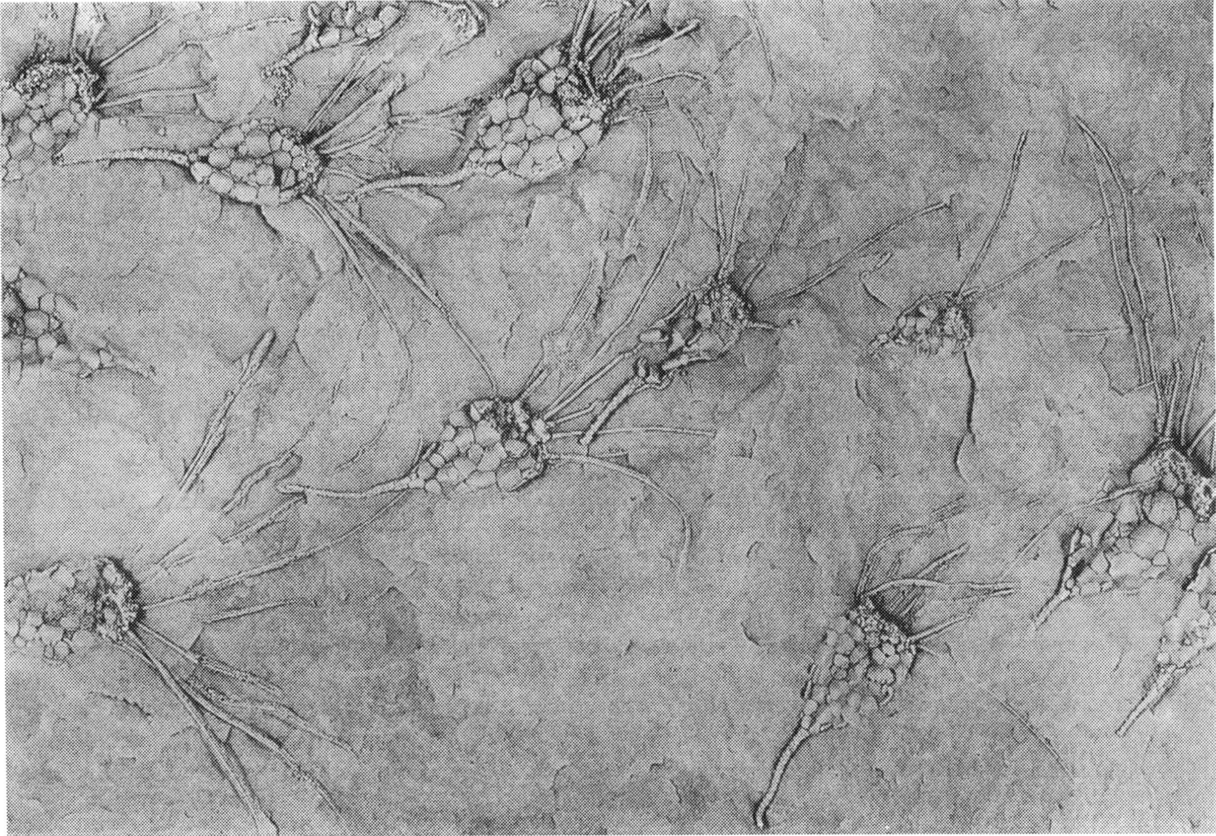

2

Fig. 2. Echinoderm obrution deposits from the Spence Shale, Wellsville Mountains, Utah. 1, Latex cast of a bedding surface with *Ctenocystis utahensis* Robison & Sprinkle (Ctenocystoidea),USNM 163253, X4.0. Preservation of these vagrant animals with articulated skeletons at oblique angles in mudstone suggests their failure to escape rapid deposition of smothering sediment. 2, Latex cast of bedding surface with aligned *Gogia* n. sp. (Eocrinoidea), X1.5. Lack of disarticulation and near-uniform orientation of these sessile animals suggests toppling and rapid burial parallel to a dominant current with most crowns down-current. Both surfaces indicate the abrupt introduction of high-energy conditions into normally quiet-water environments (cf. Brett & Baird, 1986).

ing the Wellsville Mountains site, during the *Peronopsis bonnerensis* Biochron (Fig. 1.2). At the same time, carbonate sediment was accumulating in platform environments to the south and east. Based on regional setting, lithology, and bedding features, the Spence Shale of the Wellsville Mountains appears to contain distal tempestite deposits that accumulated in a ramp environment (cf. Aigner, 1985, fig. 49). This possibility is supported by evidence of rapid burial and smothering of some epifauna (Fig. 2). It also is suggested by some infaunal fossils. Compact curling of the type shown by the annelid *Palaeoscolex* (see Fig. 6.5) is commonly induced in modern annelids by anaerobic conditions, which may develop following rapid deposition of sediment (Dean *et al.*, 1964). The unusually high diversity of trilobite genera (see Table 1) and species (Appendix) in the Spence Shale may be in part a result of downslope transport and accumulation of some trilobite remains during storm events.

Beginning in the middle Cambrian (*Ptychagnostus gibbus* Biochron) local deepening of the shelf platform produced the House embayment (compare Fig. 1.3), which is thought to have been caused by displacement along an inferred rotational fault that presently trends northeastward (Rees, 1986). The trough was asymmetrical and became deeper and wider as it crossed more than 400 kilometers of shelf. The Wheeler and Marjum formations represent sedimentary infilling along the axial part of the embayment near its present-day southern margin. Stratigraphic relationships of the two formations have been discussed by Hintze & Robison (1975) and further elaborated by Rees (1986).

The Wheeler Formation is thickest and best developed in west-central Utah. Two Lagerstätten, one in the southern Drum Mountains and one in the central House Range (Fig. 1.3), are preserved in superficially similar limestone and shale lithofacies of the upper hundred meters of the Wheeler, but they accumulated in quite different environments. Diverse trilobites in each lagerstätten represent the middle *Ptychagnostus atavus* Zone (Robison, 1984a).

Fossiliferous, platy limestone and shale deposits in the upper Wheeler Formation of the Drum Mountains are here informally called the *shallow Wheeler lagerstätte*. Rapid lithofacies changes in the area have been described and analyzed by Rees (1984). Limestone in the Lagerstätte is predominantly carbonate mudstone with variable argillaceous content. Rees identified thin beds of intraclastic and peloidal grainstone that are scattered throughout the lithofacies, and showed that some clasts contain the probable cyanobacterium *Girvanella* and the tiny mollusk *Pelagiella*. Minor stromatolitic limestone is also interbedded, which contains the probable cyanobacteria *Epiphyton, Girvanella* and *Renalcis* (Rees, 1984, fig. 23). From analysis of both regional stratigraphic relationships and local depositional sequences, Rees (1984) concluded that the platy limestone and shale lithofacies in the Drum Mountains represents deposition in a protected, low-energy, shallow-subtidal environment. The site was near the lee side of peritidal shoals along the margin of the carbonate platform on the northeastern flank of the House embayment.

A thin-bedded limestone and shale lithofacies in the upper Wheeler Formation of the House Range in places contains richly fossiliferous deposits that are here informally called the *deep Wheeler lagerstätte*. The lithofacies has been described in detail by Rees (1986), who concluded that it was deposited on a deep-shelf ramp below wave base. It differs from the limestone and shale lithofacies of the upper Wheeler in the Drum Mountains by being less calcareous, more widespread, and associated with gravity-flow deposits. Also, stromatolitic beds are absent and the limestone lacks *Girvanella*. Trilobites are commercially mined from this lithofacies at the type locality of the Wheeler Formation in the Wheeler Amphitheater of the House Range. Hundreds of thousands of specimens of *Elrathia kingii* and other species have been produced and distributed throughout the world, mainly as curiosities, but also for teaching and scientific purposes. Some of the rare specimens discussed here have been collected from spoil dumps of the mines.

Late Middle Cambrian progradation of carbonate platform deposits across the more gently inclined northeastern flank of the House embayment restricted deposits of the Marjum Formation to the central House Range (Rees, 1986). In that area (Fig. 1.3), discontinuous, thin, Lagerstätte deposits range from about 30 to 300 meters above the base of the formation and contain diverse trilobites of the *Ptychagnostus punctuosus* Zone (Robison, 1984a). As in the underlying Wheeler Formation, the Marjum Lagerstätte is preserved exclusively in the thin-bedded limestone and shale lithofacies of Rees (1986), and she interpreted it also as representing deep-ramp deposition.

Depositional environments and palaeoecology at quarry sites in the thin-bedded limestone and shale lithofacies of both the Wheeler and Marjum formations in the House Range have been analyzed by Rogers (1984). He concluded that turbidity currents and rapid burial were important factors in the exceptional preservation of fossils. It now seems likely, however, that storm conditions as well as turbidity currents were involved in sediment transport.

3. Taxa of the Spence, Wheeler, and Marjum Lagerstätten

Most taxa that have been observed in the Spence, Wheeler, and Marjum Lagerstätten of Utah are listed in the Appendix. Summary comments about taxa of various rank follow.

3.a. *Cyanobacteria and algae* (Fig. 3)

Material representing five genera of cyanobacteria (blue-green algae) or possible cyanobacteria have been recorded from the Middle Cambrian Lagerstätten of Utah. It includes calcareous microfossils of the problematic *Epiphyton, Girvanella*, and *Renalcis* of possible cyanobacterial affinity. Representative specimens have been illustrated by Rees (1984) from stromatolitic bioherms and grainstone clasts, all interbedded with carbonate mudstone of the shallow Wheeler Lagerstätte. These are common taxa in reefs of early Paleozoic age, but they were probably not normal members of the shallow Wheeler Lagerstätte. The *Girvanella*-bearing clasts were probably transported and redeposited from nearby environments.

A few tufted thalli of the megascopic noncalcareous cyanobacterium *Marpolia spissa* Walcott (Fig. 3.4) have been collected from the Spence Lagerstätte (Conway Morris & Robison, 1988). Broken, fine filaments

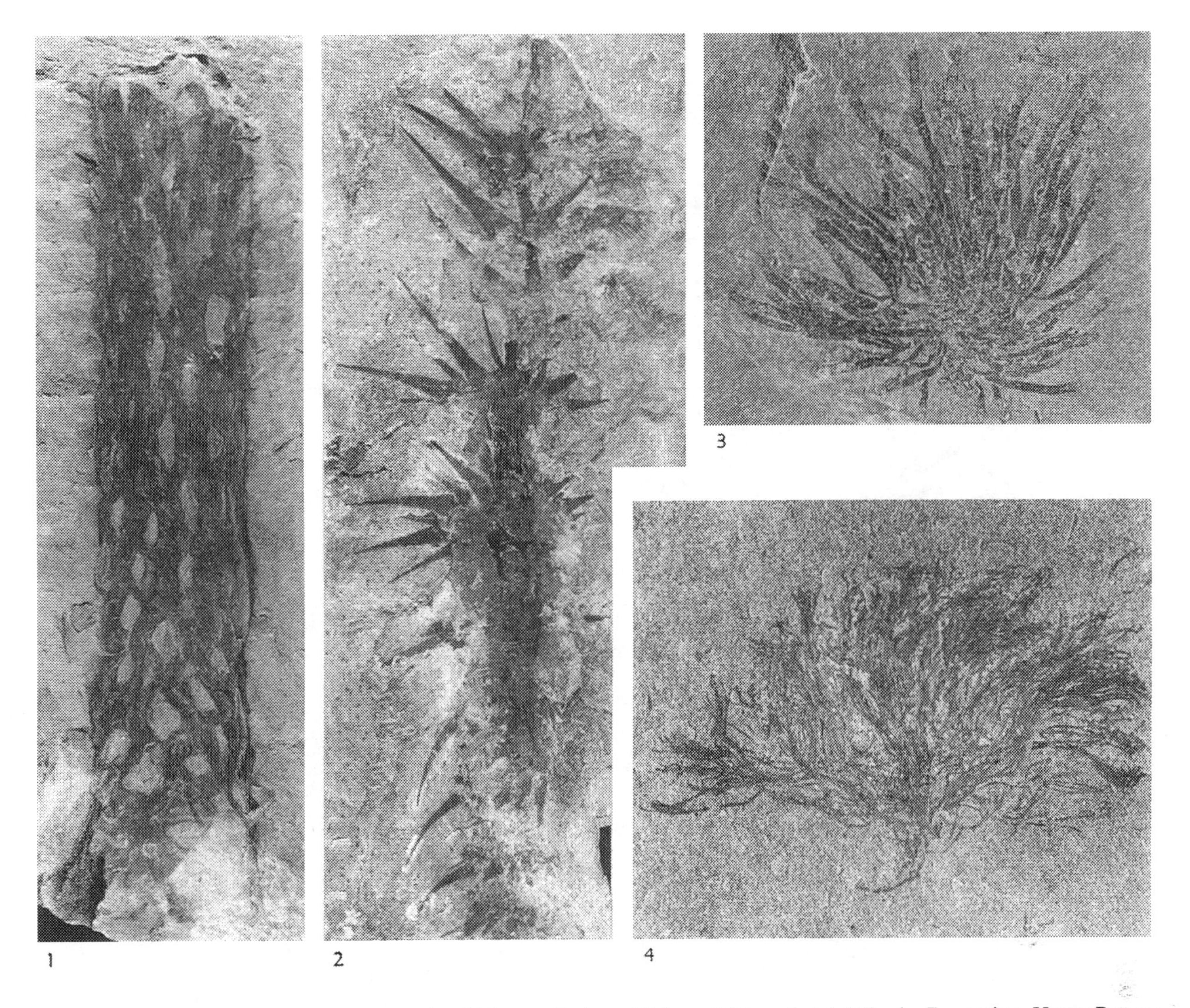

Fig. 3. Algae. 1, Fragment of compressed axis of *Margaretia dorus* Walcott (Chlorophyta); Wheeler Formation, House Range; KUMIP 127810, X1.2. 2, Holotype thallus of *Acinocricus stichus* Conway Morris & Robison (Chlorophyta?); Spence Shale, Wellsville Mountains; KUMIP 204353, X1.5. 3, Thallus of *Yuknessia simplex* Walcott (Chlorophyta?); Wheeler Formation, House Range; KUMIP 204382, X2.0. 4, Thallus of *Marpolia spissa* Walcott (Cyanobacteria); Spence Shale, Wellsville Mountains; KUMIP 111765, X2.5.

are locally abundant in the same deposit, but are easily overlooked because of their small size. One large, questionably assigned thallus is from the deep Wheeler lagerstätte. This is an abundant species in the Burgess shale, where it also occurs as masses of broken filaments (Whittington, 1985).

Small, circular, carbonaceous compressions, usually 1 to 3 millimeters in diameter, are present in all four Utah lagerstätten, and are especially abundant in some Marjum and deep Wheeler deposits (see Fig. 10.1; Robison, 1985, fig. 1). These simple fossils appear to be identical to some specimens in collections of the U.S. Museum of Natural History that are labeled in the handwriting of C. D. Walcott as *Morania fragmenta*, which he named from the Burgess shale. The taxon has been assigned to the Cyanobacteria (Walcott, 1919) but warrants further evaluation.

Three species of megascopic, noncalcareous algae have recently been documented from the Middle Cambrian Lagerstätten of Utah (Conway Morris & Robison, 1988). Two of these, *Margaretia dorus* and *Yuknessia simplex*, were originally described from the Burgess shale (Walcott, 1919), and the presence of very similar specimens in all four Utah Lagerstätten demonstrates long stratigraphic ranges and evolutionary conservatism. *M. dorus* (Fig.3.1), which is a robust tubular green alga that may have exceeded a meter in height, was widespread on the shelf around Laurentia and also is known from the Lower and Middle Cambrian of eastern Siberia.

Compared with modern algae (Bold & Wynne, 1985), the morphotypes of the Middle Cambrian algae of Utah suggest that they were adapted to intertidal or shallow subtidal environments. Because none of the specimens shows evidence of attachment at burial, and most are fragmentary, they may have been transported from shallow-water habitats.

3.b. *Microfossils*

Cambrian rocks have mostly been neglected in searches for microfossils. Recent discoveries in the Spence and Wheeler formations, however, suggest that further investigations may be productive.

An acid residue from limestone of the shallow Wheeler Lagerstätte of the Drum Mountains has yielded about 70 specimens that White (1986) described as an undetermined genus and species of radiolarian. Specimens with similar morphology and dimensions

Fig. 4. Examples of sponges. 1, *Diagonella cyathiformis* (Dawson & Hinde); Marjum Formation, House Range; KUMIP 204399, X3. 2, Holotype of *Ratcliffespongia perforata* Rigby; Marjum Formation, House Range; BYU 1482, X1. 3, Holotype of *Valospongia gigantis* Rigby; Marjum Formation, House Range; BYU 1745, X0.5. 4, Rubber cast of spicule fabric in wall of *Protospongia hicksi* Hinde; Marjum Formation, House Range; USNM 145295, X5. 5, *Vauxia gracilenta* Walcott; Spence Shale, Wellsville Mountains; KUMIP 113279, X2. 6, Bedding surface with multiple specimens of *Choia utahensis* Walcott associated with exuvia of the trilobite *Elrathia kingii*; Wheeler Formation, House Range; KUMIP 204400, X1.3. Photographs 2, 3, and 4 were provided by J. Keith Rigby.

from strata near the Middle-Upper Cambrian boundary in Australia were assigned by Bengtson (1986) to *Blastulospongia*. Bengtson questionably assigned the Australian specimens to the Radiolaria, but he noted (p. 207) that "it does not seem possible to demonstrate convincingly either a sponge or a radiolarian affinity for *Blastulospongia*."

Another acid residue from limestone in the upper Spence Shale near Malad, southern Idaho, has yielded about 30 specimens of agglutinated foraminifera representing *Ammodiscus* sp. and a new genus and new species (Lipps, 1985; 1987; 1989, personal communication).

Collections from the upper Wheeler Formation at various localities have yielded abundant sphaeromorphic acritarchs (G. D. Wood, 1986, personal communication) and rare bradoriid ostracodes (Vorwald, 1983). These all remain undescribed.

3.c. *Sponges* (Fig. 4)

The most diverse Cambrian sponge fauna in the world, including 20 genera and 34 species, has recently been documented from the Burgess shale (Rigby, 1986). A partially described sponge fauna of at least 11 genera and 20 species has been reported from the Chengjiang lagerstätte of early Cambrian age in Yunnan, China (Chen *et al.*, 1989). Less diverse but broadly similar sponge faunas are present in the four Utah Lagerstätten. To date, 14 genera and 24 species of sponges have been described from the Spence, Wheeler, and Marjum formations (Rigby, 1966, 1969, 1978, 1980, 1983; Rigby & Gutschick, 1976). Many more undescribed specimens have been collected (L.F. Gunther, 1989, personal communication), and some of these represent a few additional species (J.K. Rigby, 1989, personal communication). Half of the documented Utah genera and at least seven species are shared with the Burgess sponge fauna, demonstrating wide geographic distributions and some long stratigraphic ranges.

Recorded sponge diversity varies significantly in the Spence (3 genera, 4 species), shallow Wheeler (1 genus and species), deep Wheeler (5 genera, 8 species), and Marjum (12 genera, 19 species) lagerstätten (see Appendix). Reasons for this variation are not clear; however, differences in turbidity is suggested as a pos-

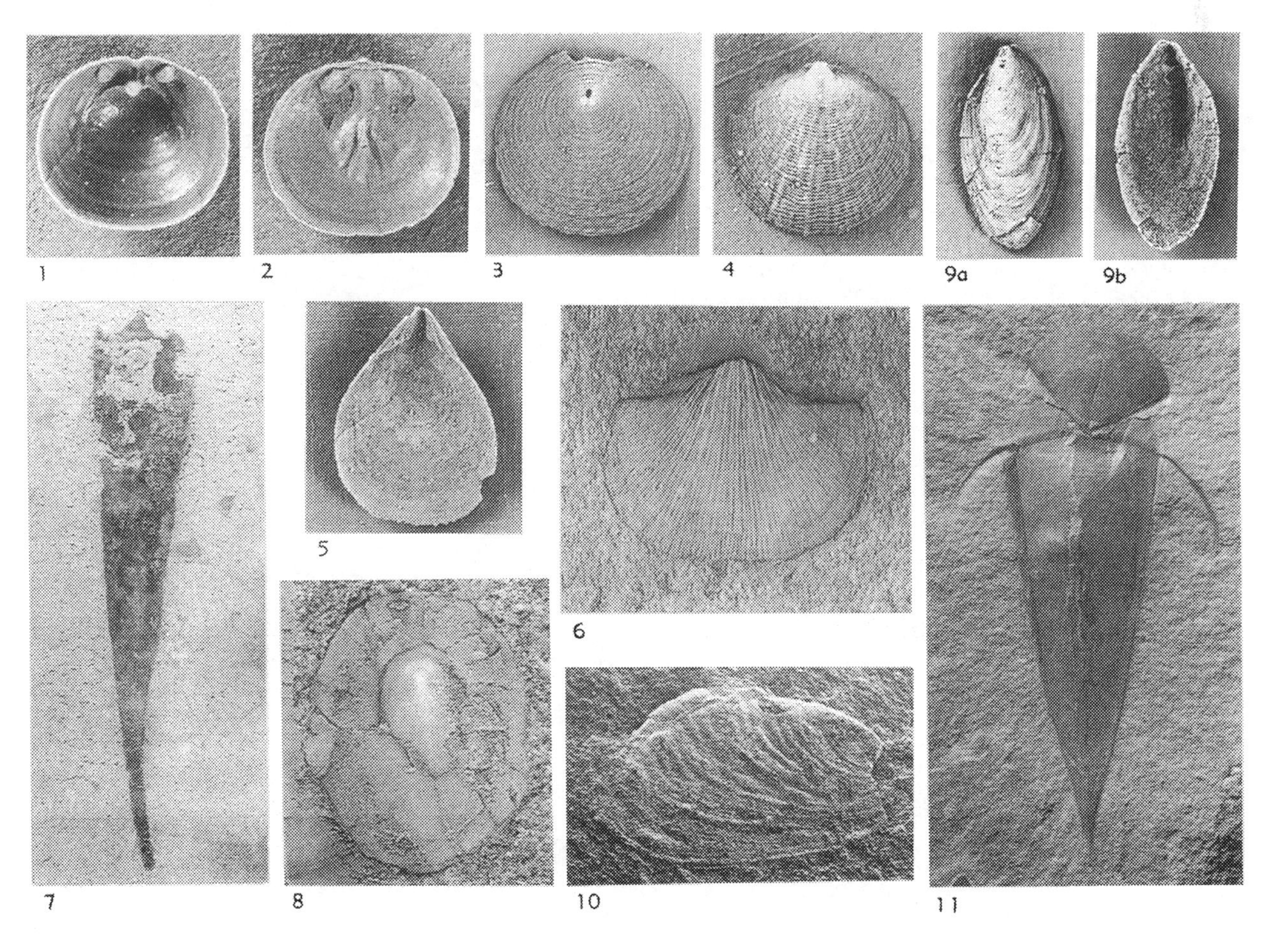

Fig. 5. Examples of brachiopods (1-6), ?cnidarians (7, 8), and mollusks (9-11). 1 and 2, Ventral and dorsal valves (interior) of *Linnarssonia ophirensis* (Walcott); Marjum Formation, House Range; USNM 141298b and 141298c, both X10. 3, Ventral valve (exterior) of *Acrothele subsidua* (White); Marjum Formation, House Range; USNM 141294a, X6.3. 4, Ventral valve (exterior) of *Micromitra modesta* (Lochman) Marjum Formation, House Range; USNM 141303a, X14. 5, Ventral valve (interior) of *Lingulella* sp.; Marjum Formation, House Range; USNM 141293a, X7.3. 6, Ventral valve (exterior) of *Diaphora*? sp.; Spence Shale, Bear River Range, Utah; KUMIP 204386, X3. 7, Tube of *Cambrorhytium major* (Walcott); Marjum Formation, House Range; KUMIP 204361, X1.5. 8, Pneumatophore of *Scenella radians* Babcock & Robison; Spence Shale, Wellsville Mountains; KUMIP 204351, X2.6. 9, Valve (a, exterior; b, interior) of *Stenothecoides elongata* (Walcott); Wheeler Formation, Drum Mountains; USNM 141308a, X4. 10, Lateral view of compressed tiny shell of *Melopegma georginensis* Runnegar & Jell; Wheeler Formation, House Range; KUMIP 204387, X16. 11, Dorsal view of *Haplophrentis reesei* Babcock & Robison showing incomplete alimentary tract on steinkern; Spence Shale, Wellsville Mountains; KUMIP 204340, X2.

sibility. In the Marjum and deep Wheeler Lagerstätten, sponges are second only to arthropods in generic diversity (see Table 1) which is the same relationship as in the Burgess shale (Whittington, 1985; Conway Morris, 1986) and the Chengjiang lagerstätte (Chen *et al.*, 1989).

3.d. *Cnidarians* (Fig. 5.7, 5.8)

Cnidarians have low generic and species diversity in Middle Cambrian deposits of both British Columbia and Utah. Some authors have considered *Scenella* to be a molluscan shell. Others have considered it to be the firm pneumatophore of a chondrophorine hydrozoan, which is the interpretation preferred here (see Babcock & Robison, 1988). Different species of *Scenella* are present in the Burgess and Spence (Fig. 5.8) biotas. Specimens are abundant on some bedding surfaces of Burgess shale but are rare in the Spence. *Cambrorhytium major*, represented mainly by tubes (Fig. 5.7), has been questionably reassigned to the Cnidaria on the basis of rare specimens with tentacular structures and other soft parts (Conway Morris & Robison, 1988). The presence of *C. major* in both the Burgess and Marjum biotas shows it to have wide geographic distribution and a long stratigraphic range.

3.e. *Brachiopods* (Fig. 5.1-6)

Brachiopod assemblages of the Burgess and Utah lagerstätten are closely similar in general aspect and diversity. They typically include a lingulide (e.g. Fig. 5.5), one or two acrotretides (e.g., Fig 5.1-3), one or two paterinides (e.g., Fig. 5.4), and rare orthides (e.g., Fig. 5.6). According to A.J. Rowell (1989, personal communication), such assemblages are common worldwide in open-shelf deposits of Middle Cambrian age.

3.f. *Mollusks* (Fig. 5.9-11)

Diversity of mollusks is low in the Burgess and Utah Lagerstätten. Distributions in Utah, except for hyoliths, tend to be patchy. Locally, some bedding surfaces in the upper Wheeler Formation may contain numerous shells of either the bivalved *Stenothecoides* (Fig. 5.9a,b) or the minute *Melopegma* (Fig. 5.10), but few other fossils.

Incomplete alimentary tracts in three specimens of *Haplophrentis reesei* (Fig. 5.11) are the first soft parts to be described from representatives of the order Hyolithida (Babcock & Robison, 1988). The specimens are from the Spence lagerstätte.

3.g. *Worms* (Fig. 6)

Remains of worms in the Burgess and Utah lagerstätten are dominated by priapulids (Conway Morris,

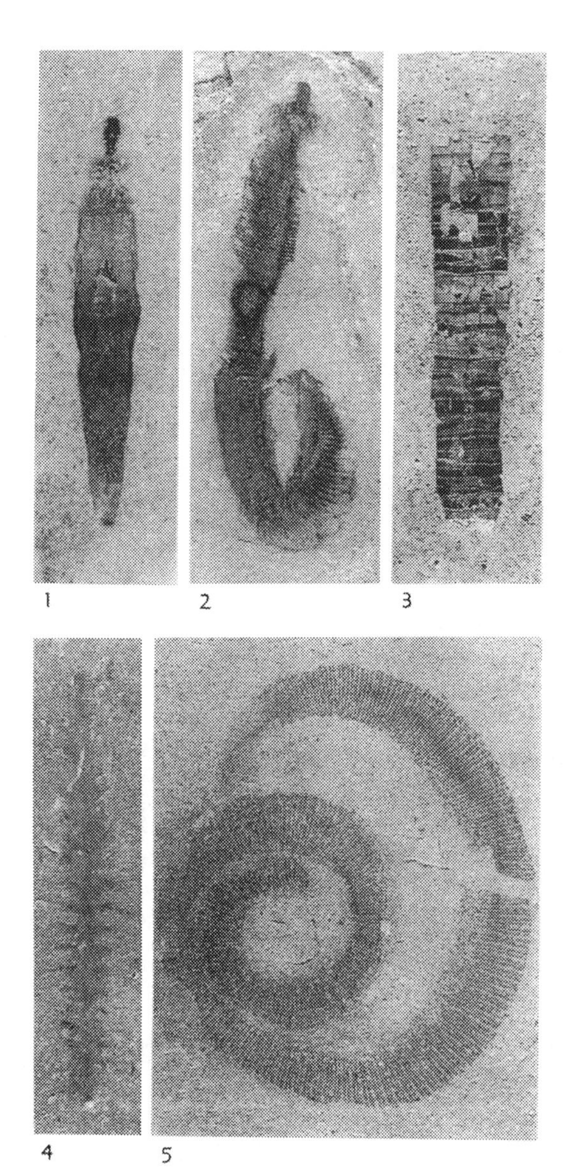

Fig. 6. Examples of miscellaneous worms. 1, Holotype tube and associated soft parts of *Selkirkia willoughbyi* Conway Morris & Robison (Priapulida); Wheeler Formation, Drum Mountains; KUMIP 204778, X3.2. 2, Soft-bodied *Ottoia prolifica* Walcott (Priapulida); Marjum Formation, House Range; KUMIP 204770, X2.7. 3, Phosphatic tube of *Hyolithellus* sp. (Pogonophora?); Marjum Formation, House Range; KUMiP 204388, X3.3. 4, Polychaete, genus and species indeterminate (Annelida); Spence Shale, Wellsville Mountains; KUMIP 204389, 2.7. 5, Soft-bodied *Palaeoscolex ratcliffei* Robison (Annelida); Spence Shale, Wellsville Mountains; KUMIP 204390, X2.6.

Fig. 7. Examples of miscellaneous arthropods. 1, Anterior half of the uniramous *Cambropodus gracilis* Robison; Wheeler Formation, Drum Mountains; KUMIP 204775, X5. 2, *Mollisonia* sp. with an undetermined acrotretide brachiopod on its trunk; Wheeler Formation, House Range; KUMIP 135149, X3. 3, Holotype of *Leanchoilia? hanceyi* Briggs & Robison; Spence Shale, Wellsville Mountains; KUMiP 374592, X1.2 (photographed under alcohol). 4, Undetermined arthropod 2 of Conway Morris & Robison (1988); Marjum Formation, House Range; KUMIP 204782, X2. 5, Holotype of *Utahcaris orion* with cluster of trilobite fragments in the alimentary tract of the head; Spence Shale, Wellsville Mountains; KUMIP 204784, X0.9. 6, Holotype of *Ecnomocaris spinosa* Conway Morris & Robison; Wheeler Formation, House Range; USNM 424114, X0.6. 7, *Pseudoarctolepis sharpi* Brooks & Caster; Wheeler Formation, House Range; X1. 8, *Branchiocaris pretiosa* (Resser) showing soft parts; Marjum Formation, House Range; KUMIP 204797, X1. 9, Right valve of *Canadaspis* cf. *perfecta*; (Walcott) Wheeler Formation, House Range; KUMIP 144401, X1.3. 10, Laterally compressed body of *Alalcomenaeus* cf. *cambricus* (Simonetta); Wheeler Formation, House Range; KUMIP 135148, X2. 11, Holotype left valve of *Tuzoia guntheri* (Robison & Richards); Marjum Formation, House Range; KUMIP 153917, X1.

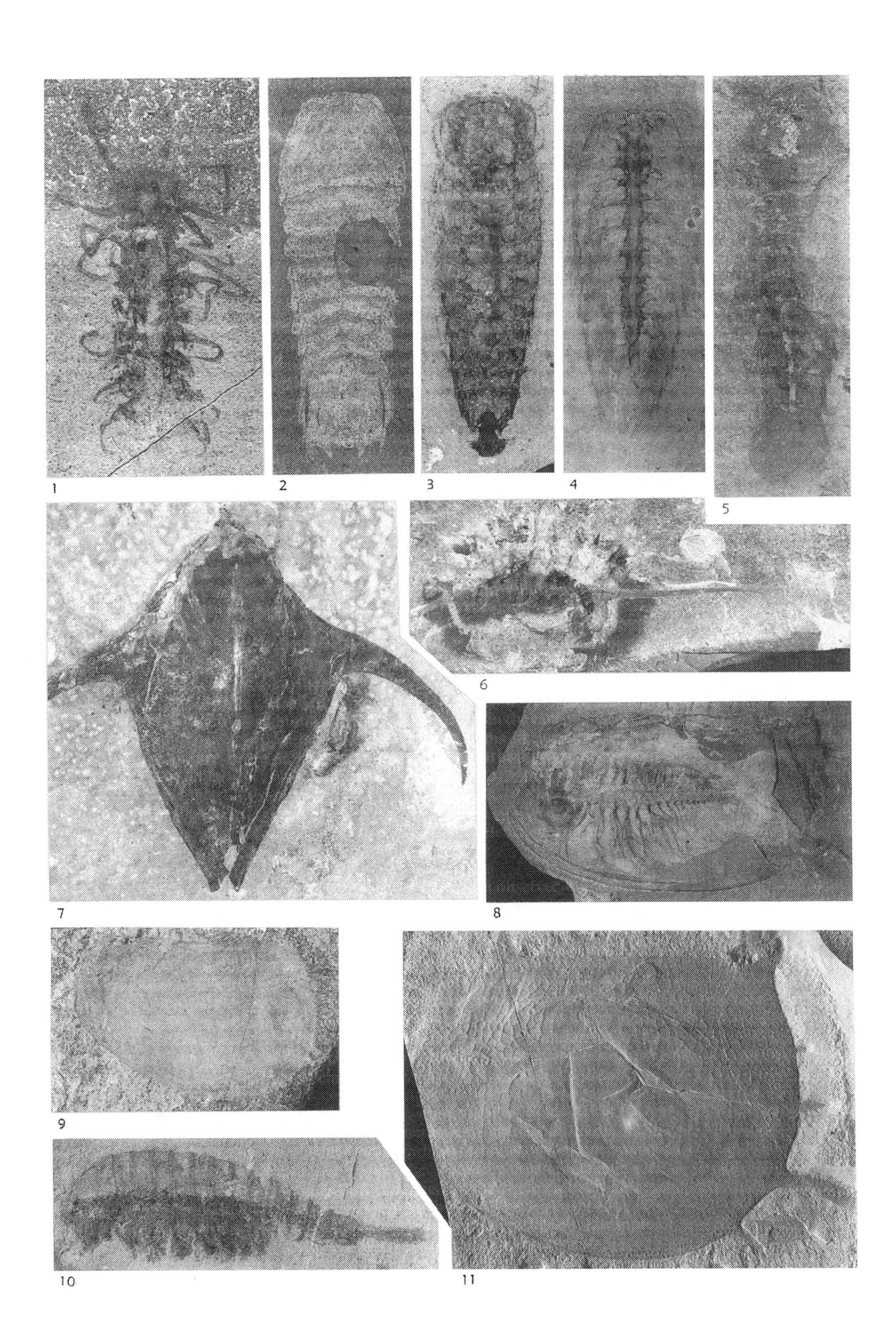
1
2
3
4
5
6
7
8
9
10
11

1977; Conway Morris & Robison, 1986, 1988), especially the burrowing *Ottoia* (Fig. 6.2) and the tube-dwelling *Selkirkia* (Fig. 6.1). Annelids (Fig. 6.4, 6.5) rank second to priapulids in diversity in both the Burgess (Conway Morris, 1979) and Spence (Robison, 1969; Conway Morris & Robison, 1988) lagerstätten. Numerous soft-bodied, vermiform specimens have been collected from the Wheeler and Marjum lagerstätten (Conway Morris & Robison, 1986, 1988) with some individuals showing vestiges of internal anatomy and parapodia-like structures. Although too poorly preserved for assignment to phylum, as indicators of diversity these vermiform specimens are counted as one indeterminate genus of worm in each lagerstätte (see Table 1). More taxa, however, may actually be represented.

3.h. *Arthropoda* (Figs. 7, 8)

As in most Cambrian faunas, arthropods clearly dominate the four Utah Lagerstätten, both in numbers of specimens and kinds. Of at least 20 nontrilobite genera (Fig. 7, Appendix), about half are also present in the Burgess shale.

Four specimens from the deep Wheeler lagerstätte in the House Range appear to represent *Alalcomenaeus*, a rare genus in the Burgess shale (Whittington, 1985), but one that has recently been found in some abundance elsewhere in the Stephen Formation (Collins *et al.*, 1983). The best preserved Wheeler specimen (Fig. 7.10) closely resembles those of *A. cambricus*, the type species. Discernible characters, however, are inadequate to confirm an assignment to that species.

Of the nontrilobite arthropods that have been recorded only from Utah Lagerstätten, *Cambropodus, Ecnomocaris*, and *Utahcaris* are especially noteworthy. *Cambropodus gracilis* (Fig. 7.1) is a marine myriapod-like animal from the shallow Wheeler lagerstätte (Robison, 1990). Its head bears three pairs of long, slender, uniramous limbs; the anterior pair being antenniform and the second and third pairs are similar to its several pairs of long, slender, trunk legs. As the oldest uniramous arthropod, *C. gracilis* may have special significance with respect to the ancestry of the terrestrial myriapods and insects. *Ecnomocaris spinosa* (Fig. 7.6) from the deep Wheeler Lagerstätte is a bizarre arthropod with several elongate spines arising from one side of the body (Conway Morris & Robison, 1988). The anterior? spine, and possibly the posterior?, is unusual in being longer than the body. Its spinosity recalls that of the enigmatic *Hallucigenia* from the Burgess shale, but the similarity is superficial. *Utahcaris orion* (Fig. 7.5), from the Spence Lagerstätte, has an elongate, multisegmented body with cephalic shield and telson of subequal size (Conway Morris & Robison, 1988). Trilobite fragments preserved in the stomach of the holotype indicate that the animal was a predator or scavenger.

Trilobite taxa that are listed in the Appendix include only those known from Lagerstätte deposits of the Spence, Wheeler, and Marjum formations (see Section 2). Specimens from the Spence Shale are mostly preserved in noncalcareous shale and all those in the Wheeler and Marjum formations are from the interbedded limestone and shale lithofacies of Rees (1986). Many more trilobite species have been described from other lithofacies in the three formations. Also, some specimens in hand from lagerstätten in the Wheeler and Marjum formations represent several new species, mostly rare, which await description. Table 1 shows a close correspondence in the total number of trilobite genera (15 or 16) that are known from the Burgess, deep Wheeler, and Marjum lagerstätten. The greater number of trilobite genera (22) in the Spence Lagerstätte may result from increased transport and accumulation of trilobite remains in more distal tempestites.

Naraoia compacta (Fig. 8.4), which lacks free thoracic segments and calcification of the exoskeleton, is one of the more unusual trilobites in the Burgess shale (Whittington, 1985) as well as all three deep-ramp Lagerstatten in Utah (Robison, 1984b, Conway Morris & Robison, 1988). Its stratigraphic range is the greatest observed for any Cambrian trilobite species. The recent description of a new species of *Naraoia* from the Chengjiang Lagerstätte of Early Cambrian age in China (Zhang & Hou, 1985) demonstrates an intercontinental distribution for the genus. The observed stratigraphic range of *Naraoia*, from the lower Lower to upper Middle Cambrian, is the greatest recorded for any Cambrian trilobite genus.

Excellent preservation of trilobite remains, especially in the deep Wheeler lagerstätte, has provided valuable information about aspects of trilobite palaeobiology. For example, Bright (1959) has documented morphologic variation in a large ontogenetic series of *Elrathia kingii*. McNamara & Rudkin (1984) have demonstrated techniques of exuviation by *Asaphiscus wheeleri*.

3.i. *Echinoderms* (Figs. 2, 9)

Of the thousands of articulated eocrinoid specimens collected from Cambrian deposits in North America, probably more than half are from the Spence, Wheeler, and Marjum Lagerstätten of Utah. Of the hundreds of articulated specimens of all other classes of Cambrian echinoderms collected in North America, probably about 60 percent are from the same Lagerstätten. The number of echinoderm genera in each of the Burgess and Utah Lagerstätten ranges from 2 to 4 (Table I, Appendix). Some classes represented in the Burgess shale (Whittington, 1985), however, are not the same as those represented in the Utah Lagerstätten.

Sprinkle (1976) has noted that many Cambrian echinoderms were gregarious. This is further demonstrated by the patchy distribution (Fig. 2) of most species listed in the Appendix. Some are represented by numerous specimens on single bedding surfaces.

The observed distribution of the carpoid *Castericystis vali* (Fig. 9.1), which is restricted to about 15 meters of beds along about 3 kilometers of outcrop in the middle of the Marjum Formation, suggests occupation of the same general area of seafloor for perhaps hundreds of thousands of years or more. Numerous articulated specimens are mostly preserved in silty partings on top of thin limestone beds or on parting surfaces in shale. The attachment of multiple juveniles to adults of *C. vali* (Fig. 9.1; Ubaghs & Robison, 1985, figs. 2.1-2, 11.3-5) suggests live burial. Alignment of specimens on some bedding surfaces (Rees, 1986, fig. 7B) indicates current influence at the time

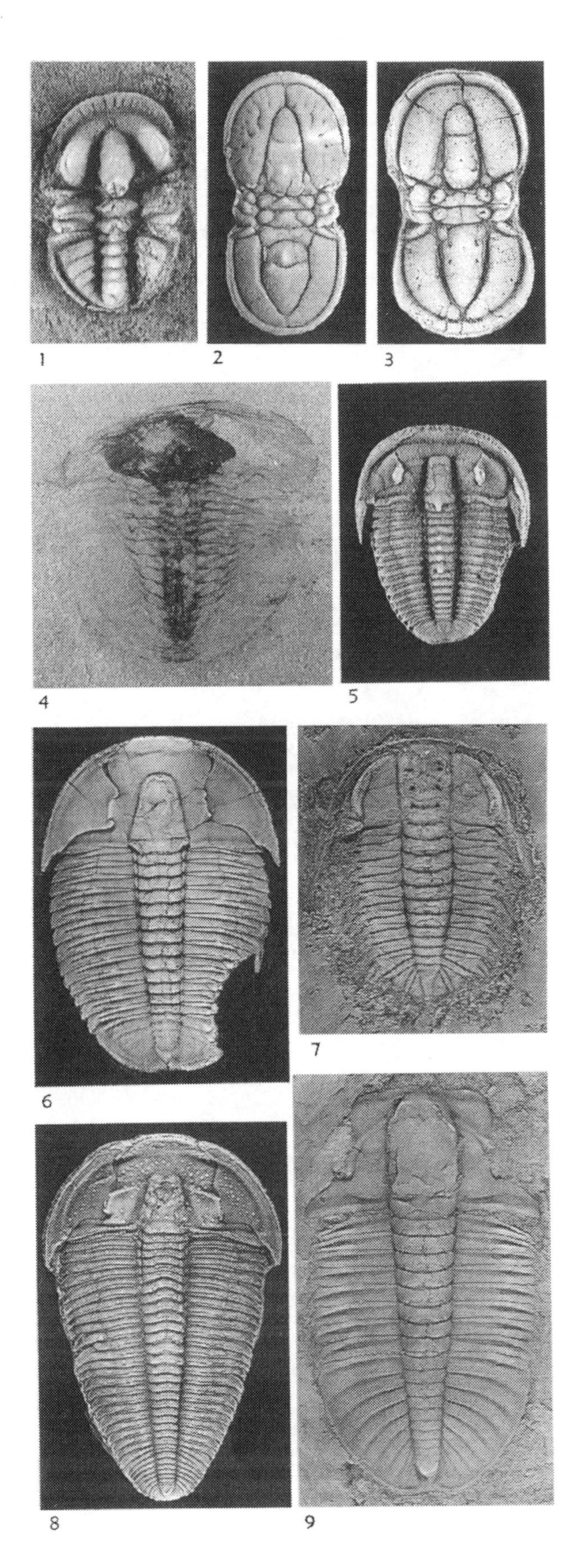

Fig. 8. Examples of trilobites. 1, *Pagetia* n. sp.; Spence Shale, Wellsville Mountains; KUMIP 204391, X8. 2, *Ptychagnostus atavus* (Tullberg); Wheeler Formation, House Range; KUMIP 153938, X4. 3, *Peronopsis interstricta* (White); Wheeler Formation, House Range; KUMIP 153983, 4.6. 4, Compression of *Naraoia compacta* Walcott showing soft parts; Wheeler Formation, House Range; USNM 424112, X2. 5, *Bolaspidella housensis* (Walcott); Wheeler Formation, House Range; KUMIP 204402, X3. 6, Dorsal exoskeleton of *Elrathia kingii* (Meek) with arcuate, healed bite mark and regenerated spine on one right pleura; Wheeler Formation, House Range; KUMIP 204773, X1.3. 7, *Oryctocephalus walcotti* Resser; Spence Shale, Two Mile Canyon, Idaho; KUMIP 204392, X3. 8, Holotype of *Alokistocare harrisi* Robison; Wheeler Formation, House Range; KUMIP 204393, X1.3. 8, Exuvium of *Ogygopsis typicalis* (Resser) Spence Shale, Wellsville Mountains; KUMIP 204394, X1.

of burial, probably during storm events. Articulated remains of other benthos are rare, except for exuviae of the trilobite *Modocia laevinucha*. Such exuviae, which are separated at the facial suture, do not represent the burial of live animals. Whether the local community consisted largely of echinoderms or whether other more active benthos, including trilobites, were present but able to escape episodic sediment deposition is unclear (compare discussion of Jurassic echinoderm preservation; Seilacher *et al.*, 1985, pp. 9, 23).

Following Conway Morris & Robison (1988), *Eldonia ludwigi* (Fig. 10.2) is here assigned as phylum undetermined rather than to the echinoderm class Holothuroidea (Durham, 1974).

3.j. *Miscellaneous animals* (Fig. 10)

Several of the more interesting problematic taxa in the Burgess shale are also represented in one or more of the four Utah Lagerstätten. Among these is the lobopodan *Aysheaia* (Fig. 10.1), represented by different eruciform species in the Burgess and deep Wheeler lagerstätten. Originally described by Walcott (1911) as a polychaete annelid, *Aysheaia* has most recently been assigned to the Arthropoda (Whittington, 1985), the Onychophora (Robison, 1985), and the Lobopodia (Dzik & Krumbiegel, 1989).

The enigmatic *Anomalocaris nathorsti* is the largest known Cambrian animal and a probable predator of debated affinities (Whittington & Briggs, 1985; Bergström, 1986, 1987; Briggs & Whittington, 1987). In Utah, specimens have been recorded from the three deep-ramp Lagerstätten (Fig. 10.3-5), but all are incomplete (Conway Morris & Robison, 1982, 1988; Briggs & Robison, 1984) and provide no additional information about affinities. Indirect evidence has been used, however, to suggest possible trophic relationships between *Anomalocaris* and other Utah fossils. Healed injuries were cited by Vorwald (1982) as evidence of a possible predator-prey relationship between *Anomalocaris* and the trilobite *Asaphiscus wheeleri* in what is here called the shallow Wheeler lagerstätte. Further evidence of predation was cited by Babcock & Robison (1989a, b), who noted that arcuate healed bite marks (Fig. 8.6) on various trilobites from all Utah Lagerstätten are similar in size and shape to the opening between the jaw plates of *Anomalocaris*. If these inferred trophic relationships are correct, it is reasonable to expect that further collecting will produce specimens of *Anomalocaris* in the shallow Wheeler lagerstätte.

Together with two other problematic genera, Bengtson & Missarzhevsky (1981) reassigned *Chancelloria* and *Wiwaxia* to the class Coeloscleritophora of open phylum assignment. Animals of this class were said to be characterized by a "composite exoskeleton,

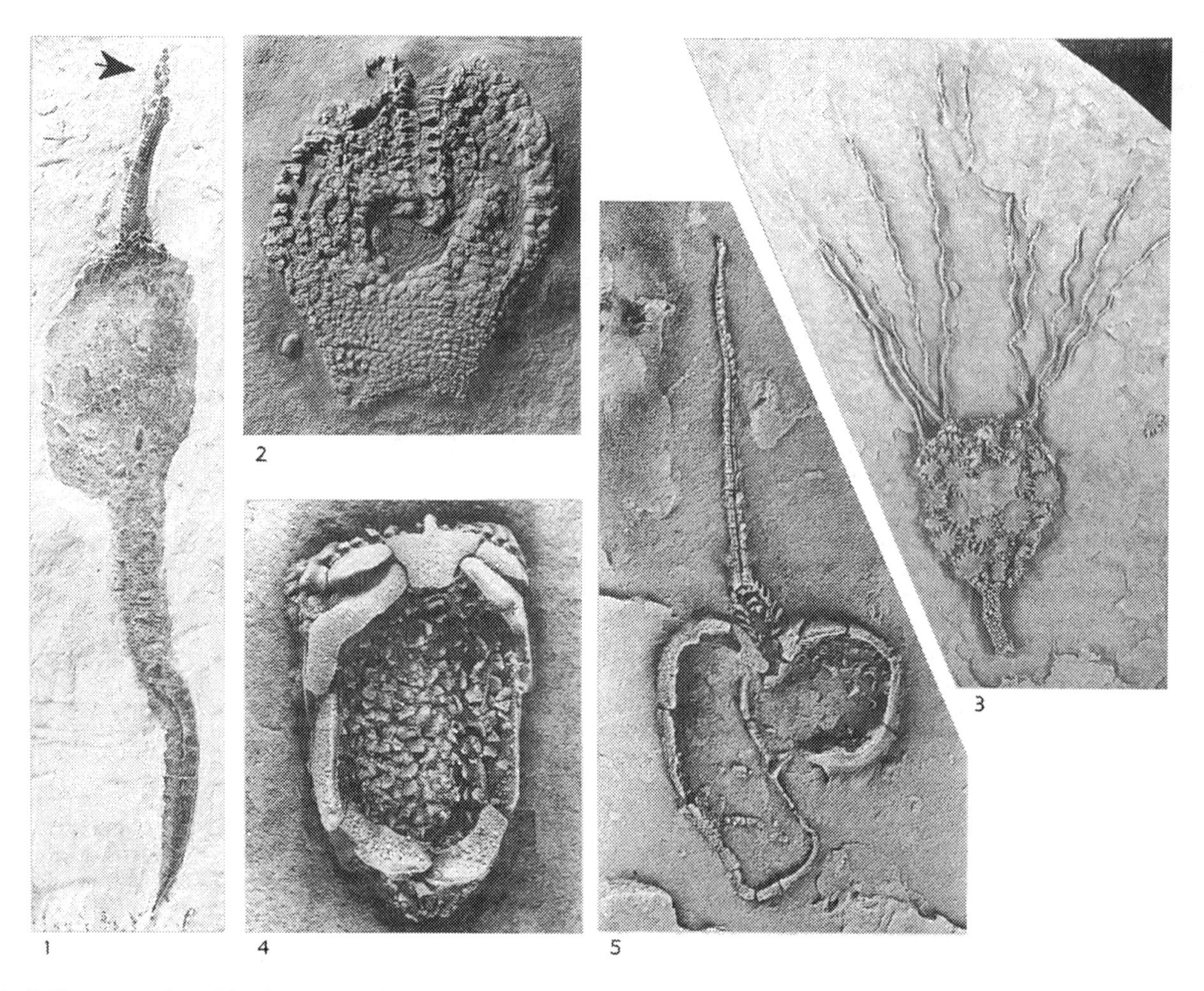

Fig. 9. Representative echinoderms. 1, Holotype adult of *Castericystis vali* Ubaghs & Robison (class Homoiostelea) with juvenile attached to arm (arrow); Marjum Formation, House Range; USNM 393391, X1.5. 2, Holotype theca of *Totiglobus? lloydi* Sprinkle (class Edrioasteroidea); Marjum Formation, House Range; USNM 172047, X3.3. 3, Holotype (latex cast) of *Gogia granulosa* Robison (class Eocrinoidea); Spence Shale, Wellsville Mountains; KUMIP 204395, X3.0. 4, Holotype theca (latex cast, dorsal surface) of *Ctenocystis utahensis* Robison & Sprinkle (class Ctenocystoidea); Spence Shale, Wellsville Mountains; USNM 163252, X7.9. 5, New species of class Stylophora (latex cast), Spence Shale, Wellsville Mountains, X3.7. Photographs 2 and 5 were provided by James Sprinkle.

the individual sclerites of which have a prominent internal cavity and a restricted basal foramen." The sessile sponge-like *Chancelloria* (Fig. 10.8) is known from rare articulated vasiform skeletons in the deep Wheeler lagerstätte (Rigby, 1978) and locally common stellate clusters of sclerites in the three Wheeler and Marjum Lagerstätten. At some Marjum localities, such clusters are preserved as black carbonaceous compressions in light-colored shale, and have been misidentified as tiny starfish by some unwary collectors. Until recently, the vagrant lepidote *Wiwaxia* has been recorded only from the Stephen Formation at a few localities in British Columbia (Conway Morris, 1985a). Rare sclerites (Fig. 10.6-7) in the Spence Lagerstätte (Conway Morris & Robison, 1988), however, have shown this unusual genus to have a much wider geographic distribution.

Rare medusiform specimens of *Eldonia ludwigi* have been found in both the Spence (Fig. 10.2) and Marjum Lagerstätten (Conway Morris & Robison, 1988), but otherwise the species has been recorded only from the Burgess shale. Some authors have considered the species to be a holothurian and others have rejected such an affinity. I presently consider its affinities to be uncertain.

3.k. *Trace fossils and coprolites* (Fig. 11)

Trace fossils are generally rare in all four Utah Lagerstätten, but may be present in modest numbers on some bedding surfaces. Most are horizontal or near-horizontal, simple burrows that are assigned to either *Planolites* (Fig. 11.2) or *Palaeophycus* as revised by Pemberton & Frey (1982).

Lobate structures assigned to *Brooksella* have most commonly been interpreted to be medusoid impressions, but have also been considered to be fecal masses, organic burrows, and inorganic compaction structures. Rare specimens of *Brooksella* from the Spence Lagerstätte (Fig. 11.4) were previously considered by Willoughby & Robison (1979) to be medusoid impressions. Following Kauffman & Fursich (1983), however, I now regard them as complex burrowing and probing structures produced by an unknown metazoan.

Rare "feather-stitch" burrows in the Spence Lagerstätte (Fig. 11.3) that were previously described as Burrow Type A (Robison, 1969) are here assigned to *Trept-*

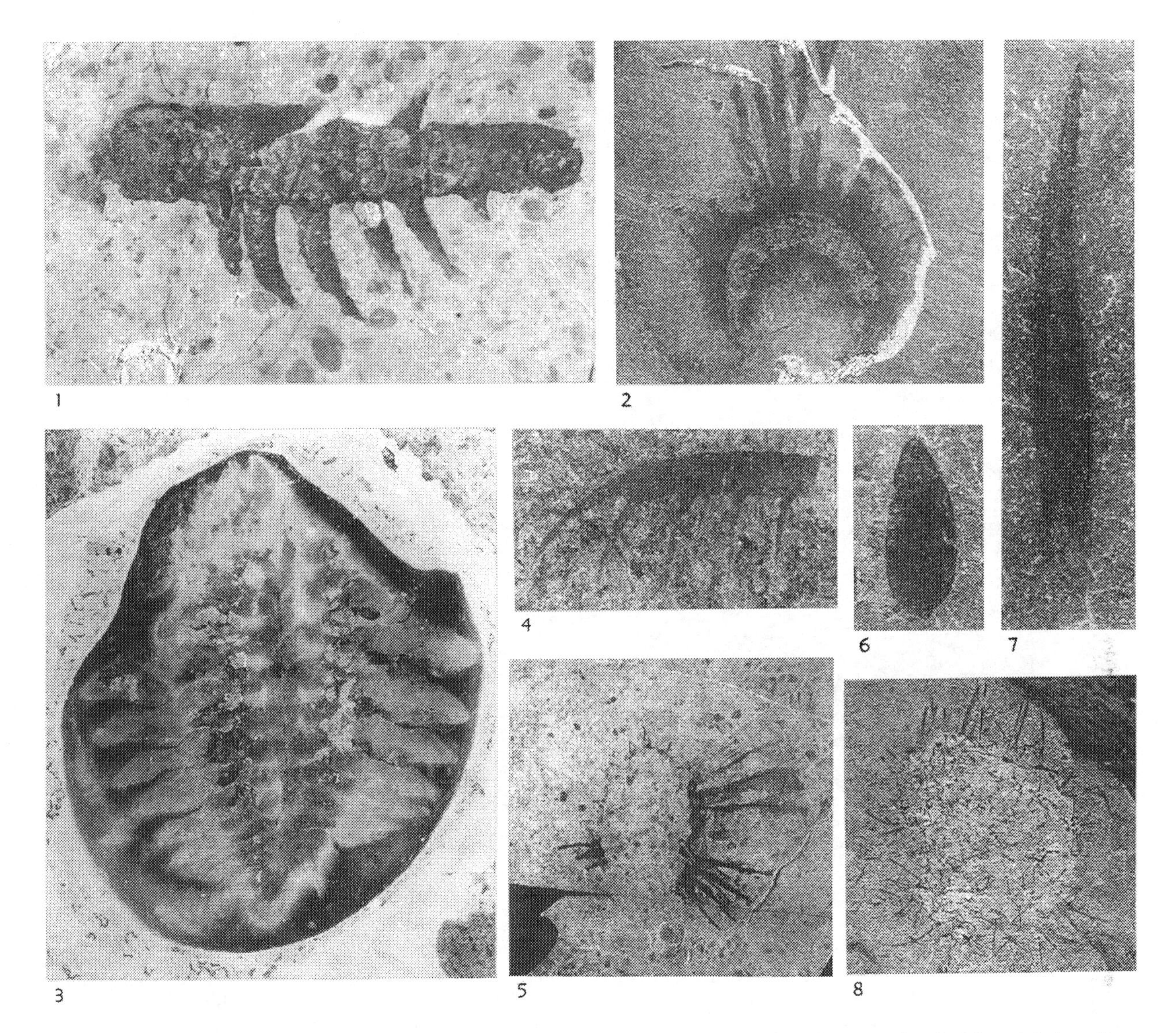

Fig. 10. Examples of miscellaneous animals. 1, Incomplete holotype (exuvium?) of *Aysheaia prolata* Robison (phylum Onychophora) on bedding surface with several *Morania fragmenta*, a questionable blue-green alga; Wheeler Formation, House Range; KUMIP 153923, X2.8. 2, Carbonaceous compression of *Eldonia ludwigi* Walcott; Spence Shale, Wellsville Mountains; KUMIP 204370, X2. 3-5, Parts of *Anomalocaris nathorsti* (Walcott); 3, posterior body, Marjum Formation, House Range, USNM 374593, Xl; 4, isolated blade of cephalic appendage, Spence Shale, Wellsville Mountains, KUMIP 204777, X3; 5, incomplete jaw, Wheeler Formation, House Range, KUMIP 153093, X1.3. 6 and 7, Isolated scale and spine of *Wixaxia* cf. *corrugata* (Matthew); Spence Shale, Wellsville Mountains; KUMIP 204367 and 204366, both X5. 8, Sclerite cluster of *Chancelloria eros* Walcott; Wheeler Formation, House Range; KUMIP 204401, X2.5.

ichnus sp. A few additional specimens have been collected by Lloyd and Val Gunther, and are in collections at the University of Kansas.

About a dozen previously undescribed specimens of *Bergaueria* (Fig. 11.5) are preserved in hyporelief on thin, dark-gray, lime-mudstone beds from the shallow Wheeler Lagerstätte of the Drum Mountains. These specimens are hemispherical, each with a shallow central depression. Height is less than the diameter, which ranges up to about 40 mm. Such structures have most commonly been interpreted as dwelling burrows of probable sea anemones (e.g., Alpert, 1973), and if correct, the Wheeler fossils represent an otherwise unknown cnidarian. They most closely resemble specimens of *Bergaueria* illustrated by Narbonne (1984, fig. 6G) from Silurian intertidal carbonates of arctic Canada, which is in accord with the shallow-water habitat inferred here.

Some healed injuries of trilobites (Fig. 8.6) from the four Utah lagerstätten (Vorwald, 1982, 1983; Babcock & Robison, 1989a, b), as well as many other Palaeozoic deposits, have been interpreted as traces of unsuccessful predation. Strong asymmetry in the distribution of sublethal predation scars suggests that predators more commonly attacked the right side of trilobites (Babcock & Robison, 1989a, b). This evidence of behavioural asymmetry, which in modern animals is correlated with asymmetry of the brain, further suggests that brain laterality had developed about 500 million years earlier than shown by previous evidence.

Unusually large, round coprolites (Fig. 11.1) are locally common in the Spence lagerstätte (Conway Morris & Robison, 1988). Some contain mixed skeletal and shell fragments of trilobites, brachiopods, and echinoderms, indicating that their producer was omnivorous and either a scavenger or predator. Moreover,

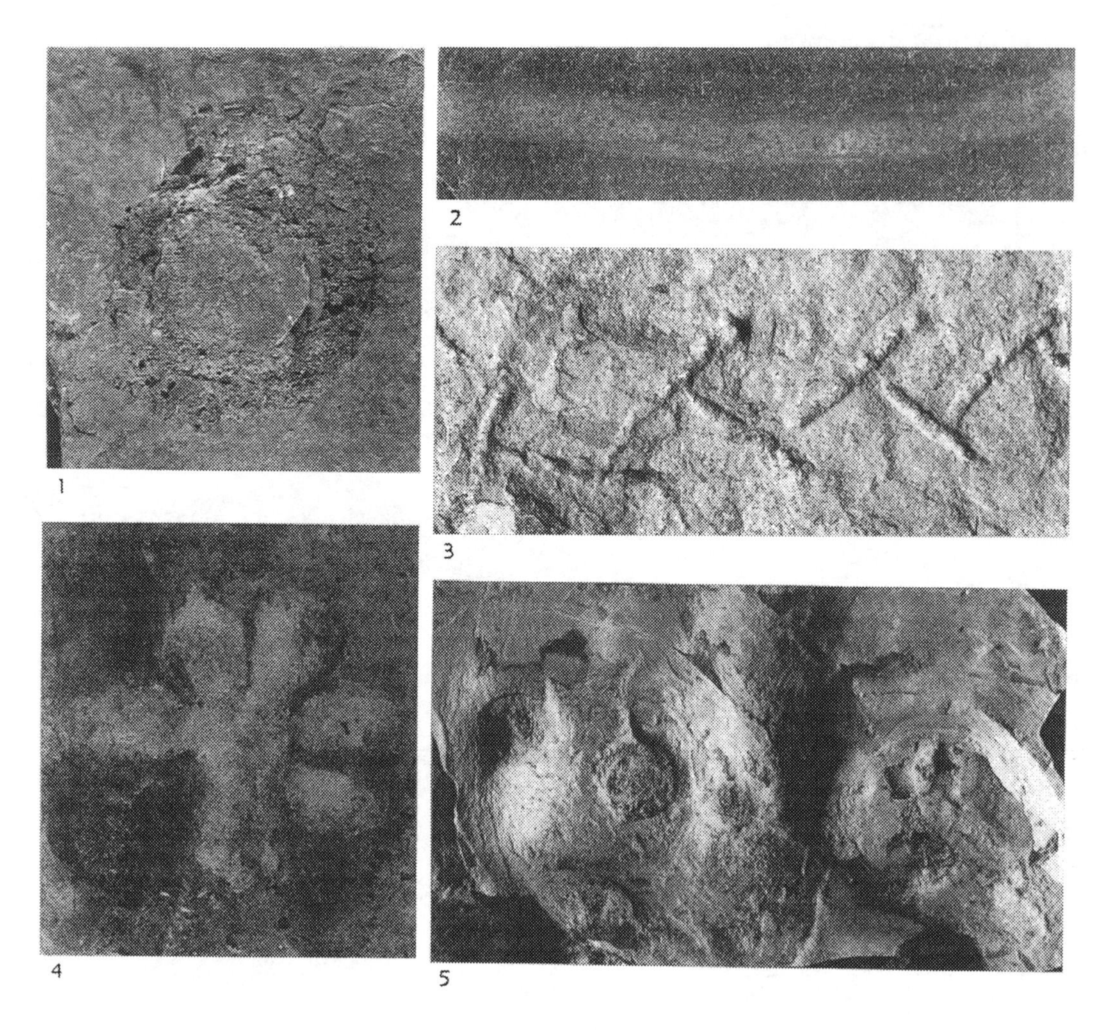

Fig. 11. Examples of coprolite (1) and trace fossils (2-5). 1, Coprolite containing skeletal and shell fragments of echinoderms, trilobites, and brachiopods; Spence Shale, Wellsville Mountains; KUMIP 204369, X1. 2, *Planolites beverleyensis* (Billings) burrow; Wheeler Formation, House Range; KUMIP 204396, X2. 3, *Treptichnus* sp., "feather-stitch" burrow; Spence Shale, Wellsville Mountains; KUMIP 204397, X1. 4, *Brooksella* sp., burrow; Spence Shale, Wellsville Mountains; BYU 1545, X1. 5, *Bergaueria* sp., preserved in hyporelief; Wheeler Formation, Drum Mountains; KUMIP 204398, X1.

their diameter, ranging up to 45 mm, suggests that the producer may have been substantially larger than previously imagined for a Cambrian animal.

4. Significance of the Utah lagerstätten

Applying the classification of fossil Lagerstätten by Seilacher *et al.* (1985), the extraordinary biotas of the Spence, Wheeler, and Marjum formations are all preserved in conservation deposits of obrution type. All accumulated in marine-shelf environments on the passive, rapidly subsiding, continental margin of equatorial Laurentia. All are preserved in fine-grained, laminated mudstone that varies greatly in carbonate content. Bedding is commonly flaggy. Fossils, as well as bioturbation, are scarce in much of each deposit. In this latter respect, they recall the classic Solnhofen Lagerstätte, of which Seilacher *et al.* (1985, p. 11) noted that "had there been no quarries, the unit (Solnhofen) would have possibly been mapped as 'non-fossiliferous' ". Scattered beds in the Utah deposits contain rare to abundant shelly fossils, usually dominated by trilobites. Rare soft-bodied or lightly skeletized fossils were mostly, if not entirely, preserved by rapid burial in fine sediment, probably during episodic storm events. Characteristic features are whole organisms that are associated with thin tempestites and show current alignment. Wave ripples are conspicuously absent. Although one Wheeler Lagerstätte accumulated in shallow water and the other on a deep ramp, their lithologies are superficially similar because of prevailing low-energy conditions in each environment.

Biotas in the four Utah Lagerstätten (Appendix) provide further evidence that many taxa that were described from the Burgess Lagerstätte (Whittington, 1985; references therein) are widespread in shelf lithofacies of Laurentia. The stratigraphic distribution of the Utah Lagerstätten, ranging from lower to upper Middle Cambrian, also provides information about evolution within major biotic groups through a substantial interval of time. In general, species of such shelly groups as trilobites, brachiopods, and echinoderms have shorter stratigraphic ranges than do many of the algae and soft-bodied or lightly skeletized animal species. A limited indication of this is seen even within one class, the Trilobita, where ordinary calcareous species rarely range beyond one or two trilobite zones, but the unusual, lightly skeletized *Naraoia compacta* ranges through most of the Middle Cambrian, and possibly from the upper Lower Cambrian. Also, as noted above (Section 3.h), *Naraoia* has the longest observed strati-

Table 1: Number of genera in major animal groups of the Burgess (Bu), Spence (Sp), shallow Wheeler (sW), deep Wheeler (dW), and Marjum (Ma) Lagerstätten, with percentages based on the total genera in each. Except for subsequent minor emendation, the numbers of Burgess genera are from Whittington (1985) and Rigby (1986); numbers for the Spence, Wheeler, and Marjum are from the Appendix.

	Bu	%	Sp	%	sW	%	dW	%	Ma	%
Arthropoda	45	40	29	55	17	50	24	53	20	40
Trilobita	(16)	(14)	(22)	(42)	(12)	(35)	(15)	(33)	(15)	(30)
Others	(29)	(26)	(7)	(13)	(5)	(21)	(9)	(20)	(5)	(10)
Porifera	20	18	3	6	1	3	5	11	12	24
Worms	13	12	5	10	4	12	1	2	4	8
Brachiopoda	6	5	5	10	5	15	5	11	6	12
Echinodermata	4	4	4	8	2	6	3	7	3	6
Cnidaria	4	4	1	2	0	0	1	2	1	2
Mollusca	1	1	2	4	4	12	3	7	2	4
Miscellaneous	20	18	3	6	1	3	3	7	2	4
Total genera	113		52		34		45		50	

graphic range by far of any Cambrian trilobite genus. Such differences in stratigraphic ranges, moreover, do not seem to result from the application of different taxonomic philosophy by different taxonomists.

Compared with the Burgess lagerstätte, recorded faunas of the three deep-ramp Lagerstätten of Utah have smaller numbers of lightly skeletized nontrilobitic arthropods, soft-bodied worms, and soft-bodied or lightly skeletized miscellaneous genera (Table 1). From available evidence, however, these numbers seem to be influenced more by taphonomic and collecting biases than by fundamental community differences. Other differences probably result from slight variation in paleogeographic settings, different geologic ages and corresponding phyletic changes, and the chance preservation and discovery of either rare taxa or taxa with patchy distributions.

The Utah biotas provide additional information about various aspects of Middle Cambrian trophic structure and animal behavior.The Spence arthropod *Utahcaris* (Fig. 7.5), with trilobite skeletal remains among its stomach contents, was either a predator or scavenger. Extraordinarily large coprolites (Fig. 11.1) containing skeletal or shell fragments of brachiopods, echinoderms, and trilobites provide evidence of a possibly unnamed omnivorous predator or scavenger in the Spence biota. Size of the coprolites further suggests that the producer may be larger than any animal previously recorded from Cambrian rocks. A gut filling in a specimen of the Spence hyolithid *Haplophrentis reesei* (Fig. 5.11), the first reported from the order Hyolithida, is similar to matrix surrounding the shell. This supports previous suggestions (e.g., Yochelson, 1961) that hyolithids were deposit feeders. Statistically significant differences in the distribution of healed bite marks on trilobites from all of the Utah Lagerstätten (e.g., Fig. 8.6), as well as other Palaeozoic deposits, have provided indirect evidence of behavioural asymmetry among trilobite predators. This also suggests that the trilobite predators had developed brain laterality.

The Burgess shale has been interpreted to represent part of a sediment apron that accumulated adjacent to a near-vertical submarine escarpment (Fritz, 1971; McIlreath, 1975, 1977). Using the uppermost occurrences of *Glossopleura* as an inferred isochronous surface, Fritz (1971) calculated that minimum water depth at the front of the escarpment exceeded 200 meters. That physiographic model has subsequently received wide acceptance (e.g., Whittington, 1985; several references therein). Based on the observed distribution patterns of the *Glossopleura* and *Bathyuriscus-Elrathina* faunas, however, Briggs & Robison (1984) suggested that the boundary between the two may have greater ecological than temporal significance, and that inferred water depths during deposition of the Burgess shale may be exaggerated. Recently, Ludvigsen (1989) has disputed Fritz's physiographic model, contending that there is no convincing evidence that a Cambrian escarpment was ever present in the area. Instead, Ludvigsen (1989, p. 58) concluded that "the biostratigraphic and sedimentologic evidence strongly suggests that the Burgess Shale was deposited on a ramp as a sequence of tempestites." If that interpretation is correct, the depositional environments of Burgess, Spence, deep Wheeler, and Marjum Lagerstätten were even more similar than has generally been thought.

Acknowledgments

I am indebted to individuals too numerous to mention, who have collected and kindly contributed rare fossils that are cited here. Exceptionally important for their diligence and generosity, however, are Lloyd and Val Gunther of Brigham City, Utah. Without their sustained efforts, the appendix of this paper would be substantially shorter. Constructive reviews of this paper were given by L.E. Babcock, Lloyd Gunther, M.N. Rees, A.J. Rowell, and James Sprinkle. This study is based on work supported by National Science Foundation grant EAR-8720333. The University of Camerino is thanked for their hospitality during this Symposium.

REFERENCES

AIGNER, T. 1985. *Storm Depositional Systems*, Berlin: Springer-Verlag.

AITKEN, J.D. 1978. Revised models for depositional grand cycles, Cambrian of the southern Rocky Mountains, Canada. *Bulletin of Canadian Petroleum Geology* **26**, 515-542.

ALPERT, S.P. 1973. *Bergaueria* Prantl (Cambrian and Ordovician), a probable actinian trace fossil. *Journal of Paleontology* **47**, 919-924.

BABCOCK, L.E. & ROBISON, R.A. 1988. Taxonomy and paleobiology of some Middle Cambrian *Scenella* (Cnidaria) and hyolithids (Mollusca) from western North America. *University of Kansas Paleontological Contributions, Paper* **121**, 1-22.

BABCOCK, L.E. & ROBISON, R.A. 1989a. Preferences of Palaeozoic predators. *Nature* **337**, 695-696.

BABCOCK, L.E. & ROBISON, R.A. 1989b. Asymmetry of predation on trilobites. *28th International Geological Congress, Abstracts* **1**, 66.

BENGTSON, S. 1986. Siliceous microfossils from the Upper Cambrian of Queensland. *Alcheringa* **10**, 195-216.

BENGTSON, S. & MISSARZHEVSKY, V.V. 1981. Coeloscleritophora -- a major group of enigmatic Cambrian metazoans. *U.S. Geological Survey, Open-file Report* **81-743**, 19-21.

BERGSTRÖM, J. 1986. *Opabinia* and *Anomalocaris*, unique Cambrian 'arthropods.' *Lethaia* **19**, 241-246.

BERGSTRÖM, J. 1987. The Cambrian *Opabinia* and *Anomalocaris*. *Lethaia* **20**, 187-188.

BOLD, H.C. & WYNNE, M.J. 1985. *Introduction to the Algae*, Englewood Cliffs, New Jersey: Prentice-Hall.

BRETT, C.E. & BAIRD, G.C. 1986. Comparative taphonomy: a key to paleoenvironmental interpretation based on fossil preservation. *Palaios* **1**, 207-227.

BRIGGS, D.E.G. & ROBISON, R.A. 1984. Exceptionally preserved nontrilobite arthropods and *Anomalocaris* from the Middle Cambrian of Utah. *University of Kansas Paleontological Contributions, Paper* **111**, 1-23.

BRIGGS, D.E.G. & WHITTINGTON, H.B. 1987. The affinities of the Cambrian animals *Anomalocaris* and *Opabinia*. *Lethaia* **20**, 185-186.

BRIGHT, R.C. 1959. A paleoecologic and biometric study of the Middle Cambrian trilobite *Elrathia kingii* (Meek). *Journal of Paleontology* **33**, 83-98.

CAMPBELL, D.P. 1974. *Biostratigraphy of the* Albertella *and* Glossopleura *zones (lower Middle Cambrian) of northern Utah and southern Idaho.* Unpublished M.S. thesis, University of Utah, 295 pp.

CHEN JUN-YUAN, HOU XIAN-GUANG & LU HAO-ZHI 1989. Lower Cambrian leptomitids (Demospongea), Chengjiang, Yunnan. *Acta Palaeontologica Sinica* **28**, 17-31.

COLLINS D., BRIGGS, D. & CONWAY MORRIS, S. 1983. New Burgess shale sites reveal Middle Cambrian faunal complex. *Science* **222**, 163-167.

CONWAY MORRIS, S. 1977. Fossil priapulid worms. *Special Papers in Palaeontology* **20**, 1-95.

CONWAY MORRIS, S. 1979. Middle Cambrian polychaetes from the Burgess shale of British Columbia. *Philosophical Transactions of the Royal Society of London* B **285**, 227-274.

CONWAY MORRIS, S. 1985a. The Middle Cambrian metazoan *Wiwaxia corrugata* (Matthew) from the Burgess shale and *Ogygopsis* shale, British Columbia, Canada. *Philosophical Transactions of the Royal Society of London* B **307**, 507-586.

CONWAY MORRIS, S. 1985b. Cambrian lagerstätten: their distribution and significance. *Philosophical Transactions of the Royal Society of London* B **311**, 49-65.

CONWAY MORRIS, S. 1986. The community structure of the Middle Cambrian phyllopod bed (Burgess shale). *Palaeontology* **29**, 423-467.

CONWAY MORRIS, S. 1989. The persistence of Burgess shale-type faunas: implications for the evolution of deeper water faunas. *Transactions of the Royal Society of Edinburgh, Earth Sciences* **80**, 271-283.

CONWAY MORRIS, S., PEEL, J.S., HIGGINS, A.K., SOPER, N.J. & DAVIS, N.C. 1987. A Burgess shale-type fauna from the Lower Cambrian of North Greenland. *Nature* **326**, 181-183.

CONWAY MORRIS, S. & ROBISON, R.A. 1982. The enigmatic medusoid *Peytoia* and a comparison of some Cambrian biotas. *Journal of Paleontology* **56**, 116-122.

CONWAY MORRIS, S. & ROBISON, R.A. 1986. Middle Cambrian priapulids and other soft-bodied fossils from Utah and Spain. *University of Kansas Paleontological Contributions, Paper* **117**, 1-22.

CONWAY MORRIS, S. & ROBISON, R.A. 1988. More soft-bodied animals and algae from the Middle Cambrian of Utah and British Columbia. *University of Kansas Paleontological Contributions, Paper* **122**, 1-48.

DEAN, D., RANKIN, J.S. & HOFFMANN, E. 1964. A note on the survival of polychaetes and amphipods in stored jars of sediment. *Journal of Paleontology* **38**, 608-609.

DURHAM, J.W. 1974. Systematic position of *Eldonia ludwigi* Walcott. *Journal of Paleontology* **48**, 750-755.

DZIK, J. & KRUMBIEGEL, G. 1989. The oldest 'onychophoran' *Xenusion*: a link connecting phyla? *Lethaia* **22**, 169-181.

FRITZ, W.H. 1971. Geologic setting of the Burgess shale. In *Proceedings of the North American Paleontological Convention*, vol. 2 (ed. E.L. Yochelson), pp. 1155-1170. Lawrence, Kansas: Allen Press.

GOULD, S.J. 1989. *Wonderful Life: The Burgess Shale and the Nature of History*, New York: Norton.

GUNTHER, L.F. & GUNTHER, V.G. 1981. Some Middle Cambrian fossils of Utah. *Brigham Young University Geology Studies* **28**, 1-81.

HINTZE, L.F. & ROBISON, R.A. 1975. Middle Cambrian stratigraphy of the House, Wah Wah, and adjacent ranges in western Utah. *Geological Society of America Bulletin* **86**, 881-891.

HOU XIAN-GUANG. 1989. Features and significance of Chengjiang fauna: exceptionally preserved soft-bodied fauna from lowest Cambrian in Yunnan, China. *28th International Geological Congress (Washington, D.C.), Abstracts* **2**, 75.

HOU XIAN-GUANG & SUN WEI-GUO. 1988. Discovery of Chengjiang fauna at Meishucun, Jinning, Yunnan. *Acta Palaeontologica Sinica* **27**, 1-12 [In Chinese, with English summary].

KAUFFMAN, E.G. & FÜRSICH, F. 1983. *Brooksella canyonensis*: a billion-year-old complex metazoan trace fossil from the Grand Canyon. *Geological Society of America, Abstracts with Programs* 15, 608.

LIPPS, J.H. 1985. Earliest foraminifera and radiolaria from North America: evolutionary and geological implications. *Geological Society of America, Abstracts with Programs* **17**(7), 644-645.

LIPPS, J.H. 1987. Cambrian foraminifera. *Geological Society of America, Abstracts with Programs* **19**(7), 747.

LUDVIGSEN, R. 1989. The Burgess shale: not in the shadow of the Cathedral escarpment. *Geoscience Canada* **16**, 51-59.

MCILREATH, I.A. 1975. Stratigraphic relationships at the western edge of the Middle Cambrian carbonate facies belt, Field, British Columbia. *Geological Survey of Canada, Paper* **75-1**(A), 557-558.

MCILREATH I.A. 1977. Accumulation of a Middle Cambrian, deepwater limestone debris apron adjacent to a vertical, submarine carbonate escarpment, southern Rocky Mountains, Canada. *Society of Economic Paleontologists and Mineralogists, Special Publication* **25**, 113-124.

MCNAMARA, K.J. & RUDKIN, D.M. 1984. Techniques of trilobite exuviation. *Lethaia* **17**, 153-173.

NARBONNE, G.M. 1984. Trace fossils in Upper Silurian tidal flat to basin slope carbonates of arctic Canada. *Journal of Paleontology* **58**, 398-415.

ORIEL, S.S. & ARMSTRONG, F.C. 1971. Uppermost Precambrian and lowest Cambrian rocks in southeastern Idaho. *U.S. Geological Survey, Professional Paper* **394**, 1-52.

PEEL, J.S. 1988. Molluscs of the Holm Dal Formation (late Middle Cambrian), central North Greenland. *Meddelelser om Groenland, Geoscience* **20**, 145-168.

PEMBERTON, S.G. & FREY, R. W. 1982. Trace fossil nomenclature and the *Planolites-Palaeophycus* dilemma. *Journal of Paleontology* **56**, 843-881.

REES, M.N. 1984. *A fault-controlled trough through a carbonate platform: Middle Cambrian House Range embayment, Utah and Nevada*. Unpublished Ph.D. dissertation, University of Kansas, 136 pp.

REES, M.N. 1986. A fault-controlled trough through a carbonate platform: the Middle Cambrian House Range embayment. *Geological Society of America Bulletin* **97**, 1054-1069.

RESSER, C.E. 1939. The Spence Shale and its fauna. *Smithsonian Miscellaneous Collections* **97**(12), 1-29.

RIGBY, J.K. 1966. *Protospongia hicksi* Hinde from the Middle Cambrian of western Utah. *Journal of Paleontology* **40**, 549-554.

RIGBY, J.K. 1969. A new Middle Cambrian hexactinellid sponge from western Utah. *Journal of Paleontology* **43**, 125-128.

RIGBY, J.K. 1978. Porifera of the Middle Cambrian Wheeler Shale, from the Wheeler Amphitheater, House Range, in western Utah. *Journal of Paleontology* **52**, 1325-1345.

RIGBY, J.K. 1980. The new Middle Cambrian sponge *Vauxia magna* from the Spence Shale of northern Utah and taxonomic position of the Vauxiidae. *Journal of Paleontology* **54**, 234-240.

RIGBY, J.K. 1983. Sponges of the Middle Cambrian Marjum Limestone from the House Range and Drum Mountains of western Millard County, Utah. *Journal of Paleontology* **57**, 240-270.

RIGBY, J.K. 1986. Sponges of the Burgess shale (Middle Cambrian), British Columbia. *Palaeontographica Canadiana* **2**, 1-105.

RIGBY, J.K. & GUTSCHICK, R.C. 1976. Two new lower Paleozoic hexactinellid sponges from Utah and Oklahoma. *Journal of Paleontology* **50**, 79-85.

ROBISON, R.A. 1960. Lower and Middle Cambrian stratigraphy of the eastern Great Basin. In *Guidebook to the Geology of East Central Nevada* (eds. J.W. Boettcher and W.W. Sloan, Jr.), pp. 43-52. Salt Lake City: Intermountain Association of Petroleum Geologists.

ROBISON, R.A. 1964. Late Middle Cambrian faunas from western Utah. *Journal of Paleontology* **38**, 510-566.

ROBISON, R.A. 1965. Middle Cambrian eocrinoids from western North America. *Journal of Paleontology* **39**, 355-364.

ROBISON, R.A. 1969. Annelids from the Middle Cambrian Spence Shale of Utah. *Journal of Paleontology* **43**, 1169-1173.

ROBISON, R.A. 1971. Additional Middle Cambrian trilobites from the Wheeler Shale of Utah. *Journal of Paleontology* **45**, 796-804.

ROBISON, R.A. 1972. Hypostoma of agnostid trilobites. *Lethaia* **5**, 239-248.

ROBISON, R.A. 1976. Middle Cambrian trilobite biostratigraphy of the Great Basin. *Brigham Young University Geology Studies* **23**(2), 93-109.

ROBISON, R.A. 1978. Origin, taxonomy, and homeomorphs of *Doryagnostus* (Cambrian Trilobita). *University of Kansas Paleontological Contributions, Paper* **91**, 1-12.

ROBISON, R.A. 1982. Some Middle Cambrian agnostoid trilobites from western North America. *Journal of Paleontology* **56**, 132-160.

ROBISON, R.A. 1984a. Cambrian Agnostida of North America and Greenland; Part I, Ptychagnostidae. *University of Kansas Paleontological Contributions, Paper* **109**, 1-59.

ROBISON, R.A. 1984b. New occurrences of the unusual trilobite *Naraoia* from the Cambrian of Idaho and Utah. *University of Kansas Paleontological Contributions, Paper* **112**, 1-8.

ROBISON, R.A. 1985. Affinities of *Aysheaia* (Onychophora), with descriptions of a new Cambrian species. *Journal of Paleontology* **59**, 226-235.

ROBISON, R.A. 1990. Earliest-known uniramous arthropod. *Nature* **343**, 163-164.

ROBISON, R.A. & CAMPBELL, D.P. 1974. A Cambrian corynexochoid trilobite with only two thoracic segments. *Lethaia* **7**, 273-282.

ROBISON, R.A. & RICHARDS, B.C. 1981. Larger bivalve arthropods from the Middle Cambrian of Utah. *University of Kansas Paleontological Contributions, Paper* **106**, 1-28.

ROBISON, R.A. & SPRINKLE, J. 1969. Ctenocystoidea: new class of primitive echinoderms. *Science* **166**, 1512-1514.

ROGERS, J.C. 1984. Depositional environments and paleoecology of two quarry sites in the Middle Cambrian Marjum and Wheeler formations, House Range, Utah. *Brigham Young University Geology Studies* **31**, 97-115.

ROWELL, A.J. 1966. Revision of some Cambrian and Ordovician inarticulate brachiopods. *University of Kansas Paleontological Contributions, Paper* **7**, 1-36.

SCHWIMMER, D.R. 1989. Taxonomy and biostratigraphic significance of some Middle Cambrian trilobites from the Conasauga Formation in western Georgia. *Journal of Paleontology* **63**, 484-494.

SCOTESE, C.R. 1986. Phanerozoic reconstructions: a new look at, the assembly of Asia. *Texas Institute for Geophysics, Technical Report* **66**, 1-54.

SEILACHER, A., REIF, W.E. & WESTPHAL, F. 1985. Sedimentological, ecological and temporal patterns of fossil lagerstätten. *Philosophical Transactions of the Royal Society of London* **B 311**, 5-23.

SPRINKLE, J. 1973. Morphology and evolution of blastozoan echinoderms. *Harvard University Museum of Comparative Zoology, Special Publication.*

SPRINKLE, J. 1976. Biostratigraphy and paleoecology of Cambrian echinoderms from the Rocky Mountains. *Brigham Young University Geology Studies* **23** (2), 61-73.

SPRINKLE, J. 1985. New edrioasteroid from the Middle Cambrian of Utah. *University of Kansas Paleontological Contributions, Paper* **116**, 1-4.

UBAGHS, G. & ROBISON, R.A. 1985. A new homoiostelean and a new eocrinoid from the Middle Cambrian of Utah. *University of Kansas Paleontological Contributions, Paper* **115**, 1-24.

UBAGHS, G. & ROBISON, R. A. 1988. Homalozoan echinoderms of the Wheeler Formation (Middle Cambrian) of western Utah. *University of Kansas Paleontological Contributions, Paper* **120**, 1-17.

VORWALD, G.R. 1982. Healed injuries in trilobites-evidence for a large Cambrian predator. *Geological Society of America, Abstracts with Programs* **14**, 639.

VORWALD G.R. 1983. *Paleontology and paleoecology of the upper Wheeler Formation (late Middle Cambrian), Drum Mountains, west- central Utah.* Unpublished M.S. thesis, University of Kansas, 176 pp.

WALCOTT, C.D. 1908. Cambrian geology and paleontology; No. 1. Nomenclature of some Cambrian Cordilleran formations. *Smithsonian Miscellaneous Collections* **53**, 1-12.

WALCOTT, C.D. 1911. Cambrian geology and paleontology II; No. 5.-Middle Cambrian annelids. *Smithsonian Miscellaneous Collections* **57 (5)**, 109-144.

WALCOTT, C.D. 1919. Cambrian geology and paleontology IV; No. 5.-Middle Cambrian algae. *Smithsonian Miscellaneous Collections* **67 (5)**, 217-260.

WALCOTT, C.D. 1920. Cambrian geology and paleontology IV; No. 6.-Middle Cambrian Spongiae. *Smithsonian Miscellaneous Collections* **67 (6)**, 261-364.

WHITE, R.D. 1986. Cambrian radiolaria from Utah. *Journal of Paleontology* **60**, 778-780.

WHITTINGTON, H.B. 1985. *The Burgess Shale.* New Haven: Yale University Press.

WHITTINGTON, H.B. & BRIGGS, D.E.G. 1985. The largest Cambrian animal, *Anomalocaris*, Burgess shale, British Columbia. *Philosphical Transactions of the Royal Society of London* **B 309**, 569-609.

WILLOUGHBY, R.H. & ROBISON, R.A. 1979. Medusoids from the Middle Cambrian of Utah. *Journal of Paleontology* **53**, 494-500.

YOCHELSON, E.L. 1961. The operculum and mode of life of *Hyolithes*. *Journal of Paleontology* **35**, 152-161.

ZHANG WEN-TANG & HOU XIAN-GUANG. 1985. Preliminary notes on the occurrence of the unusual trilobite *Naraoia* in Asia. *Acta Palaeontologica Sinica* **24**, 591-595 [In Chinese, with English abstract].

APPENDIX

Middle Cambrian species that are known or questionably known from the Spence (S), shallow Wheeler (sW), deep Wheeler (dW), and Marjum (M) Lagerstätten of Utah are indicated by an 'x' or '?', respectively, under the appropriate abbreviation. Generic names that have also been assigned to fossils from the Burgess shale of British Columbia are preceded by a cross (+) combined generic and species names from the Burgess are preceded by an asterisk (*). The listed references generally include the most recent descriptions or discussions of the taxon.

	Formations				References
	S	sW	dW	M	
Division Cyanobacteria					
**Marpolia spissa* Walcott	x				Conway Morris & Robison, 1988
Division ?Cyanobacteria					
Epiphyton sp.		x			Rees, 1984
Girvanella sp.		x			Rees, 1984
**Morania fragmenta* Walcott	x	x	x	x	Walcott, 1919; Robison, 1985
Renalcis sp.		x			Rees, 1984
Division Chlorophyta					
**Margaretia dorus* Walcott	x	x	x	x	Conway Morris & Robison, 1988
Division ?Chlorophyta					
Acinocricus stichus Conway Morris & Robison	x				Conway Morris & Robison, 1982, 1988
**Yuknessia simplex* Walcott	x	x	x	x	Conway Morris & Robison, 1988
Division unknown					
Group Acritarcha					
several species			x		undescribed
Phylum ?Actinopoda					
Class ?Radiolaria					
Blastulospongia sp.		x			White, 1986; Bengtson, 1986
Phylum Foraminifera					
Ammodiscus sp.	x				Lipps, 1985, 1987
n. gen. & n. sp.	x				undescribed
Phylum Porifera					
Class Demospongea					
**Choia carteri* Walcott			x	x	Rigby, 1978, 1983
+ *Choia hindei* (Dawson)				x	Rigby, 1983
+ *Choia utahensis* Walcott			x	x	Rigby, 1978, 1983
+ *Choia* sp.	x				undescribed
**Hamptonia bowerbanki* Walcott				x	Rigby, 1983
**Hazelia palmata* Walcott				x	Rigby, 1983
Leptomitella metta (Rigby)				x	Rigby, 1983, 1986
**Sentinella draco* Walcott			x		Rigby, 1978, 1986
**Vauxia gracilenta* Walcott	x				Fig. 4.5
+ *Vauxia magna* Rigby	x				Rigby, 1980
+ *Vauxia* sp.		x			undescribed
archaeoscyphiid, n. gen. & n. sp.				x	undescribed
Class Hexactinellida					
+ *Diagonella cyathiformis (Dawson & Hinde)*			*x*	*x*	*Rigby, 1983*
**Diagonella hindei* Walcott				x	Rigby, 1983, 1986
+ *Diagonella robisoni* Rigby			x		Rigby, 1978
+ *Diagonella* sp.				x	Rigby, 1983
Hintzesnonaia bilamina Rigby & Gutschick				x	Rigby & Gutschick, 1976, Rigby 1983
Kiwetinokia spiralis Walcott			x	x	Rigby, 1978, 1983
Kiwetinokia utahensis Walcott			x	x	Rigby, 1978
**Protospongia hicksi* Hinde	x			x	Rigby, 1966, 1986; undescribed
+ *Protospongia? elongata* Rigby				x	Rigby, 1983
+ *Protospongia fenestrata* Salter			x	x	Walcott, 1920
Ratcliffespongia perforata Rigby				x	Rigby, 1969
Testispongia venula Rigby				x	Rigby, 1983
Valospongia gigantis Rigby				x	Rigby, 1983

Phylum Cnidaria					
Class Hydrozoa					
+*Scenella radians* Babcock & Robison	x				Babcock & Robison, 1988
Class uncertain					
Cambromedusa furcula Willoughby & Robison				x	Willoughby & Robison, 1979
Phylum ?Cnidaria					
**Cambrorhytium major* (Walcott)				x	Conway Morris & Robison, 1988
Phylum Brachiopoda					
Class Inarticulata					
Acrothele affinis Resser	x				Resser, 1939
Acrothele subsidua (White)		x	x	x	Robison, 1964; Vorvald, 1983
Dictyonina sp.	x				Gunther & Gunther, 1981
+*Lingulella* sp.	x	x		x	Robison, 1964; undescribed
Linnarssonia ophirensis (Walcott)			x	x	Robison, 1964; Rowell, 1966
+*Micromitra modesta* (Lochman)	x	x	x	x	Robison, 1964; undescribed
Pegmatreta bellatula (Walcott)				x	Robison, 1964
Prototreta attenuata (Meek)		x	x	x	Robison, 1964; Vorwald, 1983
Class Articulata					
+*Nisusia* sp.		x	x		Robison, 1964; Vorwald, 1983
Diaphora? *spencei* (Walcott)	x				Resser, 1939; emendation here
Phylum Mollusca					
Class Hyolitha					
+*Haplophrentis reesei* Babcock & Robison	x				Babcock & Robison, 1988
'Hyolithes' comptus Howell	x				Babcock & Robison, 1988
'Hyolithes' idahoensis Resser	x				Babcock & Robison, 1988
'Hyolithes' spp.		x	x	x	Ubaghs & Robison, 1985; undescribed
Class Monoplacophora					
Latouchella arguta (Resser)		x	x	x	Robison, 1964; Peel, 1988
Melopegma georginensis Runnegar & Jell			x		Fig. 5.10
Pelagiella sp.		x			Vorwald, 1983; Rees, 1984
Class Stenothecoida					
Stenothecoides elongata (Walcott)		x			Robison, 1964
Phylum Priapulida					
**Ottoia prolifica* Walcott	x	?		x	Conway Morris & Robison, 1986
**Selkirkia* cf. *columbia* Conway Morris	x				Conway Morris & Robison, 1988
+*Selkirkia spencei* Resser	x				Conway Morris & Robison, 1986
+*Aelkirkia willoughbyi* Conway Norris & Robison		x	x		Conway Morris & Robison, 1986
+*Selkirkia* sp.		x			Conway Morris & Robison, 1986
Phylum ?Pogonophora					
Hyolithellus sp.	x	x		x	Fig. 6.3; Vorwald, 1983
Phylum Annelida					
Palaeoscolex ratcliffei Robison	x				Robison, 1969
polychaete, gen. & sp. indet.	x				Robison, 1969; Conway Norris, 1979
Phylum Onychophora					
+*Aysheaia prolata* Robison			x		Robison, 1985
Phylum Arthropoda					
Class Crustacea					
**Canadasnis* cf. *perfecta* Walcott	x				Robison & Richards, 1981
+*Perspicaris*? *dilatus* Robison & Richards		x	x		Robison & Richards, 1981; Vorwald, 1983
+*Persnicaris*? *ellipsopelta* Robison & Richards				x	Robison & Richards, 1981
bradoriid, genus & species undet.		x			Vorwald, 1983
Class Trilobita					
Achylysopsis sp.	x				Gunther & Gunther, 1981
Alokistocare harrisi Robison		x	x		Robison, 1971; undescribed
Alokistocare idahoense Resser	x				Campbell, 1974; Gunther & Gunther, 1981
Alokistocare laticaudum Resser	x				Resser, 1939; Campbell, 1974
Alokistocare n. sp.				x	undescribed

Asaphiscus wheeleri Meek		x	x	x	Robison, 1964
Athabaskia bithus (Walcott)	x				Campbell, 1974; Gunther & Gunther, 1981
Athabaskia wasatchensis (Resser)	x				Gunther & Gunther, 1981
Baltagnostus centerensis (Resser)		x	x	x	Robison, 1964; Schwimmer, 1989
Bathyuriscus brighamensis Resser	x				Resser, 1939; Campbell, 1974
Bathyuriscus fimbriatus Robison		x	x	x	Robison, 1964
Bathyuriscus wasatchensis (Resser)	x				Gunther & Gunther, 1981
Bolaspidella contracta Robison				x	Robison, 1964
Bolaspidella drumensis Robison		x			Roblson, 1964
Bolaspidella housensis (Walcott)			x		Robison, 1964
Bolaspidella wellsvillensis (Lochman & Duncan)			x	x	Robison, 1971; undescribed
Brachyaspidion microps Robison		x	x		Robison, 1971
Brachyaspidion sulcatum Robison		x	x		Robison, 1964
Bythicheilus typicum Resser	x				Campbell, 1974
+*Chancia evax* Walcott	x				Resser, 1939
Clavaspidella anax (Walcott)	x				Resser, 1939
Cotalagnostus sp.				x	Robison, 1964
Dorypyge n. sp.	x				Campbell, 1974; Guntner & Gunther, 1981
Ehmaniella quadrans (Hall & Whitfield)	x				Resser, 1939
+*Elrathia kingii* (Meek)		x	x		Robison, 1964
+*Elrathina spencei* (Resser)	x				Campbell, 1974
Elrathina n. sp.			x		Gunther & Gunther, 1981
Glossopleura bion (Walcott)	x				Campbell, 1974; Gunther & Gunther, 1981
Glossopleura gigantea Resser	x				Resser, 1939; Gunther & Gunther, 1981
Glossopleura n. sp.	x				Campbell, 1974
Glossopleura utahensis Resser	x				Resser, 1939
Hemirhodon amplipyge Robison			x	x	Robison, 1964; undescribed
Hypagnostus parvifrons (Linnarsson)				x	Robison, 1964
Jenkinsonia varga Robison		x	x		Robison, 1971
Kochina vestita Resser	x				Gunther & Gunther, 1981
+*Kootenia gracilis* Resser	x				Resser, 1939
+*Kootenia mendosa* Resser	x				Campbell, 1974
+*Kootenia spencei* Resser	x				Campbell, 1974; Gunther & Guntner, 1981
+*Kootenia* n. sp.		x			Gunther & Gunther, 1981
Lejopyge lundgreni (Tullberg)				*x*	*Robison, 1984a*
Lejopyge rigbyi Robison				x	Robison, 1984a
Modocia brevispina Robison		x			Robison, 1964
Modocia laevinucha Robison			x	x	Robison, 1964, 1971
Modocia typicalis Resser				x	Robison, 1964
**Naraoia compacta* Walcott	x		x	x	Robison, 1984b; Conway Morris & Robison, 1988
Ogygopsis typicalis (Resser)	x				Gunther & Gunther, 1981
+*Olenoides brighamensis* Resser	x				Resser, 1939
+*Olenoides evansi* Resser	x				Resser, 1939
+*Olenoides nevadensis* (Meek)		x	x		Robison, 1971; Vorwald, 1983
+*Olenoides wahsatchensis* (Hall & Whitfield)	x				Resser, 1939
Oryctocara geikei Walcott	x				Resser, 1939; Campbell, 1974
+*Oryctocephalus walcotti* Resser	x				Campbell, 1974
+*Pagetia clytia* Walcott	x				Resser, 1939; Campbell, 1974
+*Pagetia* cf. *fossula* Resser	x				Campbell, 1974
+*Pagetia* n. sp.	x				Campbell, 1974
+*Peronopsis bonnerensis* (Resser)	x				Campbell, 1974
+*Peronopsis brighamensis* (Resser)	x				Resser, 1939; Robison, 1978
+*Peronopsis ferox* (Tullberg)			x		Robison, 1972
+*Peronopsis interstricta* (White)		x	x	x	Robison, 1964, 1982; Vorwald, 1983
+*Peronopsis segmenta* Robison			x		Robison, 1964, 1982
+*Ptychagnostus affinis* (Brogger)				x	Robison, 1984a
+*Ptychagnostus atavus* (Tullberg)			x	x	Robison, 1982, 1984a
+*Ptychagnostus germanus* Opik			x	x	Robison, 1984a
+*Ptychagnostus michaeli* Robison				x	Robison, 1984a
+*Ptychagnostus occultatus* Opik			x	x	Robison, 1984a
+*Ptychagnostus punctuosus* (Angelin)				x	Robison, 1984a
Schmalenseeia sp.				x	undescribed
Thoracocare minuta (Resser)	x				Robison Campbell, 1974
Trymataspis depressa Robison				x	Robison, 1964
Zacanthoides grabaui Pack	x				Campbell, 1974; Gunther & Gunther, 1981

Taxon					Reference
Zacanthoides idahoensis Walcott	x				Campbell, 1974; Gunther & Gunther, 1981
Zacanthoides n. sp.	x				Gunther & Gunther, 1981
Class uncertain					
**Alalcomenaeus* cf. *cambricus* Simonetta			x		Fig. 7.10
Cambropodus gracilis Robison		x			Robison, in press
**Branchiocaris pretiosa* (Resser)	x	x		x	Robison & Richards, 1981; Briggs & Robison, 1984
Ecnomocaris spinosa Conway Morris & Robison			x		Conway Morris & Robison, 1988
+*Emeraldella*? sp. Briggs & Robison	x				Briggs & Robison, 1984
+*Leanchoilia*? *hanceyi* Briggs & Robison				x	Briggs & Robison, 1984
+*Mollisonia* sp.			x		Gunther & Gunther, 1981
**Proboscicaris agnosta* Rolfe			x		Robison & Richards, 1981
Pahvantia hastata Robison & Richards			x		Robison & Richards, 1981
Pseudoarctolepis sharpi Brooks & Caster		x	x		Robison & Richards, 1981; Briggs & Robison
+*Sidneyia*? sp. Briggs & Robison			x		Briggs & Robison, 1984
+*Tuzoia guntheri* Robison & Richards				x	Robison & Richards, 1981
+*Tuzoia*? *peterseni* Robison & Richards			x		Robison & Richards, 1981
**Tuzoia retifera* Walcott	x				Robison & Richards, 1981
Utahcaris orion Conway Morris & Robison	x				Conway Morris & Robison, 1988
Undetermined genus Gunther & Gunther [merostome-like arthropod]	x				Gunther & Gunther, 1981
Undetermined arthropod 1 Conway Morris & Robison	x				Conway Morris & Robison, 1988
Undetermined arthropod 2 Conway Morris & Robison				x	Conway Morris & Robison, 1988
Phylum Echinodermata					
Class Ctenocystoidea					
Ctenocystis colodon Ubaghs & Robison		x			Ubaghs & Robison, 1988
Ctenocystis utahensis Robison & Sprinkle	x				Robison & Sprinkle, 1969
Class Edrioasteroidea					
Totiglobus? *lloydi* Sprinkle				x	Sprinkle, 1985
Class Eocrinoidea					
+*Gogia granulosa* Robison	x				Robison, 1965; Sprinkle, 1973
+*Gogia guntheri* Sprinkle	x				Sprinkle, 1973
+*Gogia spiralis* Robison		x	x		Robison, 1965; Sprinkle, 1973
+*Gogia* n. sp.	x				Sprinkle, 1976
Marjumicystis mettae Ubaghs & Robison				x	Ubaghs & Robison, 1985
n. gen. & n. sp.	x				undescribed
Class Homoiostelea					
Castericystis vali Ubaghs & Robison				x	Ubaghs & Robison, 1985
Castericystis? *sprinklei* Ubaghs & Robison			x		Ubaghs & Robison, 1988
Class Stylophora					
Cothurnocystis? *bifida* Ubaghs & Robison			x		Ubaghs & Robison, 1988
n. gen.? & n. sp.	x				Sprinkle, 1976
Phylum and class undetermined					
**Anomalocaris nathorsti* (Walcott)	x		x	x	Briggs & Robison, 1984; Conway Morris & Robison 1982, 1988
**Chancelloria eros* Walcott			x		Rigby, 1978
+*Chancelloria pentacta* Rigby			x		Rigby, 1978
+*Chancelloria* spp.		x	x	x	Rigby, 1978; undescribed
**Eldonia ludwigi* Walcott	x			x	Conway Morris & Robison, 1988
**Wiwaxia* cf. *corrugata* (Matthew)	x				Conway Morris & Robison, 1988
Undetermined worms Conway Morris & Robison		x	x	x	Conway Morris & Robison, 1986, 1988; undescribed
Trace fossils					
Bergaueria sp.		x			Fig. 11.5
Brooksella sp.	x				Willoughby & Robloon, 1979; undescribed
Cruziana sp.	x				undescribed
Gyrophyllites? sp.	x				undescribed

Neonereites? sp.	x			undescribed
Palaeophycus heberti (Saporta)	x	x	x	Ubaghs & Robison, 1985; undescribed
Planolites beverleyensis (Billings)	x	x	x	Robison, 1969; undescribed
Rusophycus sp.	x			undescribed
Tasmanadia sp.			x	Ubaghs & Robloon, 1985
Treptichnus sp.	x		?	Robison, 1969; Ubaghs & Robison, 1985; undescribed
Coprolites	x		x	Conway Morris & Roblson, 1986, 1988

Extant "Problematica" within or near the Metazoa

Gerhard Haszprunar[1], Reinhard M. Rieger[1] & Peter Schuchert[2]

This paper is dedicated to Univ.-Prof. Dr. Wolfgang Wieser on the occasion of his 65th birthday.

Abstract

This paper attempts to summarize the literature on certain extant "Problematica" within or near the Metazoa. The groups reviewed represent taxa with only one or a few species that do or did significantly contribute to the discussion on the early radiation of the Metazoa (*Proterospongia, Trichoplax, Buddenbrockia, Salinella*) or of the Bilateria (*Xenoturbella, Lobatocerebrum, Planktosphaera*). It is concluded that the problematic status of these taxa is due to poor knowledge of either their microanatomy (in particular in cases where only the original description is available: *Buddenbrockia, Salinella, Planctosphaera*) or due to incomplete understanding of their life-cycle (*Trichoplax, Xenoturbella, Lobatocerebrum*).

Introduction

In spite of the recent advances in our understanding of Metazoan phylogeny (see e.g. Conway Morris *et al.*, 1985), most extant phyla include some representatives or subgroups with unresolved phylogenetic affinities. In many cases the relationships along phyla are also still ambiguous or controversial. For instance the "aschelminth" phyla, the Chaetognatha and the recently discovered groups at the phylum level such as Gnathostomulida (Ax, 1956), Vestimentifera (Jones, 1981) or the Loricifera (Kristensen, 1983) all pose problems of their wider affinities.

Here we want to provide brief characterisations and a listing of the literature for only a few species, respectively small subgroups, of such problematic extant Metazoa. These species are either little known - some have not been found or reinvestigated since the original descriptions — or have special significance for concepts of metazoan phylogeny, particularly with respect to the question of the origin of the Bilateria. Most of these species are of microscopic size and move by ciliary gliding, both features traditionally being viewed as plesiomorphic characters within the Bilateria (see Rieger *et al.*, this volume).

Case studies

A. *Proterospongia haeckeli* (Choanoflagellata)

Kent (1880-1882) described *Proterospongia (=Protospongia) haeckeli* from a freshwater pond in England, and he regarded it as a connecting link between the Choanoflagellata and the Porifera; the close affinity between these groups had already been assumed by James-Clark (1868). This idea has been denied by Dubuscq & Tuzet (1938) and Tuzet (1963), who regarded *Proterospongia* as a so-called resting body of a freshwater sponge. Up to now about eight species (status of the species regains problematic, cf. Ertl, 1981) of this peculiar genus, which lives in both marine or freshwater habitats, have been described in varying detail (Table 1, see also review in Leadbeater, 1983).

Proterospongia species generally build colonies consisting of a few up to 100.000 cells. These are embedded in a gelatinous matrix, and in addition are linked together by the intercrossing collars of the cells forming a hexagonal pattern. Recent observations on the marine species *Proterospongia choanojuncta* (= "*Choanoeca perplexa*"; cf. Leadbeater, 1977, 1983) revealed a rather complicated life-cyle consisting of a motile colony, which produces monocellular "minute cells" of unknown significance, and a sessile monocellular stage, the latter has been described as another genus, *Choanoeca* (*Proterospongia* has priority). In addition, the *Choanoeca*-stage may produce monocellular swarmers by inequal cell-division.

Fine-structural features of all stages of the *Proterospongia* life-cycle are characteristic of the Choanoflagellata. In addition, all life-cycle phenomena are more or less well known from other choanoflagellates, and in particular from the related genera *Sphaeroeca* and *Cladospongia* (Leadbeater, 1983). According to Leadbeater (1983), *Proterospongia* can no longer be regarded as a direct "missing link" between flagellates and sponges or as a resting body of a freshwater sponge. On the other hand, fine-structural similarities between choanoflagellates and sponge choanocytes support the assumption of a close choanoflagellate-poriferan relationship (e.g. Hibbert, 1975). The mode of ontogeny and in particular the complex life-cycle of *Proterospongia* and the other choanoflagellates might prove to be of specific significance in further discussions of the parazoan level of organization (cf. Salvini-Plawen, 1978 and Buss, 1987:40 for stimulating discussion).

B. *Trichoplax adhaerens* (Placozoa)

The Placozoa are composed of three species: *Trichoplax adhaerens* F.E. Schulze, 1883 (cultivated as clones from the Red Sea and Bermuda at the Universities of Tübingen and Bochum, FRG), *Trichoplax* sp. (an unnamed second species being cultivated in Moscow, USSR), and *Treptoplax reptans* Monticelli, 1893 (not found since its original description). Table 2 gives an overview of our current status of knowledge.

The systematic position appears now fairly well settled among the early Metazoa between the level of Parazoa [lack of muscle and nerve cells: but recently it

[1]Institut für Zoologie der Universität, Technikerstrasse 25, A-6020 Innsbruck, Austria.
[2]Institut für Zoologie der Universität Basel, Rheinsprung 9, CH-4051 Basel, Switzerland.

Table 1: Original data on Choanoflagellata of the genus *Proterospongia* (species are alphabetically arranged)

Species	Habitat	Cell/Colony	Reference
Proterospongia:			
P. choanojuncta	marine	up to 120	Leadbeater 1977, 1983
P. dybsoeensis	marine	3-10	Grøntved 1956
P. haeckeli	freshw.	6-60	Kent 1880-1882
P.h. var. *clarki*	freshw	about 30	Schiller 1953
P. lackeyi	freshw.	about 24	Lackey 1959; Bourrelly 1968
P. nana	marine	about 40	Braarud 1935
P. pedicellata	freshw.	10.000-20.000	Oxley 1884, Skuja 1937
P. skujae	freshw.	4-16	Skuja 1956; Lackey 1959; Ertl 1968, 1981
P.s. var. *gracilis*	freshw.	4-16	Skuja 1956; Francé 1897

could be shown that Trichoplax contains several dozen cells along the whole margin that give a clearly positive signal with labeled antibodies against the neuropeptide RFamid (P. Schuchert, in preparation)] and that of the coelenterate phyla (position and orientation of accessory centriole, presence of true epithelial junction complexes). This view is further corroborated by recent molecular data (e.g. Christen *et al.*, this volume).

What is still unsettled is our understanding of the life-cycle. Are the placozoans indeed suitable models for a proposed phagocytella or planuloid-like ancestor of all higher Metazoa? Or can they be regarded as neotenic (progenetic) larva of an unknown and probably extinct group with a biphasic life-cycle consisting of a microscopic larva and a sessile adult similar to sponges? Apart from binary fission Thiemann & Ruthmann (1988) recently described other modes of asexual reproduction in the form of two sorts of multicellular swarmers (hollow/solid), consisting of dorsal (outside/posterior half) and ventral (inside/anterior half) cells with fibre cells between/inside. Sexual reproduction of placozoans is known from fine-structural descriptions and movies of yolky eggs, spermatocytes and early cleavage. However, meiosis has never been observed, the eggs may be asexual spores, and the very aberrant S-cells (spermatocytes?) could possibly be interpreted as cells with another function. While present data suggest that sexual reproduction may occur, it is still to be demonstrated in detail, before the placozoans can be accepted without hesitation as organisms relevant to our understanding of early metazoan evolution.

C. *Buddenbrockia plumatellae* and *Salinella salve* (Mesozoa)

The enigmatic worm *Buddenbrockia plumatellae* Schröder, 1910 is found attached to the peritoneum, or free in the coelom of freshwater phylactolaematan Bryozoa (Entoprocta). The animal is elongate and reaches about 3.6 mm lengths as adult. Its cuboidal epidermis lacks cilia, but a basal membrane is present. The muscle system consists of four longitudinal tracts only, between these tracts rows of single cells occupy median and lateral positions. The uniformly shaped internal cells are interpreted as oocytes, the presence of male gametes near the anterior end is doubtful. Locomotion is nematode-like, cleavage resembles that of the Mesozoa or the Nematoda again. However, alimentary, excretory and in particular the nervous system are wanting, contradicting nematode affinities. On the other hand, Schröder's (1910, 1912) tentative inclusion of *Buddenbrockia* among the Mesozoa (see below) is contradicted by the presence of true muscle fibres which are unknown in that group, thus the systematic position of this species remains enigmatic (Marcus, 1941). To our knowledge there are no recent studies on this animal.

Salinella salve, which has been described by Frenzel (1892) in water from a salt mine in Argentina, is another puzzling animal for which mesozoan affinities have been claimed. According to Frenzel's (1892) original description the microscopic, planuloid-like animal consists of a single layer of cells surrounding a gastric cavity with anus and mouth, whereby the latter is surrounded by long (locomotory?, sensory?) cilia. The cells appear to be ciliated at both their internal and external surfaces, a case unique in the animal kingdom. Juveniles are supposed to be monocellular, and conjugation of individuals produce cysts. Again the proposed affinities with the Mesozoa are contradicted by the presence of an alimentary tract. To our knowledge there are no further studies on this animal (McConnaughey, 1963).

The Mesozoa (=Planuloida: Orthonectida and Dicyemida) with which *Buddenbrockia* and *Salinella* have been tentatively allied, can without doubt be called "Problematica" (McConnaughey, 1963; Lapan & Morowitz, 1975). Their primitive organisation, consisting of a multiciliated epidermis, a single or few axonal cells within the body cavity, and several specialized cells composing the anterior end of the microscopic worms sharply contrasts with the complicated life cycle which includes larva, asexual and sexual stages. The discussion whether the organisation is primarily primitive or secondarily simplified due to endoparasitism in cephalopod kidneys (Dicyemida) or body cavities of various invertebrates (Orthonectida) is still open. However, the majority of authors favour the latter hypothesis. In addition, the monophyletic origin of both major subgroups is questionable. Ultrastructural studies in the 1960s (Brescian & Fenchel, 1965, 1967; Ridley, 1968; Kozloff, 1969) could not clear up the phylogenetic position of the Mesozoa. Moreover, recent studies on rRNA-sequences (Ohama *et al.*, 1984, Hori *et al.*, 1988) suggest the Mesozoa to be an early

Table 2: Original data on Placozoa (chronological arrangement)

Species	Method	Object	Reference
Trichoplax adhaerens	LM	histology	Schulze 1883, 1891
	LM	some details	Stiasny 1903
	LM	predation	Riedl 1959
	LM	locomotion	Kuhl & Kuhl 1966
	LM	biology	Miller 1971
	LM	histology	Grell 1971a
	LM	egg, cleavage	Grell 1971b
	TEM	cytology	Grell & Benwitz 1971
	LM	egg, cleavage	Grell 1972
	movie	whole, locomotion	Grell 1973a
	movie	reproduction	Grell 1973b
	movie	eggs, cleavage	Grell 1973c
	TEM	fibre cells	Grell & Benwitz 1974a
	TEM	eggs	Grell & Benwitz 1974b
	molec.	DNA-content, chromosomes	Ruthmann & Wenderoth 1975; Ruthmann 1977
	cytol.	dis-reaggregation	Ruthmann & Terwelp 1979
	SEM	whole	Rassat & Ruthmann 1979
	molec.	DNA-content	Ruthmann *et al.* 1981
	TEM	cytology	Grell 1981; Grell & Benwitz 1981
	TEM, exp.	locomotion	Klauser & Ruppert 1981; Klauser 1982
	LM	cultivation	Schulze & Grell 1984
	LM	cultivation	Grell 1983
	LM, TEM	reproduction	Grell 1984
	cytol.	differentiation	Schwartz 1984
	TEM	ventral epithelium	Ruthmann *et al.* 1986
	TEM	fiber cells	Behrendt & Ruthmann 1986
	TEM	cytophagy	Wenderoth 1986
	TEM	feeding vacuoles	Seravin & Gerasimova 1988
	TEM, LM	swarmers	Thiemann & Ruthmann 1988
	TEM	fiber cells	Thiemann & Ruthmann 1989
	LM	fauna	Grell & Lopez-Ochoterena 1989
	LM	growth, behavior	Pearse 1989
	LM/TEM	review	Grell & Ruthmann 1990
	molec.	rRNA	Christen *et al.* (this volume)
T. sp.	SEM	adults, vagrants	Ivanov *et al.* 1980
	LM	adults	Ivanov *et al.* 1981
	TEM	adults	Ivanov *et al.* 1982
	LM	cultivation	Shubravyi 1983
	LM	biology	Okshtein 1987
	LM	cultivation	Okshtein 1988
	LM	chromosomes	Birstein 1989
T. reptans	LM	histology	Monticelli 1893, 1896

Abbreviations: cytol - methods of cell biology; exp - experimental methods; molec - methods of molecular biology; LM - light microscopy, SEM - scanning electron microscopy, TEM - transmission electron microscopy.

metazoan offshoot without close affinities to any other extant phylum.

D. *Xenoturbella bocki* (Acoelomorpha/Deuterostomia?)

Xenoturbella bocki was originally found by Sixten Bock in 1915 on muddy bottoms off the Väderöarna Islands (Skagerrak) In about 100 m depth (also later only discovered there and off Britain; cf. Reisinger, 1960). In spite of collecting a large amount at specimens and data, Sixten Bock did not publish any of his observations before his death in 1946. It was Westblad (1949) who did describe this peculiar animal based on own observations and the refound original data (27 section series, 30 drawings, 70-80 photographs) of Sixten Bock as a quite large (up to 30 mm) peculiar (hence xenos = strange), primitive turbellarian worm. This view was primarily based on the turbellarian-like appearance of the animal, the lack of an anus and the obvious hermaphroditism. In contrast, Reisinger (1960) regarded *Xenoturbella* as a neotenic larva of a coelomopore (i.e. oligomeric or archicoelamate such as Pterobranchia, Echinodermata or Enteropneusta), calling attention to the completely intraepithelial nervous system, the enteropneust-like high and complex epidermis, the spermatozoa (primitive type), and to the

Table 3: Original data on problematic Metazoa (chronological arrangement)

GROUP: Species	Method	Object	Reference
MESOZOA (???):			
Buddendrockia	LM	anatomy, histology	Schröder 1910, 1912
plumatellae	LM	anatomy, histology	Marcus 1941
Salinella salve	LM	anatomy, histology	Frenzel 1892
TURBELLARIA - ACOELOMORPHA / DEUTEROSTOMIA (???):			
Xenoturbella	LM	anatomy, histology	Westblad 1949
bocki	LM	epidermis, statocyst	Reisinger 1960
	TEM	extracellular matrix	Pedersen & Pedersen 1986
	TEM	epidermis	Franzén & Afzelius 1987
	TEM	epidermis	Pedersen & Pedersen 1988
LOBATOCEREBROMORPHA (ANNELIDA - OLIGOCHAETA??):			
Lobatocerebrum	TEM	cuticle	Rieger & Rieger 1976
psammicola,	TEM	adhesive system	Rieger & Tyler 1979
L. sp.	LM, TEM	whole, phylogeny	Rieger 1980
	TEM	body wall	Rieger 1981
	TEM	gonoduct	Rieger 1988
ENTEROPNEUSTA (??):			
Plancosphaera	LM	anatomy, histology	Spengel 1932
pelagica	LM	anatomy	Van der Horst 1936
Pl. sp. 1?	LM	whole	Damas & Stiasny 1961
Pl. sp. 2?	LM	whole	Scheltema 1970
Pl. sp. 3?	LM	whole	Hadfield & Young 1983

Abbreviations: exp - experimental methods; molec - methods of molecular biology; LM - light microscopy, SEM - scanning electron microscopy, TEM - transmission electron microscopy.

statocyst which resembles those found in certain holothurians.

Recent fine-structural studies have rekindled the old controversy. On one hand, the extracellular matrix system is of the type found in coelomates (Pedersen & Pedersen, 1986). On the other, epidermal fine-structure is different from that of Enteropneusta (Pedersen & Pedersen, 1988). Moreover, the ultrastructure of the ciliary system, in particular the tips, closely resemble conditions found in the acoelomorph (Nemertodermatida and Acoela; see Smith & Tyler, 1985) turbellarians (Franzén & Afzelius, 1987). The ciliary rootlet system apparently shows affinities to both, acoelomorphs and enteropneusts (Rieger *et al.*, 1990).

In addition, despite a common origin of the Acoelomorpha and the remaining Turbellaria respectively, the monophyly of the Platyhelminthes is difficult to demonstrate and therefore presently uncertain (Smith *et al.*, 1986; Rohde *et al.*, 1988). Indeed, aside from the turbellarian-like appearance, many structures (e.g. ciliary system, rhabdoids, frontal gland system, nervous system, excretory system, cleavage-type) of the Acoelomorpha differ considerably when compared with the remaining Turbellaria. Molecular data concerning this qestion are still wanted.

Thus, at present the systematic position of *Xenoturbella* appears intermediate between the Acoelomorpha and certain deuterostomes. Should further evidence corroborate such an unorthodox conclusion, then *Xenoturbella* will certainly become again (cf. Jägersten, 1955) a key animal for our understanding of the early radiation of the Bilateria.

E. *Lobatocerebrum psammicola* (Annelida?)

The meiobenthic turbellarian-like *Lobatocerebrum psammicola* Rieger, 1980 and another undescribed species have been reported from sandy and muddy bottoms off the North Carolina coast (USA) by Rieger (1980, 1981). A third species of this genus, mentioned by Rieger (1980), occurs in shallow subtidal sands in the Red Sea, near Eilat (Israel). Kristensen (1983) refers to yet another specimen similar to *Lobatocerebrum* from the Skagerrak. However, the life observations of this single specimen (courtesy Prof. R. Kristensen) suggest to us that it to belong to a different genus or even a more distantly related, new group.

Lobatocerebrum is a slender worm of 2-4 mm length which moves by ciliary gliding. The presence of a microvillar-type cuticle (1 μm thick) with well developed epicuticular membrane, with a grid of (collageneous?) crossing fibres (see Rieger & Rieger, 1976: Rieger, 1981) together with the complete body ciliation is unique among the Bilateria. The completely acoelomate worm has a large brain with complex lobes (hence the name) and a protonephridial excretory system. Whereas a circulatory system is lacking, the anus-hindgut system is fully developed. The animals are hermaphrodites, the elongated sperm-cells (modified type sensu Franzén 1956) are uniflagellate, eggs are entolecithal. A single male gonoduct opens mid-dorsally, a female canal system is lacking (Rieger, 1980, 1981, 1988).

As already discussed in detail by Rieger (1980), the cuticle-type, the epidermal fine-structure, the anus-hindgut system, and in particular the organisation of the male gonoduct contradict any attempt to include *Lobatocerebrum* in the Turbellaria, in spite the striking turbellariomorph appearence (see Ivanov, 1988, for a contrasting view). Based on the particular sign of these forementioned features, the genus had been interpreted originally as an aberrant family with annelid affinities (Rieger, 1980) While the incorporation of the genus in the Annelida-Oligochaeta remains questionable (see Rieger 1988) the lobatocerebrid organisation as a whole suggests a derivation by progenesis (see Rieger, 1980, 1988; Westheide, 1987) from a vermiform coelomate spiralian ancestor near the radiation of the Annelida. Indeed, *Lobatocerebrum* might represent a new phylum (Lobatocerebromorpha nom. nov.) close to other spiralian coelomate phyla such as Echiura, Sipuncula or Annelida. If so, its turbellarian-like appearance may serve as a model for possible origins of acoelomate Bilateria.

Planctosphaera

F. *Planctosphaera pelagica* (Enteropneusta?)

The paper by Spengel (1932), published years after his death, described in detail a peculiar pelagic organism of about 10 mm diameter called *Planctosphaera pelagica* from tropical North Atlantic deep-waters (250 m). Later specimens up to 28 mm in length have been found. The animal is spherical, transparent, and the surface is covered by branching, arborescent ciliary tracts. The internal anatomy is bilaterally symmetrical. An U-shaped gut, five coelomic cavities and a heart vesicle are present. Two coelomopores open near the anus. Two "internal sacs" (infundibula) of ectodermal origin arise laterally. Gonads have not been described.

In all subsequent investigations a similarity between *Planctosphaera* and the tornaria larvae of enteropneusts was noted. However, whereas Van der Horst (1936) regarded *Planctosphaera* as a neotenic larva of an (unknown or extinct) class of Hemichordata, more recent authors tend to regard the animals simply as specialized larva of a (unknown?) deep-sea enteropneust, having enlarged its ciliary bands due to its size. It is also not clear whether the different specimens found world-wide (e.g. Hadfield & Young, 1983) represent different stages of one species or (more probably) represent larvae of different species.

Thus, the enteropneust affinities of *Planctosphaera* now appear to be settled. Nevertheless, the animals (larvae?) remain problematic with respect to possible adult organization.

Conclusions

These examples demonstrate two main reasons for the enigmatic status of neontological problematica: (1) Poor knowledge of their microanatomy, particularly in cases where the animals could not be reinvestigated after their original description (e.g. *Buddenbrockia, Salinella, Planctosphaera*). (2) Lack of understanding of the complete life-cycle in forms which are quite well studied by means of electron microscopy or other modern techniques (e.g. *Trichoplax, Xenoturbella, Lobatocerebrum*). An intimate knowledge of the life history is significant, because heterochrony (e.g. paedomorphosis, progenesis, neoteny sensu Gould, 1977) is turning out to be a major driving force in macroevolution of many Metazoa (e.g. Gould, 1977; Westheide, 1987, McKinney, 1988).

Therefore we want to stimulate a more careful consideration of life-cycle processes of such problematic animals, which figure significantly in theories of metazoan origin and evolution. In our view the various life-cycle strategies (direct development, bi- or polyphasic life-cycle, asexual-sexual reproduction) have not been evaluated critically enough for models of early metazoan phylogeny (see Rieger *et al.* in this volume for further considerations).

REFERENCES

Ax, P. 1956. Die Gnathostomulida, eine rätselhafte Wurmgruppe aus dem Meeres sand. *Abhandlungen der Akademie der Wissenschaften und Literatur Mainz, mathematisch-naturwissenschaftliche Klasse* **8**, 1-32.

Behrendt, G. & Ruthmann, A. 1986. The cytoskeleton of the fiber cells of *Trichoplax adhaerens* (Placozoa). *Zoomorphology* **106**, 123-130.

Birstein, V.J. 1989. On the karyotype of *Trichoplax* sp. (Placozoa). *Biologisches Zentralblatt* **108**, 63-67.

Braarud, T. 1935. The "Ost" expedition to the Denmark Strait, 1929. II. The phytoplankton and its conditions of growth (including some qualitative data from the Arctic in 1930). *Hvalrdets skrifter* **10**, 5-173.

Brescian, J. & Fenchel, T. 1965. Studies on dicyemid Mesozoa. 1. The fine structure of the adult (the nematogen and rhombogen stage). *Videnskabelige Meddelelser Dansk naturhistorik Forening* **124**, 367-408.

Brescian, J. & Fenchel, T. 1967. Studies on dicyemid Mesosoa. 2. The fine structure of the infusiform larva. *Ophelia* **4**, 1-17.

Bourelly, P. 1968. *Les Algues d'eau douce.* Tome II. Les Algues jaunes et brunes. Paris: N. Boubee et Cie.

Buss, L.V. 1987. *The Evolution of Individuality.* Princeton: University Press.

Christen, R., Ratto, A., Baroin, A., Perasso, R., Grell, K.G. & Adoutte A. Origin of metazoans, a phylogeny deduced from sequences of the 28S ribosomal RNA (this volume).

Conway Morris, S., George, J.D., Gibson, R. & Platt, K.M. (Eds.) 1985. *The Origins and Relationships of lower Invertebrates. Systematics Association Special Volume* **28**.

Damas, D. & Stiasny, G. 1961. Les larves planctonique d'enteropneustes. *Memoirs de l'Academie Royale, Classe des Sciences* **15**, 1-68.

Duboscq, O. & Tuzet, O. 1938. La collerete des chonaocytes chez les ésponges calcaires hétérocoeles. *Comptes rendus de la Societée de Biologie* **129**, 296-298.

Ertl, M. 1968. Über das Vorkommen von *Protospongia haeckeli* in der Donau und einige Bemerkungen zur Taxonomie dieser Art. *Archiv für Protistenkunde* **111**, 18-23.

Ertl, M. 1981. Zur Taxonomie der Gattung *Proterospongia* Kent. *Archiv für Protistenkunde* **124**, 259-266.

France, R. 1897. *Der Organismus der Craspedomonaden.* Budapest.

Franzén, A. 1956. On spermiogenesis, morphology of the spermatozoon and biology of fertilization among invertebrates. *Zoologisca Bidrag fran Uppsala* **31**, 355-482.

Franzén, A. & Afzelius, B.A. 1987. The ciliated epidermis of *Xenoturbella bocki* (Platyhelminthes, Xenoturbellida), with some phylogenetic considerations. *Zoologica Scripta* **16**, 9-17.

Frenzel, J. 1892. Untersuchungen über die mikroskopische Fauna Argentiniens. *Archiv für Naturgeschichte* **58**, 66-96.

Gould, S.J. 1977. *Ontogeny and Phylogeny.* Cambridge, Massachusetts: The Belknap Press of Harvard University.

Gottschalk, C. 1971. Zur Frage stammesgeschichtlicher Beziehungen zwischen Plathelminthen und Mesozoen. *Parasitologische Schriftenreihe* **21**, 29-32.

Grell, K.G. 1971a. *Trichoplax adhaerens* F.E. Schulze und die Entstehung der Metazoen. *Naturwissenschaftliche Rundschau* **24 (4)**, 160-161.

GRELL, K.G. 1971b. Embryonalentwicklung von *Trichoplax adhaerens* F.E. Schulze. *Naturwissenschaften* **58 (11)**, 570.

GRELL, K.G. 1972. Eibildung und Furchung von *Trichoplax adhaerens* F.E. Schulze. *Zeitschrift für Morphologie der Tiere* **73**, 297-314.

GRELL, K.G. 1973a. *Trichoplax adhaerens* (Placazoa). Bewegung und Organisation. Encyclopedia cinematographica (G. Wolf ed.) E 1918, Göttingen.

GRELL, K.G. 1973b. *Trichoplax adhaerens* (Placazoa). Vermehrung. Encyclopedia cinematographica (G. Wolf ed.) E 1919, Göttingen.

GRELL, K.G. 1973c. *Trichoplax adhaerens* (Placazoa). Eizellen und Furchungsstadien. Encyclopedia cinematographica (G. Wolf ed.) E 1920, Göttingen.

GRELL, K.G. 1981. *Trichoplax adhaerens* and the origin of Metazoa. In *L'origine dei grandi phyla dei Metazoi*. pp. 107-121. Atti dei Convegni Lincei **49**.

GRELL, K.G. 1983. A new cultural method for *Trichoplax adhaerens* F.E. Schulze. *Zeitschrift für Naturforschung* **C38**, 1072.

GRELL, K.G. 1984. Reproduction of Placozoa. *Advances in Invertebrate Reproduction* **3**, 541-546.

GRELL, K.G. & BENWITZ, G. 1971. Die Ultrastruktur von *Trichoplax adhaerens* F.E. Schulze. *Cytobiology* **4**, 216-240.

GRELL, K.G. & BENWITZ, G. 1974a. Spezifische Verbindungsstrukturen der Faserzellen von *Trichoplax adhaerens* F.E. Schulze. *Zeitschrift für Naturforschung* **C29**, 790-790a.

GRELL, K.G. & BENWITZ, G. 1974b. Elektronenmikroskopische Beobachtungen über das Wachstum der Eizelle und die Bildung der "Befruchtungsmembran" von *Trichoplax adhaerens* F.E. Schulze (Placozoa). *Zeitschrift für Morphologie der Tiere* **79**, 295-310.

GRELL, K.G. & BENWITZ, G. 1981. Ergänzende Untersuchungen zur Ultrastruktur von *Trichoplax adhaerens* F.E. Schulze (Placozoa). *Zoomorphology* **98**, 47-67.

GRELL, K.G., GRUNER, H.-E. & KILIAN, E.F. 1980. Einführung. In *Lehrbuch der Speziellen Zoologie* Vol.1: *Wirbellose Tiere*. (ed. H.-E. Gruner), pp. 15-156. Jena: Gustav Fischer Verlag.

GRELL, K.G. & LOPEZ-OCHOTERENA E. 1989. *Trichoplax adhaerens* F.E. Schulze (Phylum Placzoa) in the Mexican Caribbean Sea. *Anales del Centro de Ciencias del Mar y Limnologia (Universidad Nacinal Autonomia de Mexico* **14**, 255-256.

GRELL, K.G. & RUTHMANN, A. 1990. Placozoa. In *Microscopic Anatomy of Invertebrates*, vol. 2, chapter 2 (ed. F.W. Harrison and J. Westphal), pp. (in press). New York: A. Liss.

GRONTVED, J. 1956. Planktological contributions. II. Taxonomical studies in some Danish coastal localities. *Meddelelser fra Danmarks Fiskeri- og Havundersogelser* **1(12)**, 13 pp.

HADFIELD, M. G. & YOUNG, R.E. 1983. *Planctosphaera* (Hemichordata, Enteropneusta) in the Pacific Ocean. *Marine Biology* **73**, 151-153.

HIBBERT, D. 1975. Observation on the ultrastructure of the choanoflagellate *Codosiga botrytis* (Ehr.) Saville-Kent, with special reference to the flagellar apparatus. *Journal of Cell Science* **17**, 191-219.

HORI, H.I, MUTO, A., OSAWA, S., TAKAI, M., LUE, K.-Y. & KAWAKATSU, M. 1988. Evolution of Turbellaria as deduced from 5S ribosomal RNA. *Forschritte der Zoologie* **36**, 163-167.

IVANOV, A.V. 1988. On the early evolution of the Bilateria. *Fortschritte der Zoologie* **36**, 349-352.

IVANOV, D.L., MALAKHOV, V.V., PRILEPSKY, G.V. & TZETLIN, A.B. 1982. Fine morphology and ultrastructure of a primitive multicellular organism, *Trichoplax* sp. 2. Ultrastructure of adult individuals. *Zoologicheskii Zhurnal* **61**, 645-652. [in Russian, with English summary].

IVANOV, D.L., MALAKHOV, V.V. & TZETLIN, A.B. 1980. Fine morphology and ultrastructure of a primitive multicellular organism, *Trichoplax* sp. 1. Morphology of adults and vagrants by the data of scanning electron microscopy. *Zoologicheskii Zhurnal* **59**, 1765-1767. [in Russian, sith English summary].

IVANOV, D.L., MALAKHOV, V.V. & TZETLIN, A.B. 1981. A finding of a primitive multicellular organism *Trichoplax* sp. *Zoologicheskii Zhurnal* 60, 1735-1739. [in Russian, with English summary].

JAMES-CLARK, H. 1868. On the Spongiae Ciliatae as Infusoria Flagellata; or observations on the structure, animality, and relationship of *Leucosolenia botryoides*, Bowerbank. *Annual Magazine of Natural History* **1**, 133-142, 188-215, 250-264.

JÄGERSTEN, G. 1955. On the early phylogeny of the Metazoa. The Bilaterogastraea-theory. *Zoologiski Bidrag fran Uppsala* **30**, 321-354.

JONES, M.L. 1981. *Riftia pachyptila*, new genus, new species, the vestimentiferan worm from the Galapagos Rift geothermal vents (Pogonophora). *Proceedings of the Biological Society of Washington* **93**, 1295-1313.

KENT, W.S. 1880-1882. *Manual of the Infusoria*. Vols. 1-3 London: D. Bogue.

KLAUSER, M.D. 1982. An ultrastructural and experimental study of locomotion in *Trichoplax adhaerens*. Unpublished Master Thesiis. Clemson University: Department of Zoology.

KLAUSER, M.D. & RUPPERT, E.E. 1981. Non-flagellar motility in the phylum Placozoa: Ultrastructural analysis of the terminal web of *Trichoplax adhaerens*. *American Zoologist* **21**, 1002 (abstract).

KOZLOFF, E.N. 1969. Morphology of the orthonectid *Rhopalura ophiocomae*. *Journal of Parasitology* **55**, 171-195.

KRISTENSEN, R.M. 1983. Loricifera, a new phylum with Aschelminthes characters from the meiobenthos. *Zeitschrift für zoologische Systematik und Evolutionsforschung* **21**, 163-180.

KUHL, W. & KUHL, G. 1966. Untersuchungen über das Bewegungsverhalten von *Trichoplax adhaerens* F.E. Schulze (Zeittransformation: Zeitraffung). *Zeitschrift für Morphologie und ökologie der Tiere* **56**, 417-435.

LACKEY, J.B. 1959. Morphology and biology of a new species of *Protospongia*. *Transactions of the American microscopic Society* **78**, 202-206.

LAPAN, E.A. & MOROWITZ, H. 1975. The Mesozoa. *Scientific American* **227**, 94-101.

LEADBEATER, B.S.C. 1977. Observations on the life-history and ultrastructure of the marine choanoflagellate, *Choanoeca perplexa* Ellis. *Journal of the Marine Biological Association of the U.K.* **57**, 285-301.

LEADBEATER, B.S.C. 1983. Life-history and ultrastructure of a new marine species of *Proterospongia* (Choanoflagellida). *Journal of the Marine Biological Association of the U.K.* **63**, 135-160.

MARCUS, E. 1941. Sobre Bryozoa do Brazil (in Portugese). *Boletim da facultade de Filosofia, Ciencias e Letras, Universidat de Sao Paulo* **22**, *Zoologia* **5**, 3-208.

MCCONNAUGHEY, B.H. 1963. The Mesozoa. In *The Lower Metazoa* (ed. E.C. Dougherty), pp. 151-165. Berkeley and Los Angeles: University of California Press.

MCKINNEY, M.L. (ed.) 1988 *Heterochrony in Evolution. A multidisciplinary approach. Topics in Geobiology*, 7. New York: Plenum Press.

MILLER, R.L. 1971. Observations on *Trichoplax adhaerens* Schulze. *American Zoologist* **11**, 698-699a.

MONTICELLI, F.S. 1893. *Treptoplax reptans* n.g., n.sp. *Atti dell'Accademia dei Lincei, Rendiconti* **(5) II**, 39-40.

MONTICELLI, F.S. 1896 Adelotacta zoologica. *Treptoplax reptans* Montic. *Mitteilungen der Zoologischen Station Neapel* **12**, 444-462.

OHAMA, T., KUMAZAKI, T., HORI, H. & OSAWA, S. 1984. Evolution of multicellular animals as deduced from 5S rRNA sequences: a possible early emergence of the Mesozoa. *Nucleic Acid Research* **12**, 5101-5108.

OKSHTEIN, I.L. 1987. On the biology of *Trichoplax* sp. (Placozoa). *Zoologicheskii Zhurnal* **66**, 339-347. [in Russian, with English summary].

OKSHTEIN, I.L. 1988. New method of culturing of *Trichoplax* sp. (Placozoa). *Zoologicheskii Zhurnal* **67**, 923-926. [in Russian, with English summary].

OXLEY, F. 1884. On *Protospongia pedicellata*, a new compound infusorian. *Journal of the Royal Microscopical Society* **2**, 530-532.

PEARSE, V.B. 1989. Growth and behavior of *Trichoplax adhaerens*. First record of the phylum Placozoa in Hawaii (USA). *Pacific Science* **43**, 117-121.

PEARSE, V.B. 1989. Stalking the wild placzoan: Biogeography and ecology of *Trichoplax* in the Pacific. *American Zoologist* **29 (4)**, 175A (abstract).

PEDERSEN, K.J. & PEDERSEN, L.R. 1986. Fine structural observations of the extracellular matrix (ECM) of *Xenoturbella bocki* Westblad, 1949. *Acta Zoologica (Stockholm)* **67**, 103-114.

PEDERSEN, K.J. & PEDERSEN, L.R. 1988. Ultrastructural observations on the epidermis of *Xenoturbella bocki* Westblad, 1949; with a discussion of epidermal cytoplasmic filament systems of invertebrates. *Acta Zoologica (Stockholm)* **69**, 231-246.

RASSAT, J. & RUTHMANN, A. 1979. *Trichoplax adhaerens* F.E. Schulze (Placozoa) in the scanning electron microscope. *Zoomorphologie* **93**, 59-73.

REISINGER, E. 1960. Was ist *Xenoturbella*? *Zeitschrift für wissenschaftliche Zoologie* **164**, 188-198.

RIDLEY, R.K. 1968. Electron microscopical studies of dicyemid Mesozoa: 1. vermiform stages. *Journal of Parasitology* **54**, 957-998.

RIEDL, R. 1959. Beitrage zur Kenntnis der *Rhodope veranii*. 1. Geschichte und Biologie. *Zoologischer Anzeiger* **163**, 107-122.

RIEGER, R.M. 1980. A new group of interstitial worms, Lobatocerebridae nov. fam. (Annelida), and its significance for metazoan phylogeny. *Zoomorphologie* **95**, 41-84.

RIEGER, R.M. 1981. Fine structure of the body wall, the nervous system, and the digestive system of the Lobatocerebridae Rieger (Annelida) and remarks to the organization of the gliointerstitial systems in Annelida. *Journal of Morphology* **167**, 139-165.

RIEGER, R.M., HASZPRUNAR G. & SCHUCHERT P. On the origin of the Bilateria: Traditional views and recent alternative concepts. (This volume).

RIEGER, R.M. 1988. Comparative ultrastructure and the Lobatocerebridae: Keys to understand the phylogenetic relationship of Annelida and the acoelomates. - The Ultrastructure of Polychaeta. *Microfauna Marina* **4**: 373-382.

RIEGER, R.M. & RIEGER, G.E. 1976. Fine structure of the archiannelid cuticle and remarks on the evolution of the cuticle within the Spiralia. *Acta Zoologica (Stockholm)* **57**, 53-68.

RIEGER, R.M., RIEGER, G.E., TYLER, S. & SMITH, J.P.S. III. (1990) Turbellaria. In *Microanatomy of the Invertebrates* (ed. F. Harrison). New York: A.Liss (in press).

RIEGER, R.M. & TYLER, S. 1979. The homology theorem in ultrastructural research. *American Zoologist* 19, 655-664.

ROHDE, K., WATSON, N. & CANNON, L.R.G. 1988. Ultrastructure of epidermal cilia of *Pseudactinoposthia* sp. (Plathelminthes, Acoela); implications for the phylogenetic status of the Xenoturbellida and Acoelomorpha. *Journal of submicroscopic Cytology and Patholology* **20**, 759-767.

RUTHMANN, A. 1977. Cell differentiation, DNA content and chromosomes of *Trichoplax adhaerens* F.E. Schulze. *Cytobiology* **15**, 58-64.

RUTHMANN, A., BEHRENDT, G. & WAHL, R. 1986. The ventral epithelium of *Trichoplax adhaerens* (Placozoa): Cytoskeletal structures, cell contacts and endocytosis. *Zoomorphology* **106**, 115-122.

RUTHMANN, A., GRELL, K. G. & BENWITZ, G. 1981. DNA content and fragmentation of the egg-nucleus of *Trichoplax adhaerens* F.E. Schulze. *Zeitschrift für Naturforschung* **C36**, 564-567.

RUTHMANN, A. & TERWELP, U. 1979. Disaggregation and reaggregation of cells of the primitive metazoon *Trichoplax adhaerens*. *Cell and Differentiation* **13**, 185-198.

RUTHMANN, A. & WENDEROTH, H. 1975. Der DNA-Gehalt der Zellen bei dem primitiven Metazoon *Trichoplax adhaerens* F.E. Schulze. *Cytobiology* **10**, 421-431.

SALVINI-PLAWEN, L.v. 1978. On the origin and evolution of the lower Metazoa. *Zeitschrift für zoologische Systematik und Evolutionsforschung* **16**, 40-88.

SCHELTEMA, R.S. 1970. Two new records of *Planctosphaera* larvae (Hemichordata: Planctosphaeroidea). *Marine Biology* **7**, 47-48.

SCHILLER, J. 1953. Uber neue Craspedomonaden (Choanoflagellaten). *Archiv für Hydrobiologie* **48**, 248-259.

SCHRÖDER, O. 1910. *Buddenbrockia plumatellae*, eine neue Mesozoenart aus *Plumatella repens* L. und *Pl. fungosa* Pall. *Zeitschrift für wissenschaftliche Zoologie* **96**, 525-537.

SCHRÖDER, O. 1912. Zur Kenntnis der *Buddenbrockia plumatellae* Ol. Schröder. *Zeitschrift fur wissenschaftliche Zoologie* **102**, 79-91.

SCHULZE, F.E. 1883. *Trichoplax adhaerens*, nov.gen., nov.spec. *Zoologischer Anzeiger* **6**, 92-97.

SCHULZE, F.E. 1891. Über *Trichoplax adhaerens*. *Physikalische Abhandlungen der Akademie der Wissenschaften in Berlin* **1891**, 1-23.

SCHULZE, F.E. & GRELL, K.G. 1984. A new culture method for *Trichoplax adhaerens*. *Zeitschrift für Naturforschung* **C38**, 1072.

SCHWARTZ, F.E. 1984. Das radiopolare Differenzierungsmuster bei *Trichoplax adhaerens* F.E. Schulze. *Zeitschrift für Naturforschung* **C39**, 812-832.

SERAVIN, L.N. & GERASIMOVA, Z.P. 1988. Some ultrastructural features of *Trichoplax adhaerens* (type Placozoa) feeding on dense plant substrates. *Tsitologiya* **30(10)**, 1188-1193. [In Russian, with English summary].

SHUBRAVYI, O. 1983. An aquarium with artificial sea-water for keeping and breeding a primitive multicellular organism, *Trichoplax* and other marine invertebrates. *Zoologicheskii Zhurnal* 62, 618-621. [In Russian, with English summary].

SKUJA, H. 1937. Beitrag zur Algenflora Lettlands. I. *Acta Horti botanici Universitatis latviensis* **7**, 25-86.

SKUJA, H. 1956. Taxonomische und biologische Studien über das Phytoplankton schwedischer Binnengewasser. *Nova acta Regiae Societatis scientiarum upsaliensis* **16(3)**, 404 pp.

SMITH, J.P.S. & TYLER, S. 1985. The acoel turbellarians: kingpins of metazoan evolution or a specialized offshoot? In: *The Origins and Relationships of Lower Invertebrates.* (ed. S. Conway-Morris, J.D. George, R. Gibson & H.M. Platt), *Systematic Association Special Volume* **28**, 123-142.

SMITH, J.P.S., TYLER, S. & RIEGER, R.M. 1986. Is the Turbellaria polyphyletic? *Hydrobiologica* **132**, 13-21.

SPENGEL, J.W. 1932. *Planctosphaera pelagica*. Report on the scientific results of the "Michael Sars North Atlantic Deep- Sea Expedition 1910". vol.Ç5. Bergen.

STIASNY, G. 1903. Einige histologische Details über *Trichoplax adhaerens*. *Zeitschrift für wissenschaftliche Zoologie* **75**, 430-436.

THIEMANN, M. & RUTHMANN, A. 1988. *Trichoplax adhaerens* F.E. Schulze (Placozoa): The formation of swarmers. *Zeitschrift für Naturforschung* **C43**, 955-957.

THIEMANN, M. & RUTHMANN, A. 1989. Microfilaments and microtubules in isolated fiber cells of *Trichoplax adhaerens* (Placozoa). *Zoomorphology* **109**, 89-94.

TUZET, O. 1963. The phylogeny of sponges according to embryological, histological and serological data, and their affinities with the Protozoa and the Cnidaria. In *The Lower Metazoa.* (ed. E.C. Dougherty), pp. 129-148. Berkeley & Los Angeles: University of California Press.

VAN DER HORST, C.J. 1936. *Planctosphaera* and tornaria. *Quarterly Journal of microscopical Science* **78**, 605-613.

WENDEROTH, H. 1986. Transepithelial cytophagy by *Trichoplax adhaerens* F.E. Schulze (Placozoa) feeding on yeast. *Zeitschrift für Naturforschung* **C41**, 343-347.

WESTBLAD, E. 1949. *Xenoturbella bocki* n.g., n.sp., a peculiar, primitive turbellarian type. *Arkiv für Zoologie* **1**, 11-29.

WESTHEIDE, W. 1987. Progenesis as a principle in meiofauna evolution. *Journal of Natural History* **21**, 843-854.

On the origin of the Bilateria: traditional views and recent alternative concepts

R.M. Rieger[1], G. Haszprunar[1] & P. Schuchert[2]

Dedicated to Prof. Wolfgang Wieser on the occasion of his 65th birthday.

Abstract

Microscopic dimensions, ciliary locomotion and direct development are most often considered plesiomorphic features of ancestral Bilateria. This paper argues that several lines of phenotypic evidence suggest the possibility of other assumptions for the bilaterian stem species: a biphasic life cycle combining a microscopic, pseudocoelomate/acoelomate larva, which moves by ciliary action, with a filter-feeding, coelomate adult building up colonies of macroscopic dimensions. Acoelomate/pseudocoelomate and coelomate body cavities are viewed as adaptations within the life cycle. The acoelomate and pseudocoelomate phyla accordingly are viewed as having arisen by progenesis from larvae or juveniles of such a bilaterian stem species. Data relevant for this alternative proposal are discussed. It is suggested that animal size and life history evolution (e.g. heterochronies) at the first major radiation of the Bilateria need to be critically reconsidered by neontologists and palaeontologists alike.

Introduction

The many hypotheses proposed for the origin of the Bilateria fall for the most part under two contrasting models: one that the bilaterian ancestor had an acoelomate planuloid body form and the other model an ancestor with an oligomerous coelomate plan (see e.g. Barnes, 1985; Rieger, 1985). In either case the ancestor would be derived from a stem species that was itself at the diploblastic level of tissue organization, either a diploblastic larval form or a diploblastic adult (see Turbeville & Ruppert, 1983). Recent molecular data have produced some unorthodox cladograms among the most primitive metazoans. They suggest an independent origin of the Parazoa, Placozoa, Ctenophora and Cnidaria and of the bilaterian phyla and thus postulate at least a diphyletic origin of the Metazoa from the protistan stock (see Christen *et al.*, this volume). Traditional theories often discuss the independent origin of the Parazoa and the rest of the Metazoa (see e.g. summary in Mohn, 1984), or a diphyletic origin of the Bilateria within the coelenterate level (see e.g. Grell *et al.*, 1980 p. 154; Anderson, 1981, p. 163). Except for the recent proposal in a new textbook on invertebrates (Barnes *et al.*, 1988), no phenotypic character analysis to our knowledge does support the separation of the Parazoa, Placozoa, Ctenophora and Cnidaria from all other metazoan phyla, as is suggested by the 18S RNA sequence data set.

Traditionally, microscopic dimensions (mm-size range), ciliary locomotion and direct development have been assumed as plesiomorph features in theories on the origin of the Bilateria. Many authors have correlated these features with an acoelomate, turbellariomorph organization (see Table 1). Also the molecular data support the early branching of the acoelomate Platyhelminthes within the Bilateria (Hori *et al.*, 1988; Field *et al.*, 1988; Riutort *et al.*, 1989) and thus support indirectly the plesiomorphy of small size, acoelomate body cavity, benthic mode of life, ciliary locomotion, and direct development within the Bilateria at that level of organization. Other investigators have developed models for the evolution of the Bilateria, in which small size and ciliary locomotion are found originally in conjunction with an acoelomate/pseudocoelomate design and pelagic mode of life (e.g. Nielsen & Nørrevang, 1985; Nielsen, 1985, 1987) or even with true coelomic organization and a benthic mode of life (see e.g. Remane, 1958, 1963).

The primitiveness of indirect development (= biphasic life cycle with pelagic larva and benthic adult) for the Bilateria was first elaborated as a comprehensive theory by Jägersten (1955, 1959, 1972). His concept of the evolution of the metazoan life cycle was accepted, though modified, by various subsequent authors (e.g. Grell *et al.*, 1980; Nielsen, 1985, 1987). However, in one of the most recent proposals on the bilaterian stem species by Ax (1984, 1987) a set of special arguments is again listed in favour of the primitiveness of direct development within the Bilateria. It should be mentioned here that Ax's proposal formalized most clearly the various features of the bilaterian ancestor. Larger size (cm-size range) of the bilaterian stem species has been postulated but rarely (see Table 1). Apparently it is assumed by some proponents of the archicoelomate-concept (e.g. Siewing, 1985). This is also true for Lang's Ctenophore hypothesis (see Hyman, 1951) and Gutmann's (1981, 1989) gallertoid concept, in which additional muscular locomotion and body size in the cm-range, may be assumed as significant features in the ancestral Bilateria.

Biomechanical considerations until recently have corroborated the notion that a stem species with small size and coelomic organization in the adult can hardly be regarded as plesiomorphic within the Bilateria (e.g. Clark, 1964, 1979a). However, Mettam, (1985, p. 302-303) has summarized new evidence pointing that Clark's (1964, 1979a) postulate of a functional correlation of large body-size, coelomic organization and burrowing life style can be questioned (see also Gray, 1969).

Towards a new point of view

A possible synthesis of the differences concerning

[1]Institut für Zoologie der Universität Innsbruck, Technikerstrasse 25, A-6020 Innsbruck, Austria.
[2]Institut für Zoologie der Universität Basel, Rheinsprung 9, CH-4051 Basel, Switzerland.

Table 1: Some basic assumptions on the bilaterian stem species (archetype) during the last 40 years. Authors are alphabetically arranged. Microscopic means mm-size, macroscopic means cm-size.

Dimension (Adult)	Mode of locomotion (Adult)	Mode of development	Organization of body cavity (Adult)	Reference
microscopic	ciliary	direct	acoelomate	Ax 1984, 1985, 1987
microscopic	ciliary	direct	acoelomate	Beklemishev 1969
microscopic	ciliary	direct	acoelomate	Bergström 1986
microscopic	ciliary	direct	acoelomate	Boaden 1975
microscopic	ciliary	direct?	acoelomate	Clark 1963, 1964, 1979b
macroscopic?	burrowing	direct?	gallertoid	Gutmann 1981, 1989
microscopic	ciliary	direct	acoelomate	Hadzi 1963
microscopic	ciliary	direct	acoelomate	Hand 1963
microscopic	ciliary	direct	acoelomate	Hyman 1951, 1959
microscopic	ciliary	direct	acoelomate	Ivanov 1968, 1988; Ivanova-Kazas & Ivanov 1967, 1987
microscopic	ciliary	biphasic	"Protocoeloma"	Jägersten 1955, 1959, 1972
microscopic	ciliary	direct?	archicoelomate	Marcus 1958
microscopic	ciliary	direct	acoelomate	Möhn 1984
microscopic	ciliary	direct	"Trochaea"	Nielsen 1985, 1987; Nielsen & Nørrevang 1985
microscopic	ciliary	direct	acoelomate	Reisinger 1970, 1972
microscopic	ciliary	direct?	archicoelomate	Remane 1963
microscopic	sessile	biphasic	clonal coelomate	Rieger 1986, 1988
microscopic	ciliary	direct	acoelomate	Salvini-Plawen 1978, 1980, 1982
microscopic?	hemisessile?	biphasic?	archicoelomate	Siewing 1976, 1980, 1981, 1985
microscopic	ciliary	direct	acoelomate	Steinböck 1963, 1966

original size, mode of locomotion and body organization during the first major radiation of the Metazoa has been recently proposed for the Bilateria. This was based on new evidence from histological data (see Rieger, 1986, 1988) as well as from new considerations on the evolution of reproductive strategies (see e.g. Olive, 1985) and life histories (see e.g. Jägersten, 1972). The new proposal suggests (see Rieger, 1980, 1984, 1985, 1986, 1988 for details):

1. The bilaterian stem-species had a biphasic life-cycle with an acoelomate/pseudocoelomate, short-lived larva of microscopic dimensions and with ciliary locomotion. Such a larva is envisioned to alternate with a filter-feeding, clonal coelomate adult building up colonies of macroscopic dimensions (cm-range). The evolution of the two main types of body cavity organizations (acoelomate/pseudocoelomate and coelomate) are interpreted as specialisations during the ontogeny of the stem-species.

One line of support for this critique of the traditional view (see Hadzi, 1956 for a stimulating summary) on the bilaterian stem-species comes from the rediscovery of several parallel lines of evolution within the Annelida, leading to functional acoelomates in several of small (mm size range) polychaetes (cf. Westheide, 1967; Fransen, 1980, 1988; Smith P.R. *et al.*, 1986). Progenesis is the most likely evolutionary mechanism that produced such functional acoelomates in the Annelida (cf. Gould, 1977; Westheide, 1984, 1985, 1987; Rieger, 1988 for literature). Ultrastructural studies on the body cavity of such (polyphyletic) "archiannelids" suggest a strong correlation between meiobenthic size, ciliary gliding and a trend towards an acoelomate condition in an originally coelomate group (Fransen, 1980, 1988; Rieger, 1980, 1988; Smith P.R. *et al.*, 1986).

2. In analogy to such an evolutionary process in the Annelida, the acoelomate and pseudocoelomate phyla are thought to be derived through multiple events of progenesis (sensu Gould, 1977, see also Inglis, 1985 on the new concept of "evolutionary wave"). Extant acoelomate and pseudocoelomate phyla may actually be the only remnants of the biphasic life-cycle of the early bilaterians. This assumption could explain the traditional phylogenetic pattern based on various morphological or molecular data.

The following phylogenetic interpretations of histological features and of life strategies have primarily lead to the formulation of such a synthesis concerning the organization and biology of the bilaterian stem-species:

1) Recent reviews on the histological organization of the tissues filling the body cavity of acoelomates such as the primary clades of the Platyhelminthes (Rieger, 1981, 1985; Smith, J.P.S. *et al.*, 1986), the Nemertea (Turbeville & Ruppert, 1985) and the Gnathostomulida (Sterrer *et al.*, 1985) note a significant heterogeneity of these tissues. This heterogeneity suggests that the hypothesis of a common ancestral parenchymate body plan at the level of the bilaterian stem species can be put in question (Rieger, 1986, 1988).

2) Epithelio-muscle cells, of the plesiomorphic monociliated organization, similar to that of the Cnidaria, as well as the integration of muscle cells into various types of myoepithelia have now been described in many lophophorates and in deuterostomes. Such conditions occur also in certain coelomate spiralians (see summary in Rieger & Lombardi, 1987 and Fransen, 1988 for further discussion). This allows a reconsideration of the traditional view regarding the subepithelial muscle grid of small vermiform phyla to be a plesiomorphy in the Bilateria (see recent proposal by Ax, 1984, 1987): Epithelio-muscle cells apparently are lacking entirely in the body wall musculature of acoelomates such as Turbellaria (Ehlers, 1985; Smith & Tyler, 1985; Smith J.P.S. *et al.*, 1986), Nemertea (Turbeville & Ruppert, 1983), Gnathostomulida (Lammert, 1986), and meiobenthic pseudocoelomates such as the Gastrotricha (see Rieger *et al.*, 1974; Teuchert & Lappe, 1980).

A recent study by Sarnat (1984) on the cytology of muscle cells in planarians apparently supports the notion of a paedomorphic origin of the platyhelminth muscle system within the Bilateria.

3) Since Thorson's (1946) analysis of the major types of life cycles in marine invertebrates and Franzén's (1956) discussion of the plesiomorphy of external fertilization, it was especially Jägersten (1972) who further produced data and reviewed evidence for the widespread existence of the biphasic life cycle among the metazoan phyla. These facts led Jägersten to formulate his theory on the evolution of the metazoan life cycle. The pattern of occurence of the primitive monociliated (uniciliated) cell (see Tyler, 1981; Barnes, 1985 for terminology) in bilaterian larvae is now taken as an indication for an early dichotomy (Grell *et al.*, 1980; Nielsen, 1987) in the evolution of the biphasic life. Rieger (1984, 1986, 1988) has reemphasized Jägersten's concept, accepting the plesiomorphy of a biphasic life cycle not only for the bilaterian but also for the poriferan and radiate level of organization. At this point it should be emphasized that embryological (see e.g. Lemche & Tendal, 1977; Salvini-Plawen, 1978) and ultrastructural (see e.g.Rieger, 1976) data could be taken as evidence for the view that the biphasic life cycle of the sponges is derived convergently to that of the other Metazoa.

4) It has been proposed (Rieger & Rieger, 1976, see also Rieger, 1984 for further discussion) that true cuticles (= extracellular matrix secreted by the epidermis and surrounding all or most of the body wall of the organism) originated from the epidermal glycocalyx of small, vermiform Bilateria with ciliary locomotion. This proposal is in line with the traditional view that the bilaterian stem species was a vermiform organism in the mm-size range. However, Rieger (1984) pointed out that the origin of true cuticles preceeded the bilaterian level of organization: cuticles appear to be rare in the sponges, but they are common — and sometimes structurally complex — in adult cnidarians (e.g. the episarc of certain hydrozoans). The early appearance of true cuticles as mechanical support of large (cm) organisms already at the diplobastic (radiate) level of organization is thus in agreement with the assumption of the appearance of larger adults (cm-size range) during the very early phase of metazoan evolution. Therefore, the hypothesis about the origin of true cuticles from the epidermal glycocalyx can support the traditional view (Rieger & Rieger, 1976), but can be seen equally well to be in line with the new concept discussed here.

That the traditional view of the primitiveness of microscopic size, of ciliary locomotion, and of direct development for the bilaterian stemspecies warrants re-evaluation is also apparent when considering the occurence of these features among the histologically plesiomorphic metazoan phyla (Parazoa, Placozoa, Ctenophora, Cnidaria).

1. A biphasic life-cycle with microscopic, lecithotrophic larvae and macroscopic adults (often with colonial organization), is basic for extant Parazoa as well as for extant Cnidaria. The character combination of microscopic size and ciliary locomotion is typical for the larval forms of these groups, but is exceptional and clearly derived when occurring in the adult organization (e.g. Swedmark, 1964; Clausen, 1971; Clausen & Salvini-Plawen, 1986, 1987; Thiel, 1988). If the cnidarian level of organization is considered ancestral to the Bilateria (e.g. the Planula-theory, Hyman, 1951), a progenetic origin of the turbellariomorph Bilateria appears therefore likely.

2. The Ctenophora are exceptional among the primitive Metazoa, because of their unique mode of ciliary locomotion. They succeeded in having ciliary locomotion combined with macroscopic size and direct development. Deriving the Bilateria from the level of organization of the Ctenophora via larger turbellariomorph acoelomates has been attempted in the past (see above) and is now apparently reconsidered in Gutmann's (1981, 1989) gallertoid theory. This hypothesis might have advantages for an easy explanation of the origin of the muscle grid in vermiform Bilateria. However, at present too little is known about development of circular and longitudinal muscle cells in the body wall of lower Bilateria for a firm answer on that question. Since ciliary locomotion with combplates and macroscopic size appear to be interrelated in the ctenophoran body plan (see development of macrocilia in Tamm & Tamm, 1988a, b), the derivation of uniformly ciliated, turbellariomorph Bilateria from ctenophoran-like ancestors remains unlikely, due to the lack of any intermediate forms, fossil or extant.

3. Among the most primitive Metazoa only the Placozoa (*Trichoplax*) exhibit an adult organization (Grell, 1971a, b, 1981; Grell & Ruthmann, 1990), which corroborates the traditional view that the bilaterian stem-species was small in size and moved with uniformly distributed cilia. However, our present knowledge of the sexual reproduction of *Trichoplax* does not completely exclude the possibility that this organism might be a paedomorphic (sensu Gould, 1977) larva of a yet unknown or extinct primitive metazoan (Salvini-Plawen, 1978). In fact, the occurrence of sexual reproduction is still somewhat doubtful in this group (see Haszprunar, Schuchert & Rieger, this volume). Models the stem-species of the Bilateria based on the organization of *Trichoplax* need to take this uncertainty into account.

The question whether the bilaterian stem-species was microscopic, moving by ciliary locomotion and exhibiting direct development or whether it was equipped with a biphasic life cycle with microscopic larva and clonal, macroscopic adult (cm-size range) is particularly relevant in the light of recent advances. These are our knowledge of the Precambrian metazoan fossils such as the Ediacara fauna (see e.g. Glaessner, 1984; Seilacher, 1984; Conway Morris, 1989), new concepts in life history strategies and life cycle evolution (e.g. Stearns, 1982, 1987; Sibly & Calow, 1986), recent arguments about possible evolutionary strategies of asexual versus sexual reproduction and clonal versus aclonal organization (e.g. Boardman *et al.*, 1973; Jackson *et al.*, 1985; Buss, 1987) and, finally, recent advances in evolutionary biomechanics (Alexander, 1988; Wainwright, 1988).

Future research

As one aspect in further corroborating the assumption of the origin of the acoelomate and pseudocoelomate phyla through progenesis, we think it useful to investigate by electron microscopical techniques the body cavity organization, the body wall, and the nervous system of minute (mm-size), turbellariomorph organisms within the Bilateria, for which the acoelomate organization is known to be secondary. This is the case in some interstitial Annelida (see Westheide, 1967; Fransen, 1980, 1988; Kristensen & Nørrevang, 1982; Smith P.R. *et al.*, 1986) and in the meiobenthic Gastropoda: Rhodopidae, which have lost the pericardial/ heart system (Böhmig, 1893; Riedl, 1960; Salvini-Plawen, 1990). The ultrastructural data already available on the afore mentioned examples do not contradict the new proposal [see Rieger, 1988; Fransen, 1988; Smith P.R. *et al.*, 1986 on interstitial Annelida, unpublished TEM-micrographs of *Rhodope* (G. Haszprunar) and *Helminthope* (Rhodopidae) (R.M. and G.E. Rieger)].

An excellent case-study for testing the likelihood of the new proposal on the bilaterian stem species by Rieger (1986, 1988) should be a microanatomical analysis of the dwarf-male of *Bonellia viridis*. We have already begun such investigations in studying the fine structure of the coelomic lining and the excretory system in the adult male (see Schuchert & Rieger, 1990; Schuchert, 1990). The data show, as expected from the data on the Annelida (cited above), that progenesis does not necessarily result in a complete loss of the coelomic organization (see Rieger, 1985, Fig. 1a for the histological design of this organization). The *Bonellia* male is of striking turbellariomorph shape, the entire ciliation is composed of multiciliated cells, a circulatory system is lacking (obviously correlated with the presence of secondary protonephridia in the adult (see Schuchert, 1990)), yet it exhibits a coelomate body cavity with myoepithelial organization of the gastric muscle grid and fibre-type musculature in the body wall similar in design to the Polycladida and Tricladida (see Rieger, 1985 for literature on Turbellaria). Peculiar to the *Bonellia* male, apparently among all Bilateria, is the microanatomical relationship between the sarcoplasmatic portions of all three layers of muscle cells (see Schuchert & Rieger, 1990 for details). Further studies of the indifferent larva of *Bonellia* and the differentiation of the muscle system and its relationship to the extracellular matrix are necessary to unravel this puzzle.

REFERENCES

ALEXANDER, R. McN. 1988. *Elastic Mechanisms in Animal Movement.* Cambridge: University Press.

ANDERSON, D.I. 1981. Origins and relationships among the animal phyla. *Proceedings of the Linnean Society of New South Wales* **106**, 151-166.

AX, P. 1984. *Das phylogenetische System. Systematisierung der lebenden Natur aufgrund ihrer Phylogenese.* Stuttgart: Gustav Fischer Verlag.

AX, P. 1985. The position of the Gnathostomulida and Plathyelminthes in the phylogenetic system of the Bilateria. In *The Origins and Relationships of Lower Invertebrates* (ed. S. Conway Morris, J.D. George, H.M. Platte & R. Gibson), *Systematic Association Special Volume* **28**, 168-180.

AX, P. 1987. *The Phylogenetic System. The systematization of organisms on the basis of the phylogenesis* (Translation of Ax 1984 by R.P.S. Jefferies). Chicester, New York: John Wiley & Sons.

BARNES, R.D. 1985. Current perspectives on the origins and relationships of lower invertebrates. In *The Origins and Relationships of lower Invertebrates* (ed. S. Conway Morris, J.D. George, H.M. Platt & R. Gibson), *Systematic Association Special Volume* **28**, 360-367.

BARNES, R.S.K., CALOW, P. & OLIVE, P.J.W. 1988. *The Invertebrates: a new synthesis.* Oxford: Blackwell Scientific Publications.

BEKLEMISHEV, V.N. 1969. *Principles of Comparative Anatomy of Invertebrates. Vol. 1: Promorphology; Vol. 2: Organology.* Chicago: University of Chicago Press.

BERGSTRÖM, J. 1986. Metazoan evolution - a new model. *Zoologica Scripta* **15**, 189-200.

BOADEN, P.J.S. 1975. Anaerobiosis, meiofauna and early metazoan evolution. *Zoologica Scripta* **4**, 21-24.

BOARDMAN, R.S., CHEETHAM, A.H. & OLIVER, W.A. 1973. *Animal Colonies, Development, and Function through Time.* Stroudsberg PA: Dowden, Hutchinson & Ross.

BÖHMIG, L. 1893. Zur feineren Anatomie von *Rhodope veranii* Kölliker. *Arbeiten aus dem Zoologischen Institut Graz* **5 (2)**, 33-108.

BUSS, L.W. 1987. *The Evolution of Individuality.* Princeton: University Press.

CLARK, R.B. 1963. The evolution of the coelom and metameric segmentation. In *The Lower Metazoa* (ed. E.C. Dougherty), pp. 91-107, Berkely & Los Angeles: University of California Press.

CLARK, R.B. 1964. *Dynamics in Metazoan Evolution. The origin of the coelom and segments.* Oxford: Clarendon Press.

CLARK, R.B. 1979a. Functional correlation of the coelom. *Fortschritte in der zoologischen Systematik und Evolutionsforschung* **1**, 141-149.

CLARK, R.B. 1979b. Radiation of the Metazoa. In *The Origin and Evolution of Major Invertebrate Groups* (ed. M.R. House), *Systematics Association Special Volume* **12**, 55-102.

CLAUSEN, C. 1971. Interstitial Cnidaria: Present status of their systematics and ecology. In *Proceedings of the 1st International Conference on Meiofauna* (ed. N.C. Hulings). *Smithsonian Contribution in Zoology* **76**, 1-8.

CLAUSEN, C. & SALVINI-PLAWEN, L.v. 1986. Cnidaria. In *Stygofauna Mundi* (ed. L. Botosaneanu). Leiden: Brill/Backhuys.

CONWAY MORRIS, S. 1989. Early metazoans. *Science Progress, Oxford* **73**, 81-99.

EHLERS, U. 1985. *Das phylogenetische System der Plathelminthes.* Stuttgart: Gustav Fischer Verlag.

FIELD, K.G., OLSON, G.J., LANE, D.J., GIOVANNONI, S.J., GHISELIN, M.T., RAFF, E.C., PACE, N.R. & RAFF, R.A. 1988. Molecular phylogeny of the animal kingdom. *Science* **239**, 748-753.

FRANSEN, M.E. 1980. Ultrastructure of coelomic organization in annelids. I. Archiannelids and other small polychaetes. *Zoomorphology* **95**, 235-249.

FRANSEN, M.E. 1988. Coelomic and vascular system. In *The Ulträstructure of the Polychaeta* (ed. W. Westheide & C.O. Hermans). *Microfauna Marina* **4**, 199-213.

FRANZÉN, A. 1956. On spermiogenesis, morphology of the spermatozoon and biology of fertilization among invertebrates. *Zoologiski Bidrag fran Uppsala* **31**, 355-482.

GLAESSNER, M.F. 1984. *The Dawn of Animal Life. A biohistorical study.* Cambridge: University Press.

GOULD, S.J. 1977. *Ontogeny and Phylogeny.* Cambridge, Massachusetts: Harvard University Press.

GRAY, J.S. 1969. A new species of *Saccocirrus* (Archiannelida) from the west coast of North America. *Pacific Science* **23**, 238-251.

GRELL, K.G. 1971a. Über den Ursprung der Metazoen. *Mikrokosmos* **4**, 97-102.

GRELL, K.G. 1971b. *Trichoplax adhaerens* F.E. Schulze und die Entstehung der Metazoen. *Naturwissenschaftliche Rundschau* **24**, 160-161.

GRELL, K.G. 1981. *Trichoplax adhaerens* and the origin of Metazoa. *Atti dei Convegni Lincei (Roma)* **49**, 107-121.

GRELL, K.G., GRUNER, H.E. & KILIAN, E.F. 1980. Einführung. In *Lehrbuch der Speziellen Zoologie, vol. I. Wirbellose Tiere. 1. Teil: Einführung Protozoa, Placozoa, Porifera.* (ed. Gruner, H.E.). pp. 15-156. New York: A. Liss.

GRELL, K. G. & RUTHMANN, A. 1990. Placozoa. In *Microscopic Anatomy of Invertebrates*, vol. II, (ed. F.W. Harrison & J. Westphal). New York: A. Liss (in press).

GUTMANN, W.F. 1981. Relationships between invertebrate phyla based on functional-mechanical analysis of the hydrostatic skeleton. *American Zoologist* **21 (1)**, 63-81.

GUTMANN, W.F. 1989. *Die Evolution hydraulischer Konstruktionen. Organismische Wandlung statt altdarwinistischer Anpassung.* (ed. W. Kramer), Frankfurt/Main. 200 pp.

HADZI, J. 1956. Das Kleinsein und Kleinwerden im Tierreiche. Ein weiterer Beitrag zu meiner Turbellarientheorie der Knidarien. *Proceedings of the XIV International Congress in Zoology, Copenhagen*: 154-158.

HADZI, J. 1963. *The Evolution of the Metazoa.* Oxoford: Pergamon Press.

HAND, C. 1963. The early worm: a planula. In *The Lower Metazoa.* (ed. E.C. Dougherty), pp. 33-39. Berkeley and Los Angeles: University of California Press.

HORI, H., MUTO, A., OSAWA, S., TAKAI, M., LUE, K.-Y. & KAWAKATSU, M. 1988. Evolution of Turbellaria as deduced from 5S ribosomal RNA sequences. *Fortschritte der Zoologie* **36**, 163-167.

HYMAN, L.H. 1951. *The Invertebrates. Vol. II. Platyhelminthes and Rhynchocoela, The acoelomate Bilateria II.* New York: McGraw-Hill.

HYMAN, L.H. 1959. *The Invertebrates. Vol. V. Smaller Coelomate Groups: Chaetognatha, Hemichordata, Pogonophora, Phoronida, Ectoprocta, Brachiopoda, Sipunculida, The coelomate Bilateria.* New York: McGraw-Hill.

INGLIS, W.G. 1985. Evolutionary waves: patterns in the origins of animal phyla. *Australian Journal of Zoology* **33**, 153-178.

IVANOV, A.V. 1968. *The Origin of Metazoa.* Nauka, Leningrad.

IVANOV, A.V. 1988. On the early evolution of the Bilateria. *Fortschritte der Zoologie* **36**, 349-352.

IVANOVA-KAZAS, O.M. & IVANOV, A.V. 1967. On the origin of the Metazoa and their ontogeny. (A critical essay of the Zakhvatkin's synzoospora-hypothesis). *Trudy Zoological Institute, Academy of Science, USSR* **44**, 5-25. [in Russian, English summary].

IVANOVA-KAZAS, O.M. & IVANOV, A.V. 1987. The trochaea theory and phylogenetic significance of ciliate larvae. *Marine Biology Vladivostok* **2**, 6-21. [In Russian, English summary].

JACKSON, J.B.C., BUSS, L.W. & COOK, R.E. 1985. *Population Biology and Evolution of Clonal Organisms.* New Haven: Yale University Press.

JÄGERSTEN, G. 1955. On the early phylogeny of the Metazoa. The Bilaterogastraea-theory. *Zoologiski Bidrag fran Uppsala* **30**, 321-354.

JÄGERSTEN, G. 1959. Further remarks on the early phylogeny of the Metazoa. *Zoologiski Bidrag fran Uppsala* **33**, 79-108.

JÄGERSTEN, G. 1972. *Evolution of the Metazoan Life Cycle. A comprehensive theory.* London & New York: Academic Press.

KRISTENSEN, R.M. & NØRREVANG, A. 1982. Description of *Psammodrilus aedificator* sp.n. (Polychaeta), with notes on the Arctic interstitial fauna of Disko Island, W. Greenland. *Zoologica Scripta* **11**, 265-279.

LAMMERT, V. 1986. *Vergleichende Ultrastruktur-Untersuchungen an Gnathostomuliden und die phylogenetische Bewertung Ihrer Merkmale.* Unpublished Dissertation Universität Göttingen, 217 pp.

LEMCHE H. & TENDAL O.S. 1977. An interpretation of the sex cells and the early development in sponges, with a note of the terms acrocoel and spongocoel. *Zeitschrift für zoologische Systematik und Evolutionsforschung* **15**, 241-252.

MARCUS, E. 1958. On the evolution of the animal phyla. *Review of Biology* **33 (1)**, 24-58.

METTAM, C. 1985. Functional constraints in the evolution of the Annelida. In *The Origins and Relationships of Lower Invertebrates.* (ed. S. Conway Morris, J.D. George, R. Gibson & H.M. Platt), *Systematics Association Special Volume* **28**, 297-309.

MÖHN, E. 1984. *System und Phylogenie der Lebewesen. Band 1: Physikallsche, chemische und biologische Evolution, Prokaryonta, Eukaryonta (bis Ctenophora).* Stuttgart: Schweizerbart'sche Verlagsbuchhandlung.

NIELSEN, C. 1985. Animal phylogeny in the light of the trochaea theory. *Biological Journal of the Linnean Society* **25**, 243-299.

NIELSEN, C. 1987. Structure and function of metazoan ciliary bands and their phylogenetic significance. *Acta Zoologica (Stockholm)* **68**, 205-262.

NIELSEN, C. & NØRREVANG, A. 1985. The trochaeatheory: an example of life cycle phylogeny. In *The Origins and Relationships of Lower Invertebrates.* (ed. S. Conway Morris, J.D. George, H.M. Platt & R. Gibson) *Systematics Association Special Volume* **28**, 28-41.

OLIVE, P.J.W. 1985. Covariability of reproductive traits in marine Invertebrates: Implications for the phylogeny of the lower invertebrates. In *The Origins and Relationships of Lower Invertebrates.* (ed. S. Conway Morris, J.D. George, H.M. Platt & R. Gibson). *Systematics Association Special Volume* **28**, 42-59.

REISINGER, E. 1970. Zur Problematik der Evolution der Coelomaten. *Zeitschrift für zoologische Systematik und Evolutionsforschung* **8**, 81-109.

REISINGER, E. 1972. Die Evolution des Orthogons der Spiralier und das Archicoelomatenproblem. *Zeitschrift für zoologische Systematik und Evolutionsforschung* **10**, 1-43.

REMANE, A. 1958. Zur Verwandtschaft und Ableitung der niederen Metazoen. *Verhandlungen der Deutschen Zoologischen Gesellschaft Graz* **(1957)**, 179-196.

REMANE, A. 1963. The enterocoelic origin of the coelom. In *The Lower Metazoa.* (ed. E.C. Dougherty), pp. 78-90. Berkeley and Los Angeles: University of California Press.

RIEDL, R. 1960. Beiträge zur Kenntnis der *Rhodope veranii.* 2. Entwicklung. *Zeitschrift für wissenschaftliche Zoologie* **163**, 237-316.

RIEGER, R.M. 1976. Monociliated epidermal cells in Gastrotricha: Significance for concepts of early metazoan evolution. *Zeitschrift für zoologische Systematik und Evolutionsforschung* **14**, 198-226.

RIEGER, R.M. 1980. A new group of interstitial worms, Lobatocerebridae nov.fam. (Annelida) and its significance for metazoan phylogeny. *Zoomorphology* **95**, 41-84.

RIEGER, R.M. 1981. Fine structure of the body wall, nervous system, and digestive tract In the Lobatocerebridae Rieger, and the organization of the gliointerstitial system of the Annelida. *Journal of Morphology* **167**, 139-165.

RIEGER, R.M. 1984. Evolution of the cuticle in the lower Eumetazoa. In *Biology of the Integument, Vol. I. Invertebrates.* (ed. J. Bereiter-Hahn, A.G. Matoltsy & K. S. Richards), pp. 389-399. Berlin: Springer Verlag.

RIEGER, R.M. 1985. The phylogenetic status of the acoelomate organization within the Bilateria: a histological perspective. In *The Origins and Relationships of Lower Invertebrates* (ed. S. Conway Morris, J.D. George, H.M. Platt & R. Gibson), *Systematics Association Special Volume* **28**, 101-122.

RIEGER, R.M. 1986. Über den Ursprung der Bilateria: Die Bedeutung der Ultrastrukturforschung für ein neues Verstehen der Metazoenevolution. *Verhandlungen der Deutschen Zoologischen Gesellschaft* **79**, 31-50.

RIEGER, R.M. 1988. Comparative ultrastructure and the Lobatocerebridae: Keys to understand the phylogenetic relationship of Annelida and the Acoelomates. In *The Ultrastructure of the Polychaeta.* (ed. W. Westheide & C.O. Hermans), *Microfauna Marina* **4**, 373-382.

RIEGER, R.M. & LOMBARDI, J. 1987. Ultrastructure of coelomic lining in echinoderm podia: Significance for concepts in the evolution of muscle and peritoneal cell. *Zoomorphology* **107**, 191-208.

RIEGER, R.M. & RIEGER G.E. 1976. Fine structure of the archiannelid cuticle and remarks on the evolution of the cuticle within the Spiralia. *Acta Zoologica (Stockholm)* **57**, 53-68.

RIEGER, R.M., RUPPERT, E., RIEGER, G.E. & SCHOEPFER-STERRER, CH. 1974. On the fine structure of gastrotrichs with description of *Chordodasys antennatus* sp. nov. *Zoologica Scripta* **3**, 219-237.

RIUTORT, M., FIELD, K.G., RAFF, R.A. & BAGUNA, J. 1989. Phylogeny of the phylum Platyhelminthes by 18S rRNA sequences. *Abstract of the 2nd ESEB Congress, Roma 1989*; 54.

Salvini-Plawen, L. v. 1978. On the origin and evolution of the lower Metazoa. *Zeitschrift für zoologische Systematik und Evolutionsforschung* **16**, 40-88.

Salvini-Plawen, L.v. 1980. Phylogenetischer Status und Bedeutung der mesenchymaten Bilateria. *Zoologische Jahrbücher, Abteilung Anatomie* **103**, 354-373.

Salvini-Plawen, L.v. 1982. A paedomorphic origin of the oligomerous animals? *Zoologica Scripta* **11**, 77-81.

Salvini-Plawen, L.v. 1987. Mesopsammic Cnidaria from Plymouth (with systematic notes). *Journal of the Marine Biological Association of the U.K.* **67**, 623-637.

Salvini-Plawen, L. v. 1990. The status of Rhodopidae (Gastropoda, Euthyneura). *Malacological Review, Supplement* **5**, 123-136.

Sarnat, H.B. 1984. Muscle histochemistry of the planarian *Dugesia tigrina* (Turbellaria: Tricladida). Implication in the evolution of muscle. *Transactions of the American Microscopical Society* **103 (3)**, 284-294.

Schuchert, P. 1990. Fine structural Investigations on the excretory organs of the dwarf male of *Bonellia viridis* (Echiura). *Acta Zoologica (Stockholm)* **71**, 1-4.

Schuchert, P. & Rieger, R.M. 1990. Ultrastructural observations on the dwarf male of *Bonellia viridis* (Echiura). *Acta Zoologica (Stockholm)* **71**, 1-4.

Seilacher, A. 1984. Late Precambrian Metazoa: Preservational or real Extinctions? In *Patterns of Change in earth evolution*. (ed. H.D. Holland & A.F. Trendall). pp. 159-168. Berlin: Springer Verlag.

Sibly, R.M. & Calow P. 1986. *Physiological Ecology of Animals. An Evolutionary Approach.* Oxford: Blackwell Scientific Publications.

Siewing, R. 1976. Probleme und neuere Erkenntnisse in der Großsystematik der Wirbellosen. *Verhandlungen der Deutschen Zoologischen Gesellschaft* **1976**, 59-83.

Siewing, R. 1980. Das Archicoelomatenkonzept. *Zoologische Jahrbücher, Abtellung Anatomie* **103**, 439-482.

Siewing, R. 1981. Problems and results of research on the phylogenetic origin of Coelomata. *Atti dei Convegni Lincei* 49, 123-160.

Siewing, R. 1985. *Lehrbuch der Zoologie.* Band 2: Systematik. 3. Aufl. Stuttgart: Gustav Fischer Verlag, 1107 pp.

Smith, J.P.S. III & Tyler, S. 1985. The acoel turbellarians: kingpins of metazoan evolution or a specialized offshoot? In *The Origins and Relationships of Lower Invertebrates*. (ed. S. Conway Morris, J.D. George, R. Gibson & H.M. Platt), *Systematics Association Special Volume* **28**, 123-142.

Smith, J.P.S. III, Tyler, S. & Rieger, R.M. 1986. Is the Turbellaria polyphyletic? *Hydrobiologia* **132**, 13-21.

Smith, P.R., Lombardi, J. & Rieger, R.M. 1986. Ultrastructure of the body cavity lining in a secondary acoelomate, *Microphthalmus cf. listensis* Westheide (Polychaeta: Hesionidae). *Journal of Morphology* **188**, 257-271.

Stearns, S.C. 1982. The role of development in the evolution of life histories. In *Evolution and Development*. (ed. J.T. Bonner), pp. 237-258. Berlin: Springer Verlag.

Stearns, S.C. 1987. *The Evolution of Sex and its Consequences.* Basel: Birkhäuser.

Steinböck, O. 1963. Origin and affinities of the lower Metazoa: The "acoeloid" ancestry of the Eumetazoa. In *The Lower Metazoa.* (ed. E.C. Dougherty), pp. 40-54. Berkeley: University of California Press.

Steinböck, O. 1966. Die Hofsteniiden (Turbellaria acoela). Grundsätzliches zur Evolution der Turbellaria. *Zeitschrift für zoologische Systematik und Evolutionsforschung* **4**, 58-195.

Sterrer, W., Mainitz, M. & Rieger, R.M. 1985. Gnathostomulida: enigmatic as ever. In *The Origins and Relationships of Lower Invertebrates* (ed. S. Conway Morris, J.D. George, R. Gibson & H. M. Platt), *Systematics Association Special Volume* **28**, 181- 199.

Swedmark, B. 1964. The Interstitial fauna of marine sand. *Biological Review* **39**, 1-42.

Tamm, S. & S.L. Tamm 1988a. Development of macrociliary cells in *Beroe*. 1. Actin bundles and centriole migration. *Journal of Cell Science* **89**, 67-80.

Tamm, S.L. & Tamm S. 1988b. Development of macrociliary cells in *Beroe*. 2. Formation of macrocilia. *Journal of Cell Science* **89**, 81-95.

Teuchert, G. & Lappe, A. 1980. Zum sogenannten "Pseudocoel" der Nemathelminthes. Ein Vergleich der Leibeshöhlen von mehreren Gastrotrichen. *Zoologisches Jahrbuch, Abteilung Anatomie* **10**, 424-438.

Thiel, H. 1988. Cnidaria: In *Introduction to the Study of Meiofauna*. (ed. R.P. Higgins & H. Thiel), pp. 266-272. Washington: Smithsonian Institution Press.

Thorson, G. 1946. Reproduction and larval development of Danish marine bottom Invertebrates. *Meddelelser fra Kommissionen for Danmarks Fiskeri og Havundersogelser, Serie Plancton* **4**, 1-523.

Turbeville, J.M. & Ruppert E.E. 1983. Epidermal muscles and peristaltic burrowing in *Carinoma tremaphoros* (Nemertini): correlates of effective burrowing without segmentation. *Zoomorphology* **103**, 103-120.

Turbeville, J.M. & Ruppert E.E. 1985. Comparative ultrastructure and the evolution of Nemertines. *American Zoologist* **25**, 53-71.

Tyler, S. 1981. Development of cilia in embryos of the turbellarian *Macrostomum. Hydrobiologia* **84**, 231-239.

Wainwright, S.A. 1988. *Axis and Circumference.* Cambridge, Massachusetts: Harvard University Press, 132 pp.

Westheide, W. 1967. Monographie der Gattungen *Hesionides* Friedrich und *Microphthalmus* Mesznifow (Polychaeta, Hesionidae). Ein Beitrag zur Organisation und Biologie psammobionter Polychaeten. *Zeitschrift der Morphologie der Tiere* **61**, 1-159.

Westheide, W. 1984. The concept of reproduction in polychaetes with small body size: adaptations in interstitial species. In *Polychaete Reproduction*. (ed. A. Fischer & H.-D. Pfannenstiel), *Fortschritte der Zoologie* **29**, 265-287.

Westheide, W. 1985. The systematic position of the Dinophilidae and the archiannelid problem. In *The Origins and Relationships of Lower Invertebrates.* (ed. S. Conway Morris, J.D. George, R. Gibson, H.M. Platt, *Systematics Association Special Volume* **28**, 310-326.

Westheide, W. 1987. Progenesis as a principle in meiofauna evolution. *Journal of Natural History* **21**, 843-854.

Problematic reef-building sponges

Rachel A. Wood[1]

Abstract

Several groups of formerly problematic status have played a major role in reef-building from the base of the Cambrian to the early Cretaceous. The rediscovery of strikingly similar living sponges has since provided a basis for their biological placing, and the finding of incorporated spicules within stromatoporoids, chaetetids and sphinctozoans has shown them to be sponges which bear a massive calcareous skeleton. Spicule data, however, reveal a polyphyletic origin for these groups, with representatives from different orders of both the Demospongiae and Calcarea. Calcareous skeletons have thus been independently acquired at different times in many unrelated clades. Functional and constructional analyses of archaeocyaths also strongly support a poriferan affinity for this group.

Existing lines of evidence suggest that sponges can calcify with relative ease, as non-spicular skeleton production is a simple process and only a limited number of biomineralization mechanisms are employed. Although these calcareous skeletons form the most prominent aspect of the calcified sponge fossil record, they are therefore of little value when assessing evolutionary ancestries. These formerly problematic groups can be better defined as grades of organisation, where grades are here interpreted as reflections of non-systematically related soft-tissue and aquiferous filtration system organisations, which are probably ecologically determined. This interpretation also explains the presence of many seemingly intermediate forms.

The aspiculate archaeocyaths, often characterised by stalked and cup-shaped morphologies, a solitary functional system, complex porosity, the presence of septa and a strongly developed individuality inhabited soft-substrates in often unstable environmental conditions. As they lacked encrusting morphologies, they were unable to construct rigid reef frameworks. Archaeocyaths gave rise to some stromatoporoid- and sphinctozoan-grade forms, but all perished by the end of the Cambrian. Stromatoporoids and chaetetids are functionally modular. Their skeletal integration provides mechanical strength, stability, great plasticity of growth morphology and an ability to grow indefinitely. These factors, together with the ability to encrust hard substrates and to produce rapidly built skeletons, enabled forms with a stromatoporoid organisation to construct the framework of the first extensive reefs when they appeared in large numbers in the mid-late Ordovician, and continue to dominate reef-building for much of the Palaeozoic. Sphinctozoans are solitary or pseudomodular and being small and fragile were mainly reef-dwellers and bafflers. In contrast to the steady trend of increasing integration within corals during the Phanerozoic, calcified sponges show the early appearance (mid-Ordovician) of well integrated forms. However, corals became the dominant reef-builders through most of the Mesozoic and Tertiary, presumably due to the acquisition of more porous, less dense skeletons and faster calcification rates than stromatoporoids which was aided by their symbiotic relationship with zooxanthellae. There is some evidence to suggest that modular grades of calcified sponge may be derived from forms with a solitary organisation by paedomorphosis.

1. Introduction

Sponges occupy an unusual phylogenetic position as the most primitive of metazoans, but given their minimum number of cell types, they have evolved a considerable array of morphological forms and exploited a large range of ecological habitats. Although they possess no distinct tissues, their cellular activities achieve a significant degree of integration and they are able to secrete several skeletal components including a primary organic skeleton of collagenous strands known as spongin, and siliceous or calcareous spicules. Since most living sponges bear only these two structural components, their fossil record was also thought to be restricted to such forms, which were usually found as disaggregated spicule associations. However, the rediscovery of a diverse fauna of massive calcified sponges in caves and other cryptic habitats of tropical coral reefs have shown that sponges are able to produce an additional calcified skeleton and this has renewed interest in the biology, ecology and phylogeny of related fossils. Indeed, the fossil record has proved to be extremely rich in forms which combine these three skeletal elements and many of these have been gleaned from reef-builders of previously problematic status, including representatives from the stromatoporoids, chaetetids, spinctozoans and archaeocyaths.

Unlike most problematic groups, which tend to be rare, restricted in time and represented by only a few species, these groups are widespread both stratigraphically and geographically and as major reef-builders they offer a relatively complete fossil record with numerous examples (see Figure 1). With the confirmation of archaeocyaths from the middle Cambrian of Siberia, late Cambrian of Texas and upper Cambrian of Antarctica (Debrenne *et al.*, 1986), and sphinctozoans from the early Cambrian of Australia (Pickett & Jell, 1983) and North America (Debrenne & Wood, 1990), the previously discrete ranges of these groups now overlap. Likewise, the poorly known and transitional period in stromatoporoid history from the Lower Carboniferous to Upper Jurassic, which has often been cited to maintain a systematic segregation between the Palaeozoic and Mesozoic stromatoporoids, needs reinterpretation. Rich faunas of stromatoporoids, chaetetids and sphinctozoans have been recovered from the Permian of Tunisia (Termier & Termier, 1977), the Upper Triassic of Italy and Turkey (Wendt, 1974; Dieci *et al.*, 1974a and b) and from the Upper Permian, Triassic and Lower Jurassic of North and South America (George Stanley, pers. comm.). Stratigraphic obstacles for the interrelationships of these groups have thus been removed, and we are able to consider them as a whole.

The history of pre-Cretaceous sponges is intimately related to the geological history of reef-building, with the evolution of reef-building poriferans mirroring the general pattern of skeletal reefs (Fagerstrom, 1987). Poriferans dominated reef-building during the early

[1]Department of Earth Sciences, University of Cambridge, Downing St., Cambridge CB2 3EQ, U.K.

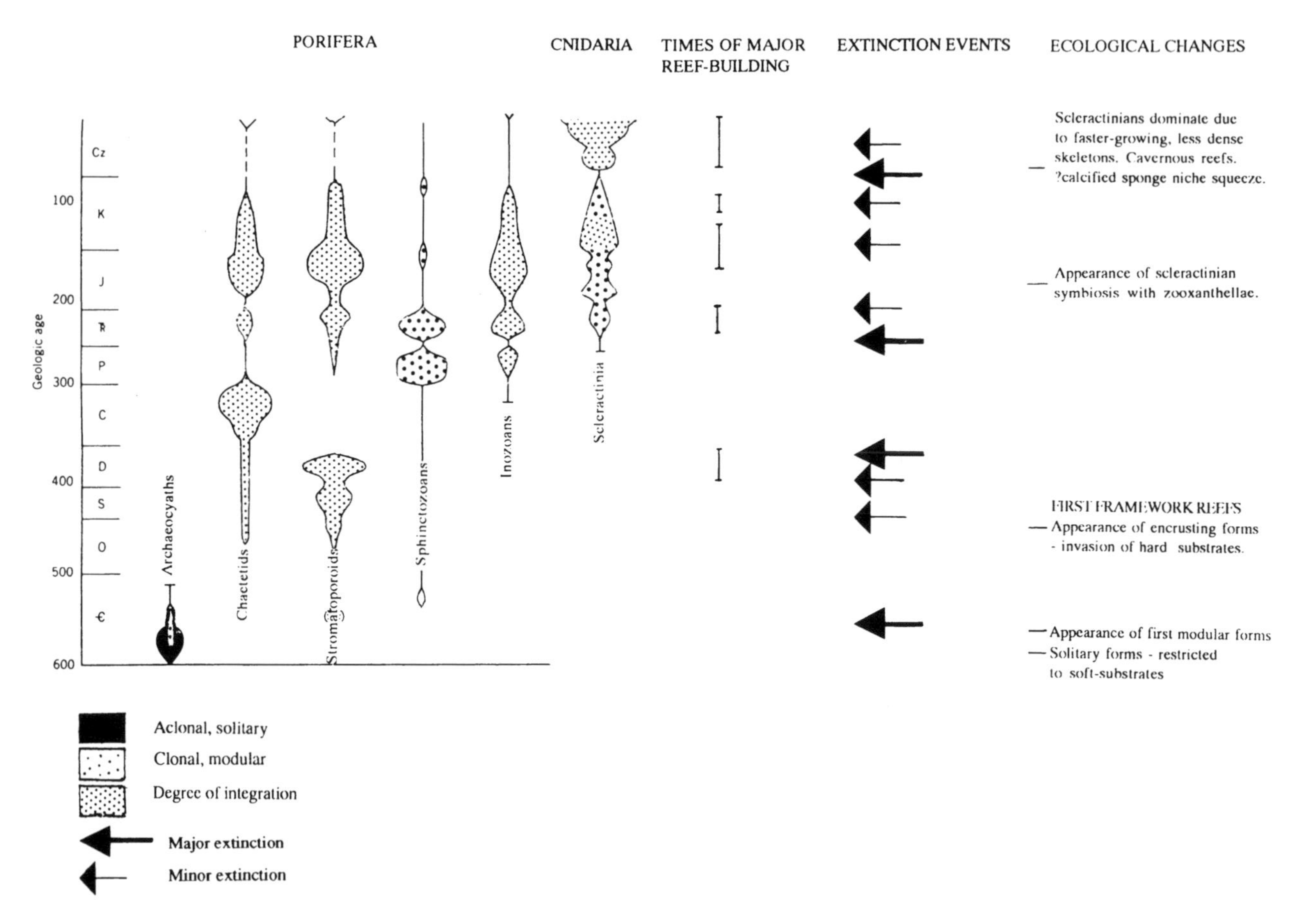

Fig. 1. The distribution of calcified reef-building sponge grades through geological time, with an indication of the major changes in ecological status and major extinction events.

Cambrian (archaeocyaths), late Carboniferous (chaetetids), and the Permian and mid-Triassic (sphinctozoans), and were of major importance during the mid-Palaeozoic and mid-late Jurassic (stromatoporoids and lithistids respectively). Episodes in which sponges were not major reef-builders correspond to global extinctions followed by a general dearth of skeletal reefs (see Figure 1). Stanley (1988) has determined a recurring pattern of reef development during much of the Phanerozoic which he interprets as reef-community reorganisation following mass extinction and eustatic sea level drop. Dispersal and diversification was facilitated by sea-level rise and an attendant increase in suitable shallow carbonate platform environments for colonisation.

Sponges have achieved the rigid skeleton required for reef-building in two ways: by the fusion of robust spicules, as found in lithistid demosponges or by the acquisition of an additional basal calcareous skeleton. Both these adaptations have convergently evolved in a large number of clades (Vacelet, 1985; Wood, 1987; Levi, in press; Vacelet, in press (b)). In addition, the ecological requirements of being simple, sessile organisms restricts them to a limited number of body plans and growth types, such as massive, encrusting or dendroid, which seem to reflect morphological designs for survival under different conditions independent of phylogenetic relationships (Jackson, 1985). This leads to further problems of convergence in the characters of the readily preservable calcareous skeleton. Firstly, by exacerbating taxonomic difficulties by presenting characters of little taxonomic value and secondly by creating the need for special conditions to preserve the vital taxonomic tools, the spicules, which being usually siliceous are thermodynamically unstable when trapped within a calcareous skeleton. Therefore, when compared to the number of representatives in the fossil record, there is a paucity of useful data.

In fact, as with the conodont animal, the key to the affinities of these previously problematic calcified sponge groups was to be found lying forgotten in museum drawers - both in terms of their living representatives, which were known since 1901 when the first example *Astrosclera* was described, and in the undiscovered but abundant presence of sufficiently well preserved fossil examples. Only since the rediscovery in the last two decades of living calcified sponges have palaeontologists been furnished with a basis for their recognition and the information derived from these relict faunas has indicated our ideas concerning related fossil groups to have been misguided. The presence of artificial taxonomic boundaries dividing these forms meant that groups were studied in isolation by specialists, thus obscuring their true affinities and leading to a lack of sophistication. Each group has been encumbered by a complex nomenclature which hindered comparison even though such nomenclatures can be easily synonymised (see Debrenne & Wood, 1990; Zhuravlev *et al.*, in press).

These groups, now recognised to be calcified sponges, seem to represent a reducing pool of Problematica. Demonstration of their poriferan affinity has considerable implications for our understanding of poriferan evolution and also confirms the major ecological role of sponges in ancient communities, especially in reefs.

2. Convergence of the calcareous skeleton

When living calcified sponges were first rediscovered many were placed in a new class, the Sclerospongiae (Hartman & Goreau, 1970). Others, found in the Mediterranean by Jean Vacelet bore calcareous spicules and were allocated variously within the Calcarea (Vacelet, 1970). However, using spicule and soft-tissue criteria, Vacelet (1985) has shown that living sclerosponges are a collection of assorted demosponges, which can be distributed within pre-existing orders without the need for the erection of a new taxon. The calcareous skeletons of these sponges are therefore features which must have evolved independently in many unrelated clades. In addition, Vacelet found that many calcified forms had very closely related non-calcified counterparts. He suggested that these forms had either evolved in parallel or had lost their calcareous skeletons in relatively recent times. As a result of this revelations, Vacelet's work was an invitation to palaeontologists to apply and test his phylogenetic proposals to the fossil record.

Assuming spicule type and arrangement to be as taxonomically significant as for living forms, the spicule types found in fossil calcified sponges have also revealed the presence of many unrelated clades. Chaetetids are proposed to be an assortment of demosponges (Vacelet, 1985; Wood & Reitner, 1988), while the sphinctozoans and stromatoporoids are assorted demosponges and calcareans (e.g. Wood, 1987; Reitner & Engeser, 1985; Reitner, 1987). The polyphyletic origin of sphinctozoans based on spicule criteria is corroborated by the clearly widespread palaeogeographic distribution and varied morphologies of early sphinctozoans, which are known from the USA, USSR and Australia in various island arc settings of the palaeopacific rim (Webby & Rigby, 1985; Debrenne & Wood, 1990).

Calcified demosponges are now known from the Orders Haplosclerida, Axinellida, Hadromerida, Choristida, Vaceletida, Keratosa and Poecilosclerida (Reitner, in press). Calcified calcareans are known from both the Orders Calcaronea and Calcinea (Vacelet, in press (b)). Their distribution within sub-classes of the Porifera is shown in Figure 2.

In the absence of much spicule information, faunas are still too poorly known to allow any detailed reconstruction of phylogeny. However, some of the genera now extant have a fossil record. *Acanthochaetetes* can be traced back to the Lower Cretaceous (Reitner & Engeser, 1987), while *Ceratoporella* first appeared in the Upper Permian (Wood, 1987). In addition, some fossil forms can be directly related to the living sphinctozoan *Vaceletia*, with a clade that extends into the Upper Triassic (Reitner & Engeser, 1985) and representatives of the Family Minchinellidae within the Order Calcinea are found from the late Cretaceous to Recent times (Reitner, 1987). Extinct clades can also be detected, such as the Family Milleporellidae within the Order Axinellida which extends from the Upper Triassic to Upper Cretaceous (Wood, 1987; Wood *et al.*, 1989).

Finks (1983) and Wendt (1980) concluded that ex-

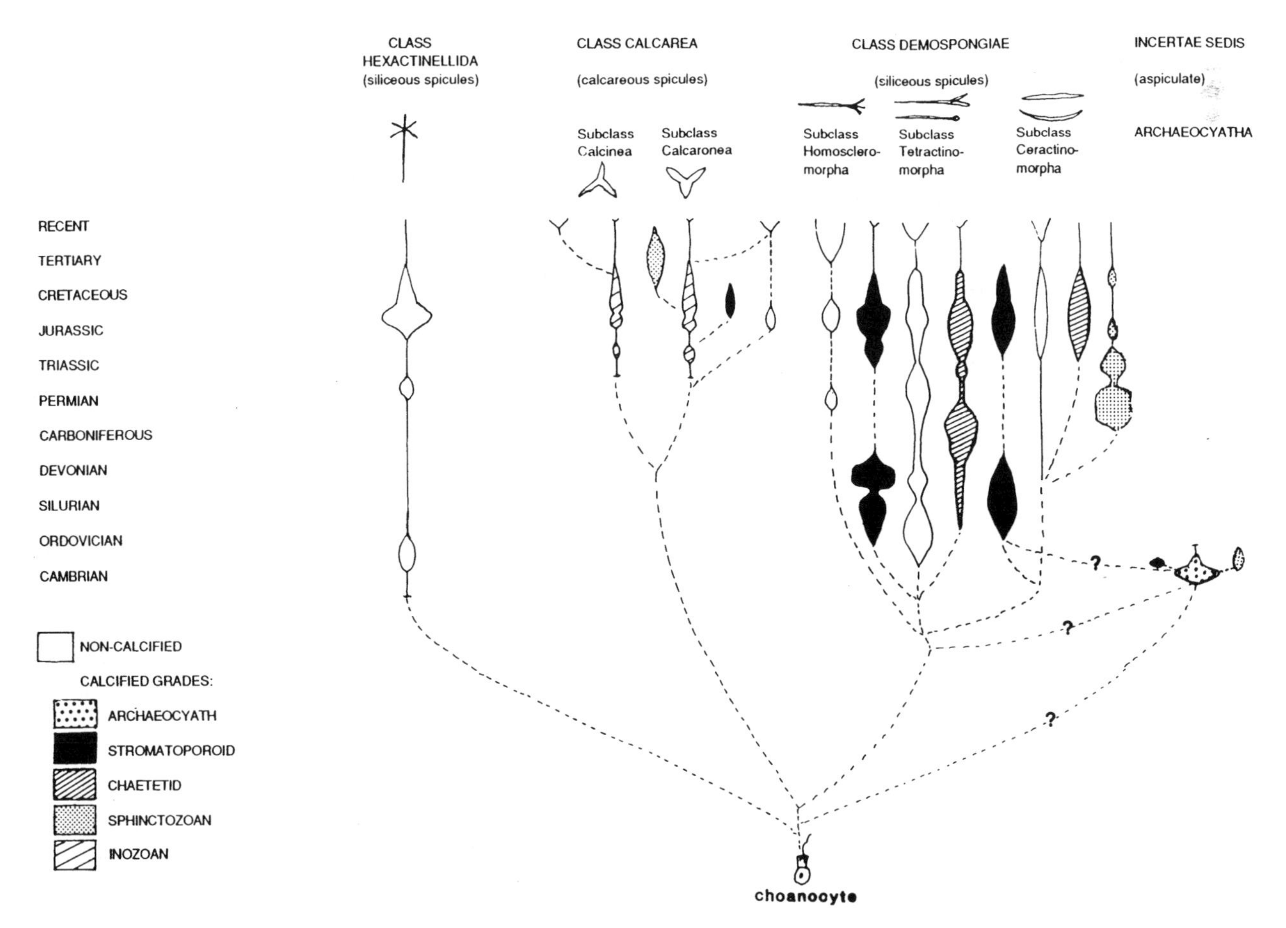

Fig. 2. The convergent development of a calcareous skeleton within sub-classes of the Porifera (from Wood 1990).

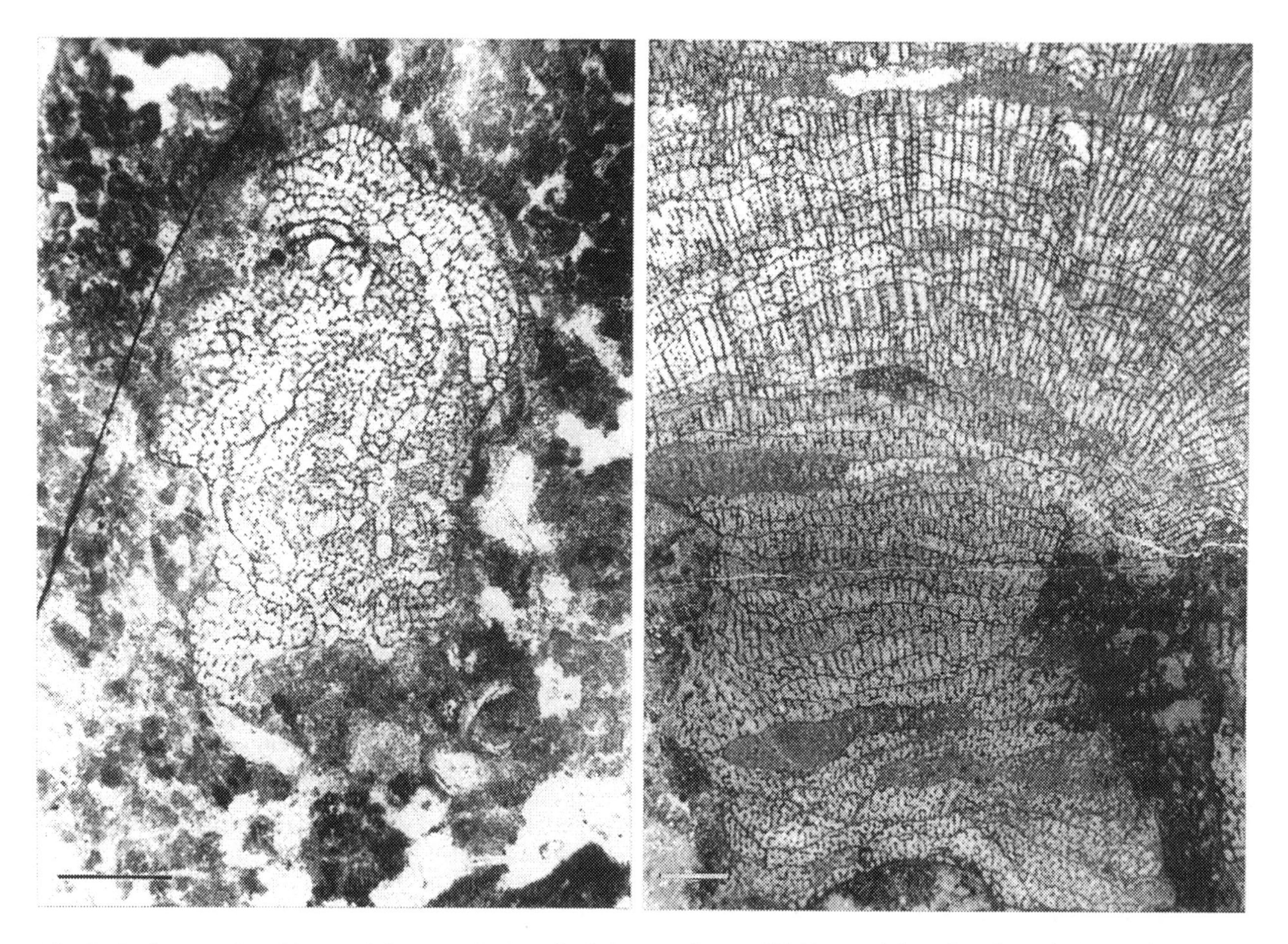

Fig. 3. Archaeocyaths with a modular organisation. Scale bars = 2 mm. (A) Tangential section through *Praeactinostroma vologdini* (Yavorsky) showing astrorhizae, characteristic of sponges with a thin veneer of soft-tissue and bidirectional filtration surfaces. Srebenka Section, Kazyr River, Eastern Sayan, USSR. Botomian. (B) Longitudinal section through *Zunicyathus grandis* (Yuan and Zhang) showing a stromatoporoid or chaetetid organisation. Note the lack of the complex ontogenetic development typical of archaeocyaths; the juvenile skeletal tissue is retained throughout adult life. Jindingshan, Zunyi Province, South China. Botomian.

cept for their sharp decline during the early Triassic, there is considerable continuity at the family and generic level through the late Palaeozoic to early Mesozoic. Other workers have also concluded that, as far as can detected, many known calcified demosponges appear to form part of long-lived and conservative clades, some surviving from the Upper Permian to the present day (Wood, 1987; Wood *et al.*, 1989; Reitner, in press).

3. Biology of archaeocyaths

Archaeocyaths are characterised by stalked, cup-shaped morphologies, the presence of septa, a complex porosity and a highly developed sense of individuality. The latter is expressed by their radial symmetry and complex reactions in response to the proximity of other individuals, where they display a hierarchy of allogenic incompatibility (Brasier, 1976). These supposed apomorphies have confounded many workers, but detailed comparative morphology and functional analysis of archaeocyaths reveals that their skeletal arrangement is entirely consistent with that of a poriferan organisation (Debrenne & Vacelet, 1984; Zhuravlev, 1985; 1989; Debrenne & Wood, 1990):

Skeletal morphology. Features that have been considered purely archaeocyathan have now been described in various calcified sponges, especially those with a thalamid organisation:

a) Thalamid arrangements comparable to that of the sphinctozoan sponges are found in coscinocyaths and capsulocyathines (Figure 5B),

b) Archaeocyath-like dictyonal networks are known from spicular demosponges (Debrenne & Vacelet, 1984),

c) Hexactinellids can form a cup similar to the tubuli of syringocnemid and erismacoscinids (Zhuravlev, 1989),

d) Septa are known from Devonian and Triassic sponges (Ott, 1974; Pickett & Rigby, 1983; Zhuravlev, 1989),

e) Some early Cambrian sphinctozoans, e.g. *Polythalamia americana* share the same pore size and organisation as archaeocyaths (Debrenne & Wood, 1990; and see Figure 5A). The pore spacing of the early Cambrian sponge ?*Jawonya tiro* also resembles that of monocyathine archaeocyaths (Kruse, 1987),

f) The distinctive multiperforate sieves which form tumuli in the early Cambrian sponge *Wagima* resemble the multi-perforate tumuli of archaeocyaths and may represent incipient outer wall development (Kruse, 1987). This feature was thought to be restricted to some archaeocyath superfamilies.

Skeletal microstructure. The microgranular microstructure of archaeocyaths has been thought to be unique (Debrenne & Vacelet, 1984). However, this fabric can be interpreted as diagenetic alteration of an original irregular microstructure comparable to that found in many fossil and living calcified sponges (Zhuravlev, 1989; Debrenne & Wood, 1990). Archaeocyaths are suggested to have mineralized via a collagenous matrix, as found in the Recent sphinctozoan *Vaceletia*, to produce an irregular microstructure of magnesium calcite mineralogy (Debrenne & Vacelet, 1984; Zhuravlev, 1989). New collagenous material would form at the uppermost parts of the skeleton, as evidenced by the frequent deformation and enhanced diagenetic susceptibility of this area.

Secondary filling tissue. The soft-tissue of living calcified sponges occupies only the upper parts of the skeleton. As a result, some form of secondary calcareous tissue is required to separate the abandoned parts of the skeleton and to support the relatively thin layer of soft-tissue. Archaeocyaths show such secondary tissue in the form of secondary thickening, stereoplasm and dissepiments. The position of these structures also thus aids reconstruction of the soft-tissue distribution.

Functional morphology and soft-tissue reconstruction. Brasier (1976) notes the absence of epibionts in the central cavity of archaeocyaths suggesting that a constant water flow was present and various experiments have shown that the archaeocyath cup was well suited to passive filtration (Balsam & Vogel, 1973; Savarese, 1988). Water would enter through the outer microporous sheath (inhalant surface), be filtered through the choanocyte chambers in the tissue-filled intervallum and exit via the inner wall pores (exhalant surface) into the central cavity. The central cavity was thus empty and homologous to the spongocoel of a sponge. The diameter of the inner-wall pores is always greater than those on the outer wall, which also supports this direction of flow. The soft-tissue would have been mainly internal and restricted to the upper parts of the inter-vallum, with abandoned parts of the skeleton being sectioned-off by filling tissue. A single-celled pinacoderm may also have covered the inner and outer walls.

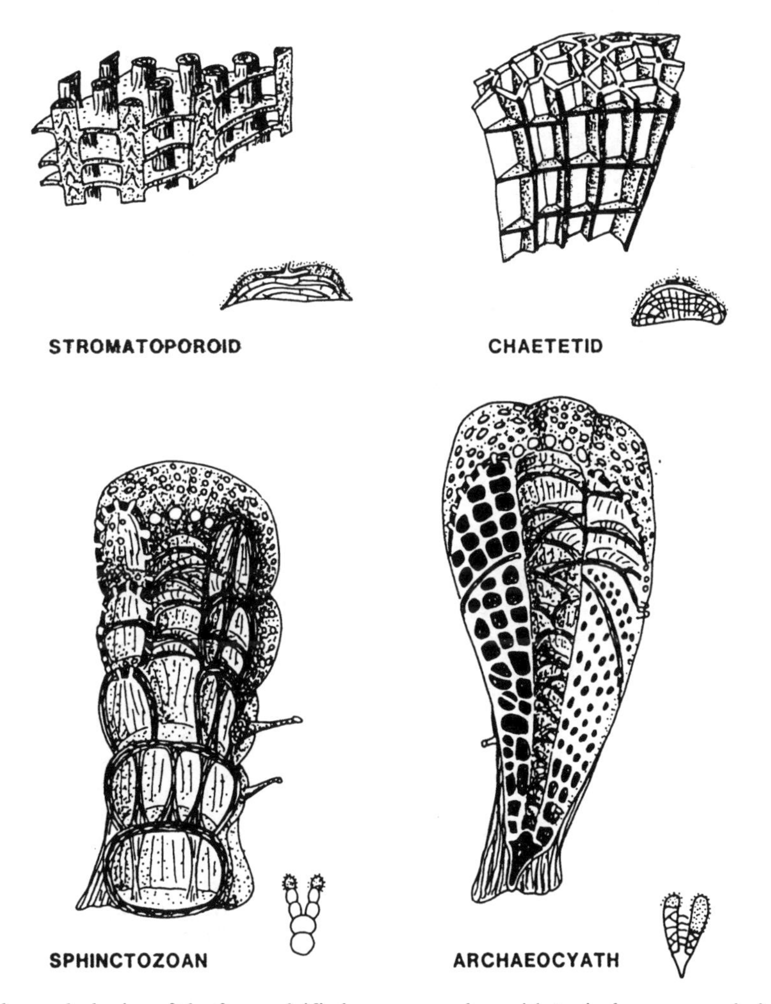

Fig. 4. Stylized internal morphologies of the four calcified sponge grades, with typical gross morphologies and inferred tissue distributions (stippled areas). (Modified from Zhuravlev *et al.*, in press).

Astrorhizal canals are found in some forms e.g. *Bottonaecyathus* and *Praeactinostroma vologdini* (Figure 3A) and are characteristic of sponges which possess only thin veneers of tissue where the upper surface contains both the inhalant and exhalant pores. Boyajian & LaBarbara (1987) have shown that these are ideal structures for active filtration. This would suggest that some archaeocyaths possessed only a thin layer of soft-tissue on their upper surfaces which supported

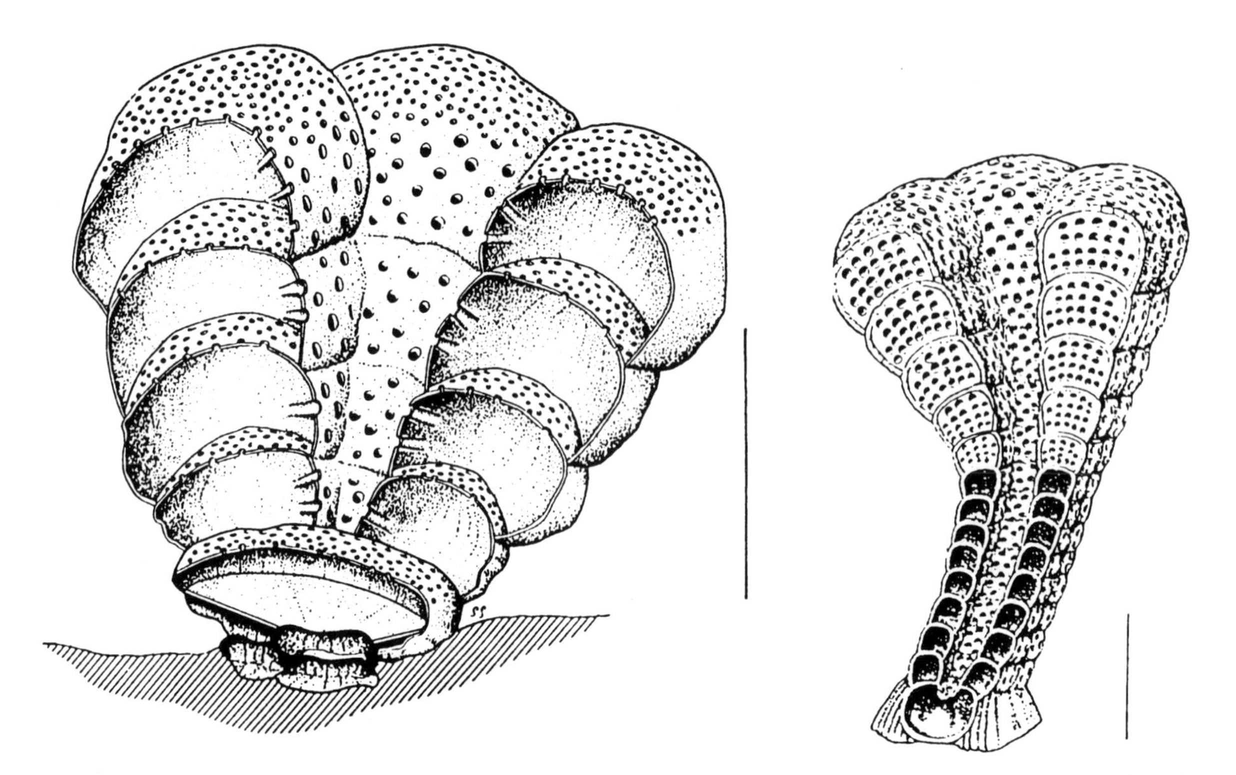

Fig. 5. Possible inter-relationship between archaeocyath and sphinctozoan sponges. (A) Generalised reconstruction of the early Cambrian (Botomian) sphinctozoan *Polythalamia americana* (from Debrenne & Wood, 1990), suggested to have arisen by paedomorphosis from coscinocyathine archaeocyaths. (B) Generalised reconstruction of a coscinocyathine archaeocyath (from Debrenne, Zhuravlev & Rozanov, 1989). Scale bars = 1mm.

a bidirectional water flow and were able to employ active filtration.

Thus we are able to demonstrate that the skeletal features of archaeocyaths are consistent with a sponge organisation: archaeocyaths were highly organised, aspiculate calcified sponges, some capable of active as well as passive filtration, which produced a skeleton by direct calcitization of an organic matrix. However, the lack of spicules does not allow any placing within the existing poriferan classification, and we must seek some further explanation for the mosaic of features not commonly found in other calcified sponges. We may then place archaeocyaths in the context of calcified sponge evolution as a whole.

4. Mineralization of the calcareous skeleton

The active scavenging of sediment grains falling upon the surface of aspicular sponges suggests a need for some form of mineralized reinforcement. Spicule formation processes are well controlled by the organism as spicules change in form and arrangement according to their position within the individual. However, Recent sponges show only four main non-spicular precipitation mechanisms (Wood, in press), and within these a variety of microstructures may be formed:
1) Direct mineralization of an extensive organic framework e.g. *Calcifibrospongia* (spherulitic microstructure) and *Vaceletia* (irregular microstructure),
2) Mineralization via a thin secretory pinacoderm and a possibly mucopolysaccharide-filled layer e.g. *Acanthochaetetes* (irregular microstructure) and *Ceratoporella* (elongate spherulitic microstructure),
3) A "passive" almost cement-like formation of orthogonal fibres e.g. *Minchinella* (calcarean), and
4) An intra-cellular mechanism of spherule production, transportation to the skeletal wall and then *in situ* syntaxial crystal growth to form the skeletal wall e.g. *Astrosclera*.

From microstructural characteristics, fossil representatives can be inferred for all four mechanisms. The distribution and form of organic tissue appears to play an important role, but its influence is variable. Calcified demosponges show a range of seemingly "biologically controlled", "biologically-induced" through to "organic matrix-mediated" mechanisms. In contrast, the role of spicules is of subsidiary importance, but they may in some cases act as nucleation sites.

The type of biomineralization mechanism, when higher taxonomic placing has been achieved by spicule means is of significance only in lower taxonomic categories. However, too little is known about the systematic distribution of mechanisms in sponges to assess the relative status of differing microstructures, but it would seem probable that "biologically controlled" precipitations e.g. intracellular spherulites, would have greater higher systematic use than "biologically-induced" ones e.g. orthogonal cement.

Aragonitic forms predominated in Permian/Triassic and Recent times, which corresponds to the proposed aragonite facilitating phases of Sandberg (1983). Mineralogy in demosponges may therefore be a response of physiochemical precipitates relative to water chemistry and may be determined by the appearance of calcification within a particular clade relative to Sandberg's scheme. Different mineralogies are found within members of the same order, which indicates that mineralogical type is not of very high taxonomic significance (Wood *et al.*, 1989; Wood, 1990 and in press).

Existing lines of evidence suggest that non-spicular

skeleton production is a simple process and that sponges can calcify with relative ease as (Wood, in press):

a) Sponges which bear calcareous skeletons have a wide systematic distribution.

b) Different microstructures and seemingly different biomineralization mechanisms are present within closely related forms e.g. both *Astrosclera* and *Ceratoporella* are agelasids, but one produces a spherulitic microstructure with an intracellular origin and the other elongate spherulites via a secretory pinacoderm. In addition, members of the same clade appear to have independently produced different calcareous mineralogies and microstructures at different times e.g. *Newellia mira* and *Euzkadiella erenoensis* (Wood *et al.*, 1989).

c) Closest relatives of Recent calcified forms are non-calcified e.g. *Spirastrella* to *Acanthochaetetes*, and *Agelas* to *Astrosclera* and *Ceratoporella*.

d) Some species appear to have a facultative ability. There are four species of the genus *Merlia* but only two produce a calcareous skeleton, and

e) The crystals precipitated by sponges do not have a unique crystal habit, trace element or isotopic composition which points to a minimal intervention of biological processes. The $\delta^{18}O$ signature of calcified demosponges all fall within the field of normal sea water, indicating that little fractionation occurs (Land, 1989). However, calcareans show heavy $\delta^{18}O$ signatures when compared to sea water (Reitner, 1989).

5. Grades of organisation

These former discrete systematic groupings, which were defined solely on calcareous skeleton arrangement, can best be redefined simply as "grades of organisation" of the calcareous skeleton (Vacelet, 1985; Wood, 1987; see Figure 4). These grades have no higher systematic value. Now that the classification of these various calcareous sponges has been placed on a more secure footing, the once puzzling range of apparent intermediates such as "archaeocyath-like sphinctozoans" and "stromatoporoid-like chaetetids" are explained. These various grades of organisation can be interpreted as reflections of non-systematically related soft-tissue and aquiferous system arrangements and are described below.

5.a Archaeocyaths

Archaeocyaths, which represent the earliest skeletal component of reefs, appeared in the Anabar-Sinyana Basin of the Siberian Platform during the early Tommotian, where the base of the Cambrian is determined by the simultaneous appearance of five archaeocyath species (Rozanov & Debrenne, 1974). Under the most recent classification (Debrenne *et al.*, 1989) these species have been placed within 3 families, suggesting that archaeocyaths already had a not inconsiderable evolutionary history as soft-bodied forms which have left no fossilizable remains. The Atabanian was a period of archaeocyathan dispersal, diversification and the appearance of new morphological features. The late Atabanian to early Bottomian saw a further doubling of archaeocyath genera and the widest range of morphologies and environmental settings. Archaeocyaths became restricted to Altay-Sayan and NW Canada during the Toyonian, with only one species remaining in Siberia by the mid-Cambrian and one species in Antarctica and Texas by the late Cambrian (Debrenne *et al.*, 1986).

Archaeocyaths are generally uni-oscular or solitary, with cup-bowl or cylindrical shapes (*cyathiform* of Finks 1983 see Fig. 4). Multi-oscular or modular forms may be hemispherical e.g. *Altaicyathus notabilis* Vologdin and *Zunyicyathus grandis* (Yuan & Zhang) (Figure 3B), patelliform e.g. *Retilamina* Debrenne, cateniform e.g. *Salairocythus (Polystillicocyathus) erbosimilis* Debrenne, or dendroid e.g. *Archaeocyathus*. The proportion of modular archaeocyaths increased during the Early Cambrian and reached 100% by the end of the Toyonian (R. Wood, A.Yu. Zhuravlev and F. Debrenne, personal observation).

In solitary forms water would enter through the outer microporous sheath, be filtered through the choanocyte chambers in the tissue-filled intervallum and exit via the inner wall pores into the central cavity. However, in hemispherical forms the soft-tissue would be present as a relatively thin surficial layer, necessitating the need for the amalgamation of inhalant and exhalant surfaces to produce a single bidirectional surface.

5.b Stromatoporoids

Forms exhibiting a stromatoporoid grade of organisation have been noted from the Botomian, in the early Cambrian (Khalfina & Yavorsky, 1967). However these forms were short-lived and became extinct by the end of the early Cambrian (Toyonian). Forms with a stromatoporoid organisation next appeared in the middle Ordovician. They became abundant and widespread in the Silurian to middle Devonian and formed extensive framework reefs. However, numbers sharply declined at the end of the Frasnian (early late Devonian) although some members of the labechoids (thought to be the most primitive order) are found in the Famennian, but these became extinct at the end of the Devonian. A second acme of stromatoporoids was present in the Middle - Upper Jurassic, when they also built reefs on a biohermal scale. The Mesozoic forms possess a fibrous microstructure absent in most Palaeozoic forms and almost certainly represent a second phase of calcareous skeleton acquisition. Stromatoporoid grade calcareans appeared in the late Palaeozoic, but only became abundant in the Jurassic and Cretaceous. Many clades appear to have become extinct by the end of the Cretaceous. *Astrosclera* (agelasid demosponge) and *Calcifibrospongia* (haplosclerid demosponge) represent living calcified demosponges of the stromatoporoid grade.

The stromatoporoid grade is characterised by a multi-oscular, "compound", "colonial" or modular aquiferous system and a layered organisation of radial and concentric skeletal elements (Figure 4). Growth morphology may be hemispherical, dendroid, columnar or encrusting. The response to damage and great plasiticity of shape indicates a well integrated organisation. The soft-tissue would have occupied the upper parts of the skeleton only. Drainage would have been superficial, producing astrorhizae, necessitating the need for a bidirectional filtration surface with excurrent canals parallel or sub-parallel to the skeletal surface. The sedimentary relationship between some laminar skeletons and internal sediment confirms the rapid growth rates of many stromatoporoids.

5.c Chaetetids

Chaetetids appeared in the Late Ordovician and continue to the present day. They were significant reef-builders during the Carboniferous and late Jurassic, but are also found on shallow-water carbonate level-bottom environments, where they were binders or encrusters. West & Clark (1984) have noted the importance of hard substrates, such as mollusc debris or lithic fragments, as basal attachment sites for columnar or hemispherical chaetetids. Living forms of the chaetetid grade are represented by *Merlia* (poecilosclerid demosponge), *Acanthochaetetes* (hadromerid demosponge) and Ceratoporellids (agelasid demosponges).

Chaetetids are also functionally modular and show similar growth morphologies to stromatoporoids, but their skeleton is composed of minute conjoined tubules, or calicles (Figure 4). They are characterised by possession of only a very thin veneer of tissue, which is mostly external to the skeleton. Considerable secondary filling tissue is often present which sections-off abandoned parts of the skeleton. Chaetetids also possessed a bidirectional filtration surface with an extremely surficial excurrent drainage system. Vacelet (in press) has suggested that the formation of tabulated calicles in demosponges may be to form protected areas to contain the totipotent archaeocytes, which are commonly found at the bases of calicles in living chaetetids.

5.d Sphinctozoans

Although sphinctozoans appeared in the early Cambrian, their record during most of the Palaeozoic is scarce. Pickett & Jell (1983) have described porate sphinctozoans from the late early Cambrian [Ordian] of New South Wales, Australia, and Debrenne & Wood (1990) have described a further form from North America. Other lower Palaeozoic sphinctozoan faunas have been documented from the Ordovician of Australia, California, and the Yukon and Alaska (Webby & Rigby, 1985; Rigby & Potter, 1986; Rigby *et al.*, 1988), the Silurian of the Canadian Arctic (de Freitas, 1987) and from the Devonian of Australia and Alaska (Pickett & Rigby, 1983; Rigby & Blodgett, 1983). They became major reef-builders during the mid-late Permian and mid-late Triassic. During these times they were bafflers, but rarely produced a rigid framestone. At other times they were present in the inter-biohermal areas. The genus *Vaceletia* represents the only known living sphinctozoan.

The spinctozoan grade is characterised by a thalamid, or chambered skeleton (Figure 4). The aquiferous units are usually arranged in a solitary organisation, but phaceloid and laminar forms are known which correspond to a pseudomodular arrangement. The living tissue would have been almost wholly internal occupying the uppermost one or two chambers, with a one-celled external pinacoderm covering. Growth would have been episodic, with the production of a new chamber requiring the migration of soft-tissue. The direction of water flow would have been the same as for archaeocyaths, with water entering small pores on the upper and outer surfaces and exiting via a central cavity.

6. The inter-relationships of calcified grades

Significant differences have been documented in the morphology, distribution, population dynamics and life histories of aclonal and clonal epibenthic invertebrates (e.g. Jackson, 1977a and b) and attempts have been made to explain the adaptive significance of these traits (e.g. Jackson, 1985). Clonal invertebrates grow and reproduce through potentially unlimited production of identical individuals, which are all ultimately derived from the same zygote. The resulting individuals are generally capable of independent existence separate from the parent organism. As much of the organism can be destroyed without the entire loss of the genotype, clonal organisms do not senesce and can attain very large sizes. In contrast, the growth of an aclonal animal is finite. Aclonal animals are all solitary, as opposed to clonal ones which are often colonial or modular, where the individuals are smaller and less complex than those in solitary forms, and repeat many times in morphological continuity. Living sponges, although best considered individuals, are often clonal and show subdivision of their aquiferous system into distinct modules. Thus some sponges may be considered solitary and possibly aclonal while others are modular and clonal (Jackson, 1985).

Table 1 presents the major morphological and ecological differences exhibited by aclonal/solitary and clonal/modular animals. The characteristics differ in many aspects, and have been correlated with adaptations to different substrate types and environmental settings (Jackson, 1985). It has been found that the relative abundance and diversity of clonal and aclonal animals increases with a variety of measures of environmental stability and predictability. Aclonal/solitary forms are adapted to soft-substrates in unstable environments whereas clonal/modular forms are adapted to hard-substrates in stable environments. Generally, modular forms have an unlimited size, life expectancy and regenerative abilities, and have superior competitive abilities on hard substrates. There is also a clear correlation between modularity, philopatry, long generation time and low fecundity.

All these differences between clonal and aclonal epibenthic invertebrates have evolved many times in unrelated higher taxa. Major groups, e.g. the Cnidaria, exhibit the entire range of aclonal and clonal life history patterns displayed among all epibenthic invertebrates known (Jackson, 1985). It is clear that these patterns must have considerable adaptive significance. These theories have been applied to the history of modularity/coloniality within the Cnidaria, and increasing modularity and colony-wide cooperation i.e. integration, has been noted throughout the Phanerozoic (Coates & Oliver, 1973). They may also be applied to explain the development and temporal distribution of the various grades of organisation within calcified sponges.

Most archaeocyaths were clearly solitary organisms. They generally bore only one functional unit, were small and showed comparatively poor regenerative abilities. They possessed only one direction of growth (at the top of the cup) and had radial symmetry. They were anchored to the substrate by means of a small holdfast, which did not allow a firm attachment to a hard substrate. Palaeoecological analysis implies that they lived anchored to soft, shifting substrate in intertidal and subtidal area of shelf seaways or edges which were subject to frequent erosive events. There is no evidence of growth into the shallow, turbulent

	SIZE	LIFE EXPECTANCY	REGENERATIVE CAPACITY	INTEGRATION	FECUNDITY
SOLITARY ACLONAL e.g. archaeocyaths	small	low	low	low philopatry	high
COLONIAL CLONAL e.g. stromatoporoids	large	long	high	high philopatry	low

	LARVAL DISPERSAL	LARVAL GROUPING	GROWTH FORM	ENVIRONMENTAL SETTING	
SOLITARY ACLONAL	long	synchronous + aggretive settlement	stalked, cup-like	soft-substrates Unstable	NON-REEF BUILDING
COLONIAL CLONAL	short	asynchronous + non-aggretive settlement	encrusting massive, branching	hard-substrates Stable	REEF-BUILDING

Table 1. Ecological characteristics of clonal/colonial forms compared to aclonal/solitary

zone. Due to their lack of encrusting morphologies and their frequently highly developed sense of individuality, they were unable to cement or overgrow other individuals and so were unable to form rigid reef frameworks. It may also be possible that the relatively small maximum size achieved by most archaeocyaths was due to the absence of an efficient mechanism to discard metabolic waste products. Most early Cambrian reefs are small, algally constructed bioherms, where archaeocyaths are often only contributory skeletal elements.

In contrast, stromatoporoids and chaetetids bear many conjoined functional units. They are clearly modular in organisation and have achieved a high degree of integration between functional units. Due to their lack of radial symmetry, they had gained many more margins for growth. This enabled them to attain very large sizes and to become gregarious. Most importantly, the presence of an encrusting morphology allowed them to attach to hard substrates and become the first framework builders.

Sphinctozoans are solitary or pseudomodular and being small and fragile were mainly reef-dwellers and bafflers.

Long-term trends may be detected within the evolution of the calcified sponge skeleton. When the archaeocyaths, sphinctozoans, stromatoporoids and chaetetids are considered as a whole, they show a range of integration from solitary forms (archaeocyaths), through pseudomodular (sphinctozoans) to fully integrated modular forms (stromatoporoids and chaetetids; see Figure 1). Skeletal integration provides mechanical strength, stability and a larger individual size. Also, integration reduces extreme incompatibility shown between individuals and allows coexisting forms to encrust. These factors, together with the ability to produce rapidly built skeletons enabled forms with a stromatoporoid organisation to construct the first extensive framework reefs when they appeared in large numbers in the mid-late Ordovician, and continue to dominate reef-building for much of the Palaeozoic.

In contrast to the steady development of coloniality within Phanerozoic coral evolution (Coates & Oliver, 1973), calcified sponges show the relatively early appearance (mid-Ordovician) of well integrated forms. However, scleractinian corals became the dominant reef-builders through most of the Mesozoic and Tertiary, presumably due to the acquisition of well integrated, less dense, more porous skeletons and faster calcification rates than stromatoporoids, which was aided by their symbiotic relationship with zooxanthellae (see Figure 1).

There is some indication that modular grades have been derived paedomorphically from solitary ones (Debrenne & Wood, 1990). The early Cambrian sphinctozoan *Polythalamia americana* (Figure 5A) is closely associated with a diverse archaeocyath assemblage and draws attention to morphological similarities between some archaeocyaths and sphinctozoans. *P. americana* shares a similar pore arrangement to many archaeocyaths and to the chambered juvenile stages of coscinocyathine archaeocyaths (Figure 5B). In addition, the small size of *P. americana* corresponds to that of the juvenile archaeocyath skeleton. This raises the possibility that *P. americana* may be derived from such archaeocyaths by paedomorphosis. In particular, the juvenile stages of the genus *Clathricoscinus*, which were erroneously described as a separate genus, *Kameschkovia*, is remarkably similar. Unlike archaeocyaths, *P. americana* is able to encrust firm substrates and algae, e.g. *Epiphyton*. This may then represent the predicted niche differences between solitary and modular forms. The driving pressure for this niche change may be the need to escape the unstable environments in which archaeocyaths lived. Substantial proportions of archaeocyath communities are often found toppled and reworked, indicating that they lived in environments which were susceptible to frequent erosive or high-energy events.

A number of Cambrian stromatoporoid-grade forms e.g. *Zunyicyathus grandis* (Figure 3B) also bear a resemblance to the undifferentiated juvenile tissue of some irregular archaeocyaths, again suggesting a paedomorphic origin. These forms do not show the sequence of complex developmental changes typical of cylindriconal, solitary archaeocyaths. An important point in the transition from solitary to modular organisation may therefore be a function of soft-tissue distribution. In some forms, juvenile tissue may have been present only as a thin layer, thus necessitating the need for a bidirectional filtration surface. The mainteinance of this surface allows the retention of mobile, near external soft-tissue, the development of a modular habit with the possibility of active filtration, and the acquisition of an encrusting habit with many directions of growth. The development of a complex, segmented and solitary skeleton requires the separation of inhalant and exhalant surfaces. Since well integrated, modular forms are characterised by bidirectional filtration surfaces, retention of this juvenile character would have been advantageous to colonisation of the hard substrate, framework building habit in more stable environmental areas.

Calcified sponges are deeply conservative, and although their calcareous skeletons are the most prominent aspect of their fossil record, they are of little value when assessing phylogeny (Wood, 1990). In fact, it has been suggested that the identification of "living fossils", such as living calcified demosponges, may simply be an artifact of our palaeontological classificatory system, which by necessity emphasizes shared hard-part morphologies which may not accurately reflect original soft-part differences (Schopf, 1984). Although sponge calcified skeletons appear to show little morphological change over considerable periods of the Phanerozoic, this may be because these simple organisms have only a limited number of possible options available in response to similar environmental or ecological constraints. The widespread convergence apparent in calcified sponges indicates considerable turnover and innovation, but meaningful rates of evolution and any real indication of the diversity they achieved cannot be assessed without great difficulty, as relatively few forms have as yet yielded spicule data and present species definitions are so dependent upon characters which are clearly of little taxonomic value (Wood, 1990).

7. Conclusions

(1) The presence of incorporated spicules within stromatoporoids, chaetetids and sphinctozoans has shown these groups to be sponges which bear a calcareous skeleton.

(2) Spicule data, however, reveals a polyphyletic origin for these groups, with representatives from several different orders of both the Demospongiae and the Calcarea. Calcareous skeletons have been independently acquired in many different clades. This confirms the artificial nature of these higher taxonomic groups, which can be better defined as grades of organisation.

(3) Grades are interpreted as reflections of non-systematically related soft-tissue and aquiferous filtration system organisations, which are probably ecologically based.

(4) Archaeocyaths were highly organised, aspiculate calcified sponges, some capable of active as well as passive filtration, which produced a skeleton by direct calcitization of an organic matrix.

(5) Existing lines of evidence suggest that sponges can calcify with relative ease, as non-spicular skeleton production is a simple process and only a limited number of biomineralization mechanisms are employed. This may be a reflection of the limited number of responses available to these organisms in response to similar ecological pressures. Although these calcareous skeletons form the most prominent aspect of the calcified sponge fossil record, they are therefore of little value when assessing evolutionary ancestries.

(6) The different strategies adopted by sessile epibenthic invertebrates to soft and hard substrates can be used to interpret the various grades of organisation found in calcified sponges.

(7) The aspiculate archaeocyaths, often characterised by the presence of septa, a complex porosity and a strongly developed individuality, were adapted to a soft-substrate in unstable environmental conditions. They also possessed a solitary functional system, which explains many of their seemingly apomorphic characters. The relatively small maximum size achieved by most archaeocyaths may have been due to the absence of an efficient mechanism to discard metabolic waste products. They were unable to encrust and reach sufficiently large sizes to build framework reefs.

(8) Stromatoporoids and chaetetids are functionally modular and are able to encrust hard substrates. Stromatoporoids were able to produce rapidly built skeletons and attain very large sizes. They thus became the first framebuilders of extensive reefs when numerous forms appeared in the mid-late Ordovician.

(9) Sphinctozoans are solitary or pseudomodular and with their small size and fragile, skeletons were mainly reef-dwellers and bafflers.

(10) A trend can be detected within the evolution of calcified sponge grades. Calcified sponge grades show a range of integration from solitary forms (archaeocyaths), through pseudomodular (sphinctozoans) to fully integrated modular forms (stromatoporoids and chaetetids). Skeletal integration provides mechanical strength, stability and a larger individual size and allows cohabiting forms to cooperate, such as mutually encrust.

(11) Some archaeocyaths have given rise to non-spiculate stromatoporoid- and sphinctozoan-grade clades via paedomorphosis, but all forms perished by the end of the Cambrian. Paedomorphosis is therefore suggested as a possible mechanism for acquisition of modular grades from forms with a solitary organisation. The retention of juvenile characters before expres-

sion of individuality in the adult and the ability to reach the large sizes offered by modularity would provide better adaptations for constructing reef frameworks.

Acknowledgements

This work was carried out under tenure of a NERC Fellowship (GT5/87/GS/1), which is gratefully acknowledged. This is Earth Sciences contribution no. 1808.

REFERENCES

BALSAM, W.L. & VOGEL, S. 1973. Water movement in archaeocyathids: evidence and implications of passive flow in models. *Journal of Palaeontology* **45**, 970-84.

BOYAJIAN, G.E. & LABARBARA, M. 1987. Biomechanical analysis of passive flow in stromatoporoids - morphological, paleoecologic and systematic implications. *Lethaia* **20**, 223-29.

BRASIER, M.D. 1976. Early Cambrian intergrowths of archaeocyathids, *Renalcis* and pseudostromatolites from South Australia. *Palaeontology* **19**, 223-45.

COATES, A.G. & OLIVER, W.A. JR. 1973. Coloniality in zooantharian corals. In *Animal colonies, developments and function through time* (eds. R.S. Boardman, A.H. Cheetham, and W.A. Oliver, Jr.), pp.3-27. Stroudsburg, PA: Dowden, Hutchinson and Ross.

DEBRENNE, F. & VACELET, J. 1984. Archaeocyatha: is the sponge model consistent with their structural organisation? *Paleontographica Americana* **54**, 358-69.

DEBRENNE, F., ROZANOV, A. YU. & WEBERS, G.F. 1986. Upper Cambrian Archaeocyatha from Antarctica. *Geological Magazine* **121**, 291- 99.

DEBRENNE, F., ZHURAVLEV, A.YU. & ROZANOV, A.YU. 1989. Pravilnye Archeosiaty [Regular archaeocyaths.] *Trudy Paleontologicheskogo Instituta* **233**, 198pp. Moskow, Nauka 1989. (in Russian).

DEBRENNE, F. & WOOD, R. 1990. A new Cambrian sphinctozoan from North America, its relationship to archaeocyaths and the nature of early sphinctozoans. *Geological Magazine*, **127**, 435-443.

DE FREITAS, T. 1987. A Silurian sphinctozoan sponge from east- central Cornwallis Island, Canadian Arctic. *Canadian Journal of Earth Sciences* **24**, 840-44.

DIECI, G., RUSSO, A. & RUSSO, F. 1974a. Nota preliminare sulla microstruttura di spugne aragonitiche del Trias medio-superiore. *Bollettino della Società paleontologica Italiana* **13**, 99-107.

DIECI, G., RUSSO, A. & RUSSO, F. 1974b. Revisione del genere *Leiospongia* d'Orbigny (Sclerospongia triassica). *Bollettino della Società paleontologica Italiana* **13**, 135-46.

FAGERSTROM, J.A. 1987. *The Evolution of Reef Communities.* New York: John Wiley & Sons.

FINKS, R.M. 1983. Pharetronida: Inozoa and Sphinctozoa. In *Sponges and Spongiomorphs. Notes for a Short Course* (eds. J.K. Rigby and C.W. Stearn), pp.59-69. University of Tennessee, Department of Geological Sciences, Studies in Geology **7**.

HARTMAN, W.D., & GOREAU, T.E. 1970. Jamaican coralline sponges: their morphology, ecology and fossil relatives. *Symposium Zoological Society London* **25**, 205-43.

JACKSON, J.B.C. 1977a. Competition on marine hard substrata: The adaptive significance of solitary and colonial strategies. *American Naturalist* **111**, 743-67.

JACKSON, J.B.C. 1977b. Habitat area, colonisation, and development of epibenthic community structure. In *Biology of Benthic Organisms* (eds. B.F. Keegan, P.O. Ceidigh and P.J.S. Boaden), pp. 349-58. Oxford: Pergamon Press.

JACKSON, J.B.C. 1985. Distribution and ecology of clonal and aclonal benthic invertebrates. In *Population Biology and Evolution of Clonal Organisms* (eds. J.B.C. Jackson, L.W. Buss, and R.E. Cook), pp. 297-355. New Haven: Yale University Press.

KHALFINA, V.K. & YAVORSKY, V.I. 1967. O drevneishikh stromatoporoideyakh [The oldest stromatoporoids]. *Paleontologishes Zeitshrift* **1967** (3), 133-36 (in Russian).

KRUSE, P. 1987. Further Australia Cambrian sphinctozoans. *Geological Magazine* **124**, 543-53.

LAND, L. 1989. The Carbon and Oxygen isotope chemistry of surficial Holocene shallow marine carbonate sediment and Quaternary limestone and dolomite. In *Handbook of Environmental Isotope Geochemistry.* (eds. P. Fritz and J.Ch. Fontes), pp. 191-216. *The Marine Environment* **3**.

LEVI, C. in press. Recent lithistids from the Norfolk Rise, Recent and Mesozoic genera. *Berliner Geowissenshaftliche Abhandlungen*.

OTT, E. 1974. *Phragmocoelia* n.g. (Sphinctozoa) ein segmentierter Kalkschwamm mit neuem Füllgewebetyp aus em Alpinen Trias. *Neues Jahrbuch für Geologie und Paläontologie Monatshefte* **1974**(2), 712-23.

PICKETT, J. & JELL, P.A. 1983. Middle Cambrian Sphinctozoa (Porifera) from New South Wales. *Memoirs of the Association of Australasian Palaeontologist* **1**, 85-92.

PICKETT, J. & RIGBY, J.K. 1983. Sponges from the early Devonian Garra Formation, New South Wales. *Journal of Paleontology* **57**, 720-41.

REITNER, J. 1987. Phylogenie und Konvergenzen bei Rezenten und Fossilen Calcarea (Porifera) mit einem kalkigen Basalskelett ("Inozoa", "Pharetronida"). *Berliner Geowissenshaftliche Abhandlungen* **86**, 87-125.

REITNER, J. 1989. Struktur, Bildung und Diagenese der Basalskelette bei rezenten Pharetroniden unter besonderer Berüruckischtigung von *Petrobiona massiliana* Vacelet & Levi, 1958 (Minchinellida, Porifera). *Berliner Geowissenshaftliche Abhandlungen* **106**, 343-83.

REITNER, J. in press. Phylogenetic aspects of calcified demosponges with consideration of calcified Hadromerida. *Berliner Geowissenschaftliche Abhandlungen.*

REITNER, J. & ENGESER, T. 1985. Revision der demospongier mit einem Thalamiden, aragonitischen Basalskelett und trabekularer Internstruktur ("Sphinctozoa" pars). *Berliner Geowissenschaftliche Abhandlungen* **60**, 151-93.

REITNER, J. & ENGESER, T. 1987. Skeletal structures and habits of Recent and fossil *Acanthochaetetes* (subclass Tetractinomorpha, Demospongiae, Porifera). *Coral Reefs* **6**, 13-18.

RIGBY, J.K. & BLODGETT, R.B. 1983. Early Middle Devonian sponges from the McGrath Quadrangle of west-central Alaska. *Journal of Paleontology* **57**, 773-86.

RIGBY, J.K. & POTTER, A.W. 1986. Ordovician sphinctozoan sponges from the eastern Klamath Mountains, northern California. *Paleontological Society Memoir* no. 20 (Journal of Paleontology **60** (4) Supplement), 1-47.

RIGBY, J.K., POTTER, A.W. & BLODGETT, R.B. 1988. Ordovician sphinctozoan sponges of Alaska and Yukon Territories. *Journal of Paleontotogy* **62**, 731-48.

ROZANOV, A. & DEBRENNE, F. 1974. Age of archaeocyathid assemblages. *American Journal of Science* **274**, 833-48.

SANDBERG, P.A. 1983. An oscillating trend in Phanerozoic nonskeletal carbonate mineralogy. *Nature* **305**, 19-22.

SAVARESE, M. 1988. Functional analysis of archaeocyathan skeletal morphology: implications for the group's paleobiology. *Geological Society of America Abstracts with Programs* **20**, A201.

SCHOPF, T.J.M. 1984. Rates of evolution and the notion of living fossils. *Annual Review of Earth and Planetary Sciences* **12**, 192-245.

STANLEY, G.D. 1988. The History of Early Mesozoic reef communities: A three step process. *Palaios* **3**, 170-83.

TERMIER, H. & TERMIER, G. 1977. Monographie paléontologique des affleurements Permiens du Djebel Tebaga (Sudtunisien). *Palaeontographica Abteilung A* **156**, 1-109.

VACELET, J. 1970. Les éponges Pharetronides actuelles. *Zoological Society of London Symposium* **25**, 189-204.

VACELET, J. 1985. Coralline sponges and the evolution of the Porifera. *Systematics Association Special Publication* **28**, 1-13.

VACELET, J. in press. The storage cells of calcified relict sponges. *Proceedings of the Third International Conference on the Biology of Sponges.*

VACELET, J. in press. The Recent Calcarea with a reinforced skeleton ("Pharetronid" sponges). *Berliner Geowissenshaftliche Abhandlungen.*

WEBBY, B.D. & RIGBY, J.K. 1985. Ordovician sphinctozoan sponges from central New South Wales. *Alcheringa* **9**, 209-20.

WENDT, J. 1974. Der Skelettbau aragonitischen Kalkschamme aus dem Alpinen Obertrias. *Neues Jahrbuch für Geologie und Paläontologie Monatshefte* **1974**, 498-511.

WENDT, J. 1980. Calcareous sponges - development through time. In *Living and Fossil sponges - notes for a short course.* (eds. W.D. Hartman *et al.*), pp. 169-78. *Sedimenta* 7.

WEST, R.R. & CLARKE, G.R. 1984. Paleobiology and biological affinities of Paleozoic chaetetids. *Palaeontographica Americana* **54**, 337-48.

WOOD, R. 1987. Biology and revised systematics of some late Mesozoic stromatoporoids. *Special Papers in Palaeontology* **37**, 1-89.

WOOD, R. 1990. Reef-building sponges. *American Scientist* **78**, 224-35.

WOOD, R. in press. Non-spicular biomineralization in demosponges. *Berliner Geowissenschaft Abhandlungen.*

WOOD, R. & REITNER, J. 1988. The Upper Cretaceous "chaetetid" demosponge *Stromatoaxinella irregularis* nov. gen. (MICHELIN) and its systematic implications. *Neues Jahrbuch für Geologie und Paläontologie, Abhandlungen* **177**, 213-24.

WOOD, R., REITNER, J. & WEST, R.R. 1989. Systematics and phylogenetic implications of the haplosclerid stromatoporoid *Newellia mira* nov. gen. *Lethaia* **22**, 85-93.

ZHURAVLEV A. YU. 1985. Sovremennye arkheotsiaty? [Recent archaeocyaths?] In *Problematiki pozdnego dokembriya paleozoya* (eds. B.S. Sokolov and I.T. Zhuravleva), pp. 24- 34. *Trudy Instituta Geologii i Geofiziki, Sibirskoe Otdelenie* **632** (in Russian).

ZHURAVLEV, A. YU. 1989. Poriferan aspects of archaeocyathan skeletal function. *Proceedings of the Sixth Conference on Fossil Cnidaria. Brisbane, Australia. Memoir Association of Australian Palaeontologists* **8**, 387-99.

ZHURAVLEV, A. YU., DEBRENNE, F. & WOOD, R.A. in press. A synonymised nomenclature for calcified sponges. *Geological Magazine.*

Beach and laboratory experiments with the jellyfish *Aurelia* and remarks on some fossil "medusoid" traces

David L. Bruton[1]

Abstract

Beach and laboratory experiments with fresh living jellyfish have shown how casts and impressions can be obtained from both fine and coarse sand. Animals thrown up onto the beach above high-water mark, dehydrate quickly leaving well defined impressions of the underside. Windblown sand can also cover the upper surface, become impregnated with decomposing fluids and the result is a cast of the upper surface. On a falling tide, stranded specimens can be covered quickly by sand thrown up by waves. Collapse structures result during decomposition. Dehydrated animals stranded on a fine silt flat leave delicate traces of the outline, genital organs and radial tubes. Jellyfish can only form moulds and potential fossils if stranded and allowed to dry. Dead animals sinking to the sea floor leave no identifiable impressions. These basic facts cast doubts on the "medusoid" traces from the Late Precambrian.

1. Introduction

Impressions of soft-bodied organisms in Late Precambrian rocks include those with a rounded outline and radial symmetry commonly referred to as "medusoids" (Wade, 1972). They are numerous in the Ediacara faunas first described from South Australia (see Glaessner, 1984 and refs.), but are now known from more than 20 localities all over the world. Traditionally, the Ediacara fossils are thought to be an assemblage of soft-bodied metazoan organisms (Cnidaria, annelids or arthropods) representing forerunners of modern phyla (Glaessner, 1984; Fedonkin, 1985) or, more radically, a unique set of metazoans, the Vendozoa (Seilacher, 1984, 1989). I agree with Seilacher that the preservational data provided by Glaessner & Wade (1966) and by Wade (1960, 1972) do not support the idea that the "medusoids" are Cnidaria and that the Ediacara sedimentary environments were not the sites where jellyfish sank quietly to the bottom nor beaches where they could strand leaving preservable impressions. My ideas are based on beach and laboratory studies with living jellyfish which, when deposited on a beach by waves, dry out and leave traces which have the potential of becoming fossilised. These traces are not like any Ediacara medusoids.

2. Beach observations

During the summer and early autumn, large numbers of jellyfish appear in the North Sea (Linke, 1956) and also invade the coastal waters of southern Norway. Of several species, two are most common, the transparent *Aurelia aurita* (Linnaeus, 1758) and the brown *Cyania capillata* (Linnaeus, 1758). The latter, among the largest invertebrates known with a diameter of up to 2m, prefers relatively cold water and extends far north to East Greenland, Jan Mayen, Svalbard and the Barents Sea. *A. aurita* has a more southerly distribution and is especially common in areas of low salinity. Concentrations of both species are also high immediately following periods of strong off-shore wind when warmer surface water is blown seawards and the jellyfish appear with the colder upwelling water.

From mid-June to mid-September 1977, hundreds of thousands of *A. aurita* were observed in parts of the inner Oslofjord and Lid (1979, p. 132) states that in Mossesundet "They formed a more or less continuous carpet near the surface and in many places formed such thick clumps that they reduced the speed of a boat with a small outboard motor" (transl.). Similar concentrations were observed offshore from the bay at Huk, Bygdøy, Oslo, whilst ashore many had been thrown up onto the sandy beach. The beach is made up of coarse sand layed out each summer and becomes mixed with local shale cobbles from the underlying bedrock. Tide marks can be distinguished by seaweed or by tracing lines of tidal debris. Figure 1A shows the positions of three high tide marks; T4 being the line of the high water spring, T3, the high tide mark on the evening of Saturday September 10., T2, the high tide mark at 4am the following day and T1 the low water mark 5hrs 17mins (i.e. 09.17hrs) later. All the following observations were made around this time.

No jellyfish were observed between the high water spring line and T3, but between the latter and T2, the high area of beach covered by the tide on the Saturday, were large specimens of *C. capillata*. Seven individuals were noted, all in an advanced state of decay, this having taken place during Saturday with a midday maximum temperature of 10.7 degrees C and six hours of sunshine. Parts of the pellicle were still visible mainly as a contracted brown slime. Where the animal had been deposited dorsal side up, a raised central area of firm sand occurred in the region of the mouth and radiating ridges corresponding to the tentacles (Fig. ID). This sand mould was firm to touch having been impregnated by fluid from the decomposing body. These impressions of the ventral surface of the bell had obviously survived the incoming Saturday evening tide. Animals deposited dorsal side down left only a faint and featureless trace. On Sunday morning's falling tide (Fig. 1A), large numbers of *A. aurita* (also a few *C. capillata*), were being deposited on the beach in concentrations up to 122 individuals per square metre and an average *Aurelia* diameter of 85mm (n = 50). As the tide fell, individuals were stranded in lines along the waters edge (Fig. 1B), and because of he 10 degree dip of the beach, the landward edge of the animal became stranded first whilst the seaward

[1]Paleontologisk Museum, Sars Gate 1, 0562 Oslo 5, Norway.

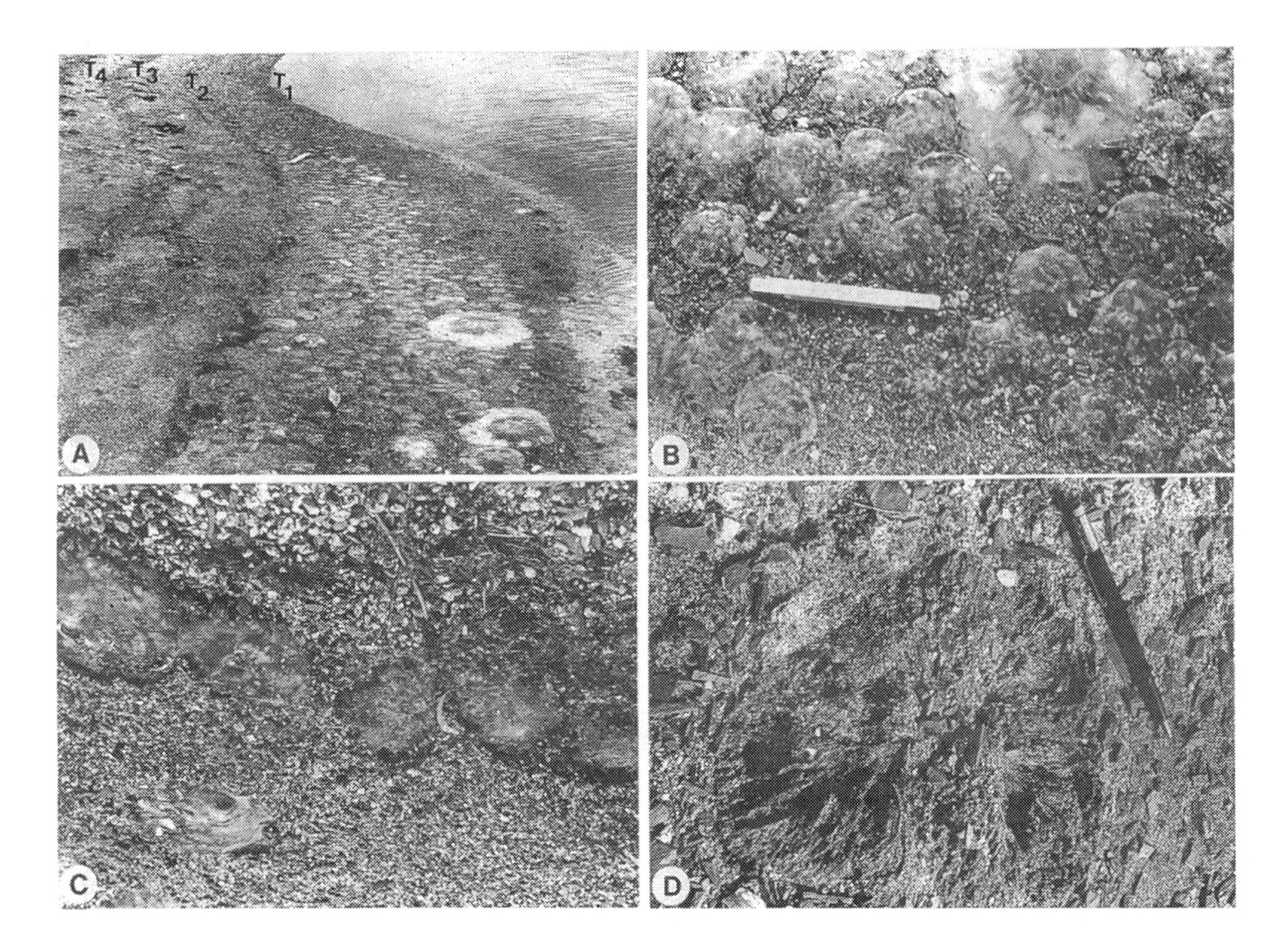

Fig. 1. Stranded specimens of *Aurelia aurita* and *Cyania capillata* (the larger), on beach at Huk, Bygdøy, Oslofjord. Photographs by author.
A. The observed beach area between high tide mark T2 and low water mark, T1, Sunday September 11, 1977.
B. Detail of observed area showing *A. aurita* stranded in lines parallel to tide mark at that time. Scale = 24cm.
C. *A. aurita* stranded at dead low water. Note partial burial by winnowed gravel.
D. *Cyania capillata*, mould in sand of ventral surface showing radial depressions of tentacles and raised central area of mouth. Beach area between T3 and T2, Saturday 10, September 1977. Pencil scale = 14cm.

edge remained afloat. Small waves and ripples repeatedly covered the stranded animals and eventually caused them to become totally or partially covered with sand or winnowed gravel at the time of dead low water (Fig. 1C). By midday, the earliest stranded animals were drying out and firm sand was already compressed beneath them. The succeeding incoming tide at 16.03hrs, aided by a fresh on-shore breeze, made up on its morning mark and the entire observed area was covered with water. By Monday morning (one high tide later), the jelly fish had been covered with sand. A similar situation had occurred earlier around the spring tide mark west of the observed area. Here, several *A. aurita*, all dorsal side up (Fig. 2A), had been exposed to the air for several days.

Their sand covered pellicle had dried out and formed a stiff, semi-transparent disc. When lifted off, this disc bore well defined traces of the radiating food canals, the central mouth and the gonad areas also preserved on the sand mould below (Figs. 2B,C). Both mould and disc were firm to touch and had survived a heavy rainfall as shown by surrounding rain drop impressions (Fig. 2A).

3. Laboratory experiments

In an attempt to replicate some of the beach observations described above, laboratory experiments were performed using living *Aurelia* and two different substrates. One substrate was a sample of sand from the beach at Bygdøy, the other from the Glomma river chosen for its fineness. Grain size tests were made following the techniques of Visher (1969) and raw data are available on request. Here it is sufficient to say that the Bygdøy sand is an immature arkosic arenite with no fraction finer than 125 μ while the river sand is moderately well sorted with an almost symmetrical grain size ranging from c. 3.9 μ to 50 μ. Sand substrates were prepared in watertight wooden or plastic boxes and covered with seawater. Two specimens of *A. aurita* (approximately 10cm diameter), were floated into a box containing the coarse sand and allowed to settle onto the substrate as the water was slowly drained off through a plug hole in the bottom of the box. The box was then left at room temperature (22 degrees C) for a period of four days. After this time dessication was complete and clear traces of the circular outline were

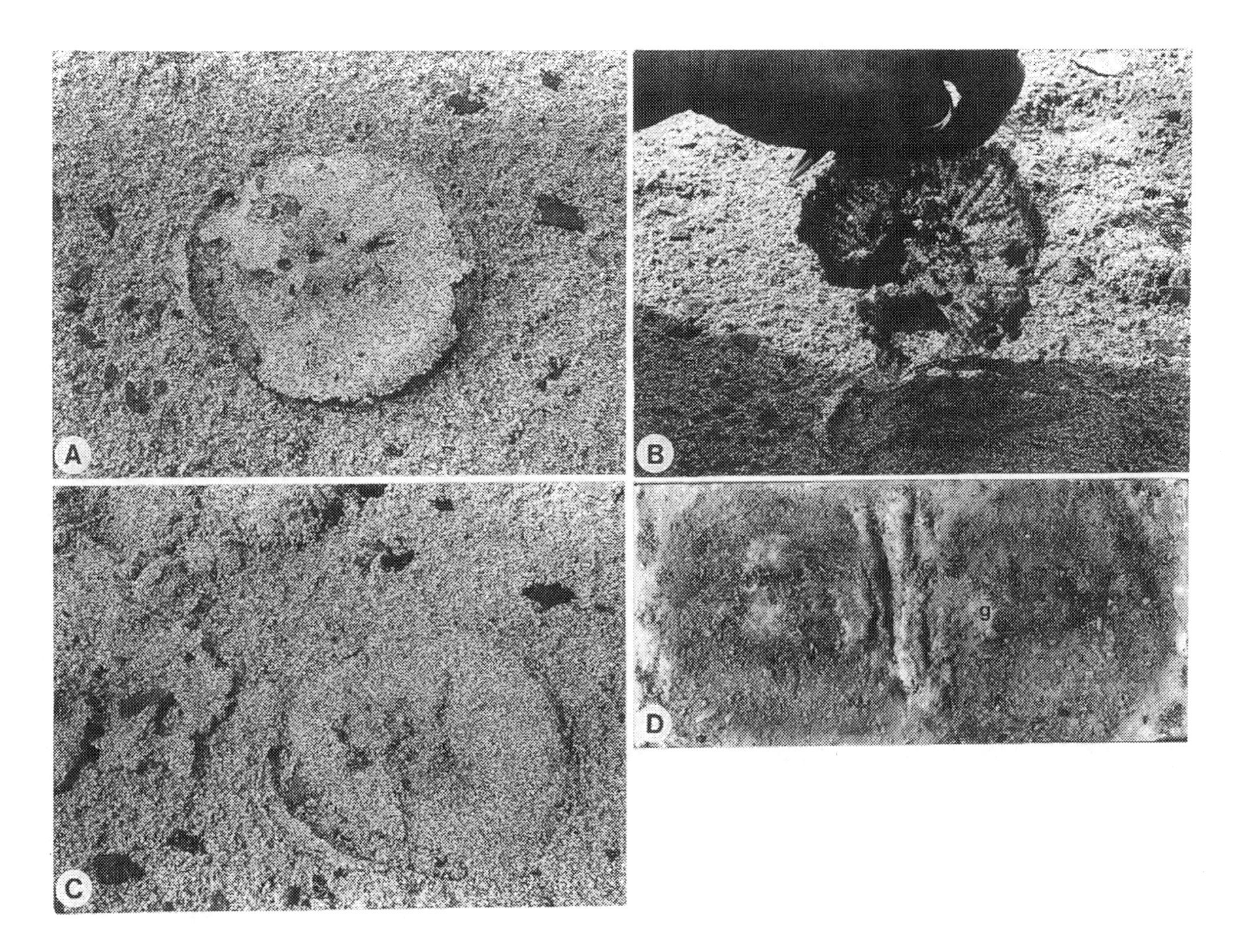

Fig. 2 A-C. Sand moulds of *Aurelia aurita* near high water spring tide mark, T4, on beach at Huk, Bygdøy, Oslofjord. Photographs by author.
A. Contracted pellicle, dried and sand covered.
B. Pellicle removed as stiff, semi-transparent disc showing radial food canals and central area.
C. Sand mould of ventral surface of umbrella showing same features as on disc (B). Both disc and mould could be potential fossils. Diameter 11cm.
D. Epoxy resin replica of coarse sand substrate showing outline of two dried out *A. aurita*. Note raised gonad area (g).

present together with faint outlines of the radial canals and gonad tracts (Fig. 2D). These features were much better preserved after desiccation of a large specimen (20cm diam.) which was allowed to settle on the finer sand substrate (Fig. 3B). In both experiments little or no relief was present in the region of the mouth. As observed on the beach, the substrates were hardened by seeping body fluids before dessication was complete and the resulting moulds could be cast directly without damage using silicone rubber (Type 533, from Wacker Chemicals U.K. Ltd., Bridge Street, Walton-on-Thames, Surrey KT 12 1AS, UK). The extent of natural hardening of the substrate is shown by the detail picked up by the silicone rubber and portrayed on positive reliefs in polyester resin (Jotun 4780, from Jotun Group A/S, 3200 Sandefjord, Norway), with a 1' methyl ethyl ketone (MEK) peroxide catalyst (Figs. 3A, C).

In the above experiments the animals were allowed to settle slowly whilst the supporting water was gradually tapped. As the water depth decreased, the animals became more agitated causing clouds of finer sediment to rise beneath the bell and this sediment finally became the casting medium for the ventral surface features. When in contact with the substrate slight contractions of the bell margin caused deepening and accentuation of the circular outline (Fig. 3A). At no time during these experiments did the animals attain inverted positions and all touched bottom with the oral surface down.

A third experiment was performed whereby a specimen was tossed, to land oral side down, onto a dry sandy substrate, covered with fine sand and allowed to dry out over a period of several days at 30 degrees C (Fig. 3D).

Subsidence of the bell occurred quite quickly with a hole appearing around the area of the mouth. Finally the entire structure collapsed and cracks appeared. Shrinkage caused an irregular outline with turned up edges but the entire sand structure was sufficiently strong to be cast using silicone rubber.

4. Discussion

Schafer (1941, 1972) has discussed the fossilisation potential of jellyfish and many of his observations have

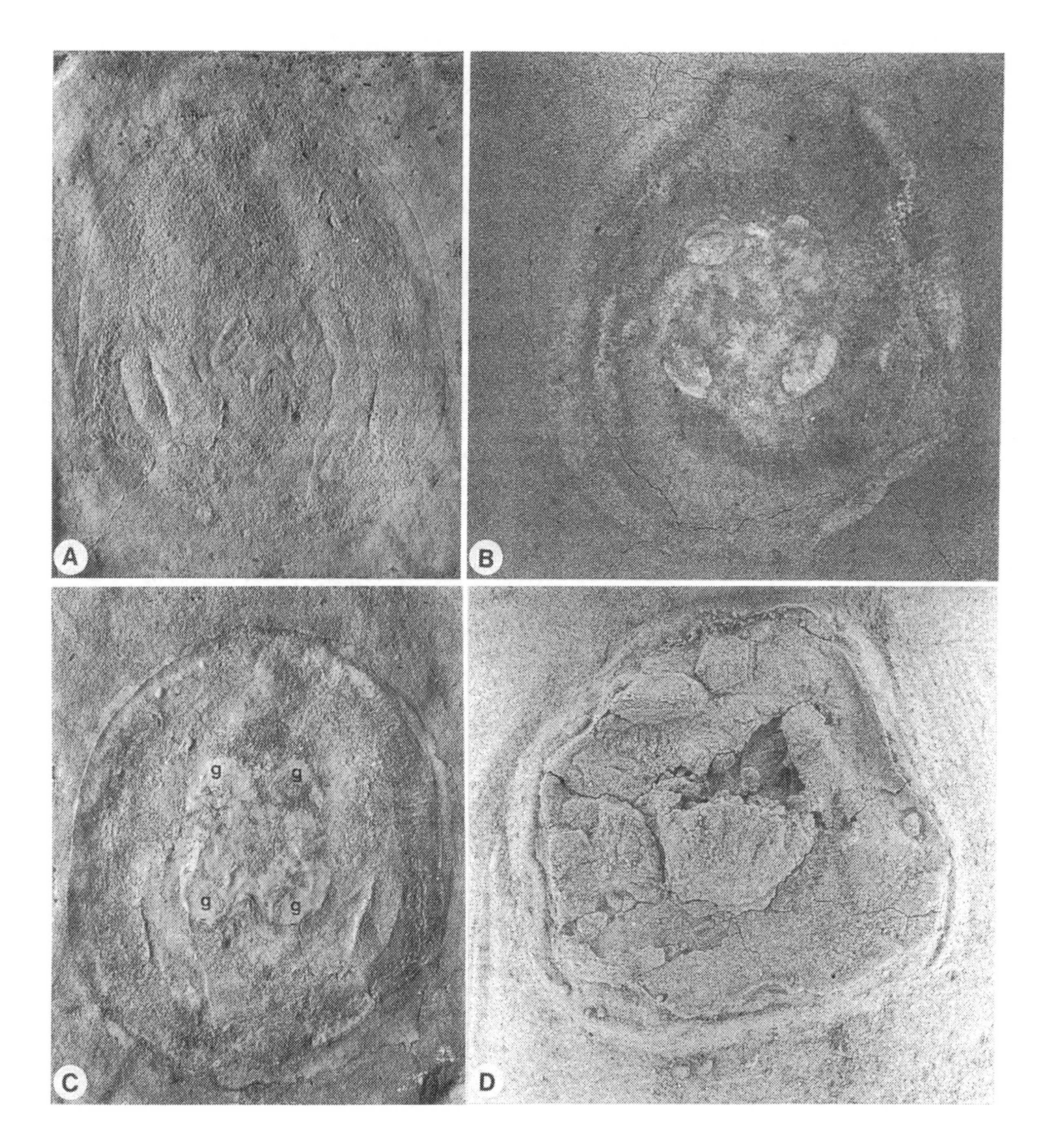

Fig. 3. Laboratory experiments with *Aurelia aurita* on fine sand substrate. Replicas by Aa. Jensen, photographs by P.Aas.
A. Silicone rubber mould of dried out specimen (B)
B. Dried out specimen showing well defined outline, radial canals and gonad tracts. Diameter 20cm
C. Epoxy resin cast from latex rubber mould (A) of specimen (B). Note surface detail especially of gonad tracts (g).
D. Dried out dorsal sand covered surface of *A. aurita*. Note collapse around central area, cracks in pellicle and shrinkage causing irregular outline. Diameter approx. 14 cm.

been confirmed here, the only difference being that I have used living specimens. Jellyfish can only be preserved as moulds and it is the interpretation of these which is important in interpreting the Ediacara "medusoids". The latter belong to what Wade (1968) called "non-resistant", meaning that they decayed before the covering sediments were compacted. The resulting fossils are therefore casts of decayed or compressed surfaces. The casts, in positive relief, occur on the base of sandstones which originally filled depressions of a clay or silt substrate. Wade (1968, p. 265) maintains that the majority of Ediacara "medusoids" are preserved exumbrellar (= dorsal) side down and that the casts are of this and other surfaces. However, in my view, the detail preserved is not that of an organism that has decayed nor would collapse of an inverted jellyfish produce such consistent shapes. My experiments have shown that collapse results in damage of the dorsal surface and distortion of the margin. Furthermore, studies of fresh living specimens of *Aurelia* have shown that it is extremely difficult to get the animals to turn over and almost all stranded specimens were dorsal side up. Those few that were upside down may well have been dead before grounding and turned in the swell. They left little or no casts in the sand. Wade (1968, p. 265), experimented with "nearly dead specimens" of modern jellyfish when trying to see how these inverted when water depth and agitation was changed. They apparently turned easily when water depth exceeded more than about 2/3 of their diameter.

Because jelly fish have nearly the density of sea water, they can only form moulds if stranded and allowed to dry, whilst dead animals sinking to the sea floor, retain so much buoyancy that they make no identifiable impression on the bottom (Schafer, 1972). These basic facts are, in themselves, enough to cast doubts on the Ediacara "medusoid" interpretations. Sedimentological data and palaeoenvironmental studies of the fossiliferous Ediacara Member is equivocal (see Goldring & Curnow, 1967; Jenkins *et al.*, 1983), but it is generally agreed that shallow water environments were present and at no time were the animals deposited on a beach (Glaessner, 1984, p. 49; Wade, 1968, p. 265).

What the Ediacara "medusoids" were is still not clear and Mary Wade's interpretaion of the fossils still remains ingenious. However, her starting point was that they were jellyfish and perhaps now we should look at other possibilities including an animal with a convex ventral surface anchored into the sediment. Seilacher's recent observations are pertinent but he does not interpret the observed morphology and the mode of preservation.

Acknowledgements

I wish to thank my wife, Anne, for help in the field and laboratory and for discussions throughout all phases of this work. Aage Jensen kindly made the rubber moulds and epoxy casts and Per Aas photographed them. Nicola Möller generously analysed the sand samples used. Discussions with Drs M.A. Fedonkin and W.G. Sun are gratefully acknowledged. I am also glad to acknowledge the hospitality of the University of Camerino and of the other agencies who sponsored this symposium.

REFERENCES

FEDONKIN, M.A. 1985. Precambrian metazoans: the problems of preservation, systematics and evolution. *Philosophical Transactions of the Royal Society of London* **B 311**, 27-45.

GLAESSNER, M.F. 1984. *The Dawn of Life. A Biohistorical Study. Cambridge: University Press.*

GLAESSNER, M.F. & WADE, M. 1966. The Late Precambrian fossils from Ediacara, South Australia. *Palaeontology* **9**, 599-620.

GOLDRING, R. & CURNOW, C.N. 1967. The stratigraphy and facies of the late Precambrian at Ediacara, South Australia. *Journal of the Geological Society of Australia* **14**, 195-214.

JENKINS, R.J.F., FORD, C.H. & GEHLING, J.G. 1903. The Ediacara Member of the Rawnsley Quartzite: the context of the Ediacara assemblage (Late Precambrian, Flinders Ranges). *Journal of the Geological Society of Australia* 30, 101-119.

LINKE, O. 1956. Quallen-Spülsäume. Ökologische Voraussetzung und aktuogeologische Ausdeutung. *Natur und Volk* **86**, 119-127.

LID, G. 1979. Forekomsten av stormaneter langs norskekysten. *Fauna* 32, 129-136.

SCHÄFER, W. 1941. Fossilisations-Bedingungen von Quallen und Laichen. *Senckenbergiana* **25**, 459-466.

SCHÄFER, W. 1972. *Ecology and Palaeoecology of marine environments.* (Transl. Oertel, I., ed., Craig, G.Y.). Oliver & Boyd, Edinburgh.

SEILACHER, A. 1984. Late Precambrian and Early Cambrian Metazoa: Preservational or real extinctions? In *Patterns of Change in Earth Evolution, (Dahlem Konferenzen)*, (eds. H.D. Holand & A.F. Trendall, pp. 159-168, Heidelberg: Springer Verlag.

SEILACHER, A. 1989. Vendozoa: Organismic construction in the Proterozoic biosphere. *Lethaia*, **22**, 229-239.

VISHER, G.S. 1969. Grain size distributions and epositional processes. *Journal of Sedimentary Petrology*, **39**, 1074-1106.

WADE, M. 1968. Preservation of soft-bodied animals in Precambrian sandstones at Ediacara, South Australia, *Lethaia*, **1**, 238-267.

WADE, M. 1972. Hydrozoa and Scyphozoa and other medusoids from the Precambrian Ediacara Fauna, South Australia. *Palaeontology*, **15**, 197-225.

Early Cambrian medusiform fossils from Chengjiang, Yunnan, China

Sun Weiguo[1]

Abstract

Heliomedusa Sun & Hou is no longer regarded as a hydrozoan medusoid as initially described (Sun & Hou, 1987). A re-examination (see also Conway Morris & Robison, 1988, p.39) indicates that it is actually an inarticulate brachiopod with two small circular delicate valves and unusually well-preserved soft-parts, including marginal setae, mantle canals and lophophore filaments, etc. *Rotadiscus* Sun & Hou seems a porpitid chondrophorine, which is essentially preserved with a large annularly chambered disc. It differs from other fossils and the living porpitids in possessing a trilete central structure on the assumed oral side. *Rotadiscus* is potentially significant in providing evidence for the prolonged conservative evolutionary history of Chondrophorina. Both *Stellostomites* Sun & Hou and *Yunnanomedusa* Sun & Hou are probably scyphozoan medusae (Sun & Hou, 1987) judging from their moderately large discoid umbrellas, distinct annular muscle bands and radial canals, which are simple unbranched in *Stellostomites* and trifurcately branched in *Yunnanomedusa*. However, comparison of *Stellostomites* with the Middle Cambrian *Eldonia* Walcott, which has been interpreted as an umbrella-bearing holothurian (Durham, 1974) may cast doubt on the supposed scyphozoan affinities of *Stellostomites* and possibly *Yunnanomedusa* as well (see also Conway Morris & Robison, 1988).

REFERENCES

Conway Morris, S. & Robison, R.A. 1988. More soft-bodied animals and algae from the Middle Cambrian of Utah and British Columbia. *The University of Kansas Paleontological Contributions, Paper* **122**, 1-48.

Durham, J.W. 1974. Systematic position of *Eldonia ludwigi* Walcott. *Journal of Paleontology* **48**, 750-755.

Sun Weiguo & Hou Xianguang 1987. Early Cambrian medusae from Chengjiang, Yunnan, China. *Acta Palaeontologica Sinica* **26**, 257-271. [In Chinese, with English abstract].

[1]Nanjing Institute of Geology and Palaeontology, Academia Sinica, Nanjing, China 210008.

The enigma of conulariid affinities

Loren E. Babcock[1]

Abstract

The conulariids are an extinct group of organisms whose phylogenetic affinities have been debated for more than 160 years. New or recently published evidence shows that the group has been used as a taxonomic wastebasket, and that most hypotheses of affinity have been based on organisms that have no demonstrable close relationship to *Conularia*. Misinterpretations about preserved morphology and disagreements about morphological terminology have further limited attempts to resolve the question of their affinities.

As now understood, the Conulariida Babcock & Feldmann, 1986a, is a monophyletic lineage of Bilateria having no obvious relationship to any extant phylum. They secreted an elongate, four-sided pyramidal exoskeleton having a phosphatic composition and bilateral symmetry. The exoskeleton was strengthened in part by transverse thickenings called rods, which have a pairwise arrangement on each face. Faces are separated by distinct longitudinal invaginations at each corner. Except perhaps as larvae, conulariids were sessile, being attached by a flexible, chitinoid? stalk. Evidence about soft-part morphology indicates that they may have achieved a triploblastic, coelomic? grade of organization. The conulariids are considered to be an independent, extinct phylum.

1. Introduction

Few organisms, living or extinct, have inspired as many hypotheses of affinity as have the conulariids. Statements about the phylogenetic relationships of these extinct organisms have been published for more than 160 years. In that time, they have been suggested to be mollusks, 'worms,' cnidarians, conodont animals, invertebrate chordates, vertebrate relatives, and an independent animal phylum.

Discovering the zoological relationships of conulariids has for many years been hindered by a variety of problems, including problems of definition, a lack or misinterpretation of preserved morphology, and use of either inconsistent morphological terminology or terms that connote a presumed affinity. Recent increased interest in conulariids has produced much new information about their hard- and soft-part morphology, and older interpretations about their affinities have been reevaluated. It is now clear that the word "conulariid" has been used as a wastebasket taxonomic term and that less than 60 percent of genera recently assigned as 'conulariids' (He, 1987; Hergarten, 1988) can be accommodated without reservation in the group. Furthermore, it has become evident that most hypotheses of affinity were primarily based on organisms having no demonstrated relationship with *Conularia* Miller in Sowerby, 1820, the nominate genus of the group.

The purpose of this paper is to summarize and evaluate hypothesized relationships of conulariids. Current information about their morphology and paleobiology is discussed, and cluster analysis is used to help clarify relationships among putative conulariids. Although their ancestry remains obscure, increasing evidence about *Conularia* and its close relatives indicates that they are a group of Bilateria having no obvious relationship to any phylum of living organisms. Convergent characters and lack of some critical anatomical information, however, contribute to the enigmatic, and still somewhat ambiguous, nature of conulariids.

2. Hypothesized relationships

Interpretations about the affinities of conulariids have greatly changed as information about their morphology gradually accumulated and as concepts of included organisms changed. Because of a former common belief that they must fit into a living higher taxonomic group, they have been considered to be representatives of an extant phylum or class by almost all workers until the late 1960's, and by some until the present.

The word "conulariid" has become a wastebasket term for a large number of relatively simple pyramidal, conical, or tubular fossils of variable composition (Babcock & Feldmann, 1986c; Feldmann & Babcock, 1986). Most published phylogenetic inferences have been based primarily on organisms that differ substantially in skeletal morphology from *Conularia* and that are probably not closely related. To avoid confusion in this paper, organisms referred to as conulariids are ones having a body plan conforming in fundamental characters to *Conularia* (discussed below). Such organisms are here interpreted to represent a monophyletic (holophyletic) group. Where fossils have been classified in a demonstrably polyphyletic group, the informal name is enclosed in single quotes, 'conulariid'.

Morphological terminology for conulariids mostly follows that of Babcock & Feldmann (1986a). Use of the term "carina" follows Babcock *et al.* (1987b).

2.a. *MOLLUSKS*

From their initial scientific description (Miller in Sowerby, 1820) until the late 1930's, conulariids were accepted as mollusks by most authorities. The early association of *Conularia* with the cephalopods (e.g., Eichwald, 1840; Vanuxem, 1842) seems to have been based on its general resemblance to orthoconic nautiloids and the presence of apical walls. Indeed, one species referred by Sowerby (1820) to *Conularia* was later shown (Barrande, 1867) to be an orthoconic nautiloid.

The interpretation of *Conularia* as a cephalopod,

[1]Department of Geology, The University of Kansas, Lawrence, Kansas 66045, U.S.A. and Kansas Geological Survey, 1930 Constant Avenue, The University of Kansas, Lawrence, Kansas 66047, U.S.A. Present address: Department of Geology and Mineralogy, The Ohio State University, 125 South Oval Mall, Columbus, Ohio 43210, U.S.A.

which was common until the 1840's and accepted by some (e.g., Geinitz, 1853; Ihering, 1881) until much later, was largely supplanted by Archiac & Verneuil's (1842) placement of the genus in the Pteropoda. The problematic genera *Tentaculites* Schlotheim, 1820, and *Hyolithes* Eichwald, 1840, were also reassigned to the Pteropoda. Reasons for the new assignments were not stated, but the proposed relationship was adopted by most contemporary paleontologists (e.g., Sandberger, 1847; McCoy, 1852; Barrande, 1867; Hall, 1879; Holm, 1893), and published in textbooks until the 1930's. Lindström (1884), who favoured an affinity of conulariids with pteropods, discussed the similarities in external morphology between the shells of some species of present-day pteropods and *Conularia*. He hypothesized that the "internal longitudinal septa" (carinae *sensu* Babcock *et al.*, 1987b) were homologous with structures of similar appearance in present-day pteropods. Wiman (1895) illustrated line drawings of a conulariid in cross section showing elongate bifurcate "septa" extending inward from the midlines that seemed to support Lindström's (1884) argument of homology.

Other workers who favoured an affinity of *Conularia* with mollusks but who were reluctant to place the genus either in the Cephalopoda or the Pteropoda include Pelseneer (1889), who regarded it as a lineage of opisthobranch gastropods other than pteropods, and Neumayer (1879), Slater (1907), and Grabau & Shimer (1910), who regarded the genus as representative of an independent molluscan group.

Walcott (1885) erected the family Conulariidae within the Pteropoda to include *Conularia*, as well as *Palaenigma* Walcott, 1885, which is an unrecognizable conulariid genus based on badly weathered specimens, and *Matthevia* Walcott, 1885, which is now generally considered to be a polyplacophoran mollusk (Runnegar *et al.*, 1979).

The name Conularida was first used (Waagen, 1891) for mollusks that were "certainly not pelagic shells". The group consisted of three families: 1) Conulariidae; 2) Thecidae, which included hyoliths; and 3) Tentaculitidae, which included *Tentaculites*. Miller & Gurley (1896) redefined Conularida as an order of mollusks with "pelagic shells having lime-phosphate" to include taxa in Walcott's Conulariidae and Miller & Gurley's (1896) Enchostomidae. The Enchostomidae is based upon *Enchostoma* Miller & Gurley, 1896, a possible worm (Fisher, 1962). The new order included pyramidal shells having cross sections that varied by species "from square and subquadrate, to octagonal and somewhat rounded". This broad definition permitted the inclusion of a variety of problematic organisms, both by Miller and Gurley, and later by others (see especially Termier & Termier, 1950). Grabau & Shimer (1910) raised the Conularida to class rank within the Mollusca.

2.b. *'WORMS'*

Ruedemann (1896a, 1896b) reported the discovery of sessile 'conulariids' from the Ordovician of New York and suggested that they are 'worms.' His figures showed small individuals seemingly budding from larger ones, and attached by basal attachment disks. Examination of Ruedemann's extant specimens in the New York State Museum (Albany, New York) indicated that his figures are probably composites. The specimens are not *Conularia* as Ruedemann reported, but *Sphenothallus* Hall, 1847 (see Feldmann *et al.*, 1986; Bodenbender *et al.*, 1989), which is now regarded as a 'worm' tube by most authorities (e.g., Schmidt & Teichmüller, 1958; Mason & Yochelson, 1985; Fauchald *et al.*, 1986; Feldmann *et al.*, 1986; Bodenbender *et al.*, 1989; but see Cox *et al.*, 1986).

Other authors who supported an interpretation of 'conulariids' as 'worms' include Weller (1925) and Lalicker & Moore (1952).

2.c. *CNIDARIANS*

Kiderlen (1937) published stimulating ideas about the functional morphology and phylogenetic relationships of 'conulariids' that dominated thinking about the group until the late 1960's, and that have continued to influence discussions to the present. His suggestion that 'conulariids' are scyphozoan cnidarians was based primarily on an interpretation of radial symmetry that was strengthened by Wiman's (1895) figures of "septa" in *Eoconularia loculata* and Ruedemann's (1896a, 1896b) figures of small "*Conularia*" (= *Sphenothallus*) specimens budding from larger ones. Other evidence of a radial symmetry came from supposed pathologic conulariids having two or three sides (Kiderlen, 1937). Kiderlen reconstructed 'conulariids' with tentacles and interpreted their life cycle as consisting of an initial attached polypoid phase, followed by a severing of the attached end, and development of a free-living medusa.

A cnidarian affinity was supported (Knight, 1937) by the suggested relationship of the problematic fossil *Conchopeltis* Walcott, 1876, with conulariids. Knight wrote that *Conchopeltis* bore a "striking agreement" in morphology with *Conularia*, and he reassigned *Conchopeltis* from the patellid gastropods to Walcott's (1885) Conulariidae, which he suggested should be placed in the phylum Coelenterata. Later, Moore & Harrington (1956a) figured a specimen of *Conchopeltis* having preserved tentacles, which reinforced the claim that it was a cnidarian.

Moore & Harrington (1956a) erected the monordinal subclass Conulata (class Scyphozoa) to embrace the order Conularida (respelled Conulariida) Miller & Gurley, 1896, which had been expanded to receive a variety of problematic fossils. They also used the subordinal name Conulariina Miller & Gurley, 1896 (*nomen translatum*) for the first time. It included the same taxa as their subclass Conulata and the order Conularida. Moore & Harrington (1956a) also suggested that 'conulates' formed their skeletons not only of calcium phosphate (phosphorite or collophane; Bouček & Ulrich, 1929), but also of chitin or a chitinophosphatic substance.

Chapman (1966) reviewed contemporary information about 'conulariids' and concluded that they were ancestral to the scyphozoan *Scyphistoma*. About the same time, Werner (1966, 1967, 1969, 1973) suggested that the living *Stephanoscyphus* (a provisional generic name for the polypoid stage of various genera of coronatid scyphozoans; Chapman & Werner, 1972) represents a primitive scyphozoan link with 'conulariids,' an interpretation that was followed by some authors (e.g., Grasshoff, 1984; Robson, 1985) but challenged by Kozłowski (1968) and Scrutton (1979).

Glaessner (1971) figured a specimen of the late Precambrian species *Conomedusites lobatus* Glaessner & Wade, 1966, having putative tentacles and considered it to be related to *Conchopeltis*. The suborders Conchopeltina and Conulariina were raised to ordinal rank and the subclass Conulata was raised to class rank within the Cnidaria. Glaessner regarded 'conulariids' as being less closely related to scyphozoans than thought by Chapman (1966) and Werner (1966, 1967, 1969, 1973), and as characteristically polypoid, not transitional to medusae as previously thought (Kiderlen, 1937; Moore & Harrington, 1956a).

Bischoff (1973, 1978) described and figured some small tubular and conical phosphatic fossils of Palaeozoic age and interpreted them as 'conulariids'. It was suggested (Bischoff, 1973) that some conodont elements were secreted by certain 'conulariids'. Later, Bischoff (1978) erected a suborder of 'conulariids', the Circoconulariina, for fossils showing internal septa that have possible homology with the septa of living cnidarians.

The cnidarian hypothesis of affinity was strongly challenged during the 1980's, particularly by Oliver (1984) and Babcock & Feldmann (1986a, 1986c). Nearly all of Kiderlen's (1937) interpretations of conulariid paleobiology had been shown to be in error (see section 4) and some were based on organisms that have been subsequently removed from taxonomic association with conulariids. Organisms that had been used to support a cnidarian affinity, but which are doubtfully related to *Conularia*, are discussed in section 3.a.1.

Renewed support for a cnidarian interpretation during the 1980's was based primarily on studies of uncharacteristic 'conulariids'. *Sphenothallus*, which is thought by most authors to be a 'worm' (Schmidt & Teichmuller, 1958; Mason & Yochelson, 1985; Fauchald *et al.*, 1986; Feldmann *et al.*, 1986; Bodenbender *et al.*, 1989), has been interpreted by some (Cox *et al.*, 1986) to be a cnidarian and closely related to 'conulariids.' Van Iten (1987) asserted that 'conulariids' reproduced through asexual budding, and that apical walls were secreted to close the exoskeleton when individuals broke free of their apical attachments to become free-living. Van Iten (1988) also compared the corner and midline areas of 'conulariids' with presumed homologous structures in three present-day taxa, and stated that 'conulariids' show more homologies with scyphozoans than with either mollusks or annelids.

2.d. *INVERTEBRATE CHORDATES OR VERTEBRATES*

Twice it has been suggested that conulariids are either invertebrate chordates or vertebrate relatives. Termier & Termier (1949, 1953) suggested that they were related to the "stomocordes", a deuterostome group that includes hemichordates, graptolites, and pterobranchs. Aspects of the internal and external morphology, particularly the division of the interior into chambers by apical walls and the supposed fusellary tissue of the skeleton, were cited as support for that interpretation. Steul (1984), who examined specimens showing pyritized soft parts, interpreted conulariids as vertebrate relatives on the basis of their internal bilateral symmetry, the characteristic phosphatic composition of the exoskeleton, a supposed segmented axial skeleton with segmentally arranged muscle bunches, and supposed photosensitive organs.

2.e. *INDEPENDENT PHYLUM*

A number of authors, including Sinclair (1948), Lalicker & Moore (1952), Kozłowski (1968), and Mortin (1985) suggested the possibility that 'conulariids' represent an independent, extinct phylum. Babcock & Feldmann (1986a, 1986c) recognized that Miller & Gurley's (1896) order Conularida was a taxonomic wastebasket. Based partially on new morphological evidence about *Conularia* and *Paraconularia* Sinclair, 1940a, they (Babcock & Feldmann, 1986a) introduced the phylum Conulariida on the argument that authentic conulariids have a fundamentally different morphology from that of any known animal, making any interpretation of a close phylogenetic relationship very unlikely. Only fossils having a demonstrable morphological similarity and probable phylogenetic relationship with *Conularia* were included in the phylum (Babcock & Feldmann, 1986a, 1986b, 1986c; Feldmann & Babcock, 1986; Babcock *et al.*, 1987a).

McMenamin (1987) cited similarities of conulariids with some frondose organisms of the late Precambrian Ediacaran fauna and suggested that frondose organisms gave rise to the phylum.

3. What is a conulariid?

Much confusion about conulariid affinities has resulted from the long historical use of the group as a taxonomic wastebasket that at various times has included many elongate pyramidal, conical, or tubular fossils of phosphatic, chitinous, chitinophosphatic, or even calcareous composition. The problem has been exacerbated by the common practice of adding the suffix *-conularia* to new generic names for putative conulariids. This suffix implies an affinity with *Conularia* that may not always be justified. Differing interpretations through history about what constitutes a conulariid have greatly contributed to misunderstandings about inferred homologies with other organisms.

Recent morphological analyses of *Conularia* and *Paraconularia* (Babcock & Feldmann, 1986a; Feldmann & Babcock, 1986; Babcock *et al.*, 1987a, 1987b; Babcock, 1990) permit a less ambiguous definition of conulariids than was previously possible. As now understood, the Conulariida includes only those fossils having a four-sided elongate pyramidal exoskeleton of phosphatic composition and weak bilateral symmetry (Figure 1). A slight curvature, which is most noticeable in the apical region, indicates that skeletal growth was allometric. The exoskeleton is multilayered, and an organic material may have separated phosphatic laminae. Rods, which are slender, arched, transversely oriented, thickened structures in the integument, probably served to reinforce the exoskeleton. Externally, they produce narrow ridges. After death, the rods are somewhat more resistant to decomposition than the rest of the exoskeleton. Rather than being continuous across each face, they are arranged in pairs that meet at a longitudinal midline. Commonly, the midline is thickened into a low internal carina. Alternatively, a pair of carinae flanking the midline may be present (Sinclair, 1940b), but bifid septa are unproven in any authentic conulariid. The rods articulate in longitudinal invaginations in the corner regions. These corner grooves also probably added structural strength to the exoskeleton. Species for which the apical morphology is known had a closed apex (Figure 2) and were attached by a flexible, perhaps chitinoid, stalk that partially sheathed the apical end. Some species secreted one or more smooth, transverse, internal walls near the apical end. Available evidence of soft parts (Babcock & Feldmann, 1986a, 1986b; see also Steul, 1984) indicates the presence of

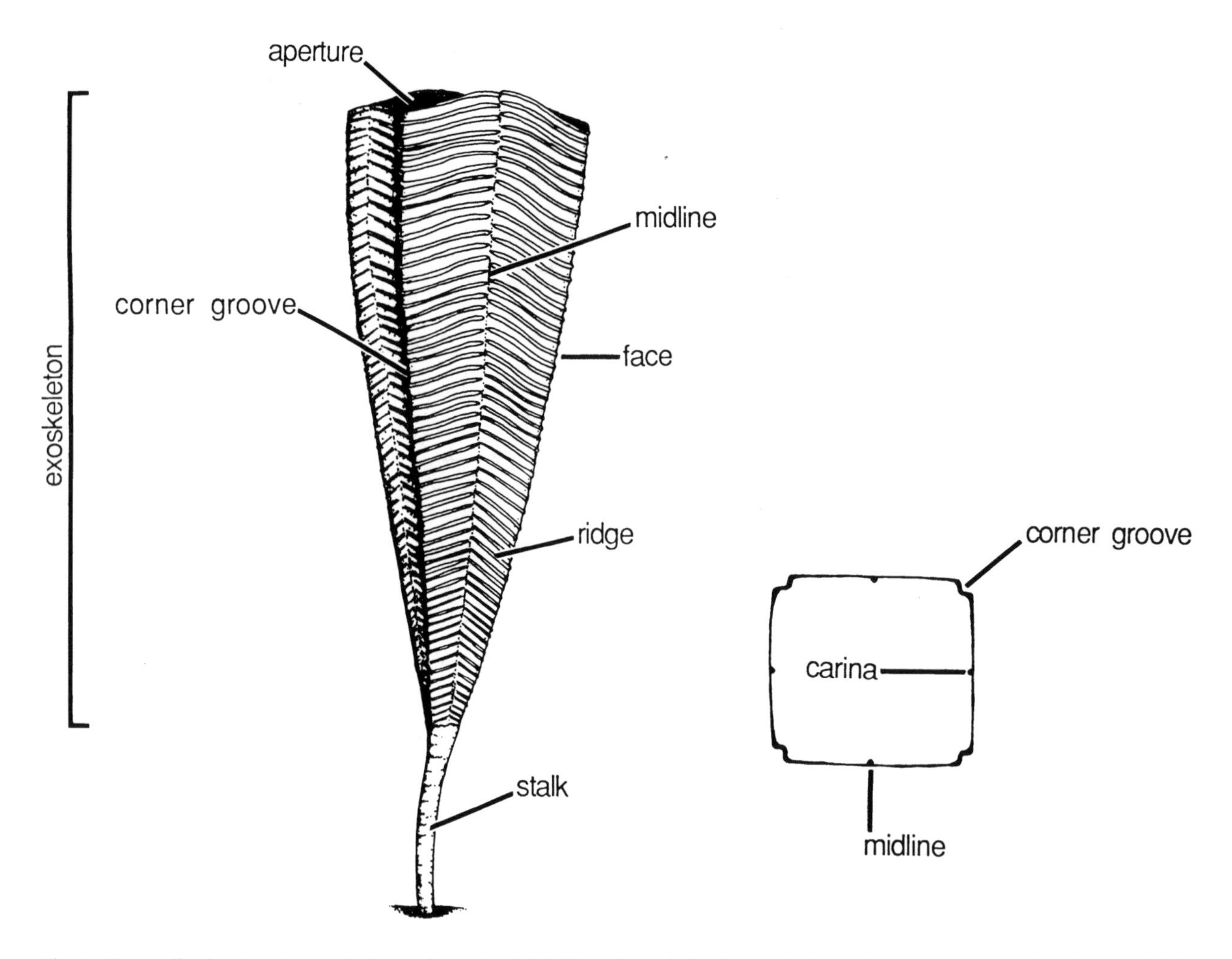

Fig. 1. Generalized external morphology of conulariid (left) and exoskeletal morphology shown in cross-sectional view (right), based on *Paraconularia*.

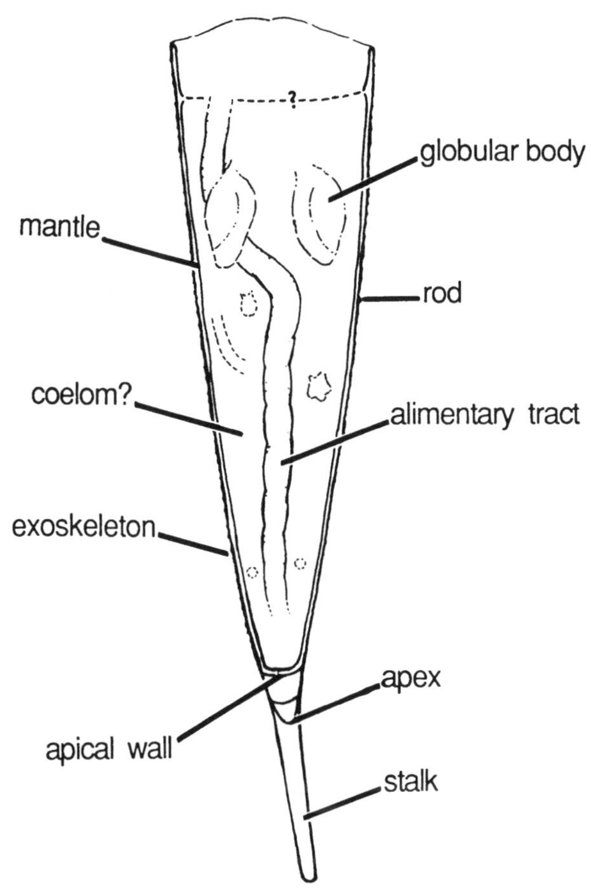

Fig. 2. Generalized conulariid in exsagittal section showing interpretive relationships of the exoskeleton, stalk, and soft parts, based on *Conularia* and *Paraconularia*.

a straight (blind?) alimentary tract and several globular-shaped bodies, some of which are paired (Fig. 2). A mantle or mantle-like tissue that surrounded the soft parts and secreted the exoskeleton has also been inferred (Babcock, 1990).

3.a. *CLUSTER ANALYSIS*

Cluster analysis was used to quantitatively explore patterns of morphology among 'conulates' or 'conulariids' on the assumption that morphological similarity or dissimilarity may reflect, in a general way, relative phylogenetic distance. Genera representing all three suborders assigned to Miller & Gurley's (1896) order Conularida (subclass Conulata, class Scyphozoa) by Hergarten (1988) were included in the analysis. They are the suborders Conchopeltina, Conulariina, and Circoconulariina. Although excluded by Hergarten (1988) from the 'conulariids', *Sphenothallus* was included in the analysis because it has historically been considered to be a 'conulariid' or closely related to them (Ruedemann, 1896a, 1896b; Moore & Harrington, 1956b; Cox *et al.*, 1986). *Conchopeltis* was similarly included in the analysis mostly for historical reasons although Oliver (1984) concluded that it is unrelated to conulariids. Conodont elements placed by Bischoff (1973) in the 'conulariid' suborder Conodontina, however, were not considered in the analysis.

Binary (presence-absence) data were compiled and analyzed for 40 characters preserved in 25 genera (Table 1) that have been described as 'conulariids'. The included genera are known from relatively complete material but where a character state was unknown, it was so coded. Computations were performed using the program Systat, which ignores character states that are coded as "unknown". Coefficients of Euclidean distance were computed for the data matrix in Q mode. In Q-mode analyses, objects (genera) are related to each other on the basis of their attributes (characters). A dendrogram (Figure 3) was constructed from the distance matrix using unweighted pair group arithmetic means (UWPG).

The dendrogram (Figure 3) shows a straightforward division of 'conulates' into two primary clusters: 1) an assemblage of low-level clusters that were collectively linked at 0.39 (labelled "conulariids" in Figure 3); and 2) an assemblage of clusters of variable levels that were collectively linked at 0.59 (labelled "non-conulariids" in Figure 3) The two primary clusters were linked at 0.64 but separated for illustration. Qualitative morphological studies (e. g., Babcock & Feldmann, 1986c; Feldmann & Babcock, 1986; Babcock *et al.*, 1987a) have shown that genera in the non-conulariid primary cluster have only a superficial similarity to *Conularia* and characters homologous with those in *Conularia* cannot be demonstrated. *Conularia* and other fossils that can be related by presumably homologous structures are in the conulariid cluster. The general conclusion is that the groups Conulata Moore & Harrington, 1956b, Conularida Miller & Gurley, 1896, and Conulariina Miller & Gurley, 1896, are polyphyletic.

Among the genera analyzed, the only ones here considered to be authentic conulariids are those grouped in the primary cluster containing *Conularia* Miller in Sowerby, 1820. In addition to *Conularia*, they include *Archaeoconularia* Bouček, 1939, *Climacoconus* Sinclair, 1952, *Eoconularia* Sinclair, 1943, *Exoconularia* Sinclair, 1952, *Glyptoconularia* Sinclair, 1952, *Metaconularia* Foerste, 1928, *Notoconularia* Thomas, 1969, *Paraconularia* Sinclair, 1940a, *Pseudoconularia* Bouček, 1939, *Reticulaconularia* Babcock & Feldmann, 1986b, and *Tasmanoconularia* Parfrey, 1982. All other genera considered in the analysis show a high level of morphological dissimilarity from the Conulariida and are here excluded from that group because a close relationship cannot be demonstrated.

3.a.1. *Discussion of organisms excluded from the Conulariida*

Thirteen genera in the analysis formed a primary cluster of heterogeneous fossils at a high level of dissimilarity (0.64) with organisms considered to be authentic conulariids, which were collectively linked at 0.39. Few of these problematic organisms can be confidently placed in higher taxa, and some may represent extinct phyla. With few exceptions, these taxa seem to be very distantly related to one another.

Conchopeltis Walcott, 1876, and *Conomedusites* Glaessner & Wade, 1966, are probable cnidarians. The interpretation of *Conchopeltis* as a cnidarian is based on several morphological characters (Oliver, 1984), the most convincing of which is the presence of preserved tentacles in one specimen (Oliver, 1984, pl. 1, figs. 1, 2). Previously, that specimen was used in support of the notion that 'conulariids' were tentacle-bearing cnidarians (Moore & Harrington, 1956a). Criteria for considering *Conomedusites* to be a cnidarian were published by Glaessner (1971), and include the presence of possible tentacles. *Conchopeltis* and *Conomedusites* were linked at a dissimilarity value of 0.33. This is considered to be a meaningful cluster of relatively similar organisms.

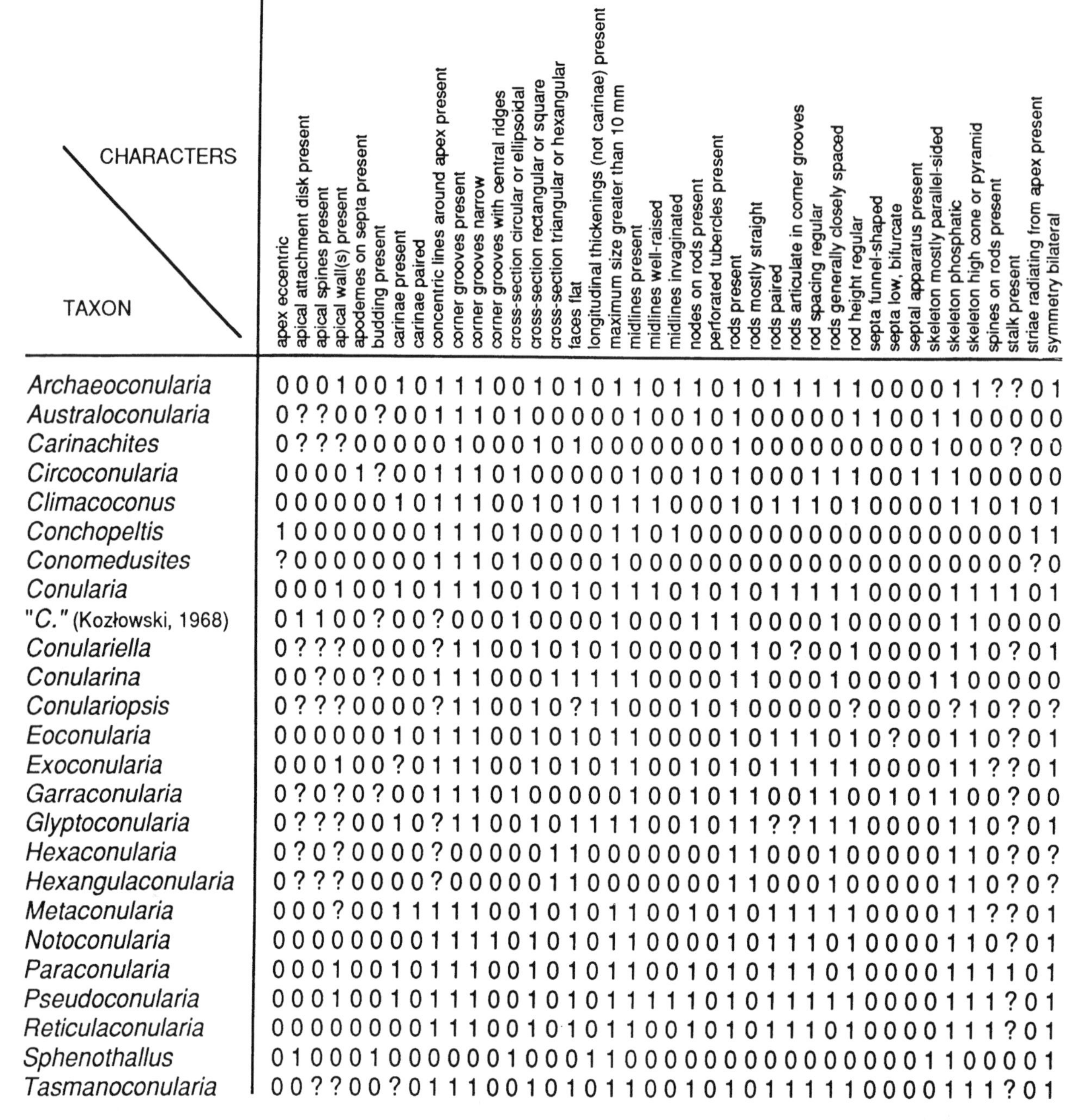

TAXON \ CHARACTERS	apex eccentric	apical attachment disk present	apical spines present	apical wall(s) present	apodemes on septa present	budding present	carinae present	carinae paired	concentric lines around apex present	corner grooves present	corner grooves narrow	corner grooves with central ridges	cross-section circular or ellipsoidal	cross-section rectangular or square	cross-section triangular or hexangular	faces flat	longitudinal thickenings (not carinae) present	maximum size greater than 10 mm	midlines present	midlines well-raised	midlines invaginated	nodes on rods present	perforated tubercles present	rods present	rods mostly straight	rods paired	rods articulate in corner grooves	rod spacing regular	rods generally closely spaced	rod height regular	septa funnel-shaped	septa low, bifurcate	septal apparatus present	skeleton mostly parallel-sided	skeleton phosphatic	skeleton high cone or pyramid	spines on rods present	stalk present	striae radiating from apex present	symmetry bilateral
Archaeoconularia	0	0	0	1	0	0	1	0	1	1	1	0	0	1	0	1	0	1	1	0	1	1	0	1	0	1	1	1	1	1	0	0	0	0	1	1	?	?	0	1
Australoconularia	0	?	?	0	0	?	0	0	1	1	1	0	1	0	0	0	0	0	1	0	0	1	0	1	0	0	0	0	0	1	1	0	0	1	1	0	0	0	0	0
Carinachites	0	?	?	?	0	0	0	0	0	1	0	0	0	1	0	1	0	0	0	0	0	0	0	1	0	0	0	0	0	0	0	0	0	1	0	0	0	?	0	0
Circoconularia	0	0	0	0	1	?	0	0	1	1	1	0	1	0	0	0	0	0	1	0	0	1	0	1	0	0	0	1	1	1	0	0	1	1	1	0	0	0	0	0
Climacoconus	0	0	0	0	0	0	1	0	1	1	1	0	0	1	0	1	0	1	1	1	0	0	0	1	0	1	1	1	0	1	0	0	0	0	1	1	0	1	0	1
Conchopeltis	1	0	0	0	0	0	0	0	1	1	1	0	1	0	0	0	0	1	1	0	1	0	0	0	0	0	0	0	0	0	0	0	0	0	0	0	0	0	1	1
Conomedusites	?	0	0	0	0	0	0	0	1	1	1	0	1	0	0	0	0	1	0	0	0	0	0	0	0	0	0	0	0	0	0	0	0	0	0	0	0	0	?	0
Conularia	0	0	0	1	0	0	1	0	1	1	1	0	0	1	0	1	0	1	1	1	0	1	0	1	0	1	1	1	1	1	0	0	0	0	1	1	1	1	0	1
"*C.*" (Kozłowski, 1968)	0	1	1	0	0	?	0	0	?	0	0	0	1	0	0	0	0	1	0	0	0	1	1	1	0	0	0	0	1	0	0	0	0	0	1	1	0	0	0	0
Conulariella	0	?	?	?	0	0	0	0	?	1	1	0	0	1	0	1	0	1	0	0	0	0	0	1	1	0	?	0	0	1	0	0	0	0	1	1	0	?	0	1
Conularina	0	0	?	0	0	?	0	0	1	1	1	0	0	0	1	1	1	1	1	0	0	0	0	1	1	0	0	0	1	0	0	0	0	1	1	0	0	0	0	0
Conulariopsis	0	?	?	?	0	0	0	0	?	1	1	0	0	1	0	?	1	1	0	0	0	1	0	1	0	0	0	0	0	?	0	0	0	0	?	1	0	?	0	?
Eoconularia	0	0	0	0	0	0	1	0	1	1	1	0	0	1	0	1	0	1	1	0	0	0	0	1	0	1	1	1	0	1	0	?	0	0	1	1	0	?	0	1
Exoconularia	0	0	0	1	0	0	?	0	1	1	1	0	0	1	0	1	0	1	1	0	0	1	0	1	0	1	1	1	1	1	0	0	0	0	1	1	?	?	0	1
Garraconularia	0	?	0	?	0	?	0	0	1	1	1	0	1	0	0	0	0	0	1	0	0	1	0	1	1	0	0	1	1	0	0	1	0	1	1	0	0	?	0	0
Glyptoconularia	0	?	?	?	0	0	1	0	?	1	1	0	0	1	0	1	1	1	1	0	0	1	0	1	1	?	?	1	1	1	0	0	0	0	1	1	0	?	0	1
Hexaconularia	0	?	0	?	0	0	0	0	?	0	0	0	0	0	1	1	0	0	0	0	0	0	0	1	1	0	0	0	1	0	0	0	0	0	1	1	0	?	0	?
Hexangulaconularia	0	?	?	?	0	0	0	0	?	0	0	0	0	0	1	1	0	0	0	0	0	0	0	1	1	0	0	0	1	0	0	0	0	0	1	1	0	?	0	?
Metaconularia	0	0	0	?	0	0	1	1	1	1	1	0	0	1	0	1	0	1	1	0	0	1	0	1	0	1	1	1	1	1	0	0	0	0	1	1	?	?	0	1
Notoconularia	0	0	0	0	0	0	0	0	1	1	1	1	0	1	0	1	0	1	1	0	0	0	0	1	0	1	1	1	0	1	0	0	0	0	1	1	0	?	0	1
Paraconularia	0	0	0	1	0	0	1	0	1	1	1	0	0	1	0	1	0	1	1	0	0	1	0	1	0	1	1	1	0	1	0	0	0	0	1	1	1	1	0	1
Pseudoconularia	0	0	0	1	0	0	1	0	1	1	1	0	0	1	0	1	0	1	1	1	1	1	0	1	0	1	1	1	1	1	0	0	0	0	1	1	1	?	0	1
Reticulaconularia	0	0	0	0	0	0	0	0	1	1	1	0	0	1	0	1	0	1	1	0	0	1	0	1	0	1	1	1	0	1	0	0	0	0	1	1	1	?	0	1
Sphenothallus	0	1	0	0	0	1	0	0	0	0	0	0	1	0	0	0	1	1	0	0	0	0	0	0	0	0	0	0	0	0	0	0	0	1	1	0	0	0	0	1
Tasmanoconularia	0	0	?	?	0	0	?	0	1	1	1	0	0	1	0	1	0	1	1	0	0	1	0	1	0	1	1	1	1	1	0	0	0	0	1	1	1	?	0	1

Table 1. Skeletal characters of putative conulariids used in cluster analysis. 1 = presence of character; 0 = absence of character; ? = insufficient data to determine whether character is present or not. Specimens referred by Kozłowski (1968) to *Conularia* spp. are designated as "C".

Sphenothallus Hall, 1847, a phosphatic tube having a pair of longitudinal rods, is considered to be a 'worm' by most authorities (e.g., Mason & Yochelson, 1986; Fauchald *et al.*, 1986) except Cox *et al.* (1986), who regarded it as a cnidarian. Specimens showing pyritized soft parts (Fauchald *et al.*, 1986) indicate the presence of paired tentacles but segmentation, which would suggest an annelidan affinity, is not evident. Budding and attachment of juveniles by apical disks in *Sphenothallus* (Ruedemann, 1896a, 1896b; see also Feldmann *et al.*, 1986) previously supported Kiderlen's (1937) view that 'conulariids' had an attached phase during the early part of the life cycle.

Specimens that Kozłowski (1968) figured and described as *Conularia* spp. are certainly not Conularia, and cluster analysis indicates that they are not conulariids. These fossils are phosphatic but they lack flat faces, bilateral symmetry, paired rods, corner grooves, and other characters of conulariids. Also, they were interpreted as having been secreted internally, in contrast to conulariid skeletons, which were secreted externally (Babcock *et al.*, 1987a; Babcock, 1990). The specimens have hollow spines that extend from the apical end, a character that is not known in any authentic conulariid. Bischoff (1978) interpreted the apical spines as evidence of budding using stolons, which supported his view that 'conulariids' were cnidarians. Kozłowski's "*Conularia*" and *Sphenothallus* were linked at 0.50. Clusters containing conchopeltids (*Conchopeltis* + *Conomedusites*) and *Sphenothallus* + Kozłowski's (1968) "*Conularia*" were linked at 0.59. Taxonomically, conchopeltids and *Sphenothallus* are

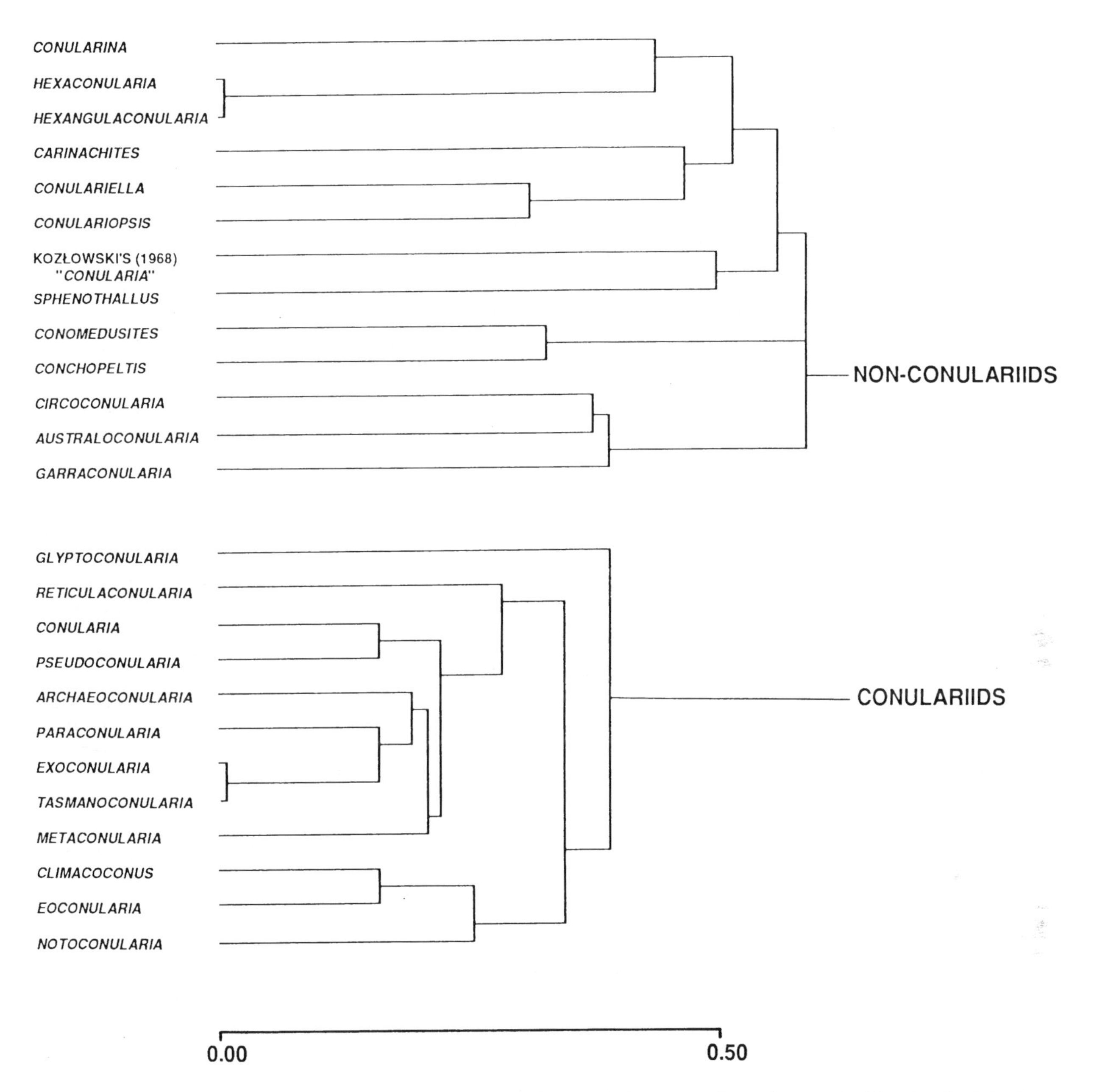

Fig. 3. Dendrogram formed by Q-mode clustering of preserved skeletal characters of putative conulariids. Meaningful primary clusters are labelled as conulariids or non-conulariids depending on their degree of morphological similarity to, and interpreted phylogenetic relationship with, *Conularia*. The distance metric is Euclidean distance.

differentiated at the phylum level by most authorities.

The circoconulariids (*Australoconularia* Bischoff, 1978, *Circoconularia* Bischoff, 1978, and *Garraconularia* Bischoff, 1978) are very small phosphatic fossils having circular to elliptical cross-sections, an apparent radial symmetry, and well-developed internal septa. They were linked at the relatively low value of 0.39, which suggests that they may represent a monophyletic group. Their affinities are unknown. A radial symmetry suggests a possible relationship with cnidarians or ctenophores. The phosphatic composition of the skeleton, however, weighs against such an interpretation because skeletal phosphate is not present in any appreciable amount in any known 'coelenterate'. Circoconulariids are possibly representatives of an extinct group that had a radial body plan convergent with that of the cnidarians. Previously, the septa of circoconulariids were interpreted as skeletal supports for mesenteries, and were thought to be homologous with the septa of living scyphozoans (Bischoff, 1978). That evidence, in turn, was used to support then-current arguments (e.g., Kiderlen, 1937; Moore & Harrington, 1956a) for including 'conulariids' in the Cnidaria.

Carinachites Qian, 1977, is a small shelly fossil of unknown affinities known only in Lower Cambrian rocks. It is a four-sided tube having deeply invaginated corner grooves and almost no taper over most of its observed length. Some specimens appear to indicate that the tube was flexible and that ridges across the faces are spurious structures. Remains of *Carinachites* may be secondarily rather than originally phosphatic.

Conulariella Bouček, 1928, and *Conulariopsis* Sugiyama, 1942, are elongate pyramidal fossils of unknown affinities. *Conulariella* is a rapidly expanding pyramidal tube having a very elongate rectangular cross section and single ridges across each of the faces. *Conulariopsis* is a pyramidal fossil that has widely spaced transverse and longitudinal subdivisions across the surface. The subdivisions are not regularly arranged and do not appear to be related to significant thickenings of the integument. Their origin could be taphonomic.

Hexaconularia Yang *et al.*, 1983, and *Hexangulaconularia* He, 1983, are Early Cambrian phosphatic tubes having six sides and ridges across the faces that resemble rods. The possible exoskeletal thickenings are not paired or arched on the faces. Whether or not they are bilaterally symmetrical has not been determined. The tubes are one to two orders of magnitude smaller than undoubted conulariids from post-Cambrian rocks. Their affinities cannot be ascertained without additional study. *Hexaconularia* and *Hexangulaconularia* were linked at 0.01, which supports the suggestion (Brasier & Singh, 1987) that differences between *Hexaconularia* and *Hexangulaconularia* may be preservational.

Conularina Sinclair, 1942, is an elongate phosphatic tube that has a triradial symmetry. Its morphology was discussed by Babcock *et al.* (1987a), who considered it to be an *incertae sedis.*

4. Phylogeny of the Conulariida

Attempts to resolve the phylogenetic relationships of the Conulariida are based on relatively limited information about skeletal morphology and secretion, and about soft-part morphology. Available anatomical evidence allows little to be interpreted beyond their grade of organization. The group is regarded as an extinct clade that shows a substantial conservatism of skeletal morphology over a known geologic range from Early Ordovician to Late Triassic.

The early history of conulariids is unclear. They may have originated in the late Precambrian but convincing fossil evidence is lacking. It has been suggested (McMenamin, 1987) that their four-sided exoskeleton developed during the Cambrian by the fusion of two or more frondose organisms similar to those from Ediacaran faunas. That hypothesis is based in part on the erroneous notions that conulariids had apical attachment disks and a radial symmetry. The few similarities that exist between conulariids and frondose fossils of the Ediacaran fauna are more likely due to convergences than homologies. Such putative conulariids as *Hexaconularia* and *Hexangulaconularia* have been described from Lower Cambrian rocks but they require further study before they can be confidently assigned to the Conulariida. The cluster analysis suggests that their skeletons are structurally different from post-Cambrian conulariids. It is possible that these fossils represent an early group of conulariids or conulariid relatives that did not survive into the Ordovician. Intermediates between them and undoubted post-Cambrian conulariids, however, are not known. It is also possible that conulariids did not mineralize their skeleton before the Early Ordovician and that *Hexaconularia* and *Hexangulaconularia* acquired similar skeletal forms through convergence.

It is doubtful, as was commonly assumed until the late 1930's, that conulariids are an extinct molluscan group. Among much evidence for rejecting this hypothesis are the lack of skeletal phosphate (Yochelson, 1975) and rods in any mollusk. Such organisms as hyoliths (Runnegar *et al.*, 1975; Marek & Yochelson, 1976; Runnegar, 1980) and tentaculitids (Yochelson & Lindemann, 1986), whose inclusion in groups with conulariids once tended to support a molluscan affinity, have since been removed from taxonomic association with *Conularia.*

Almost all arguments supporting a cnidarian hypothesis of affinity of the 'conulariids' (e.g., Kiderlen, 1937; Moore & Harrington, 1956a; Van Iten, 1988) are now untenable because *Conchopeltis, Sphenothallus,* circoconulariids, and Kozłowski's (1968) "*Conularia*" should be excluded from the Conulariida. Furthermore, most characters of authentic conulariids that were used to support the cnidarian hypothesis have been shown to be incorrectly interpreted or of dubious value in formulating statements of homology. Apertural "flaps" or "lappets" were taphonomically produced (Babcock & Feldmann, 1986a, 1986c), not structures that either closed the aperture to protect the soft parts or flapped during swimming (Kiderlen, 1937). Apical walls are simple internal partitions (Babcock & Feldmann, 1986a, 1986c; Babcock, 1988) rather than scar tissue secreted when a supposed medusa separated from its attached apical end (Kiderlen, 1937; Moore & Harrington, 1956a; Van Iten, 1987). Specimens of *Conularia* preserving only two or three faces (Kiderlen, 1937) instead of the normal four were reinterpreted as having been crushed (Babcock *et al.*, 1987a).

The proposed homology of large, bifurcate "septa" extending inward from the midlines of *Eoconularia loculata* (Wiman, 1895) with the interradial septa of a group of living scyphozoan cnidarians, the Stauromedusae, is weak. The "septa" of *Eoconularia,* assuming they are not spurious structures (see Sinclair, 1948; Babcock & Feldmann, 1986a, 1986c), would be composed of hard skeletal tissue secreted by a mantle or mantle-like tissue (see Figure 2) whereas the septa of stauromedusans are infoldings of ectodermal tissue (Hyman, 1940). Furthermore, there are no longitudinal invaginations in the perradial positions of stauromedusans that correspond to the corner grooves (Figure 1) of conulariids.

Additional support for the idea that conulariids are not cnidarians comes from evidence of their soft parts and functional morphology. X-radiographs of specimens from the celebrated Hunsrück Slate (Steul, 1984) of Devonian age show no evidence of tentacles although other soft parts (see Figure 2) are preserved. Furthermore, a mantle or mantle-like tissue, which was present in conulariids (Babcock, 1990) is unknown in any cnidarian. Functional considerations that argue against the interpretation (Kiderlen, 1937; Moore & Harrington, 1956a; Van Iten, 1987) that conulariids developed into free-swimming medusae or medusoid-like forms are the following: 1) Calcium apatite (collophane), which has a specific gravity of 3.15 to 3.20 (Hurlbut & Klein, 1977), is too dense to form the skeleton of a large nektonic organism having a low surface area to volume ratio. Some conulariids reached lengths of 20 cm or more (Fletcher, 1938; Lamont, 1946; Babcock & Feldmann, 1986a, 1986b; Babcock *et al.*, 1987b) and probably could not have achieved neutral buoyancy in water unless most of their body mass were composed of lipids. 2) The long, slender pyramidal shape is hydrodynamically unstable for a nektonic organism. More stable designs would include lower, broader, or rounded shapes. 3) The exoskeleton was rigid or only slightly flexible. The rods would have strengthened the skeleton to some degree but, more importantly, the corner grooves would have prevented it from pulsating like the bell of a scyphozoan cnidarian. 4) If apical walls were secreted to close the apical ends of medusae, no more than one should be present in any individual specimen. Individuals preserving more than one apical wall, however, have been well-documented (e.g., Slater, 1907; Sinclair, 1948; Brood, 1986).

Conulariids were evidently sessile, benthic organisms except as larvae, assuming they had a larval phase. They lived attached by a flexible, possibly chitinoid, stalk (Figure 1) that partially sheathed the apical end and that had no observed connection to the central cavity. How it was secreted is unknown. Apical attachment disks or spinose projections are not known in any conulariid. Monospecific associations of large and small individuals in clusters (e.g., Babcock & Feldmann, 1986a, figs. 4.1, 4.4; 1986b, figs. 24.1, 24.3), without any evidence of connections between them indicates that conulariids were gregarious (Babcock & Feldmann, 1986a). In the absence of any evidence for asexual reproduction, only a sexual habit can be inferred at present.

A bilateral symmetry, which is reflected in both hard- (Babcock & Feldmann, 1986a, 1986c) and soft-part (Steul, 1984; Babcock & Feldmann, 1986a, 1986c) morphology indicates that conulariids belong in the Bilateria. A triploblastic grade of organization is suggested by the production of pearls (Babcock, 1990) in some species. Among present-day organisms, pearls develop only in some triploblastic taxa. Conulariids had a predominantly straight (Babcock & Feldmann, 1986a), perhaps blind, alimentary tract. The presence of globular bodies, which are probable organs (Steul, 1985; Babcock & Feldmann, 1986a, 1986b), between the gut and the inner exoskeletal wall can be taken as evidence of a possible coelom (Figure 2).

Conulariids seemed to have achieved a grade of evolution similar to the Protostomia, Lophophorata, or Deuterostomia but there is no available evidence that supports a close phylogenetic relationship with any phylum within any of these groups. Unique, derived characters of conulariids include: 1) development of a four-sided elongate pyramidal exoskeleton; 2) mineralization of the exoskeleton with calcium phosphate; 3) strengthening of the exoskeleton with slender, curved transverse thickenings (rods); 4) development of paired sets of rods on each of the four faces of the exoskeleton; 5) development of longitudinally invaginated corners; and 6) acquisition of a stalk for attachment. Smooth, transverse internal partitions (apical walls) and carinae are not present in all species and may have resulted from homoplasy among lineages within the Conulariida.

Some of the characters considered to be synapomorphic among conulariids are present in other animals but there is no satisfactory evidence of homology with them. Convergence is thought to be responsible for the development of a phosphatic skeleton in inarticulate brachiopods, some arthropods, chordates, conulariids, and other problematic fossils such as *Sphenothallus* and circoconulariids. Similarly, the ability to thicken exoskeletal laminae into rods may have developed independently in conulariids, circoconulariids, *Hexaconularia*, and other taxa. Whether or not the presence of a mantle or mantle-like tissue was acquired in the Conulariida independently of the Mollusca, Brachiopoda, and Tunicata (Chordata) has yet to be resolved. Conulariids may have affinities with chordates although the evidence of fusellar tissue in the exoskeleton (Termier & Termier, 1949, 1953) or a segmented skeleton, segmentally arranged muscle bunches, and photosensitive organs (Steul, 1984) is dubious.

Much remains to be learned about the soft-part anatomy of conulariids. The inner surfaces of midlines, for example, are commonly thickened into single or, rarely, double carinae. Exoskeleton in the corner regions is also thickened. The carinae, and perhaps the corner thickenings, may have strengthened the exoskeleton, served as muscle attachment surfaces, or had some other function. The function of paired rods on each face and whether or not they had any relationship to soft parts is not known. Resolution of these and other questions may eventually improve understanding of conulariid affinities.

5. Summary

The subclass Conulata Moore & Harrington, 1956a, order Conularida Miller & Gurley, 1896, and suborder Conulariina Miller & Gurley, 1896, were used as taxonomic wastebaskets for a variety of problematic fossils found in rocks of late Precambrian through Mesozoic age. The primary characters that these fossils have in common are pyramidal, conical, or tubular skeletons of relatively simple construction, commonly with external ridges. Their compositions could be phosphatic, chitinophosphatic, chitinous, or even calcareous.

The Conulariida Babcock and Feldmann, 1986a, is considered to be a relatively small monophyletic lineage that secreted an elongate, four-sided pyramidal exoskeleton having a phosphatic composition and bilateral symmetry. The exoskeleton was strengthened primarily by rod pairs that meet at a midline on each face and articulated in a corner groove between two faces. From what is known about the conulariid body plan, they had no obvious relationship to any living phylum, and are therefore considered to be an independent, extinct phylum of Bilateria. They probably achieved a triploblastic level of tissue organization and may have possessed a coelom. The sister group of the conulariids is not known. Based on available evidence, that group could have had a protostome, lophophorate, or deuterostome organization.

Acknowledgements

Information about small shelly fossils of Early Cambrian age has come partially from discussions with S. Bengtson and S. Conway Morris. S. Conway Morris also provided some specimens for study. C.G. Maples assisted with computing and reviewed this paper. This paper was also reviewed by K.E. Evans and R.A. Robison. Research was supported in part by the Erasmus Haworth fund, administered by The University of Kansas. I am grateful to the University of Camerino and the Italian National Research Council who sponsored my attendance at this symposium.

REFERENCES

ARCHIAC, E. D' & VERNEUIL, E. DE. 1842. On the fossils of the older deposits of the Rhenish provinces, preceded by a general survey of the fauna of the Palaeozoic rocks, and followed by a tabular list of the organic remains of the Devonian System in Europe. *Geological Society of London, Transactions, Series* **2, 6(2)**, 303-410.

BABCOCK, L.E. 1988. New Permian conulariids from Devon Island, Canada. *Journal of Paleontology* **62**, 615-617.

BABCOCK, L.E. 1990. Conulariid pearls. In *Evolutionary Paleobiology of Behavior and Coevolution* (ed. Boucot, A.J.), pp. 68-71. Amsterdam: Elsevier Science Press.

BABCOCK, L.E. & FELDMANN, R.M. 1986a. Devonian and Mississippian conulariids of North America. Part A. General description and *Conularia. Annals of Carnegie Museum* **55**, 349- 410.

BABCOCK, L.E. & FELDMANN, R.M. 1986b. Devonian and Mississippian conulariids of North America. Part B. *Paraconularia, Reticulaconularia*, new genus, and organisms rejected from Conulariida. *Annals of Carnegie Museum* **55**, 411-479.

BABCOCK, L.E. & FELDMANN, R.M. 1986c. The phylum Conulariida. In *Problematic Fossil Taxa* (eds. A. Hoffman and M. H. Nitecki), pp. 135-147. New York: Oxford University Press.

BABCOCK, L.E., FELDMANN, R.M. & WILSON, M.T. 1987a. Teratology and pathology of some Paleozoic conulariids. *Lethaia* **20**, 93-105.

BABCOCK, L.E., FELDMANN, R.M., WILSON, M.T. & SUÁREZ-RIGLOS, M. 1987b. Devonian conulariids of Bolivia. *National Geographic Research* **3**, 210-231.

BARRANDE, J. 1867. *Système silurien du centre de la Boheme: Vol. 3: Classe des Mollusques, Ordre des Pteropodes*. Prague and Paris: published by the author.

BISCHOFF, G.C.O. 1973. On the nature of the conodont animal. *Geologica et Palaeontologica* **7**, 147-174.

BISCHOFF, G.C.O. 1978. Internal structures of conulariid tests and their functional significance, with special reference to Circoconulariina n. suborder (Cnidaria, Scyphozoa). *Senckenbergiana lethaea* **59**, 275-327.

BODENBENDER, B.E., WILSON, M.A. & PALMER, T.J. 1989. Paleoecology of *Sphenothallus* on an Upper Ordovician hardground. *Lethaia* **22**, 217-225.

BOUČEK, B. 1928. Revise českych paleozoických Konularii. České Akademie Ved a Umeni v Praze, Trida II. Palaeontographica Bohemiae **XI**, 1-108. Prague: Nakladem České Akademie Ved a Umění.

BOUČEK, B. 1939. Conularida. In *Handbuch der Paläozoologie*, **2A** (ed. O.H. Schindewolf), pp. A113-A131. Berlin: Gebruder Borntraeger.

BOUČEK, B. & ULRICH, F. 1929. O skorapce rodu *Conularia* Miller. *Statniho geologického ustavu Ceskoslovenské Republicky, Vesnik* **5(2/3)**, 1-25.

BRASIER, M.D. & SINGH, P. 1987. Microfossils and Precambrian-Cambrian boundary stratigraphy at Maldeota, Lesser Himalaya. *Geological Magazine* **124**, 323-345.

BROOD, K. 1986. Conulaires. In *Le Groupe de Liévin. Pridoli Lochkovien de l'Artois (N. France)* (ed. Racheboeuf, P.R.), pp. 171-174. *Biostratigraphie du Paléozoique* **3**.

CHAPMAN, D.M. 1966. Evolution of the scyphistoma. In *The Cnidaria and Their Evolution* (ed. Rees, W.J.), pp. 51-75. London: Academic Press.

CHAPMAN, D.M. & WERNER, B. 1972. Structure of a solitary and a colonial species of *Stephanoscyphus* (Scyphozoa, Coronatae) with observations on periderm repair. *Helgoländer wissenschaftliche Meersuntersuchungen* **23**, 393-421.

COX, R.S., VAN ITEN, H. & MAPES, R.H 1986. The problematic *Sphenothallus*: annelid or cnidarian? *Geological Society of America, Abstracts with Programs* **18**, 573.

EICHWALD, C.E. D' 1840. Über das silurische Schichtensystems in Esthland. *Zeitschrift für Natur-und Heilkunde der K. Medicinisch-chirurgische Akademie St. Petersburg* **1/2**, 1-210.

FAUCHALD, K., STÜRMER, W. & YOCHELSON, E.L. 1986. *Sphenothallus* "Vermes" in the Early Devonian Hunsrück Slate, West Germany. *Paläontologische Zeitschrift* **60**, 57-64

FELDMANN, R.M. & BABCOCK, L.E. 1986. Exceptionally preserved conulariids from Ohio-reinterpretation of their anatomy. *National Geographic Research* **2**, 464-472.

FELDMANN, R.M., HANNIBAL, J.T. & BABCOCK, L.E. 1986. Fossil worms from the Devonian of North America (*Sphenothallus*) and Burma ("Vermes") previously identified as phyllocarid arthropods. *Journal of Paleontology* **60**, 341-346.

FISHER, D.W. 1962. Small conoidal fossils of uncertain affinities. In *Treatise on Invertebrate Paleontology: Part W, Miscellanea* (ed. Moore, R.C.), pp. W98-W143. New York and Lawrence: Geological Society of America and University of Kansas.

FLETCHER, H.O. 1938. A revision of the Australian Conulariae. *Australian Museum, Records* **20**, 235-255.

FOERSTE, A.F. 1928. American Arctic and related cephalopods. *Denison University Bulletin* **38(2)**, *Scientific Laboratories, Journal* **23(1-2)**, 1-110.

GEINITZ, H.B. 1853. *Conularia hollebeni* Gein. aus dem unteren Zechstein von Ilmenau. *Deutsche geologische Gesellschaft, Zeitschrift* **5**, 465-466.

GLAESSNER, M.F. 1971. The genus *Conomedusites* Glaessner & Wade and the diversification of the Cnidaria. *Paläontologische Zeitschrift* **45**, 7-17.

GLAESSNER, M.F. & WADE, M. 1966. The Late Precambrian fossils from Ediacara, South Africa. *Palaeontology* **9**, 599-628.

GRABAU, A.W. & SHIMER, H.W. 1910. *North American Index Fossils: Invertebrates. Volume II.* New York: A.G. Seiler & Company.

GRASSHOFF, M. 1984. Cnidarian phylogeny - a biomechanical approach. *Palaeontographica Americana* no. **54**, 127-135.

HALL, J. 1847. *Palaeontology of New-York. Volume I. Containing Descriptions of the Organic Remains of the Lower Division of the New-York System. (Equivalent to the Lower Silurian Rocks of Europe).* Albany: C. Benthuysen.

HALL, J. 1879. *Geological Survey of the State of New York. Palaeontology: Vol. V. Part II. Containing Descriptions of the Gasteropoda. Pteropoda and Cephalopoda of the Upper Helderberg, Hamilton, Portage and Chemung Groups*. Albany: Charles Van Benthuysen & Sons.

HE TING-GUI, 1983. (Description of fossils *pars*). In *The Sinian-Cambrian Boundary of China* (ed. Xing Yusheng *et al.*) *Bulletin of the Institute of Geology. Chinese Academy of Sciences*, **10**. (In Chinese).

HE TING-GUI. 1987. (Early Cambrian conulariids from Yangtze Platform and their early evolution.) *Journal of Chengdu College of Geology* **14**, 7-18. (In Chinese).

HERGARTEN, B. 1988. Conularien in Deutschland. *Aufschluss* **39**, 321-256.

HOLM, G. 1893. Sveriges Kambrisk-Siluriska Hyolithidae och Conulariidae. *Sveriges Geologiska Undersökning, Afhandlinger och uppsatser*, Series C, **112**, 1-172.

HURLBUT, C.S., JR. & KLEIN, C. 1977. *Manual of Mineralogy (After James D. Dana). 19th Edition*. New York: John Wiley & Sons.

HYMAN, L.H. 1940. *The Invertebrates: Protozoa Through Ctenophora*. New York and London: McGraw-Hill.

IHERING, H. VON, 1881. Die Aptychen als Beweismittel für Dibranchiaten-Natur der Ammoniten. *Neues Jahrbuch für Mineralogie, Geologie und Paläontologie, Jahrgang 1881* **1**, 44-92.

KIDERLEN, H. 1937. Die Conularien. Über Bau und Leben der ersten Scyphozoa. *Neues Jahrbuch für Mineralogie, Geologie und Paläontologie*, **77(B)**, 113-169.

KNIGHT, J.B. 1937. *Conchopeltis* Walcott, an Ordovician genus of the Conulariida. *Journal of Paleontology* **11**, 186-188.

KOZŁOWSKI, R. 1968. Nouvelles observations sur les Conulaires. *Acta Palaeontologica Polonica* **13**, 497-535.

LALICKER, C.G. & MOORE, R. C. 1952. Annelids and other worms.In *Invertebrate Fossils* (eds. R.C. Moore, C.G. Lalicker & A.G. Fischer), pp. 452-462. New York: McGraw-Hill.

LAMONT, A. 1946. Largest British *Conularia. Quarry Managers' Journal* **29**, 569-570.

LINDSTRÖM, G. 1884. On the Silurian Gastropoda and Pteropoda of Gotland. *Kongliga Svenska Vetenskaps-Akademiens, Handlingar* **19(6)**, 1-250.

MAREK, L. & YOCHELSON, E.L. 1976. Aspects of the biology of Hyolitha (Mollusca). *Lethaia* **9**, 65-82.

MASON, C. & YOCHELSON, E.L. 1985. Some tubular fossils (*Sphenothallus*: "Vermes") from the middle and late Paleozoic of the United States. *Journal of Paleontology* **59**, 85-95.

MCCOY, F. 1852. Systematic description of the British Palaeozoic fossils in the Geological Museum of the University of Cambridge. In *Synopsis of the Classification of the British Palaeozoic Rocks* (ed. A. Sedgwick), pp. 407-661. Cambridge: University Press.

McMenamin, M.A.S. 1987. The fate of the Ediacaran fauna, the nature of conulariids, and the basal Paleozoic predator revolution. *Geological Society of America, Abstracts with Programs* **19**, 29.

Miller, S.A. & Gurley, W.F.E. 1896. New species of Palaeozoic invertebrates from Illinois and other states. *Illinois State Museum of Natural History. Bulletin* **11**, 50 pp.

Moore, R.C. & Harrington, H.J. 1956a. Scyphozoa. In *Treatise on Invertebrate Paleontology: Part F: Coelenterata* (ed. R.C. Moore), pp. F27-F38. New York and Lawrence: Geological Society of America and University of Kansas.

Moore, R.C. & Harrington, H.J. 1956b. Conulata. In *Treatise on Invertebrate Paleontology: Part F: Coelenterata* (ed. R.C. Moore), pp. F54-F66. New York and Lawrence: Geological Society of America and University of Kansas.

Mortin, J. 1985. The shell structure and zoological affinities of conulariids. *Palaeontological Association Annual Conference Abstracts*, pp. 12-13. Aberystwyth: University College of Wales.

Neumayer, M. 1879. Zur Kenntniss der Fauna des untersten Lias in den Nordalpen. *Kaiserlich-königlichen geologischen Reichsanstalt. Abhandlungen* **7(5)**, 1-46.

Oliver, W.A., Jr. 1984. *Conchopeltis*: its affinities and significance. *Palaeontographica Americana no.* **54**, 141-147.

Parfrey, S.M. 1982. Palaeozoic conulariids from Tasmania. *Alcheringa* **6**, 69-75.

Pelseneer, P. 1889. Sur un nouveau *Conularia* du Carbonifère et sur les prétendus "Pteropodes" primaires. *Société belge de Géologie, de Paléontologie et d'Hydrologie, Mémoires* **3**, 124-136.

Qian Yi, 1977. (Hyolitha and some problematica from the Lower Cambrian Meishucun Stage in Central and S.W. China.) *Acta Palaeontologica Sinica* **16**, 255-278. (In Chinese.).

Robson, E.A. 1985. Speculations on coelenterates. In *The Origins and Relationships of Lower Invertebrates* (eds. S. Conway Morris, J.D. George, J. Gibson & H.M. Platt), pp. 60-77. *Systematics Association Special Volume* **28**. Oxford: Clarendon Press.

Ruedemann, R. 1896a. Note on the discovery of a sessile *Conularia*. Article I. *American Geologist* **17**, 158-165.

Ruedemann, R. 1896b. Note on the discovery of a sessile *Conularia*. Article II. *American Geologist* **18**, 65-71.

Runnegar, B. 1980. Hyolitha: status of the phylum. *Lethaia* **13**, 21-25.

Runnegar, B., Pojeta, J. jr., Morris, N.J., Taylor, J.D., Taylor, M.E. & McClung, G. 1975. Biology of the Hyolitha. *Lethaia* **8**, 181-191.

Runnegar, B., Pojeta, J. jr., Taylor, M.E. & Collins, D. 1979. New species of the Cambrian and Ordovician chitons *Matthevia* and *Chelodes* from Wisconsin and Queensland: evidence for the early history of polyplacophoran mollusks. *Journal of Paleontology* **53**, 1374-1394.

Sandberger, G. 1847. Die Flossenfüsser oder Pteropoda der ersten Erdbildungs-Epoche. *Conularia* und *Coleoprion*. *Neues Jahrbuch für Mineralogie, Geologie und Paläontologie, Jahrgang* 1845, 379-402.

Schlotheim, E.F. von, 1820. *Die Petrefactenkunde auf ihrem jetzigen Standpunkte durch die Beschreibung seiner Sammlung versteinester und fossiler Uberreste des Thier- und Pflanzenreichs der Vorwalt erlänt.* Gotha: Inder Becker'schen Buchandlung.

Schmidt, W. & Teichmüller, M. 1958. Neue Funde von *Sphenothallus* auf dem westeuropäischen Festland, insbesondere in Belgien, und erganzende Beobachtungen zur Gattung *Sphenothallus*. *Association pour l'étude de la Paléontologie et de la Stratigraphie Houillere, Publication* **33**, 1-34.

Scrutton, C. T. 1979. Early fossil cnidarians. In *The Origin of Major Invertebrate Groups* (ed. M.R. House), pp. 161-207. *Systematics Association Special Volume* **12**. London: Academic Press.

Sinclair, G.W. 1940a. The genotype of *Conularia*. *Canadian Field-Naturalist* **54**, 72-74.

Sinclair, G.W. 1940b. A discussion of the genus *Metaconularia* with descriptions of new species. *Royal Society of Canada, Section IV. Transactions, Series 3*, **34**, 101-121.

Sinclair, G.W. 1942. The Chazy Conularida and their congeners. *Annals of Carnegie Museum* **29**, 219-240.

Sinclair, G.W. 1943. Notes on the genera *Archaeoconularia* Boucek and *Eoconularia*, new genus. *Royal Society of Canada, Proceedings, Series 3*, **37**, 122.

Sinclair, G.W. 1948. *The Biology of the Conularida.* Unpublished Ph.D. thesis. 442 pp. Montreal: McGill University.

Sinclair, G.W. 1952. A classification of the Conularida. *Fieldiana. Geology* **10**, 135-145.

Slater, I.L. 1907. *A Monograph of the British Conulariae*. London: Palaeontographical Society, London. 41 pp.

Sowerby, J. 1820 [dated 1821]. *The Mineral Conchology of Great Britain; or Coloured Figures and Descriptions of those Remains of Testaceous Animals or Shells, which have been Preserved at Various Times and Depths in the Earth.* Volume 3, Part 46. pp. 1-194. London: W. Arding Co.

Steul, H. 1984. Die systematische Stellung der Conularien. *Giessener Geologische Schriften* **37**, 1-117.

Sugiyama, T. 1942. Studies on the Japanese Conularida. *Geological Society of Japan, Journal* **49**, 390-399.

Termier, G. & Termier, H. 1950. Paléontologie marocaine. Tome II. Invertébrés de l'Ere Primaire. Fascicule IV: Annélides, Arthropodes, Échinodermes, Conularides et Graptolithes. *Maroc, Service géologique. Notes et Memoires No.* **79**, pp. 1- 279. (Also issued as: *Actualitées scientifiques et industrielles*, No. 1095, pp. 1-279. Paris: Hermann & Cie.

Termier, H. & Termier, G. 1949. Position systematique et biologie des Conulaires. *Revue scientifique*, 86, No. 3300, 711-722.

Termier, H. & Termier, G. 1953. Les Conularides. *Traité de Paléontologie: Vol. III: Onychophores, Arthropodes, Echinodermes, Stomocordés* (ed. J. Piveteau), pp. 1006-1013. Paris: Masson & Cie.

Thomas, G.A. 1969. *Notoconularia*, a new conularid genus from the Permian of eastern Australia. *Journal of Paleontology* **43**, 1283-1290.

Van Iten, H. 1987. The mode of life of the Conulariida and its implications for conulariid affinities. *Geological Society of America, Abstracts with Programs* **19**, 876.

Van Iten, H. 1988. Morphology and phylogenetic significance of conulariid corners/midlines. *Geological Society of America, Abstracts with Programs* **20**, 393.

Vanuxem, L. 1842. *Geology of New-York. Part 3. Survey of the Third Geological District*. Albany: C. Van Benthuysen.

Waagen, W. 1891. Salt Range fossils. *Geological Results. Geological Survey of India, Memoirs, Palaeontologia Indica, Series 13*, **4**, 89-242.

Walcott, C.D. 1876. Descriptions of new species of fossils from the Trenton Limestone. *New York State Museum of Natural History, 28th Annual Report*, pp. 93-97.

Walcott, C.D. 1885. Note on some Paleozoic pteropods. *American Journal of Science* **30**, 17-21.

Weller, S. 1925. A new type of Silurian worm. *Journal of Geology* **33**, 540-544.

Werner, B. 1966. *Stephanoscyphus*, (Scyphozoa Coronatae) und siene direkte Abstammung von den fossilen Conulata. *Helgoländer Wissenschaftliche Meeresuntersuchungen* **13**, 317- 347.

Werner, B. 1967. *Stephanoscyphus* Allman (Scyphozoa Coronatae), ein rezenter Vertreter der Conulata? *Paläontologische Zeitschrift* **41**, 137-153.

Werner, B. 1969. Neue Beitraege zur Evolution der Scyphozoa und Cnidaria. *I Simposio Internacional de Zoofilogenia*, pp. 223-244. Salamanca: University of Salamanca.

Werner, B. 1973. New investigations on systematics and evolution of the class Scyphozoa and the phylum Cnidaria. *Seto Marine Biology Laboratory Publications* **20**, 35-61.

Wiman, C. 1895. Palaeontologische Notizen. 1 und 2. *University of Upsala, Geological Institution, Bulletin* **2** (1894-1895), 109-117.

Yang Xian-He, He Yuan-Xiang & Deng Shou-He. 1983. (On the Sinian-Cambrian boundary and the small shelly fossil assemblages in Nanjiang area, Sichuan.) *Bulletin of the Chengdu Institute of Geology and Mining Research. Chinese Academy of Geological Sciences* **4**, 91-105. (In Chinese.).

Yochelson, E.L. 1975. Discussion of Early Cambrian "molluscs." *Journal of the Geological Society of London* **131**, 661-662.

Yochelson, E.L. & Lindemann, R.H. 1986. Considerations on systematic placement of the styliolines (*incertae sedis:* Devonian). In *Problematic Fossil Taxa* (eds. A. Hoffman & M. H. Nitecki), pp. 45-58. New York: Oxford University Press.

Evolutionary affinities of conulariids

Heyo Van Iten[1]

Abstract

The hypothesis that conulariids were closely related to scyphozoan cnidarians, once widely regarded as the most likely interpretation of conulariid affinities, has recently been questioned. However, challenges to this hypothesis are not supported by results of detailed microstructural and anatomical comparisons. Such comparisons confirm that steeply pyramidal, generally four-sided conulariid tests show numerous similarities to scyphozoan thecae and soft parts. These similarities, many of them uniquely shared by conulariids and scyphozoans, suggest that conulariids were a group of septate cnidarians most closely related to scyphozoans.

Based in part on analysis of relic conulariid soft parts, some authors claim that conulariids could not have been cnidarians. However, relic soft parts documented by Steul (1984) in specimens of *Conularia* from the Hunsrück Slate (Early Devonian, Germany), and possible soft part relics in the holotype of *Eoconularia amoena* (Sinclair), show evidence of homology to soft parts of scyphozoan polyps undergoing polydisc strobilation.

Introduction

Previous comparisons of conulariids and scyphozoan cnidarians have revealed intriguing similarities in gross morphology, thecal microstructure and growth, and mode of life (e.g., Kiderlen, 1937; Moore & Harrington, 1956a; Chapman, 1966; Werner, 1966, 1967a, 1971a; Bischoff, 1978). These similarities have been interpreted as evidence that conulariids were scyphozoans (e.g., Kiderlen, 1937; Bouček, 1939; Moore & Harrington, 1956a, b; Werner, 1966, 1967a, 1971a; Bischoff, 1978) or that they were more closely related to scyphozoans than to any other currently recognized taxon of comparable rank (Van Iten, in press).

These interpretations of conulariid affinities have been questioned or rejected by several authors (e.g., Termier and Termier, 1949, 1953; Kozłowski, 1968; Oliver, 1984; Steul, 1984; Mortin, 1985; Feldmann & Babcock, 1986; Babcock & Feldmann, 1986a, b; Oliver & Coates, 1987). With the exceptions of Termier & Termier (1949, 1953) and Steul (1984), however, none of these authors has proposed alternative hypotheses of affinity involving conulariids and non-cnidarian taxa. Rather, dissatisfaction with the hypothesis of a scyphozoan affinity for conulariids has generally been based on contentions that similarities between conulariids and scyphozoans are superficial, and/or that conulariids and scyphozoans are characterized by fundamental dissimilarities.

The present contribution seeks to evaluate conflicting interpretations of conulariid affinities by addressing three critical areas of disagreement. Two of these concern the interpretation of the anatomy of the steeply pyramidal, generally four-sided conulariid test. Based on analyses of the gross anatomy, microstructure, and growth of this test, proponents of a scyphozoan affinity for conulariids propose (1) that the midline of each of the conulariid test's four sides (faces) was the site of a gastric mesentery, or septum, homologous to the four gastric septa of scyphozoans (e.g., Kiderlen, 1937; Werner, 1966, 1967a, 1971a, Bischoff, 1978); and (2) that the conulariid test is a mineralized, ectodermally secreted theca, homologous to the theca of polypoid scyphozoans of the order Coronatida (e.g., Werner, 1966, 1967a, 1971a). Opponents of these interpretations claim that similarities thought to corroborate the first hypothesis of homology are superficial or based on misinterpretation of features observed in the test cavity of certain conulariids (e.g., Kozłowski, 1968; Oliver, 1984; Babcock & Feldmann, 1986a, b; Oliver & Coates, 1987), and that the second hypothesis of homology is precluded by fundamental dissimilarities in test (thecal) composition and microstructure (e.g., Kozłowski, 1968; Feldmann and Babcock, 1986). As will be discussed below, similarities between conulariids and scyphozoans are more numerous and detailed than hitherto realized, and claims that conulariids and scyphozoans are fundamentally dissimilar are not supported by available anatomical and microstructural evidence.

The third area of disagreement centres on the interpretation of possible relics of conulariid soft parts. Such features have recently been documented by Steul (1984) and Babcock & Feldmann (1986a, c), who maintain that conulariids exhibited a higher level of anatomical organization than that shown by scyphozoans or other cnidarians. As will be discussed in this paper, comparison of Steul's (1984) relics to scyphozoan soft parts, and analysis of a possible additional occurrence of relic conulariid soft parts, discovered by the present author and documented here, suggest that features reported by Steul (1984) can be homologized to soft parts of scyphozoans.

Conulariid Corners and Midlines

ANATOMICAL BACKGROUND

General -- Proponents of a scyphozoan affinity for conulariids have emphasized similarities between the corners and midlines of the conulariid test and anatomical features located at the scyphozoan perradii and interradii, respectively. These similarities have been interpreted as indicating that conulariids possessed an enteron, with a single opening at the animal's apertural end and with four radially disposed, longitudinal endodermal infoldings, or septa, one at each of the four midlines.

[1]Museum of Paleontology, The University of Michigan, Ann Arbor, MI 48109, U.S.A.

Scyphozoans -- Except in medusae of the orders Rhizostomatida and Semaeostomatida, the enteron of all scyphozoans is subdivided by radially disposed, longitudinal septa (Hyman, 1940). Scyphozoans usually exhibit four such structures, but aberrant individuals produce three, five, or six septa. Where four septa are present, the septa lie in one of two mutually perpendicular planes of symmetry, called the interradii. The angles formed by the interradii are bisected by two additional symmetry planes, the perradii, which intersect the corners of the rectangular mouth. Each septum is penetrated adorally by a deep, ectoderm-lined, funnel-shaped invagination, the peristomial funnel, that causes the septum to bulge laterally. In some members of the order Stauromedusida (a group of exclusively polypoid scyphozoans), each of the two sides of the septum exhibits a sheet-like longitudinal outgrowth, called a claustrum, that is fused along most of its length with a claustrum of the laterally adjacent septum. The septum's free or adaxial edge exhibits two symmetrically arranged, longitudinal rows, or phacellae, of nematocyst-laden gastric filaments. Nearer the septum's fixed, or abaxial edge is a pair of longitudinal, entodermal gonads. Situated still closer to the septum's abaxial edge is a longitudinal, ectodermal retractor muscle. In most scyphozoans, apparently, this muscle consists of a single tissue bundle; however, in the stauromedusans *Craterolophus* and *Lucernaria*, the septal muscle is adorally bifurcate (Antipa, 1892; Gross, 1900). In *Lucernaria*, the paired bundles originate near the polyp's oral end and diverge at about 20 degrees; in *Craterolophus*, they originate nearer the basal end, at or near the apex of the peristomial funnel, and diverge at about 1 to 2 degrees.

Polyps of the order Coronatida are sheathed in a chitinous, steeply conical theca that in some species exhibits internal structures at the interradii and perradii (e.g., Werner, 1966, 1967a, b, 1970, 1971a, b, 1974, 1979, 1983; Chapman & Werner, 1972). The theca itself is built of numerous, extremely thin (12 nanometers) lamellae that are grouped in two layers. In species exhibiting internal thecal structures, the perradii and interradii are sites of a single file or series of discrete, thorn-like longitudinal invaginations of the theca's inner layer. These invaginations, here designated internal thecal projections, are also arranged in whorls, with each whorl consisting of a set of four perradial and four interradial projections. Projections at the perradii are consistently larger (longer, wider, higher) than projections at the interradii, with the perradial projections often extending well over half-way to the theca's longitudinal axis. In most species, both sets of projections are smooth; in other species, however, the perradial projections exhibit small pustules.

Conulariids -- Conulariids are characterized by an apatitic, finely lamellar, steeply pyramidal test that usually exhibits four sides, or faces. *Conularina triangulata* (Raymond) has only three faces (Sinclair, 1942, 1948), and Babcock *et al.* (1987a) have documented a specimen of *Paraconularia* that exhibits six faces. As seen in transverse section, non-distorted, four-sided tests may be trapezoidal, square, or rectangular. In nearly all conulariids, the corners of the test are furrowed by a longitudinal groove, and in some conulariids the midline of each face is also grooved or is marked by a longitudinal ridge or fold.

Corners and/or midlines of some or all species of at least nine of the 21 currently recognized conulariid genera are sites of localized inflection and thickening of the innermost test lamellae (Van Iten, in press; this paper, Figure 1). Corners of some or all species of at least six genera — *Archaeoconularia* Bouček, *Climacoconus* Sinclair, *Conularina* Sinclair, *Eoconularia* Sinclair, *Glyptoconularia* Sinclair, and *Paraconularia* Sinclair — possess a smooth, broadly rounded or keel-like internal carina. In some taxa (e.g., certain species of *Climacoconus*), this carina projects up to about one-third of the way to the centre of the test cavity. Although corner carinae have previously been characterized as continuous structures (e.g., Sinclair, 1948), examination of etched specimens (Van Iten, in press) revealed that in some *Climacoconus*, corner carinae are seriated, consisting of discrete segments that exhibit a gently arched longitudinal profile and range from about 4 to 5 millimeters long.

A substantially more varied array of internal structures occurs at conulariid midlines. Among the structures discovered thus far are: (1) a low, smooth, broadly rounded or keel-like carina (present in some species of *Archaeoconularia, Climacoconus, Conularia, Ctenoconularia,* and *Paraconularia*; Ulrich, 1892; Knod, 1908; Bouček, 1939; Sinclair, 1948; Babcock *et al.*, 1987a,b; Babcock, 1988; Van Iten, in press); (2) a pair of carinae, with members of each pair situated on opposite sides of the midline proper and gradually diverging toward the aperture (present in *Metaconularia* and *Conularina narrawayi* Sinclair; Holm, 1893; Sinclair, 1940, 1948; Van Iten, in press); (3) a pair of broad, very low thickenings, likewise with members of each pair situated on opposite sides of the midline proper and gradually diverging toward the aperture (present in other North American species of *Conularina*; Van Iten, in press); (4) a low carina that is adaxially bifid (present in *Conularia splendida* Billings; Van Iten, in press); (5) a high, adaxially bifid carina that exhibits a more or less Y-shaped transverse cross section (present in *Eoconularia loculata* (Wiman); Wiman, 1895); (6) a high, non-bifid carina that projects about one-third of the way to the opposite midline and is covered by small pustules (present in specimens of *Paraconularia?* sp. from the Devonian of Australia; Bischoff, 1978); (7) a series of low, short, narrow ridges running along the midline proper (present in *Archaeoconularia membranacea* (Ringueberg); Van Iten, in press); and (8) paired series of short, closely spaced, I-shaped ridges, with members of each pair of series situated on opposite sides of the midline proper and gradually diverging toward the aperture (present in certain *Climacoconus*; Van Iten, in press).

Some species of *Archaeoconularia, Climacoconus, Conularina, Eoconularia,* and *Paraconularia* exhibit internal structures both at the corners and the midlines (Knod, 1908; Bouček, 1939; Van Iten, in press; this paper, Figure 1). In nearly all of these taxa, internal structures at the corners are consistently higher than internal structures at the midlines (Van Iten, in press).

Among the most striking internal test structures are the four high, adaxially bifid carinae documented by Wiman (1895) at the midlines of *Eoconularia loculata* (Wiman). The existence of these features has recently been questioned (Babcock & Feldmann, 1986a, b), due in part to the fact that identical structures have not been found in other conulariids. Unfortunately, the two

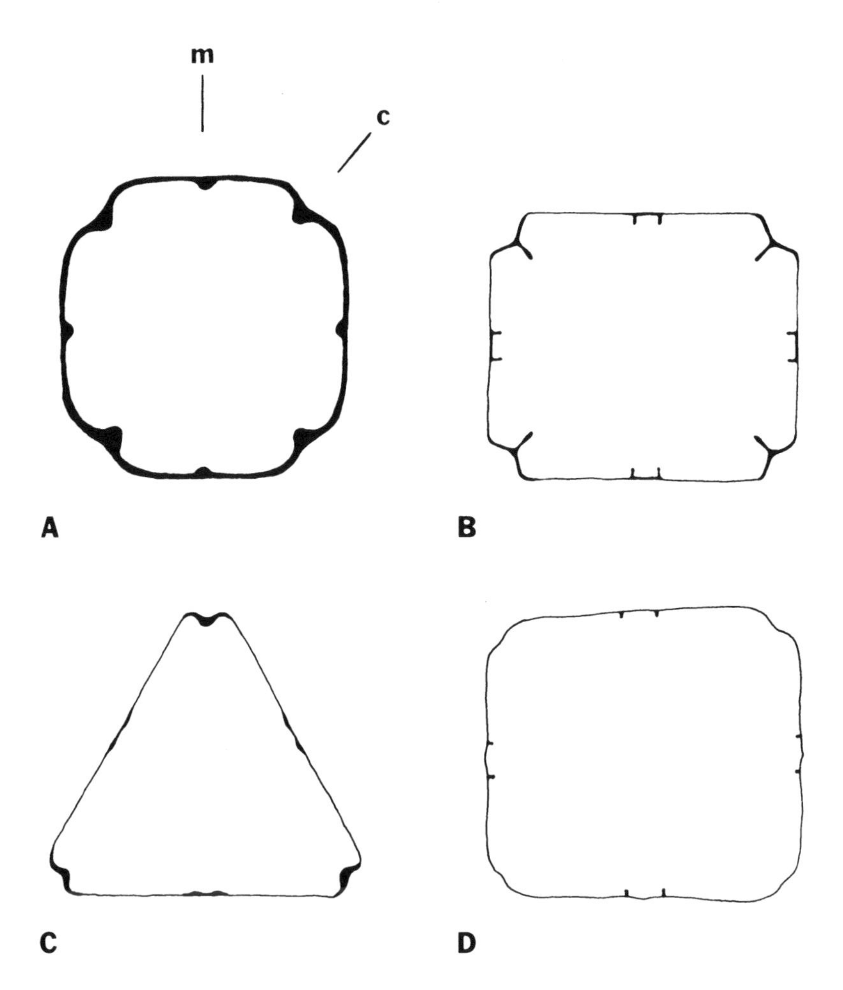

Fig. 1A-D. Transverse cross sections through some of the conulariids known to exhibit internal test structures at their corners (c) and/or midlines (m). A, *Paraconularia crustula* (White) (Pennsylvanian, midcontinental United States; corners and midlines both with single carina; x5); B, *Climacoconus* sp. (Middle and Late Ordovician, Iowa and Minnesota, USA; corners with single carina, midlines with paired, seriated carinae; x7); C, *Conularina triangulata* (Raymond) (Middle Ordovician, Quebec, Canada; corners with single carina, midlines with paired, low thickenings; x7); D, *Metaconularia* Foerste (Ordovician and Silurian, worldwide; corners non-thickened, midlines with paired carinae; x1.8). Figure 1D taken from Holm, 1893, Plate 6, Figure 39. All other drawings based on specimens in collections of the University of Michigan (Ann Arbor), the State University of Iowa (Iowa City), or the Geological Survey of Canada (Ottawa).

E. loculata specimens figured by Wiman have been lost or misplaced (Babcock and Feldmann, 1986a, b). Nevertheless, as shown in Wiman's line drawings of transverse sections through one of these specimens, reproduced here in Figure 2, the four Y - shaped features depicted at the midlines are built of thickened, inwardly deflected lamellae that continue into the faces. Since this is exactly what one sees in carinae of other conulariids, it seems wisest to accept Wiman's drawing of the internal midline anatomy of *E. loculata* as accurate. This conclusion is reinforced by the presence of a low bifid carina at midlines of *Conularia splendida* (Billings), a fairly common species in the lower Maquoketa Formation (Late Ordovician) of northeast Iowa, USA (Figure 3).

COMPARISONS OF CONULARIIDS AND SCYPHOZOANS

The hypothesis that conulariid midlines were sites of four gastric septa, homologous to the four gastric septa of scyphozoans, was originally based in large part on comparisons of high, adaxially bifid midline carinae of *Eoconularia loculata* (Wiman) to claustra-bearing septa of stauromedusans such as *Craterolophus* (Kiderlen, 1937). As documented by Kiderlen (1937; see also Moore & Harrington, 1956a), both sets of structures are identical in number and arrangement, and show similarities in size (both absolute and relative to the body, or test cavity) and cross-sectional form. Moreover, no currently known non-scyphozoan group exhibits sort-part or test structures that more closely resemble *E. loculata* midline carinae than do stauromedusan septa. Coupled with microstructural evidence indicating that the conulariid test was an ectodermal derivative produced by soft tissues lining its inner surface (e.g., Bischoff, 1978; see also below), this suggests that high, adaxially bifid midline carinae, while not directly homologous to stauromedusan septa, were covered in life by soft tissue structures that were homologous to stauromedusan septa.

Midline carinae in other conulariids also exhibit interesting similarities to stauromedusan septa. As noted above in connection with details of septal anatomy, the stauromedusan septal muscle consists either of a single tissue bundle, or is adorally bifurcate. Similarly, conulariids whose midlines are internally carinate exhibit either a single carina, or a pair of carinae, comparable in spacing and angle of divergence to paired septal muscle strands of *Craterolophus*. Given that conulariid carinae and scyphozoan septal muscles are

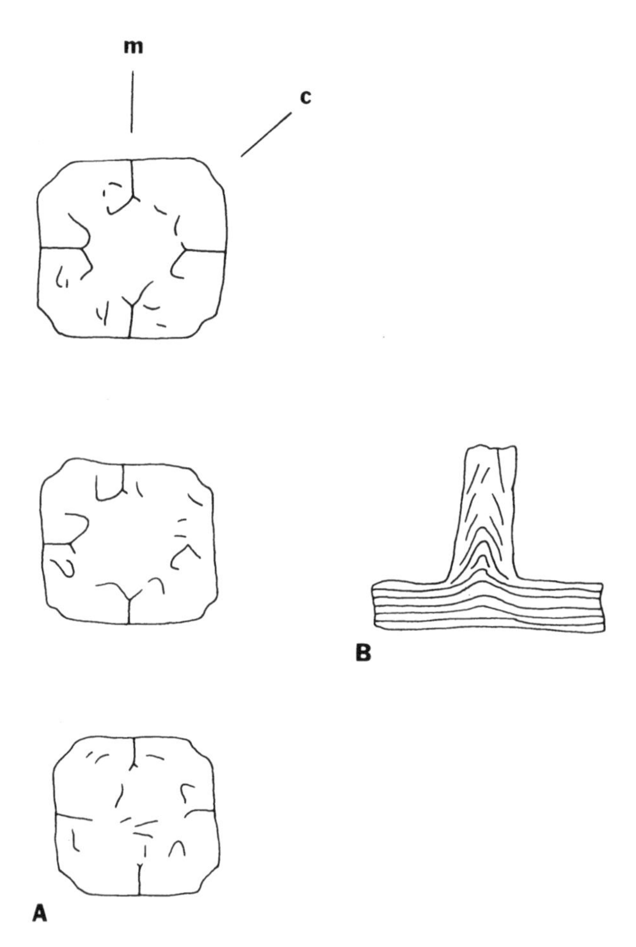

Fig. 2. Wiman's (1895) line drawings of three transverse sections through a specimen of *Eoconularia loculata* (Wiman), showing (A) the presence of a high, adaxially bifid internal carina at each of the four midlines, and (B) details of the organization of test lamellae in the base of one of the carinae. The uppernost cross section in Figure 2A is about 2.1 millimeters wide.

conulariid carinae and scyphozoan septal muscles are both ectodermal derivatives, these similarities strengthen the hypothesis that the organization of soft tissues located at conulariid midlines and scyphozoan interradii was fundamentally similar. More specifically (and as suggested earlier by Bischoff, 1978), the fact that internal carinae at conulariid midlines are single or paired can be interpreted as evidence of the former presence of a septal retractor muscle that, like stauromedusan retractor muscles, could be either single or adorally bifurcate.

Similarities between internal carinae at conulariid corners and midlines and internal thecal projections at the perradii and interradii of coronatid scyphozoans are also important. Both sets of structures consist of tetramerally arranged, longitudinally elongate inflections of inner test or thecal lamellae, and may be smooth or covered by small pustules. Just as coronatid perradial projections are consistently larger (higher, broader, longer) than projections at the interradii, in nearly all conulariids whose corners and midlines are both carinate, internal structures at the corners, interpreted as perradial in position, are consistently higher than internal structures at the midlines. Although internal corner and midline structures of conulariids are generally continuous, internal structures at midlines of *Archaeoconularia membranacea* (Ringueberg), and at corners or midlines of some *Climacoconus*, are seriated. Moreover, seriated internal structures in *Climacoconus* are organized in whorls (currently available specimens of *A. membranacea* are too incomplete to determine if their internal midline structures also are arranged in whorls), and, like coronatid perradial projections, individual segments of seriated *Climacoconus* corner structures are several times longer and higher than segments of seriated midline structures.

Possession of internal test or thecal structures having the characteristics just summarized is a feature uniquely shared by certain conulariids and coronatid scyphozoans. Together with similarities in test (thecal) microstructure and growth (see below), similarities between internal test (thecal) structures not only reinforce the hypothesis that conulariid midlines were sites of gastric septa, but they also suggest that the conulariid test and coronatid theca are themselves homologous (see below).

Finally, just as some conulariids exhibit three or six faces, and not four, aberrant scyphozoans produce three, five, or six septa. The only other metazoans resembling conulariids and known to exhibit this ar-

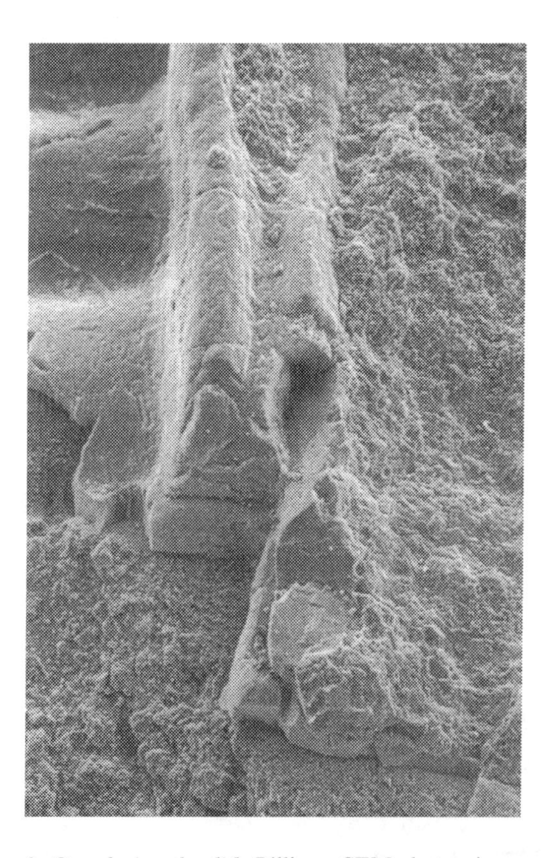

Fig. 3. *Conularia splendida* Billings. SEM photomicrograph of a portion of the interior of one midline, showing the presence there of a low, longitudinally bifid carina, similar to the high bifid carina seen at midlines of *E. loculata* (x100). The specimen is in collections of the Burke Museum, University of Washington, Seattle.

ray of alternate symmetries are hydrozoan cnidarians (which, for example, show variation in the number of primary branches in the medusa's radial canal system; Werner, 1973).

To summarize, conulariid corners and midlines exhibit numerous similarities to anatomical structures located, respectively, at the scyphozoan perradii and interradii. Except for similarities in symmetry patterns (which are also shared with hydrozoan cnidarians), all of the similarities outlined above are uniquely shared by conulariids and scyphozoans, and there is no evidence of greater similarity between conulariid corner and midline anatomy and the test or soft-part anatomy of any non-scyphozoan group. The relatively complex character of similarities between conulariids and scyphozoans makes it difficult to dismiss these similarities as superficial or due to convergence. Rather, these similarities suggest that conulariid midlines were sites of four gastric septa, homologous to the four gastric septa of scyphozoans.

The Conulariid Test

BACKGROUND

General - Advocates of a scyphozoan affinity for conulariids interpret the conulariid test as a mineralized, ectodermally secreted theca, similar to non-mineralized thecae produced by polyps of certain scyphozoans. Based on comparisons with thecal of coronatid scyphozoans, Werner (1966) proposed that the conulariid test and coronatid theca are homologous. In his opinion (Werner, 1966), coronatids inherited their theca from a conulariid ancestor, through rounding of the corners and loss of apatitic mineralization.

Coronatid theca -- Additional aspects of the anatomy and growth of the steeply conical, chitinous coronatid theca have been discussed by Werner (1966, 1967a,b, 1970, 1971a,b, 1974, 1979, 1983), and Chapman and Werner (1972). Briefly, the theca completely covers the side of the calyx, which grows up to eight centimeters long (Werner, 1970). The theca terminates adorally in a small, disc-like holdfast, attached in life to rocks or other hard substrates. Its numerous, twelve-nanometer-thick lamellae parallel the outer thecal surface and are arranged in two layers: a thin, outer layer that is crossed by fine, closely spaced transverse and longitudinal corrugations; and a thicker, non-corrugated, inner layer. The outer, corrugated layer ranges from about 2 to 4 microns thick and is secreted by soft tissues overlapping the apertural margin. Growth of the outer layer evidently involves extension of the apertural edge of existing lamellae (Van Iten, in press). The inner layer, locally almost 20 microns thick, is secreted by ectoderm of the side of the calyx, which accretes whole lamellae to the theca's inner surface. Especially near the aperture, where the inner layer is thin or absent, the theca is highly pliable. Animals whose theca has been punctured or severed repair such damage by secreting a thin patch of non-ornamented lamellae to the inner thecal surface.

Conulariid test -- As noted above in connection with conulariid corner and midline anatomy, the steeply pyramidal conulariid test is apatitic, finely lamellar, and usually exhibits four faces. Tests collected thus far range from less than two centimeters long to nearly 30 centimeters long. The largest known tests (e.g., *Metaconularia papillata* (Hall); Middle Ordovician, New York) are incomplete, and some of them originally measured at least 50 centimeters long (Van Iten, in prep.). Although tests of some conulariids (e.g., *Metaconularia calderi* Sinclair; Middle Ordovician, Ontario) are locally as much as 2 or 3 millimeters thick, many specimens, including some that were well over 20 centimeters long in life, are only a few tens of microns thick and show evidence (e.g., plication or wrinkling) of having been flexible.

In some conulariids, the apertural end of each face projects beyond the ends of the corners, forming a broadly rounded or triangular extension, called a lappet (e.g., Moore & Harrington, 1956a, b). Lappets are often folded over the aperture, in such a manner as to indicate that conulariids were capable of bending the lappets in life (e.g., Ulrich, 1892; Bouček, 1928; Reed, 1933; Kowalski, 1935). Although the opposite, or apical end of the test is usually not preserved, Kozłowski (1968) documented several specimens of *Conularia* that terminate adapically in a minute (less than one millimeter wide), inverted-cup-like structure that probably was part of a holdfast (see also below). Inside this structure, the test cavity exhibits several more or less globular, mutually contiguous chambers (see also Termier & Termier, 1949, 1953). Some conulariid specimens, now missing a substantial portion of their apical end, terminate in a smooth, finely lamellar, generally outwardly convex transverse wall, called the schott, that also extends along the inside of the faces, toward the aperture (Sinclair, 1948; Van Iten, in press). Conulariids usually exhibit a single such structure, though several specimens have been found that exhibit one or two additional schotts within the test cavity. Analysis of patterns of occurrence of schott-bearing conulariids with respect to sedimentary facies (Van Iten, 1987, in press) suggested that the schott represents a cicatrix produced when the animal was severed near its apex by currents.

Faces of nearly all conulariids are crossed by closely spaced, adaperturally arching transverse corrugations (the crests of which are generally referred to as tranverse ribs), or they are covered by minute, closely spaced tubercles, arranged in gently arched tranverse rows and longitudinal files (e.g., Sinclair, 1948). In corrugated taxa, the spaces between the tranverse ribs may be more or less smooth (e.g., *Climacoconus*), or they may be longitudinally corrugated (e.g., *Conularia*). Transverse ribs are commonly offset at the midlines, and in some taxa (e.g., *Climacoconus, Paraconularia*) they are consistently offset and deflected toward the aperture at the corners.

SEM analysis of polished sections through representatives of six conulariid genera (Van Iten, in press) revealed that the conulariid test is built of numerous, very thin (1-5 microns), alternately dense and vacuity-rich lamellae that parallel the outer test surface. No evidence of proteinaceous lamellae, reported by Babcock & Feldmann (1986a), was detected. As seen in longitudinal section, conulariid tests are thicker at transverse ribs and tubercles than at sites between these features. This is due to localized thickening of individual test lamellae. Faces of some conulariids exhibit healed punctures, sustained below the apertural margin (e.g. Babcock *et al.*, 1987a; Van Iten, in press).

As in the case of severance of the test, mentioned above, such injuries were repaired by deposition of non-ornamented lamellae on the test's inner surface. Damage repair involving deposition of lamellae on the test's outer surface has not been observed, indicating that the test was most likely an external skeleton. Analysis of healed injuries also suggests that living conulariids were characterized by two distinct thecal layers (Van Iten, in press): a mineralized, inner layer, composed of lamellae that were accreted to the test's inner surface by ectoderm of the body wall; and a non-mineralized, outer layer, composed of lamellae that were secreted by soft tissue overlapping the apertural margin. Growth of the outer, non-mineralized layer involved extension of the apertural edge of existing lamellae and proceeded until the animal reached a certain age or size, at which point the animal began to thicken its theca by producing the inner, mineralized layer.

DISCUSSION

The conulariid test and coronatid theca exhibit similarities in gross anatomy, microstructure, and mode of growth and damage repair. Some of these similarities, for example the presence of two sets of tetramerally arranged, continuous or seriated longitudinal inflections of the innermost lamellae, are uniquely shared by conulariids and coronatids. Although certain details of the anatomy and growth of the conulariid test are not yet known or fully understood, currently available microstructual and anatomical evidence suggests that the conulariid test is most similar to the theca of coronatid scyphozoans.

Opponents of the hypothesis that the conulariid test and coronatid theca are homologous have pointed to differences between these two structures that they regard as fundamental. Among these differences are the presence of strong, apatitic mineralization in the conulariid test versus the absence of mineralization in the coronatid theca, and the steeply pyramidal shape of the conulariid test versus the steeply conical shape of the coronatid theca. These differences have been interpreted by Kozłowski (1968) as posing serious difficulties for the hypothesis that conulariids and coronatid scyphozoans were closely related. However, similarities in details of thecal microstructure, gross anatomy, and growth, outlined in this paper, suggest that differences in shape and degree of mineralization can be explained as reflecting divergence of conulariids and coronatids from a thecate common ancestor possessing internal thecal projections at the interradii and perradii. According to this interpretation, conulariids and coronatids lacking internal thecal projections lost these structures secondarily.

Other proposed objections to the hypothesis that the conulariid test and coronatid theca are homologous include the following two claims: (1) that the conulariid test was secreted internally, not externally as in the case of the coronatid theca (Kozłowski, 1968); and (2) that thickening of transversely corrugated conulariid tests at sites of transverse ribs is due to the presence within the ribs of a discrete, non-laminated transverse rod, a structure unique to conulariids and with no homologue in scyphozoans or any other group (Feldmann & Babcock, 1986; Babcock and Feldmann, 1986a,b). The first of these claims, that conulariid tests were secreted internally, is not supported by microstructural evidence, which as noted above indicates that test-secreting soft tissues probably were not present on the test's exterior surface (as would have been the case were the test secreted internally). Microstructural data also refute the second claim, namely that thickening of corrugated conulariid tests at transverse ribs is due to the presence of discrete rods. As shown by SEM analysis of polished sections, thickening at sites of transverse ribs is actually due to localized thickening of individual test lamellae (Figure 4). Although the distinction between lamellae within the core of tranverse ribs is often obscured by accentuated mineralization, lamellae can still be traced through such ribs, indicating that they are part of a single, continuously laminated structure.

Finally, it should be noted that Termier & Termier (1949, 1953), writing before the publication of comparisons of conulariids and coronatid scyphozoans, rejected a scyphozoan affinity for conulariids based in part on interpretation of the significance of offset of conulariid tranverse ribs at corners and midlines. In their opinion, such offset indicates that tranverse ribs are comparable to fusellar half-rings of pterobranchs. However, microstructral analysis (Van Iten, 1989) has yielded no evidence that traverse ribs are structurally discrete, as are fusellar half-rings, and thus transverse ribs are probably better regarded as a kind of divaricate ornament.

To summarize, similarities between the conulariid test and the theca of coronatid scyphozoans have been interpreted as indicating that these two structures are homologous. Opponents of this hypothesis have yet to show that the conulariid test is more similar (and therefore more likely to be homologous) to the test or theca of some non-scyphozoan group. Claims that the conulariid test and coronatid theca are fundamentally dissimilar are not supported by comparative microstructural and anatomical analyses. Instead, such analyses suggest that observed differences can be explained as due to divergence of conulariids and coronatids from a thecate common ancestor.

Relic Conulariid Soft Parts

BACKGROUND

Previous discoveries -- Possible relics of conulariid soft parts have recently been documented by Steul (1984) and Babcock & Feldmann (1986a,c). Steul (1984), using X-radiography to examine pyritized specimens of *Conularia* from the Hunsrück Slate (Early Devonian), Germany, discovered dense, localized, more or less regular aggregations of pyritic material in matrix filling the test cavity. Among the features documented by Steul (1984), in at least four specimens, is a slender, elongate band of segment-like aggregations, apparently coincident with the test's longitudinal axis. Steul (1984) refers to this feature as the axial element. At or near its apertural end, the axial element is associated with aggregations of pyritic material extending across the test cavity, at right angles to the axial element. Interestingly, Steul's (1984) photographs show no clear evidence of circumoral tentacles, featured in reconstructions of the conulariid organism made by Kiderlen (1937).

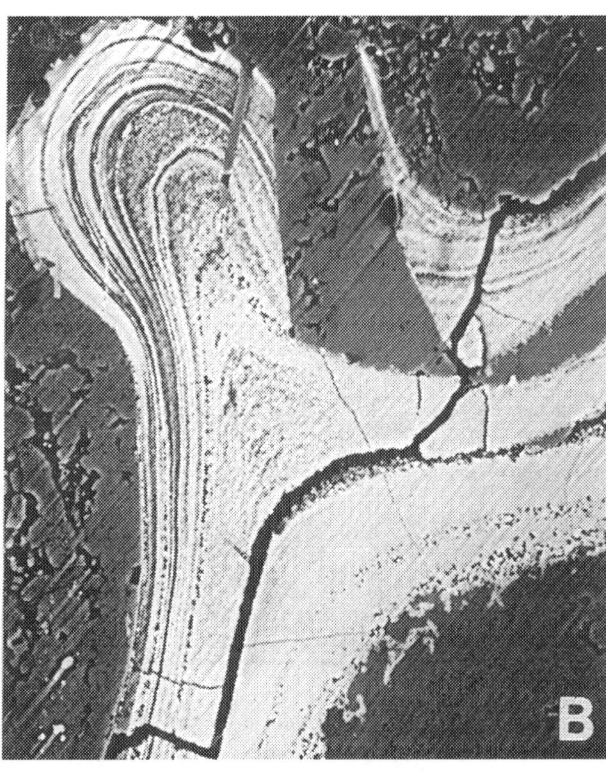

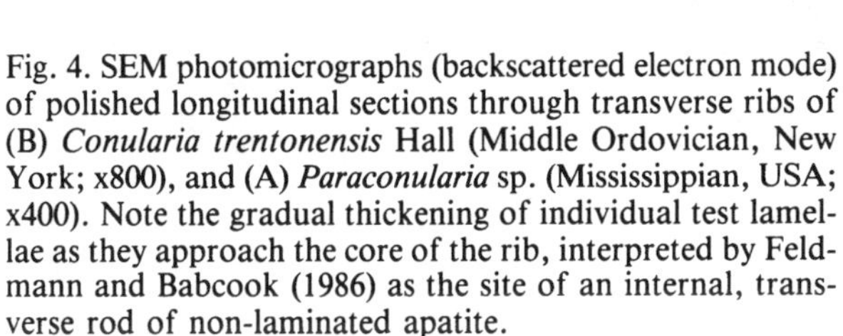

Fig. 4. SEM photomicrographs (backscattered electron mode) of polished longitudinal sections through transverse ribs of (B) *Conularia trentonensis* Hall (Middle Ordovician, New York; x800), and (A) *Paraconularia* sp. (Mississippian, USA; x400). Note the gradual thickening of individual test lamellae as they approach the core of the rib, interpreted by Feldmann and Babcook (1986) as the site of an internal, transverse rod of non-laminated apatite.

Steul (1984) interprets the axial element as remains of an internal, axially situated support structure, with components homologous to the spinal muscles, vertebrae, and notochord of vertebrates; she interprets the aggregations of pyritic material extending laterally from the axial element's apertural end as relics of a pair of eyes. Although these interpretations have been challenged by Babcock and Feldmann (1986b; see also below), the fact that the features in question occur in several specimens, apparently at the same site within the test cavity, suggests that, whatever their homology, they are relics of original anatomical structures.

Babcock and Feldmann (1986b, p. 365), examining two specimens of *Paraconularia subulata* (Hall) (Early Mississippian, USA) preserved in fine-grained, phosphatic concretions, discuss what they describe as a "single elongate tube that extends the length of the central cavity and a large globular structure near the aperture". In their opinion (Babcock and Feldmann, 1986b), the elongate tube corresponds to Steul's (1984) axial element. In contrast to Steul (1984), however, Babcock and Feldmann (1986b) interpret the axial element as a relic intestine or alimentary tract.

Additional material -- Inspection of the micrite-filled test cavity of the holotype and only known specimen of *Eoconularia amoena* Sinclair, a conulariid from the Trenton Group (Middle Ordovician) of Quebec, Canada (Sinclair, 1944, 1948), revealed the presence of a small, centrally located circular feature at the specimen's apical end, which has been broken about 2 centimeters above the apex (Figure 5A). The circular feature lies in a plane oriented approximately perpendicular to the specimens's long axis. It is about 0.25 millimeters in diameter and consists mostly of extremely fine-grained (less than one micron) micrite that exhibits a regular fabric resembling concentric lamination (Figure 5B). Disposed about the circular feature's centre, in three of four quadrants defined by planes containing the test's midlines, is a set of sub-triangular regions of relatively coarse micrite showing a centripetal fabric (Figure 5B). The outer margin of the circular feature is scalloped (Figure 5B). The exact number of lobes along this margin is difficilt to determine but appears to be at least 13 and possibly 16. Four lobes are present in the southwest quadrant (Figure 5B), suggesting that four lobes may also have been present in each of the other four quadrants.

Additional work is needed to establish whether or not the circular feature extends along the test cavity's longitudinal axis, but the morphology and orientation of this feature relative to the conulariid's corners and midlines suggest that it is not a diagenetic artifact or the remains of some non-conulariid fossil (e.g., a crinoid ossicle or stick-like bryozoan) preserved inside the conulariid. For these reasons, the circular feature is here provisionally interpreted as a relic of something originally present in the conulariid test cavity, possibly a transverse section through Steul's (1984) axial element.

DISCUSSION

Steul's (1984) interpretation of the axial element as remains of an internal support structure, comparable to the vertebral column of vertebrate chordates, is based in part on analysis of a single feature (Steul, 1984, Figure 17) resembling an axial element and that, like vertebrate bone, is characterized by phosphatic mineralization (as demonstrated by x-ray fluorecence). However, Steul (1984) presents no evidence of phosphatic mineralization in any of the other features she interprets as axial elements, and comparison of her photographs of these features (Steul, 1984, Figures 11a, 12a, 14a, 15a, 16a) to known pyritized soft parts in

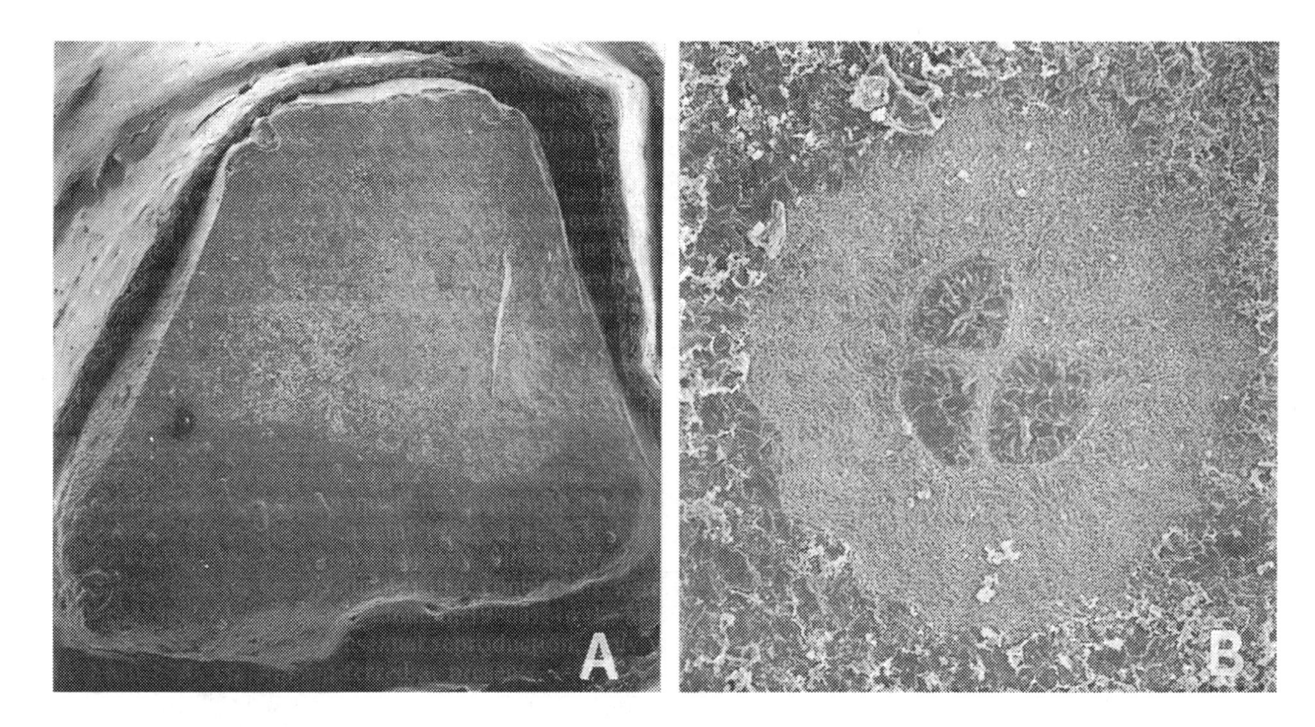

Fig. 5A-B. SEM photomicrographs of possible relic soft parts in the test cavity of the holotype of *Eoconularia amoena* (Sinclair). A, photomicrograph of the specimen's broken apical end, showing the central location (arrow) of the circular feature (x30); B, close-up of the circular feature, showing its scalloped outer margin and the three sub-triangular regions of relatively coarse micrite disposed about its centre (x250). The specimen is housed in collections of the Royal Ontario Museum, Toronto, Canada (ROM 23297).

other Hunsrück fossils (e.g., Stürmer and Bergström, 1973; Fauchald *et al.*, 1986) suggests that they consist of pyrite only.

An alternative interpretation of the axial element is suggested by comparisons of this feature to scyphozoan polyps undergoing polydisc strobilation, a process of medusa formation characteristic of many scyphozoans and documented in some detail by Buchsbaum (1948), Werner (1967b, 1974), and Werner & Hentschel (1983). During this process, the polyp's septa and tentacles degenerate, and the bulk of its soft body is transformed into a slender, axially situated series of incipient medusae, or ephyrae. As suggested for example by comparison of Steul's (1984) Figures 10a, 11a, 12a, 13a, and 14a to Werner's (1967b) Figure 6, and Steul's (1984) Figures 15a and 16a to Werner's (1967b) Figure 13, features cited by Steul (1984) as characteristic of *Conularia* axial elements (save phosphatic mineralization, which as argued above probably is not a feature of *Conularia* axial elements) are matched by similar features in strobilating coronatids. Coupled with similarities between scyphozoans and the conulariid test, this suggests that the axial element could be interpreted as a string of conulariid ephyrae, attached to the inside of the test by means of relic soft tissue interpreted by Steul (1984) as remains of a pair of eyes. This interpretation accounts for the apparent absence of circumoral tentacles in Hunsrück *Conularia* (Steul, 1984), thought by Steul (1984) and Babcock and Feldmann (1986b) to constitute a serious problem for hypotheses of a scyphozoan affinity for conulariids. As noted above, scyphozoan polyps undergoing strobilation lack tentacles.

Additional evidence of homology to soft parts of strobilating scyphozoans may be provided by the circular feature preserved inside the test cavity of *Eoconularia amoena* Sinclair (Figure 5). This feature, interpreted above as originally belonging to the conulariid, exhibits evidence of tetrameral symmetry and appears to be situated in the same position as the axial element in Hunsrück *Conularia.* The geometry of the circular feature is in poor agreement with transverse sections through the vertebrate notochord and vertebral column (see for example Jollie, 1962), interpreted by Steul (1984) as homologues of the conulariid axial element, and it is also in poor agreement with transverse sections through vertebrate and invertebrate alimentary tracts (see for example Barnes, 1987), interpreted by Babcock and Feldmann (1986b) as possible axial element homologues. The circular feature agrees in several respects with the anatomy of scyphozoan ephyrae and medusae, particularly medusae of the Order Samaeostomatida (Figure 6). The circular feature's scalloped outer margin could be interpreted as comparable to the lappet-bearing scyphomedusan bell margin, and the sub-triangular regions disposed in three of four quadrants about the circular feature's centre could be interpreted as relics of a set of four gonads (one of which has not been preserved or is absent in the plane of section), similar in position to the gonads of semaeostome medusae.

That conulariid soft parts could be preserved as calcite is further supported by the recent discovery of cnidarian polyps preserved as calcite in a specimen of the tabulate coral *Favosites* from the Jupiter Formation (Lower Silurian) of Anticosti Island, Quebec (Copper, 1985). Many of the polyps exhibit beautifully preserved circumoral tentacles and have had their enteron filled with argillaceous sediment. Exactly how the polyps came to be preserved as calcite is not yet clear Copper (1985, p. 144) suggests that this unusual form of preservation was facilitated by the presence of algal

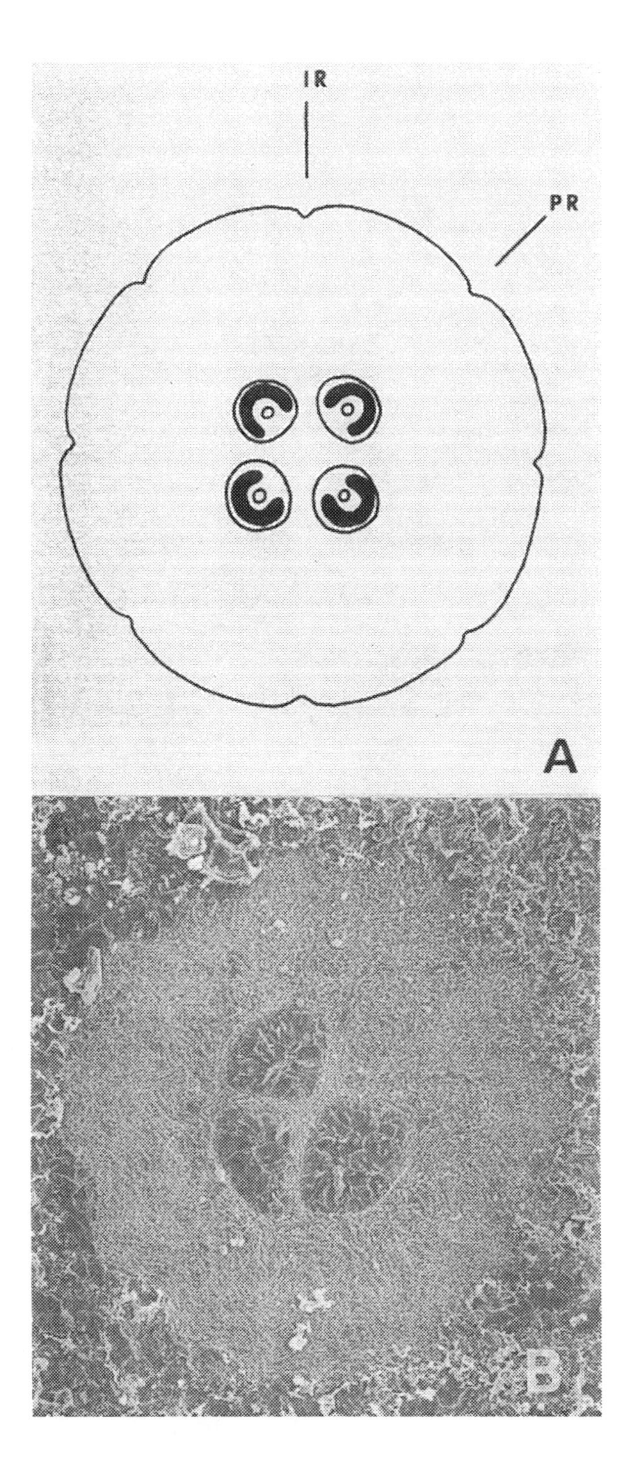

Fig. 6. Comparison of (A) a medusa of the extant, semaeostome scyphozoan *Aurelia* and (B) the circular feature in the test cavity of *Eoconularia amoena* Sinclair. (Figure 6A modified from Barnes, 1987, Figure 5-27B).

symbionts that "may have initiated precipitation of calcite nuclei or spicules inside the polyp[s]". The mode of preservation of the circular feature in *Eoconularia amoena* is also unclear. One possibility, casting of conulariid soft tissues by lime mud sediment and/or calcite cement, is suggested by results of Walcott's (1898) taphonomic experiments on extant scyphozoan medusae. Walcott (1898) demonstrated that the digestive cavity and gonads of *Aurelia* medusae buried in wet plaster are readily cast, with outstanding preservation of their original three-dimensional geometry. In light of published discussions of the mechanical properties of jellyfish membranes (e.g., Walcott, 1898), it is conceivable that, shortly after the death of the *E. amoena*, one or more tears or punctures developed in the external skin or in the lining of the digestive cavity or gonads of one of its ephyrae, allowing fine sediment to enter the ephyra as its original watery contents leaked out.

Needless to say, the interpretations of the circular feature outlined above are in need of further testing. One approach is to see how far the circular feature extends into the matrix. If it turns out that the circular feature does extend into the matrix, and that subtriangular structures occur in discrete groups of four members arranged in series along the specimen's long axis, then it will be difficult to argue against the interpretations suggested here.

Additional Evidence

Palaeontologists have discovered several additional lines of evidence that are consistent with or corroborate interpretations of homology advocated above. Conulariid specimens discussed by Sinclair (1948), Finks (1955, 1960), Rooke & Carew (1983), Babcock *et al.* (1987b), and Van Iten (in press) provide more or less direct evidence that conulariids, like scyphozoan polyps, were sessile organisms, attached to shell material or directly to the sea floor at their apical end. Interestingly, conulariids sometimes attached to and were overgrown by calcareous sponges (Finks, 1955, 1960; Van Iten, in press), a life relationship exhibited by extant sponges and polyps of certain coronatid scyphozoans (e.g., Werner, 1970). Conulariids preserved in fine-grained sediments often occur in monospecific clusters, in some cases with the individuals arranged in radial fashion and with their apical ends pointing toward a common centre (e.g., Slater, 1907; Ruedemann, 1925; Sinclair, 1948; Babcock & Feldmann, 1986a). One of the *Conularia* apices documented by Kozłowski (1968) exhibits features suggesting that conulariid clusters represent clonally budded colonies, similar to clonal polyp colonies of certain scyphozoans. Finally, the known geologic range of conulariids, from Middle Ordovician to Triassic (Babcock & Feldmann, 1986b), lies well within the range of scyphozoans, which have a fossil record extending as far back as the Vendian (Glaessner, 1984).

The Question of Relationship

Proponents of hypotheses of homology advocated in this paper interpret conulariids as extinct members of the phylum Cnidaria. Relationships among extant cnidarians have been subject to conflicting interpretations (e.g., Glaessner, 1971; Hand, 1959; Hadži, 1963; Werner, 1973; Salvini-Plawen, 1978; Grasshoff, 1984), and at present there appears to be no consensus on this problem. Still, the presence of four gastric septa in scyphozoans and conulariids, coupled with evidence suggesting that conulariids, like scyphozoans, were characterized by medusa formation involving polydisc strobilation, indicates that conulariids and scyphozoans may be more closely related to each other than either group is to anthozoan or hydrozoan cnidarians. Cladistic analysis of relationships among conulari-

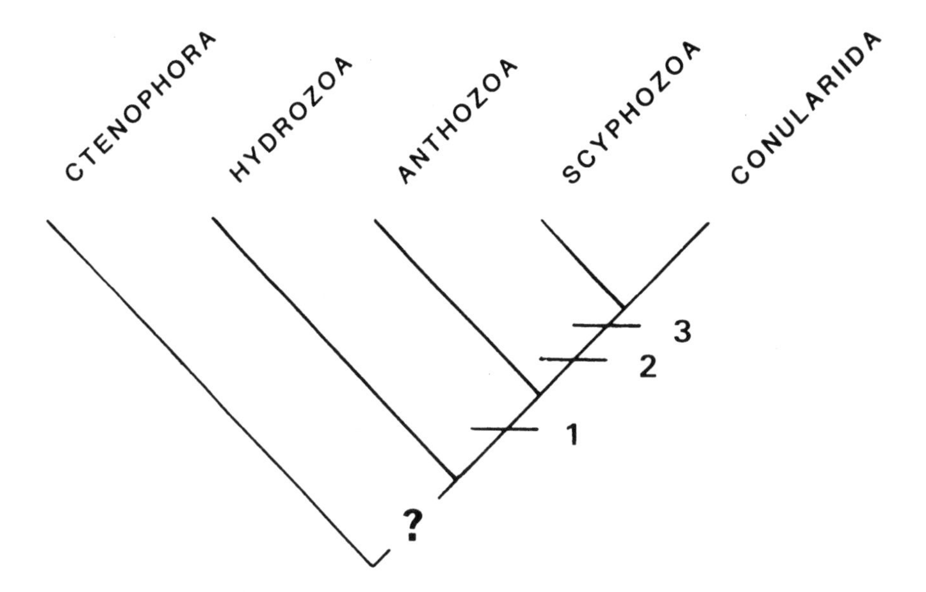

Fig. 7. Cladrogram summarizing the interpretation of phylogenetic relationships among conulariids and extant cnidarians favoured by the present author. Putative synapomorphies discussed in the text are as follows: (1) enteron septate; (2) enteron with four septa; (3) production of medusae through strobilation. The phylum Ctenophora is interpreted as the most likely candidate for an extant nearest relative of Cnidaria.

ids and extant cnidarians (Van Iten, 1989), based on scoring of over 20 anatomical and life history characters obtained from Hyman (1940) and Barnes (1987), revealed that the most parsimonious hypothesis of relationships (i.e., the one requiring the smallest number of transitions in character states) is that conulariids and scyphozoans are nearest relatives, and that together they and anthozoan cnidarians, which are also septate, form a monophyletic taxon within Cnidaria that excludes hydrozoan cnidarians, which are non-septate (Figure 7). This interpretation of relationships among extant cnidarians is controversial, but if corroborated by analysis of additional characters (e.g., biochemical or molecular characters) it will have profound implications for discussions of the evolutionary history of this peculiar group of metazoans.

Acknowledgements

I thank Simon Conway Morris (Cambridge University, Cambridge, England) and Alberto M. Simonetta (University of Camerino, Camerino, Italy) for inviting me to participate in the Camerino Symposium on problematic fossil taxa and the origins of Metazoa. I also thank my former professors and graduate student colleagues in the Museum of Paleontology, University of Michigan, for helpful discussions of conulariid paleobiology and affinities. Much of the monetary support for this work was provided by funds from NSF Research Grant BNS 8521097 to Daniel C. Fisher (Museum of Paleontology, The University of Michigan, Ann Arbor). Preparation of illustrations for this paper was done with the assistance of my wife, Tatiana Van Iten. Finally, I am grateful to David Rudkin (Royal Ontario Museum, Toronto, Canada) for permission to examine the holotype of *Eoconularia amoena* (Sinclair).

REFERENCES

ANTIPA, G. 1892. Die Lucernariden der Bremer Expedition nach Ostspitzbergen im Jahre 1889. *Zoologische Jahrbücher, Abteilung für Systematik, Geographie, und Biologie der Tiere* 3, 377-398.

BABCOCK, L.E. 1988. New Permian conulariids from Devon Island, Canada. *Journal of Paleontology* **62**, 615-617.

BABCOCK, L.E. & FELDMANN, R.M. 1986a. The phylum Conulariida. In *Problematic Fossil Taxa* (ed. A. Hoffman and M. H. Nitecki), pp. 135-147. New York: Oxford University Press.

BABCOCK, L.E. & FELDMANN, R.M. 1986b. Devonian and Mississippian conulariids of North America. Part A. General description and *Conularia. Annals of Carnegie Museum* **55**, 349- 410.

BABCOCK, L.E. & FELDMANN, R.M. 1986c. Devonian and Mississippian conulariids of North America. Part 8. *Paraconularia, Reticulaconularia*, new genus, and organisms rejected from Conulariida. *Annals of Carnegie Museum* **55**, 411-479.

BABCOCK, L.E., FELDMANN, R.M. & WILSON, M.T. 1987a. Teratology and pathology of some Paleozoic conulariids. *Lethaia* **20**, 93-105.

BABCOCK, L.E., FELDMANN, R.M., WILSON, M.T. & SUÁREZ-RIGLOS, M. 1987b. Devonian conulariids of Bolivia. *National Geographic Research* 3, 210-231.

BARNES, R.D. 1987. *Invertebrate Zoology*, 5th ed., Philadelphia: Saunders College Publishing, 893 pp.

BISCHOFF, G.C.O. 1978. Internal structures of conulariid tests and their functional significance, with special reference to Circoconulariina n. suborder (Cnidaria, Scyphozoa). *Senckenbergiana lethaea* **59**, 275-327.

BOUČEK, B. 1928. Revise Českych paleozoických Konularii. České Akademie Ved a Umění v Praze, Trida II. Palaeontographica Bohemiae **XI**, 1-108. Prague: Nakladem České Akademie Ved a Umění.

BOUČEK, B. 1939. Conularida. In *Handbuch der Paläozoologie, 2A* (ed. O.H. Schindewolf), pp. A113-A131. Berlin: Gebruder Borntraeger.

BUCHSBAUM, R. *Animals without backbones*, revised ed. Chicago: University of Chicago Press, 405 pp.

CHAPMAN, D.M. 1966. Evolution of the scyphistoma. In *The Cnidaria and Their Evolution* (ed. W.J. Rees), pp. 51-75. London: Academic Press.

CHAPMAN, D.M. & WERNER, B. 1972. Structure of a solitary and a colonial species of *Stephanoscyphus* (Scyphozoa, Coronatae) with observations on periderm repair. *Helgoländer wissenschaftliche Meersuntersuchungen* **23**, 393-421.

COPPER, P. Fossilized polyps in 430-Myr-old *Favosites* corals. *Nature* **316 (2)**, 142-144.

Fauchald, K., Stürmer, W. & Yochelson, E.L. 1986. *Sphenothallus* "Vermes" in the Early Devonian Hunsrück Slate, West Germany. *Paläontologische Zeitschrift* **60**, 57-64.

Feldmann, R.M. & Babcock, L.E. 1986. Exceptionally preserved conulariids from Ohio-- reinterpretation of their anatomy. *National Geoaraphic Research* **2 (4)**, 464-472.

Finks, R.M. 1955. *Conularia* in a sponge from the west Texas Permian. *Journal of Paleontology* **29 (5)**, 831-836.

Finks, R.M. 1960. Late Paleozoic sponge faunas of the Texas region. The siliceous sponges. *Bulletin of the American Museum of Natural History* **120**, 7-160.

Glaessner, M. F. 1971. The genus *Conomedusites* Glaessner & Wade and the diversification of the Cnidaria. *Paläontologische Zeitschrift* **45**, 7-17.

Glaessner, M.F. 1984. *The dawn of Animal Life. A Biohistorical Study*. Cambridge: University Press, 244 pp.

Grasshoff, M. 1984. Cnidarian phylogeny - a biomechanical approach. *Palaeontographica Americana* **54**, 127-135.

Gross, J. 1900. Zur Anatomie der Lucernariden. *Jenaische Zeitschrift für Naturwissenschaft* **34**,611-622.

Hadži, J. 1963. The evolution of the Metazoa. *International Series of Monographs of Pure and Applied Biology, Division of Zoology* **16**,1-499.

Hand, G. 1959. On the origin and phylogeny of coelenterates. *Systematic Zoology* **8 (4)**, 191-202.

Holm, G. 1893. Sveriges Kambrisk-Siluriska Hyolithidae och Conulariidae. *Sveriges Geologiska Undersökning, Afhandlinger och uppsatser, Series C, 112*, 1-172.

Hyman, L. H. 1940. *The Invertebrates: Protozoa Through Ctenophora*. Vol. I, pp. 365-696. New York,London: McGraw-Hill.

Jollie, M. 1962. *Chordate Morphology*. New York: Reinhold Publishing Corporation, 478 pp.

Kiderlen, H. 1937. Die Conularien. Über Bau und Leben der ersten Scyphozoa. *Neues Jahrbuch für Mineralogie, Geologie und Paläontologie*, **77(B)**, 113-169.

Knod, R. 1908. Devonische Faunen Boliviens. In *Beiträge zur Geologie und Paläontologie von Sudamerika, XIV*, (ed. G. Steinmann). *Neues Jahrbuch für Mineralogie, Geologie und Paläontologie, Abteilung* **B 77**, 113-169.

Kowalski, J. 1935. Les Conulaires. Quelques observations sur leur structure anatomique. *Société des sciences naturèlles de l'Ouest France, Bulletin*, Série 5, 5, 281-293.

Kozłowski, R. 1968. Nouvelles observations sur les *Conularides. Acta Palaeontologica Polonica* **13**, 497-535.

Moore, R.C. & Harrington, H.J. 1956a. Scyphozoa. In *Treatise on Invertebrate Paleontology: Part F: Coelenterata* (ed. R.C. Moore), pp. F27-F38. New York and Lawrence: Geological Society of America and University of Kansas.

Moore, R.C. & Harrington, H.J. 1956b. Conulata. In *Treatise on Invertebrate Paleontology: Part F: Coelenterata* (ed. R.C. Moore), pp. F54-F66. New York and Lawrence: Geological Society of America and University of Kansas.

Mortin, J. 1985. The shell structure and zoological affinities of conulariids. *Palaeontological Association Annual Conference Abstracts*, pp. 12-13. Aberystwyth: University College of Wales.

Oliver, W.A., Jr. 1984. *Conchopeltis*: its affinities and significance. *Palaeontographica Americana no.* **54**, 141-147.

Oliver, W.A. & Coates, A.G. 1987. Phylum Cnidaria. In *Fossil Invertebrates* (ed. R.S. Boardman, A.H. Cheetam & A.J. Rowell), pp. 140-193. Palo Alto: Blackwell Scientific Publications.

Reed, F.R.C. 1933. Some new Ordovician species of *Conularia* from Girvan. *Geological Magazine* (n.S.) **70**, 354-358.

Rooke, H.G. & Carew, J.L. 1983. New light on a poorly understood fossil group. *Geological Society of America, Abstracts with Programs* **15 (2)**, 53.

Ruedemann, R. 1925. Some Silurian (Ontarian) Faunas of New York. *New York State Museum Bulletin* **265**, 5-84.

Salvini-Plawen, L. von, 1978. On the origin and evolution of the lower Metazoa. *Zeitschrift für zoologische Systematik und Evolutionsforschung* **16**, 40-88.

Sinclair, G.W. 1940. A discussion of the genus *Metaconularia* with descriptions of new species. *Royal Society of Canada, Section IV. Transactions, Series 3*, **38**, 87-95.

Sinclair, G.W. 1942. The Chazy Conularida and their congeners. *Annals of Carnegie Museum* **29**, 219-240.

Sinclair, G.W. 1944. Notes on the genera *Archaeoconularia* and *Eoconularia. Royal Society of Canada, Section IV, Transactions*, Series **3**, **34**, 101-121.

Sinclair, G.W. 1948. *The Biology of the Conularida*. Unpublished Ph. D. thesis. 442 pp. Montreal: McGill University.

Slater, I.L. 1907. *A Monograph of the British Conulariae*. London: Palaeontographical Society, London. 41 pp.

Steul, H. 1984. Die systematische Stellung der Conularien. *Giessener Geologische Schriften* **37**, 1-117.

Stürmer, W. & Bergström, J. New discoveries on trilobites by X-rays. *Paläontologische Zeitschrift* **47** (1/2), 104-141.

Termier, H. & Termier, G. 1949. Position systématique et biologie des Conulaires. *Revue scientifique*, 86, No. 3300, 711-722.

Termier, H. & Termier, G. 1953. Les Conularides. *Traité de Paléontologie: Vol. III: Onychophores, Arthropodes, Echinodermes, Stomocordés* (ed. J. Piveteau), pp. 1006-1013. Paris: Masson & Cie.

Ulrich, A. 1892. Paläozoische Versteinerungen aus Bolivien. in Beiträge zur Geologie und Palaeontologie von Sudamerika, (ed. G. Steinmann) *Neues Jahrbuch für Mineralogie, Geologie und Paläontologie* **8**, 5-116.

Van Iten, H. 1987. The mode of life of the Conulariida and its implications for conulariid affinities. *Geological Society of America. Abstracts with Programs* **19**, 876.

Van Iten, H. 1989. *Anatomy, life history, and evolutionary affinities of conulariids*. Unpublished Ph.D. thesis. Ann Arbor: University of Michigan, 215 pp.

Walcott, C.D. 1898. Fossil Medusae. *United States Geological Survey, Monograph* **30**, 1-201.

Werner, B. 1966. *Stephanoscyphus*, (Scyphozoa, Coronatae) und siene direkte Abstammung von den fossilen Conulata. *Helgoländer Wissenschaftliche Meeresuntersuchungen* **13**, 317- 347.

Werner, B. 1967a. *Stephanoscyphus* Allman (Scyphozoa, Coronatae), ein rezenter Vertreter der Conulata? *Paläontologische Zeitschrift* **41**, 137-153.

Werner, B. 1967b. Morphologie, Systematik und Lebensgeschichte von *Stephanoscyphus* (Scyphozoa, Coronatae) sowie seine Bedeutung für die Evolution des Scyphozoa. *Zoologischer Anzeiger, Supplement* **30**, 297-319.

Werner, B. 1970. Contribution to the evolution of the genus *Stephanoscyphus* (Scyphozoa, Coronatae) and ecology and regeneration qualities of *Stephanoscyphus racemosus. Publications of the Seto Marine Biology Laboratory* **18**, 1-20.

Werner, B. 1971a. Neue Beiträge zur Evolution der Scyphozoa und Cnidaria. *Acta Salmanticensia. Ciencias* **36**, 223-244.

Werner, B. 1971b. *Stephanoscyphus planulophorus* n. spec., ein neuer Scyphopolyp mit einem neuen Entwicklungsmodus. *Helgoländer wissenschaftliche Meeresuntersuchungen* **22**, 120-140.

Werner, B. 1973. New investigations on systematics and evolution of the class Scyphozoa and the phylum Cnidaria. *Seto Marine Biology Laboratory Publications* **20**, 35-61.

Werner, B. 1974. *Stephanoscyphus eumedusoides* n. spec., ein Höhlenpolyp mit einem neuen Entwicklungsmodus. *Helgoländer wissenschaftliche Meeresuntersuchungen* **26**, 434-463.

Werner, B. 1979. Coloniality in the Scyphozoa: Cnidaria. in *Biology and systematics of colonial organisms* (ed. G. Larwood & B.R. Rosen) *Systematics Association Special Volume* **11**, 81-103. London: Academic Press.

Werner, B. 1983. Weitere Untersuchungen zur Morphologie, Verbreitung und Ökologie von *Stephanoscyphus planulophorus* (Scyphozoa, Coronata). *Helgoländer wissenschaftliche Meeresuntersuchungen* **36**, 119-135.

Werner, B. & Hentschel, J. 1983. Apogamous life cycle of *Stephanoscyphus planulophorus. Marine Biology* **74**, 301-304.

Wiman, C. 1895. Palaeontologische Notizen. 1 und 2. *University of Upsala. Geological Institution. Bulletin* **2 (1894-1895)**, 109-117.

Functional morphology of the Class Helcionelloida nov., and the early evolution of the Mollusca

John S. Peel[1]

Abstract

Functional analysis of apertural structures indicates that the Early-Middle Cambrian molluscs here assigned to a new class, Class Helcionelloida, were untorted molluscs with endogastrically coiled shells, not exogastric as previously supposed. The Class Tergomya (previously a Sub-class) is established to accomodate the exogastrically coiled tryblidiaceans which form a major Cambrian-Recent lineage of untorted molluscs alongside the helcionelloids. The term Class Monoplacophora is abandoned as a taxon for untorted molluscs on account of its embracive and varied earlier usage; tergomyans, helcionelloids and representatives of several other supposed Molluscan classes have been grouped frequently within the Monoplacophora.

Helcionelloids are considered to be the ancestors of the Class Rostroconchia and members of both classes may show homologous morphological adaptations to an infaunal mode of life. In consequence, the Subphylum Diasoma (classes Bivalvia, Rostroconchia and Scaphopoda) is considered to be polyphyletic, since bivalves are generally considered to be derived from the exogastric Tergomya or tergomyan-like molluscs. The earliest cephalopods may have been derived from the similarly coiled endogastric helcionelloids instead of from exogastric molluscs, casting doubt also upon the Cyrtosoma (classes Monoplacophora, Gastropoda, Cephalopoda) as a monophyletic group.

Eotebenna viviannae sp. nov. is described from the Middle Cambrian of Denmark.

Introduction

Much of the interest in the early fossil history of the shelled molluscs during the last fifteen years stems from the elegant model of molluscan evolution proposed by Bruce Runnegar and John Pojeta in 1974, and developed in a series of subsequent papers (Pojeta, 1980; Pojeta & Runnegar, 1976; Runnegar, 1978, 1983; Runnegar & Pojeta, 1985). While not universally accepted (cf. Yochelson, 1978, 1979; Wingstrand, 1985), the model proved popular by presenting an integrated palaeontological perspective complementing discussions based on a neontological standpoint (e.g., Stasek, 1972; Salwini-Plawen, 1981, 1985).

Runnegar & Pojeta (1974) viewed the univalved and bivalved molluscs as belonging to two sub-phyla (fig. 1). In the Cyrtosoma, which includes the classes Monoplacophora, Gastropoda and Cephalopoda, the shell is usually univalved and the gut bent into a U-shape or twisted on account of torsion. The gut is essentially straight in the Diasoma, to which sub-phylum the classes Rostroconchia, Bivalvia and Scaphopoda were assigned by Runnegar & Pojeta (1974).

The Sub-phylum Diasoma of Runnegar & Pojeta (1974) was derived from a group of univalves collectively termed helcionellaceans and assigned to the Class Monoplacophora. Runnegar & Pojeta included a variety of molluscs within this class in addition to traditionally accepted monoplacophorans such as the tryblidiaceans *Pilina* Koken, 1925 (fig. 2A) and *Tryblidium* Lindström, 1880 (fig. 2B, C). Helcionellaceans were typified in the model of Runnegar & Pojeta by the genus *Latouchella* Cobbold, 1921 (figs 3, 4). They are a group of Early and Middle Cambrian molluscs named after *Helcionella* Grabau & Shimer, 1909 (fig. 5) which is relatively more rapidly expanding and less strongly coiled than *Latouchella*. In the helcionellaceans the shell is usually bilaterally symmetrical and most commonly curved through about half to one and a half whorls. In tryblidiaceans the bilaterally symmetrical shell is usually only slightly coiled and spoon or cap-shaped.

Runnegar & Pojeta (1974) followed contemporary authors in considering the helcionellaceans to be untorted molluscs contrary to the practice of Knight & Yochelson (1960) who placed them within the archaeogastropod Prosobranchia (Class Gastropoda). As with tryblidiaceans such as *Pilina* (fig. 2A), *Tryblidium* (fig. 2B, C) and the living *Neopilina*, Runnegar & Pojeta considered the helcionellaceans to be exogastrically coiled, i.e., the apex of the coiled shell was located anteriorly and the shell expanded posteriorly. Thus, the generally concave sub-apical surface was located anterior of the apex while the convex supra-apical surface was posterior (fig. 4). Water currents were considered to enter the mantle cavity anteriorly, above the head, and pass along either side of the body over laterally disposed respiratory structures prior to posterior exhalation (Pojeta & Runnegar, 1976).

It is the main thesis of the present paper that helcionellaceans represent a major lineage of untorted univalved molluscs distinct from the tryblidiaceans. Helcionellaceans are here considered to be endogastrically coiled (fig. 4), as suggested by Yochelson (1978), Geyer (1986) and Peel & Yochelson (1987). This conclusion is supported by the interpretation of a number of morphological features of the helcionellacean shell in terms of functional adaptation. It is proposed that the shell apex was located posteriorly and that the shell expanded anteriorly, as is the case with the gastropods. Unlike gastropods, helcionellaceans were untorted. The mantle cavity in helcionellaceans was probably located posteriorly, essentially beneath the sub-apical surface. Water currents usually entered the mantle cavity laterally, passing over postero-laterally arranged gills prior to leaving the mantle cavity as a single, median posterior stream (fig. 4).

The present re-interpretation of the functional morphology of helcionellaceans is significant in terms of molluscan evolution and systematics. As distinct lineages of untorted molluscs, the tryblidiaceans and helcionellaceans are recognised herein as separate classes, defined below. Helcionellaceans are assigned to a new Class Helcionelloida. The tryblidiaceans are placed

[1]Geological Survey of Greenland, Øster Voldgade 10, DK-1350 Copenhagen K, Denmark

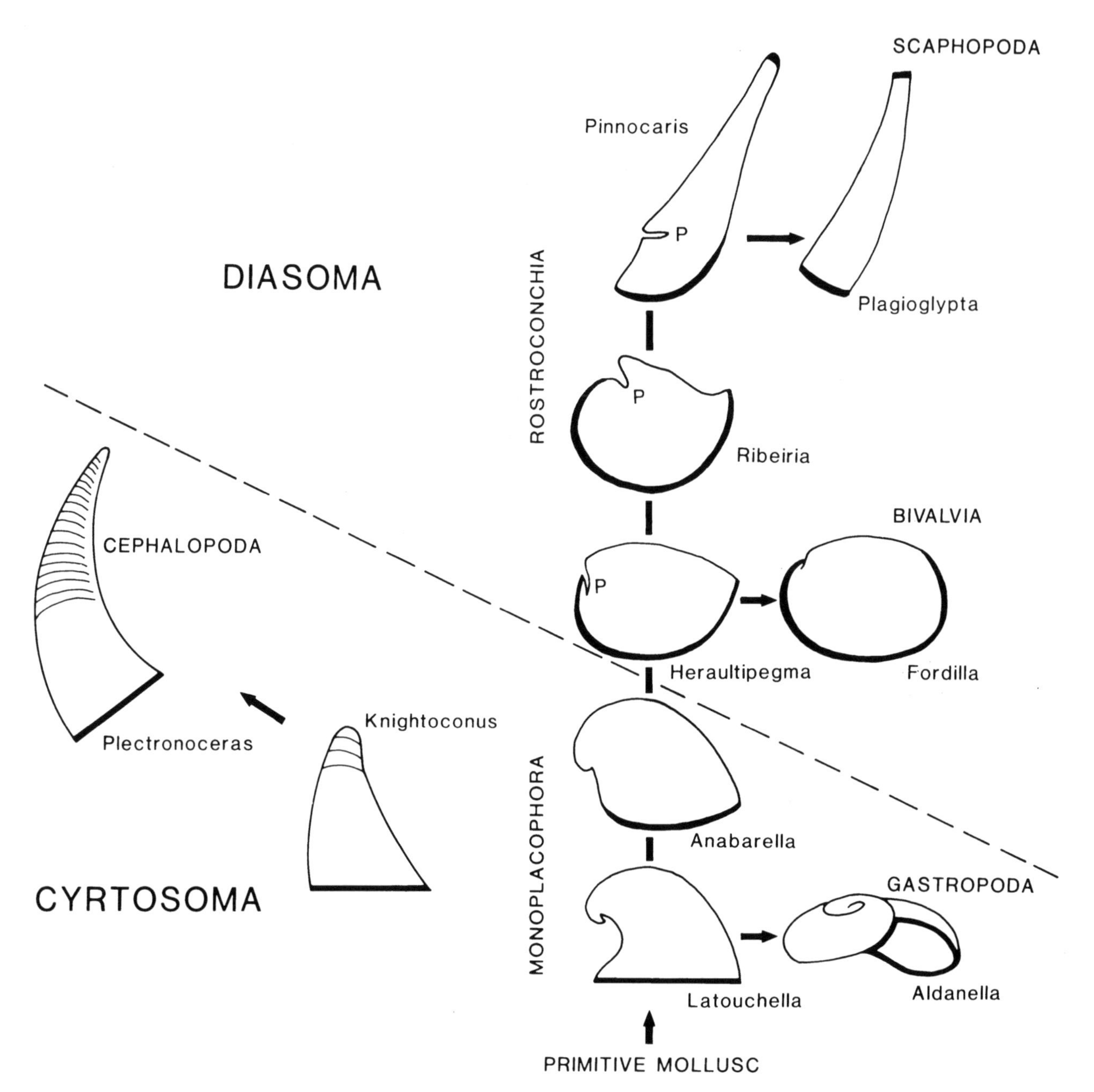

Fig. 1. Origin of the univalved and bivalved molluscs according to Runnegar & Pojeta (1974). Constituent classes of the Sub-phylum Diasoma (Rostroconchia, Bivalvia and Scaphopoda) are derived from the Class Monoplacophora. The Sub-phylum Cyrtosoma comprises the classes Monoplacophora, Gastropoda and Cephalopoda. Thick black lines indicate open apertures; p, pegma (redrawn from Runnegar & Pojeta, 1974, fig. 4). Note that Runnegar & Pojeta considered the illustrated monoplacophorans *Latouchella* and *Anabarella* to be exogastric, i.e., with the shell coiled clock-wise with anterior to the left as illustrated, while *Knightoconus* and its descendant cephalopods are endogastric, coiling anti-clockwise with anterior to the left (cf. fig. 4).

within a Class Tergomya, based upon a sub-class of the same name described by Horny (1965a, b). The Class Monoplacophora is abandoned as a formal term on account of its varied usage. In the embracive sense of Runnegar & Jell (1976) and Runnegar & Pojeta (1985), the Class Monoplacophora includes both the Tergomya and Helcionelloida, together with representatives of other molluscan classes.

The Helcionelloida is considered to be the direct ancestor of the Class Rostroconchia and members of this latter class are also interpreted here as endogastric molluscs. It follows, therefore, that the model of diasome evolution proposed by Runnegar & Pojeta is no longer tenable and two parallel lineages are recognised within the former Diasoma: Tergomya - Bivalvia, and Helcionelloida - Rostroconchia. The scaphopods may have been derived from either lineage but are considered here to be descendants of the rostroconchs.

The Cephalopoda may have been derived from the endogastric helcionelloids and not from pseudo-endogastric molluscs (Tergomya) such as *Knightoconus*, as proposed by Yochelson *et al.* (1973). Acceptance of this proposal, together with the widely accepted notion of derivation of the Gastropoda from the Tergomya or tergomyan-like forms, suggests that the Cyrtosoma is also not monophyletic.

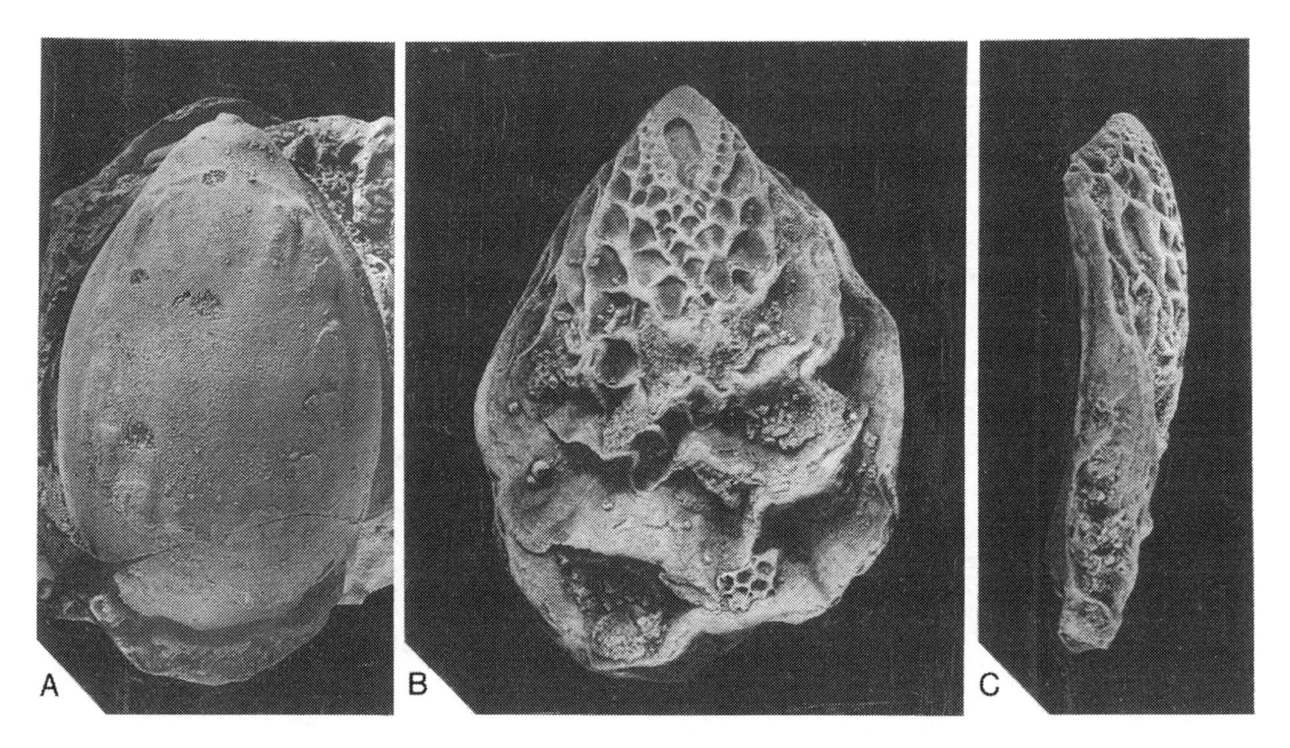

Fig. 2. A, *Pilina cheyennica* Peel, 1977, Late Ordovician, Chimneyhill Limestone, near Ada, Oklahoma, U.S.A.; YPM 74, holotype, x 1.6. Dorsal view of internal mould of tryblidiacean tergomyan oriented with anterior at top, showing the paired, raised muscle scars and the small anterior tubercle representing the earliest growth stages. B, C, *Tryblidium reticulatum* Lindström, 1880, Silurian, Högklint Beds, Lauterhorn, Fårö, Gotland, Sweden; MGUH 16.469, x 3. Dorsal and lateral views of tryblidiacean tergomyan with anterior at top, showing the unusual irregular lamellose ornamentation.

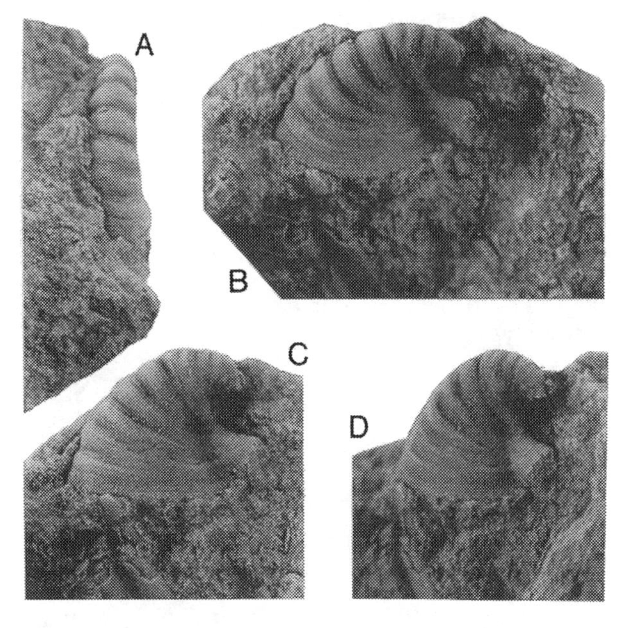

Fig. 3. *Latouchella holmdalense* Peel, 1988, late Middle Cambrian, Holm Dal Formation, Peary Land, central North Greenland; MGUH 18.678 from GGU collection 225535, holotype, x 10. A, antero-dorsal view showing the high degree of lateral compression. B-D, lateral, dorso-lateral and antero-lateral views showing the prominent comarginal ornamentation, shallowly proscyrt lateral apertural margins, and fold with resultant emargination in the sub-apical surface.

Status of the Class Monoplacophora

The simplicity of the model of molluscan evolution presented by Runnegar & Pojeta (1974) depends largely on the adoption of a wide concept for the stem-group Class Monoplacophora; these authors considered the Rostroconchia to be the only extinct molluscan class. The complex relationships within the morass of early untorted molluscs are obscured by this embracive use of a Class Monoplacophora and the term becomes too generalised for satisfactory employment in phylogeny. Apart from traditionally recognised monoplacophoran taxa such as *Pilina* (fig. 2A), *Tryblidium* (fig. 2B, C) and *Neopilina*, Runnegar & Pojeta (1985) formally assigned the helcionellids, hypseloconids, bellerophontids, archinacellids, pelagiellids, tuarangiids and cyrtonellids to the class. The Stenothecoida, considered a separate class of bivalved molluscs by Yochelson (1968, 1969, 1978) and Aksarina (1968; under the name Probivalvia), was dismissed as a group of bivalved monoplacophorans, although without formal placement. While most of these groups share a basic untorted anatomy (except for the torted bellerophontids, archinacellids and possibly cyrtonellids) and univalved shell (except for the bivalved stenothecoidans), they clearly represent a broad spectrum of adaptations in the earliest Palaeozoic history of molluscan evolution.

In the *Treatise Invertebrate Paleontology* Knight & Yochelson (1960) included within the Class Monoplacophora three orders which are now assigned by many workers to three distinct molluscan classes. Most of their tryblidioideans are Tergomya (in the sense defined below, generally equivalent to the Tryblidiida of Wingstand, 1985, the Tergomya of Horny, 1965a, b and the restricted Monoplacophora of Harper & Rollins, 1982). Their archinacelloideans and the cyrtonellacean tryblidioideans are gastropods according to Harper & Rollins (1982), Yochelson (1988) and Peel (1989b), while cambridioideans are stenothecoidans.

Horny (1965a, b) proposed two sub-classes within

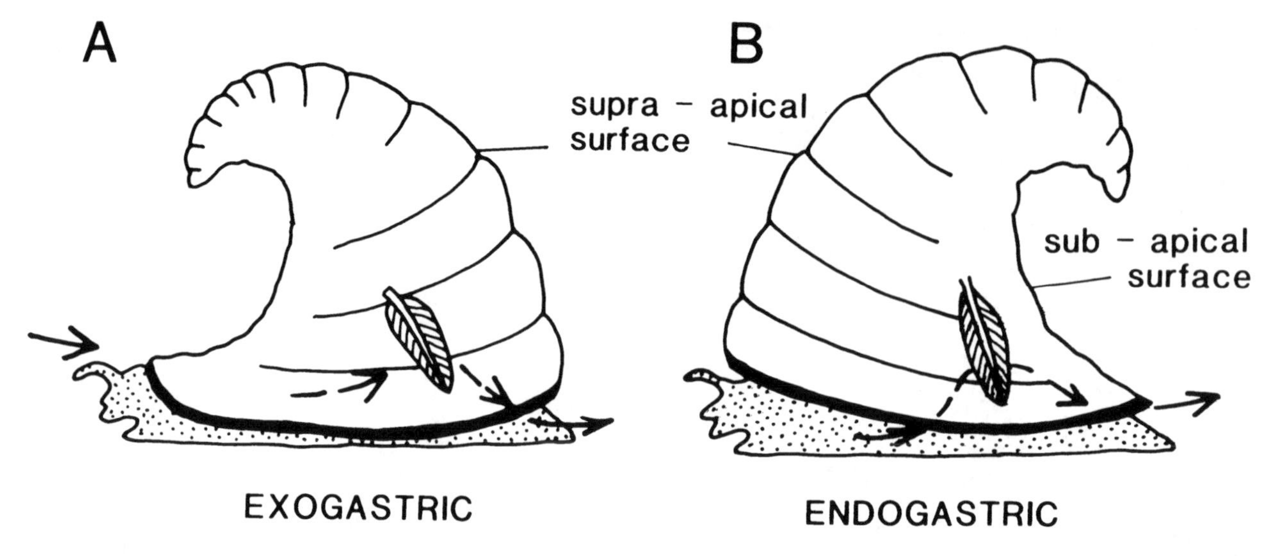

Fig. 4. Alternative reconstructions of the helcionelloid *Latouchella*. A, exogastric coiling with inhalant water currents (arrows) entering the shell anteriorly, passing over the lateral gills prior to postero-lateral exhalation; this is the reconstruction favoured by Runnegar & Pojeta but rejected here. B, endogastric coiling with postero-lateral inhalant currents and a median posterior exhalant stream; this reconstruction forms the basis for the Class Helcionelloida proposed herein. Thick black lines, apertural margins.

the Monoplacophora, one of which (Tergomya) corresponds to the traditionally recognised tryblidiacean lineage. A second sub-class (Cyclomya) included the archinacelloideans and the more strongly coiled cyrtonellacean tryblidioideans of Knight & Yochelson (1960).

Harper & Rollins (1982) rigorously reviewed the significance of characters employed by earlier workers in attempting to delimit the untorted monoplacophorans from the torted gastropods. They restricted the Class Monoplacophora to the tryblidiacean lineage, the Tergomya of Horny (1965a, b; equivalent to the Tryblidiida of Wingstrand, 1985) while Horny's cyclomyans were considered to be gastropods. By direct statement or implication, Harper & Rollins (1982) excluded from their restricted Monoplacophora ten families of supposedly untorted molluscs placed in three orders within the Class Monoplacophora of Runnegar & Jell (1976) and Runnegar & Pojeta (1985).

The recommendation of Wingstrand (1985) and others concerning abandonment of the term Class Monoplacophora is endorsed, although the informal term monoplacophore for untorted univalves in general may still prove useful.

It is proposed to elevate the Sub-class Tergomya of Horny (1965a, b) to a full class, defined below, to essentially replace Monoplacophora in the restricted sense of Harper & Rollins (1982). Tryblidiida has been used by several authors (e.g., Wingstrand, 1985; Haszprunar, 1988) at a similar systematic level to the use of Tergomya here, but without formal definition. It has also been widely used, however, as an order attributed to Lemche (1957) by many authors (e.g. Horny 1965a, b; Runnegar & Jell, 1976; Runnegar & Pojeta, 1985), and this usage is preferred here.

HELCIONELLOID AUTOECOLOGY

A number of morphological features in the group of Cambrian molluscs which are here assigned to the new Class Helcionelloida support the reconstruction of an essentially posterior mantle cavity in an endogastric shell. In *Helcionella* itself (fig. 5) the shell is low and in plan view ovoid, weakly reminiscent of tryblidlacean tergomyans such as *Pilina* and *Neopilina* (fig. 2). While considered to be endogastric (contra the exogastric tryblidiaceans) *Helcionella* may have possessed a mantle cavity extending along the lateral surfaces. With the increased coiling and lateral compression typical of *Latouchella* and other helcionelloids, the mantle cavity became concentrated in the posterior portion of the shell in similar fashion to the gastropods and cephalopods, due to life in the narrow cone-like shell. This reduction of the mantle cavity and probably also a concomitant reduction in the number of pairs of gills was also probably present in tergomyans of the Superfamily Hypseloconellacea.

YOCHELCIONELLA AND THE FUNCTION OF THE SNORKEL

Species of *Yochelcionella* are described from Lower and Middle Cambrian strata throughout the world and may vary in form from strongly curved and laterally compressed (such as *Yochelcionella americana* Runnegar & Pojeta, 1980; see also Peel, 1988a) to tall and slender cones, e.g., *Yochelcionella ostentata* Runnegar & Jell, 1976. All species exhibit the characteristic snorkel on the sub-apical surface (figs 5, 6C-E). This tube-like extension of the shell is usually relatively short although Runnegar & Jell (1980, fig. 1) figured a specimen of *Yochelcionella cyrano* Runnegar & Pojeta, 1974 from the Middle Cambrian of Australia in which the snorkel was longer than the preserved height of the helcionelloid shell.

Pojeta & Runnegar (1976, fig. 9) discussed the function of the snorkel in *Yochelcionella* indicating four possible reconstructions (fig. 7). In compliance with their earlier evolutionary model (Runnegar & Pojeta, 1974), they concluded that the shell of *Yochelcionella* was exogastric, with the snorkel located anteriorly, and that the tube served as a conduit for water enter-

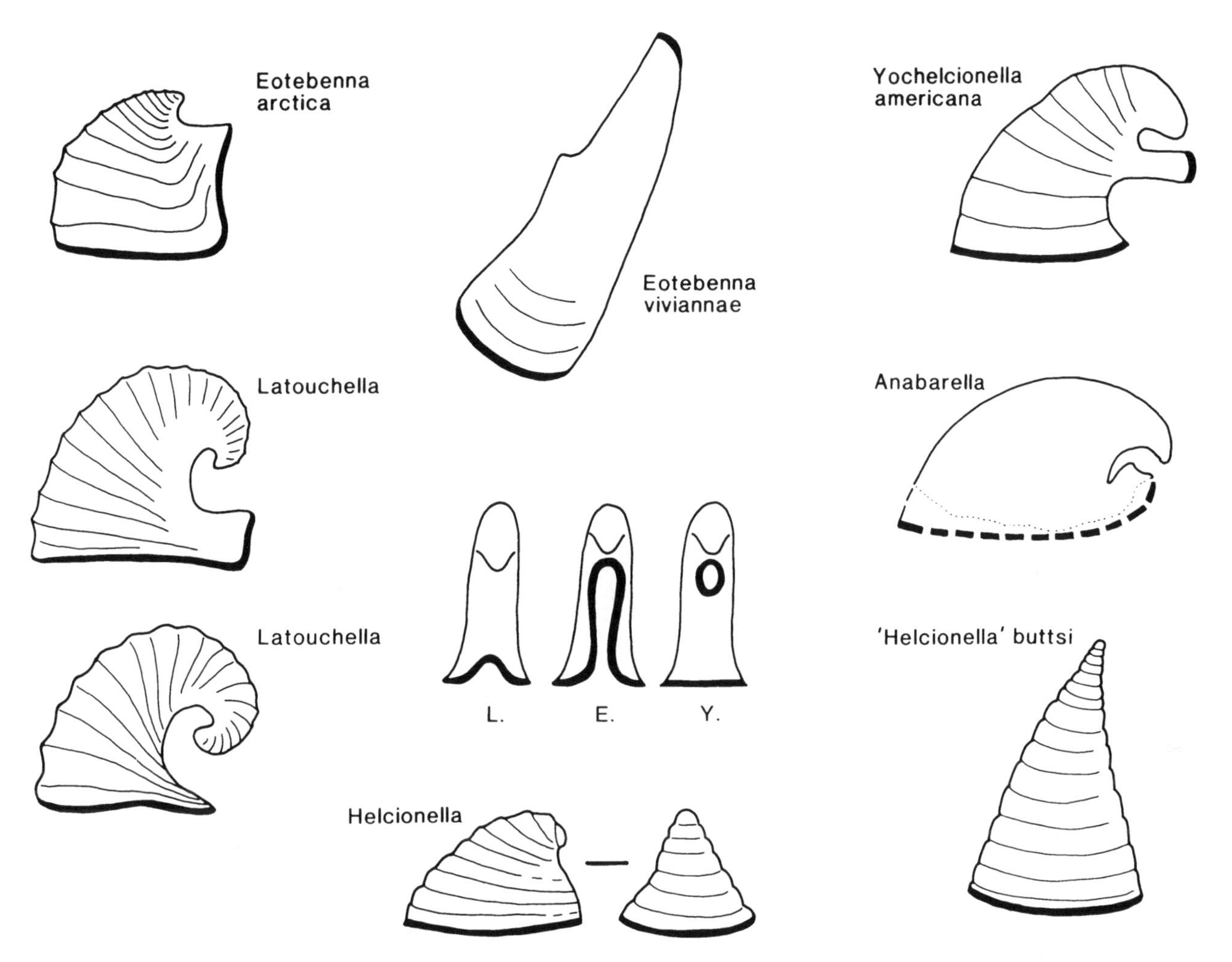

Fig. 5. Class Helcionelloida, morphological series showing the development of exhalant structures in the sub-apical wall. All specimens are drawn in lateral view with the anterior to the left; *Helcionella* is also shown in anterior view. The three schematic drawings in the centre show the extent of the emarginations in *Latouchella, Eotebenna arctica* and the snorkel in *Yochelcionella. Anabarella* is drawn from an internal mould figured by Runnegar (1983, fig. 4A); the preserved margin is shown by the finely stippled line and the inferred aperture by the heavy broken line. Thick black lines indicate apertural margins.

ing the mantle cavity (fig. 7D). After entering the mantle cavity, the inhalant current divided prior to passing over the respiratory surfaces which they considered to be located laterally. The exhalant streams left the mantle cavity along each postero-lateral surface. This interpretation was challenged by Yochelson (1978), Geyer (1986) and Peel & Yochelson (1987: see also Berg-Madsen & Peel, 1987) who considered that the snorkel housed the exhalant stream from the mantle cavity and was located posteriorly in an endogastrically coiled shell (fig. 7E). Inhalant currents entered the endogastric shell postero-laterally and united prior to expulsion through the snorkel as a single median, posterior stream.

In the reconstruction favoured here (fig. 7E), the mantle cavity in *Yochelcionella* is essentially restricted to the posterior, reflecting physical confinement in the narrow, cone-shaped shell. Oxygen-rich water enters the mantle cavity postero-laterally along two relatively wide surfaces, achieving maximum supplies for respiration. The force of expulsion of the exhalant current is increased by the unification of the two lateral streams and its concentration in a narrow conduit, the snorkel. As a result, oxygen-depleted water is carried away from the shell.

Employment of the snorkel for the inhalant current, as suggested by Pojeta & Runnegar (fig. 7D), would restrict intake of oxygenated water to a single narrow stream. However, in their reconstruction, the mantle cavity occurs all around the shell aperture and the gills are considered to be placed laterally in the often laterally compressed shell. Consequently, the single narrow inhalant stream would be diminished further by being divided into streams passing along each side of the mollusc. In the model of Pojeta & Runnegar the narrowness of the snorkel thus inhibits the supply of oxygenated water to the mantle cavity. In the favoured reconstruction (fig. 7E), the constriction provided by the snorkel is interpreted as an adaptation to carry oxygen-depleted water away from the mantle cavity.

OELANDIA

This Middle Cambrian helcionelloid, originally described by Westergård (1936) from Sweden, has been redescribed on the basis of new material by Peel & Yochelson (1987) who also noted species from China and North Africa (originally referred to *Latouchella* by Yu & Ning, 1985 and Geyer, 1986). Peel & Yochelson described a short tubular structure on the sub-apical wall of *Oelandia* (figs 7F, 6A-B) which they compared

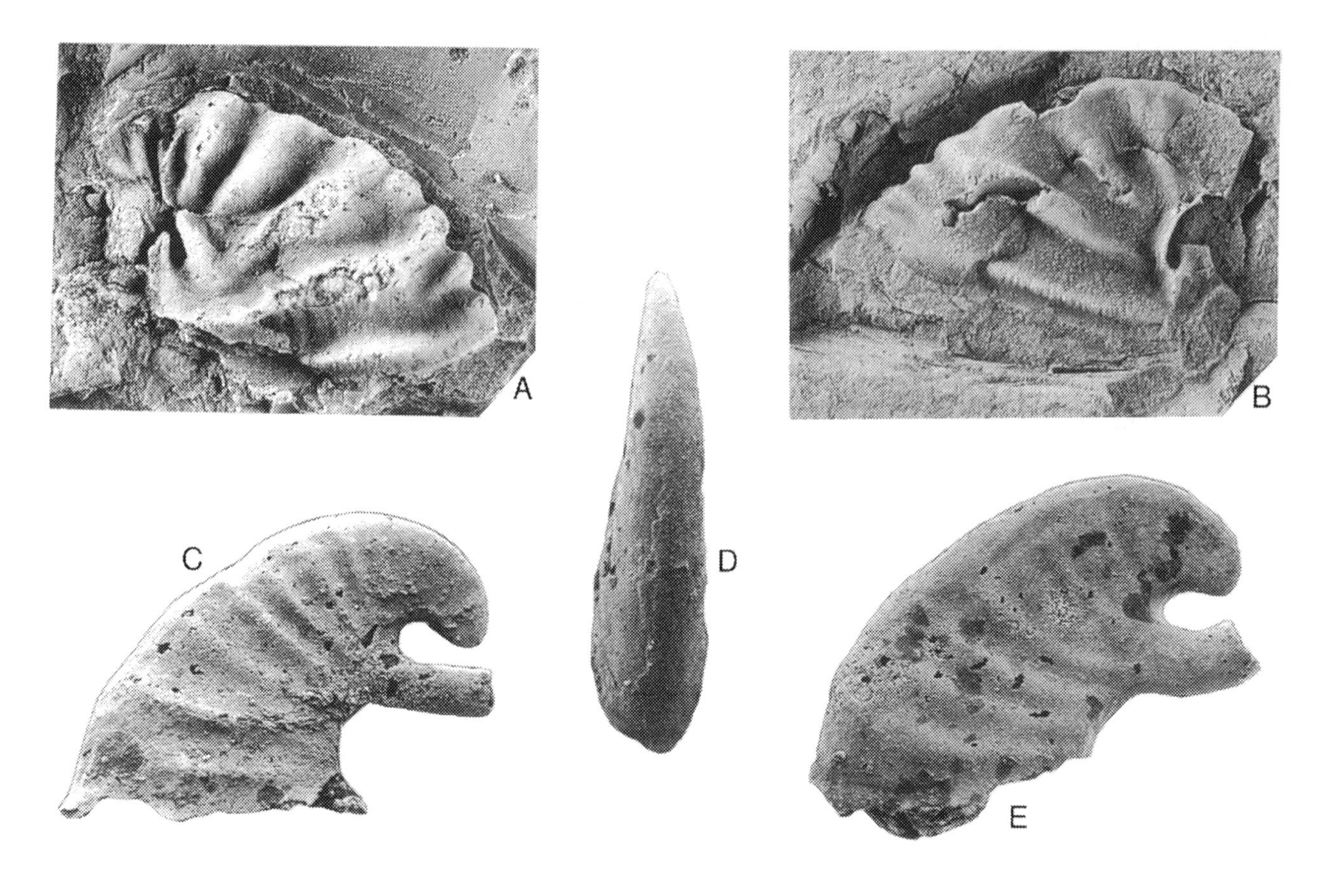

Fig. 6. *Yochelcionella* and *Oelandia*. A, B, *Oelandia pauciplicata* Westergård, 1936, Middle Cambrian, Öland, Sweden, internal moulds in lateral aspect, x 6. A, LO 5758t, right side with snorkel on sub-apical surface, showing the vertical snorkel which is closed in the specimen as preserved. B, LO 5757t, left side with anterior to left. Water currents entering the shell laterally were initially expelled through the posterior snorkel and subsequently through a median fold located posterior to the snorkel. In both specimens the prominent comarginal rugae are essentially restricted to one side of the shell; traces of rugae from the other side interdigitate with rugae from the opposing side in the median dorsal area. C-E, *Yochelcionella americana* Runnegar & Pojeta, 1980, late Early Cambrian, Forteau Formation, Gros Morne, western Newfoundland, Canada, lateral views of internal moulds with anterior to left, x 45. C, GSC 85865. D, GSC 85863, dorsal view. E, GSC 85862. Note the prominent snorkel on the sub-apical surface of the strongly coiled shell.

to the snorkel of *Yochelcionella*: Geyer (1986, pl. 3, fig. 37) has illustrated a similar structure in *Oelandia comma* (Geyer, 1986). This tube rises perpendicular from the plane of the aperture toward the apex and is closed at its adapical termination. Peel & Yochelson (1987) suggested that closure of the tube indicated abandonment of the snorkel with increased growth. Its function as the locus of the exhalant current from the mantle cavity may have been taken over by the median fold in the aperture below the apex. The snorkel may also have been developed periodically during ontogeny, as are the spines and anterior siphon of some gastropods, but no described material preserves more than a single tube.

Oelandia is unusual among helcionelloids on account of its lack of bilateral symmetry in shell ornamentation. The prominent comarginal plications develop alternately on each lateral area and terminate near the mid-dorsum. Peel & Yochelson (1987) could find no explanation for this structure which forms the base for recognition of the genus.

EOTEBENNA

Eotebenna was originally described from the Middle Cambrian of Australia where it occurs together with *Yochelcionella* and a variety of other molluscs (Runnegar & Jell, 1976). Peel (1989a) has described a Lower Cambrian species from North Greenland (figs. 5, 8, 9) and a late Middle Cambrian species is described below from Bornholm, Denmark (*Eotebenna viviannae* sp. nov., figs 5, 9, 19). As with *Yochelcionella, Eotebenna* is characterised by the development of a snorkel but it differs from the former genus in that the small perforation (or trema) at the distal end of the snorkel is usually connected to the relatively larger shell aperture by a narrow slit. The snorkel is thus less tube-like than in *Yochelcionella*. and more closely resembles a deep fold in the sub-apical apertural margin.

As with *Yochelcionella* the snorkel in *Eotebenna* is considered to lie posteriorly and to serve as a conduit for the exhalant current from the mantle cavity. The shell aperture is antero-ventral, which is the orientation proposed for the aperture in rostroconchs by Pojeta & Runnegar (1976) and Pojeta (1980; 1987) although their interpretation of *Eotebenna* would place the aperture postero-ventrally. Similar separation of an antero-ventral aperture (principally for extension of the foot) and a posterior entrance to the mantle cavity is also seen in bivalves and discosorid cephalopods (cf. Dzik, 1981).

A morphological series can be recognised from *Latouchella* to the late Early Cambrian *Eotebenna arctica* Peel, 1989a, through the early Middle Cambrian *E. papilio* Runnegar & Jell, 1976 and the medial Middle Cambrian *E. pontifex* to the late Middle Cambrian *E. viviannae* sp. nov. (fig. 9). The series is charac-

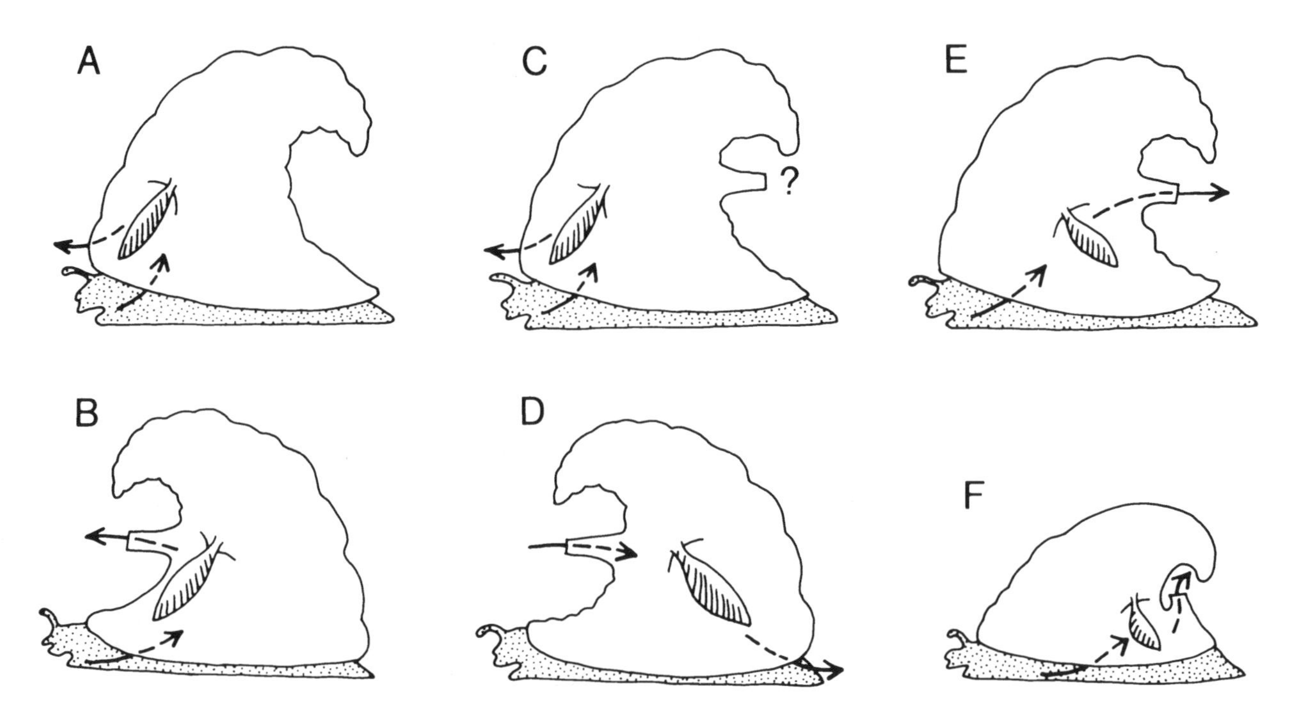

Fig. 7. Possible reconstructions of helclonelloids based on *Latouchella* (A), *Yochelcionella* (B-E) and *Oelandia* (F). A, *Latouchella* restored as a torted endogastric mollusc gastropod as suggested by Knight et al. (1960). Antero-lateral inhalant currents pass over the gills prior to exhalation in a single median, anterior stream. B, *Yochelcionella* interpreted as an exogastric gastropod. Water currents enter the mantle cavity antero-laterally and are exhaled through the anterior snorkel; gastropods, however, are typically endogastric. C, *Yochelcionella* interpreted as an endogastrlc gastropod. Water currents enter the mantle cavity antero-laterally and are expelled anteriorly; the posterior snorkel has no obvious function in terms of respiration. D, *Yochelcionella* interpreted as an exogastric untorted mollusc (same shell orientation as Tergomya) in which the the single inhalant current enters the mantle cavity through the snorkel and divides to pass over the laterally disposed gills prior to postero-lateral exhalation. This is the reconstruction favoured by Runnegar & Pojeta (1974 and subsequent papers) which is rejected here. E, *Yochelcionella* interpreted as an endogastric untorted mollusc (helcionelloid) with water currents entering the mantle cavity laterally and passing over the gills prior to expulsion as a single median stream through the posterior snorkel. This is the reconstruction favoured by Geyer (1986) and Peel & Yochelson (1987) and adopted herein. F, *Oelandia* interpreted as an endogastric helcionelloid, compare with E (from Pojeta & Runnegar, 1976 modified by Peel & Yochelson, 1987).

terised by increasing separation of the deepest part of the snorkel-emargination from the aperture, manifested in elongation of the shell as a result of pronounced allometry. In the Early Cambrian *E. arctica* the shell is equidimensional in lateral aspect and the deepest part of the slit-like emargination in the sub-apical wall lies 'below' the apex (figs 5, 8, 9). In *E. papilio* from the early Middle Cambrian, a more tube-like snorkel is developed, connected to the aperture by a narrow slit, although the opening of the snorkel is still mainly below the apex (Runnegar & Jell, 1976, fig. 11B, 12-14). In the late Middle Cambrian type species *E. pontifex* the snorkel is more fully developed and cylindrical in form; its open end is raised high above the apex. In *E. viviannae* sp. nov., from the late Middle Cambrian, elevation of the snorkel opening above the apex achieves maximum development and the shell is several times higher than long (figs 5, 9, 19). In addition, the narrow slit connecting the aperture with the trema at the tip of the snorkel is frequently closed.

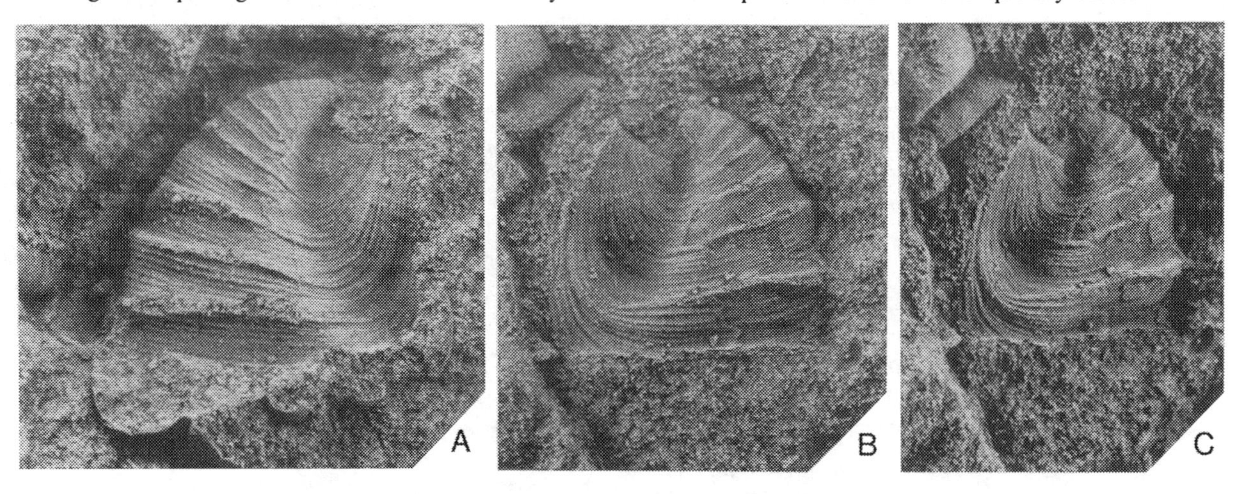

Fig. 8. *Eotebenna arctica* Peel, 1989, late Early Cambrian, Henson Gletscher Formation, south-east Freuchen Land, central North Greenland, GGU collection 315109, x 10. A, MGUH 18.702, lateral view with anterior to left. B, C, 18.701, lateral and oblique lateral views with anterior to right. Note the prominent sail-like snorkel-fold on the sub-apical surface.

The morphological series is interpreted in terms of increasing penetration of the sediment-water interface by successively younger species of *Eotebenna*. Early species lived at the sediment-water interface with only slight penetration, but the end-member *E. viviannae* sp. nov. was largely infaunal, with only the shell posterior protruding from the sediment (fig. 10). *E. viviannae* sp. nov. is bivalve-like in its form, but differs in the direction of coiling of the shell and in being univalve.

The proposed semi-infaunal mode of life of *Eotebenna viviannae* sp. nov. probably caused modification of the pattern of circulation of mantle water currents from that envisaged in older species of the genus and most species of *Yochelcionella* (an exception, the strongly coiled *Y. americana* may be interpreted in a similar manner to *E. viviannae* sp. nov., fig. 10). As with scaphopods, and many bivalves and rostroconchs, the snorkel may have served both inhalant and exhalant functions since the antero-ventral aperture was presumably buried within the bottom sediment. Water supply to the mantle cavity may have been achieved with simultaneous inhalation and exhalation through separate siphons, as in bivalves, but it is perhaps more likely that *E. viviannae* employed alternating inhalation and exhalation, as described in scaphopods (Yonge & Thompson, 1976, p. 232). When restored in the manner suggested herein, *Eotebenna viviannae* sp. nov. parallels the mode of life inferred for many rostroconchs by Pojeta & Runnegar (1976) and Pojeta (1987; see fig. 10 and discussion below). Many rostroconchs have small posterior and rostral openings above the sediment surface (cf. Runnegar & Pojeta, 1976; Pojeta, 1987) which also probably maintained separation of the inhalant and exhalant streams.

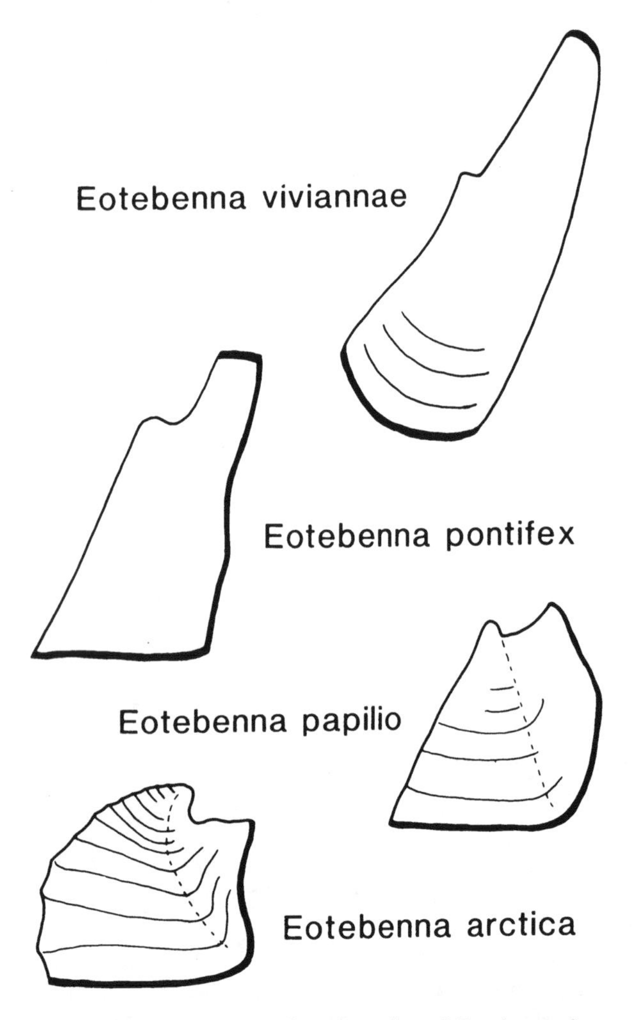

Fig. 9. Morphological series of species of *Eotebenna* from the Early Cambrian *Eotebenna arctica* to the late Middle Cambrian *E. viviannae* sp. nov. All species are oriented in lateral view with the anterior to the left; thick black lines indicate the extent of shell openings. For discussion see text.

EUREKAPEGMA

This Middle Cambrian helcionelloid from New Zealand (figs 10, 11) is characterised by extreme lateral compression, apertural margins convex away from the apex when viewed laterally, and by the presence of an internal plate (zygion) connecting opposing lateral areas beneath the apex (MacKinnon, 1985). It resembles *Eotebenna viviannae* sp. nov., to which it is more closely compared below. While the zygion might be fancifully interpreted as "pegma-like", MacKinnon (1985, fig. 6) considered it to be a support for internal musculature. He restored *Eurekapegma* as semi-infaunal, with the sub-apical surface within the sediment (fig. 11A). Following Runnegar & Pojeta (1974), this surface was considered anterior, although MacKinnon proposed that the mantle cavity was located posteriorly, near the supra-apical margin, and not concentric around the margins of the shell aperture as postulated by Pojeta & Runnegar (1976) in *Yochelcionella*.

It is proposed here that the plate-like zygion partially delimits the mantle cavity in the sub-apical surface. Thus, *Eurekapegma* is restored here with the opposite orientation to that suggested by MacKinnon (1985), namely with the posterior mantle cavity below the sub-apical surface (fig. 11B). *Eurekapegma* is therefore closely homologous to *Eotebenna viviannae* sp. nov. and a morphological descendant of *Mellopegma* Runnegar & Jell, 1976. The shield-like, convex lateral margins also suggest a semi-infaunal mode of life. The close juxtaposition between the adapertural termination of the zygion and the almost conjoined apertural margins indicates that water currents probably entered and left the shell on the sub-apical side of the zygion. This parallels the mantle cavity structure proposed for *E. viviannae* sp. nov. and many rostroconchs and bivalves where the larger antero-ventral gape was buried in sediment (fig. 10).

INTERNAL LONGITUDINAL RIDGES IN LATOUCHELLA

The presence of longitudinal ridges on the interior of the sub-apical surface of *Latouchella* was reported by Robison (1964) in silicified material from the Middle Cambrian of the western U.S.A. Similarly preserved material from the Middle Cambrian of Australia was described by Runnegar & Jell (1976) and the present discussion is based on some of this material (fig. 12). On internal moulds of the type described here from the Middle Cambrian of Greenland (fig. 13) the ridges are preserved in the form of deep grooves. These grooves are not present at the earliest growth stages, although subsequent interior thickening of the shell may have obscured their presence. The grooves are symmetrically disposed about the plane of symmetry; as few as two and as many as six have been observed. In the latter case, the inner pair originates nearer to the shell apex

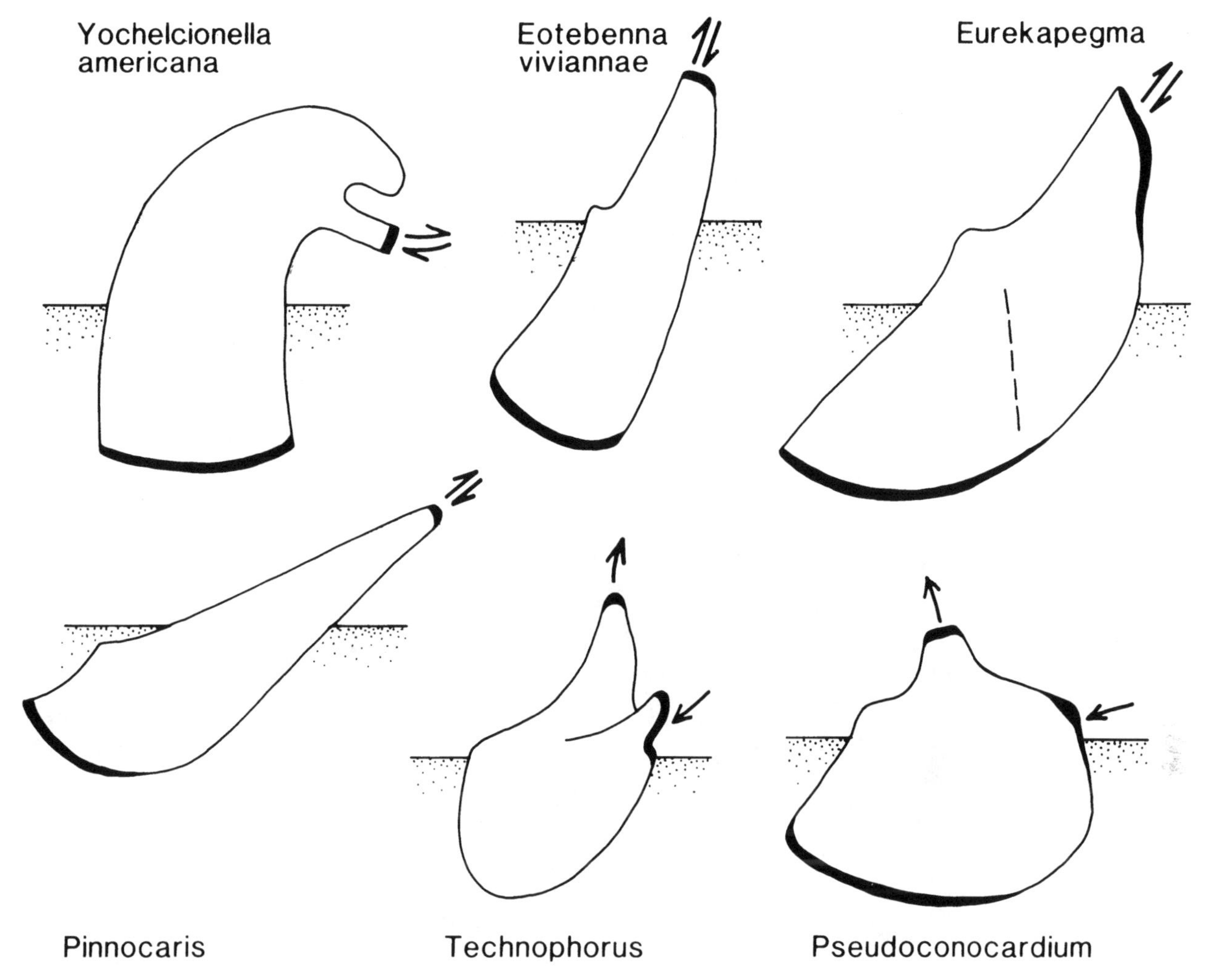

Fig. 10. Reconstructions of helcionelloids (*Yochelcionella americana, Eotebenna viviannae* sp. nov., *Eurekapegma* and rostroconchs (*Pinnocaris, Technophorus, Pseudoconocardium*), based on Pojeta & Runnegar, 1976) showing semi-infaunal mode of life. Thick black lines indicate the margins of shell gapes and apertures; arrows indicate inhalant and exhalant water currents while the zygion occurring internally in *Eurekapegma* is illustrated by a broken line. All shells are similarly oriented with anterior to the left. The large shell gape occupied by the foot (except in *Technophorus*) is located antero-ventrally within the sediment, indicating that respiration was accomplished via the postero-dorsal shell opening(s).

while the outer pairs commence more distally, although this may not be a result of simple ontogenetic increase in the number of ridges. Individual ridges are not necessarily entire but may be periodically interupted (fig. 13A-C). Silicified material indicates that the ridges terminate just within the aperture; it also clearly demonstrates the cross-sectional shape of the structures as well-rounded ridges (fig. 12) with a tendency to be T-shaped in cross-section (fig. 12E).

The ridges are considered to reflect folds in the overlying soft-tissue of the mantle cavity wall which served to separate inhalant water currents from an exhalant stream. Inhalant streams would have been located laterally, on either side of the median exhalant stream, producing a mantle cavity configuration reminiscent of that described in the bellerophontacean *Plectonotus* by Peel (1974; 1984; see also Knight, 1952). In *Plectonotus*, interpreted by Peel (1974) as a torted mollusc (i.e. gastropod), conspicuous trilobation of the dorsal area (fig. 14) is taken to reflect effective separation of the two lateral inhalant currents from the median exhalant current. In many large specimens of *Plectonotus* the degree of trilobation is so intense that internal moulds are marked by deep spiral channels separating the three dorsal lobes (Peel, 1974).

Trilobation in *Plectonotus* and relatives is inter-

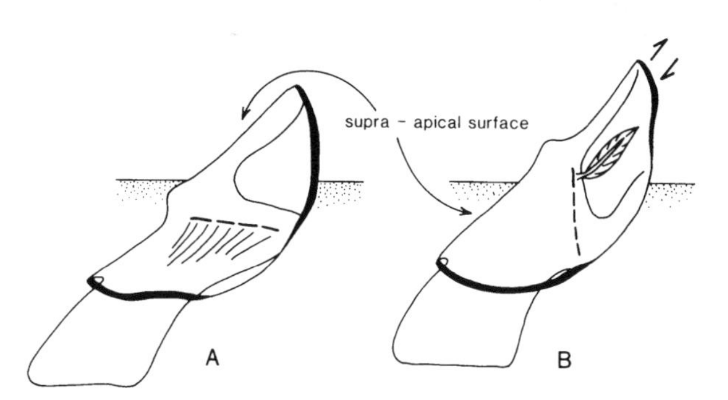

Fig. 11. Reconstructions of the mode of life of *Eurekapegma*. A, after MacKinnon (1985); the exogastric *Eurekapegma* is oriented with the sub-apical surface within the sediment. The zygion (dashed line) provides muscle attachment for the foot. The postero-dorsal mantle cavity is located beneath the supra-apical surface. B, interpretation favoured herein; the endogastric *Eurekapegma* is oriented with the supra-apical surface within the sediment and with the postero-dorsal mantle cavity below the sub-apical surface. Thick black lines indicate shell gapes; arrows indicate inhalant and exhalant water currents.

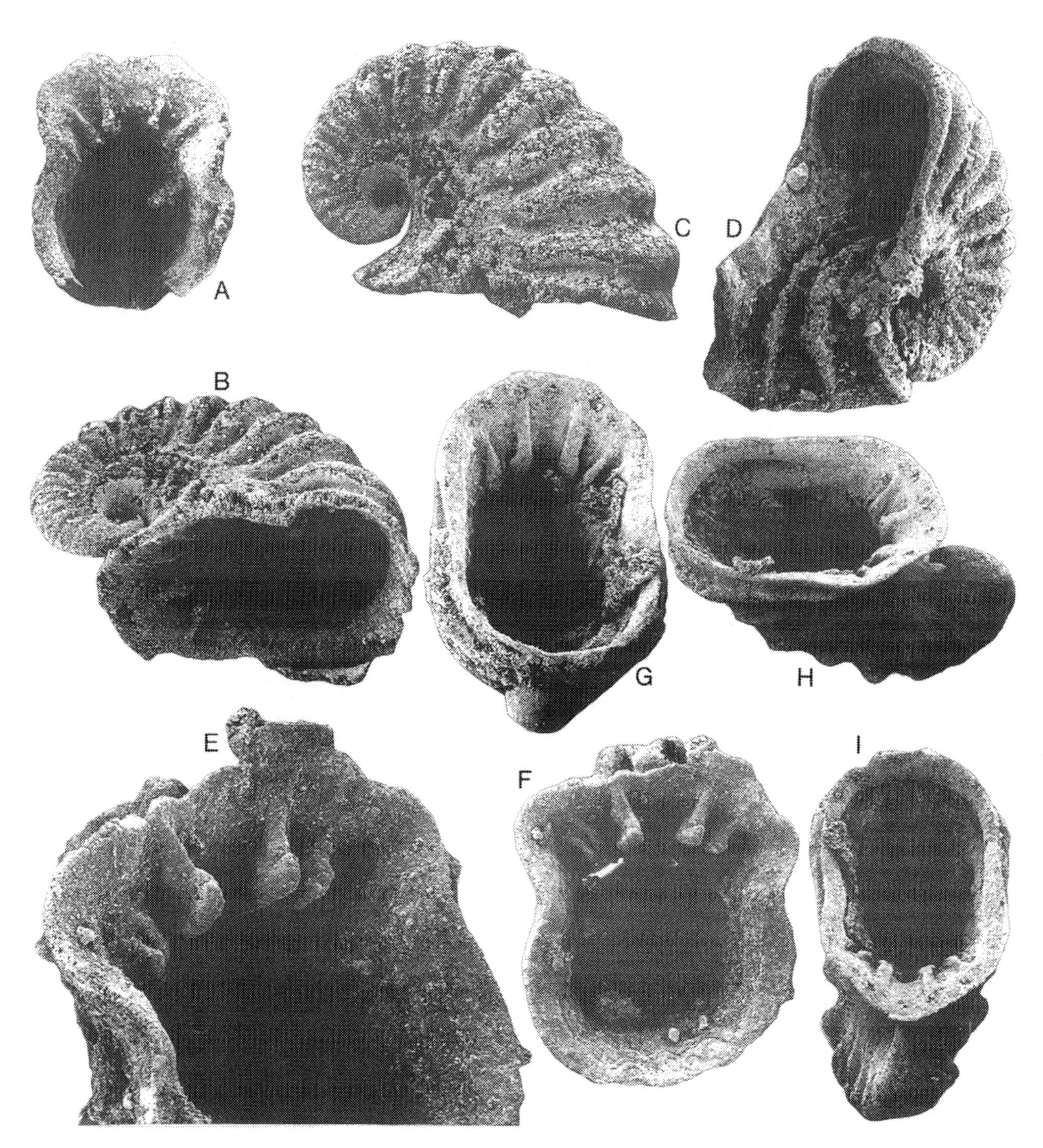

Fig. 12. Internal ridges in the helcionelloid *Latouchella* controlling water currents in the posterior mantle cavity. A-I, *Latouchella merino* Runnegar & Jell, 1976, Middle Cambrian, Coonigan Formation, New South Wales, Australia, Australian National University Bulk Collection 10352 (for details see Runnegar & Jell, 1976, p. 110), silica replicas showing well-developed internal ridges on the sub-apical wall. A, MGUH 19.556, apertural view showing two pairs of ridges terminating within the keyhole-shaped aperture, x 10. B, C, MGUH 19.557, lateral and aperturo-lateral views showing two pairs of ridges within the aperture. Note that the prominent rugae on the shell exterior are restricted to dorso-lateral areas of the shell; umbilico-lateral areas are smooth. The constriction in the aperture producing the key-hole shape (see A, E, F) coincides with the termination of the comarginal rugae, as does the shallow sinus in each lateral margin (see I), x 10. D, MGUH 19.558, oblique apertural view of broken specimen showing ridges on the interior of the sub-apical wall extending deep into the shell interior, x 10. E, F, MGUH 19.559, apertural views showing the key-hole shape of the aperture and the bulbous thickening of the free ends of the two pairs of ridges on the sub-apical wall, x 16 and x 10, respectively. G-I, MGUH 19.560, apertural views showing two pairs of ridges. Note the shallow sinus in the lower right margin of the aperture in H, x 8.

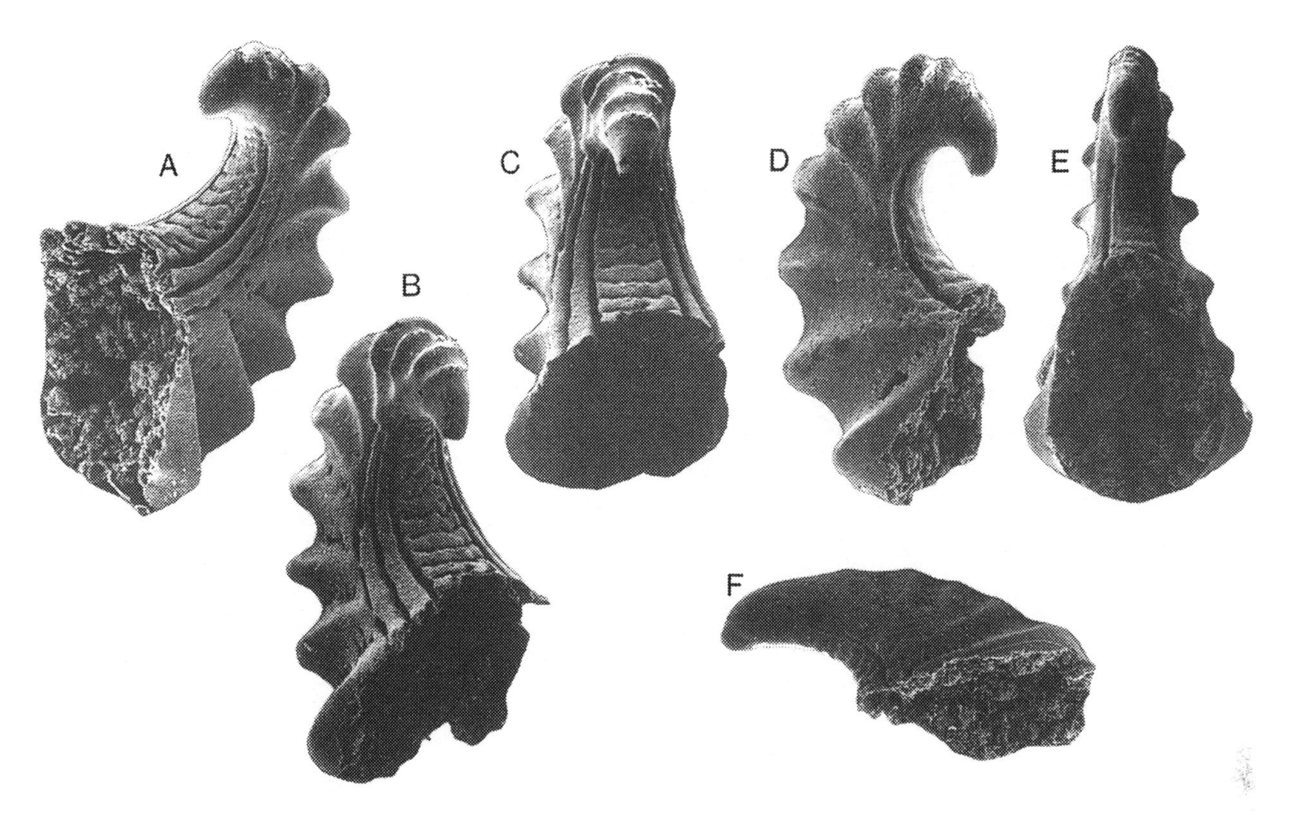

Fig. 13. Internal moulds of the helcionelloid *Latouchella* showing furrows produced by ridges on the shell interior. The ridges control water currents in the posterior mantle cavity. A-F, *Latouchella* spp., late Middle Cambrian, Henson Gletscher Formation, Løndal, Peary Land, central North Greenland, GGU collection 271718, internal moulds, x 30. A-C, MGUH 19.561, oblique posterior views of the sub-apical surface showing three pairs of ridges preserved as furrows in the surface of the internal mould. Note that the outer pair is discontinuous. D, E, MGUH 19.562, slightly oblique lateral view and posterior view showing a single pair of grooves on the sub-apical surface of the internal mould representing ridges on the shell interior. F, MGUH 19.563, oblique lateral view of a second species of *Latouchella* showing one of a pair of grooves on the internal mould corresponding to a ridge on the shell interior.

preted in terms of a pair of laterally disposed gills, by analogy with primitive gastropods (Knight, 1952; Peel, 1974, 1984). The presence of as many as three pairs of folds in *Latouchella* might suggest that a corresponding number of gills was present. However, the narrowness of the shell and presumably restricted size of the mantle cavity argue against this interpretation, although multiple gills may be present in the small posterior mantle cavity of polyplacophorans (cf. Smith, 1960; Yonge, 1960).

The illustrated silicified specimens of *Latouchella* show no trace of a median emargination in the apertural margin of the sub-apical surface, although such a feature is conspicuous in many helcionelloids (Runnegar & Jell, 1976) and also in the plectonotiform gastropods described by Peel (1974). Specimens with well-preserved apertural margins may show shallow lateral emarginations associated with a tendency for the aperture to become keyhole-shaped (fig. 12A, E, H). The lateral impression of the apertural margins producing the narrower waist in this keyhole-shape in apertural view corresponds to the sudden termination of the prominent comarginal rugae of the supra-apical surface against the essentially unornamented sub-apical surface when the shell is viewed in lateral perspective (fig. 12C). The keyhole-shape of the aperture suggests location of the mantle cavity in the area delimited by the narrow waist and the apertural margin of the sub-apical surface. It can be interpreted in terms of incipient separation of an antero-ventral aperture from the more posteriorly located mantle cavity (see discussion of *Yochelcionella, Eotebenna* and *Eurekapegma* above).

Runnegar & Jell (1976, p. 127) suggested that the ridges in *Latouchella* formed an anterior inhalant channel in an exogastrically oriented shell. Presumably this would require that the inhalant current divided as it passed over the respiratory surfaces and that exhalant streams left the mantle cavity along each lateral surface (cf. Pojeta & Runnegar, 1976, fig. 9B; see also fig. 4). This is the opposite orientation to that accepted here where inhalant currents are considered to have entered the endogastric shell laterally and to have united prior to expulsion as a single median, posterior stream. The ridges reflect mantle cavity folds which helped to separate these lateral inhalant currents from the posterior exhalant stream. As in *Yochelcionella* discussed above, oxygen-rich water would enter the mantle cavity along two relatively wide surfaces, achieving maximum supplies for respiration. The rate of expulsion of the exhalant current would be increased by the unification of the two lateral inhalant streams, thus carrying oxygen-depleted water away from the shell.

Interpretation of the internal ridges as supports for shell musculature is rejected both on account of their positive relief within the shell interior and their persistence to the apertural margin. Muscle scars would be impressed within the shell and raised on the internal mould (cf. Horny, 1965a; Peel, 1972, 1976, 1977, 1980, 1982). Longitudinal structures are often associated with muscle scars in bellerophonts and gastropods (cf. Peel, 1977, 1982) but these show little morphological similarity to the prominent internal ridges described here.

Fig. 14. *Plectonotus boucoti* Peel, 1974, Silurian, near Llandeilo, Wales; NMW 39.180 G14, x 3. Antero-dorsal view of this bilaterally symmetrical, planispirally coiled bellerophontacean gastropod showing the prominent dorsal trilobation and the deep sinus culminating in a short slit which generates a broad, band-like selenizone on the median dorsal lobe. Water currents entered the mantle cavity antero-laterally beneath the lateral lobes, and passed over the single pair of gills prior to exhalation via the median emargination. Trilobation assisted separation of the inhalant and exhalant streams within the mantle cavity.

Another possible interpretation is that the ridges may have controlled mantle folds in association with retraction of soft parts into the shell. However, on account of the rapid rate of expansion of the shell and the planar aperture suited to clamping against the substrate, there is little to suggest that these helcionelloids were capable of substantial retraction into the shell.

Signor & Kat (1984) noted a high degree of correlation between columellar folds and the burrowing habit in living high-spired gastropods, enabling the recognition of burrowing in high-spired gastropods at least as old as Silurian (Peel, 1984). While the internal ridges in *Latouchella* are morphologically reminiscent of columellar ridges there is little in their disposition or in the form of the shell in *Latouchella* to suggest that they served the muscle-control function documented by Signor & Kat (1984).

Many living pulmonate gastropods severely constrict the shell aperture with series of lamellae or palatal folds which are often interpreted as serving a defensive role; they may also strengthen the outer lip against breakage by predators. The form of these folds may be reminiscent of the ridges described here in *Latouchella* where the distribution of the ridges, however, is far too restricted to serve the same function as in pulmonates.

The present interpretation of the internal ridges as being associated with the control of water currents entering and leaving the mantle cavity locates this respiratory area below the sub-apical surface. The posterior position of the mantle cavity, deduced from the supposed median exhalant stream, supports the interpretation of helcionelloids as endogastrically coiled untorted molluscs.

PROTOWENELLA

Protowenella was first described from the Middle Cambrian of Australia by Runnegar & Jell (1976) and discussed in detail by Berg-Madsen & Peel (1978) on the basis of material of similar age from Denmark (fig. 15). Subsequently, it has been reported from strata of Early and Middle Cambrian age from many parts of the world (e.g. Peel, 1979; MacKinnon, 1985; Geyer, 1986). The small, globose shell is coiled through a little more than a whorl and its shape prompted Runnegar & Jell (1976) to place it as an intermediate between helcionelloids and the bellerophontiform molluscs in their Order Bellerophontida of the Monoplacophora. Berg-Madsen & Peel (1978) could not accept wholesale interpretation of all bellerophontiform molluscs as monoplacophores, considering many to be gastropods. However, functional interpretation of emarginations located on the umbilico-lateral surfaces suggested that *Protowenella* was untorted and Berg-Madsen & Peel (1978) assigned it to the Monoplacophora.

Protowenella is now transferred to the Class Helcionelloida and the lateral emarginations taken to mark inhalant water currents are considered comparable to the emarginations noted above in *Latouchella*. *Protowenella* is thus considered to be endogastric and not exogastric as earlier suggested by Runnegar & Jell (1976) and Berg-Madsen & Peel (1978). Australian and Danish specimens assigned to *Protowenella* compare closely to *Perssuakiella* described from the latest Middle Cambrian of central North Greenland (Peel, 1988b), which itself resembles the early growth stages of *Helcionella*. In this context, it should be noted that *Protowenella*-like morphologies might result from preservation of the early growth stages of different helcionelloids.

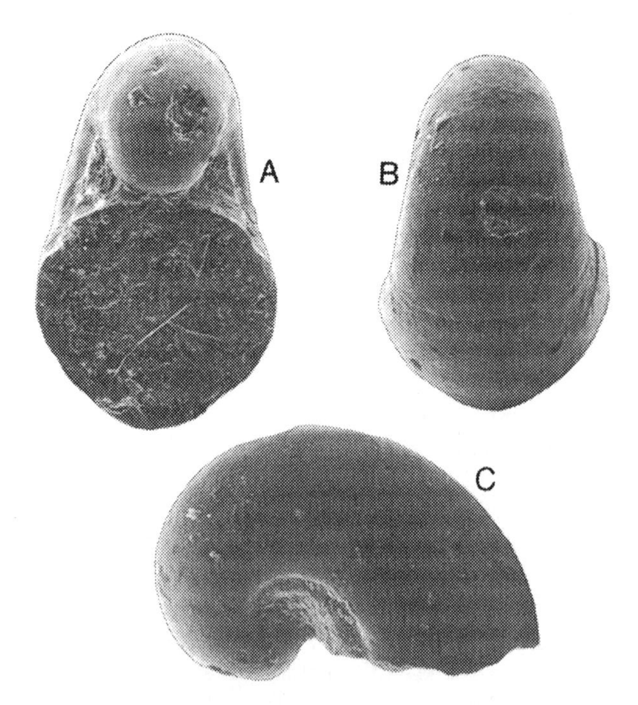

Fig. 15. *Protowenella flemingi* Runnegar & Jell, 1976, late Middle Cambrian, Kalby Clay, Bornholm, Denmark; MGUH 13.991, internal mould, x 85. A, lateral view with anterior to right, showing the prominent circumbilical fold. B, slightly oblique antero-dorsal view showing traces of transverse ornamentation. C, apertural view with posterior at top and anterior at base, showing the prominent circumbilical folds.

Tichkaella Geyer, 1986 from the Middle Cambrian of Morocco has similar folds on the sub-apical wall which may be interpreted as the loci of inhalant currents (Geyer, 1986, pl. 4, fig. 52c). *Tichkaella* resembles *Protowenella* on account of its convex dorsal profile but is much more loosely coiled.

Rostroconchs and diasome evoution

The hypothesis of Runnegar & Pojeta (1974) that rostroconchs were derived from helcionelloids and themselves gave rise to the Class Bivalvia (fig. 1) naturally draws heavily on morphological information offered by the bivalves as the only extant group within the lineage (scaphopods are widely acknowledged as being related to bivalves or rostroconchs, as also suggested by Runnegar & Pojeta (1974), but their high degree of morphological adaptation divorces them from the present discussion). An anterior apex and coiling toward the rear are a unifying theme within the concept Diasoma and helcionelloids were homologised with the exogastric tryblidiacean tergomyans by Runnegar & Pojeta in similar orientation to the majority of bivalves. As an intermediate link, rostroconchs are assumed to have had a similar original orientation. Detailed and well illustrated accounts of rostroconch morphology and evolution are given by Pojeta & Runnegar (1976), Pojeta et al. (1977), Runnegar (1978) and Pojeta (1985; 1987).

Many rostroconchs posses a tubular extension of the shell reminiscent of the helcionelloid snorkel (fig. 10). By analogy with scaphopods and bivalves this rostrum was interpreted by Runnegar & Pojeta as serving a respiratory function and was located posteriorly. In contrast, the snorkel of *Yochelcionella* and *Eotebenna* was considered by Runnegar & Pojeta to perform a similar function, but it was placed anteriorly (fig. 7D). Several rostroconchs display similar morphological adaptations to advanced species of *Eotebenna* (e.g. *E. viviannae* sp. nov.) and are interpreted also as having lived partially infaunally (fig. 10). However, the model of Runnegar & Pojeta requires that the posterior protrudes from the sediment in rostroconchs while the anterior protrudes in Eotebenna (see fig. 11 and discussion of *Eurekapegma*).

Runnegar & Pojeta commented that many rostroconchs have a large gape, interpreted as antero-ventral by comparison with living bivalves, and a smaller posterior gape often forming the tip of the rostrum. Similar gapes are present in *E. viviannae* sp. nov. but the model of Runnegar & Pojeta requires that the larger gape is postero-ventral and the smaller opening at the tip of the snorkel is antero-dorsal.

In the current helcionelloid reconstruction, the larger gape of *E. viviannae* sp. nov. and morphologically similar taxa, and the snorkel opening are both interpreted to be in conformity with rostroconchs (fig. 10). Thus, the snorkel of *Yochelcionella* and *Eotebenna* and the rostrum of rostroconchs are considered posterior. The large gape through which the foot gained contact with the sediment is considered antero-ventral in both rostroconchs and *Eotebenna*. Similar modes of life are inferred for *Eotebenna viviannae* sp. nov. and many rostroconchs on the basis of homologous functional adaptations (fig. 10).

A transverse strengthening bar, the pegma, is typical of rostroconchs and its appearence in the diasome lineage demarcates members of the class from the ancestral helcionelloids (cf. Pojeta, 1985, p. 302; see fig. 1). By reference to the common supposedly posterior extension of the rostroconch shell, the pegma is generally considered to lie anterior to the apex. While the presence of the pegma is by definition characteristic of rostroconchs, a number of helcionelloids preserve structures interpreted by Runnegar & Pojeta as "pegma-like". These structures are preserved on the sub-apical surface of the helcionelloid shell and comparison with the pegma of rostroconchs is a major element in the interpretation of this surface as anterior in the model of Runnegar & Pojeta. However, the supposed homology between many of these structures and the rostroconch pegma is not convincing or is the subject of debate.

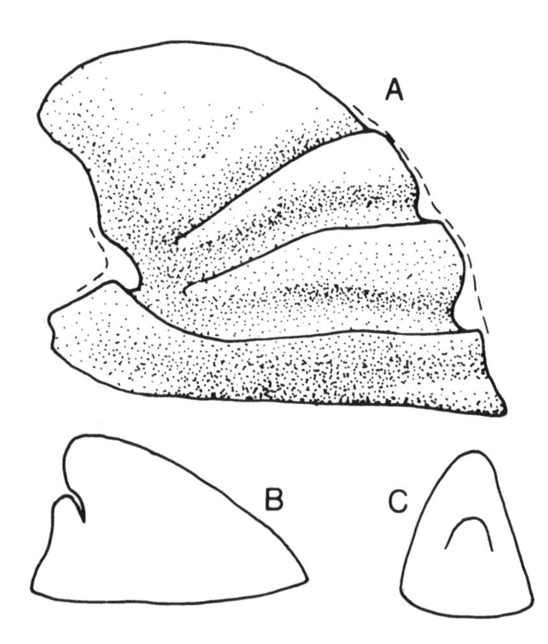

Fig. 16. Pegma-like structures in *Latouchella*? and *Enigmaconus*. A, lateral view of internal mould of *Latouchella*?, Early Cambrian, Parara Limestone of South Australia showing prominent constrictions produced by varix-like thickening on the shell interior. The pegma-like notch below the apex represents part of a continuous comarginal thickening of the shell interior. The original outline of the shell exterior surface is represented by a broken line (drawn from Runnegar, 1985, fig. 1A). B, C, *Enigmaconus* MacKinnon, 1985, Middle Cambrian, New Zealand, sketches based on the holotype, a largely exfoliated internal mould, as illustrated by MacKinnon (1985) showing the pegma-like notch in lateral and posterior views.

MacKinnon (1985) refuted the supposed presence of a pegma in the Lower Cambrian *Heraultipegma* which Runnegar & Pojeta considered to be the oldest rostroconch (fig. 1). He considered the structure to be produced by phosphatisation around the margin of the sub-apical surface leaving an impression of the shell edge. Kerber (1988) considered *Heraultipegma* to be an arthropod (Crustacea, Branchiopoda, ?Conchostraca), returning in part to the previously expressed opinions of Cobbold (1935) and Müller (1975). He synonymised *Heraultipegma* with *Watsonella* Grabau, 1900. Landing (1989) also considered *Heraultipegma* to be a junior synonym of *Watsonella* although he followed Runnegar & Pojeta (1976) in considering *Watsonella* to be a rostroconch.

Sudden changes in growth parameters on the sub-apical surface such as those described below in *Eotebenna viviannae* sp. nov. can produce notches in this surface reminiscent of the rostroconch pegma. A deep notch in the internal mould of a species of *Anabarella* from the Cambrian of Australia illustrated by Runnegar (1983, fig. 4A) and Runnegar & Pojeta (1985, fig. 20A; fig. 5 herein) is reminiscent of a similar structure illustrated by Geyer (1986, pl. 3, figs 35-42) in *Oelandia comma* from the Middle Cambrian of Spain. *Anabarella simesi* MacKinnon, 1985 from the Middle Cambrian of New Zealand exhibits a less strongly incised notch more closely comparable to other helcionelliods. The notch in the Australian species reflects the development of a sub-apical exhalant emargination in this unusually strongly coiled form and can be compared with similar apertural folds developed in *Eotebenna arctica* (figs 5, 8) and *Oelandia pauciplicata* (fig. 6).

Enigmaconus MacKinnon, 1985, from the Middle Cambrian of New Zealand, preserves a more convincing pegma-like structure on the sub-apical surface (fig. 16), but the quite different morphology of the relatively broad, cone-shaped *Enigmaconus* and the strongly laterally compressed early rostroconchs makes any proposed equivalence in function between the respective transverse bars tenuous.

Runnegar (1985, fig. 1A-E) illustrated an internal mould from the Lower Cambrian of South Australia as *Latouchella*? n. sp. with pronounced constrictions produced by varix-like thickenings of the shell interior. A deep cleft below the apex resembles the impression of a pegma when the shell is viewed laterally (fig. 16). However, the constriction producing the cleft is completely continuous around the shell and can not be compared to the rostroconch pegma.

Transverse furrows on internal moulds of *Merismoconcha* Yu, 1979 occur on the broadly convex, apparently supra-apical surface and may resemble a pegma-like structure (cf. Kerber, 1988, fig. 19). Kerber (1988) relegated the supposed Class Merismoconchia to a family of the helcionelloids, but the status of this group of problematic Lower Cambrian fossils remains unresolved (cf. Qian & Bengtson, 1989; Yu, 1989).

The derivation of the rostronchs from the helcionelloids proposed by Runnegar & Pojeta (1974) is accepted as a general model, although the opinions expressed by MacKinnon (1985) and Kerber (1988) concerning the non-rostroconch status of *Heraultipegma* merit close investigation. Rostroconchs, however, are considered to be originally endogastric, in similar fashion to helcionelloids, and not exogastric (as are tryblidiacean tergomyans or bivalves), as proposed by Runnegar & Pojeta (1974, and later references). This derivation of rostroconchs from endogastric helcionelloids accepts most of the rostroconch morphological and functional analyses described by Runnegar & Pojeta but allows similar structures in both groups to be Interpreted as performing the same function, in the same orientation (fig. 10). Neither is the analogy between rostroconch and bivalve functional morphology weakened since the relative orientation of members of these two classes in life remains as proposed by Runnegar & Pojeta.

The Sub-phylum Diasoma of Runnegar & Pojeta thus comprises two distinct but parallel lineages and can not be maintained as a monophyletic group. Bivalves were not derived from helcionelloids but are considered to be descendants of early tryblidiacean tergomyans or an exogastric tryblidlacean-like mollusc. Rostroconchs evolved from endogastric helcionelloids, becoming a group of largely infaunal or semi-infaunal molluscs analagous to the bivalves. Despite their long geological record (Cambrian-Permian), the rostroconch solution to infaunal life proved less long-lived than that employed by the bivalves and members of the latter class survived the Rostroconchia to become increasingly abundant during the post-Palaeozoic. Scaphopods may represent the last specialised remnant of the helcionelloid-rostroconch lineage or may be an early off-shoot of the Bivalvia.

The origin of the cephalopods

Yochelson et al. (1973) derived the first cephalopods from tall monoplacophorous molluscs with septate early growth stages, by the subsequent development of a siphuncle perforating the abundant septa. They based their model on the Late Cambrian *Knightoconus* Yochelson, Flower & Webers, 1973. In which the apparently endogastric shell bears some morphological similarity to the earliest cephalopod *Plectronoceras* Ulrich & Foerste, 1933, described from the Late Cambrian of China (fig. 17). *Knightoconus* is a member of the group of supposedly untorted molluscs termed Hypseloconellacea by Stinchcomb (1986) and is distinguished from the nominate genus *Hypseloconus* Berkey, 1898 mainly on account of its internal multiseptation.

Stinchcomb (1980) compared muscle scars in *Hypseloconus* with similar muscle scar patterns in tryblidiaceans such as *Pilina* suggesting that *Hypseloconus* was indeed endogastric, as deduced by Yochelson *et al.* (1973). Tryblidiaceans, however, are exogastrically coiled and any phylogenetic relationship between them and the apparently endogastric hypseloconellaceans would require a fundamental change in direction of coiling. The discrepancy was resolved by Webers et al. (in press) who described early growth stages of *Knightoconus* which are clearly exogastric. With subsequent allometric growth *Knightoconus* changes its direction of coiling to become endogastric in the adult, acquiring a pseudo-endogastric shell-form similar to that of *Plectronoceras* (fig. 18).

Runnegar & Jell (1976, p. 125) speculated that *Hypseloconus* could acquire a tall cone because of its endogastric coiling and that endogastric coiling "must have been the most important single character for the production of the Cephalopoda, for it would allow for the development of buoyancy tanks above the body mass". By analogy to species of *Yochelcionella* Runnegar & Jell (1976) anticipated the discovery of Webers et al. (in press) concerning the ontogenetic change in the direction of coiling of *Hypseloconus* noting how *Y. ostentata* (fig. 17) from the Middle Cambrian of Australia changes coiling direction during ontogeny to acquire a tall shell. They suggested that this ontogenetic change in curvature developed in response to a need to elevate the snorkel and that the curvature was retained after the snorkel was lost. It is implicit in this argumentation that Runnegar & Jell considered *Hypseloconus* and *Knightoconus* to be derived from *Yochelcionella* although in the present context they are widely separated.

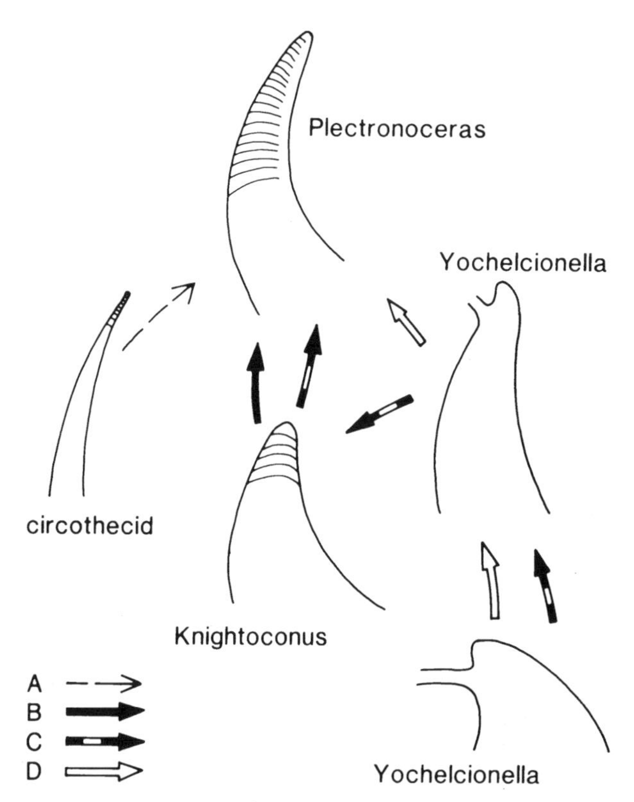

Fig. 17. Theories concerning the origin of the earliest cephalopode (*Plectronoceras*) A, Dzik (1981) suggested that cephalopods were derived from slender, planktonic monoplacophorous molluscs currently classified among the circothecid hyoliths. B, Yochelson et al. (1973) derived cephalopods from endogastric hypseloconellacean Tergomya such as *Knightoconus*. D, Runnegar & Jell (1976) considered that tall, slender, secondarily endogastric species of *Yochelcionella* gave rise to endogastric hypseloconellaceans such as *Knightoconus* which, following Yochelson *et al.*, were ancestral to the cephalopods. C, Pojeta (1980, 1987) proposed that tall, slender and secondarily endogastric species of *Yochelcionella* (such as *Y. ostentata*) gave rise to cephalopods by conversion of the snorkel into a siphuncle. The prefered theory that cephalopods were derived from endogastric helcionelloids is illustrated in fig. 18.

In the restoration of *Y. ostentata* favoured by Runnegar & Jell, coiling is originally exogastric and subsequently endogastric, and the pseudo-endogastric shell form is thus directly analagous to *Hypseloconus* and *Knightoconus* (fig. 17). *Yochelcionella ostentata* is here considered to be endogastric in its earlier growth stages and to assume exogastric coiling in its adult stage, i.e., the shell is pseudo-exogastric (fig. 18). Following the reasoning of Runnegar & Jell quoted above, both pseudo-exogastric and pseudo-endogastric shell forms permit the acquisition of tall, slender cones with a potential for the development of buoyancy above the body mass. The same result, however, can be achieved without reversal of coiling direction in shell forms with a low spiral angle or with a straight cone (e.g. some species of *Obtusoconus* Yu, 1979, see also *"Helcionella" buttsi* Resser, 1938, fig. 5). In a straight cone, an orthocline apertural margin would maintain the centre of gravity in a central position above the aperture. Development of an opisthocline apertural margin would locate the centre of gravity in a similar position in a shell (either exogastric or endogastric) with a constant, but low spiral angle. The earliest cephalopods are endogastric or straight but it is not known if the array of growth forms reflects isometry, allometry or examples of both.

In the case of *Plectronoceras*, the oldest cephalopod, the endogastric shell-form could be derived from the allometrically coiled, pseudo-endogastric *Knightoconus* or from an isometrically coiled, endogastric helcionelloid (fig. 18). In both cases, the deep and narrow, posteriorly located mantle cavity may have contributed to the differentiation of the tissue strand which ultimately developed into the relatively wide siphuncle of the earliest cephalopods. Snorkel development in *Yochelcionella* and morphologically similar species of *Eotebenna* demonstrates that the mantle cavity extended almost to the apex and a similar configuration can be anticipated in some hypseloconellaceans (fig. 18). Progressive adapertural deposition of septa around the adapical termination of such a mantle fold to close off early growth stages of a slender shell could thus produce the fore-runner to the siphuncle.

Chen & Teichert (1983) rejected the theory of Yochelson et al. (1973) that cephalopods were derived from *Knightoconus* noting that it is the development of the sipho and not the presence of septa which characterises the Cephalopoda. Septa may be expected in any relatively narrow conical shell and are widely developed in a variety of molluscs. Their presence in circothecid hyoliths even prompted Dzik (1981) to suggest that slender, planktonic monoplacophorans currently assigned to this group may have given rise to the cephalopods (fig. 17). While Chen & Teichert (1983) are undoubtedly correct in their assertion concerning the relative importance of the sipho and septation in the recognition of the cephalopods, the role of septation or the ability to produce septa should not be dismissed out of hand. It can be argued that it is the combination of septa with the sipho which provides the flotation mechanism which was so instantaneously successful in the Late Cambrian (Chen & Teichert, 1983). It is

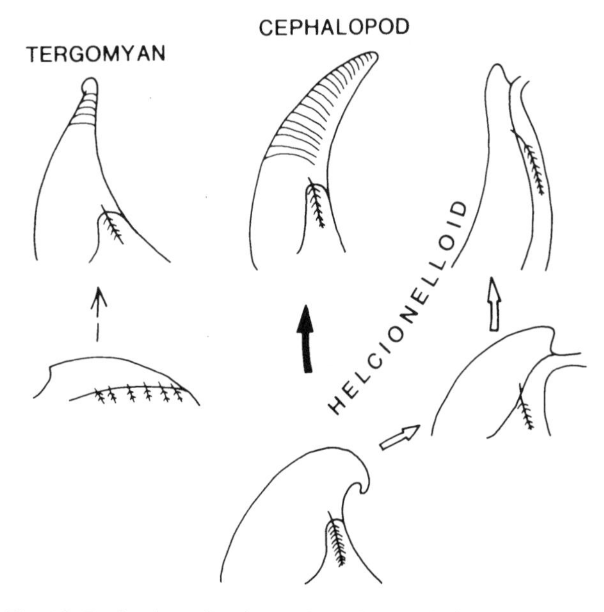

Fig. 18. Derivation of endogastric cephalopods from endogastric helcionelloids compared with the evolution of morphologically similar shell forms within the Tergomya and the helcionelloid *Yochelcionella*. All specimens are similarly orientated, with anterior to the left and posterior to the right. The mantle cavity and gills are illustrated schematically.

perhaps more correct to state that it is the development of calcareous septation in association with the sipho which permits recognition of the flotation mechanism in the geological record.

Harper & Rollins (1982) also rejected the model of Yochelson et al. (1973), considering *Knightoconus* to be a cyclomyan and not an untorted tergomyan (cf. Horny 1965a, b). Since Harper & Rollins (1982) believed the Cyclomya to be gastropods, acceptance of the theory of Yochelson *et al.* (1973) would require derivation of the untorted cephalopods from the torted Gastropoda and not from some untorted mollusc. This unacceptable derivation is avoided here since *Knightoconus* is considered to be a specialised tergomyan, hence untorted.

While slender species of *Yochelcionella* such as *Yochelcionella ostentata* may offer an important analogy to the "pre-cephalopod" it is unlikely that they were direct ancestors of *Plectronoceras* as proposed by Pojeta (1980; see also 1987 and fig. 17 herein). The reconstruction of *Y. ostentata* favoured by Pojeta & Runnegar (1976) involves a mantle cavity developed all around the shell aperture with anterior snorkel and laterally disposed gills. Development of the snorkel into a slphuncle would place it on the opposite side of the shell to that seen in *Plectronoceras* while much of the rest of the mantle cavity would be postero-laterally disposed. If the most adapical part of the elongate mantle cavity in the reconstruction of *Y. ostentata* favoured here did ultimately differentiate into a siphuncle, this would be located beneath the convex surface (in lateral view) and not beneath the concave surface, as is the case in *Plectronoceras* (fig. 18).

Cephalopods differentiate into a variety of endogastric and exogastric forms, varying from straight to shallowly curved and tightly coiled. However, the dominance of endogastric shells in the earliest fossil record of the class (Chen & Teichert, 1983) suggests derivation from an endogastric ancestor. For this reason, it is proposed that cephalopods may have been derived from the endogastric helcionelloids and not from exogastric tryblidiacean tergomyans.

The tall cones of *Yochelcionella ostentata* and similar species, *Knightoconous* and other hypseloconellaceans, and the earliest cephalopods are viewed as broadly similar morphological responses to possibly analogous environmental adaptations. Slender *Yochelcionella* are known from the Lower and Middle Cambrian and rarely approach 5 mm in height. The hypseloconellaceans appear in the late Middle Cambrian and have more massive shells up to ten times or more as large. The first *Plectronoceras* from the medial Upper Cambrian are less than 1 cm tall (Yochelson et al., 1973). There is sufficient variety both between and within these groupings to suggest that no single mode of life is involved.

Runnegar & Pojeta (1974) proposed the Subphylum Cyrtosoma to include the classes Gastropoda and Cephalopoda, and their broadly defined Monoplacophora. Wingstrand (1985) argued that the "Monoplacophora", as a stem-group to the other conchiferan molluscs, should be removed from the Cyrtosoma but he considered the remaining Cephalopoda and Gastropoda to form a monophyletic group. Acceptance of the derivation of cephalopods from the endogastric helcionelloids suggests that the Cyrtosoma may be polyphyletic, since gastropods are probably derived from exogastric tergomyan-like untorted molluscs. Similarities between cephalopods and gastropods, such as the well developed head, adanally restricted mantle cavity and frequently tightly coiled endogastric shell, may in part represent convergence on account of life within such a narrow cone.

SYSTEMATIC PALAEONTOLOGY

No synthesis of molluscan evolution is attempted in this paper since the status of many supposed monoplacophorous molluscs from the Early and Middle Cambrian remains problematic. Neither is the scope of the here defined classes Tergomya and Helcionelloida discussed in detail.

Class Tergomya

(ex. Sub-class Tergomya Horny 1965a, b)

Diagnosis: Generally bilaterally symmetrical molluscs in which the calcareous shell is usually planispirally coiled through about half a whorl. The shell is often cap-shaped or spoon-shaped, usually low, with an anterior apex which may vary from sub-central to overhanging the anterior margin. The aperture is generally planar but may be shallowly arched in lateral view. Muscle scars on the shell interior are grouped into a ring on the dorsal, supra-apical surface; the apex lies outside of this ring. Ornamentation consists of radial and/or comarginal elements; prominent rugae are only infrequently developed.

Discussion: The concept of the Class Tergomya is one of untorted univalved molluscs which are coiled exogastrically. Thus, the sub-apical surface is considered to lie anteriorly while the supra-apical surface is posterior. The Class Tergomya as proposed here corresponds quite closely to the usage of the undefined term Tryblidiida by Wingstrand (1985) and a full discussion of the anatomy of living species assigned to the class is given by Wingstand (1985).

The Class Tergomya corresponds to the Sub-class Tergomya of Horny (1965 a, b) and to the restricted Monoplacophora of Harper & Rollins (1982), with one exception; the Superfamily Hypseloconellacea is referred to the Tergomya although it is emphatically stated that the status of many of the genera placed here by Stinchcomb (1986) is left in abeyance. The hypseloconellaceans *Hypseloconus* and *Knightoconus* are at this time considered to be tergomyans in which the development of an unusually tall shell was associated with an ontogenetic change in the direction of coiling, as discussed above.

The best known tergomyan lineage is undoubtedly the Tryblidlacea which can be traced back with considerable certainty from the present day *Neopilina* and its relatives to the Late Cambrian *Proplina* Kobayashi, 1933; a variety of older univalves may even extend the antiquity of this record. Tryblidiaceans are often considered to be the ancestral group for other conchiferous molluscs or, at least, the close descendants of the ancestral group. Thus, Wingstrand (1985, p. 52) considered the tryblidiacean lineage (referred to as Tryblidiida) to be essentially a sister-group for the other Conchifera, an opinion also attributed by Wingstrand to Lauterbach (1983a, b) who used the term Neopilinida. As noted by Wingstrand (1985, p. 54) "the real troubles come when the pattern of side branches from a tryblidian-like stem is discussed...No wonder...that details are missing on this point in most phylogenetical diagrams".

Class Helcionelloida nov.

Diagnosis: Generally bllaterally symmetrical univalves in which the calcareous shell is usually planispirally coiled through about a half to one and a half whorls; the whorls may be in contact or open coiled and are often laterally compressed. The aperture is commonly planar, without re-entrants, but the sub-apical surface may develop a median sinus which may be deep and slit-like or even trematose, with a single perforation at the end of an elongate tube termed the snorkel. In some forms the lateral areas of the aperture may become prosocyrt, extended into weak lateral shields and producing broad emarginations in both the supra-apical and sub-apical surfaces astride the plane of symmetry. Ornamentation may include both comarginal and spiral elements; prominent comarginal rugae are a feature of many forms.

Discussion: The concept of the Class Helcionelloida is one of untorted molluscs which are coiled endogastrically (fig. 4B).

Helcionelloids are distinguished from tergomyans such as *Neopilina* and *Tryblidium* by their opposite direction of coiling. Members of both classes are untorted but the shell in the Tergomya is exogastric (cf. fig. 4A). Helcionelloid shells are usually more strongly coiled and more laterally compressed than tergomyan shells, the latter being mainly lower and with a much more rapid rate of whorl expansion. However, the tall, slowly expanding and only slightly coiled shells of the hypseloconellacean Tergomya provide an important exception to this generalisation.

Helcionelloids are endogastric, as is the earliest cephalopod *Plectronoceras*, although later cephalopods show a variety of coiling forms developed in response to their wide ecological diversification. The siphuncle and perforated septa clearly separate cephalopods from helcionelloids, although imperforate septa may occur in members of the latter class. Gastropods and helcionelloids are both endogastrically coiled, although the former are distinguished by having undergone torsion. A few bilaterally symmetrical and planispirally coiled (or or almost so) gastropod shells may be confused with helcionelloids or tergomyans (cf. Yochelson, 1988; Peel, 1989b). The complex of 'bellerophontiform molluscs' includes representatives of the Gastropoda and the Helcionelloida, and possibly other molluscan groups (Berg-Madsen & Peel, 1978).

Members of the Class Paragastropoda Linsley & Kier, 1984 are untorted and endogastric but can be distinguished from helcionelloids by their conispiral coiling.

Runnegar & Pojeta (1 985, p. 50) divided their Superfamily Helcionellacea Wenz, 1938 into the families Scenellidae, Helcionellidae, Yochelcionellidae, Stenothecidae and Hypseloconidae; the constituent genera were listed by Runnegar & Jell, 1976, 1980). Of these families, the Hypseloconidae is considered to be tergomyan rather than helcionelloid (cf. Stinchcomb, 1986; Peel, 1988b).

Genus *Eotebenna* Runnegar & Jell, 1976

Type species: *Eotebenna pontifex* Runnegar & Jell, 1976 from the Currant Bush Limestone (Middle Cambrian) of Queensland, Australia.

Emended diagnosis: Laterally compressed helcionelloids in which the perforation at the termination of the often sail-like snorkel-fold is usually connected to the aperture by a narrow slit; less frequently the lateral margins of the slit curve inward in later growth stages to meet, and eventually fuse, at the plane of symmetry. The shell varies from equidimensional with a sail-like snorkel-fold on the sub-apical area (*Eotebenna papilio* Runnegar & Jell, 1976 and *E. arctica* Peel, 1989a; see figs 5, 8, 9) to greatly elongated, with the sub-apical and supra-apical areas extended abapically so that their respective median dorsal areas are sub-parallel and near perpendicular to the aperture (*E. pontifex* and *E. viviannae* sp. nov.; figs 5, 9, 19). The aperture exclusive of the slit and snorkel opening varies from about as long as the shell height in *E. papilio* and *E. arctica* to about one third of the maximum dimension in *E. viviannae* sp. nov. The perforation varies from a simple slit-like extension of the aperture in the Early Cambrian *E. arctica* from Greenland to a sub-circular trema in the type species. In *E. viviannae* sp. nov. a small gape is present at the abapical extremity of the sub-apical surface.

Discussion: *Eotebenna pontifex* and *E. viviannae* sp. nov. represent a continuation of the morphological trend from *Latouchella* to *Eotebenna arctica* and *E. papilio* whereby a deep, slit-like emargination is formed in the sub-apical wall (figs 5, 9). In *E. viviannae* sp. nov. the slit closes along most of its length in many specimens to leave a single narrow perforation or trema at the apertural margin of the sub-apical surface. In *E. pontifex* the trema is proportionally larger and the snorkel more cylindrical. The result is homologous to the development of the snorkel in *Yochelcionella* and the rostral perforations in rostroconchs (fig. 10).

Eotebenna viviannae sp. nov.
Fig. 19

Derivation of name: For Vivianne Berg-Madsen who collected the type suite and kindly made it available for description.

Description: A species of *Eotebenna* Runnegar & Jell, 1976 known only from phosphatic moulds of the shell interior. The univalved shell is strongly elongated with the snorkel extending perpendicular to the aperture well beyond the apex. The shell is wedge-shaped in lateral view, converging at approximately 20 degrees from the aperture toward the most distal part of the snorkel. The earliest growth stage is preserved as an almost pyramidal elevation above the dorsal line but the direction of shell coiling remains evident despite the considerable morphological change during ontogeny. The apex may be pointed or bluntly rounded and weakly pustulose. After the initial growth stage shell coiling departs from the original spiral to produce an almost straight, slowly expanding cone. The supra-apical surface is thus shallowly concave, becoming less so as the apertural margin is approached. The sub-apical surface is initially concave, immediately below the apex but, by way of an abrupt expansion away from the original coiling spiral, it rapidly becomes sub-parallel to the supra-apical surface, forming the adapical side of the large snorkel. The abapical side of the snorkel is the locus of a narrow slit connecting the perforation at the most distal part of the snorkel with the

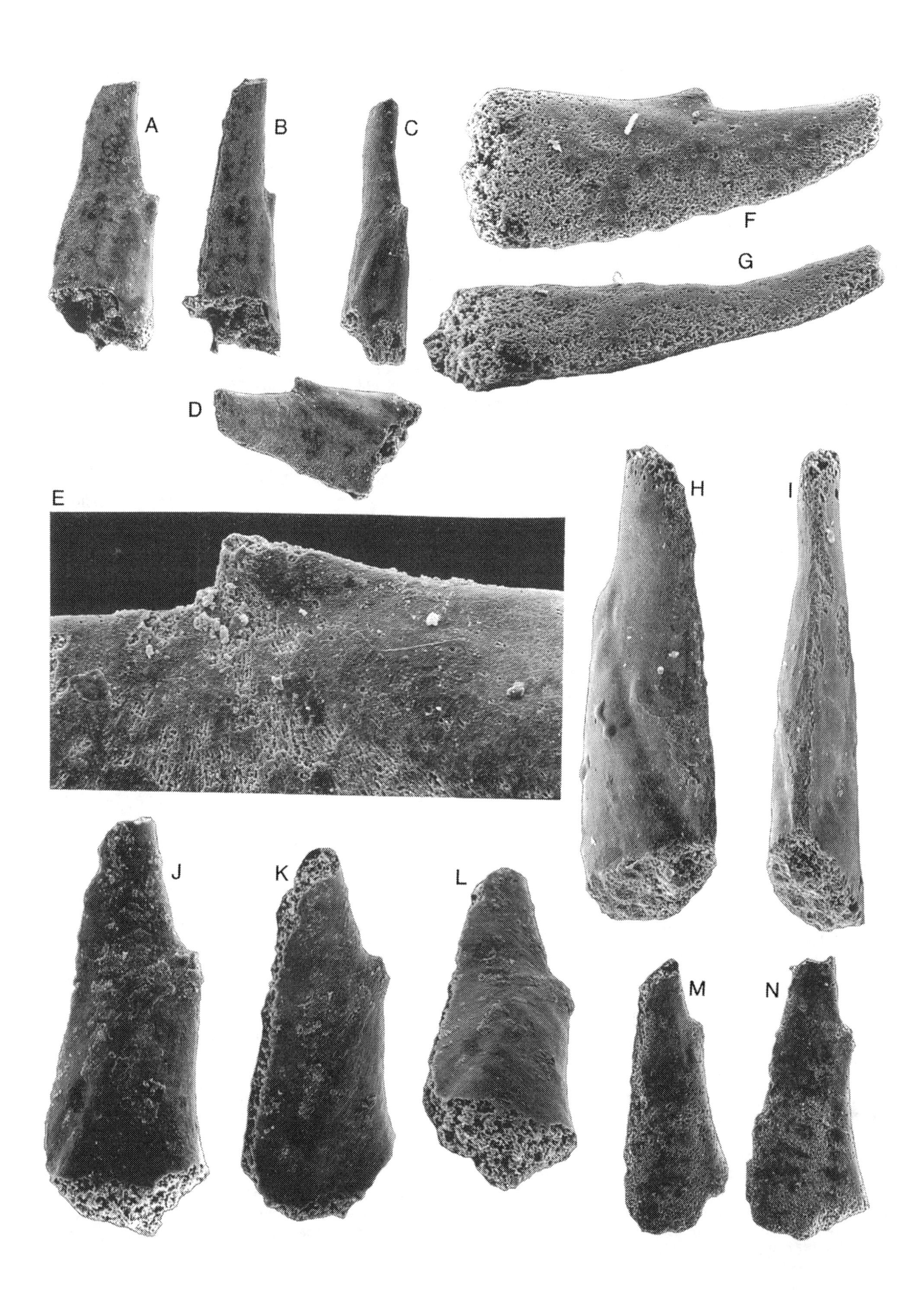
A
B
C
D
E
F
G
H
I
J
K
L
M
N

Fig. 19. *Eotebenna viviannae* sp. nov., late Middle Cambrian, Andrarum Limestone, Bornholm, Denmark, phosphatic internal moulds. A-E, MGUH 19.564, holotype; A, lateral view with anterior to right, aperture down and tip of the snorkel at top, x 50; B, same but with specimen rotated to show line of contact between opposing lateral areas connecting the aperture, below with the snorkel gape, above, x 50; C, same but rotated to show the rounded dorsal surfaces of both the sub-apical snorkel (above the apex in the illustration) and the supra-apical surface (below the apex). An oblique fold traversing the specimen from the apex toward the posterior margin of the aperture (lower left) delimits the snorkel (above) from the main body of the shell (below) 50; D, view of the snorkel showing its terminal narrow gape, x 50; E, the apex with the specimen. In similar orientation to D. Note the abrupt departure of the snorkel (left) from the sub-apical surface and the gradual change in direction of coiling of the supra-apical surface (right), x 250. F, G, MGUH 19.565, lateral views but with the antero-ventral aperture at the left to emphasise the bivalve-like and rostroconch-like form of the wedge-shaped shell. In G the internal mould is rotated to show the line of contact between the lateral areas of the shell, x 100. H, I, MGUH 19.566, slightly oblique lateral and posterior views with the aperture below and the gape at the tip of the snorkel, above. Note the fold in H delimiting the snorkel (right and above) from the main body of the shell. The irregular surface of the internal mould at the line of contact of the two incurved lateral areas in I may indicate that the slit connecting the aperture with the gape at the end of the snorkel was not completely closed, x 100. J-L, MGUH 19.567, x 100; J, lateral view with the posterior to the left; K, same but rotated to show the line of contact between opposing lateral areas and the posterior gape; L, oblique aperturo-lateral view. M, N, MGUH 19.568, oblique lateral and lateral views, x 50.

aperture in some specimens; in other specimens the sides of this slit are closed, isolating the aperture from the terminal perforation (trema). The aperture is ovoid in cross-section, somewhat longer than wide and with its total length about a third of the total height; its lateral margins are shallowly convex, producing weak lateral shields. The trema is tear-shaped in cross-section, uniformly rounded at the adapical side of the snorkel but narrowing with passage into the slit along the abapical margin of the snorkel.

Shell ornamentation is not known. Internal moulds carry traces of coarse rugae parallel to the apertural margin. A broad carina extends across the lateral areas from near the apex toward the aperture in several specimens, delimiting the snorkel from the rest of the shell.

Discussion: *Eotebenna viviannae* sp. nov. is distinguished from most other described helcionelloids by its pronounced allometry, producing the greatly elongate, wedge-shaped shell (fig. 5). *Eurekapegma* MacKinnon, 1985, from the late Middle Cambrian of New Zealand may also show considerable lateral compression and elongation but the apertural margin is convex in lateral aspect to produce a semicircular shape in contrast to the wedge-like form of *Eotebenna* (cf. MacKinnon, 1985, fig. 3K and figs 10, 11 herein). As in *Eotebenna viviannae* sp. nov., the aperture of *Eurekapegma* is also restricted medially to a narrow slit joining prominent gapes at the abapical extremities of the sub-apical and supra-apical surfaces. *Eurekapegma* is characterised by a prominent plate-like structure, the zygion, which joins the lateral surfaces below the apex. No similar structure is known in *Eotebenna*.

Eotebenna viviannae sp. nov. closely resembles *Eotebenna pontifex* Runnegar & Jell, 1976 from the Middle Cambrian of Australia in terms of the subparallel nature of the sub-apical and supra-apical surfaces and the massive, sail-like snorkel. *E. pontifex* is relatively less elongate than *E. viviannae* sp. nov. and the snorkel is more cylindrical in cross-section. The aperture in the type species is about half as long as the total height of the shell while in *E. viviannae* sp. nov. it may be only about a quarter of the height.

Eotebenna papilio and *E. arctica* (figs 8, 9) have more widely expanding shells with less prominent sail-like snorkel-folds; the tip of the snorkel does not extend beyond the apex, as is the case in *E. pontifex* and *E. viviannae* sp. nov.

Tall and narrow species of *Yochelcionella* such as *Y. ostentata* Runnegar & Jell, 1976, resemble *E. pontifex* and *E. viviannae* sp. nov. in the way in which the termination of the snorkel is elevated above the apex in lateral view. The snorkel in *Y. ostentata* differs, however, in being restricted to an ontogenetically earlier stage and its terminal perforation is not connected to the aperture by a slit, as in *Eotebenna* (although this may be closed in *E. viviannae* sp. nov.).

Acknowledgements

Peter Jell (Queensland Museum) is thanked for his kindness in donating specimens of *Latouchella merino* Runnegar & Jell, 1976 for study. Vivianne Berg-Madsen (Uppsala) kindly allowed description of the type suite of *Eotebenna viviannae*. Simon Conway Morris, Ellis L. Yochelson, Robert M. Linsley and M. Paul Smith are thanked for comments to an earlier version of the manuscript. Institutions are denoted by the following abbreviations: GGU, Geological Survey of Greenland, Copenhagen, Denmark; GSC, Geological Survey of Canada, Ottawa, Canada; LO, Palaeontological type collections of the University of Lund, Sweden. MGUH, Geological Museum, Copenhagen, Denmark; NMW, National Museum of Wales, Cardiff, Wales, U.K.; YPM, Yale Peabody Museum, New Haven, Conn., U.S.A. This paper is published with permission from the Director, Geological Survey of Greenland.

REFERENCES

Aksarina, N. A. 1968. [*New data on the geology and guide fossils of western Siberia*]. vol. 3 pp. 77-86, Tomsk: University of Tomsk, (in Russian).

Berg-Madsen, V. & Peel, J.S. 1978. Middle Cambrian monoplacophorans from Bornholm and Australia, and the systematic position of the bellerophontiform molluscs. *Lethaia* **11**, 113-125.

Berg-Madsen, V. & Peel, J.S. 1987. *Yochelcionella* (Mollusca) from the late Middle Cambrian of Bornholm, Denmark. *Bulletin of the Geological Society of Denmark* **36**, 259-261.

Cobbold, E. S. 1935. Cambrian faunas from Hérault, France. *Annals and Magazine of Natural History* **16**, 25-49.

Chen Jun-Yuan & Teichert, C. 1983. Cambrian cephalopods. *Geology* **11**, 647-650.

Dzik, J. 1981. Origin of the Cephalopoda. *Acta Palaeontologica Polonica* **26**, 161-191.

Geyer, G. 1986. Mittelkambrische Mollusken aus Marokko und Spanien. *Senckenbergiana lethaea* **67**, 55-118.

Harper, J.A. & Rollins, H.B. 1982. Recognition of Monoplacophora and Gastropoda in the fossil record: a functional morphological look at the bellerophontid controvesy. *Proceedings of third North American Paleontological Convention. Vol. I,* pp. 227-232.

HASZPRUNAR, G. 1988. On the origin and evolution of major gastropod groups, with special reference to the Streptoneura. *Journal of Molluscan Studies* **54**, 367-441.

HORNY, R.J. 1965a. *Cyrtolites* Conrad, 1838 and its position among the Monoplacophora (Mollusca). *Sborník národnihi Musea i Praze* **21 (2)**, 57-70.

HORNY, R. J. 1965b. On the systematical position of *Cyrtolites* Conrad, 1838 (Mollusca). *Časopis národniho Musea i Praze* **134**, 8-10.

KERBER, M. 1988. Mikrofossilien aus Unterkambrischen Gesteinen der Montagne Noire, Frankreich. *Palaeontographica* **A 202**, 127-203.

KNIGHT, J.B. 1952. Primitive fossil gastropods and their bearing on gastropod classification. *Smithsonian Miscellaneous Collections* **114** (13), 1-55.

KNIGHT, J.B. & YOCHELSON, E.L. 1960. Monoplacophora. In *Treatise on Invertebrate Paleontology. Part 1, Mollusca 1,* (ed. R.C. Moore), pp. 177-184. Lawrence: University of Kansas Press and Boulder: Geological Society of America.

LANDING, E. 1989. Paleoecology and distribution of the Early Cambrian rostroconch *Watsonella crosbyi* Grabau. *Journal of Paleontology* **63**, 566-573.

LEMCHE, H. 1957. A new living deep-sea mollusc of the Cambro-Devonian Class Monoplacophora. *Nature* **179**, 413-416.

MACKINNON, D.J. 1985. New Zealand late Middle Cambrian Molluscs and the origin of Rostroconchia and Bivalvia. *Alcheringa* **9**, 65-81.

Müller, K. J. 1975. *"Heraultia" varensalensis* Cobbold (Crustacea) aus dem unteren Kambrium, der älteste Fall von Geschlechtsdimorphismus. *Paläontologisches Zeitschrift* **49**, 168-180.

PEEL, J.S. 1972. Observations on some Lower Palaeozoic tremanotiform Bellerophontacea (Gastropoda) from North America. *Palaeontology* **15**, 412-422.

PEEL, J.S. 1974. Systematics, ontogeny and functional morphology of Silurian trilobed bellerophontacean gastropods. *Bulletin of the Geological Society of Denmark* **23**, 231-264.

PEEL, J.S. 1976. Musculature and systematic position of *Megalomphala taenia* (Bellerophontacea, Gastropoda) from the Silurian of Gotland. *Bulletin of the Geological Society of Denmark* **25**, 49-55.

PEEL, J.S. 1977. Relationship and internal structure of a new *Pilina* (Monoplacophora) from the Late Ordovician of Oklahoma. *Journal of Paleontology* **51**, 116-122.

PEEL, J.S. 1980. A new Silurian retractile monoplacophoran and the origin of the gastropods. *Proceedings of the Geologists' Association* **91**, 91-97.

PEEL, J.S. 1982. Muscle scars in *Bellerophon recticostatus* (Mollusca) from the Carboniferous of Ireland. *Journal of Paleontology* **56**, 1307-1310.

PEEL, J.S. 1984. Autecology of Silurian gastropods and monoplacophorans. *Special Papers in Palaeontology* **32**, 165-182.

PEEL, J.S. 1988a. *Yochelcionella americana* (Mollusca) from the Lower Cambrian of Newfoundland. *Canadian Journal of Earth Sciences* **24**, 2328-2330.

PEEL, J.S. 1988b. Molluscs of the Holm Dal Formation (late Middle Cambrian), central North Greenland. *Meddelelser om Grønland Geoscience* **20**, 145-168.

PEEL, J.S. 1989a. A Lower Cambrian *Eotebenna* (Mollusca) from Arctic North America. *Canadian Journal of Earth Sciences* **26**, 1501-1503.

PEEL, J.S. 1989b. Morphology and systematic position of *Tryblidium canadense* Whiteaves, 1884 (Mollusca) from the Silurian of North America. *Bulletin of the Geological Society of Denmark* **38**, 43-51.

PEEL, J.S. & YOCHELSON, E.L. 1987. New information on *Oelandia* (Mollusca) from the Middle Cambrian of Sweden. *Bulletin of the Geological Society of Denmark* **36**, 263-273.

POJETA, J., JR. 1980. Molluscan Phylogeny. *Tulane Studies in Geology and Paleontology* **16**, 55-80.

POJETA, J., JR. 1987. Class Rostroconchia. In *Fossil Invertebrates* (ed. R.S. Boardman, A.H. Cheetham & A.J. Rowell), pp. 358-380. Palo Alto: Blackwell Scientific Publications.

POJETA, J., JR. & RUNNEGAR, B. 1976. The paleontology of rostroconch molluscs and the early history of the Phylum Mollusca. *U.S. Geological Survey Professional Paper* **968**, 1-88.

QIAN YI & BENGTSON, S. 1989. Palaeontology and biostratigraphy of the Early Cambrian Meishucunian Stage in Yunnan Province, South China. *Fossils and Strata* **24**, 1-156.

ROBISON, R.A. 1964. Late Middle Cambrian faunas from western Utah. *Journal of Paleontology* **38**, 510-566.

RUNNEGAR, B. 1978. Origin and evolution of the Class Rostroconchia. *Philosophical Transactions of the Royal Society of London* **B 284**, 319-333.

RUNNEGAR, B. 1983. Molluscan phylogeny revisited. *Memoir of the Association of Australasian Palaeontologists* **1**, 121-144.

RUNNEGAR, B. 1985. Shell microstructure of Cambrian molluscs replicated by phosphate. *Alcheringa* **9**, 245-257.

RUNNEGAR, B. & JELL, P.A. 1976. Australian Middle Cambrian molluscs and their bearing on early molluscan evolution. *Alcheringa* **1**, 109-138.

RUNNEGAR, B. & JELL, P.A. 1980. Australian Middle Cambrian molluscs: corrections and additions. *Alcheringa* **4**, 111-113.

RUNNEGAR, B. & POJETA, J., JR. 1974. Molluscan phylogeny: the paleontological viewpoint. *Science* **186**, 311-317.

RUNNEGAR, B. & POJETA, J., JR. 1980. The monoplacophoran mollusk *Yochelcionella* identified from the Lower Cambrian of Pennsylvania. *Journal of Paleontology* **54**, 635-636.

RUNNEGAR, B. & POJETA, J., JR. 1985. Origin and diversification of the Mollusca. In *The Mollusca*, Vol. 10, Evolution, pp. 1- 57. Orlando: Academic Press.

SALVINI-PLAVEN, L. VON, 1981. On the origin and evolution of the Mollusca. In *Origine dei grandi Phyla dei Metazoi. Atti dei Convegni Lincei* **49**, 235-293. *Accademia Nazionale dei Lincei.*

SALVINI-PLAWEN, L. VON, 1985. Early evolution and the primitive groups. In *The Mollusca*, Vol.10, Evolution, pp. 59-150, Orlando: Academic Press.

SIGNOR, P.W. & KAT, P.W. 1984. Functional significance of columellar folds in turritelliform gastropods. *Journal of Paleontology* **58**, 210-216.

SMITH, A.G. 1960. Amphineura. In *Treatise on Invertebrate Paleontology. Part I, Mollusca 1,* pp. 141-176, (ed. R.C. Moore), Lawrence: University of Kansas Press and Boulder: Geological Society of America.

STASEK, C.R. 1972. The molluscan framework. In *Chemical Zoology* pp. 1-44, (ed. M. Florkin and B.T. Scheer), New York: Academic Press.

STINCHOMB, B.L. 1980. New information on Late Cambrian Monoplacophora *Hypseloconus* and *Shelbyoceras* (Mollusca). *Journal of Paleontology* **54**, 45-49.

STINCHOMB, B.L. 1986. New Monoplacophora (Mollusca) from Late Cambrian and Early Ordovician of Missouri. *Journal of Paleontology* **60**, 606-626.

WEBERS, G.F., POJETA, J., JR. & YOCHELSON, E.L. in press. Cambrian Mollusca from the Minaret Formation, Ellsworth Mountains, West Antarctica. In *Geology and Paleontology of the Ellsworth Mountains, West Antarctica* (ed. G.F. Webers, C. Craddock & J. Splettstoesser), *Memoir of the Geological Society of America* **170**.

WESTERGÅRD, A.H. 1936. *Paradoxides oelandicus* beds of öland with the account of a diamond boring through the Cambrian at Mossberga. *Sveriges Geologiska Undersökning* **C, 394**, 1-66.

WINGSTRAND, K.G. 1985. On the anatomy and relationships of Recent Monoplacophora. *Galathaea Report* **16**, 7-94.

YOCHELSON, E.L. 1968. Stenothecoida, a proposed new class of Cambrian Mollusca. *Abstracts of the International Paleontological Union Prague, Czechoslovakia*, p. 34 (abstract).

YOCHELSON, E.L. 1969. Stenothecoida, a proposed new class of Cambrian Mollusca. *Lethaia* **2**, 49-62.

YOCHELSON, E.L. 1978. An alternative approach to the interpretation of the phylogeny of ancient mollusks. *Malacologia* **17**, 165-191.

YOCHELSON, E.L. 1979: Early radiation of Mollusca and mollusc-like groups. In *The Origin of the Major Invertebrate* (ed. M. R. House), *Systematic Association Special Volume* **12**, 323-358.

YOCHELSON, E.L. 1988. A new genus of Patellacea (Gastropoda) from the Middle Ordovician of Utah: the oldest known example of the superfamily. *Memoir of the New Mexico Bureau of Mines & Mineral Resources* **44**, 195-200.

YOCHELSON, E.L., FLOWER, R.H. & WEBERS, G.F. 1973. The bearing of the new Late Cambrian monoplacophoran genus *Knightoconus* upon the origin of the Cephalopoda. *Lethaia* **6**, 275-310.

YONGE, C.M. 1960. General characters of Mollusca. In *Treatise on*

Invertebrate Paleontology, Part I, Mollusca 1, (ed. R.C. Moore) pp. 13-136, Lawrence: University of Kansas Press and Boulder Geological Society of America.

YONGE, C.M. & THOMPSON, T.E. 1976. *Living Marine Molluscs*. London: Collins, 288 pp.

YU WEN 1989. Did the shelled mollusks evolve from univalved to multivalved forms or vice versa, In *Developments in Geoscience, Contribution to 28th International Geological Congress, 1989, Washington D.C., U.S.A.* pp. 235-244. Beijing: Chinese Academy of Sciences, Science Press.

YU WEN & NING HUI 1985. Two Cambrian monoplacophorans from Borohore, Xinjiang. *Acta Palaeontologica Sinica* **24**, 47-50.

The arthropods of the Lower Cambrian Chengjiang fauna, with relationships and evolutionary significance

Hou Xian-guang[1] & Jan Bergström[2]

Abstract

The arthropods described from the Chengjiang fauna include 3 genera of trilobites, 5 genera of small bradoriids with bivalved carapaces, 7 genera with larger (> 10 mm long in adults) bivalved carapaces, 9 genera with non-mineralized exoskeleton, and the lobopod genus *Luolishania*. Most of the genera are monospecific. The specimens originally referred to the genus *Mononotella* are now considered to belong to *Chuandianella* gen. nov., and the specimens originally referred to as *Perspicaris*? sp. are here named *Perspicaris laevigata* sp. nov. Three new arthropod taxa are reported for the first time from this fauna. They are *Jiucunella paulula* gen. et sp. nov., *Chengjiangocaris longiformis* gen. et sp. nov. and *Anomalocaris canadensis* Whiteaves.

Many of the arthropods of the Lower Cambrian Chengjiang fauna are either identical with or notably similar to arthropods of the Middle Cambrian Burgess Shale. The relationships between arthropods of the two faunas are discussed.

Introduction

Due to the bias towards the preservation of mineralised parts as fossils, almost nothing has been known about arthropods with non-mineralized exoskeletons from the Early Cambrian, a time which may be considered particularly interesting in the evolution of arthropods. In 1984, a unique soft-bodied fauna, the Chengjiang fauna was found in the Lower Cambrian at Chengjiang, Yunnan Province, China (Zhang and Hou, 1985; Hou, 1987a). This fauna partly fills in a critical gap close to the base of the metazoan fossil record.

The Chengjiang fauna occurs in the basal and lower parts of the trilobite Zone with *Eoredlichia-Wutingaspis* of the Lower Cambrian Qiongzhusi (Chiungchussu) Formation, located just above the trilobite *Parabadiella* which represents the oldest trilobite in China and in certain other parts of the world (Luo *et al.*, 1984: p. 118). Within the approximately 50 m thick formation there are more than 10 beds yielding soft-bodied fossils (Hou, 1987a), including numerous arthropods without mineralized exoskeletons.

Rb-Sr isotopic age determination on the black shale of the lower part of the Qiongzhusi formation yielded 579.7 ± 8.2 Ma and 587 ± 17 Ma (Luo *et al.*, 1984, p. 122). These data would indicate an age of approximately 570 million years for the Chengjiang fauna which, however, would make it correlate with the base of the Tommotian. Anyway, it is considerably older than the Burgess Shale fauna from British Columbia. This fact is most notable in view of the great similarity between the two faunas.

Just as in the case with the Burgess fauna, the arthropods of the Chengjiang fauna dominate the fauna with regard both to number of specimens and to number of genera and species. Being the oldest fossil record of a varied arthropod fauna, the Chengjiang arthropods provide us with most valuable material for an elucidation of the earliest arthropod evolution and of the ecology of the oldest arthropod community.

[1]Nanjing Institute of Geology and Palaeontology, Academia Sinica, Nanjing, People's Republic of China.
[2]Swedish Museum of Natural History, Stockholm, Sweden.

Arthropods with a bivalved carapace

Arthropod carapaces are most common in the Chengjiang fauna. In particular, the tiny shells (from 1 mm to 6 mm in length) referred to the Bradoriida are extraordinarily abundant, representing more than 80% of the total number of individuals in the whole fauna. However, no appendages or soft tissues are preserved. Of the 14 species (belonging to 10 genera) of bradoriids in the Qiongzhusi Formation, 8 species (6 genera) were found in association with soft-bodied fossils. The latter bradoriids are *Kunmingella maotianshanensis* Huo *et al.*, 1983 (Pl. 1, figs. 3, 4), *K. dadiyakouensis* Huo and Shu, 1985, *K. venusta* Huo and Shu, 1985, *Tsunyiella zhijinensis* Yin, 1978; *Kunyangella cheni* Huo, 1965, *Liangshanella* sp., *Yaoyingella* sp. (see Huo and Shu, 1985), and a new bradoriid *Jiucunella paulula* gen. et sp. nov. (Pl. 1, fig. 5). Their tiny carapaces are preserved either in clusters (Pl. 1, fig. 4) or scattered. Most of them are preserved with the carapace articulated along the hinge line and spread out (Pl. 1, figs. 3, 4). Similarly preserved tiny bradoriids are also known from southewest China.

The new bradoriid *Jiucunella paulula* gen. et sp. nov. (Pl. 1, fig. 5) has a very small bivalved carapace with distinct marginal rim and furrow. A ridge extends from the antero-dorsal margin near the antero-dorsal angle to the ventral area and forms a node in the ventral area. It resembles species of *Zhenpingella* Li, 1975 from a corresponding stratigraphic level in Zhenping, southern Shaanxi in the presence of a ventral node and the very small carapace, but differs in the absence of antero-dorsal and dorsal nodes and in the presence of the ridge joining the anterodorsal margin and ventral node.

Notes to Plates

All the specimens illustrated here were collected from Chengjiang, Yunnan Province. The specimens marked with Field No. M2, M3 are from the western slope, Field No. Cf4, Cf5, Cf6 from the northwestern slope of Maotianshan (the Maotian Hill), respectively, while the specimens marked with Field No. DJ1 are from the hill near the Dapotou village, 3 km SW of the Maotian Hill. All of them are kept in the Nanjing Institute of Geology and Palaeontology, Academia Sinica.

Plate 1

The large (adult length of carapace 10-70 mm) bivalved arthropods are one of the main components of the Chengjiang fauna. Most of them are preserved with only carapaces, a few with "soft parts".

The valves formerly doubtfully referred to *Perspicaris* (Hou, 1987c) are subelliptical in outline, 12 mm in maximum length, with straight hinge margin (Pl.2, fig. 7). They have a slightly upwardly curved hinge margin (Pl. 2, fig. 8; Hou, 1987c, Pl. 3, figs. 6, 7), appearing to result from compaction, and a smooth surface but without antero- and posterodorsal processes. More material has been collected recently. This material indicates that the animal is distinct from the similar Burgess form at least on the species level, and it is therefore named *Perspicaris*? *laevigata* sp. nov.

Mononotella Ulrich and Bassler, 1931 is a monospecific genus from the upper Lower Cambrian (Hanfordian) of New Brunswick. It is characterised by its subfusiform outline and carapace without dorsal hinge. The specimens from the Qiongzhusi Formation assigned to *Mononotella* and placed in the Bradoriida by Li (1975) and Huo and Shu (1985) possess a larger carapace (14 mm in maximum length) covered with reticulate ornament, and they have a convex hinge margin. They differ, therefore, both from *Mononotella* from New Brunswick and from the minute carapaces from southwest China and are better placed in *Chuandianella* gen. nov. This genus is probably not a true bradoriid. At least seven species have been established for this kind of carapace (see Huo and Shu, 1985), most of which were based on deformed specimens and in fact all represent a single species, *i.e. Chuandianella ovata* (Li, 1975) (Pl. 2, figs. 5, 6).

Isoxys is widespread in the Lower and Middle Cambrian in different parts of the world. In the Chengjiang fauna this genus includes two species, viz. *I. auritus* and *I. paradoxus* (see Hou, 1987c), which are similar to the Burgess Shale species *I. acutangulus* and *I. longissimus* (see Simonetta and Delle Cave, 1975) respectively. Since *I. auritus* is numerous while *I. paradoxus* is very rare, they do not seem to represent sexual dimorphs of a single species. In *I. paradoxus* a large specimen shows that the valve can reach a length of 57 mm between the spine bases (Pl. 2, fig. 4). *Isoxys* was also found e.g. in the *Palaeolenus* Zone of the Lower Cambrian Canglangpu Formation in Wuding, eastern Yunnan, and in the approximately correlative Emu Bay Shale of Kangaroo Island, South Australia (Glaessner, 1979), and in the Buen Formation of North Greenland (Conway Morris *et al.*, 1987).

Branchiocaris? *yunnanensis* Hou 1987c greatly resembles species of *Tuzoia* in having a large (up to 60 mm long) carapace, subcircular carapace outline, short antero- and postero-dorsal processes and a reticulated ornament, but differs from *Tuzoia* in lacking marginal spines. The species is distinguished from the Burgess Shale species *Branchiocaris pretiosa* (Resser, 1929) in the absence of a distinctly pitted margin and in the presence of a reticulate ornament. This species may not belong to either *Tuzoia* nor *Branchiocaris*, but it is possible that it is more closely related to the former than to the latter. About one hundred specimens of *B.*? *yunnanensis* have so far been collected, none with anything more than the carapace preserved.

Combinivalvula chengjiangensis Hou, 1987 shows a similarity to *Branchiocaris* ?*yunnanensis* in having short antero- and postero-dorsal processes, but differs in the presence of an ankylosis between the carapace valves in the anterior part of the hinge. Therefore the carapace valves could neither open nor close completely.

Vetulicola cuneatus Hou, 1987 is a bivalved arthropod with abdomen and tail preserved (Pl. 1, figs 1, 2). The carapace is large, up to 70 mm long, with gently convex dorsal and ventral margins and a long groove along the middle of each valve. The overall appearance including the posture of the tail, indicates that a large cuneate spine is positioned on the posterodorsal margin of each valve. The spine in *V. cuneatus* may correspond to spines along the dorsal margin in *Tuzoia*. More than one hundred specimens demonstrate that the phyllocarid-like abdomen and tail is regularly preserved projecting behind the carapace, while the anterior part of the body enclosed between the valves does not appear to be preserved. This presumably indicates weaker sclerotization inside the valves, as is also commonly the case in extant crustaceans with a carapace. The common preservation of the projecting rear part of the body also suggests that forms like *Isoxys, Branchiocaris*? *yunnanensis*, and the supposed bradoriids probably had the body completely enclosed by the carapace, as no projecting parts are ever found.

At Jinning, 40 km away from Chengjiang and at a corresponding stratigraphic level there was found an incomplete specimen tentatively assigned to *Odaraia* and described under the name of *Odaraia*? *eurypetala* (Hou and Sun, 1988). It has 13 trunk segments and a telson bearing two, or possibly three, large blades. This seems to represent part of a body behind a carapace. This species is distinguished from the Burgess Shale species *Odaraia alata* Walcott, 1912 mainly through the elongate telson and the presence of a boundary groove between telson body and blades. It is distinguished from the Burgess Shale species *Waptia fieldensis* Walcott, 1912 mainly through the much shorter trunk segments.

Thus the bivalved arthropods (including *Com-*

Plate 1

Figs. 1, 2. *Vetulicola cuneatus* Hou.

1. Counterpart of Pl. V, fig. 1 illustrated by Hou (1987c), lateral aspect, showing 7 abdominal segments and a tail after preparation, x1.5. Field No. Cf4, Cat. No. 110819.

2. Part, posterior portion of closed carapace, lateral aspect, showing 6 abdominal segments and a tail, x2. Field No. M2, Cat. No. 110820.

Figs. 3, 4. *Kunmingella maotianshanensis* Huo.

3. Carapace valves outspread on the bedding plane along hinge margin, parallel aspect, x10. Field No. M3, Cat. No. 110821. 4. Cluster of carapaces with outspread valves preserved on single surface, x1.8. Field No. M3, Cat. No. 110822.

Fig. 5. *Jiucunella paulula* gen. et. sp. nov. 5a. Carapace valves opened along hinge margin, parallel aspect, holotype. 5b. Left valve. 5c. Valves. x15. Field No. M2, Cat. No. 110823, 110824 and 110825.

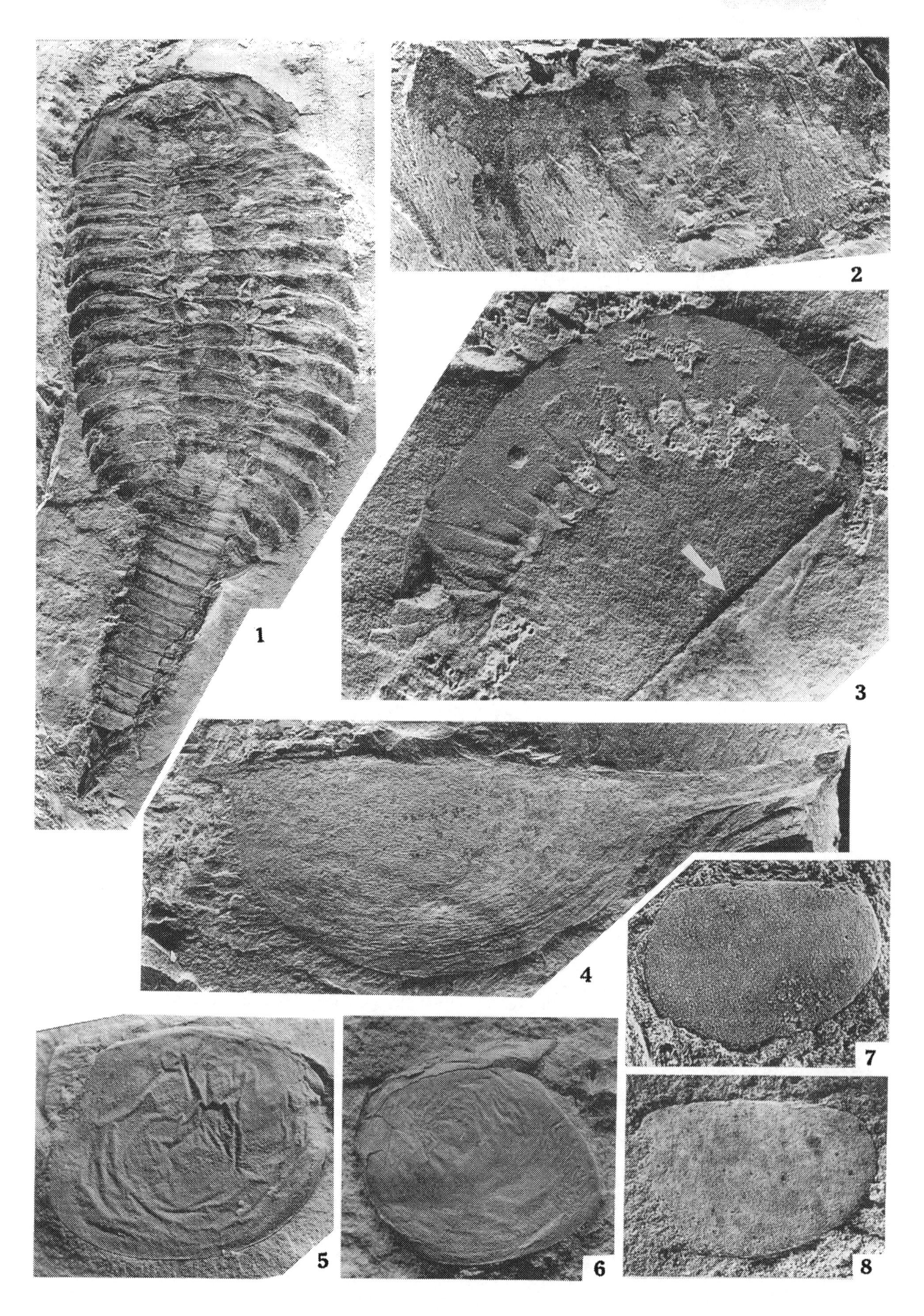
1
2
3
4
5
6
7
8

binivalvula with valves partly fused) in the Chengjiang fauna may include at least four morphological groups, viz.:

1) Tiny carapaces regarded as bradoriids, *i.e.* possible ostracodes. Bradoriids have also been extracted with dilute acid from the Middle and Upper Cambrian limestones in southwest China (Shu De-gan, personal communication).

2) Large carapaces, such as those of *Isoxys* and *Branchiocaris*? *yunnanensis*, probably representing arthropods with a trunk completely enclosed within the carapace.

3) Arthropods with the rear part of the body extending behind the carapace in a phyllocarid-like fashion: *Vetulicola cuneatus*.

4) *Odaraia*? *eurypetala* with terminal flattened blades.

Among the bivalved arthropods of the Chengjiang fauna, some may be crustaceans, while others may represent taxa of ranks equivalent to the Crustacea.

Arthropods without carapace and with a non-mineralised exoskeleton

Cambrian fossils are strongly dominated by trilobites, which were readily preserved due to the strong calcification of their dorsal exoskeleton. In the Chengjiang fauna, however, trilobites are sparse in number of specimens and species, being represented only by rare specimens of *Yunnanocephalus yunnanensis* (Mansuy, 1912), *Eoredlichia intermedia* (Lu, 1941) and *Kuanyangia pustulosa* (Lu, 1941). Only the calcified parts of the exoskeleton have been found so far.

Most common among the forms with non-mineralised exoskeletons are the two species of *Naraoi*a, *N. longicaudata* and *N. spinos*a Zhang and Hou, 1985. They compare well with the two Burgess Shale species *N. compacta* Walcott, 1912 and *N. spinifer* Walcott, 1931, respectively. The beautifully preserved Chengjiang specimens promise to yield additional information of importance for determining the evolutionary and taxonomic position of the genus.

Alalcomenaeus? *illecebrosus* Hou, 1987 is the third most abundant species in this group of arthropods. The specimens are preserved in dorsal, oblique and lateral compressions. The better-preserved specimens indicate that the great appendage of the head (like that in *Leanchoilia superlata* Walcott, 1912) possesses three long flagella (Pl. 3, fig. 2). The animal has 11 rather than 12 trunk segments, each bearing a pair of biramous appendages (Pl. 3, fig. 1). The single specimen in dorsal compression clearly shows the two separate cephalic segments (Pl. 3, fig. 3), each possessing a pair of biramous appendages. *A.*? *illecebrosus* has a great similarity to the Burgess Shale species *A. cambricus* Simonetta, 1970, and is probably closely related also to *Leanchoilia superlata* Walcott, 1912 and *Actaeus armatus* Simonetta, 1970.

Jianfengia multisegmentalis Hou, 1987 (Pl. 3, fig. 4) is another form which is beautifully preserved, but is very rare in the fauna. It has a head, 22 trunk segments and a telson. This species bears a similarity to *Yohoia tenuis* Walcott, 1912 only in the great appendage carrying a group of spines at the tip. This resemblance is probably only superficial. *J. multisegmentalis* and *A.*? *illecebrosus* resemble each other in the pair of great appendages on the head and in a series of similar biramous appendages on the head and trunk regions.

Kuamaia lata Hou, 1987, *Rhombicalvaria acantha* Hou, 1987 *Retifacies abnormalis* Hou *et al.*, 1989 and *Fuxianhuia protensa* Hou, 1987 all have a wide exoskeleton with a length exceeding 8 cm. *K. lata* and *R. acantha* are similar to the Burgess Shale species *Helmetia expansa* Walcott, 1918 in the tail shield and especially in the presence of a possible labrum projecting beyond the anterior margin of the cephalic shield. The exoskeleton of *Retifacies abnormalis* has an ornament in the shape of a polygonal network. It has a cephalic shield, 10 trunk tergites and an elliptical tail shield. A pair of antennae project beyond the anterior margin of the cephalic shield. *Fuxianhuia protensa* has a head with a pair of stalked eyes projecting beyond the cephalic shield, 31 trunk tergites (20 anterior ones with long pleura, 11 narrow posterior ones with very short pleura), and a short telson. An approximately 10 cm long specimen shows a pair of short antennae placed at the outer side of the stalked eyes (Pl. 2, fig. 1).

In contrast to the last four taxa, *Acanthomeridion serratum* Hou *et al.*, 1989 and *Urokodia aequalis* Hou *et al.* 1989 have a parallel-sided, narrow body

Plate 2
Fig. 1. *Fuxianhuia protensa* Hou. Counterpart, parallel aspect, x1.5. Field No. M2, Cat. No. 110826.

Figs. 2, 3. *Anomalocaris canadensis* Whiteaves.
2. Part, lateral aspect, x7. Field No. Dj1. Cat. No. 110827.
3. Part, lateral aspect, x5. Field No. M2. Cat. No. 110828. An incomplete specimen of *Isoxys auritus* denoted by arrows in bottom right.

Fig. 4. *Isoxys paradoxus* Hou. Left valve, x1.5. Field No. Cf 4, Cat. No. 110829.

Figs. 5, 6. *Chuandianella ovata* (Li).
5. Closed carapace valves, lateral aspect, anterior toward right, showing reticular ornament on the surface, x7. Field No. M3, Cat. No. 110830.
6. Carapace valves opened along upward curved hinge margin, lateral aspect, anterior toward left, x7. Field No. Cf6, Cat. No. 110831.

Figs. 7, 8. *Perspicaris*? *laevigata* sp. nov.
7. Right valve, holotype, x4. Field No. M2, Cat. No. 110832.
8. Right valves, x4. Field No. M3, Cat. No. 110833. 26

Plate 3

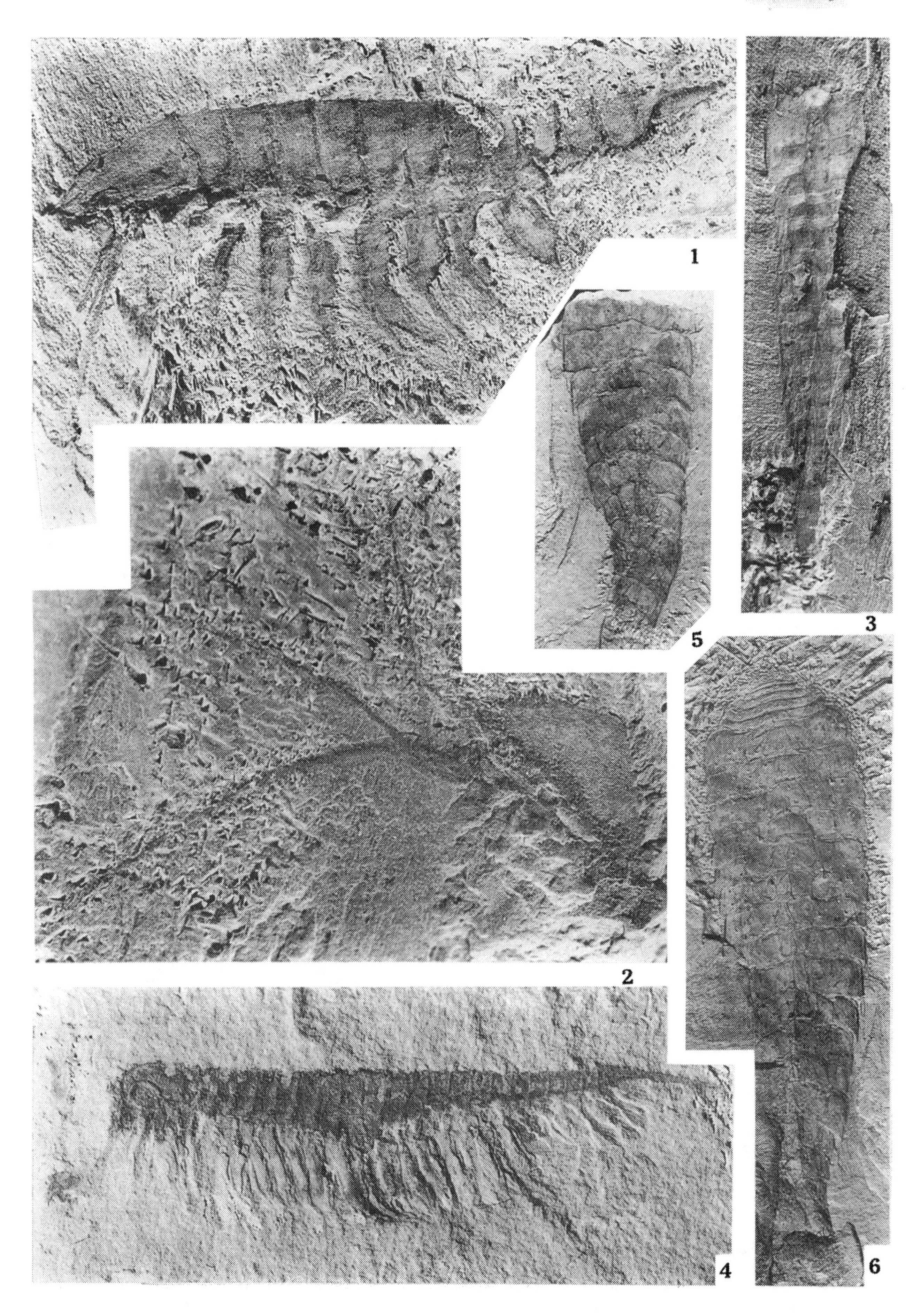

shape. *A. serratum* has cephalic shield, 10 tergites and a pair of flabellate posterior spines. The lateral extremities of the tergites are spinose and equally developed in the first 6 tergites, but become successively elongate in the posterior ones, and form exceedingly long lateral spines in the 9th one, so that the lateral extremities of the 9th tergite extend posteriorly far beyond the telson.

In *Urokodia aequalis* Hou *et al.*, 1989 the cephalic and tail shields resemble each other in side and morphological features, bearing a pair of anterior or posterior spines and pairs of lateral spines respectively. The 14 trunk tergites have almost the same width as length. *U. aequalis* is similar to *Mollisonia symmetrica* Walcott, 1912 from the Burgess Shale, while the other species have no close counterpart known from any part of the world. None of them can be accommodated within any extant arthropod group.

The lobopodan animal Luolishania

Among all fossils in the Burgess Shale, the more or less soft-bodied *Aysheaia pedunculata* Walcott, 1911 is particularly interesting because of its similarity to living Tardigrada and Onychophora, and to some degree also Myriapoda. Another species, *A. prolata* Robison, 1985, was recently described from the Middle Cambrian Wheeler Formation of western Utah. Although the latter species is poorly preserved, it can be stated to differ from *A. pedunculata* through the elongation of its cephalic region and lobopods (Robison, 1985).

A marine animal comparable to *Aysheaia* is *Luolishania longicruris* Hou and Chen, 1989 recently found in the Chengjiang fauna. Despite a general similarity, *Luolishania* differs in a number of respects:

1) The number of lobopod pairs is different.

2) The segmentation is not expressed as annulations, only as transverse rows consisting of three nodes.

3) There is a tagmosis. Five anterior trunk segments increase in length rearwards, with the number of intrasegmental annulations increasing from 3 to 5; trunk segments are of subequal length and have probably 7 annulations each; 6 posterior trunk segments decrease in length rearwards, and the number of annulations decrease from 5 to 3, less in the last segment. In *Aysheaia* the number of annulations is the same throughout the body length.

4) The *Luolishania* lobopod is slenderer, lacks spines and has a larger number of annulations.

5) The body projects beyond the last pair of lobopods. *Xenusion auerswaldae* from the earliest Cambrian of the Baltic area was recently restudied in detail by Dzik and Krumbiegel (1989), after a second specimen was found. The authors conclude that "*Xenusion* differs from *Aysheaia* in lacking specialized appendages, in the presence of a long proboscis and in the strongly heteronomous annulation of the body". *Luolishania* is distinguished from *Xenusion* at least by the transverse row of three rather than two nodes in the trunk-region for each lobopod pair, the more evidently heteronomous segments of the body and the regular change in number of the annulations in the trunk segments.

Luolishania and *Xenusion*, both from the earliest Cambrian, are geographically widely separated. The differences between them as well as between them and *Aysheaia* suggest that the evolutionary radiation of marine lobopods was well under way already at that early date.

A new arthropod of unknown affinities

Chengjiangocaris longiformis gen. et sp. nov. (Pl. 3, figs. 5, 6), is a new taxon represented by part and counterpart of one specimen in the Chengjiang fauna. It has at least 22 segments with an unusual type of tagmosis. The first tergites are much shorter than the successive ones. The whole set narrows into a blunt anterior angle. It is therefore probable that these tergites represent the head tagma of the animal.

The successive 17 abdominal tergites are of subequal length but decrease in width backwards. The body is terminated abruptly; apparently the telson is missing.

The organisation is unique and shows no evidence of any close relationship with any other group. The somewhat eurypterid-like outline may be misleading. Information on the appendages is needed for a placement in a major group.

Anomalocaris canadensis

Several isolated specimens of the appendage of *Anomalocaris canadensis* Whiteaves, 1892 have been found associated with the above mentioned arthropods at Chengjiang (Pl. 2, figs. 2, 3). The appendage tapers distally and consists of 13 segments, each bearing a pair of elongate ventral spines. Each ventral spine possesses a pair of small lateral spines diverging from the middle of each ventral spine. The first segment, attached proximally to the body, is longer than the others and bends upwards (Pl. 2, fig. 2). The 13th segment carries two terminal claws. The dorsal margin of the more distal segments shows the presence of dorsal spines.

The specimens assigned to *Anomalocaris canadensis* in the Chengjiang fauna appear to be indistinguish-

Plate 3
Figs. 1-3. *Alalcomenaeus*? *illecebrosus* Hou.
1. Counterpart, lateral aspect, x6 Field No. M2, Cat. No. 110834
2. Counterpart, lateral aspect, x10. Field No. M2, Cat. No. 110835.
3. Counterpart, parallel aspect, showing the two unfused cephalic segments and a paddle-shaped telson, x6. Field No. M2, Cat. No. 110836.

Fig. 4. *Jianfengia multisegmentalis* Hou. Counterpart, lateral aspect, x7. Field No. M2, Cat. No. 110123.

Figs. 5, 6. *Chengjiangocaris longiformis* gen. et sp. nov.
Part and counterpart, parallel aspect, x1.5. Field No. Cf5, Cat. No. 110837.

able from those from North America in morphology and structure. There is no evidence for referring the Chinese material to a new species. The occurrence in the Chengjiang fauna extends the known geographical range of this taxon from North America and its stratigraphic range downwards.

Anomalocaris has recently been considered unrelated to normal arthropods, and either regarded as representing a separate trend of arthropodization (Bergström, 1986, 1987; Dzik & Lendzion, 1988) or as a non-arthropod (Whittington & Briggs, 1985; Briggs and Whittington, 1987). The difference between the two views is semantic. The present material of *Anomalocaris canadensis* does not provide any new information useful for the discussion on its affinities.

Conclusion

The non-trilobite arthropods of the Chengjiang fauna represent a wide range of morphologies, surprisingly so with respect to the early Cambrian age. This shows that the radiation of at least some arthropod groups had advanced far already at this early date. It is notable that even with this earliest Cambrian fauna it is extremely difficult to trace the origin of the Arthropoda as a whole. In fact, the Chengjiang arthropods are not much different from the Burgess Shale arthropods from an evolutionary point of view despite their distinctly older age, and does not bring us much closer to a solution, at least not from a morphologic point of view. Part of its importance may lie in the possible implication that the initial radiation was extremely rapid and morphologically far-reaching. It is uncertain if uniramian forms are represented, if the lobopodan *Luolishania* is not counted with that category.

Formal descriptions of new taxa

JIUCUNELLA gen. nov.
Derivation of name: From Jiucun, the village northeast of the Maotian Hill (the Maotianshan).
Diagnosis: Bradoriid genus with marginal furrow and anteroventral ridge.
Remarks: *Jiucunella* gen. nov. is similar to *Zhenpingella* Li, 1975 in the outline and size of carapace and in having a ventral node, but differs from *Zhenpingella* in lacking antero-dorsal and dorsal nodes, and in having an antero-ventral ridge.
Type species: *J. paulula* sp. nov.

Jiucunella paulula sp. nov.
Pl. 1., fig. 5
Derivation of name: From Latin diminutive of *paulus*, little, referring to its small size.
Diagnosis: Bradoriid with a bivalved carapace not exceeding 2 mm in maximum length, subelliptical in outline, with a moderate retral swing. Marginal rim and furrow distinct. A ridge extends from the antero-dorsal margin near the antero-dorsal angle ventrally to form a node in the ventral area.
Holotype: Nanjing Institute of Geology and Palaeontology, Academia Sinica, Cat. no. 110823.

PERSPICARIS? *LAEVIGATA* sp.nov.
Pl. 2, figs 7, 8
Derivation of name: From Latin *laevigatus*, smooth, referring to the smooth carapace surface.
Diagnosis: *Perspicaris*-like species with subelliptical carapace, valves reaching 12 mm in maximum length, tapering and rounded anteriorly, expanding posteroventrally, without antero- and postero-dorsal processes.
Remarks: The valves, formerly doubtfully referred to *Perspicaris* (Hou, 1987c), are subelliptical in outline, 12 mm in maximum length, with straight (Pl. 2, fig. 7) to slightly upwardly curved hinge margin (Pl. 2, fig. 8; see also Hou, 1987c, Pl. 3, figs 6, 7), the latter appearing to result from compaction, and smooth surface, but without antero- and postero-dorsal processes. More material has been collected recently. This material indicates that the animal is distinct from the similar Burgess forms *P. dictynna* and *P. recondita* (cf. Briggs, 1977), and from the Utah forms *P.*? *ellipsopelta* and *P.*? *dilata* (cf. Robison and Richards, 1981) at least on the species level, having more rounded posterodorsal corners. It is therefore assigned to a new species.
Holotype: Nanjing Institute of Geology and Palaeontology, Academia Sinica, Cat. no. 110832.

CHUANDIANELLA gen. nov.
Derivation of name: From Chuan and Dian, the abbreviations for the Sichuan and Yunnan Provinces respectively, where the new form was found.
Diagnosis: Bradoriid? with bivalved carapace of large size, more than 10 mm in maximum length, subcircular in outline, hinge margin curved upward, with puncto-reticulate surface; ornament marginal rim very narrow, without marginal furrow.
Remarks: *Chuandianella* gen. nov. differs from *Mononotella* Ulrich and Bassler, 1931 in the subcircular carapace outline and in having a hinged bivalved carapace.
Type species: *Mononotella ovata* Li, 1975 (illustrated here as Pl. 2, figs 5, 6).

CHENGJIANGOCARIS gen. nov.
Derivation of name: From the type locality Chengjiang in Yunnan.
Diagnosis: Arthropod with anterior tapering tagma with notably short tergites and posterior tagma with long tergites: tergites provided with pleura.
Type species: *C. longiformis* sp. nov.

C. longiformis sp. nov.
Pl. 3, figs. 5, 6
Derivation of name: refers to the elongate body shape.
Diagnosis: Anterior tagma with 5 tergites, the whole series of which narrows into a blunt anterior angle. The successive 17 tergites of subequal length, but decrease in width back-wards.
Remarks: Posterior end of body missing in the single individual known.
Holotype: Nanjing Institute of Geology and Palaeontology, Academia Sinica, Cat. no. 110837.

Acknowledgements

This study is part of the studies of the Chengjiang fauna supported by the Academia Sinica Foundation grant no. 862004. Cooperation was facilitated by a travel grant to JB from the Swedish Natural Science Foundation. The photographs were produced by Mao Ji-

Liang and Deng Dong-xing, both of whom are gratefully acknowledged by the authors. We are also grateful to the University of Camerino and the other Agencies who sponsored the Symposium for their hospitality and for meeting travel expenses.

REFERENCES

BERGSTRÖM, J. 1986. *Opabinia* and *Anomalocaris*, unique Cambrian 'arthropods'. *Lethaia* **19**, 241-246.

BERGSTRÖM, J. 1987. The Cambrian *Opabinia* and *Anomalocaris*. *Lethaia*, **20**, 187-188.

BRIGGS, D.E.G. 1977. Bivalved arthropods from the Cambrian Burgess Shale of British Columbia. *Palaeontology* **20**, 67-72.

BRIGGS, D.E.G. & WHITTINGTON, H.B. 1987. The affinities of the Cambrian animals *Anomalocaris* and *Opabinia*. *Lethaia* **20**, 185-186.

CONWAY MORRIS, S., PEEL, J., HIGGINS, A.K., SOPER, N.J. & DAVIS, N.C. 1987. A Burgess Shale-like fauna from the Lower Cambrian of North Greenland. *Nature* **326**, 181-188.

DZIK, J. & KRUMBIEGEL, G. 1989. The oldest 'onychophoran' *Xenusion*: a link connecting phyla? *Lethaia* **22**, 169-181.

DZIK, J. & LENDZION, K. 1988. The oldest arthropods of the East European Platform. *Lethaia* **21**, 29-38.

GLAESSNER, M.F. 1979. Lower Cambrian Crustacea and annelid worms from Kangaroo Island, South Australia. *Alcheringa* **3**, 21-31.

HOU XIAN-GUANG 1987a. Two new arthropods from Lower Cambrian, Chengjiang, eastern Yunnan. *Acta Palaeontologica Sinica* **26 (3)**, 236-256. [in Chinese, with English abstract].

HOU XIAN-GUANG 1987b. Three new large arthropods from Lower Cambrian, Chengjiang, eastern Yunnan. *Acta Palaeontologica Sinica* **26 (3)**, 272-285 [in Chinese, with English abstract].

HOU XIAN-GUANG 1987c. Early Cambrian large bivalved arthropods form Chengjiang, eastern Yunnan. *Acta Palaeontologica Sinica* **26 (3)**, 286-298 [in Chinese, with English abstract].

HOU XIAN-GUANG & CHEN JUN-YUAN 1989. Early Cambrian arthropod- annelid intermediate sea animal, *Luolishania* gen. nov. from Chengjiang, Yunnan. *Acta Palaeontologica Sinica* **28 (2)**, 221-227 [in Chinese, with English abstract].

HOU XIAN-GUANG, CHEN JUN-YUAN & LU HAO-ZHI 1989. Early Cambrian new arthropods from Chengjiang, Yunnan. *Acta Palaeontologica Sinica* **28 (1)**, 42-57 [in Chinese, with English abstract].

HOU XIAN-GUANG & SUN WEI-GUO 1988. Discovery of Chengjiang fauna at Meichucun, Jinning, Yunnan. *Acta Palaeontologica Sinica* **27 (1)**, 1-12 [in Chinese, with English abstract].

HUO SHI-CHENG 1965. Additional notes on Lower Cambrian Archaeostraca from Shensi and Yunnan. *Acta Palaeontologica Sinica* **13**, 291-307.

HUO SHI-CHENG & SHU DEGAN 1985. *Cambrian Bradoriida of South China*. Xian: Publishing House of Northwest University, 251 pp.

HUO SHI-CHENG, SHU DEGAN, ZHANG XI-GUANG, CUI ZHI-LIN & TONG HAO-WEN 1983. Notes on Cambrian bradoriids from Shaanxi, Yunnan, Sichuan Guizhou, Hubei and Guangdong. *Journal of Northwest University* **13**, 56-75.

LI YU-WEN (LEE YUWEN) 1975. Cambrian ostracodes and their new knowledge from Sichuan, Yunnan and Shaanxi. *Stratigraphy and Palaeontology Papers* **(2)**, 37-72. Beijing: Geological Publishing House.

LU YAN-HAO 1940. On the ontogeny and phylogeny of *Redlichia intermedia* Lu (sp. nov.). *Bulletin of the Geological Society of China* **20**, 333-342.

LU YAN-HAO 1941. Lower Cambrian stratigraphy and trilobite fauna of Kunming, Yunnan. *Bulletin of the Geological Society of China* **21**, 71-90.

LUO HUI-LIN, JIANG ZHI-WEN, SONG XUE-LIANG, OU-YANG LIN, XING YU-SHENG, LIU GUI-ZHI, ZHANG SHI-SHAN & TAO YON-GHE 1984. *Sinian-Cambrian boundary stratotype section at Meishucun, Jinning, Yunnan, China*. Yunnan: People's Publishing House, 154 pp.

MANSUY, H. 1912. Etude geologique du Yunnan Oriental. *Mémoires du Service géologique de l'Indo-Chine* **1 (2)**, 1-146.

RESSER, C.E. 1929. New Lower and Middle Cambrian Crustacea. *Proceedings of the U.S.National Museum* **76 (9)**, 1-18.

ROBISON, R.A. 1985. Affinities of *Aysheaia* (Onychophora), with description of a new Cambrian species. *Journal of Paleontology* **59**, 226-235.

ROBISON, R.A. & RICHARDS, B.C. 1981. Larger bivalve arthropods from the Middle Cambrian of Utah. *Paleontological Contributions of the University of Kansas* **106**, 1-19.

SIMONETTA, A. 1970. Studies on non-trilobite arthropods of the Burgess Shale (Middle Cambrian) *Palaeontographia Italica* **66 (n.s. 36)**, 35-45.

SIMONETTA, A. & DELLE CAVE, L. 1975. The Cambrian non trilobite arthropods from the Burgess Shale of British Columbia. A study of their comparative morphology, taxonomy and evolutionary significance. *Palaeontographia Italica* **69 (n.s. 39)**, 1-37.

ULRICH, E.O. & BASSLER, R.S. 1931. Cambrian bivalved Crustacea of the Order Conchostraca. *Proceedings of the U.S. National Museum* **78 (4)**, 1-130.

WALCOTT, C.D. 1911. Cambrian Geology and paleontology. II. Middle Cambrian annelids. *Smithsonian Miscellaneous Collections* **57 (5)**, 109-144.

WALCOTT, C.D. 1912. Cambrian geology and paleontology. II. Middle Cambrian Branchiopoda, Malacostraca, Trilobita and Merostomata. *Smithsonian Miscellaneous Collections* **57 (6)**, 145-228.

WALCOTT, C.D. 1918. Geological explorations in the Canadian Rockies. *Smithsonian Miscellaneous Collections* **68 (12)**, 4-20.

WALCOTT, C.D. (ed C.E.Resser) 1931. Addenda to descriptions of Burgess Shale fossils. *Smithsonian Miscellaneous Collections* **85**, 1-46.

WHITEAVES, J.F. 1892. Description of a new genus and species of phyllocarid crustacean from Middle Cambrian of Mount Stephen, British Columbia. *Canadian Record of Science* **5**, 205-208.

WHITTINGTON, H.B. & BRIGGS, D.E.G. 1985. The largest Cambrian animal, *Anomalocaris*, Burgess Shale, British Columbia. *Philosophical Transactions of the Royal Society of London* **B 309**, 569-609.

YIN GONZENG 1978. Arthropoda: In *Palaeontological Atlas of Southwest China*, Volume of Guizhou 1 (ed. Work Team of Stratigraphy and Palaeontology of Guizhou), pp. 383-385. Beijing: Geological Publishing House.

ZHANG WEN-TANG & HOU XIAN-GUANG 1978. Preliminary notes on the occurrence of the unusual trilobite *Naraoia* in Asia. *Acta Palaeontologica Sinica* **24 (6)**, 591-595 [in Chinese, with English abstract].

Early Palaeozoic Arthropods and problems of arthropod phylogeny; with some notes on taxa of doubtful affinities

L. Delle Cave[1] & A.M. Simonetta[2]

Abstract

The present paper, after considering the general morphological and evolutionary principles that we shall take as fundamental for the interpretation of the fossil, basically lower Palaeozoic, Arthropods, discusses seriatim the morphology and possible evolutionary significance of these animals and suggests some plausible phylogenetic arrangements of them. Section IV of this paper considers a group of segmented animals which clearly are neither Arthropods nor Annelids and discusses their relationships with both these major phyla. It is argued that, just as for Arthropods, also at this higher level the traditional classifications are inadequate to convey the complexities of the Palaeozoic radiations.

INTRODUCTION

The affinities of the Arthropods, both among themselves and with other phyla of segmented invertebrates, have been the subject of long and extensive debate. Several authors, following Tiegs & Manton (1958, see also Manton 1977,1979) have considered the Arthropoda as a composite, paraphyletic group, which should be split into Phyla Crustacea, Chelicerata and Uniramia, the last comprising the Insecta, Myriapoda and Onychophora. About ten years ago, after we had pointed to the links existing between the Cambrian *Aysheaia*, the Onychophora and the Tardigrada (Simonetta,1975; Simonetta & Delle Cave, 1975, Delle Cave & Simonetta, 1975, for a dissenting opinion see Robison, 1985), Manton (1979) and Anderson (1979) have added, with some qualifications, the Tardigrada to the Uniramia.

Various contributions have focused attention on the possible significance of several lower and middle Palaeozoic taxa for the understanding of the early arthropod evolution (for bibliography see Schram, 1978; Simonetta & Delle Cave, 1981, 1982; McKenzie, 1983; Briggs, 1983, 1985; Briggs & Whittington, 1985; Robison, 1985). Moreover, the recent discovery of a number of extremely puzzling fossils has considerably complicated the issue.

This contribution aims at a brief and comprehensive assessment of this evidence and its possible significance. The first two sections of this paper deal with critical issues in the interpretation of morphological evidence. This must be quite clear when considering unquestionable Arthropods, but it is equally significant when dealing with several other puzzling animals which position should, at least tentatively, be assessed within the framework of the "Articulata". The third section is a brief systematic discussion of Palaeozoic Arthropod taxa, with occasional references to younger taxa.

Section IV discusses other puzzling articulate groups, and their relationships with subsequent higher taxa. Finally some general conclusions are addressed in Section V.

To deal succintly with the evolution of Palaeozoic Arthropods is no easy task when we consider their extraordinary variety over this long period and the difficulties in interpreting specimens which are usually incomplete or otherwise poorly preserved.

SECTION I

Preliminary principles

General problems of morphology and taxinomy

The problem of the general significance of Palaeozoic Arthropods has been discussed by various authors, not only in the light of different principles, but sometimes even without a clear statement as to the general assumptions made and to the taxinomic and evolutionary principles followed. The attitude of the majority of scholars has been to try to fit the various fossils into traditional taxinomic categories, often by broadening their usual definition as much as needed to accommodate the new animals so described. This is, I think, an objectionable approach if we seek to build a phylogenetic taxinomy.

Indeed any phylogenetic discussion starting from a taxinomic definition of categories is intrinsically invalid as it logically begs the issue. We must first establish, if possible, the phyletic relationships between the various animals known and then see whether previously established taxa correspond with identified branches of our phyletic tree. If they do not we establish new taxinomic categories and, if necessary, abandon previously accepted ones.

As a preliminary to the arguments which follow we contend that the principles of Hennigian cladism have been falsified and can be largely discounted (Scott-Ram, 1990; Simonetta, 1983, 1987, 1988 and in press). We do not consider, therefore, that there is any reason to assume *a priori* that the ancestor of taxon "A" will be such as to be included into a higher category comprehensive of both "A" and its ancestor. We hold that taxa, being concepts, do not have a material existence (whether they can be treated as individuals is controversial; we think that they must be considered as "logical operators", but that is not directly relevant here). Populations are the embodiment of genetic pools. This statement applies at least to groups such as the Arthropods which are usually gonocoric, evolve by more or less gradual steps (cf. Simonetta, 1987, in press). Accordingly any taxon "A" must derive from a "pre-A" taxon of the same rank.

In fact morphological features are very often coordinated and evolve together, so that they cannot be

[1]Dipartimento di Scienze della Terra, Università di Firenze, Via La Pira 4, 50121 Firenze, Italy.
[2]Dipartimento di Biologia Molecolare, Cellulare e Animale, Università di Camerino, Via Camerini 2, 62100 Camerino (MC), Italy.

graduated in a series of successive and separate evolutionary events, and even if we pick one feature out of a series so coordinated as "eponymous" of a group, the same basic opportunities and similar selective pressures will often allow for the invention of very similar structures. This is especially true during the early phases of evolution, when niche differentiation must necessarily have been less developed than nowadays (cf. Valentine & Erwin, 1987; Simonetta, 1987, 1988a and in press). From consideration of the known mechanisms of evolution we hold that, while a nested taxinomy is a technical necessity, a nested phylogeny is a meaningless expression.

Modern mathematical techniques of analysis do not give very significant results where fossils are concerned, and especially when animals of different ages are compared. They are usually based on binary logic (for instance, the various methods of multivariate analysis) and there are basic theoretical limitations to the application of binary logic to problems where time is a factor (Simonetta, 1987, 1988a). A combined approach using modal logics, trend analysis, "Fuzzy sets" and "game theory" seems most promising, but all these are still in their infancy in this context. It follows from the previous considerations that there are no hard and fast rules on which to rely, but this does not mean that there are no general principles to follow. We may add that for any discussion on morphological evolution, it is irrelevant whether any change occurred by a long and gradual process or whether it was attained in a few generations. The current argument on "punctuated equilibria", therefore, is irrelevant for our purposes.

A correct procedure requires that we first place each known animal in a plausible morphologic relationship to the other known animals; it is quite possible that the available evidence may be consistent with more than one placement. If then, on evidence other than purely morphologic, we may rule out some alternatives, so much the better. Nevertheless, in spite of claims by biochemists and geneticists, their evidence, though it may be significant, must usually be considered as auxilary to morphology (Simonetta, 1987, 1988).

Once we are satisfied as to the (approximate) position of the various animals in a phylogenetic reconstruction, we may superimpose on our phyletic tree a formal taxinomy. It is arbitrary whether we use a vertical or a horizontal classifications, or whether we amplify or restrict the definition of any taxon. The only trouble is that to find a formal classification suitable for an approximate description of phyletic relationships appears to be increasingly difficult, so that most taxinomists conveniently ignore whatever groups fail to fit into their scheme. For example, the taxinomic arrangement proposed by Schram (1978) omits the Cycloidea, the Euthycarcinida, *Sidneyia*, etc.

So when we try to use fossil evidence for the assessment of the origin of the living Arthropod taxa it must be granted that:

a) the rarity of fossil Arthropods in comparison with their presumed numbers at any time, and b) the intervals of time and space separating the available samples make it extremely unlikely that we ever have in our sample not even an actual ancestor of a later taxon[1], let alone a close relative. On the other hand, if we consider broader categories, such as stating that Family or Order X can be derived from Genus or Family Y, much depends on our broader or narrower definition of the taxa concerned. For instance, if our definition of the "Phyllocarida" includes the existence of swimmerets on all the abdominal segments, then genera such as *Canadaspis*, *Waptia* etc. are necessarily excluded from the Phyllocarida.

The criteria of assessment of morphological affinities

Let us now consider some general principles for the assessment of the morphologic and phylogenetic affinities of early Palaeozoic Arthropods.

All changes in the structure of animals can be studied by methods of *analysis situs*, in other words as topological transformations. The size and shape of any structure may change by relative increase or decrease of its constituent parts. This may, in turn, bring about a shift in the relative positions of yet other parts.

Long discussions may be found across a wide spectrum of papers on the distinction between homologous, homeomorphic and analogous structures. Analysis of the use of the term "homology" by the different classical authors bears out that two non-exclusive concepts were the basis for the usage of this term.

Even before evolution was generally accepted, authors such as Owen, Goethe, Cuvier etc, whether they used precisely this term or not, conceived of homologous or truly comparable structures when they had reason to believe that they represented specific variations of a general archetype. This latter concept is "a basically identical structural plan that express itself by adaptive modifications in different species[1]".

Evolutionary biology superimposed on this concept the idea that the specific differences observable between homologous structures had been acquired through adaptive evolution from a common ancestor, whose structures were such as to be liable to the observed modifications. Basically the two concepts were synthesized by Gegenbauer and his school using the "Bauplan" concept of Cuvierian-Goethian origin. The concept may thus be formalized: "Two or more structures are considered to be homologous when they may be shown to be topological homeomorphs, that is when they may be derived by a direct topological transformation of one into the other. Alternatively they may be all derived by topological transformations of a single prototype by following diverging patterns of change, and there is, at the same time, sufficient evidence to show that the topological transformations implied represent the lineages of descent through which the genetic pool determining the structures considered was actually inherited during its progressive modifications".

It appears that in this field there are two sources of misunderstanding:

a) is a semantic point. Several authors use the term "homeomorph" incorrectly, equating it with "having a similiar structure and function". "Homeomorph" has a strict geometrical significance, and should not be used in the later sense. Such an instance may be found in McKenzie (1983), where he mantains that the mandibles of the Crustacea and of the "Uniramia" are homeomorphs, but not homologous. It is in fact exactly the other way round, as we shall see below.

b) two structures may be homologous but not homeomorphic or, more rarely homeomorphic, but not homologous. Moreover, two structures may be both analogous and homologous, depending on the point of view from which we consider them.

Let us clear these points by some examples:

1) the ectoderm, mesoderm and endoderm in the various classes of Vertebrates are certainly homologous, and that is clearly shown by their subsequent development and by hystochemical evidence; but their early development is so different in the various classes and sometimes even between orders of the same class, as it happens in the Mammalia, that it is extremely difficult to envisage them as topologically homeomorphs[(1)] at these early stages.

2) The mandibles of Crustaceans and Insects are homologues: both are the second postoral appendage and each can be independently derived from a serially corresponding pair of appendages structurally similar to the limbs of Trilobites. They are not, however, homeomorphs, because as Manton has shown, in the Insects and at least some Myriapods the mandibles correspond to the whole of the Trilobite limb[(2)] (apart from the vanished outer branch)[(3)], while the Crustacean mandible corresponds only to the coxa of the trilobitic limb.

3) The pectoral fins of fish (*sensu latissimo*), the paddles of Icthyosaurs, Plesiosaurs, Cetaceans, the flippers of Penguins, are homologous in that they represent the thoracic paired appendages of all Vertebrates (except Agnatha, and possibly of Acanthodii and Placoderms), but their various external or internal similarities are almost all due to convergent adaptations and henceforth all these structures are analogous, and only some of them may be considered as homeomorphs.

4) the secondary palates of Crocodiles and of Mammals are strictly homeomorphic and also their embryological development (as seen in the Eusuchia) is strictly comparable, but nevertheless they have a different morphologic significance, i.e. they are homeomorphic, but not homologous, as they have evolved independently as a response to different functional adaptations.

It appears therefore that, while the general concept of homology is clear, in practice the assessment of homologies cannot depend on any single piece of evidence, but rather on a balanced assessment of all the evidence available: morphological, embryological and palaeontological evidence are all significant, but all may be misleading. Moreover, homology may be more or less close: the homology of the wings of the various species of Birds is extremely close, while that between the wings of Birds, Bats and Pterosaurs is not closer than that with any other pectoral limb of any other tetrapod, but it is closer than that with the pectoral fins of fish.

SECTION II

MORPHOLOGICAL PRINCIPLES

General morphology

The morphology of the cephalon, postcephalic segments and the development of cephalisation

We must first stress that the frequent usage, especially by palaeontologists, of the term "somite" when they actually mean "body segment" is not only incorrect, but may lead to considerable misunderstanding.

Somites are the embryonic segments, or rather the mesodermic blocks, which give rise to most of the muscular and skeletal tissues with a connective function. That embryonic somites coincide with the adult segmentation is a frequent occurrence in invertebrates, but it is far from being a general rule, as somites often either merge or split, so that the adult or larval segments may well be made of material belonging to different somites.

The only cephalic structure which is actually capable of giving rise to the various morphologies that occur in the cranial region of the stock comprising all later Arthropods is a "cephalon": composed of the epistomium, the ocular and antennular segments, this is exemplified by *Sidneyia* and typical nauplii. Genae were probably primitively present, but, I guess, were less developed than in Trilobita.

This cephalic structure must have been followed caudally by a set of somewhat trilobite-like segments; these must have had the following differences to those of typical Trilobites: i) non calcified terga, ii) non trilobate appearance, as the longitudinal furrows separating the rachis from the pleural lobes were certainly lacking, iii) uniformly low vaulted transverse section, iv) the last opisthosomal segments were not fused to form a "pygidium" (this last feature is not diagnostic, as some primitive Trilobites have free pygidial segments).

Such a stucture is close enough to that proposed by Hessler & Newmann (1975) for their Ur-Crustacean (Fig. 1). The reconstruction by Hessler & Newmann, however, allows for some post-antennular segments to be already incorporated into the cephalon and for a long series of post-cephalic segments.

We have repeatedly mantained that the number of free segments in the most primitive Arthropods was probably the minimum necessary for efficient metachronal locomotion and for filter-feeding, that is between 6 and 10 segments (Fig. 1).

Cephalization and the development of the carapace

According to our scheme, therefore, the most primitive stage in the evolution of the cephalic region is found in *Sidneyia, Fuxianhuia*, and *Chengjiangocaris*. The morphology of these genera will be discussed below (p. 205, Fig. 8). In this context, however, it is significant that Bruton's (1981) description of *Sidneyia*, while showing that the big raptorial appendages, that I and others had credited to *Sidneyia* do not in fact belong to this genus (they belong to *Anomalocaris*, cpr. p. 235), has confirmed that the cephalon had no oral appendages. It consisted, therefore, of only a large labrum(=epistomium) and the ocular and antennal (=antennular) segments. The pleural lobes of the first two postcephalic segments are slightly different from those of the truly thoracic ones, and Bruton (1981) considers that the first pairs of legs had no exopodite. We think, when considering other rather primitive Arthropods, that the transition between appendages with a reduced exopodite and the more caudal ones, that are provided with a fully developed respiratory exopodite was probably gradual.

Most Cambrian genera in which appendages are known have three pairs of postoral appendages in the cephalon. They are, however, genera with only one postoral cephalic appendage (e.g. *Canadaspis, Marrella*), two (e.g. *Burgessia, Leanchoilia*), five (*Emeraldella*). Moreover, not only there are no obvious phyletic affinities between genera having the same number of cephalized segments, but in addition while the postor-

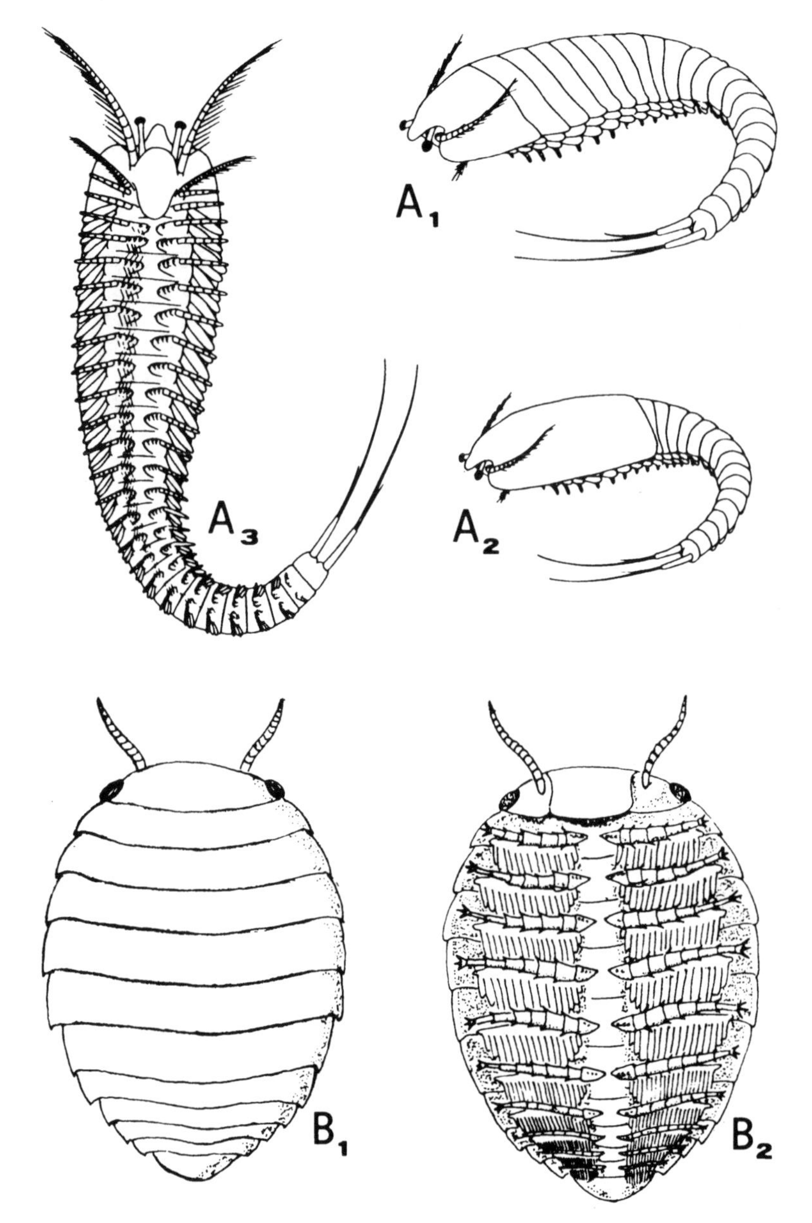

Fig. 1. A1-3: Newmann's idea of the possible Ur-crustacean. B1-2: our idea of the most primitive true Arthropod.

al appendages are, sometimes, very specialized (e.g. *Leanchoilia*), in other instances they are indistinguishable from the postcephalic appendages.

Thus cephalization of the legs must have been independent process evolved in many lineages, and also implies that even when fusion occurred it was not a constant process. The systematic account of the various taxa in section III will show that during the Palaeozoic cephalisation followed a number of different pathways. An important discussion of the problems of cephalisation in the Arthropods is provided by Lauterbach (1973).

In instances where cephalization involved the simple fusion (or rather non-separation) of segments, such as in the Trilobita, Xiphosura etc., including some Crustacean orders, we must consider two morphological alternatives. There are instances, such as the Trilobita, in which development is anamorphic, through a series of larval stages. Other, such as in the Isopoda, Amphipoda etc., show direct development, and it may be that an evolutionary phase of their past with a more or less *Mysis*-like carapace, has vanished. We see difficulties to admit it, but a thorough discussion of the Peracarida is beyond the scope of this paper.

The development and evolution of carapaces offer (Fig. 2) more complex problems. There are two basic types of carapace: a postmandibular carapace, as exemplified in the Cladocera, and the more usual mandibular carapace. An exhaustive discussion of the development and morphological significance of carapaces is to be found in Lauterbach (1974).

The postmandibular carapace arises as paired expansions of the terga of one or more fused postcephalic segments, leaving the head entirely free (Fig. 2).

The mandibular carapace is much more common. Here, a tergal fold grows backwards, outwards (and, occasionally, forwards) from the tergum of the mandibular segment. From this basic developmental type emerge secondary possibilities: i) the carapace develops without fusing with the terga of the postmandibular segments, so that, whether single or bivalve, the carapace lies above the postmandibular, free segments. ii) the carapace gradually fuses with the subsequent segments. In this latter instance usually at each successive

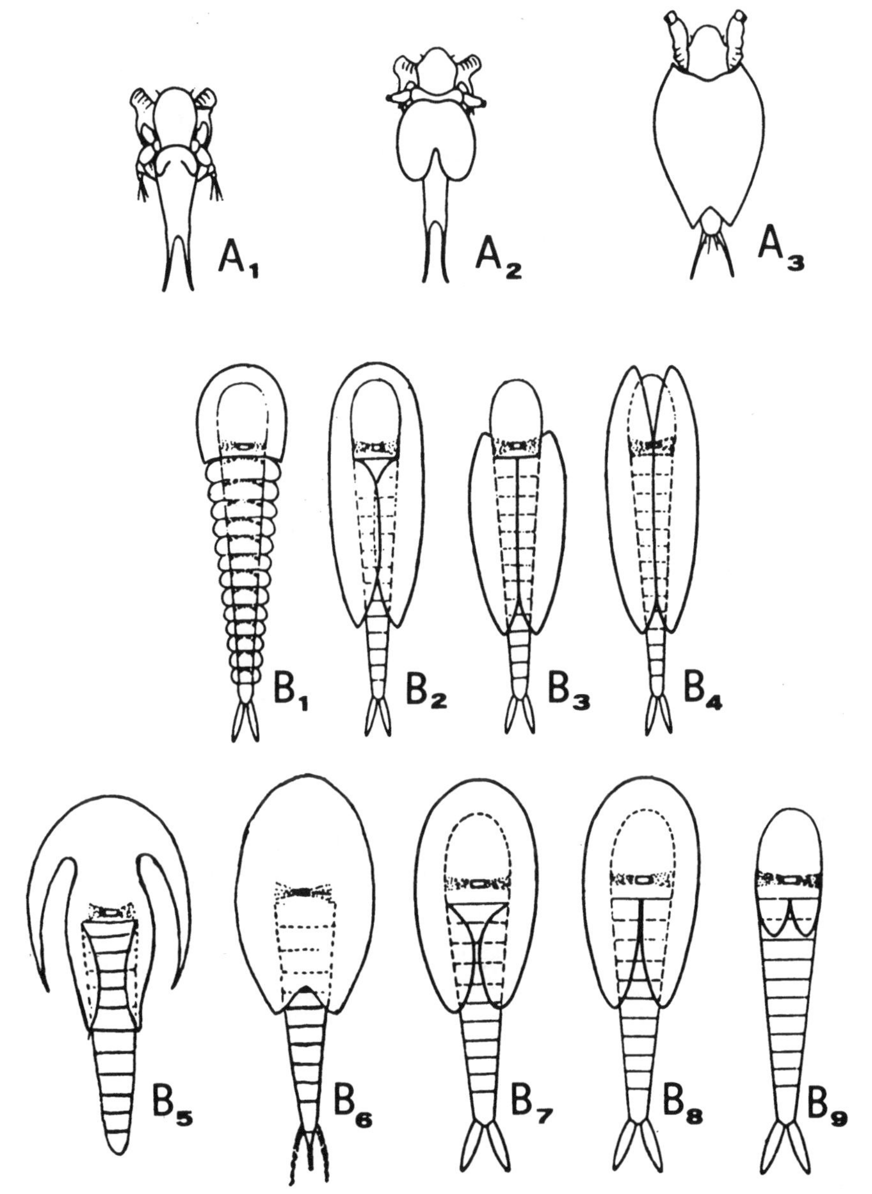

Fig. 2. A1-3: developmental stages of the crustacean carapace (after Lauterbach). B1: primitive arthropod condition without carapace, B2-9: different patterns of arthropod carapaces, position of adductor muscle in 2nd or 3rd postoral segment evidenced.

moult the dorsal cuticle of one segment and the corresponding ventral cuticle of the carapace retract, so that the insides of the body and of the carapace merge and thus the morphological dorsal region of the thorax gradually becomes directly covered by the carapace, *i.e.* by the tergum of the mandibular segment.

These simple mechanisms for carapace development probably imply that at least mandibular carapaces may have developed independently and repeatedly. The development of a carapace necessarily implies the complete reduction of the pleural lobes of the segments so incorporated or covered, as their various functions are taken over by the carapace. The complete loss of the pleural lobes brings about rearrangements in the musculature and ligaments of the cephalized segments.

Though this implies considerable taxinomic revision, secondary redevelopment of lost pleural lobes is so unlikely that it can be discounted. It follows that animals with typically developed pleural lobes on all their postcephalic segments, *e.g.* the Syncarida, cannot have evolved from animals provided with a carapace, such as the Phyllocarida.

Telson: Although the term telson has been used very loosely by various authors, the only correct usage is with reference to postanal structures: the anus opens at the posterior limit of the last complete metamere, and all structures deriving from dorsal embryonic material, developing beyond the last pair of somites, whether movable or not *are* a *telson*. Unless we agree therefore to change the morphological definition of Telson, *pace* Whittington and many others, the spines of the Xiphosurans, *Molaria, Habelia, Emeraldella*, the fixed supra-anal plate of *Sidneyia* etc. by definition are telsons. The *furcae* of many Crustaceans, Insects and other Arthropods are a difficult problem because for most of them we do not know their precise embryological development. They may develop from segmental material homologous with the anlagen of the legs, in which case they are "legs". Alternatively they may derive from dorsal material, in which case they should be considered as "telsons". A typical situation occurs in the living Notostracans where all have a furca, but

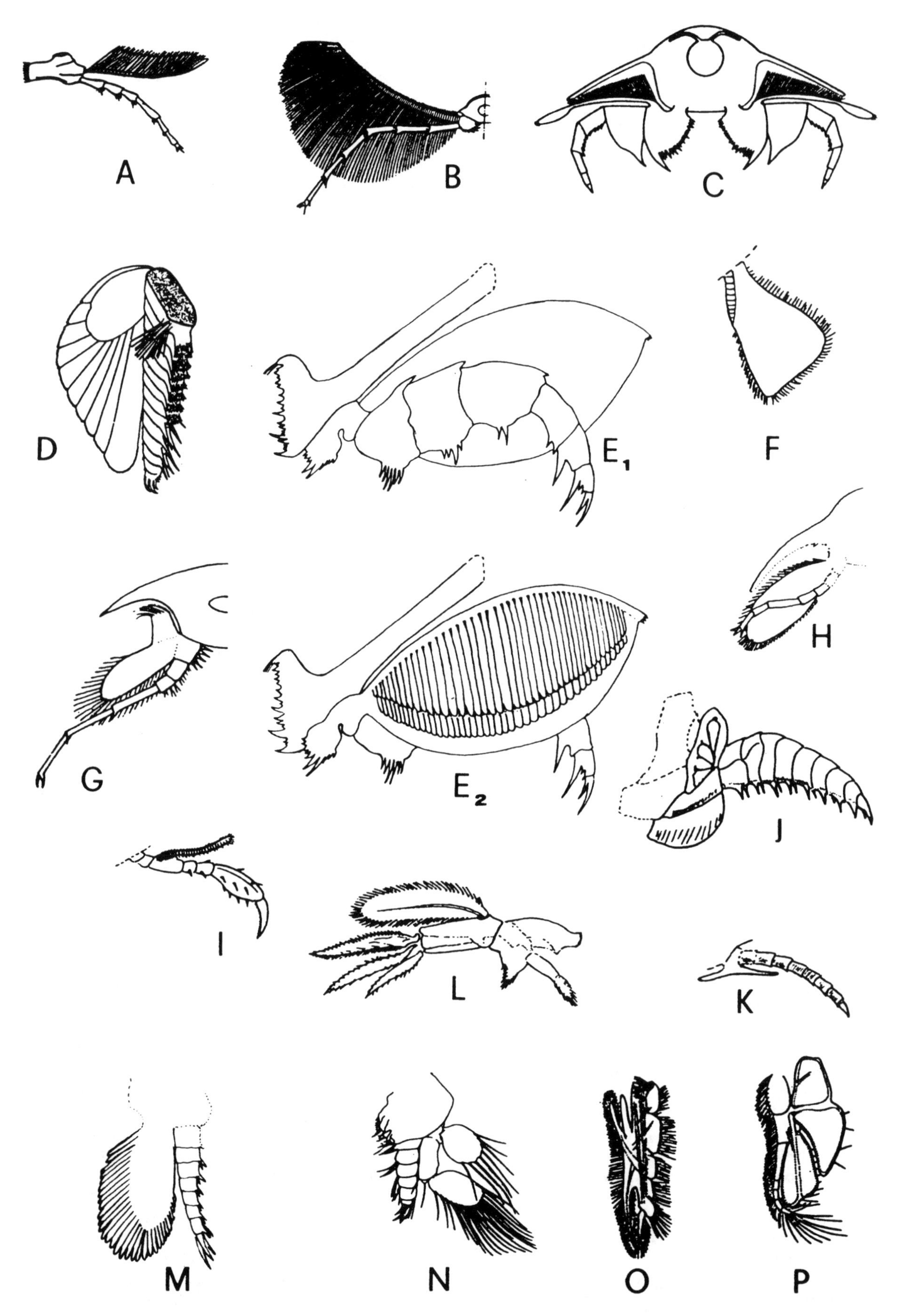

Fig. 3. Different patterns of arthropod walking legs, which may all be derived from a "trilobitic" pattern, A: the trilobite *Triarthrus*, B: *Marrella*, C: *Naraoia*, D: *Canadaspis*, E: anterior and posterior view of *Sidneyia*, F: *Protocaris* = *Branchiocaris*, G: *Emeraldella*, H: *Molaria*, I: the cyclid *Halicyne*, J: *Arthropleura*, K: *Eoarthropleura*, L: the kazacharthran *Jeanrogerium*, M: Leanchoilia, N: the living Cephalocarid *Hutchinsonella*, O: the living Conchostracan *Cyzicus*, P: the living *Nebalia*.

while in *Triops* there is no supra-anal plate, in *Lepidurus* there is a small telson plate overhanging the anus and nothing is precisely known of the embryology of the furca.

The general structure of the Arthropod limb

The fact that all the known postoral appendages of Cambrian Arthropods conform to a basically uniform pattern has led several Authors (cf. Hessler & Newman, 1975; Simonetta, 1975; Simonetta & Delle Cave, 1975, 1981; Bergström, 1979, etc.) (Fig. 3) to assume that the most primitive appendage for all Arthropods conformed with the so called "trilobitic pattern". This consisted of a basal coxa, probably provided with a weakly developed endite, from which branched an exopod with a basically respiratory and natatory function and a well segmented endopod made of subequal podomeres gently bending downwards without any specialized flexure joint. Movements of the legs may have been primitively limited to promotor-remotor movements at the body-coxa hinge and to adduction-abduction movements of the endopodite. The arthropod-like raptorial appendages of the pseudoarthropod *Anomalocaris* show the kinds of movements possible at an early stage of the evolution of articulated appendages.

The problem of the evolution of the mouthparts of the Arthropods must now be briefly reconsidered.

The papers by Tiegs & Manton (1958) and Manton (cfr. 1977, 1979) alone have provided enough evidence to assume, at least as a working hypothesis, that the jaws of Insects and Myriapods, taken as a group, and those of the Chelicerata and of the Crustacea cannot have the same evolutionary history. There is no doubt that their work has effectively disposed of the old concept opposing the "Antennata" (or "Mandibulata") to the "Chelicerata". However Tiegs and Manton's concepts are open to criticism on the following points:

i) Their entire argument about the "whole leg jaw" of the Uniramians largely depends on evidence from the Onychophora. By now there is little doubt that the Onychophorans are only distantly related to the Arthropods, so the evidence provided by the Onychophora is irrelevant to the problem.

ii) They never discuss the possibility that all the various types of arthropod mouthparts may be derived from appendages of a trilobitic type.

iii) They assume that the Crustacea are a natural group, which is doubtful, and therefore their generalization of evidence obtained from only a few groups requires verification.

The following is offered as a working hypothesis. There is no evidence against it, but it certainly requires appropriate investigations.

In all primitive arthropod groups the mouth is primitively rotated ventrally and faces either downwards or, more often, backwards. During the development of the trochophore of Annelids the mouth is primitively at right angles to the longitudinal axis of the trochophore and also of later stages. It is only when the larva settles on the sea bottom that the mouth of Annelids is rotated forwards.

The feeding mechanisms of the nauplii have never been properly investigated, but adults show one of several alternatives:

i) they can evolve as filter-feeders, and there is little doubt that this developed in a basically benthic phase of evolution, with animals crawling on a very loose substratum. Situations such as those of living Anostracans appear to be specialized. During filter-feeding the particles are moved primarily cranially along a ventral food-groove towards a backward-facing mouth. The results of the studies on living filter-feeders should be applied to those Trilobita with known mouthparts.

At the same time a careful study should be made of the various patterns of cephalic apodemes in advanced Trilobita as a possible judge of the degree of specialization of their cephalic appendages. It must be remembered that, while some Trilobita had cephalic appendages hardly different from the thoracic ones, some (e.g. *Phacops*) had cephalic appendages with very large gnathobases. Pioneer work along these lines has been done by Bergström (1969).

ii) They can feed on large sessile or slow-moving organisms. In such cases the animal can crawl on the prey, be it plant or animal, and grasp it with several legs. Thence it may either lower its body until the cephalic gnathobases contact and chew the food, or it may tear and pick up the food using the tips of its legs.

In either case the cephalic appendages must: a) develop specialized endites, and b) the grasping ability of the postcoxal section of the leg (endopodite) will be enhanced by any development of endites, spines, hairs etc., and by an increase in its flexibility. The development of major flexure joints will be an additional advantage and all these developments will probably have been attained independently in different groups.

The main joints of living Arthropods are morphologically and functionally different in different taxa, while both in the Cambrian and somewhat later times most species, even when clearly advanced towards the morphology of later Arthropods (e.g. *Odaraia, Emeraldella, Paleoisopus*, etc) had not yet developed any main joint. Nevertheless legs with well defined main flexures occurred in early animals, such as *Sidneyia* or *Cheloniellon*, which were for some other features, such as cephalization, exceedingly primitive.

This sort of piece-meal evolution is a feature common at certain stages of evolution in several groups (cf. for instance the Triassic evolution of the Therapsida). I suspect (Simonetta, in press) that in the story of each group there is at least one phase during which the group explores the possibilities of an open, poorly differentiated, and very wide ecological niche. In these circumstances advanced characters are acquired repeatedly and more or less "at random", and it is only gradually that they get into neat "packages" of coordinated structures that clearly identify the typical members of each evolutionary line.

Coming back to the possible patterns of evolution of the oral appendages, it is clear that animals feeding on large items of food have no need for a food-groove, while they had a clear advantage if a certain number of pairs of legs were either crowded around the mouth or close to it.

iii) If the animals were predators on comparatively fast moving prey, they would need some sort of grasping or stabbing appendages directed frontwards. Rotation and displacement of the ocular and antennal segments in front of the mouth must have been one of the earliest adaptations of the Arthropoda, both as an answer to the need to explore the area in front of the animal and to solve the problems posed by the backward ex-

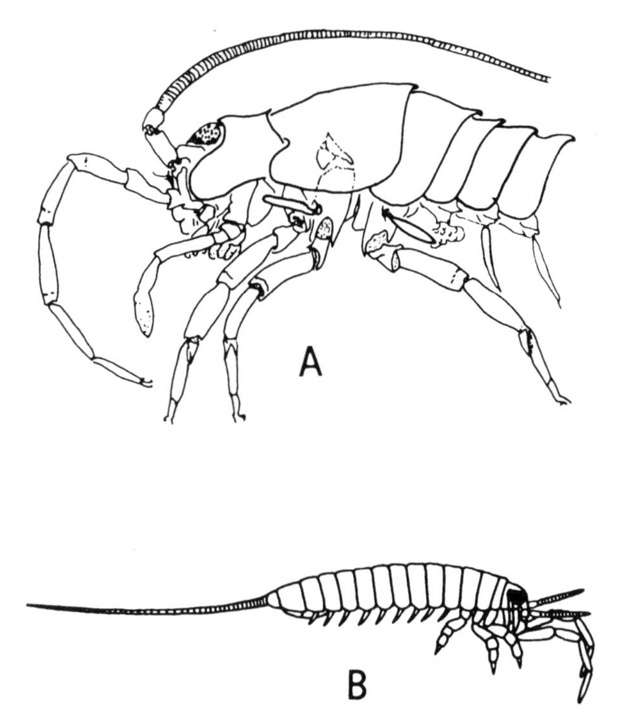

Fig. 4. The living Thysanuran *Petrobius* showing rudimentary outer branch on thoracic legs 2 and 3, primitive maxilla and labium and rudimentary abdominal appendages. The Lower Carboniferous Monuran *Dasyleptus*.

tension of the labrum (= epistomium). At this stage either the first antennae or the first pair of postoral appendages of predators had to be modified into a pair of grasping organs. If the first postoral appendages were involved, they had to be rotated to extend in front of the cephalon. Their rotation and their migration in front of the mouth (actually a complex movement involving the backwards flexing and migration of the "Anlage" of the labrum) is usual in ontogeny and may or may not reflect phylogeny. We surmise that, once the the appendages are moved to the front of the mouth or rotated to a paraxial position, they are bound either to lose their outer branch or, if retained to adopt a specialized configuration. Moreover, it must be considered that, as soon as the cephalic appendages begin to specialize for grasping and handling large food, they cannot retain the metachronal movements that are essential for the functioning of the trilobite type of leg. Reduction or specialization of the outer branch of the legs follows as a necessary consequence of the cephalization of postoral segments and of their appendages.

Can we further consider the possibility that the appendages of all moderns Arthropods have evolved from trilobitic appendages? As far as the "uniramian" appendages are concerned we must consider:

i) There is little doubt (cf. Manton, 1977) that the peculiar features of the uniramian limbs of the Pauropoda, Diplopoda, Symphyla and Chilopoda evolved independently from a very primitive structural situation, and that the same holds for the group formed by the Collembola, Diplura and Protura. The Thysanura and Pterygota share several basic common features, but must have evolved independently since their forerunners moved into a terrestrial habitat. In addition some Thysanurans (e.g. *Petrobius*) have clearly biramous appendages (Fig. 4), a point we shall return to below.

ii) the evolution of chelate appendages presents several problems. Apparently it requires only the reduction of the terminal podomere of the leg, changing it into a movable spine. Usually but not always it has a denticulate or cutting edge opposable to a similar marginal spine of the penultimate article of the leg, this marginal spine being developed as the fixed "finger" on which the movable one works.

Things are not as simple as that and it is well known that the classical 2-3-4 segmented chelicera is the result of the fusions of different podomeres (probably 5 in Scorpions) and it is not dependent on the loss of the distal ones. It follows that any subchelate first pair of postoral appendages may have evolved independently into true chelicerae and that, therefore, chelate mouthparts may have evolved independently in various lineages.

The number of chelate appendages is variable among the "Chelicerata": from one, the chelicerae, to five (in the living Xiphosura), while their number is not known in most fossil species.

It follows from these considerations that the Pantopoda (= Pycnogonida) may be an entirely isolated group, and that the relationships between the Xiphosura and Scorpiones to the remaining "Arachnida" may be very distant (cf. p. 211).

iii) In Manton's papers there is almost no discussion of the possible relationships of the "uniramian" appendages with the appendages of fossil groups.

iv) Contrary to Tiegs and Manton's tenets not all Uniramians have uniramian limbs: two pairs of thoracic limbs of some Thysanurans (e.g. *Petrobius*) have a rudimentary exopodite (Fig. 4) which closely resembles the "flabellum" of the last pair of walking legs of the Xiphosura (curiously Manton repeatedly figured these appendages without comment).

v) the legs of the Arthropleurida are very close to those of the Trilobita, apart from the loss of the exopodite and the complicated sclerotized attachment to the pleural region, which is presumably an adaptation for providing leverage in a terrestrial habitat.

Let us now turn to a consideration of the possible evolution of the "crustacean" appendages from a "trilobitic" appendage, the latter assumed to be primitive. The crucial problem is that the exopodite of the Crustacea typically arises from the second podomere, while the outer branch of the trilobitic appendage arised from the first podomere. Can the two be equated? Duplication of podomeres is possible, as shown in Diplopoda, Amblypygi etc. The postcephalic appendages of *Emeraldella* provide, on the other hand, an example of an appendage (Fig. 3) with an exopod partly divided into an upper and lower branch, much like that of the Phyllopods and Phyllocarids (Fig. 3). Anyway *Emeraldella* offers such a mixture of features that it will deserve a separate discussion (p. 207) At this point we can only stress the possibility that it provides an example of a morphological stage in the evolution of a crustacean type of biramous appendage from a trilobitic one.

The fundamental tenet by Manton that all stages of evolution must have been functionally efficient is certainly true in principle, but it may suffer exceptions in animals developing through metamorphosis. So, for instance, Manton comments on filter feeders among Crustaceans and argues that filter-feeding in those with backwards directed limbs as against those with forward directed ones must have evolved independently. She is most probably right. However if, for the sake of argument, we suppose that the forward directed leg derives from a backwards directed one, we may suppose either that the "non functional" stage with outwards directed limbs corresponds with non-feeding adults, as it is the case with some Crustacea, or that the switch was acquired as a sudden change and correspond with the acquisition of a metamorphosis (why this came about is another matter).

Several Arthropods have bifid and even trifid first antennae, but the branching is never from the basal podomere, and these branches never have a respiratory or natatory function: when the first antennae are used for swimming they are always unbranched.

Embryological and larval development

Until fairly recently the *nauplius* larva was known only in the Crustacea (using this term in its traditional meaning), where it is clearly primitive. Most Crustaceans, however, have a more or less abbreviated development, the *nauplius* stage tending to be lost.Fortey & Morris (1978) have described a *nauplius*-like stage in the Trilobita. Their interpretation of the material has been questioned by some Authors (see Schram, 1986). Nevertheless, we are inclined to accept Fortey & Morris *phaselus* larva as a trilobite larva, corresponding with the crustacean *nauplius*. If this is correct, it would follow that most probably a *nauplius*-like stage must have been a primitive feature in the development of all aquatic Arthropopds followed by anamorphic development. Such an assumption would allow for the interpretation of the Pantopod *protonymphon* as an early metanauplius-like stage in an animal which has lost the preoral antennae.

The loss of the preoral antennae and the consequent modifications of the protocerebron are irrelevant for our present argument, as they may well have occurred independently in the Pantopoda and in the other "chelicerate" lineages, just as it did in other groups, *e.g.* Protura (see Dallai, this volume). Anyway there is insufficient evidence to decide whether the chelicerae of the Pantopoda are really homologous with those of the other Chelicerates.

We shall spend here but a few words on the significance of embryological evidence for our problems. The only point that we shall raise here is the assumed relevance of the evidence from fate maps in Annelids and "Uniramians" for phylogenetic reconstructions.

That this evidence is irrelevant has been argued by several Authors, while other have made much of it. As a matter of fact if one wants to consider, as Schram does, Anderson's theses as significant he must be prepared to ignore the biomolecular evidence. The cuticle of the Annelids and of the Arthropods are structurally and chemically so different that Arthropods can not be derived from any Annelid stock, while the Annelids on one side and the Arthropods on the other form two clearly homogeneous biomolecular stocks, with minor groups which can be attached to either of them (cpr. p. 239 for a more detailed discussion)[6].

SECTION III

THE ARTHROPODA: SYSTEMATIC ACCOUNT

Trilobita

For our discussion of Trilobites the following points may be considered as relevant:

i) Apart from the probable *nauplius*-stage, whose exoskeleton was probably unmineralized, all subsequent ontogenetic stages have a chitino-calcitic exoskeleton of the terga and epistomium, while the remaining ventral structures have a purely organic cuticle.

ii) The terga have always a clear differentiation between rachis and pleural lobes (in the cephalon "genae").

iii) There is always a doublure on the terga.

iv) The cephalon and postcephalic terga may change their shape and ornamentation considerably during ontogeny, but although these changes are discontinuous (due to moults), their morphology is topologically stable.

v) The cephalon is composed by an acral, preoral lobe, forming the epistomium (= labrum) and almost certainly some of the other anteriormost structures which do not extend into the genae. These are followed by the antennal and ocular segments. The cephalon, moreover, always includes either three (e.g. *Olenoides, Triarthrus*) or four (e.g. *Rhenops*) leg-bearing segments.

The cephalic appendages are variable. In some instances (e.g. *Triarthrus*) they are only slightly differentiated from the postcephalic walking limbs, but in others (e.g. *Phacops*) they have strongly differentiated gnathobases. In other genera, though the appendages themselves are unknown, either the morphology of the cephalic apodemes (cf. *e.g. Illaenus, Dysplanus, Stenopareia, Chasmops, Pharostomus*), the epistomium (e.g. *Hypodicranotus*), or the rare occurrence of a calcified metastomium, imply a considerable differentiation of the cephalic appendages.

vi) Apart from the simple uniramous antennae, the appendages, when known are of a rather uniform pattern, though considerable variation must have occurred in the morphology of the outer branch.

vii) The postcephalic region is usually, but not always, divided into a "thorax" of between two and over 44 free segments, and a pygidium of fused segments. However in some primitive Olenellids there is a more or less clear subdivision between "thoracic" and "abdominal" segments. The latter corresponds to a large number of free pygidial elements, the pygidium itself being reduced to a small telson plate with perhaps one attached segment. In other taxa (e.g. *Triarthrus*) the small pygidial legs largely outnumber the apparent number of pygidial segments. The tergum of the pygidium shows metameric ornamentation, but this may not correspond with the boundaries of the fused segments.

As is the case in the cephalic appendages, sometimes the pygidial appendages are hardly different from those of the thorax, except that they become progressively smaller towards the anus. However, there are instances (e.g. *Rhenops*) where the pygidial appendages are markedly smaller that their thoracic counterparts, the transition being an abrupt one, located either between the thorax and pygidium or between pygidium and postpygidium.

viii) The free thoracic segments usually become

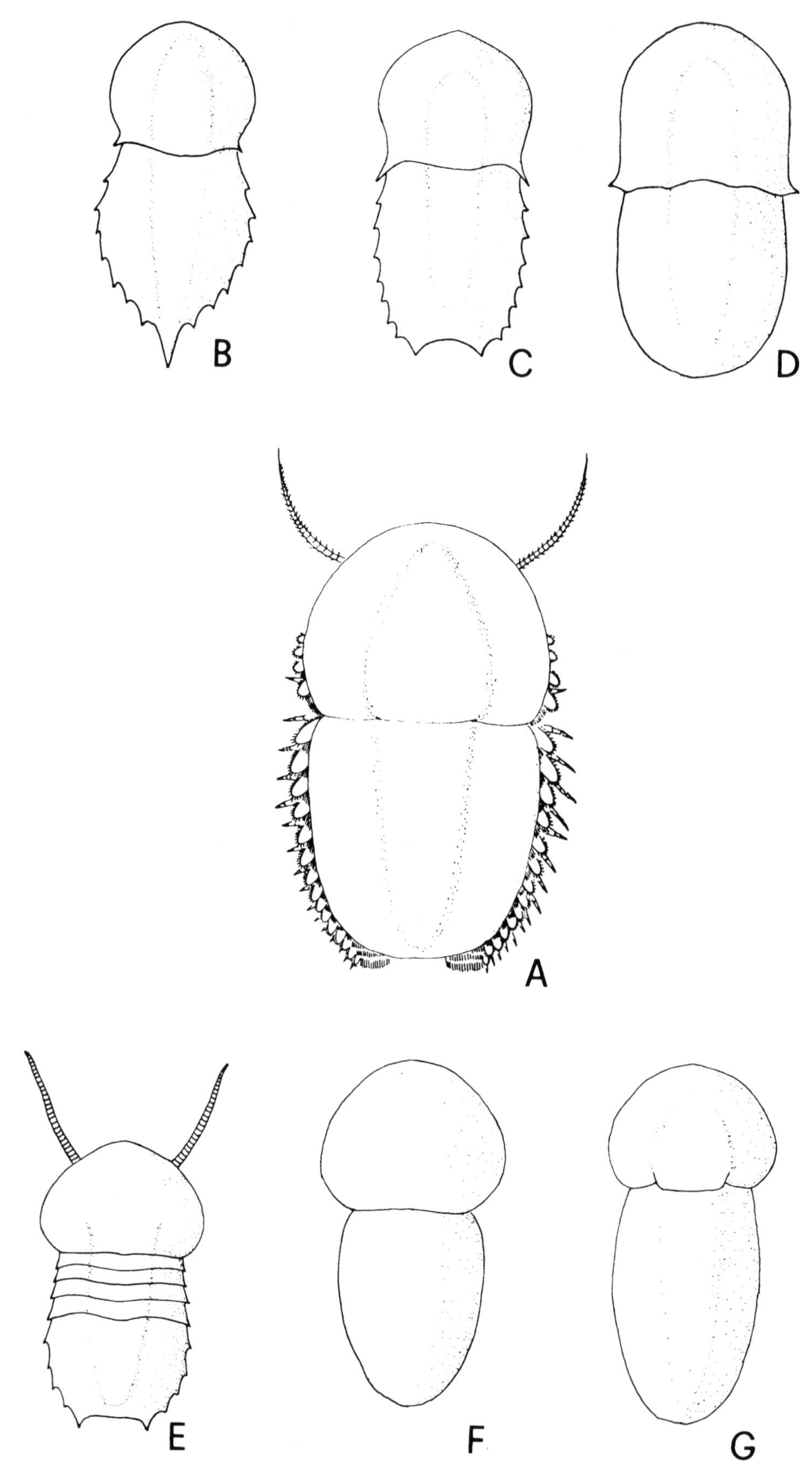

Fig. 5. The Nektaspida, A: *Naraoia compacta*, B: *N. spinifera*, C: *N. spinosa*, D: *N. halia*, E: *Liwia plana*, F: *N. pammon*, G: *N. longicaudata*.

detached from the pygidium at each successive moult, but it is significant that in some species the number of segments apparently incorporated into the pygidium is greater during development than in the adult instar. The extreme of such a condition is found in *Ceraurinella typa* Cooper, where during the whole larval development the pygidial plate grows without giving rise to any free segment. One consequence of this is that the larva appears to be formed by two tagmata: cephalon and pygidium until with the last larval moult all the thoracic segments become free from one another at the same moment.

ix) Among the Trilobita only some groups are capable of enrollment, while the others relied for defence on the ability to cling, limpet-like, to the substratum.

Nektaspida (Fig. 5)

Only the genus *Naraoia* can be referred with certainty to this order. Although we agree with Dzik & Lendzion (1988) that *Liwia* might be included also in this taxon, we consider it more likely that it should be considered closer to such genera as *Retifacies*(Fig. 6). *Orientella* (type species *O. rotundata*) and *Maritimella* (type species *M. rara*) have been described by Repina & Okuneva (1969) and referred to the Nektaspida. Robison (1984) considers them as probable pseudofossils. Dzik & Lendzion (1989), while reaffirming that these specimens are genuine, judge that the true characters of these fossils are obscured by their peculiar preservation. Therefore judgement on these two taxa must be deferred. Anyway *Orientella*, to judge from the admittedly poor figures, is strikingly similar to the well preserved *Naraoia longicaudata* Zhang & Hou, 1985, from the early lower Cambrian of Yunnan. On present records, therefore, the Nektaspida range from the lower Cambrian to the upper middle Cambrian. *Naraoia*, apart from the above discussed fossils, includes the following nominal species: *N. compacta* and *N. spinifera*[7], described by Walcott, *N. pammon* and *N. halia* described by ourselves (Simonetta & Delle Cave, 1975), *N. longicaudata* and *N.spinosa* described by Zhang & Hou. Whittington (1977a) while admitting the features that we deem warrant specific recognition of *N.halia*, considers them to reflect sexual dimorphism. The recently described Chinese material supports his contention, as some specimens show the posterolateral spines of the cephalon that we considered diagnostic of *N. halia*; however Whittington is in error when he states that the spines at the posterior margin of the cephalon were considered by us as distinctive also of *N. pammon*, which indeed we described as having an almost perfectly round cephalon. Probably Whittington misunderstood our descriptions.

Although *N.spinosa* is extremely close to *N.spinifera*, the differences mentioned by Zhang & Hou are real. If we consider Fig. 6, which incorporates all the published evidence, *Naraoia* is found to differ from all typical Trilobita, with which Whittington, followed by most authors want to group it, by the following relevant features:

Trilobita	Nektaspida
Terga calcified	terga uncalcified
Axial furrows clearly separating rachis from pleural lobes	Axial furrows absent, rachis barely distinguishable from pleural lobes.
at least 2 free thoracic segments	No free thoracic segments in *Naraoia* (but 4 in *Liwia* if it belongs here)

Taken together these features are certainly highly significant and Whittington (1977a) himself remarks that, in order to include the Nektaspida into the Trilobita, as he does, a considerable extension of the concept of Trilobite is required. Indeed the Nektaspida appear to be derived from the same stock which gave rise to the Trilobita, but to have branched off before the Trilobita acquired all their distinctive features and particularly before acquisition of a mineralized exoskeleton (Tab. I).

Other genera of possible trilobite affinities (Fig. 6-7)

The following genera are discussed here: *Retifacies, Nathorstia, Tontoia, Tegopelte, Kumaia, Rhombicalvaria, Helmetia, Mollisonia* and *Urokodia.*

Tontoia Walcott, 1912 (Fig. 6)

This monotypic genus is based on the Middle Cambrian *Tontoia kwaguntenensis* Walcott 1912; but only the tergal structures are known.Whittington, 1985, suggests that *Tontoia* should be considered as a *nomen dubium*, on account of the poor preservation of the type and only known specimen. We do not agree with him, as, having seen specimen, we consider that it provides enough evidence to consider it as validly described, and we tentatively group it with *Tegopelte*, as it appears that, like in this genus, each of its terga correspond with polisomites.

Tegopelte Simonetta & Delle Cave, 1975 (Fig. 6)

Tegopelte gigas is the type species of this monospecific genus. It is one of the largest Cambrian Arthropods, reaching the length of 270 mm. The recent redescription by Whittington (1985) confirms and completes our original one. Whittington prepared both the known specimens, and in describing the inner branch of the legs, confirmed our expectations (Simonetta & Delle Cave, 1975) that the complete leg was trilobitic type (it is apparent from Whittington's text that he misunderstood both our text and figure). There is no doubt that the 5 terga of *Tegopelte* correspond to polysomites. A pair of simple antennae and a Trilobite-like labrum also appear to be present.

A distant relationship to the Trilobita is probable, but it must be stressed that neither *Tegopelte* nor *Tontoia* have either a calcified exoskeleton or any trace of trilobation. Moreover no Trilobite has thoracic polysomites, while in *Tegopelte* each tergum corresponds with 3-4 pairs of legs. The relationship of these genera to the Nektaspida may have been somewhat closer. The total number of appendages is higher in *Tegopelte*, probably 33, than in *Naraoia*, but this may not be significant as *N. longicaudata* may have had as many as 25 pairs of legs. Nevertheless, *Naraoia* has only one

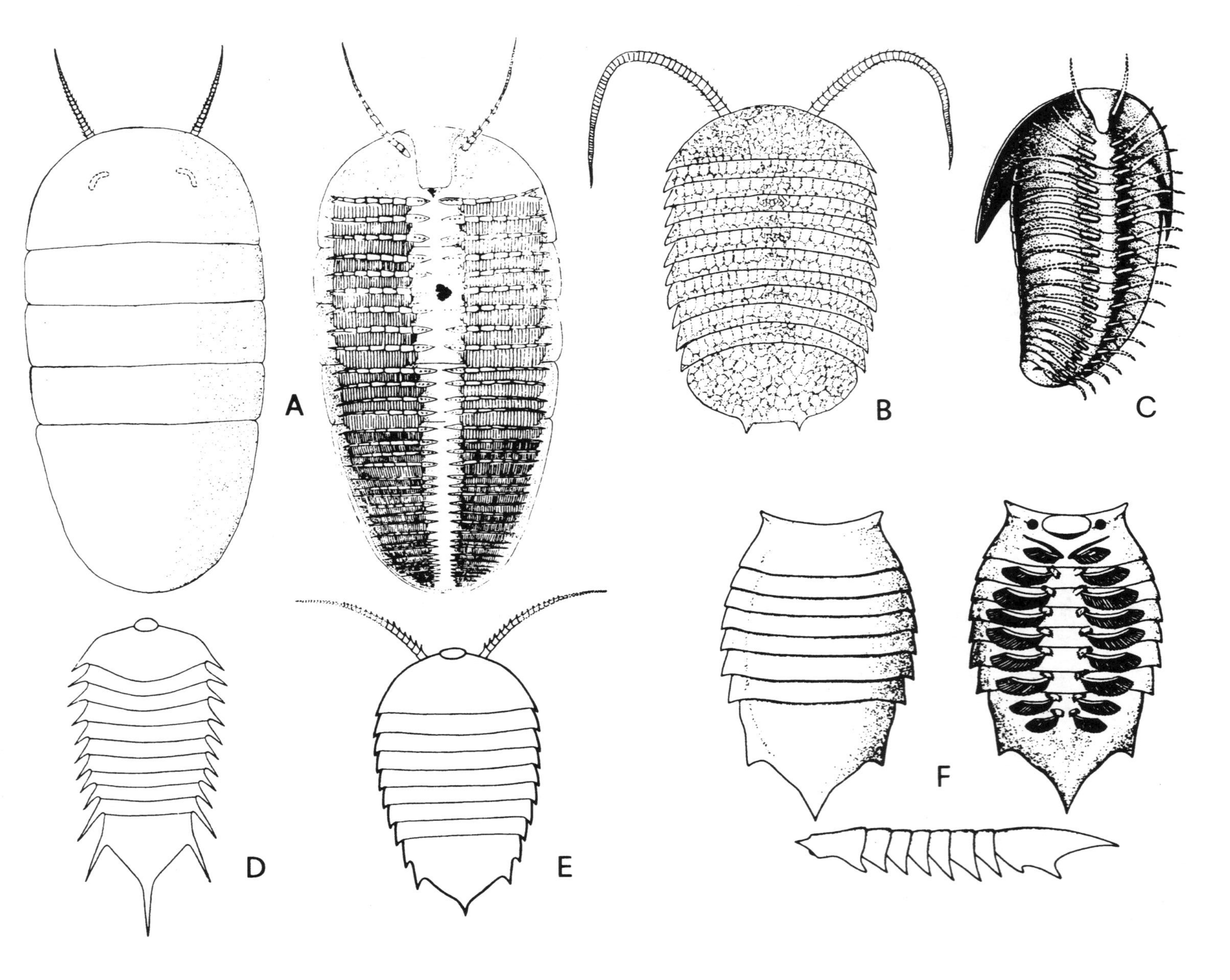

Fig. 6. A: *Tegopelte*, B: *Retifacies*, C: *Nathorstia*, D: *Rhombicalvaria*, E: *Kumaia*, F: *Helmetia*.

postcephalic tagmon and, assuming that *Orientella* and *Maritimella* were correctly identified and described, their tiny free thoracic segments cannot have been polysomites.

Nathorstia Walcott, 1912 (Fig. 6)

Nathorstia transitans, the only species of the genus, is based on a single specimen which has often been considered to be a freshly moulted trilobite, fossilized before its terga became calcified. This view has been most recently maintained by Whittington (1980). When one of us (A.S.) examined the type (and only specimen) many years ago he thought that there was sufficient evidence to warrant its separation, particularly as the cephalon appeared to have only two postoral appendages. As we could only reexamine our photographs, we can only say that, from their evidence, *Nathorstia* appears to be closer to *Retifacies* or *Liwia* than to the Trilobita. A fresh examination of the material is certainly needed.

Retifacies Hou, Chen & Lu, 1989 (Fig. 6)

A monotypic genus, based on *Retifacies abnormalis* Hou, Chen & Lu from the early lower Cambrian of China, is morphologically (apart from *Nathorstia*) the most trilobite-like of this group of taxa. However it is sharply separated from the Trilobita by its unmineralized, non-trilobate, sculptured terga. For Cambrian times it was a comparatively large animal, apparently reaching about 75 mm. This may well be the functional reason of its reticulated terga, if they had a strengthening function. In a previous paragraph (p. 195) we said that on available evidence it is not possible to decide whether the genus *Liwia* should be considered to be closer to *Naraoia* or to the *Retifacies-Kumaia* group.

Kumaia Hou, 1987 (incl. **Rhombicalvaria** Hou, 1987) (Fig. 6)

There is no doubt that by ordinary standards *Kumaia lata* and *Rhombicalvaria acantha*, both described by Hou (1987) from the early lower Cambrian of China, should be separated at generic level, however, for reasons discussed elsewhere (Simonetta, 1987, 1988, in press), we think that Cambrian taxa should "in principle" be considered of broader morphologic scope than their later equivalents. We suggest that it might be practical to merge these two genera. Anyway we propose that *Retifacies* → *Kumaia* → *Rhombicalvaria* → *Helmetia* form a morphologic series (we do not think that the different numbers of thoracic segments is of much significance). The occurrence both in *Kumaia* and in *Rhombicalvaria* of a small oval structure, figured by Hou at the foreend of the cephalon appears to correspond with the larger oval structure in *Helmetia* that we have tentatively considered as the labrum. If this suggestion is granted, then *Helmetia* appears much less isolated and its relationship to the Trilobita may be tentatively considered as possible.

Helmetia Walcott, 1917 (Fig. 6)

Helmetia expansa Walcott, 1917 is apparently related to the previously described genera. A vague and probably superficial resemblance exists also with *Corcorania trispinosa* Jell, 1980 from the lower Ordovician of Australia.

What can be made of *Helmetia* is illustrated here (Fig. 6). As a comment we must add that we now consider that the cephalon may have been actually composed of two separate terga. Accordingly the first segment may have had either one pair of postoral appendages or perhaps none. Furthermore the total number of segments may have been 8 instead of 6 as in our reconstruction of 1975.

The oval structure at the anterior of some specimens is a bit strange as a labrum, but we can not see what else it could be; a corresponding structure apparently existed in the lower Cambrian *Kumaia* and *Rhombicalvaria*. The occurrence of compound eyes is debatable, as the faint paired ventral structures may well have been something else.

Urokodia Hou, Chen & Lu, 1989 (Fig. 7)

We agree with Hou, Chen & Lu that this genus is closely comparable with *Mollisonia*. We do not think that either the cephalic and pygidial spination or the different number of segments deserve much attention. The first is just a matter of degree of development, and the second is known to be a character which is sometimes variable even within the same arthropod genus.

Mollisonia Walcott 1912 (Fig. 7)

As far as *Mollisonia symmetrica* is concerned, we reiterate (Simonetta & Delle Cave, 1975) that*Mollisonia* is a duplicate of *Thelxiope* minus the telson spine. However the new discoveries discussed in this section suggest serious doubts concerning the affinity of *Mollisonia* to the Emeraldellida. This, the probably related *Urokodia*, and perhaps *Serracaris*, may well all belong to a primary radiation of the same stock which led to the development of primitive, but typical Trilobites. This concept is incorporated in the tentative phylogeny proposed in Table I.

Corcorania Jell, 1980 (Fig. 7)

This monospecific genus (type *Corcorania trispinosa* Jell, 1980) is from the lower Ordovician of Australia, and thus is much later than *Helmetia* to which it has a superficial resemblance. However, *Corcorania* had apparently some cephalic appendages comprising a large oval podomere provided with few big spines: no trace of such appendages is found in *Helmetia*, though, if *Corcorania*'s podomeres are coxae, they may represent a late specialization.

Serracaris Briggs, 1978 (Fig. 7)

The lower Cambrian *Serracaris lineata* (Resser & Howell, 1938) is again an "orphan" genus. Its affinities are really beyond speculation. The recently discovered, highly specialized, phreatic and strange Isopod

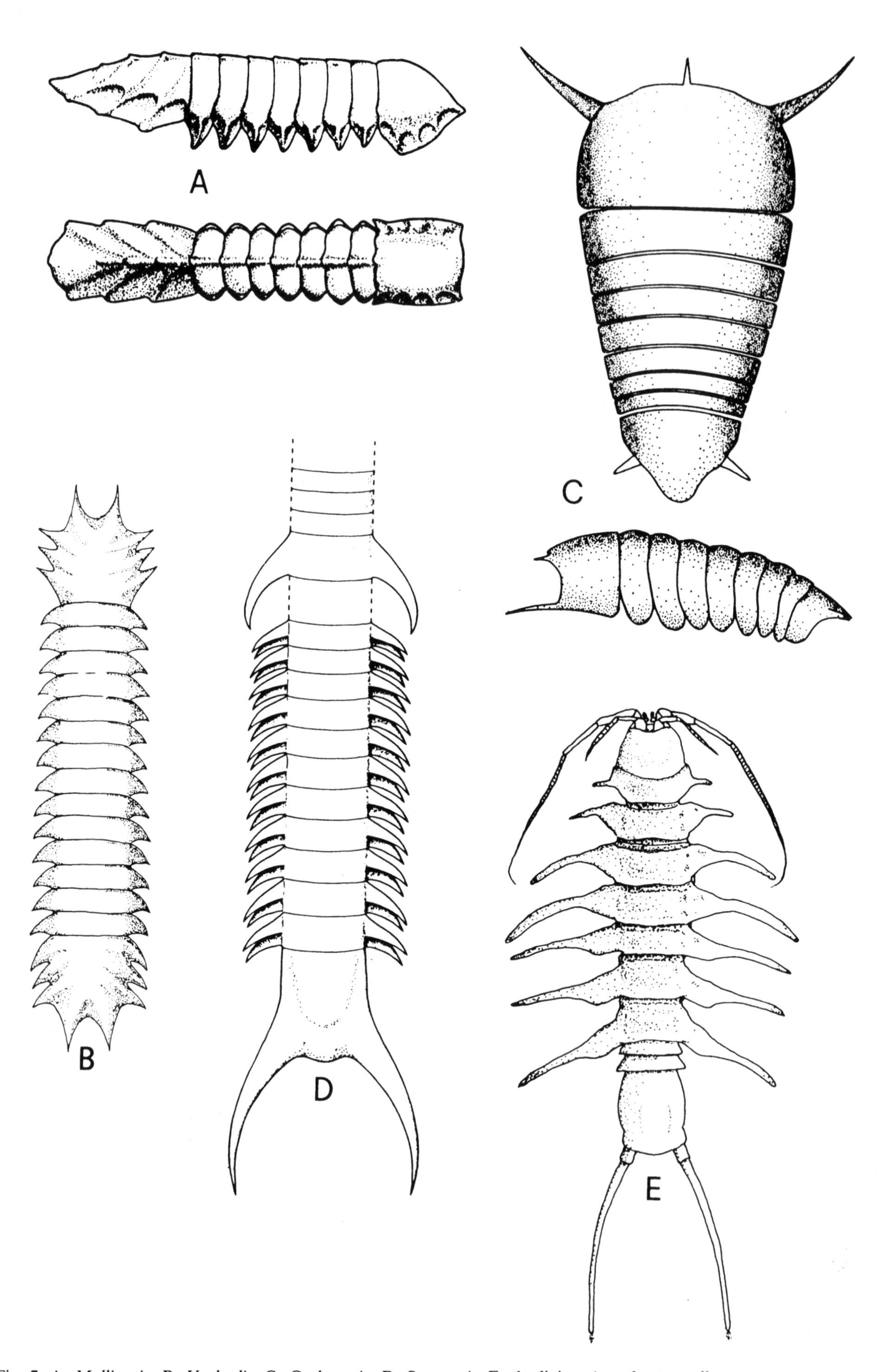

Fig. 7. A: *Mollisonia*, B: *Urokodia*, C: *Corkorania*, D: *Serracaris*, E: the living *Acanthastenasellus*.

Tab.I

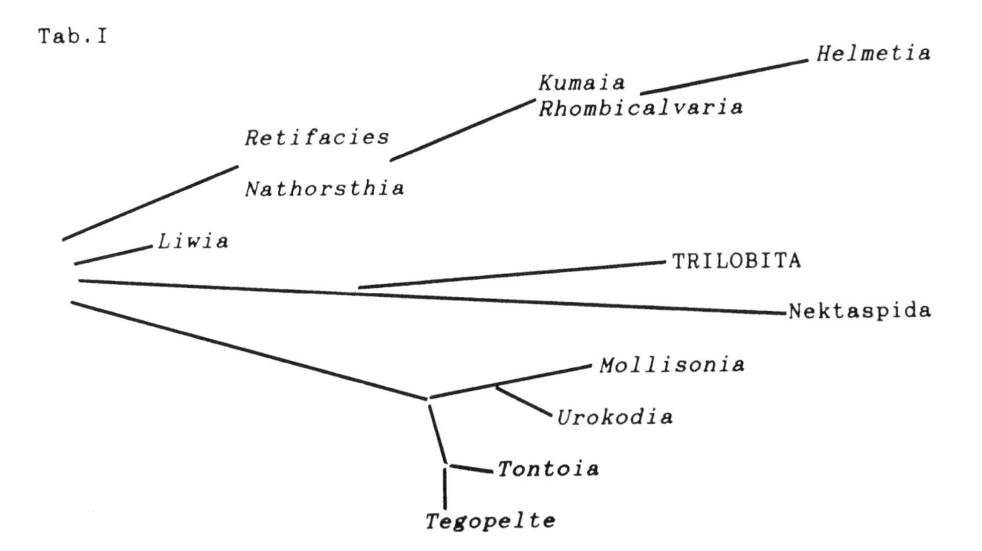

Acanthastenasellus forficuloides Chelazzi & Messana, 1985 (Fig.7), though obviously unrelated to *Serracaris* may vaguely suggest the living appearance of the Cambrian animal.

Discussion

The Nektaspida have been tentatively considered as Trilobites by Whittington. As we do not agree a discussion of these genera may be attempted here. A tentative phylogenetic reconstruction might be as follows: There was an early radiation of the "trilobite stock" late in the Precambrian, and while the branch leading to the true Trilobites acquired a basically olenelloid structure and mineralized terga before radiating into the three or four trilobite orders that occur at the base of the Cambrian, another branch evolved into the Nektaspida. This latter group has a primitive morphology of the terga, but advanced cephalic appendages. A third branch, having evolved polysomites, became the Tontoida.

Nathorstia, if it is not a freshly moulted Trilobite, and the *Retifacies, Kumaia, Rhombicalvaria* group may well be another branch (possibly including *Liwia*, if we prefer to separate it from the Nektaspida), in turn leading to the *Helmetia* complex. The group *Mollisonia-Urokodia* and possibly *Serracaris* may represent yet another branch of this radiation. As and when the legs of these animals are found, so it will be possible to verify whether they are somehow related to *Cambropodus*.

Limulava Walcott, 1912 (Fig. 8)

This order was established by Walcott (1912) with *Sidneyia* as its type genus. This genus remains, so far, the only one certainly referable to this order. Except for minor details we follow here the revised description by Bruton (1981). The points where we dissent concern the telson and uropods, as we consider it more probable that the anus opened under the anal plate near its caudal margin, rather than as a longitudinal slit in the anal plate (as suggested by Bruton). This is simply because it would be more "normal" in an Arthropod and Bruton (1981) himself agrees that there is no direct evidence as to the location of the anus. In addition, there is no doubt that the gut extended to the last segment. We also still consider it possible that the lobed portion of the uropods morphologically was the outer branch of the appendage, while the thickened outer, marginal portion may be either the reduced inner branch of the walking leg or the "flap" of the outer branch, to which the gill lamellae are attached. Under this latter interpretation the uropods are homologous with only the outer branches of normal legs, the inner branch having entirely vanished. We must, however, add that the unlobed telson of *Sanctacaris*, with its thickened outer margin like that of *Sidneyia*, will support Bruton's interpretation, if inclusion of *Sanctacaris* in the Limulavida is granted.

So far as the cephalic and thoracic appendages are concerned, while Bruton has conclusively shown that there are serial differences among the various pairs of legs in the development of the endites and spines, we consider it probable that the transition was a gradual one. Although it is probable that the first pair of postoral appendages had an outer branch either reduced or absent, we would be much surprised if the set of appendages was sharply differentiated into a fore-set devoid of an outer branch and a posterior one with fully developed respiratory branch.

Another minor point of disagreement between ourselves and Bruton concerns the functioning of the legs. Bruton, considering the remarkable overall resemblance between the legs of *Sidneyia* and those of the Xiphosura, supposes that their movements were approximately the same. We think that due consideration should be given to the fact that in the Xiphosura the legs are grouped around the downward facing mouth. In *Sidneyia*, instead, the legs are arranged in a row and the mouth, facing downwards and backwards, opens adjacent to the fore-end of the row of legs. This implies that the food, which consisted of small trilobites, hyolithids and probably soft-bodied animals, was grasped by the legs and then moved forwards while being chewed by the gnathobases.

As a whole Bruton's (1981) work has greatly improved Simonetta's (1963) reconstruction. It has also substantiated most of our former conclusions, except for the substitution of more or less normal walking legs for the big specialized appendages, which have been shown by Whittington & Briggs (1985) to belong to *Anomalocaris*. Bruton (1981) regards *Sidneyia* as in

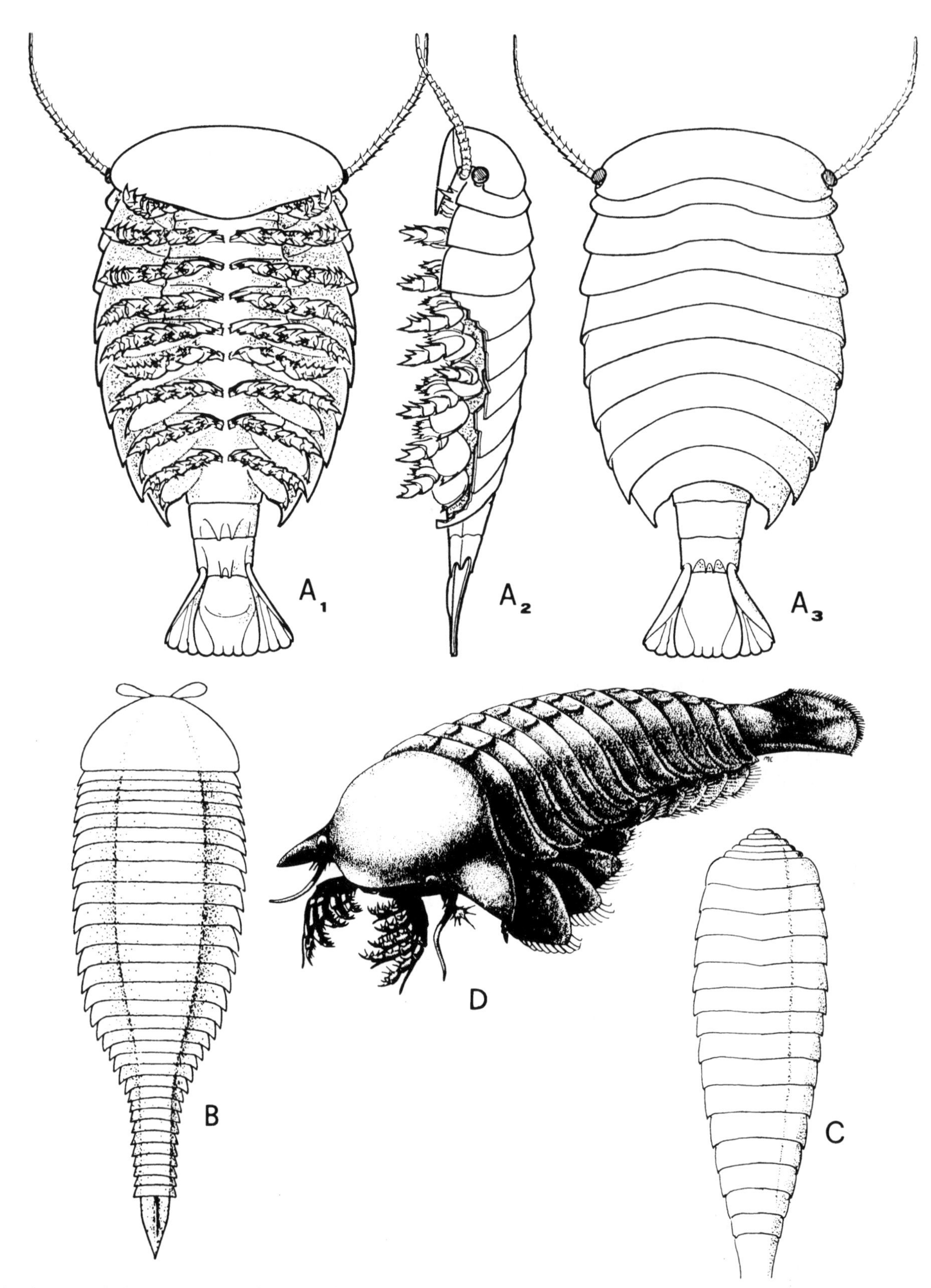

Fig. 8. A1-3: *Sidneyia*, B: *Fuxianhuia*, C: *Chengjiangocaris*, D: *Sanctacaris*.

some way related to the ancestry of the Xiphosura, though he does not formally include it into this taxon. This we consider most unlikely. First, he considers the possibility that the chelicerae are specialized preoral appendages. However, as shown by their embryological development, there is no doubt that the chelicerae are the first pair of postoral appendages and that the preoral antennae have completely vanished in the Xiphosura, except for a trace of their former existence in the prosencephalon of *Limulus*. Second, the cephalization of *Sidneyia* is minimal: Bruton (1981) has confirmed Simonetta's (1962) contention that only the antennal (= antennular) and ocular segments are fused, while all the other segments that in the Trilobita, Chelicerata, Crustacea and "uniramians" are incorporated in the cephalic tagmon, are here free and show only the slightest hint of specialization. Third, the tagmosis of the body is entirely different from that of the Xiphosura. If we accept Bruton's idea that the first four pairs of legs differed sharply from the following ones (an assumption for which Bruton admits there is no evidence), than the tagmosis of *Sidneyia* could be compared with that of the Diploaspida and Chasmataspida (fig. 11). Simple consideration of the different number of segments involved in each tagma of these different orders plainly shows that the similarity is due to superficial convergence.

Again, the comparative morphology of the appendages of the Xiphosura, Eurypterida and Scorpions, clearly shows that they cannot have been derived from appendages such as those of *Sidneyia*. They must derive from much more primitive, more trilobite-like appendages. The resemblance between the legs of *Sidneyia* and those of the Xiphosura is rather superficial, as the main joint is between podomeres 5 and 6 in *Sidneyia* and between podomeres 3 and 4 in the Xiphosura.

Finally we do not agree with Bruton's (1981) decision to leave *Sidneyia* "in the air" from the taxinomic standpoint. A genus may well be *incertae sedis*, within the broad lines of one taxon, but it must belong somewhere. Therefore, as the order Limulava was validly established for *Sidneyia*, either one considers the order to be a valid taxon and then proceeds to leave it *incertae sedis* within a Class, or if one considers that the order is not valid, then one must synonymize it and include *Sidneyia*, albeit tentatively into some other established order.

Fuxianhuia Hou,1987 (Fig. 8)

Fuxianhuia protensa Hou, 1987, from the early lower Cambrian of China, is the only species referred to this genus. We tentatively include it among the Limulava. Accepting the reconstruction proposed by Hou (1987), we would just add that, unless the published photographs are misleading, there is little doubt that the cephalon included only three segments, the foremost being apparently the labial one. The cephalon appears, therefore, to be close to that of *Sidneyia*, while the remainder of the body, even if much more elongated and with considerably more segments, approaches the outline of that of *Sidneyia*. It is plausible that here we see a primitive taxon within the Limulava, with the post-thoracic (abdomimal) segments morphologically very close to the thoracic ones, and so much less specialized than in *Sidneyia*.

The telson, represented in *Fuxianhuia* by a lanceolate plate, also is not incompatible with that of *Sidneyia*, which simply has a more truncated outline. We would dare to guess that there were no uropods and that the last pair of appendages was morphologically scarcely different from the preceding ones, a fact not unexpected in a much more primitive member of the Limulava.

The only feature which poses some problems are the paired cephalic appendages considered by Huo (1987) to be stalked eyes. They could well be eyes, but if so, we have here a major difference with *Sidneyia*. One must remember, however, that within the recent peracarid Crustaceans, the Isopoda and Anphipoda have sessile eyes, while the Mysidacea have pedunculate eyes.

Chengjiangocaris Hou & Bergström (this volume) (Fig. 8).

This monotypic genus (Hou & Bergström, this volume) is clearly more primitive than *Fuxianhuia*. Differentiation of the body in tagmata is limited and the most notable character is the head region, composed by 5 entirely separate segments that increase in size from front to back. The telson is missing and the last segments of the body are incompletely preserved; our drawing provides a tentative reconstruction. The completely segmented cephalic region of this animal corresponds in morphological development with *Sidneyia*, but has two extra segments, thus having a complement of segments comparable to at least some Emeraldellids. So far as the evidence goes, we consider *Chengjiangocaris* as a representative of a radiation leading to *Sanctacaris* and to the Emeraldellids.

Sanctacaris Collins & Briggs, 1988 (Fig. 8)

Sanctacaris uncata Collins & Briggs, 1988 may, perhaps, be considered as a side branch of the Emeraldellids. Quite apart from the similarity of the general proportions of the body, which have no morphologic significance, the following points deserve to be stressed: i) the overall similarity of the cephalic appendages with those of *Emeraldella*; ii) the morphology of the telson plate, which is close to that of *Sidneyia* (Bruton, 1981); iii) the apparent lack of the preoral antennae. In this respect *Sanctacaris* appears to be more specialized than any Emeraldellid, while the specialized, antenna-like outer branch of the cephalic appendages and the specialization of the 5th cephalic leg are unique features. However everything considered we are inclined to consider *Sanctacaris* as in some ways a side branch, intermediate between *Sidneyia* and the Emeraldellids. To consider *Sanctacaris* a Chelicerate is certainly wrong and is the result of a formalistic approach. The so-called synapomorphies of *Sanctacaris* and the Chelicerata are clearly convergences.

Utahcaris orion Conway Morris & Robison, 1988, from the Middle Cambrian of Utah is based on a single specimen. It appears to be closely related to *Sanctacaris* and we suggest, pending the availability of better specimens, that it should be considered as a junior synonym of *Sanctacaris uncata*.

Discussion

As a whole, while the lower Cambrian genera tentatively grouped in the Limulava appear to be possibly the most primitive Arthropods known, at least as far as tagmosis is concerned, the middle Cambrian ones show the different picture of advanced characters within a primitive framework. As a whole the morphology of *Sidneyia* is particularly interesting, as it shows and admirable mixture of exceedingly primitive and quite advanced features, several of them paralleling those of quite divergent classes. *Sanctacaris* tells the same story, as it mimics some advanced features of later taxa.

Emeraldellida Simonetta & Delle Cave, 1975 (Fig. 9)

It has been argued that the genera grouped here under this label are sufficiently disparate to make it doubtful whether their similarities are anything but the result of convergence and that, in any event there is insufficient evidence as to whether they are related with other taxa such as the Aglaspida, Strabopida etc.

To have positive evidence for an hypothesis is obviously one thing and to have nothing against it is quite another. Nevertheless, it has always been admitted both in pure logic and in science that, so far as there is no positive evidence against a given hypothesis, this is to be given some credit. To conclude this discussion, we will merely show that the Emeraldellida (as defined here) may be monophyletic and that animals referable to this group may have eventually evolved towards a more and more Xiphosurid-like condition.

Thelxiope Simonetta & Delle Cave, 1975 (Fig. 9)

We have no evidence to add to our original description of *Thelxiope palaeothalassia*. We must say, however, that by comparison with *Habelia*, which is clearly a more primitive member of this group, it is probable that the cephalon of this genus only included two postantennal segments. Simonetta's (1964) reconstruction has been modified accordingly.

Ecnomocaris Conway Morris & Robison 1988 (Fig. 9) This is again a monotypic genus and its authors rightly comment on its affinity with *Thelxiope*. That it is a specifically different animal we agree, but our attitude concerning monotypic genera and on the scope of Palaeozoic taxa (Simonetta, in press) lead us to suggest that it may be useful to include the new animal in *Thelxiope*.

Habelia Walcott, 1916 (Fig. 9)

We have modified Simonettàs reconstruction of *Habelia optata* Walcott, 1916 in consideration of the data published by Whittington (1981). The only aspect of which we remain unconvinced in Whittington's reconstruction is the bent telson. We persist in our belief that usually it was straight and not bent at a joint. *Habelia brevicauda* does not require any special comment.

In spite of its more pronounced ornamentation we consider *Habelia* an animal more primitive than *Molaria*.

Molaria Walcott, 1912 (Fig. 9)

We accept here the revised reconstruction by Whittington (1981). It is certainly much improved compared with ours, though the only morphologically significant difference is in the number of cephalic appendages, 4 instead of 5, as Simonetta had supposed. However Whittington may not have not fully understood my (A.M.S.) description, as I clearly said that the first pair of appendages was tactile organ, that I thought it probable that it was postoral and biramous, but that I was not sure and that alternatively it could be a simple preoral antenna, given that the presence of an outer branch could not be observed.

Whittington (1981) considers the telson to be multisegmented; of this we are not entirely convinced. However, even if Whittington (1981) is right, this may not be particularly significant. Take just one example: the telson of the scorpions is a rigid and strong sting, while in the whip-scorpions (Uropygi) it is a long, delicate, multiarticulated filament.

Sarotrocercus Whittington, 1981 (Fig. 9)

There is nothing to be added to Whittington's (1981) description, except that, although these cannot be seen on the fossils, we consider incredible that the appendages were formed by the fringed outer branch only. We are confident that, if better specimens are found, they will show walking legs very much like those of *Molaria* or *Habelia*.

Emeraldella Walcott 1912 (Fig. 9)

There is nothing to add to the descriptions by ourselves (Simonetta, 1964, Simonetta & Delle Cave, 1975) and by Bruton & Whittington (1983).

Aglaspida, Strabopida and related genera: ***Borkgrevinkium, Triopus, Bekwithia, Palaeomerus, Bunaia, Khankaspis*** (Fig. 10)

As the appendages of most of these genera are unknown there is no positive evidence on which to separate these genera from either the Emeraldellids or from any of the chelicerate genera discussed below.

These genera are especially significant as they show that through the upper Cambrian and Ordovician there was a continuous radiation of taxa more or less intermediate between the middle Cambrian Emeraldellids and the late Ordovician-Silurian Eurypterids, Xiphosurans etc.

Strabops Beecher, 1901 (Fig. 10)

The upper Cambrian *Strabops thacheri* is most notable among the early protochelicerates as it had acquired a mineralized exoskeleton, had apparently fully developed walking legs only on the prosoma, and perhaps had acquired chelate appendages. The Ordovician *Neostrabops martini* Caster & Macke, 1952, and the Beckwithiidae are probably close relatives.

Aglaspididae Miller 1887 (Fig. 10)

The only feature that separates *Aglaspis*, the best known genus of this family, from the Emeraldellidae is its phosphatic exoskeleton. This is a relevant feature, but should not obscure the significance of the overall similarity of all these taxa. That no evidence has been found of the outer branch of the legs is obviously the result of their poor preservation. An animal the size of *Aglaspis*, somewhat larger than *Emeraldella* and with a mineralized exoskeleton must necessarily have had some respiratory apparatus. This cannot possibly have been anything other than some sort of respiratory exopod.

Discussion

We may now ask ourselves whether the genera *Emeraldella, Habelia, Thelxiope, Molaria* and *Sarotrocercus* are interrelated and whether they are related to Aglaspida, Strabopida etc, and so eventually to the Xiphosura, Eurypterida and Scorpiones.

Although the morphological evidence is very scanty for several genera, the answer to both questions is positive:
i) the structure of the body in all these forms is sufficiently uniform; in some there is a tendency to differentiate the opisthosoma into two distinct regions. This is, however, of possibly little significance, and apparently evolved independently in the various groups.
ii) none of the Cambrian genera for which appendages are known had developed chelicerae, but some at least had developed a grasping ability by bending the distal part of the fore-legs against the spiny endites of the proximal segments. Thus, the known appendages are sufficiently primitive to be the topological forerunners of appendages such as those of the Xiphosura, Eurypterida, and Scorpiones.

The occurrence of a phosphatized exoskeleton in the Aglaspididae is sufficient evidence to place *Aglaspis* on an evolutionary sideline of the Emeraldellid ensemble. Other genera will probably prove to belong to other such sidelines, but neither *Aglaspis* nor any other of the genera discussed in this section may be excluded from this broad group.
iii) the occurrence, first mentioned by myself and later confirmed by Bruton & Whittington (1983), of anal valves in *Emeraldella* (as we have seen the case for *Habelia* is controversial) as in *Aglaspis* and in Eurypterida is important. Taken together with the fact that what is known for most genera of the "emeraldellid" ensemble does not allow us to say whether anal plates occurred or not, they are apparently of considerable significance, in as much as such structures are unknown in any other group.
iv) *Emeraldella*, because of its long antennae, must be considered as the most primitive member of the group (N.B. as already mentioned the first appendages of *Aglaspis* were apparently antennae). The antennae are considerably reduced in *Molaria* and *Habelia*.

All authors who have discussed these genera and ruled out the "emeraldellids" from the ancestry of at least part of the Chelicerata, have failed to discuss the possible morphology of a "non-chelicerate" ancestor. Had they done it, they would have been compelled by comparative anatomy to imagine an animal indistinguishable from a genuine "emeraldellid". In fact such an animal must have had the following features: it must have had

1) long preoral simple antennae.
2) an unmineralized, non trilobate, exoskeleton.
3) a mode of life as a predator with some sort of grasping legs liable to evolve into true chelicerae.
4) shown some tendency to the loss or specialization of the posterior pairs of legs.
5) a telson, and this may have been more or less spiny.

Having thus briefly discussed the possible significance of the "Emeraldellids" in the evolution of some, at least, of the Chelicerata, we must consider whether they may also be related to the ancestry of the "Uniramians" *sensu* Manton.

That the Tardigrada, Onychophora and *Aysheaia* have nothing to do with the "uniramian" arthropods is now widely agreeded (p. 234). Thence, if we examine with an unbiased eye the bulk of evidence assembled by Tiegs & Manton, we find that they shattered the old (but often questioned) hypothesis of the origin of the Insect-Myriapod ensemble from the Crustacea, but they never seriously discussed the possibility that "uniramians" evolved from animals with unspecialized trilobitic appendages.

Indeed until very late in Manton's life, the cephalon of the Trilobita and of the other Cambrian Arthropods was so poorly known that even the number of cephalic appendages was usually wrongly reported. We have already made several remarks on the primitive morphology of Arthropods and the "Emeraldellids" may well be considered as likely candidates for the stem-group from which also evolved some or all of the "uniramian" arthropods evolved.

Chasmataspis Caster & Brooks 1956, **Diploaspis** Størmer (Fig. 11)

These two genera are clearly isolated branches originating along with the better known Xiphosurids etc. from the broad group of genera discussed immediately above.

The Chasmataspidida are represented by the single lower Ordovician species *Chasmataspis laurencii* Caster & Brooks, 1956. Our reconstruction (Simonetta & Delle Cave, 1981) is obviously tentative, based as it is on published evidence. The walking legs are very incompletely known: we assumed them to be basically similar, but to be progressively longer caudally in order to balance the animal when walking. The eurypterid *Mixopterus kiaeri* Størmer, though of much larger size, has a shape recalling that of *Chasmataspis* and the proportions of the legs have been based on this indirect evidence. As is clear from the figure, the tagmosis of *Chasmataspis* may be compared with the tagmosis of the later, Devonian, *Diploaspis*, because in both the opisthosoma is subdivided into a pre- and postabdomen, with the latter consisting of 9 segments. The number of segments incorporated in the preabdomen is not known, but our tentative suggestion (Simonetta & Delle Cave, 1981) of three is still valid. It is possible that this similarity in tagmosis is significant and that the two monotypic orders belong to a single main branch of the xiphosuroid stock. Anyway the number of postcephalic segments in these animals represents an increase in at least three both on the number of segments of the xiphosurids proper and on the emeraldellids.

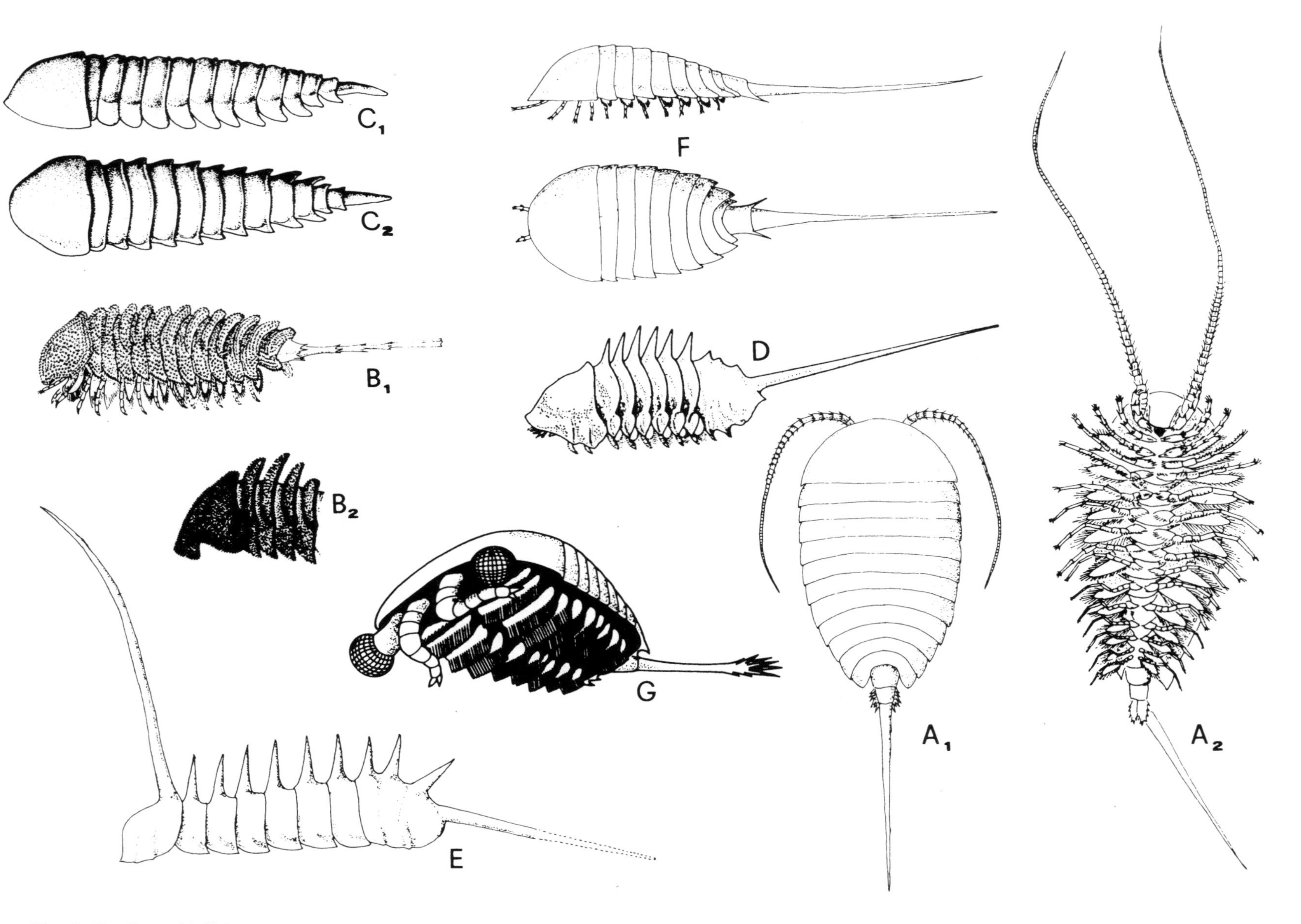

Fig. 9. The Emeraldellids, A1-2 *Emeraldella*, B1-2 the two morphotypes of *Habelia optata*, C1-2 *Habelia brevicauda*, D: *Thelxiope*, E: *Ecnomocaris*, F: *Molaria*, G: *Sarotrocercus*.

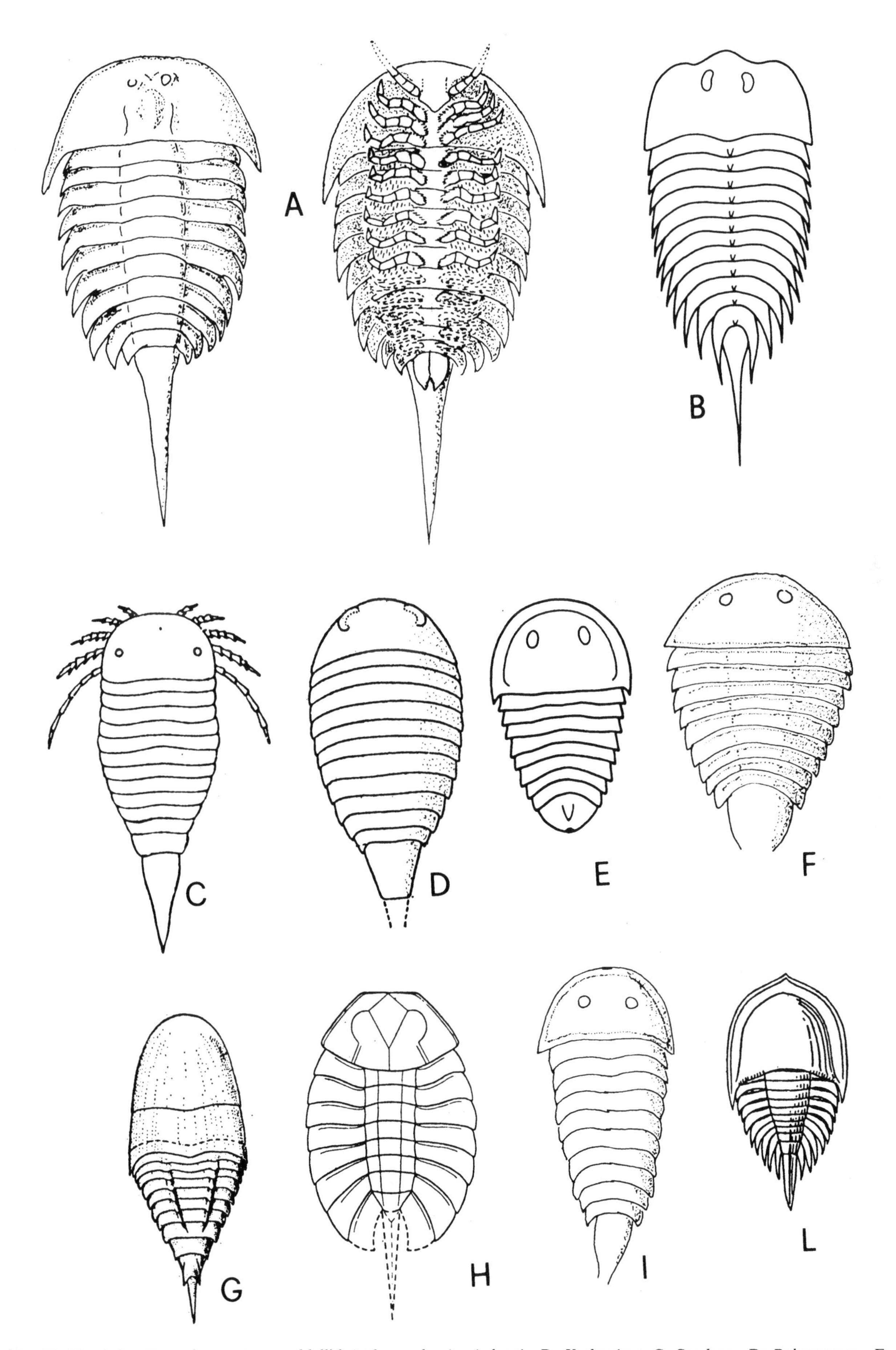

Fig. 10. Cambrian-Devonian post-emeraldellid Arthropods, A: *Aglaspis*, B: *Kodymirus*, C: *Strabops*, D: *Palaeomerus*, E: *Beckwithiatypa*, F: *Beckwithia? dambikensis*, G: *Borchgrevinkium*, H: *Triopus*, L: *Paleoniscus*.

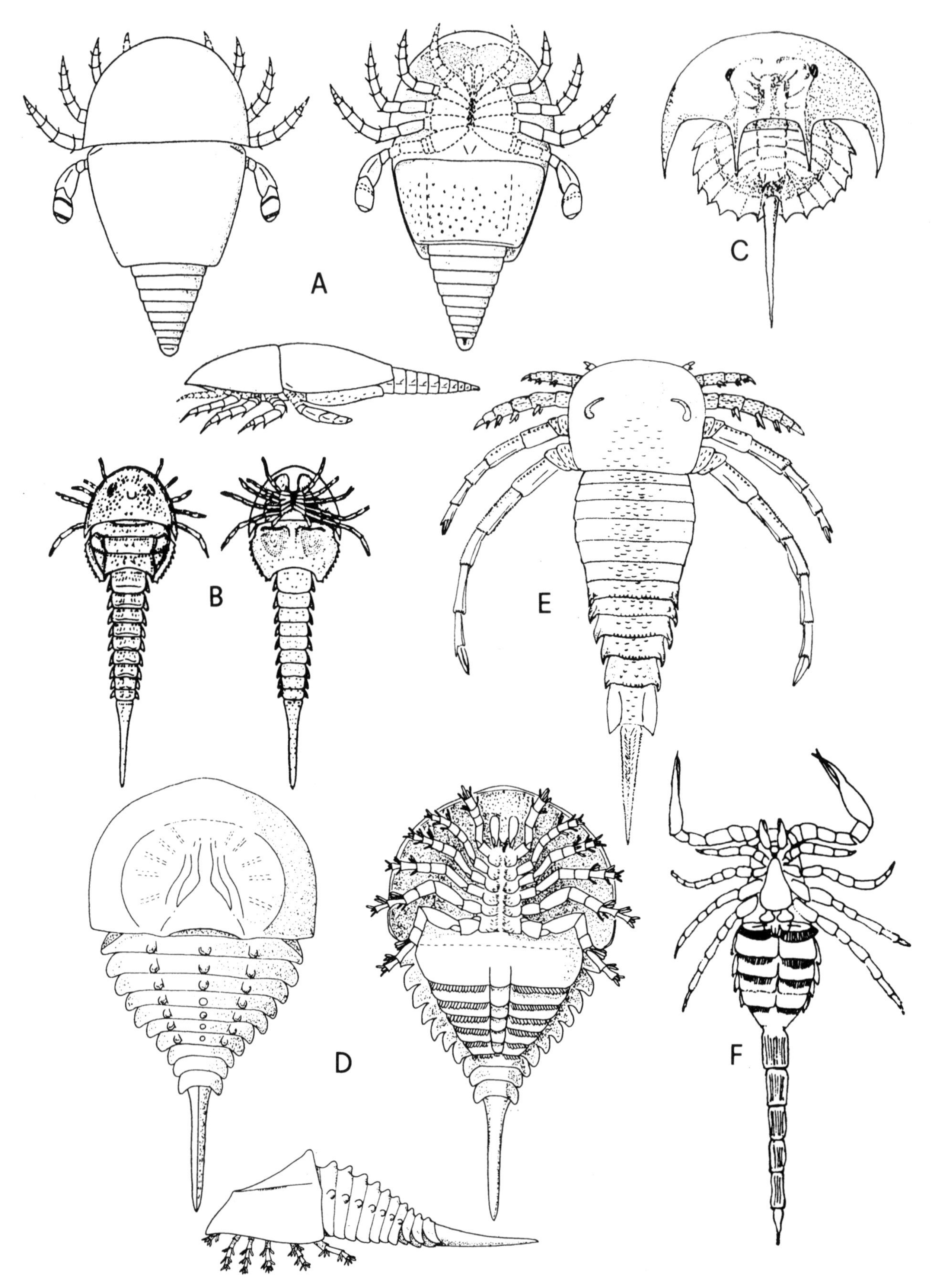

Fig. 11. A: *Diploaspis*, B: *Chasmataspis*, C: the Xiphosurid *Euproops*, D: the Sinzyphosuran *Weinbergina*, E: The Eurypterid *Parastylonurus*, F: *Waeringoscorpio*.

In both the Diploaspida and the Chasmataspidida the respiratory apparatus is limited to the preabdomen and while in *Diploaspis* we have a simple situation recalling that of the Xiphosurans, *Chasmataspis* apparently had a pair of pouches opening by some sort of slits. Such an apparatus, being almost completely closed, certainly could favour life under comparatively dry conditions. Under water it could function only if there was some sort of efficient pumping mechanism. This may, indeed, have existed, as the preabdomen certainly incorporated more than one segment (3 or 4 being the most likely number) and, therefore, a modified system of intersegmental muscles and tendons may have been evolved to act as a pump. It is equally possible that *Chasmataspis* was an intertidal dweller, foraging at low tide on stranded organisms. To suppose that there may have been Ordovician "strandloopers" does not seem now, after the discovery of truly terrestrial Arachnids in the lower Silurian as far fetched as when we suggested it in 1981.

Synziphosura, Xiphosura (Fig. 11)

These two orders are now fairly well understood and clearly represent two parallel branches issuing from a more generalized ancestor. Several of the Cambro-Ordovician "merostomoids" could be close to such an ancestor, but they are still too poorly known.

Eurypterida, Scorpiones (Fig. 11)

For the purposes of this paper these two orders do not require any special discussion. Their morphology is well understood, apart from such poorly known and marginal forms as the aberrant Eurypterid *Cyrtoctenus* Størmer & Waterston, 1968.

Beklemishev (1969) argued that if the relationship of the Devonian and Silurian marine Scorpions with the modern terrestrial ones had not been obvious, the "Palaeophonids" (or, better, Holosterni and Lobosterni) would have been certainly classified within the Eurypterida. Though there is no question as to the close relationship between these two groups, we do not go as far as Beklemishev. This is because the scorpionid lineage has some evolutionary trends which are not shared by the Eurypterids: 1) the Scorpions do not show any trace of anal valves, a primitive character that the Eurypterida share with at least some Emeraldellids; 2) the respiratory appendages appear to represent two diverging evolutionary trends, as it is apparent if one compares the Eurypterid respiratory apparatus (*e.g.* Waterston 1975), with the long branchial filaments of *Waeringoscorpio hefteri* (Størmer, 1970). The development of a preoral chamber is a purely scorpionid development and is apparently related to the acquisition of terrestrial habits.

It seems that the Scorpions may have branched off some small, primitive eurypterid stock, and while Eurypterids increased in size at an early stage, thereby precluding themselves the possibility to colonize terrestrial habitats beyond the intertidal fringe, hampered as they were by a gill system ineffective for breathing air and, at least for the larger species, too heavy a load for the appendages to support on land. The smaller kinds of Scorpions (giant aquatic scorpions up to one meter were exceptions) were better preadapted to a terrestrial habitat and even then aquatic, branchiate scorpions persisted into the Jurassic.

What is important in this context is that their origin is apparently more obscure than that of the xiphosuran-synziphosuran stock. Yet there is no morphologic problem in deriving them from some emeraldellid or "merostomoid" *sensu lato* having transverse postcephalic exopods, arranged in the same way as Whittington depictses them in *Sarotrocercus*.

Concluding remarks on Scorpiones, Eurypterida, Xiphosura, Synziphosura, Diploaspida, Chasmataspida, Cyrtoctenus

While the Eurypterida have been traditionally placed close to the Xiphosura, the Scorpions have usually been considered to be rather primitive Arachnids. This assumption appears to be quite questionable in the light of recent discoveries.

Devonian and Silurian Scorpions had well developed gill fringes on the upper side of the ventral plates that, instead of being movable sternites, as they had usually been considered, must now be considered as modified opisthosomal limbs covering the real, unsclerotized sternal surface. In spite of Kjellesvig-Waering (1986) rather confused discussion, the morphologic and embryologic evidence is perfectly compatible with the essentially homology of lung-books with gill fringes, and with the incorporation of the originally free-hanging abdominal plates into the body wall.

Moreover the aquatic Scorpions lacked the specialized gnathobases of the first and second walking legs, which appear to have evolved as an adaptation to external digestion in a terrestrial habitat. The transitional lower Devonian *Branchioscorpio* is apparently the first to have evolved them, though still retaining a primitive type of respiratory apparatus. The homology of the pecten of Scorpions with the outer branch of a "trilobitic" appendage has long been admitted and fringed, comb-like outer branches of appendages occurred in the gigantic, Hibbertopteroid Eurypterid *Cyrtoctenus*.

Thus the morphologic gap between the Scorpiones and the Eurypterid-Xiphosurid stock has narrowed considerably, while the discovery of several orders of living Arachnids (Oribatid Acari, a probable Araneid, etc.) together with extinct terrestrial orders (Trigonotarbi) in the lower and middle Devonian, has correspondingly widened the gap between the Scorpions and other Arachnids, which are now known to antedate the earliest Carboniferous terrestrial Scorpions by millions of years.

It now appears that the Scorpions, the Eurypterida, the Xiphosura, Synziphosura, *Diploaspis* and *Chasmataspis* have independently evolved from more primitive forms having the following features:
1) a prosoma of between 6 and 8 postoral metameres; 2) preoral antennae either reduced or absent; 3) an opisthosoma which may tentatively be supposed to have had 12-13 segments, not counting the telson[8]. Indeed, though an increase in the number of opisthosomal segments cannot be ruled out, there being no absolute criterion to decide the primitive number of opisthosomal segments, a reduction in the number of apparent metameres is intrinsically more probable.

Tab. II

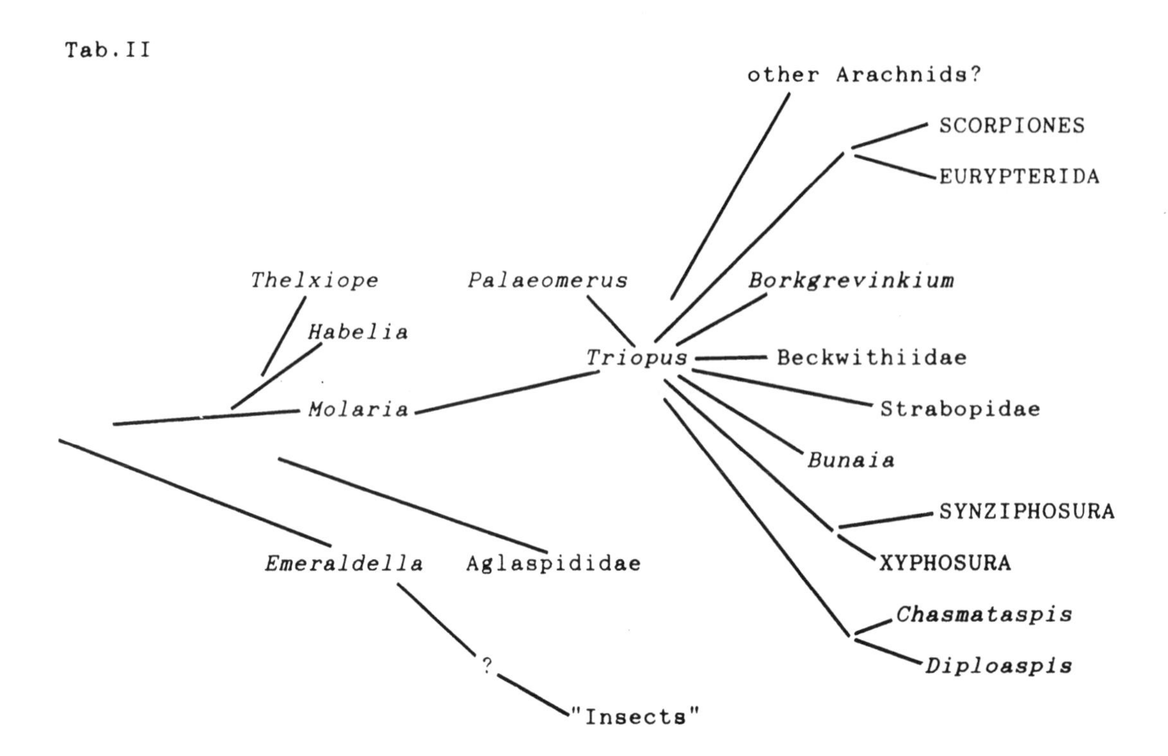

As all the post-Cambrian Chelicerata have their appendages strongly differentiated in two groups (the prosomal ones as grasping, feeding and walking organs, with the outer branch more or less completely reduced, while the opisthosomal ones are specialized for swimming and/or as sensory and respiratory organs, with the inner branch reduced), it follows that their Cambrian ancestors must have had a full set of unspecialized or poorly specialized appendages of trilobitic type.

If we now turn to consider the relationships of the Scorpions and of their relatives (Xiphosura, Eurypterida etc.) to the other traditional Chelicerata we must consider that: 1) as we have said, there are unquestionable Trigonotarbi, Pseudoscorpions, Oribatid mites Arthropleurids, associated with Collembola and a Chilopod, in different localities of the Lower and Middle Devonian, all considerably older than the Hunsrück Shale, which jelded *Waeringoscorpio*. Moreover, the Gilboa arachnid, insect, chilopod fossils, actually antedate by several millions years the Rhynie Chert with its mixed, terrestrial and aquatic fauna, which includes the oldest known Palaeophonid. It now appears that all lobosternal Scorpions were aquatic and that no terrestrial scorpion is known before the lower Pennsylvanian, though, obviously, they may have existed long before.

Taking the available evidence at face value, it appears probable that the Scorpions moved from the aquatic to the terrestrial habitat many million years after the other terrestrial Arachnids, the Centipedes and the Apterygote insects. It thus appears that there is little to commend the traditional grouping of the Scorpions within the Arachnida.

If we assume a 12-segmented opisthosoma for terrestrial Chelicerata (the actual segmentation is Palpigradi 11, Uropygi 11, Amblypygi 12 , Ricinuleida and Acarina obscure, Pseudoscorpionida 12, Solifugae 10, Araneida 12, Opiliones 9), then we see a basic correspondence with that of the Xiphosurid-Eurypterid stock, and this may indicate a true relationship, but their possible common ancestors must be looked for back in the Ordovician.

The possible relationships of the various taxa as we presently see them are tabulated in the dendrogram Tab. II.

Other Arachnida (Fig. 12)

We believe that the whole problem of the phylogenetic relationships between the various Arachnid orders needs to be entirely reexamined. A fresh study of the relevant material, especially the Carboniferous fossils, is badly needed. It must be remembered that we are still almost entirely dependent on Petrunkevich's contribution (1955). In addition we doubt that Petrunkevich work (for instance on the very interesting *Plesiosiro madeleyi* Pocock, 1911) may prove to be as incomplete and, occasionally, unreliable, as were some of Walcott's accounts of the Burgess shale fossils.

A very important question, both from the morphologic and taxinomic standpoint, is that of the homology of the opisthosomal sterna. We have seen that the lung-bearing "sterna" of the Scorpions are now known to be modified appendages. Does this hold also for the other orders of the traditional Chelicerata? If this proves to be the case, then the argument for an essential monophyly of the Chelicerata could be strengthened, but embryological data on the development of the respiratory apparatus in the various Arachnids are both old and limited to just a few examples taken among the Araneae. It will be particularly important to establish whether the "sterna" covering the "lungs" of those Arachnids so equipped are really sterna or rather, as it is the case in the Scorpions, are only modified appendages.

Equally interesting will be to clear up the problems relating to the rotation of the anus: Architarbi (but not Heterotarbi), Trigonotarbi, Antracomarti, Araneae and others have a ventral anus. Others have an apparently more primitive terminal one, but new investigations are needed to ascertain the true morphologic significance of these differences. Again the origin and development of tagmosis in the various orders is still far from being adequately studied. We would not be surprised if a fresh approach demonstrated that, at least the Acarina are entirely independent from the other Arachnids and that possibly they should be split into two main stocks.

Acarina

The Acarina pose a challenging problem (see also Bernini, this volume). They not only include a number of freshwater genera, but also the only true marine "Arachnids", with even some deep sea genera. Acarologists usually consider the marine species as specialized, but as primitive Oribatids occur in the lower Devonian, it seems worth while to consider the possibility that Mites reached their present basic structure either as dwellers of the intertidal zone or in shallow, possibly brackish waters, thence moving both up into dry lands and down into the sea.

Pantopoda (Fig. 12)

The Lower Devonian *Palaeoisopus problematicus* Broili,1928 is clearly a primitive Pycnogonid (= Pantopoda), while the contemporary *Palaeopantopus* and *Palaeolethea* are much more advanced. Inclusion of the Pantopoda among the Chelicerata has often been doubted. Our present knowledge of fossil Pantopods does not support such a conclusion in the sense that, while the "Chelicerata" appear to be a questionable assemblage, the Pantopod larvae seem to be comparable (if they are comparable at all) only with certain developmental stages of primitive mites.

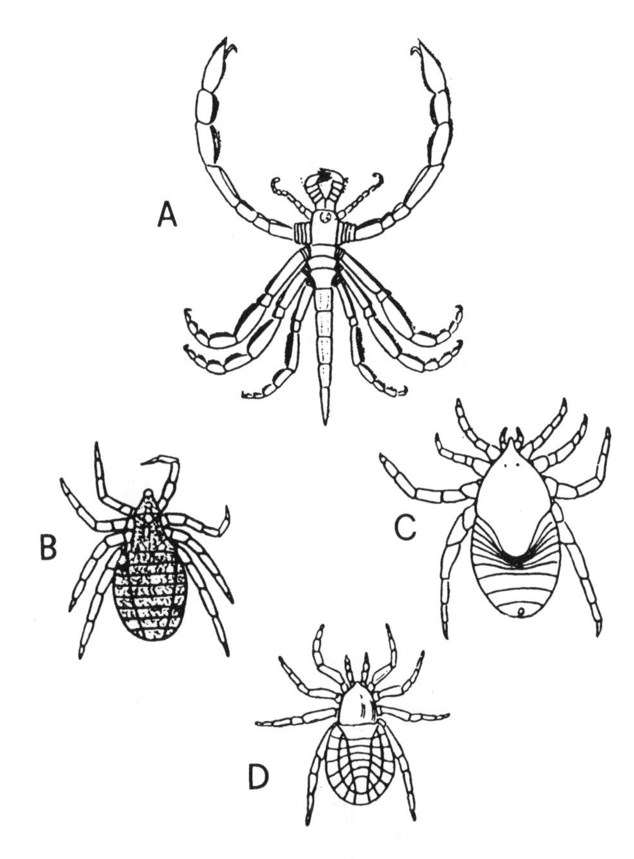

Fig. 12. A: The primitive Pantopod *Palaeoisopus*, B: *Trigonomartus*, C: *Architarbus*, D: *Cryptomartus*.

General discussion on Chelicerata

Our examination of all the taxa included in this section shows that:
i) a common origin for all the traditional Chelicerata is possible, provided that their roots are looked for in a non-chelicerate stock, morphologically like the emeraldellids *sensu lato*.
ii) both the Pycnogonida (Pantopoda) and the non-scorpionid arachnids probably derive independently from protochelicerate stocks different from those that gave rise to the Scorpions and the Eurypterids.

The following diagram is proposed as a working hypothesis (Tab. II).

Arthropleurida, Myriapoda, Insecta and possibly related forms

We have already alluded to the possibility that the search for the ancestor of Insects and Myriapods be made among the middle Cambrian "Merostomoids". It must, however, be stressed that while in the Trilobita sutures are evident in the cephalon and are prominent in the Insects, the "Merostomoids" have no cephalic sutures and sutures are equally lacking among the "Myriapods". It might, therefore, be argued that the Insects derive from animals with incompletely fused cephalic segments. As the study of the radiations of the Insects and Myriapods during the Palaeozoic are too specialized fields to be dealt with within the limits of this paper, I shall limit myself to a brief discussion of the Arthropleurida and two peculiar "Myriapod-like" animals of dubious affinity. One is a Silurian animal (Mikulic, Briggs & Kluessendorf 1985), but is still unnamed; the second, *Cambropodus*, has been described by Robison (1990) and we shall just make a brief comment on both of them.

Arthropleurida and the new Silurian animal (Fig. 13)

Nothing new has been published recently on either *Arthropleura* Jordan & Meyer or *Eoarthropleura* Stormer, and discussion of their systematic position depends on recent descriptions of other animals. These have not clarified the problem. Any of a number of different taxa may be a reasonable morphologic ancestor for these animals: a trilobite-like animal such as *Retifacies*, a "merostomoid" like *Emeraldella*, even a very primitive and rather generalized animal such as *Alalcomenaeus* may be morphologic forerunners of the Arthropleurids. *Eoarthropleura* is particularly significant as it is clearly the more primitive, of the two genera and may well have been marine. The new Silurian animal (Mikulich, Briggs & Kluessendorf 1985) from Wisconsin has not yet been properly described and the available photograph does not allow for proper comments. We may only suggest that, when described in

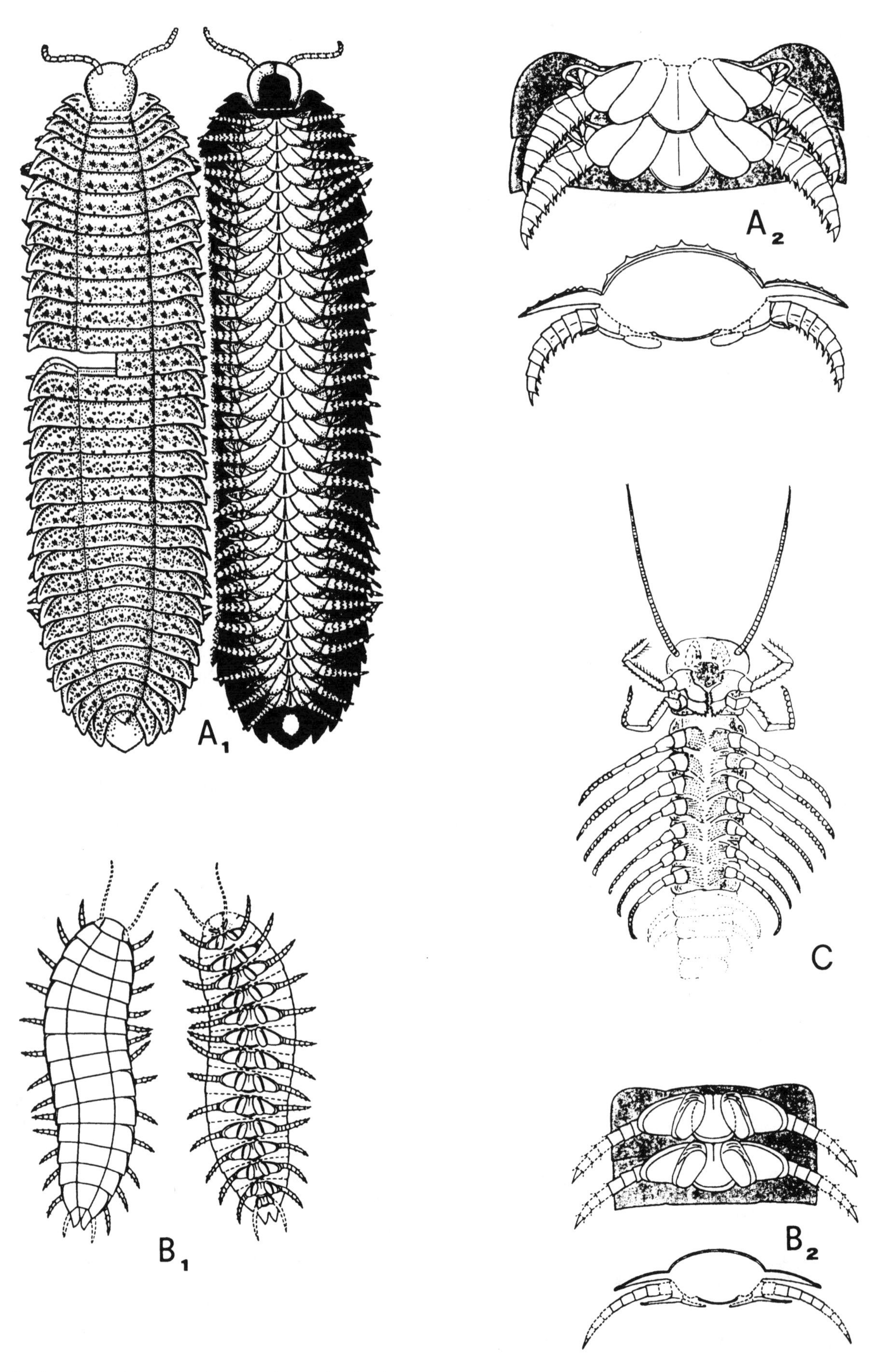

Fig 13. A: *Arthropleura*, B: *Eoarthropleura*, C: *Cambropodus*.

detail, the animal should be compared especially with *Eoarthropleura* (Fig. 13).

Cambropodus (Fig. 13)

The only known specimen of*Cambropodus gracilis* Robison 1990, is split by the cleavage plan through the body, but it is the ventral structures that are better visible. There are, at most faint and dubious indications of the outline of the terga. The legs, with their expanded 2-3 proximal articles (if our interpretation of the photograph kindly supplied by Prof. Robison is correct) followed by at least 3 long and thin articles are most peculiar and vaguely recall the long and extremely thin legs of some benthic crustaceans adapted to move on extremely loose sediments. The head includes with certainty only two postoral pairs of appendages, which were, apparently, not much specialized for feeding. There may be an indication, very doubtful, of a small pair of appendages in front of them. If the animal has some sort of delicate branchial tufts either on the pleural side of the body or on the dorsal side of the coxal region, such as those of many modern Crustacea, it is impossible to say. We have tentatively listed this genus here for lack of better evidence. Comparison with the Arthropleurida, and especially with the Devonian *Eoarthropleura* is perhaps possible. The fact that Cambropodus has long thin appendages, while the Arthropleurida have short, strong ones may or may not be significant, as extreme variations in this character occur, for instance, among the Decapoda and the Chilopoda. We agree that *Cambropodus* may be the first piece of evidence for the early evolution of the "uniramian" leg, but, as we have seen some insects still have a biramous appendages, the Cambrian genus may, at most, be an indication of an early differentiation of a "myriapod-like" lineage.

Insecta

As far as Insects are concerned the large assortment of Palaeozoic Insects shows that by the end of the Permian the Insect fauna had reached an almost modern stage of evolution, with many of the living orders already in existence. The fossils do not provide any evidence on the moot point of the inclusion of the Entognathous Apterigotes within the Insecta. The Devonian Collembolan *Rhyniella praecursor* ,1926 was probably a shore inhabitant which does not tell us much.

So far as a plausible morphologic forerunner from which the Ectognathous insects may be derived, *Emeraldella* is the best candidate. To turn *Emeraldella* into a plausible insect ancestor, it is necessary the reduction of the branchial outer branch of the legs (which even in some living Thysanurans is not completely lost, e.g. *Petrobius*), the loss of the first pair of postoral appendages, and the loss of a true telson (the published evidence is still inconclusive as to the precise morphologic significance of the various "anal filaments", cerci etc.) (Fig. 4).

Cycloidea (Fig. 14)

Nothing of significance seems to have been published on the Cycloidea in recent years.

About seven genera are included, but apparently only the tergum has been described for almost all of them. It is impossible therefore to judge which may be the most primitive of them and what were their evolutionary tendencies.

Their shortened body and vaulted tergum explains the radial arrangement of their appendages, while the wide sternal plate shows that the locomotory appendages cannot have had gnathobases. Feeding must have been the work of specialized mouthparts, which are unfortunately obscure as they are largely covered by the tergum and sternal plate which form a sort of preoral vestible. Their only crustacean-like feature is found, perhaps, in *Cyclus*, which has a sort of furca. This is, however, of very little morphologic significance. The existence of antennae makes it very difficult to suppose any relationship with either classic Arachnids or Pantopods. The abdomen must be considered to be entirely absent in at least *Halicyne* and extremely reduced or lacking in the other genera. In this feature they are as specialized as living Pantopods or Cirripedes! Relationships to the Branchiura, claimed by Hopwood (1925) can be excluded and the supposed existence of "suckers" is at least doubtful.

The fact that they are very difficult to place in any classification combined with their comparative rarity, instead of prompting fresh collecting and new investigations, has had the result that they are usually ignored in any phylogenetic discussion.

Pending new investigations the Cycloidea should be considered an entirely isolated branch of the Arthropods, of class rank.

Euthycarcinida Gall & Grauvogel, 1964 (Fig. 15)

This is also very much of an "orphan" group. Recently (Schram & Rolfe, 1982; Schneider, 1983) have added some significant new evidence to their hitherto very poorly known anatomy. The new data, however, provide somewhat contradictory evidence on the crucial cephalic region, and lend themselves to somewhat different interpretations of the cephalic structure. If we bear in mind the corresponding structures in *Sidneyia*, we may see that it is perfectly possible to consider the structures described by Schram & Rolfe (1982) as evidence of an extremely primitive head formed by two elements: the first comprising the labrum (their "head plate") and a tergum covering the basis of the eyes and antennae, and a postoral diplosomite corresponding with their sternites 1 and 2, in which appendages are still unknown. Schneider's (1983) interpretation points to a much more advanced head, made of possibly 4 segments entirely fused. In any event we wish to call the attention to the striking similarity of the larval stages of the Euthycarcinida with some of the "Xiphosuroid" animals.

We have seen that the concept of a uniramian appendage as maintained by Manton (e.g. 1979) is untenable. The appendages of the Euthycarcinids must be interpreted, therefore, independently from preconceived ideas. Their functional interpretation is easier than a morphologic one. If we suppose that these animals (which are comparatively small) moved about the bottom somewhat like living Notostracans, a multijointed, fringed appendage may have been function-

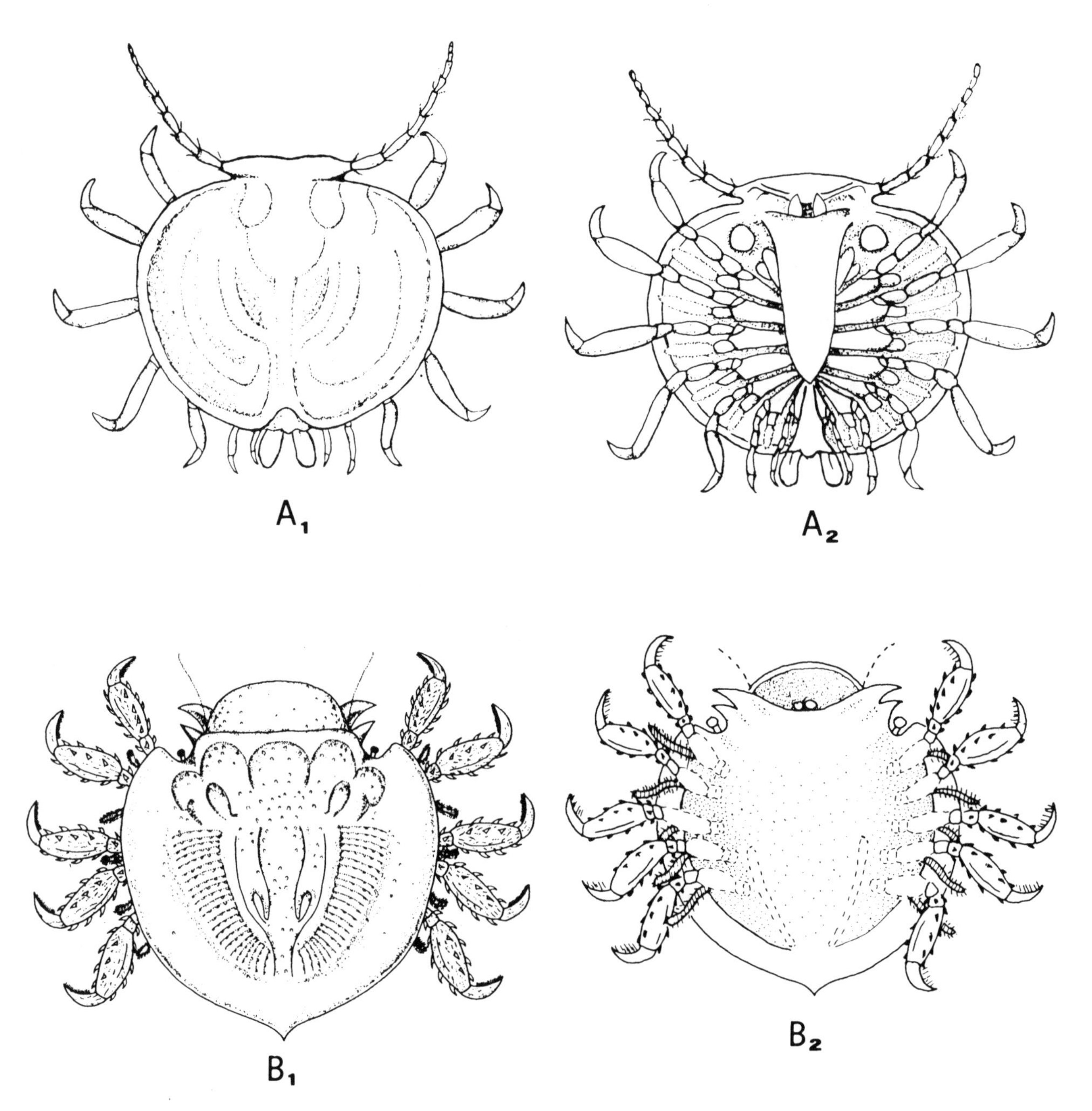

Fig. 14. The Cycloidea, A: *Cyclus*, B: *Halicyne*.

ally acceptable. However their wide sternites require a feeding mechanism very different from that of phyllopods. Just as in the Kazacarthra, their very sporadic occurrence (though known from different continents, they are rare and of extremely local distribution) points to very peculiar ecological requirements.

We are now inclined to doubt the interpretation which we proposed previously (Delle Cave & Simonetta,1980,1981). Although not excluding the possibility of a remote affinity with the Emeraldellida, we prefer to consider, at the moment that the Euthycarcinida are an entirely independent Arthropod branch.

Origin of traditional crustaceans: Preliminary remarks

We must say that we shall not propose a formal taxinomy of these groups, but rather suggest a working hypothesis of their possible phyletic relationships. We feel this may be possible only by a drastic extension of the definition of "Crustacean", which may be unacceptable to many carcinologists. For this reason consistently we refer only to "crustacea".

Starobogatov (1986) and Schram (1986) have proposed the most recent formal arrangements.

We think it necessary to discuss separately the origin of those "crustaceans" which are provided with a carapace, as against those which are either devoid of one or have apparently acquired a carapace as a late specialization. The term "carapace" has been used too loosely in arthropod anatomy. The strict usage, which we shall follow, is linguistically wrong. Indeed it is unfortunate that the term "carapace" is currently used for structures shielding the animals, while a correct linguistic usage[9] would have help to focus our attention on significant morphologic features. Although we shall use further on "scutate" and "loricate" as provisional terms to indicate some phyletic assemblages, we shall continue the traditional usage of the term "carapace", but in the strict sense defined by Lauterbach (1974) and

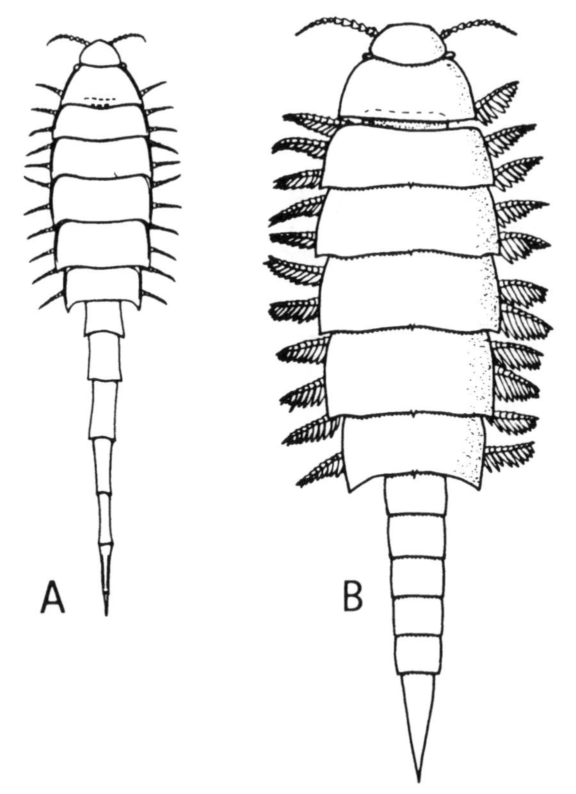

Fig. 15. Examples of Euthycarcinids, A: *Kottixerxes gerem*, B: *K. gloriosus.*

Hessler & Newman (1975): the expanded tergum of the third postoral segment covering the neighbouring postcephalic segments, which at least primitively remain free. The carapace may also extend cranially covering more or less completely the cephalic structures or fusing with them. The carapace may be undivided, but as it is primarily formed by the expanded pleurotergites of the last cephalic segment, it is generally formed by two valves. The postcephalic segments covered by the carapace preserve, in the more primitive condition, their terga. However when their appendages become specialized as maxillipedes, they tend to fuse dorsally with the carapace. Thus we have a condition, particularly in adults, where several segments may be completely incorporated in the head and the carapace fold appears to grow from the last fused thoracic segment. In such a comparatively primitive group as the Mysidacea we find situations ranging from genera in which the carapace covers several free thoracic segments to genera in which there is practically no carapace fold left, because all the thoracic segments have become fused with the head.

The functional significance of the carapace is quite different in the various groups, and sometimes even within the same order. It is, therefore, apparent that a carapace may have been independently evolved several times. The development of a carapace allows for the evolution of different kinds of branchial chambers, for specialized ways of filter feeding, for rigidity of part or all of the thorax and thus for better leverage to specialized mouthparts or walking legs, etc.. The development of a carapace is invariably accompanied by the diverging specialization of those segments that are encased or shielded by the carapace as against those which are left uncovered.

In many groups the abdominal sclerites tend to become ring-shaped and the appendages become either swimmerets or tend to be reduced and may be even lost altogether. In several instances, such as the Ostracoda and the Cirripeds the whole abdomen (and even the thorax) may be extremely reduced so as to be effectively withdrawn within the carapace. On the other hand in all the segments covered by the carapace, the pleurotergites, which have lost their function, are either reduced or lost.

It is, therefore, apparent that the development of a carapace involves a number of correlated changes. Even if, under certain circumstances such as the adoption of a mesopsammic habitat, the carapace may become reduced or lost, it is practically impossible that the lost pleurotergites or legs may re-evolve. So compare two living orders: the Cephalocarida and Mystacocarida. The first have unspecialized thoracic segments all of which are provided with pleurotergites. The Cephalocarids must be considered as animals whose ancestors were never provided with a carapace. In contrast the Mystacocarida might be descended from animals which had this structure and then lost it when adapting to a mesopsammic habitat.

If we accept this premise together with the assumptions that i) cephalization of segments and specialization of the mouthparts were progressive events which were independent, but parallel and occurred sometime later than the acquisition of other crustacean features, and ii) that the biramous crustacean leg derives from a trilobitomorph appendage, there follow several important consequences.

On our assumption ii) we only add to our previous discussion (p. 195) that as the coxal region is usually the most poorly known part of the appendage of crustacean-like species, we do not know whether any of them had subdivided the trilobitomorph coxa into coxa and basis. This subdivision however does not occur in the early larval stages of many modern crustaceans.

As far as our assumption i) is concerned, we feel that our traditional concept of crustaceans as animals provided with two pairs of antennae and three pairs of mouth appendages must now be abandoned. From an evolutionary standpoint, we see that the larval development of the Crustacea involves the addition of segments to the nauplius. In principle, the appendages grow more and more specialized with every moult, and also in a cranio-caudal direction. In the living Cephalocarida there are only three pairs (2nd antennae, mandible and 1st maxillae), while other groups have between three to six pairs of appendages specialized as mouthparts. The fact that the Cambrian ancestors of modern groups may have had less than two pairs of specialized buccal appendages should not be surprizing.

If our hypothesis is accepted, it follows that comparisons with taxa which do not fit into a traditional definition of "Crustacea" are permissible.

The loricate stock

We group under this informal name those crustaceans and crustacean-like animals that do not have a carapace, and that are clearly descended from animals equally devoid of this structure. We have chosen this term with reference to the fact that their armour recalls the segmented body armour, the "lorica" of the imperial Roman legionaries.

Alalcomenaeus Simonetta, 1970 (Fig. 16)

Alalcomenaeus cambricus Simonetta 1970, and *Alalcomenaeus illecebrosus* Hou 1987, are such exceedingly primitive arthropods that almost any arthropod ultimately could be derived from them. Their morphology is illustrated in Fig. 16. It is worth discussing the "telson". Practically all the published specimens are laterally compressed and appear to terminate with two spatulate cerci. However an incidental comment in Collins (1986), the Chinese specimens, and the specimens of *Acanthomeridion* (see below), show that the last body unit was a plate flanked laterally by large flaps which originally would have sloped downwards and backwards. We cannot exclude the possibility that, if the anus was at the end of this plate, it would not be a true telson (see Section II) and the "flaps" would be normal if elongated pleural lobes. We prefer, however, to consider the whole structure as a telson, the anus opening just in front of it. Such a telson is exceedingly primitive, as it would be patterned exactly on the body terga. In itself this is a perfectly satisfactory hypothesis, as the telson is basically formed by dorso-lateral material and would be expected to develop in a tergum-like structure in its most primitive form. The importance of *Alalcomenaeus* is that it is such a primitive animal that it may be a morphological forerunner for almost any "crustacean" or at least those "crustaceans", such as the Anostraca, Syncarida etc, which appear to derive directly from animals primitively devoid of carapace. We must, however, add that judging from some good photographs of Chinese material and of two specimens brough to Italy by our Chinese colleagues, we suspect that the Chenjang material includes two animals of approximately the same size, one being an *Alalcomenaeus*, the other a small and primitive species of *Leanchoilia*.

Acanthomeridion Hou, Chen & Lu, 1989 (Fig. 16)

Acanthomeridion serratum is the only representative of this genus. Its authors provide only a dorsal reconstruction, but judging from the photographs we are tempted to suggest the oblique reconstruction of the posterior end of the body as figured in Fig. 16.

Nothing being known of the appendages, any allocation of this genus is almost a shot in the dark. We are nevertheless tempted to place it provisionally rather close to *Alalcomenaeus*.

Leanchoilida (Fig. 16)

We tentatively include here both *Leanchoilia* and *Actaeus*, as we accept the reconstructions by Whittington (1981) and by Bruton & Whittington (1983). These warrant the inclusion of these two genera in the same order.

The problem of the number of species to be included in the genus *Leanchoilia* is irrelevant for the purposes of this paper. We have no objection to having the species which we described considered as synonyms, though we wish to point that our *Leanchoilia protogonia* Simonetta, 1970 appears to be closer to *Leanchoilia? hancyi* Briggs & Robison, 1987 than to *Leanchoilia superlata* Walcott, 1912. As stated before we consider that those Chinese specimens that show the first appendages extending into three long flagelli probably belong to *Leanchoilia* and look very much like our *L. amphiction*.. We hope that a reexamination of the specimens available may clarify this point.

Actaeus Simonetta, 1970 (Fig. 16)

The morphology of the first appendages, as clarified by Whittington (1981), plainly links *Actaeus* with *Leanchoilia*. This genus, however is much more advanced in the differentiation of its tagmata. Its large eyes are a further point of differentiation from *Leanchoilia*. Both *Leanchoilia* and *Actaeus* may well be derived from such an extremely primitive forms as *Alalcomenaeus*.

Unnamed Silurian animal (Fig. 17)

Mikulic, Briggs & Kluessendorf (1985) have given a preliminar description of this animal. We do not see why in their discussion they did not compare it to *Actaeus*, to which it seems to be clearly related.

Yohoida (Fig. 17)

In our view there is no doubt that *Jianfengia multisegmentalis* Hou, 1987 is a primitive member of the Yohoida. The body of the animal is similar to *Yohoia tenuis* Walcott, 1912, but both the cephalic tagmon and the last segments are less specialized and closer to those of such animals as *Alalcomenaeus*. The first cephalic appendage has the same outline of that of *Yohoia*, but again it is less specialized, lacking the terminal movable spine and having instead a bunch of simple, immovable small spines.

While the other appendages are built on the same pattern, there is no loss of the abdominal legs, which all bear similar appendages. As far as *Yohoia* is concerned there is little to add to what I have previously written. Whittington (1974) has reiterated his opinion that there is not sufficient evidence on the fossils either for what we supposed to be the labrum or epistome or for a secondary branching of the inner branch of the legs. It is possible that we may have misunderstood some features on some specimens, but comparative anatomy demands the existence of a labrum, even if this was not the structure that we so interpreted. Even accepting Whittington's reconstruction rather than ours, the Yohoida stand as the most "crustacean-like" of the leanchoilid-yohoid ensemble.

Lipostraca, Anostraca (Fig. 18)

The rather close relationship between the Lipostraca and the Anostraca has never been doubted. What appears more doubtful is where to look for their origins. The fact that in some Anostraca the first antennae (antennules) of the males are modified as a grasp-

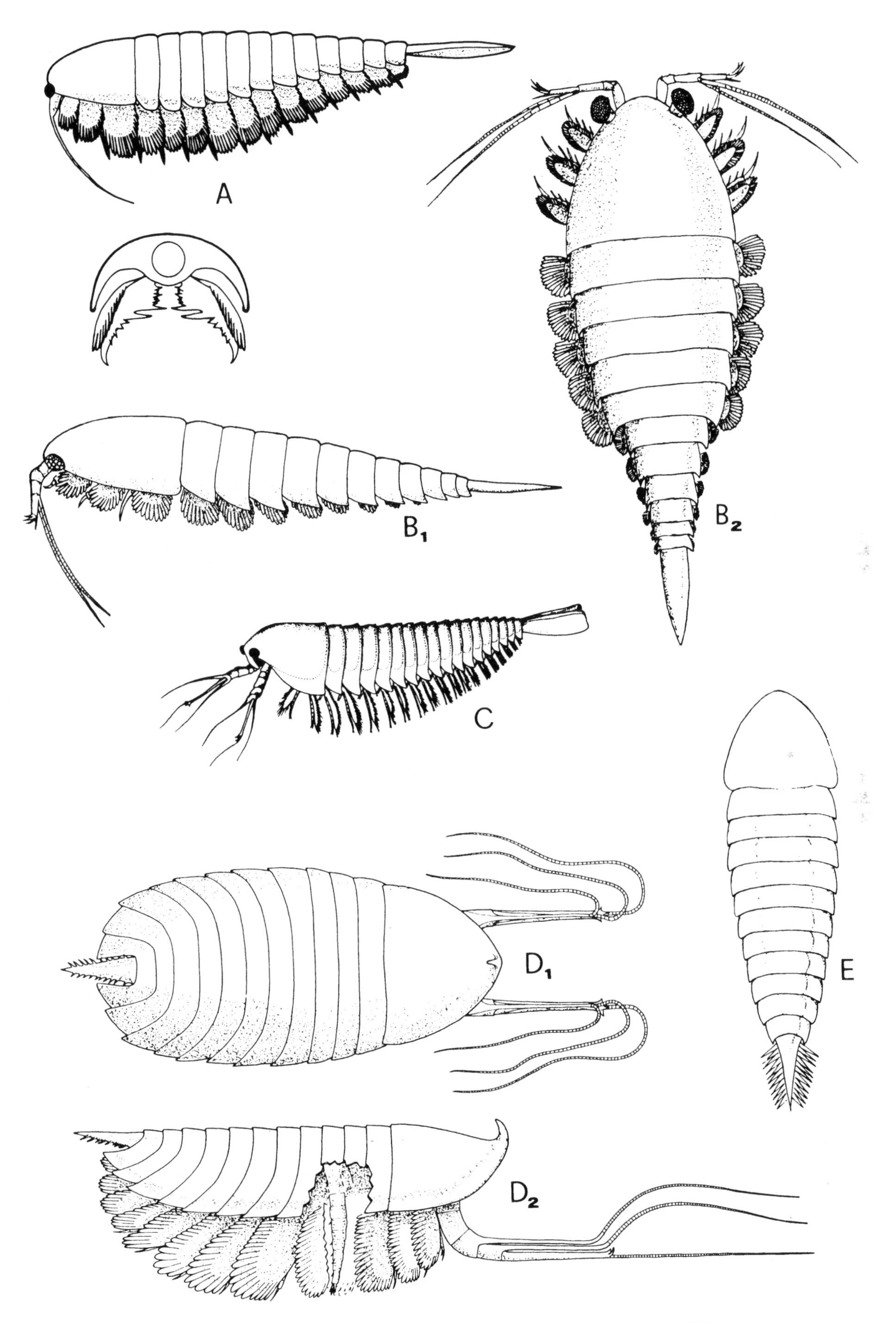

Fig. 16. A: *Alalcomeneus cambricus*, B: *Actaeus*, C: *Leanchoilia amphiction*, which we deem a valid taxon close to the Lower Cambrian Chinese Leanchoilid, D: *L. superlata*, D: *L. persephone*.

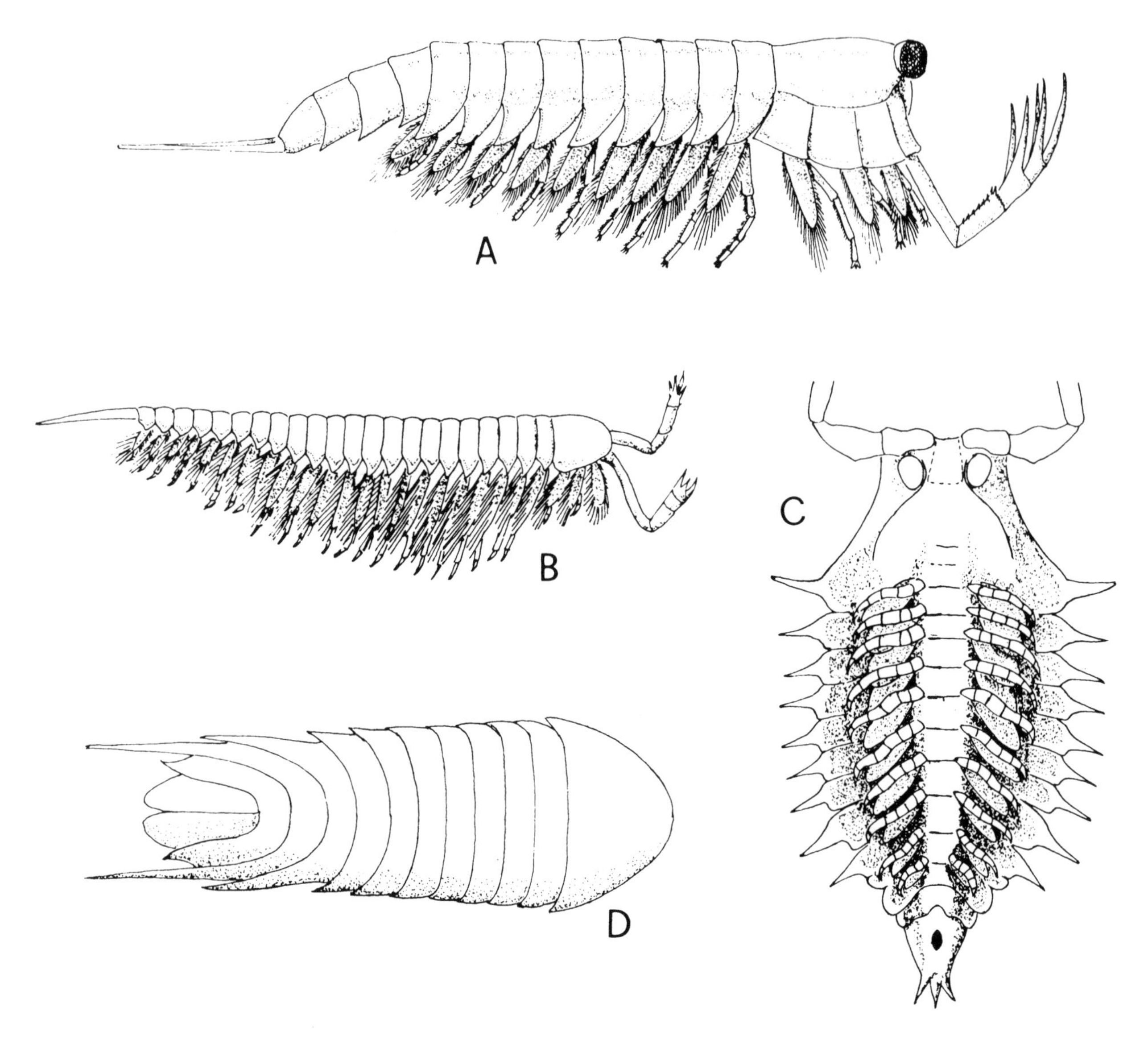

Fig. 17. A: *Yohoia*, B: *Jianfengia*, C: the unnamed animal of Mikulic et al., D: *Acanthomerion*.

ing organ is significant only to the extent that it shows that even the antennules may, occasionally, evolve as grasping organs. They are, however, structurally very different from the raptorial antennae of both the Leanchoilida and the Yohoiida.

Tesnusocarida, Nectiopoda (Fig, 18)

The Carboniferous *Tesnusocaris goldichi* Brooks (1955) and the living Nectiopoda have been grouped by Schram (1986) in his class Remipedia. This is a suggestion that stands about an equal chance of either being right or wrong. If we accept the morphologic principles expounded in Section I of this paper, it is clear that the characters common to the Nectiopoda and *Tesnusocaris* may all be derived from those of a primitive Arthropod such as *Alalcomenaeus*. In contrast, each one of them, taken separately, is known to have been evolved independently several times in the history of Arthropods: the "maxillization" of three pairs of legs (their details being as yet not entirely clear in *Tesnusocaris*), evolution of the antennae and antennulae, simplification of the inner branch of a biramous "trilobitic" leg, or of even much more advanced appendages (we can not see why Bergström (1979) did consider *Tesnusocaris* to be a "Uniramian"). Therefore, even if we are to use Hennigian principles, it is impossible to decide whether or not we have here synapomorphies. The only positive thing that may be said is that the two orders belong to that vast array of "crustaceans" which do not derive from ancestors provided with a carapace.

Cheloniellida (Fig. 19)

The single species, *Cheloniellon calmani* Broili, 1932, is usually considered as a completely isolated genus. However, *Cheloniellon* may not be so aberrant as usually considered. It could be either related to the loricate "crustacean" stock or be rather distantly linked with those taxa that we have listed as of possible Trilobite affinity (p. 199).

The long antennae are primitive, while the five pairs of specialized post-oral appendages are interesting as they are attached to at least two separate segments. Stürmer & Bergström (1978) attribute two terga to the head and it seems that but three pairs of ap-

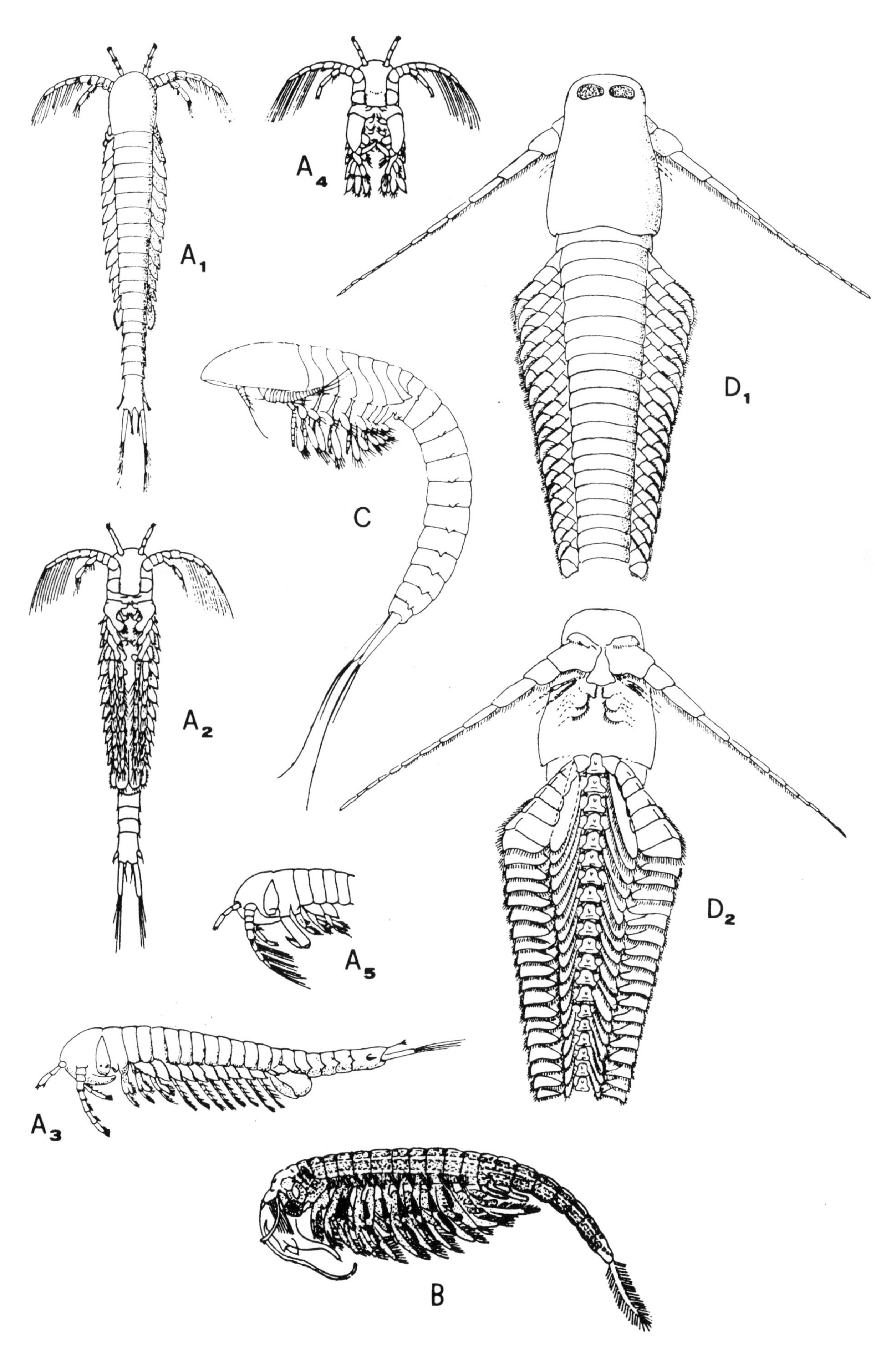

Fig. 18. A: *Lepidocaris* (A4-5 males), B: the living Anostracan *Chirocephalus*, C: the living Cephalocarid *Hutchinsonella*, D: *Tesnusocaris* (new reconstruction following Schram).

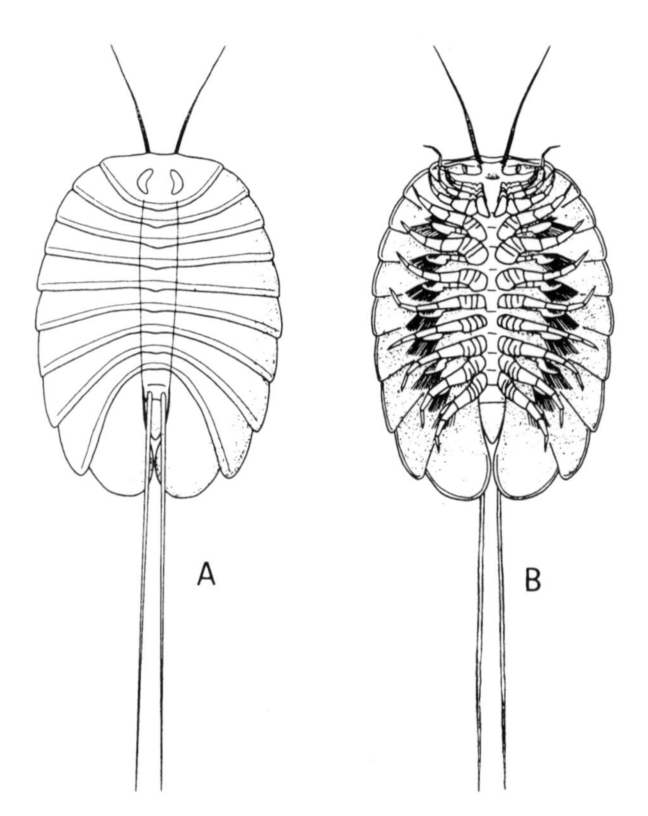

Fig. 19. *Cheloniellon.*

pendages correspond with the first tergum: the preoral antennae, and the first and second post-oral appendages. The first post-oral appendage being specialized and quite different from the following ones, and the second represents the first pair of mouth parts. The following three pairs of maxillipedes may well have all been attached to a single segment (as believed by Bergström), or to two, if the last one is attached to the segment covered by the third tergum.

That the head segments may be only partly fused or even completely free would not, by itself, debar *Cheloniellon* from being related to some "crustacean" stock. The features that have, so far, prompted the consigning of *Cheloniellon* in the "trash basket" of *incertae sedis* are i) the assumption that as it has "trilobitic appendages", this removes it to the distant galaxy of the "trilobitomorphs" and ii) that any animal belonging to the Crustacean lineage would either have only three pairs of post-oral appendages specialized as jaws, or have some "thoracic" segments fully fused to the head.

We still believe that specialization of the appendages and development of tagmosis were largely two independent processes in many lineages during the Palaeozoic, and that the position of *Cheloniellon* (and of the possibly related Ordovician genus *Duslia*) should be reassessed. *Cheloniellon* might possibly prove to be a late and somewhat evolved survivor of an early radiation of the stock leading to the "crustaceans" which either never developed a carapace or did it late in their evolutionary history.

Other "loricate crustaceans"

We shall not discuss here in any detail several other "Loricate" Crustaceans, including the Syncarida etc.. There are certainly a number of problems. For instance, if we accept that in these groups the postcephalic structure of the segments precludes their derivation from "scutate" crustaceans, not only must the traditional notion of Malacostraca be exploded (Simonetta & Delle Cave, 1978, 1981), but it becomes difficult to explain, such groups as the Peracarida, as either we credit some of them with very late acquisition of a carapace, or we must assume convergent evolution of the oostegites. We shall follow Schram (1969), who removed the Archaeostraca and the Leptostraca from the Malacostraca, though this implies a late and independent development of the carapace in different lineages of Malacostracans. In addition a revision of the original material of the obscure Devonian genus *Oxyuropoda* is clearly needed. The "undetermined Arthropod 2" of Conway Morris & Robison (1988) with apparently only 2 postoral segments in the cephalic tagmon also deserves further investigation. We tentatively consider it as close to *Alalcomeneus.*

The "scutate group"

We informally use the term "scutate" for those "crustaceans" which during the Palaeozoic developed either a univalve or a bivalve carapace. Schram (1969, 1986), Briggs & Whittington (1985), and Briggs (1977, 1978, 1983) have at least tentatively considered the opportunity of grouping the "scutate crustaceans". However, while Schram (1986) rightly, we think, concluded that the Archaeostraca and Leptostraca had to be removed from the Malacostraca, Whittington and Briggs (1978, 1981b) disregarded, however, the implications in some Burgess Shale taxa of the extremely primitive morphology of the cephalic and thoracic appendages, of the loss of the pleurotergites and of the loss of the abdominal appendages, so forcing them into the Malacostraca in the traditional sense.

Though we consider that general morphological considerations and the fossil evidence allow for alternative arrangements (e.g. instance some groups such as the Decapoda may be morphologically derived from primitive scutate ancestors similar to an archaeostracan provided with normal appendages on all abdominal segments), we have largely followed Schram (1986) for the arrangement of the post-Palaeozoic higher taxa.

Ostracoda (Fig. 20)

Ostracoda are recorded frome the lower Cambrian, though we doubt that the Archaeocopida are strict Ostracods. This is because their valves, the only known structures, lack the more diagnostic features of the Ostracod carapace: the muscle scar is in a different position, the carapace itself is poorly calcified, the valves are incompletely separated, etc. The Archaeocopida may eventually prove to be the true ancestors of classical Ostracods.

The Phosphatocopina in turn might be better considered as a separate order within a superorder including also the Ostracoda.

Isoxyda (Fig. 20)

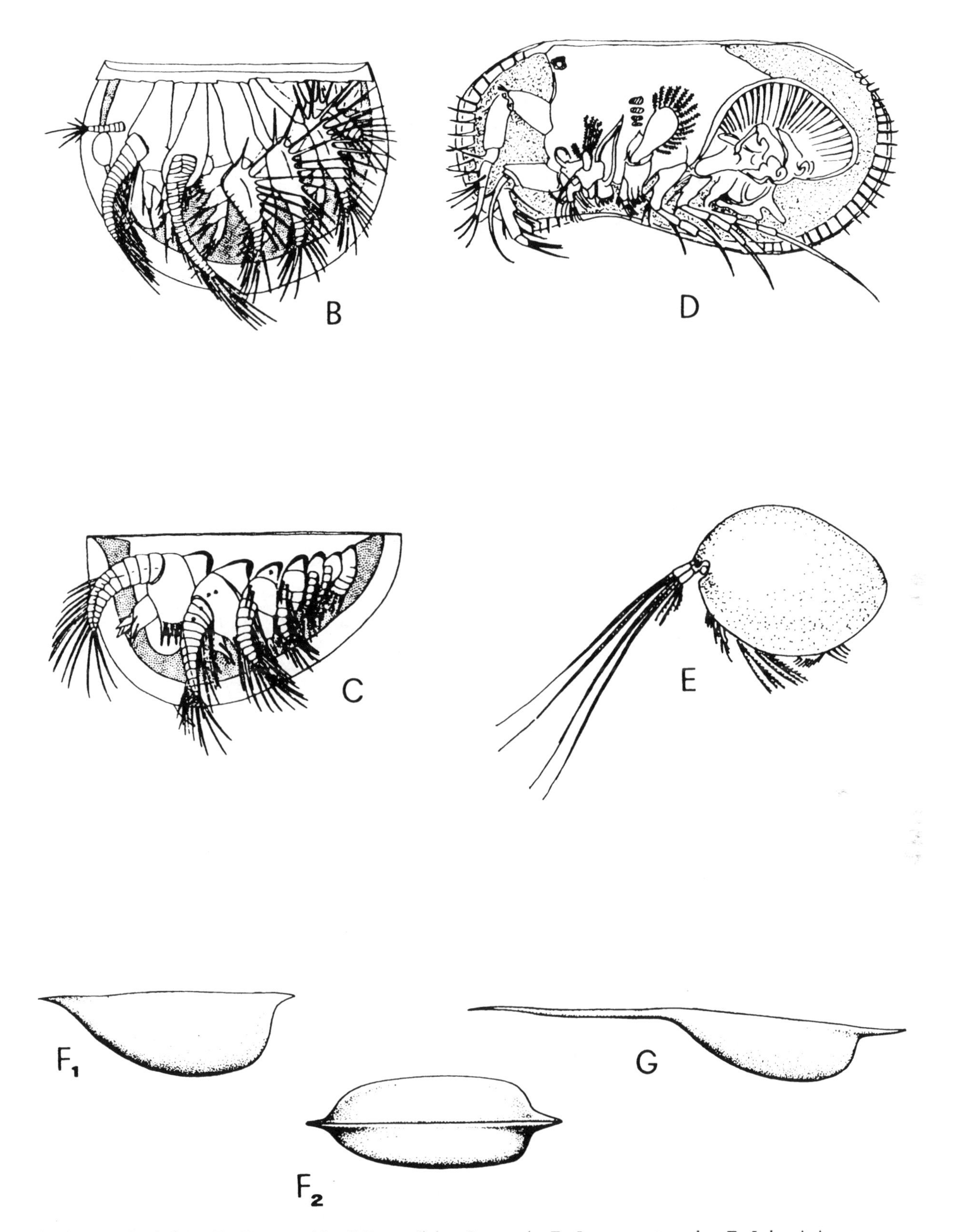

Fig. 20. A: *Hessladona*, B: *Vestrogothia*, C-D two living Ostracods, E: *Isoxys acutangulus*, F: *I. longissimus.*

The Isoxyda are listed here for mere lack of evidence as to their position. Though a widespread and at times abundant group, the only thing that may be said of their morphology is that their body was completely enclosed by the carapace.

Protocarida, Conchostraca (Fig. 21)

Protocaris is here considered to include also the genus *Branchiocaris* as we consider that differences which would warrant generic or even higher distinction as measured with the yardstick of modern carcinology, were in the Cambrian at most of specific level. One of us (Simonetta, 1987,1988, in press) has given reasons for this assumption, which are in agreement with arguments, by Valentine & Ervin (1987).

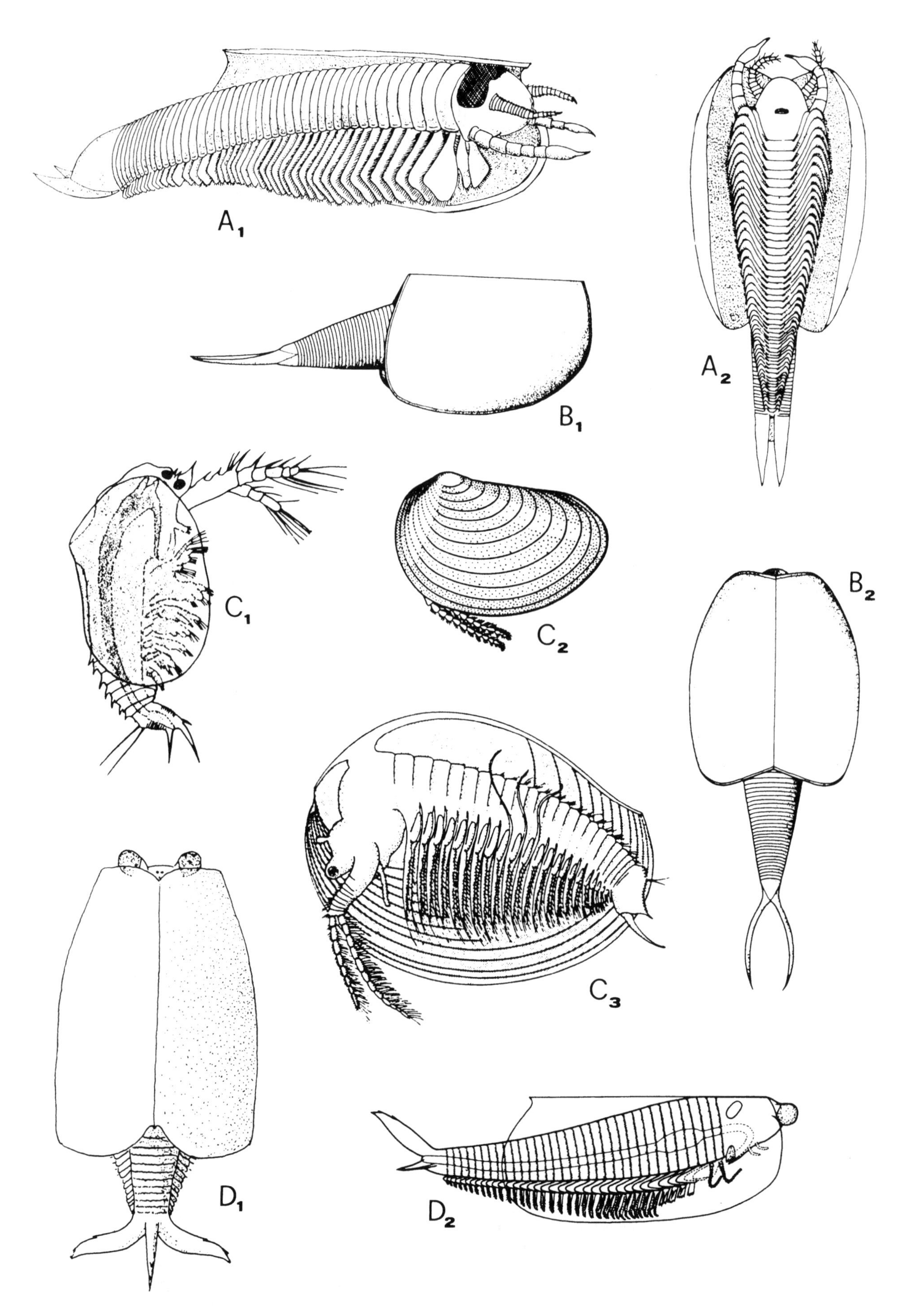

Fig. 21. A: *Protocaris (= Branchiocaris) pretiosa*, B: *Protocaris marshi*, C: living Conchostracan (C1 juv. *Cyzicus*, C2-3 *Estheria*), D: *Odaraia alata*.

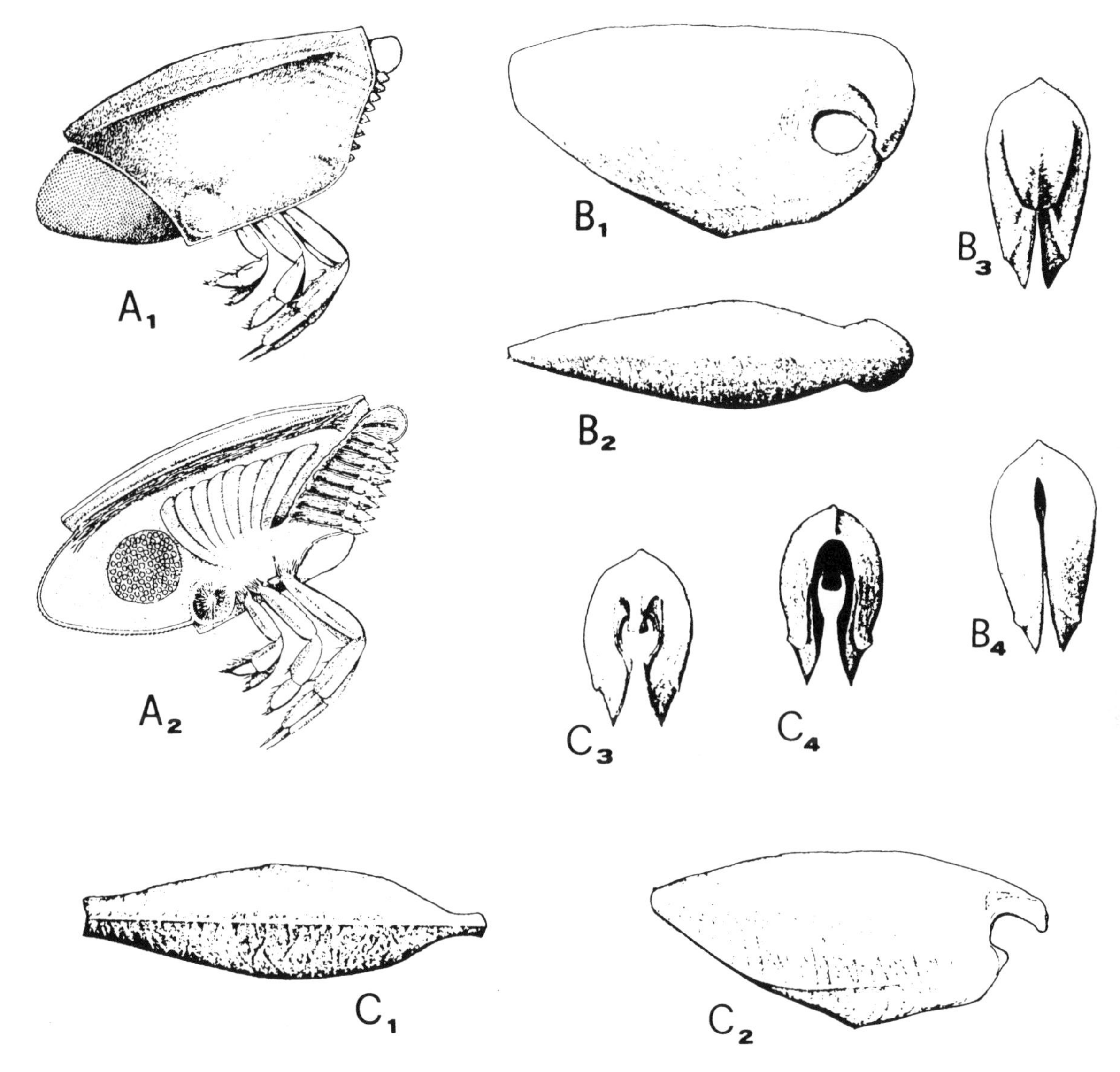

Fig. 24. A: *Ostenocaris*, B-C: examples of Concavicarids.

are a number of Arthropods which have one or more pairs of spines, filaments etc, on the sides of the anus. These are not true uropods, but have their origin from a dorsal anlage just as with the typical telson. So a "trifid" telson is not very surprising.

As for the position of *Odaraia* with respect to the other "scutata", our argument is essentially the same as for *Protocaris*: the progress of cephalization and specialization of the cephalic appendages during the Cambrian proceeded in a variety of directions and different tempos. To argue on the basis of their cephalic appendages that some scutate arthropods "were Crustaceans" and others "were not", assumes that only one group of Cambrian animals was on the path to acquiring the complement of head appendages traditionally considered typical of the Crustacea. In the light of available evidence this appears most improbable.

Other scutate genera (Fig. 22)

A number of genera, such as *Vetulicola, Proboscicaris, Papiliocaris, Tuzoia* etc. defy all systematic placement for lack of evidence other than their carapace (apart from *Vetulicola cuneatus* Hou, 1987 and for *Ohiocaris wycoffi* Rolfe, 1962). The evidence of *Vetulicola* and of the later, possibly related, genera such as *Pephricaris* point to the probability that at least *Tuzoia, Papiliocaris, Ohiocaris* and *Carnarvonia* had a rather long and, perhaps, slim abdomen protruding well beyond the carapace. This would make them superficially like large Hymenocarids or Archaeostracans. We tentatively, therefore, keep them close to the group of the Hymenocaridida etc.

Unnamed Silurian bivalve animal (Fig. 23)

Mikulich, Briggs & Kluessendorf (1985) gave a preliminary description of a curious animal, without naming it. The published evidence is insufficient to gather a reasonably reliable idea of the animals, but some features are clearly significant: the carapace is clearly thoracic, rather than cephalic, a feature which clearly sets it apart from all the genera discussed so far. The only arthropods with which this animal can be compared on account of its large free head and protruding abdomen are the Cladocera, especially such forms as e.g. *Leptodora* (Fig. 23). Obviously, until more is known about this new Silurian animal any such comparison remain very tentative.

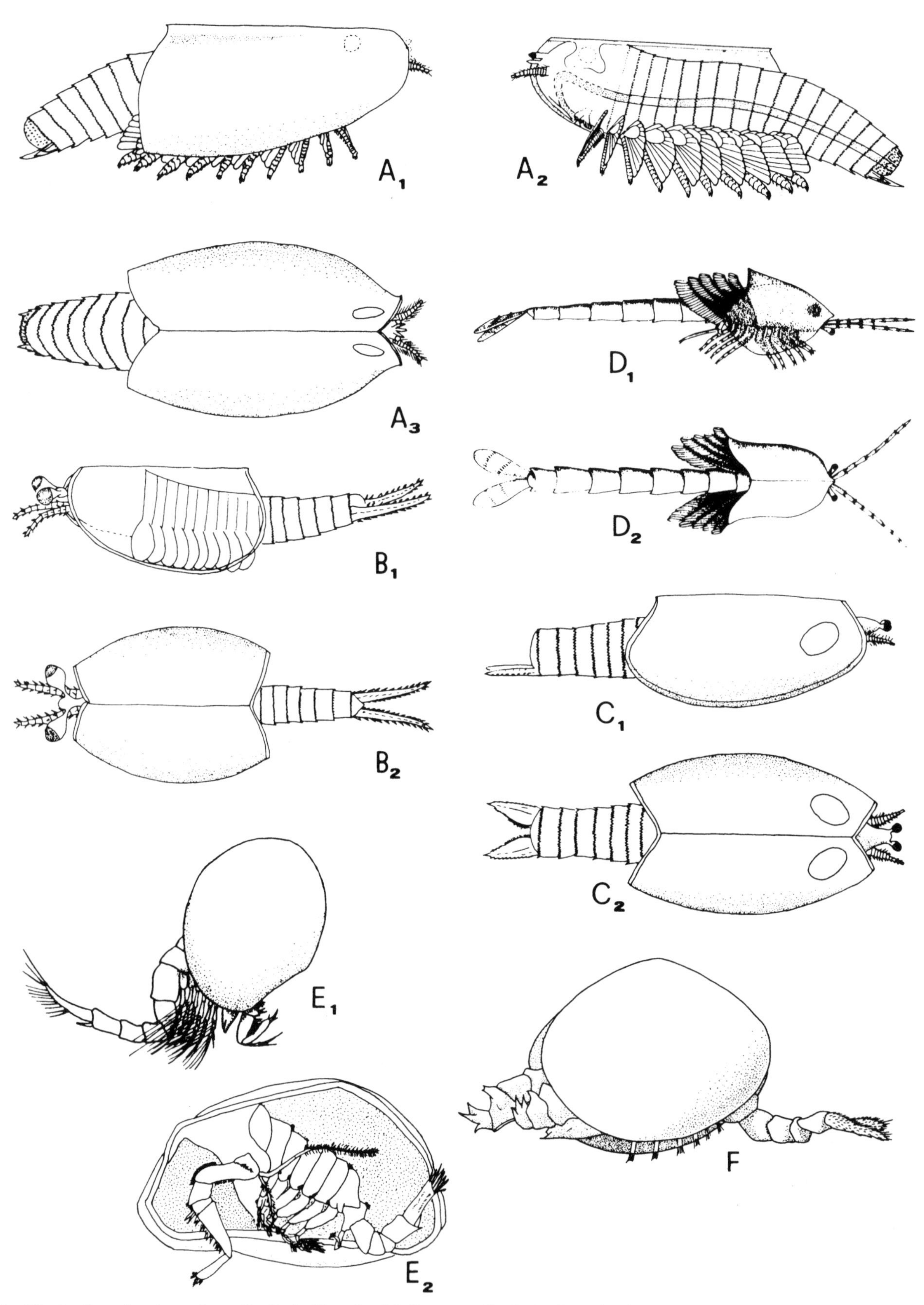

Fig. 25. A: *Canadaspis perfecta*, B: *C. (= Perspicaris) dictynna*, C: *C. (= Perspicaris) recondita*, D: *Waptia*, E: the living *Synagoga*, F: the living *Ascothorax*.

Hoplostraca, Douglasicaris (Fig. 23)

The two genera *Kellibrooksia* and *Sairocaris* are mentioned here only in order to stress that, on the evidence available, there is no special reason to consider them particularly close to the Archaeostraca. As far as *Douglasicaris collinsi* Caster & Brooks, 1956 is concerned we consider that as their only significant difference with respect to *Waptia* is in its cerci, this can hardly be considered to be of great significance. Pending the discovery of further evidence, it should be tentatively considered as a late representative of the Hymenostracans.

Unnamed "shrimp" of Collins (1986) (Fig. 23)

Pending a proper description of this animal the only published photograph is insufficient for an assessmment of its affinities. We have listed it only for the sake of completeness.

Concavicarida, Thylacocephala (Fig. 24)

There is by now fair agreement as to the affinity of the Thylacocephala, of the Concavicarida and of the unnamed species described by Mikulic, Briggs & Kluessendorf (1985). This latter form shows definitely that this group existed at least since the Silurian. It is, therefore, possible that some of the Cambrian or Ordovician Arthropods known only by their valves represent ancestral taxa.

Waptida, Hymenostraca, Cirripedia (Fig. 25)

The phylogenetic position of the genera discussed in this section is open to alternative hypotheses. We propose that all genera derive from a common ancestor, which was provided with a normal set of trilobitic appendages with a fringed outer branch. It is a reasonable assumption, therefore, to group them together, though loss of the abdominal appendages may well be an independent specialization in different lineages. Anyway such loss bars them from being grouped with the Phyllocarida and Leptostraca. They both have specialized abdominal legs, or at least retain some of them in a vestigial form. This does not mean that there may not be a more distant relationships between all these groups, as has been argued in the section on *Protocaris* and *Odaraia*. We still do not see why the genus *Canadaspis* should not be included in the Hymenostraca, and we still consider *Perspicaris* to be very close to this group. *Plenocaris* Whittington, 1974, should be included in *Waptia*. We have already said that on the available evidence *Douglasicaris* should be tentatively included in this group.

We must now consider whether there is a case for establishing a relationship between these taxa and later animals. Obviously, any later taxon having abdominal legs must be ruled out. Therefore comparison is possible only with the Cirripedia, and more precisely with the Ascothoracica. Whether *Synagoga* (which has at least two free living ectoparasitic species) and the few allied genera are actually Cirripedes has been questioned by some authors. Among other reasons, this is because their nauplii lack the typical Cirripede "horns". If comparison is limited to those features which are not specialized in connection with *Synagoga's* parasitic habits, its features can be compared one by one with those of *Canadaspis, Perspicaris* or *Waptia*. We cannot see why McKenzie(1982) rejected the hypothesis of a relationship.

The case for the Thoracica is more complex. If the Ascothoracica are Cirripedes, then we should assume a phylogeny in which the ancestors of *Synagoga* took up parasitic habits, but retained most of their primitive characters. In contrast the ancestors of the Thoracica became fixed to the substratum, perhaps by an in-

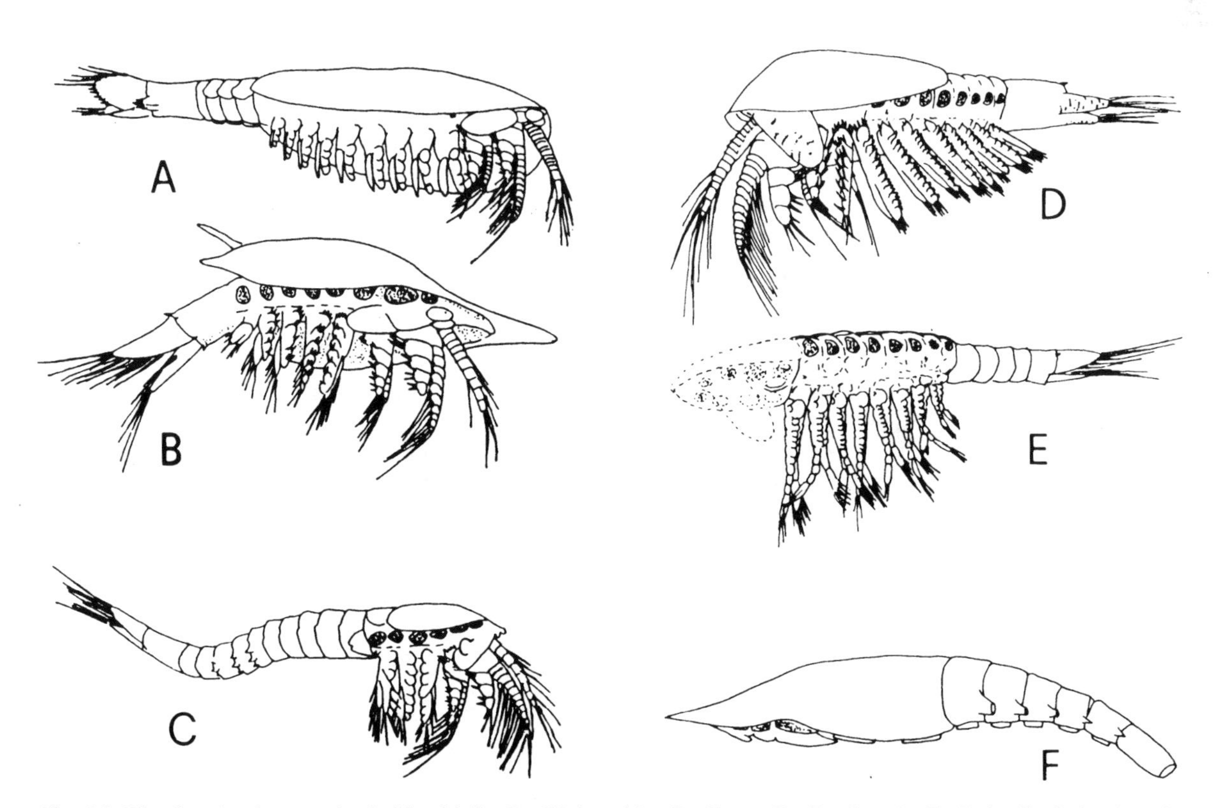

Fig. 26. The Scanian larvae, A: *Rehbachiella*, B: *Walossekia*, C: *Skara*, D: *Bredacaris*, E: *Dala*, F: *Oelandocaris*.

termediate stage, when they were foresic on marine swimming animals (as it is the case of barnacles living on Cetaceans), and specialized by drastically reducing the abdomen, etc.

Collins & Rudkin (1981) have described a supposed stalked barnacle from the Burgess shale (*Priscansermarinus barnetti*). So far as the admittedly poor preservation goes, Collins & Rudkin might be right. However, it is surprizing to have already in the middle Cambrian an advanced Lepadomorph. Considering the number of problematic taxa in the Palaeozoic, we have some reservations on the allocation of *Priscansermarinus* to the Thoracica.

An alternative hypothesis is that while the reduction of calcified plates in such Cirripedes as *Conchoderma* and *Heterolepas* is clearly secondary, it was a primitive feature in *Priscansermarinus*. This is a possibility, as it is clear that Arthropods with calcified teguments acquired them independently. So there is no inherent difficulty in admitting that the most primitive Lepadomorphs lacked differentiated calcified plates. Yet we feel reluctant to accept this genus as a Lepadomorph, and await better specimens that may demand a reassessment of its position.

The Scanian larvae (Fig. 26)

The recent discovery of several tiny Arthropods in the upper Cambrian of Sweden (*Dala peirlertae, Bredocaris admirabilis, Walossekia quinquespinosa, Rehbachiella kinnecullensis, Skara annulata, Oelandocaris oelandica*) (Müller, 1983) with perfectly preserved appendages, requires some discussion.

Müller (1983), after a brief mention of the possibility that these were larval stages of otherwise unknown animals, concludes that they are probably adults or, in two instances, that both larvae and adults are represented in his sample. In contrast Schram (1986) lists several sound reasons to consider them larvae. We entirely concur. As few crustacean-like animals are known from the Upper Cambrian, it is inherently improbable that these animals belong to otherwise known taxa. However, if we consider the probable larval appearance of animals such as *Waptia, Canadapis*, etc. (Fig. 25), it is easy to visualize larvae that must have had a morphology quite close to that of the Scanian animals.

It is well known that during development a number of Crustaceans pass through stages in which their appendages differ considerably from those of the adults and, more precisely, that when they first appear in an instar, the appendages are much more "phyllopod-like" than in the adult animals. Moreover cephalization of metameres and specialization of the appendages as mouthparts is often a gradual process to the extent that even in the Decapoda the carapace fuses very gradually with the postcephalic segments (cpr. p. 192-193) and a cephalothorax of up to 16 segments is only gradually formed.

To us the crucial problem for the interpretation of the Scanian fossils is not whether they should be formally incorporated into previously described taxa (there is a number of living Crustacean larvae provided with generic and specific names for lack of sufficient evidence for identification of any corresponding adults), but whether they should or should not be incorporated within known taxa, even orders or families, already known. Inspection of Figs. 21-25 shows that this is plausible for several of them. We must beware, however, as there is a large number of bivalve Arthropods from Cambrian strata for which next to nothing is known of their appendages: several (or all) of the Orsten larvae may well related to such poorly known taxa.

Archaeostraca, Leptostraca (Fig. 27)

The close relationship between the Phyllocarida and the Leptostraca is now beyond dispute. The group may well go back to the Cambrian if *Ohiocaris* and *Carnarvonia* should be proved to belong here. However, we do not agree with the inclusion of *Canadaspis, Waptia* etc. We agree that they may be related, but the Archaeostraca and their descendents still preserve small abdominal swimmerets and therefore must be considered as a distinct branch.

Are the Malacostraca a natural taxon ? General problems of the "Crustacea"

Zoologists have grown accustomed to the breakdown of the old concept "Entomostraca". They may even be ready to accept the dismantling of the "Euphyllopoda". They may not be so happy about the splitting of the Malacostraca. The problem of the validity of the Malacostraca hangs on the significance of tagmosis.

There is no doubt that the living Leptostraca and the Palaeozoic Phyllocarida are so closely related that it is debatable whether ordinal separation is warranted. On the other hand the Phyllocarida are a varied assemblage and they are clearly much more ancient than it is usually thought. *Ohiocaris* and *Carnarvonia* are, probably, synonymous, so dating back to the middle Cambrian. We expect that when better known *Tuzoia, Hurdia, Proboscicaris, Teles* etc. will prove not to be true Phyllocarids, but to span the morphologic gap between the Phyllocarida and other archaic bivalve proto-crustaceans.

It is difficult, therefore, to see how the Phyllocarida and Leptostraca fit in the phylogeny of the syncarid "malacostracans". They could be related, albeit distantly, to several other "malacostracan" groups such as Decapoda, Euphausiacea etc. This conclusion was reached by Schram (1969), but has not yet gained universal acceptance.

By the late Palaeozoic the arthropod fauna is basically modern, even if it still includes a few taxa of difficult allocation, which are clearly the last survivors of older radiations (e.g. *Cheloniellon, Oxyuropoda*, the Cycloidea etc.). Many of the extant orders already occur by the end of the Palaeozoic, and among these several peracarid orders are worth remembering (Cumacea, Isopoda, Tanaidacea).

The Devonian *Lepidocaris* had no carapace but well developed pleural lobes (pleurotergites). Its close relation to the Anostraca appears well grounded, so we think that the ancestors of the Anostraca are not to be sought among the Cambrian proto-crustaceans with a carapace. The case for the Mystacocarida is debatable. They may well have lost the carapace as an adaptation to a mesopsammic habitat.

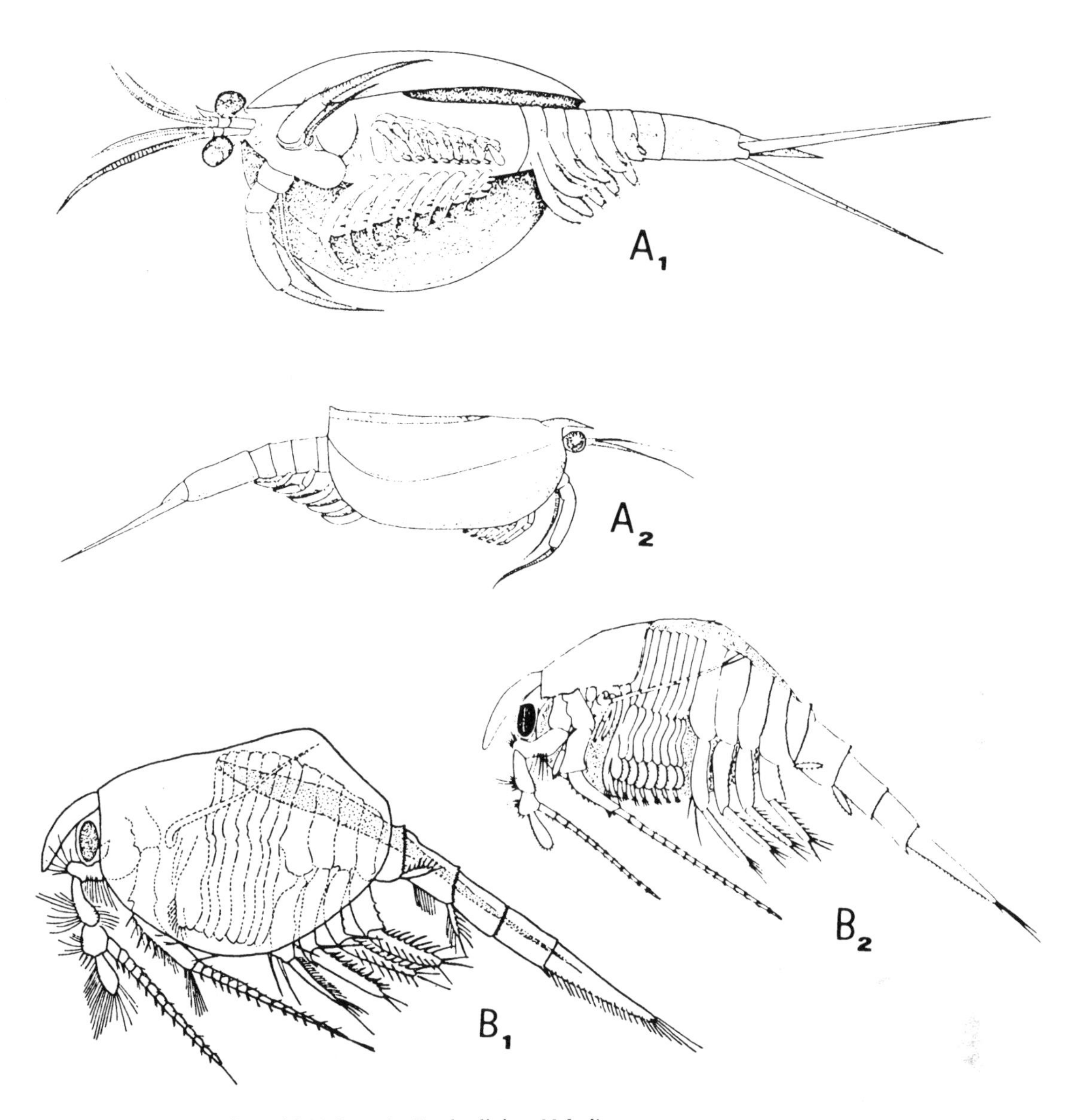

Fig. 27. A: The Phyllocarid *Nahecaris*, B: the living *Nebalia*.

The discussion of the "proto-crustacea" and of their possible relations to later Crustaceans has led us to challenge, at least by implication, the idea that the crustaceans with a carapace derived from orthodox Crustaceans not so equipped (and vice versa). That animals with a carapace ultimately derive from some animals devoid of it is obvious. But we have ample evidence that for the origin of carapace-bearing proto-crustaceans, we have to look for a lower Cambrian taxon which had not yet reached a typically crustacean level of organization.

The difficulty in assessing the phylogeny of the Peracarida stems from the fact that they include both orders provided with a carapace (Cumacea, Mysidacea, Spelaeogriphacea) and those which show a simple fusion of thoracic segments with the head (Amphipoda, Isopoda etc.). Both among the Peracarida and other Malacostraca we can follow the evolution of the carapace from a fold covering the thoracic segments, which have more or less reduced terga (e.g. in the Permian Pygaspida), to a carapace completely fused with the tergal region of the thoracic segments. It is clear that parallel evolution was rampant in these structures, but its elucidation is difficult. To take an interim stand we suggest that the Peracarida may be monophyletic with some lines developing early a carapace.

Marrella, Mimetaster, Burgessia, Acercostraca (Figs. 28-29)

Little can be added to the discussion of these taxa in Simonetta & Delle Cave (1981), as almost no new evidence has been subsequantly added.

Apart from the prosoma of *Marrella* and *Mimetaster*, these two genera appear to be closely comparable. This is, however, on the evidence of primitive characters, which are found in other groups and so are of little significance. In the head both the differences and similarities are difficult to assess. The spinose cephalic shield may well be of no significance as the spines are quite different both in number and in structure. *Mimetaster* has apparently fixed pedunculate eyes, while compound eyes are certainly absent in *Marrella*. The morphology of the first and second antennae is only superficially comparable, and may well be the result of convergence. Nevertheless we still tentatively group *Marrella* and *Mimetaster* as two branches of a single stock.

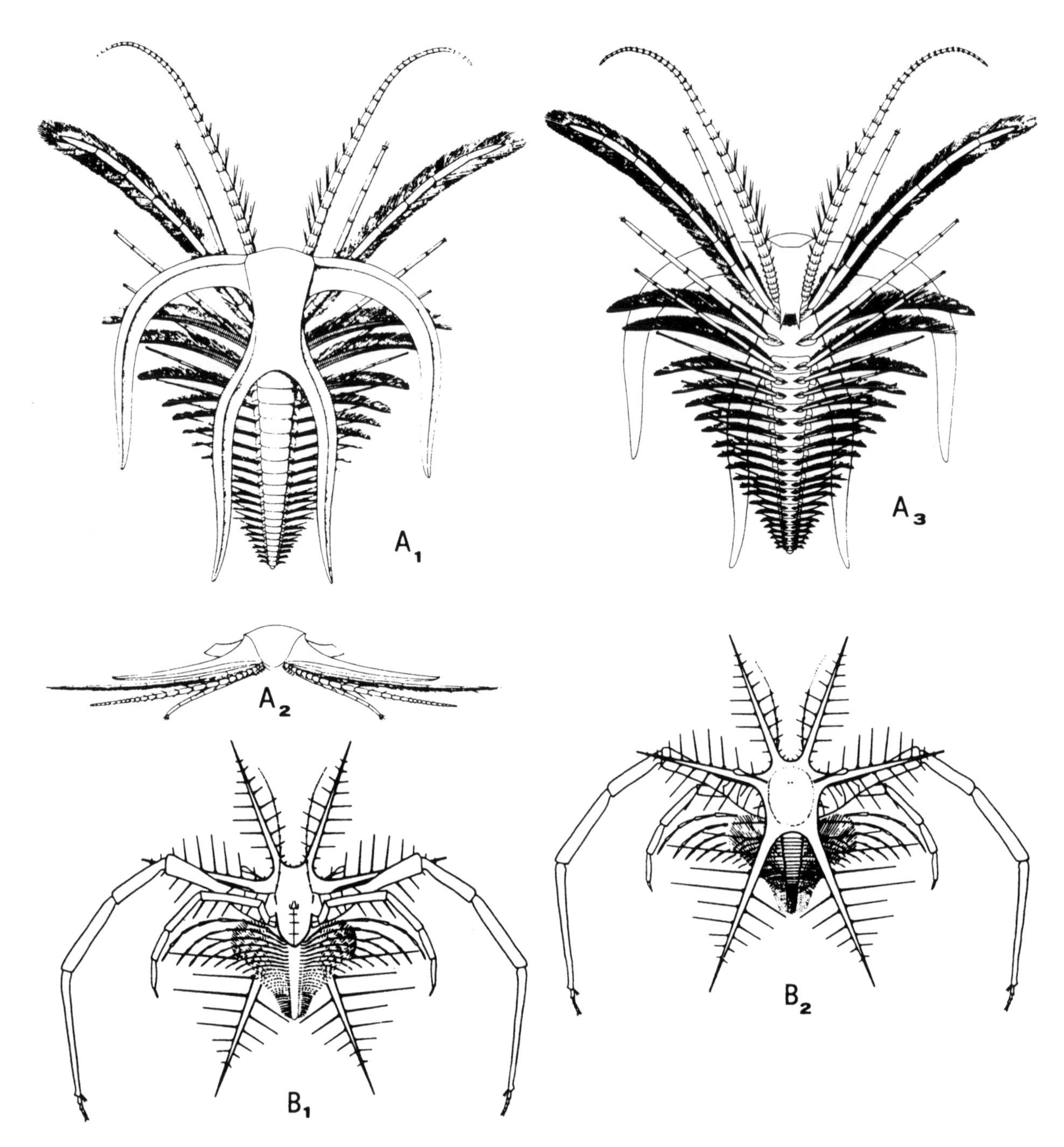

Fig. 28. A: *Marrella*, B: *Mimetaster*.

The positions of *Burgessia* and *Vachonisia* in connection with the origin of the Notostraca is an open question, but it is commonly admitted that the features by which the Anostraca, Concostraca and Notostraca were once grouped together (crustacean tagmosis of the head and phyllopodia) are simply convergent (pseudosynapomorphies). *Burgessia* and *Vachonisia* appear as reasonably primitive members of a group from which the living Notostracans (some of them having a telsonic plate comparable with the terminal spine of *Burgessia*) may have evolved. It must be emphasized that while in *Burgessia* the carapace did not fuse with any of the postcephalic segments, such evidence is lacking in *Vachonisia*.

Kazacharthra (Fig. 30)

The only new evidence concerning the Kazacharthra is the probability that a new and unnamed animal from the Lower Silurian of Wisconsin (Mikulic, Briggs & Kluessendorf, 1985) belongs within this group. Mikilic *et al.* (1985) comment on the overall similarity of the new animal to the Kazacharthra, but doubt real phyletic affinity. However, we do not see any specific evidence against there being a genuine relationship.

General conclusion

It is worth stressing as a general conclusion that even if some of the Palaeozoic Arthropods were more or less the ecological equivalents of later taxa, this equivalence must be considered as only a very rough approximation. This is because they belonged to com-

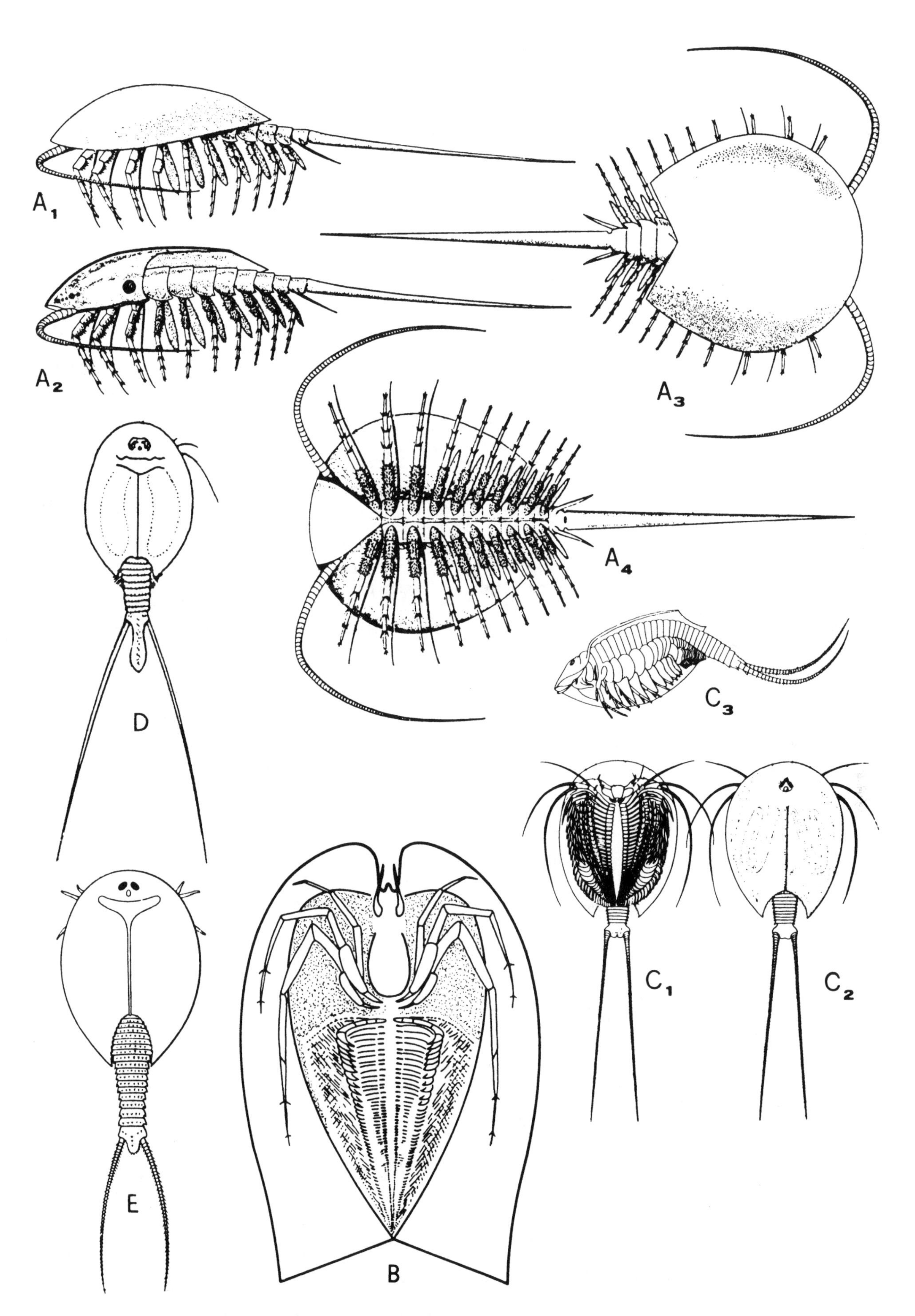

Fig. 29. A: *Burgessia*, B: *Vachonisia*, C: the living Notostracan *Triops* and D and E two species of the living *Lepidurus* to show variability of telsonic spine in Notostracans.

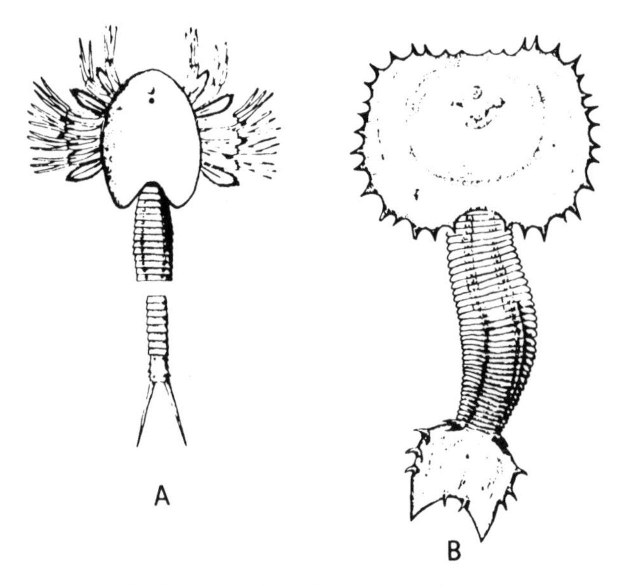

Fig. 30. The Kazacharthra, A: *Jeanrogerium*, B: *Panacanthocaris*.

munities composed of very different and much less specialized taxa than those of the present day (cpr.Simonetta, in press).

On the other hand we deplore the tendency of many authors not to name formally the animals that they describe or to allocate them to the mandatory categories established by the Rules of Nomenclature. It makes it more difficult to discuss them and it is a temptation to any later author to usurp the place of the actual discoverers. If an author is satisfied that the specimens that he describes belong to a new taxon he should name it. Any reservations are simply handled by the usual qualifications of cf. of aff.

THE OTHER ARTICULATE TAXA

Anomalocaris and its significance (Fig. 31)

Anomalocaris as a taxon based only on the segmented appendages has been known for almost century (*A. canadensis* Whiteaves was described in 1892), but just how strange an animal it was has been recognized only recently. We noticed almost all the best specimens while sorting the material for the study of Arthropods when in Washington in 1960. Later, when denied the possibility of borrowing the specimens, one of us (A.M.S.) photographed them in 1970. Those were used for a first brief communication to the congress of the Unione Zoologica Italiana in 1981.

In the meantime Whittington had noticed the same specimens and had discovered an additional one in the new collection made by the Canadian Geological Survey. Thus it happened that when we met in 1980, we discovered that we were busy describing the same animal.

We agreeded that while we were to present the preliminary description (Simonetta & Delle Cave, 1982), Whittington was to give the full description and name the animal (at that stage neither of us had realized that we were studying *Anomalocaris*). Working on photographs, and with a brief re-examination of the material during an afternoon visit to Whittington in Cambridge, we recognized the main features of the animal, including the existance of the large spiny first appendages and of some small spiny structures around the mouth. It was, however, Whittington & Briggs(1985) who, by appropriate preparation were able to show that the nominal genera *Peytoia* and *Anomalocaris* were parts of this animal. Curiously Whittington & Briggs (1985) practically do not make any comment of our paper, barely mentioning it.

The present reconstruction (Fig. 31) basically follows the account by Whittington & Briggs, 1985 except in the following points:

1) The great lobate appendages are fringed. Whittington & Briggs (1985) noted this fringe (which was confirmed by Briggs & Robison, 1984), but considered that it might be either a true fringe or was due to strengthening structures within a continuous flap. The latter is consistent with the type of locomotion that they envisaged for *Anomalocaris*: a sort of underwater flight. However, the ribs of annulated material, so evident on the ventral side of some specimens, are clearly associated with the flaps. While in the head region they are interrupted along the midline, in the postcephalic segments they run across the body. It is not clear whether this structure is distally associated with the dorsal, respiratory structure or with the ventral natatory appendage. Either way this continuity in the postcephalic region points to the appendages having been continuous across the body, probably as a narrow flap or plica. This suggests in turn that the appendages were less inclined to the horizontal than in the reconstruction by Whittington & Briggs (1985), and may be consistent with some sort of fringe.

In spite of their very arthropod-like first appendages, the buccal structures and the other appendages prevent the inclusion of *Anomalocaris* within the Arthropoda. Nevertheless the fact that its appendages are clearly ventro-lateral, rather than lateral, as in Annelida, and that their fossilization is so close to true Arthropods suggests a close structural and chemical affinity of the teguments of both Arthropods and *Anomalocaris*. We must add that we find the evidence inconclusive concerning the existence of either a smooth tergal cuticle (as proposed by Whittington) or of delicate serial terga (our inclination).

We consider *Cassubia* Lendzion, 1977, as a synonym of *Anomalocaris*, though represented by a very small specimen in comparison with the usual Anomalocarids. Not having seen the specimen we cannot definitely commit ourselves, but judging from the figures we believe the fossil to show a body of 11 (+) segments, traces of the cephalic tagmon and a "great appendage" of 8-9 podomeres.

Aysheaia, Onychophora and Tardigrada (Fig. 32)

The discovery of a new of *Aysheaia*-like animal (*Luolishania*) in the Chinese Lower Cambrian, the description of *Aysheaia prolata* Robison, 1985 from the middle Cambrian Wheeler formation of W.Utah, and of new Canadian material slightly older than the Burgess shale, add to the taxinomic roster of this group and increase its temporal range. They do not add however, very much to our morphological knowledge. The most notable new evidence, in this regard, is that in the Lower Cambrian *Luolishania* the body protrudes

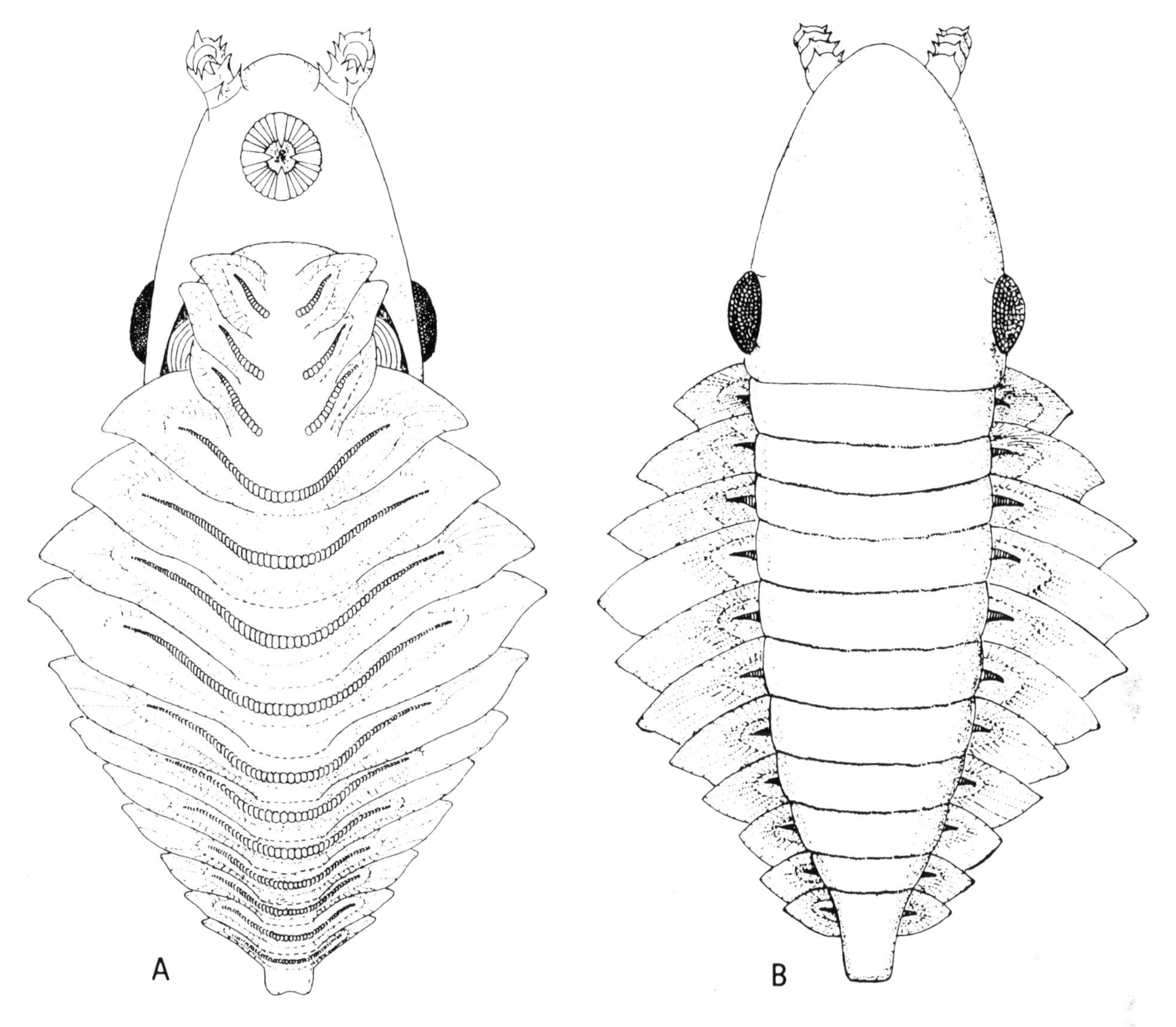

Fig. 31. *Anomalocaris.*

between the last pair of legs, just as in living Onychophorans. This substantiates the claim by Robison (1985) for inclusion of *Aysheaia* among the Onychophorans. It questions, however, his assumption that this is a specialization useful in terrestrial habitats.

Current opinion on the possible relationships of *Aysheaia* with Arthropods, Onychophorans and Tardigrada is devided on two main issues:
i) is *Aysheaia* more closely related to the Onychophora or the Tardigrada?
ii) Are the "Lobopoda" to be included with the Myriapoda and Insecta in a Phylum Uniramia?

Various aspects of the second point have already been discussed in the introductory section of this paper (see also below). Here it is sufficient to add that ultrastructural and biochemical studies (cf. Baccetti & Rosati, 1971) have shown that the cuticles of both the Onychophorans and Tardigrada show affinities to the Arthropod cuticle. They represent, however, a much simpler stage in the evolution of this kind of cuticle, the onychophoran cuticle being somewhat more arthropod-like than that of the Tardigrada. This may well be related to their very different sizes and to the different environments they inhabit.

The relative affinities of *Aysheaia*, to the Onychophora and Tardigrada requires further anatomical and embryological investigations.

The following alternatives must be considered:
1) the sensory tentacles of the Onychophora are truly preoral and thence may be homologous with the acral tentacles of the Polychaeta (which often also have circumoral tentacles)
2) the primitive "head" of the Lobopod-Arthropod stock is formed by an epistomial lobe, the only truly preoral structure, and two fused postoral segments. The latter consist of ocular and antennal (= antennular) segments, both antennae and eyes being rotated in front of the mouth at a very early stage in development. Later in evolution one or more segments may be incorporated in the head, while the mouth may rotate ventrally. In this case the antennae of the Onychophora may be homologous with the first antennae of Arthropods and the specialized first pair of appendages of *Aysheaia*, which have moved only midway in the scheme of dorsal rotation.

As for the Tardigrada, it is now commonly assumed that the marine Parastigarctidae, which show somewhat diverse cephalic appendages (fig. 31) are among the most primitive. Their embryology is unknown, and neither do we know whether the

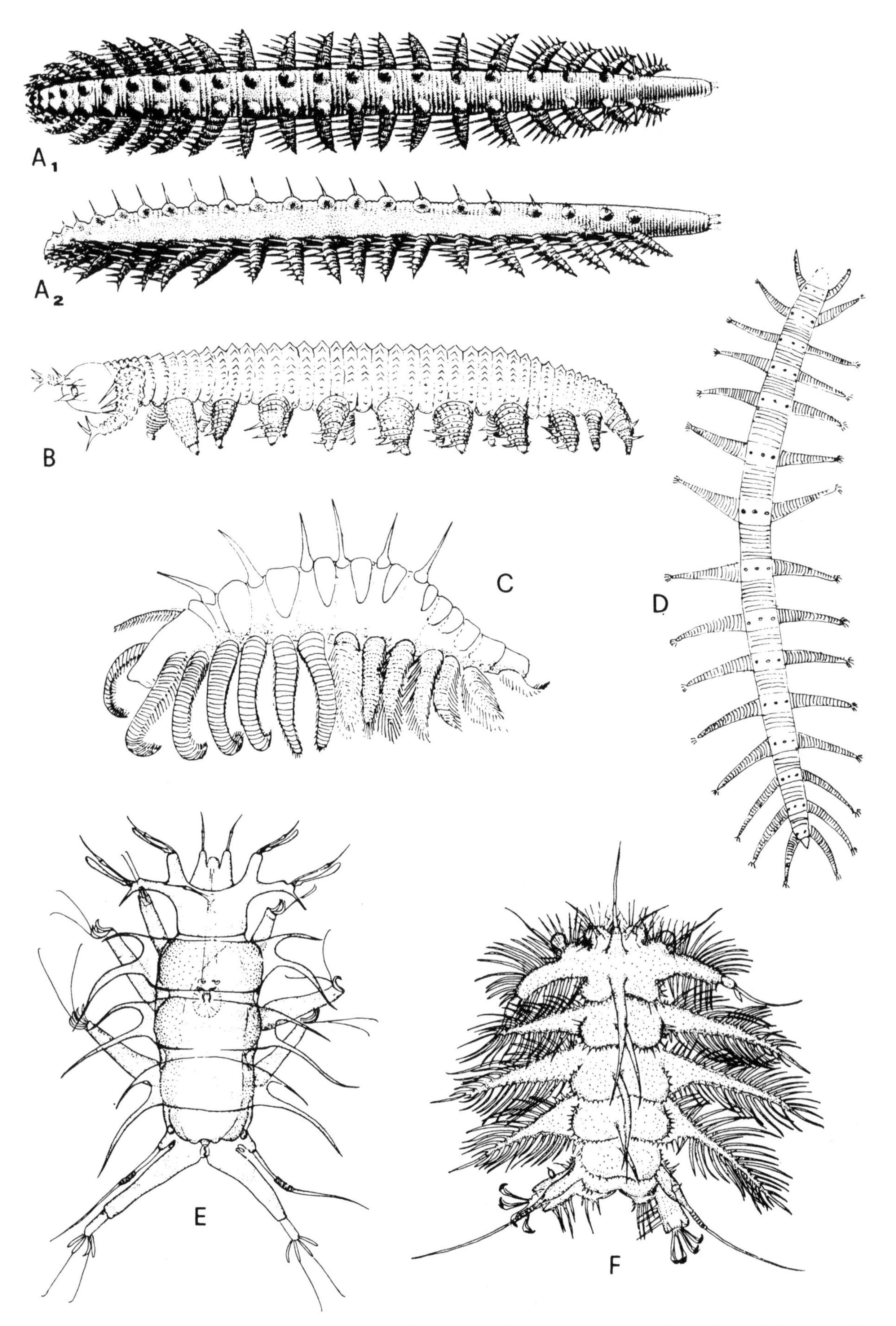

Fig. 32. A: *Xenusion*, B: *Aysheaia*, C: unnamed animal figured by Collins from the Burgess Shale, D: *Luolishania*, E-F Two living Tardigradan Parastygarthids (E: *Parastygarctus*, F: *Neostygarctus*).

"cephalic" dorsal appendages have any musculature comparable with that of the legs. Should it be so, than the appendages of the Parastigarctidae might be comparable with the specialized legs of *Aysheaia*. We could then consider a hypothetical ancestor with a series of circumoral papillae, one or more cephalic cirri of acral origin, planktonic ciliated larva, 3-4 postoral metameres and anamorphic development. From this two lines may have arisen: line A: *Aysheaia*→*Parastigarctus*→normal Tardigrada, and line B: an *Aysheaia* like animal, but retaining one pair of cirri→*Peripatus*.

The primitive ciliated larva must have been lost at the beginning of the radiation, if not before, because of the development of the chitinous cuticle.

As for the problem of the different mechanisms of mesoderm formation in Tardigrada and Onychophora, we may consider the possibility that the tardigradan enterocelic mesoderm may be a specialized condition.

Robison (1985) argues that the terminal mouth of *Aysheaia* is not significant as Tardigrades feeding on algae have a ventral mouth. Moreover he proposes that lack of antennae and mandibles is equally insignificant, as we know nothing of the pharynx of *Aysheaia* and the absence of mandibles in Tardigrada is linked with a complex pharynx. The anus on the "tail" in Onychophorans is equally considered by him as not significant as it is considered an adaptive feature of little significance in phylogenetic discussion. Robison then concludes that *Aysheaia* is an Onychophoran (Protonychophora).

We agree with this conclusion, but feel the problem should be turned the other way round. When we discuss whether two groups of animals are related, we must ask ourselves (cf. section I): can we envisage a set of topological transformations to bridge the gap between A and B?. Then to establish the position of a third taxon C we must consider: Does C correspond with one such stage of topological transformation, or is it better regarded as a third transformation-branch of this set? Stated this way, we concede that practically in all its known characters *Aysheaia* stands on the transformation set joining Tardigrada and Onychophora, albeit somewhat closer to the latter group. Therefore *Aysheaia* supports our contention that Onychophorans and Tardigradans are really related.

Xenusion and Collins' monster (Fig. 32)

Dzik (this volume) has proposed a reconstruction of *Xenusion pompecki* (Fig. 32) This shows some remarkable similarities to *Aysheaia*. There are also differences, especially in the well developped but sparse dorsal spines implanted on rather large sclerites, and in the fringed appearance of the appendages. Both features strongly recall a new animal from the Burgess shale recently figured by Collins (1986). It is obviously rash to discuss the affinities of this latter animal on the evidence of a single photograph, so the following remains tentative.

We suppose that the dorsal spines are correctly identified and not outstretched right legs protruding beyond the body. The legs, with their double set of inner, delicate spines, are suggestive of a specialization on the basic pattern of the first pair of appendages of *Aysheaia*. However the first appendages of a small specimen of *Anomalocaris* would also look much the same. At the front end of the animal Collins' (1986) photograph appears to show a structure conceivably comparable with the circumoral structures of *Aysheaia*, while the caudal half of the animal seems to show some sort of "pleural-like" folds or of matted fringes of the appendages. These, if pleural folds, would recall, *Anomalocaris*, if fringes, may be like those of *Xenusion*. All taken, it seems worth considering the possibility that this animal is close to *Xenusion*.This genus may prove to belong to the lineage leading to *Aysheaia* and the Onychophora-Tardigrada ensemble, with Collins' animal representing yet another branch of the basic radiation of α-chitinate animals.

Facivermis Hou & Chen 1989 (Fig. 33)

In addition to the paper by Hou & Chen (1989), these authors kindly provided some photographs. We feel that Hou & Chen have not given a sufficient reasons as to why they believe the appendages to be dorsal. The appendages do not really look like cirri. If we turn the animal, as reconstructed, upside down, its forepart looks very much like an Onychophoran or *Aysheaia*. We should, therefore, consider the possibility that *Facivermis* may be a burrowing or semiburrowing Lobopod.

Microdyction Chen, Hou & Lu, 1989 (Fig. 33)

Microdictyon sinicum Chen, Hou & Lu, 1989 is also problematic. Not having seen the original material, we rely on the original description and a colour photograph kindly supplied by Chen & Huo. As a working hypothesis we consider the possibility that this might be distantly related to the lobopods. Its tegument, instead of developing towards a more or less uniformly chitinized cuticle as in the Onychophora (and to some extent, *Aysheaia*) in order to strengthen itself and to afford leverage for the segmental muscles, developed its peculiar sclerites. This anticipates the development of sclerites in the Tardigrada and possibly the dorsal sclerites of *Xenusion*. The most puzzling feature of *Microdictyon* is the apparent absence of any kind of grasping or adhesive organ at the tip of its podia. These, anyway, are not comparable with Annelid parapodia.

Opabinia Walcott, 1912 (Fig. 33)

This genus, described by Walcott (1912), is monospecific (*O. regalis* Walcott, 1912)[10].

Having assumed that the animal was an Arthropod, this was the one genus for which my (Simonetta, 1970) reconstruction was grossly wrong. Here we follow the reconstruction by Whittington (1975). Apart from the gut, which was a simple straight tube, all the known features are shown in Fig. 32. We concur with Whittington (1975) that any resemblance with the Carboniferous animal *Tullimonstrum gregarium* is superficial, and the new reconstruction of *Tullimonstrum* by Beall (this volume) substantiate this.

Should we admit that *Opabinia* is closer to the Arthropoda than Annelida on account of its apparent

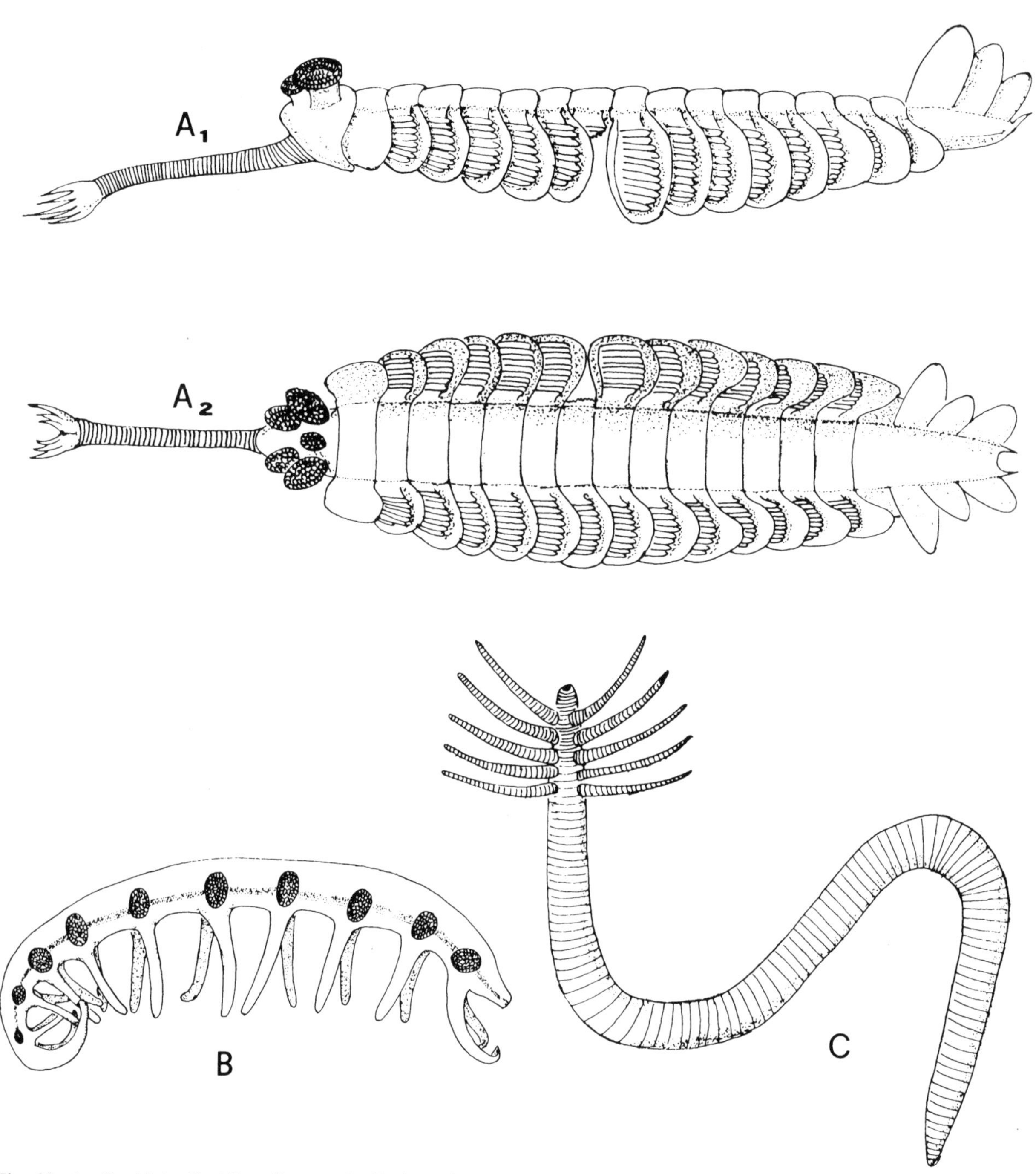

Fig. 33. A: *Opabinia*, B: *Microdictyon*, C: *Facivermis*.

cuticular compositions, then this animal would belong somewhere in the early radiation of the "Articulata". It would deserve "phylum-status" in same way as the Arthropoda, Lobopoda and *Anomalocaris*.

This strange animal fits well in the array of animals which we are discussing: A) its "flaps" are clearly latero-ventral projections as are the "flaps" of*Anomalocaris* and the appendages of the Arthropods, Onychophorans etc. and not lateral as the parapodia; B) the dorsal gills are topologically exactly in the same position as the dorsal gills of*Anomalocaris* and, maybe, of the outer branch of the trilobite leg. Unfortunately we will never have the crucial embryological data; C) the caudal flaps are apparently serially homologous with the preceding ones; D) the large, semistalked eyes recall those of *Anomalocaris*, subdivision of the eyes and central dorsal eyes are well known in a number of assorted arthropods, where they developped independently a number of times. We surmize that when, if ever, specimens of *Opabinia* showing clearly the mouth details will be discovered, the mouth will be found to be equipped with some sort of grasping-chewing armour, though not necessarily of the same type as that of *Anomalocaris*. Simply the "proboscis" is obviously a grasping organ and may have brought the food to the mouth, but a buccal armour to ingest and possibly chew the food was unavoidable.

As there is little doubt that the lateral lobes, even if their mobility was limited, were used to swim (that the animal was a swimmer is also shown by the three pairs of "tail" lobes which must have been used to stabilize the animal and prevent roll).

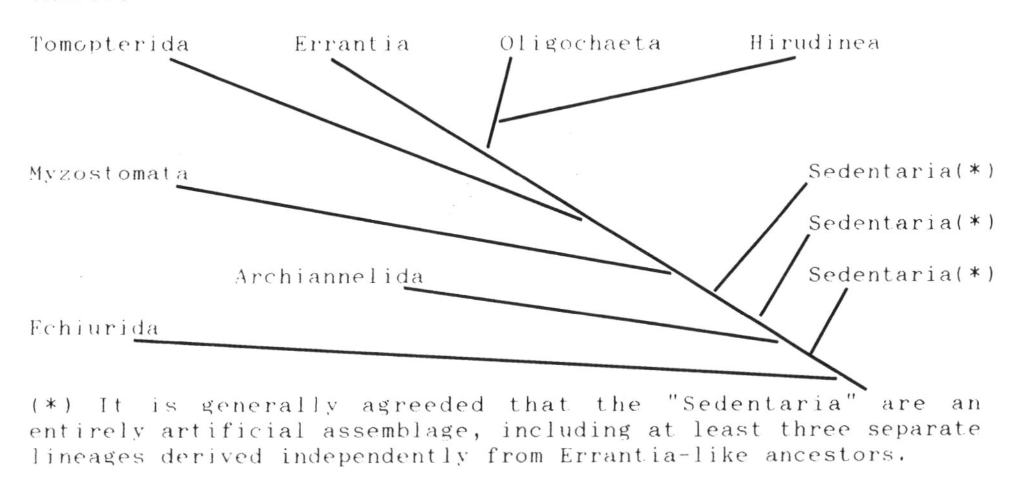

(*)It is generally agreeded that the "Sedentaria" are an entirely artificial assemblage, including at least three separate lineages derived independently from Polychete-like ancestors.

We also accept the criticisms by Briggs & Whittington (1987) concerning the reconstruction proposed by Bergström (1986).

Annelida

As a result of both recent discoveries and of revisions of former reconstructions, we now know of a number of Arthropods which do not fit into traditional higher taxa. Much the same situation obtains with other "Articulata" (Annelida, etc.), and they deserve some attention on account of the light they may throw on the early radiation of this "superphylum".

Even if we disregard the identification of a possible late Precambrian Echiurid from Namibia (*Protoechiurus edmondsi* Glaessner, 1979) as doubtful, the available comparative and embryological evidence is sufficient to establish a close relationship between Annelida and Echiurida. On the basis of presently available evidence the evolution of the Annelid stock may be diagrammatically figured as follows in Table III.

The position of the supposed Precambrian polychaetes is debatable (Conway Morris, 1985). They will not be considered here. The fact that annelid architecture rests on collagen, the synthesis of which would have been difficult in an oxygen-poor environment (see Towe, 1970) makes it likely that the members of the phylum must have remained small transparent animals well into the lower Cambrian (Simonetta, 1976). Some true Polychaetes are known from the middle Cambrian (e.g. *Canadia spinosa*, cf. Conway Morris, 1979). By that time they must have evolved their distinctive tegument.

One supposed Cambrian Annelid, *Myoscolex ateles* (Glaessner, 1979 b) has calcium phosphate "chaetae", which alone pose a very serious objection to its inclusion in the Annelida.

Strangely most authors who advocate descent of the Arthropods (or of some of them) from annelid ancestors (e.g. Anderson, 1973, Manton, 1977, 1979, Whittington, 1977, Schram, 1986) do not realize that the tegument in these two phyla is so different that any direct relationship of the two phyla is ruled out.

The annelid cuticle is basically collagenous, that is a protein. Sclerotines, which are so important in the Arthropods are unknown in the Annelids. Chitin in the Annelids is limited to aciculae and setae, which are chitino-proteic. Moreover the annelid chitins are β-chitins, quite different from the α-chitin of Arthropods. The protein component of aciculae and setae in Annelids is collagen rather than sclerotin.

Collagen is a scarce component in any part of the Arthropods and anyway does not enter the composition of the cuticle which is basically α-chitin (a polysaccharide) and sclerotin (a protein).

Onychophora and Tardigrada have a cuticle (cf. Baccetti & Rosati, 1971) that is structurally and chemically rather akin to that of the Arthropods (indeed it has been considered as a primitive evolutionary stage of the typical arthropod cuticle) and has no affinity with that of Annelids.

It is thus clear that, while the basic structural plan of the Annelids and Arthropods is sufficiently similar to justify the belief that both share a common ancestry, yet the basic differences in their teguments force us to the conclusion that both must have been derived from minute animals of no more than 6-10 metameres and with an indifferentiated tegument.

Annelida and Echiurida evolved a basically collagenous cuticle and around the chaetae and aciculae eventually evolved the parapodia and the peculiar associated muscular system. Meanwhile the stock leading to the Arthropods and related groups evolved quite independently the ability to produce α-chitin, sclerotin and resilin. The fact that α- and β-chitins on one side and that collagen, sclerotin and resilin on the other belong to closely related families of molecules is significant both in confirming the basic affinities of these two groups of phyla and of the impossibility that the Arthropods and related groups derived from Annelid ancestors.

The acquisition of a cuticle rich in α-chitin must have been closely related to the loss of the trochophore larva. Indeed aquatic Annelids generally preserve at least some areas of ciliary naked tegument, while external cilia have vanished from all members of the Arthropods-Onychophoran-etc. group.

Spriggina Glaessner, 1958

There are two alternatives concerning this animal: it can be cnsidered to be a "Vendomid", one of the "quilted animals" (cfr. Seilacher, 1984), or it can be an articulate. We do not have a definite opinion; the

Tab.IV

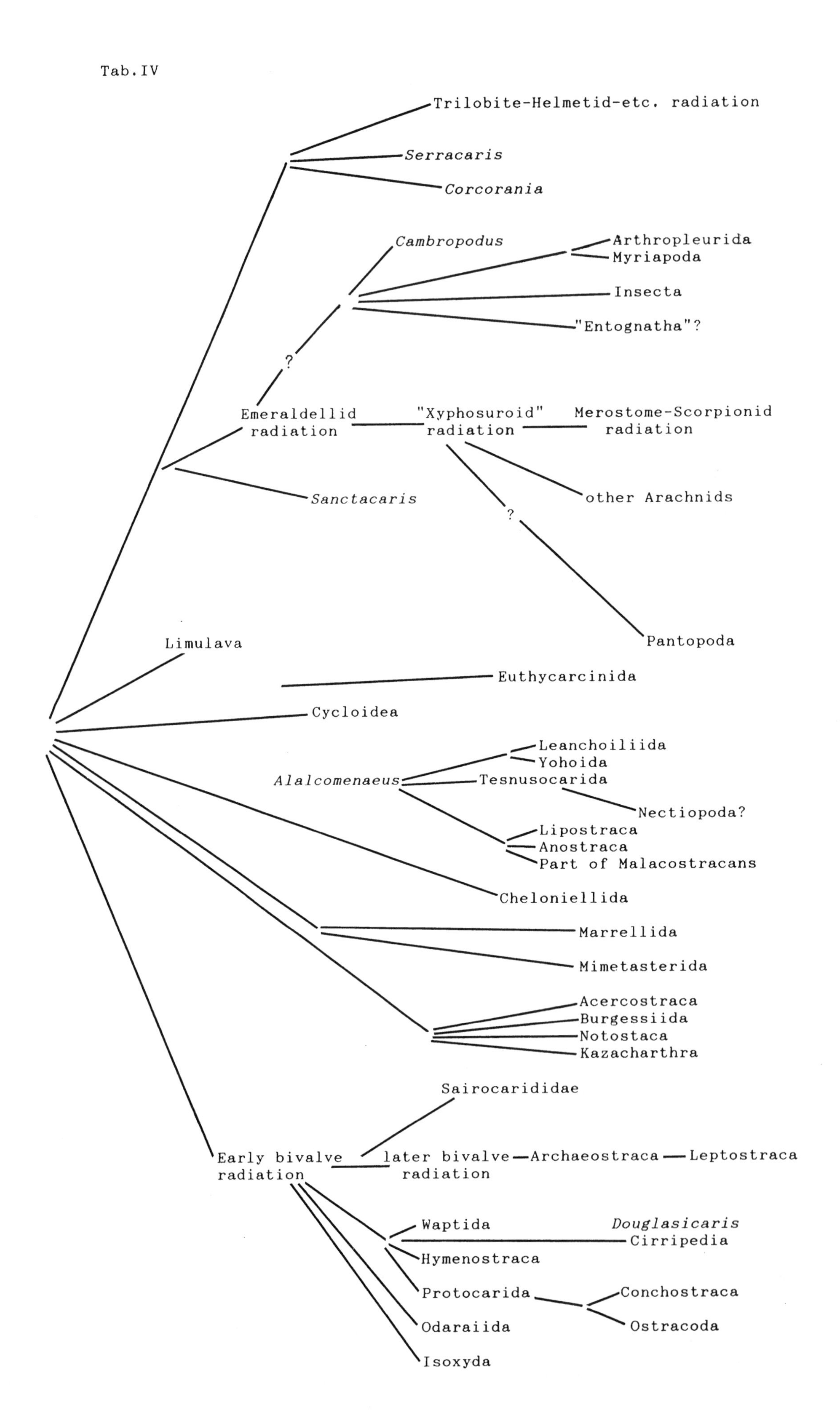

Trilobite-Helmetid-etc. radiation
Serracaris
Corcorania
Cambropodus
Arthropleurida
Myriapoda
Insecta
"Entognatha"?
?
Emeraldellid radiation
"Xyphosuroid" radiation
Merostome-Scorpionid radiation
Sanctacaris
other Arachnids
?
Pantopoda
Limulava
Euthycarcinida
Cycloidea
Leanchoiliida
Yohoida
Alalcomenaeus
Tesnusocarida
Nectiopoda?
Lipostraca
Anostraca
Part of Malacostracans
Cheloniellida
Marrellida
Mimetasterida
Acercostraca
Burgessiida
Notostaca
Kazacharthra
Sairocarididae
Early bivalve radiation
later bivalve radiation
Archaeostraca
Leptostraca
Waptida
Douglasicaris
Cirripedia
Hymenostraca
Protocarida
Conchostraca
Odaraiida
Ostracoda
Isoxyda

following lines have been written on the assumption that it was a segmented "worm".

The position of *Spriggina* must be considered in the light of what we have said on the Annelids. Quite apart from the reconstruction adopted, in order to ascertain the affinities of *Spriggina* ideally we need know the structure and composition of its cuticle.

Of the available reconstructions, and we prefer that by Birket-Smith (1981), all are compatible both with a greater affinity to the Annelids or with what I would informally term "the β-chitinous articulates". As there is no chance to obtain any evidence on this point, from *Spriggina*, we can only tentatively allocate it to a position as a separate phylum within the superphylum Articulata.

Hallucigenia Conway Morris, 1977

As *Hallucigenia* appears to be metameric, it has to be listed in this context. However its structure is so peculiar that we strongly doubt that is should ever rank within the Articulata . If there is indeed a relationship between the Mollusca and Articulata, as has often been suggested, then *Hallucigenia*, together with *Wiwaxia*, Molluscs and the Articulates, could represent as many branches of equal rank, that arose before the radiation of the Articulates themselves.

V. CONCLUSIONS

Most of the problems on the evolutionary significance of the animals described in this paper have been discussed in the framework of descriptions of the various groups of taxa. There is, however a more general set of conclusions.

While the traditional recognition of fundamental affinities in the "Bauplan" of the Annelids, Arthropods and Lobopods stands unchallenged, basic differences in ultrastructure and chemical composition of the cuticles of the Annelids versus the Arthropods-Lobopods, force us to conclude that the "Articulate" stock must have divided into at least two branches at a very early stage, just when it was about to acquire a cuticle. One branch is now represented by the Annelids (and by their close relatives the Echiurida and Myzostomata), the other by the Arthropoda and Lobopoda.

Several more branches of this early radiation are now known from the fossil record: Anomalocarida, *Opabinia, Facivermes*, Sprigginida. Their precise interrelationships and their affinities respectively with the Annelid and Arthropod branches can not presently be assessed, though it looks as if there may have been a fan-like radiation with intermediate morphologies linking what are now the two surviving stocks.

The fact that the supposed Polychete *Myoscolex ateles* Glaessner, 1979, has "chaetae" composed of calcium phosphate, shows that external appearance of worm-like animals may not give sufficient clues for the correct assessment of their phylogenetic position, especially when we deal with early Palaeozoic animals.

If we want to sort the early members of the superphylum into a reasonable scheme, we must try to study their structure in as much detail as the fossils allow and, when organic remains are available, try biochemical methods as well.

Let us turn to the next level in the evolution of the main articulate phyla, and more precisely to the Arthropoda. While we consider that the monophyletic origin of the true Arthropoda (and the related Lobopoda and probably Anomalocarida) can be vindicated, both on morphologic and biochemical evidence. The basic taxinomy and phyletic relationships of the Arthropod lineages require a reassessment. Updating the dendrograms that we have previously proposed, the phylogeny on table IV appears to be the best presently possible.

It is immediately apparent that in order not to multiply the higher taxa (such as classes) which would lead to a masking of phyletic relationships between different groups, so it is necessary to amend drastically the definition of the traditional classes. Nevertheless, even with the greatest parsimony, some new classes are needed. The splitting of the Crustacea appears to be unavoidable, and perhaps also the Insecta. The introduction of such a revised taxinomy will certainly meet with resistance, both because of innate conservatism and of the difficulty in defining various taxa. This aspect of the problems prompts us to avoid latin names. Should our proposed phylogeny meet with approval, taxinomical formalization to meet the requirements of the International Rules of Nomenclature is straightforward.

Our proposed dendrogram may also be questioned on the grounds that it is not dichotomous. We reiterate that there are sound biological and eco-geographical reasons, which have been already been rehearsed by several authors and that one of us (Simonetta in press) discusses elsewhere in detail, that holds dichotomies to be rare events in evolution. Whenever one is tempted to see a dichotomous branching in an evolutionary reconstruction, it would be wise to examine carefully all possible alternatives before accepting it.

NOTE

[1]The term is used out of convenience: a) a specimen is only a possible ancestor of a later population, not of a taxon, which is a conceptual asemblage of populations.

[2]It is usueless to go into the pedigree of this concept, sufficit to mention that it may easily be followed back to Aristotle and quite possibly to geometrical ideas of similarity and correspondence of Pythagorean origin.

[3]At least in terms of classical topology, in terms of differential topology it may, with considerable difficulties, be conceivable.

[4]Actually the argument on the "uniramian whole-leg jaw" is based on evidence from the Onychophora (which can be shown not to be "Uniramians") and on the jaws of the Myriapoda, which relationship to the Insecta is debatable. As a whole a generalisation based on somewhat flimsy evidence.

[5]The elimination by reduction of a strucutural element introduces a discontinuity in the topological transformation, which makes it difficult to admit homeomorphy in terms of classical topology. It is a "catastrophe" *sensu* Thom (1980).

[6]That the evidence from early ontogenetic stages in sometimes completely irrelavant is well known to vertebrate morphologists, it is, perhaps less so to invertebrate anatomists.

(7)Whittington is wrong in preserving the original spelling: the generic name is feminine, Walcott's use of the masculine is apparently a simple mistake.

(8)The number is 7 in living Xiphosurans, the first being partly incorporated in the prosoma; probably 4 + 9 in *Chasmataspis*; 9 free and 4? fused in *Diploaspis*, 12 (7 + 5) 12 (7 + 5) in the Eurypterida.

(9)The term carapace derives from the Spanish "carapacho", a term of obscure origin which properly means the shell of Tortoises. Thus it properly means a body armour and never a shield.

(10)We have not seen either the material or the original description by Miroschmikof & Kwarzow of *Opabinia norilica* and we can not, therefore, have any opinion on it.

SELECTED BIBLIOGRAPHY

ANDERSON, D.T. 1973. *Embryology and Phylogeny of Annelids and Arthropods.* Oxford: Pergamon Press.

ANDERSON, D.T. 1979. Embryos, Fate Maps, and the Phylogeny of Arthropods. In *Arthropod Phylogeny.* (ed. A.P. Gupta), pp. 59-135. New York: Von Norstand Reinhold Co.

BACCETTI B. & ROSATI F. 1971: Electron microscopy on Tardigrades. III. The integument. *Journal of Ultrastructure Research,* **54**, 214-243.

BEKLEMISHEV, W.N. 1969. *Principles of comparative anatomy of Invertebrates* (English translation of Russian edition of 1969), 2 Vols, Edinburgh: Oliver and Boyd.

BERGSTRÖM, J. 1969. Remarks on the appendages of trilobites. *Lethaia* **2**, 395-414.

BERGSTRÖM, J. 1979. Morphology of fossil arthropods as a guide to phylogenetic relationships, In *Arthropod phylogeny* (ed. A.P. Gupta), pp.3-56. New York: Van Nostrand Reinhold Co.

BERGSTRÖM, J. 1980. Morphology and systematics of early arthropods. Naturwissenshaftlicher Verein zu Hamburg, Abhandlungen **23**, 7-42.

BERGSTRÖM, J. 1986. *Opabinia* and *Anomalocaris*, unique Cambrian 'arthropods'. *Lethaia* **19**, 241-246.

BERGSTRÖM, J. 1987. The Cambrian *Opabinia* and *Anomalocaris. Lethaia* **20**, 187-188.

BERGSTRÖM, J., STÜRMER, W. & WINTER G. 1980. *Palaeoisopus, Palaeopantopus* and *Palaeothea*, pycnogonid arthropods from the Lower Devonian Hunsrück Shale, West Germany. *Paläontologische Zeitschrift* **54**, 7-54.

BERNINI, F. 1991. Fossil Acarida, contribution of palaeontological data to acarid evolutionary history. In *The early evolution of Metazoa and the significance of problematic taxa* (ed. S. Conway Morris & A.M. Simonetta), pp. 253-262. Camerino: Cambridge University Press.

BIRKET-SMITH, S.J.R. 1981. A reconstruction of the Pre-Cambrian *Spriggina. Zoologische Jahrbucher, Anatomie* **105**: 237-258.

BRIGGS, D.E.G. 1976. The arthropod *Branchiocaris* n. gen., Middle Cambrian, Burgess Shale, British Columbia. *Geological Survey of Canada Bulletin* **264**, 1-29.

BRIGGS, D.E.G. 1977. Bivalved arthropods from the Cambrian Burgess shale of British Columbia. *Palaeontology* **20**, 595-621.

BRIGGS, D.E.G. 1978. The morphology, mode of life, and affinities of *Canadaspis perfecta* (Crustacea, Phyllocarida), Middle Cambrian, Burgess Shale, British Columbia. *Philosophical Transactions of the Royal Society, London* **B 281**, 439-487.

BRIGGS, D.E.G. 1979. *Anomalocaris*, the largest known Cambrian Arthropod. *Palaeontology* **22**, 631-664.

BRIGGS, D.E.G. 1981a. The arthropod *Odaraia alata* Walcott, Middle Cambrian, Burgess Shale, British Columbia. *Philosophical Transactions of the Royal Society, London* **B 291**, 541-585.

BRIGGS, D.E.G. 1981b. Relationships of arthropods from the Burgess shale and other Cambrian sequences. *Open file Report* 81-743, *U.S. Geological Survey*, 38-41.

BRIGGS, D.E.G. 1983. Affinities and early evolution of the Crustacea: the evidence of the Cambrian fossils. In *Crustacean Phylogeny* (ed. Schram, F.), **1**, 1-22. Rotterdam: A.A. Balkema.

BRIGGS, D.E.G. 1985a. Les prémiers arthropodes. *La Recherche* **16**, 340-349.

BRIGGS, D.E.G. 1985b. Arthropod paleobiology. *Paleobiology*, **11 (4)**, 361-367.

BRIGGS, D.G.E., BRUTON, D.L. & WHITTINGTON, H.B. 1979. Appendages of the arthropod *Aglaspis spinifer* (Upper Cambrian, Wisconsin) and their significance. *Palaeontology* **22**, 167-180.

BRIGGS D.E.G. & COLLINS, D. 1988. A Middle Cambrian Chelicerate from Mount Stephen, British Columbia. *Palaeontology* **31**, 779-798.

BRIGGS, D.G.E. & FORTEY R.A. 1989. The early radiation and relationships of the major arthropod groups. *Science* **246**, 241-243.

BRIGGS, D.E.G. & ROBISON, R.A. 1984. Exceptionally preserved nontrilobite arthropods and *Anomalocaris* from the Middle Cambrian of Utah. *University of Kansas Paleontological Contributions, Paper* **111**, 1-23.

BRIGGS, D.E.G. & WHITTINGTON, H.B. 1985. Modes of life of arthropods from the Burgess shale, British Columbia. *Transactions of the Royal Society of Edinburgh*, **76**, 149-160.

BRUTON, D.L. 1981. The arthropod *Sidneya inexpectans*, Middle Cambrian, Burgess shale, British Columbia. *Philosophical Transactions of the Royal Society of London*, **B295**, 619-656.

BRUTON, D.L. & WHITTINGTON, H.B. 1983. *Emeraldella* and *Leanchoilia*, two arthropods from the Burgess shale, Middle Cambrian, British Columbia. *Philosophical Transactions of the Royal Society of London* **B 300**, 553-585.

CHELAZZI, L. & MESSANA, G. 1985. *Acanthastenasellus forficuloides* n. gen n. sp., a stenasellid isopod (Asellota) from Somalian phreatic layer. *Monitore zoologico italiano*, N.S. **20 suppl.**, 43-54.

COLLINS, D. 1986. Paradise revisited. *Rotunda* **19**, 30-39.

COLLINS, D. & RUDKIN, D. 1981. *Priscansermarinus barnetti*, a probable lepadomorph barnacle from the Middle Cambrian Burgess Shale of British Columbia. *Journal of Palaeontology*, **55, 5**, 1006-1015.

CONWAY MORRIS, S., 1977. A new metazoan from the Cambrian Burgess Shale, British Columbia. *Palaeontology* **20**, 623-640.

CONWAY MORRIS, S. 1985a. Non-skeletalized lower invertebrate fossils: a review. In *The origins and relationships of lower invertebrates.*(eds. S. Conway-Morris, J.D. George, R. Gibson & H.M. Platt), *Systematics Association Special Volume*, **28**, 343-359.

CONWAY MORRIS, S. 1985b. The Ediacaran biota and early metazoan evolution. *Geological Magazine* **122**, 77-81.

CONWAY MORRIS, S. & ROBISON, R.A. 1988. More soft-bodied animals and algae from the Middle Cambrian of Utah and British Columbia. *University of Kansas Paleontological Contributions, Paper* **122**, 1-48.

DELLE CAVE, L. & SIMONETTA A.M. 1975. Notes on the morphology and taxonomic position of *Aysheaia* (Onychophora?) and of *Skania* (undetermined phylum). *Monitore Zoologico Italiano, N.S.* **9**, 67-81.

DZIK, J. & LENDZION, K. 1988. The oldest arthropods of the East European Platform. *Lethaia*, **21(1)**, 29-38.

DZIK, J. 1991. Is fossil evidence consistent with traditional views of the early metazoan phylogeny? In *The early evolution of Metazoa and the significance of problematic taxa* (ed. S. Conway Morris & A.M. Simonetta), pp. 47-56. Camerino: Cambridge University Press.

FORTEY, R.A. & MORRIS S.F. 1978. Discovery of nauplius-like trilobite larva. *Paleontology*, **21(4)**, 823-833.

GLAESSNER, M.F. 1979. Lower Cambrian Crustacea and annelid worms from Kangaroo Island, South Australia. *Alcheringa* **3**, 21-31.

GLAESSNER, M.F. 1979. An echiurid worm from the late Precambrian. *Lethaia* **12**, 121-124.

HESSLER, R.R. & NEWMANN, W.A. 1975. A trilobitomorph origin for the Crustacea. *Fossils and Strata* **4**, 437-459.

HOPWOOD, A.T. 1925. On the family Cyclidae Packard. *Geological Magazine* **62**, 289-309.

HOU, XIANG-GUANG 1987. Two new Arthropods from Lower Cambrian, Chengjiang, eastern Yunnan. *Acta Palaeontologica Sinica* **26**, 236-256.

HOU XIAN-GUANG 1987a. Two new arthropods from Lower Cambrian, Chengjiang, Eastern Yunnan. *Acta Palaeontologica Sinica* **26**, 236-256.

HOU XIAN-GUANG 1987b. Three new large arthropods from Lower Cambrian, Chengjiang, Eastern Yunnan. *Acta Palaeontologica Sinica* **26**, 272-285.

HOU XIAN-GUANG 1987c. Early Cambrian large bivalved arthropods from Chengjiang, Eastern Yunnan. *Acta Palaeontologica Sinica* **26**, 286-298.

HOU XIAN-GUANG, CHEN JUN-YUAN & LUO HAO-ZHI 1989. Early Cambrian new arthropods from Chengjiang, Eastern Yunnan. *Acta Palaeontologica Sinica* **28**, 42-57.

HOU, XIAN-GUANG, CHEN, JUN-YUAN & LUO HAO-ZHI 1989. Early Cambrian new arthropods from Chengjiang, Yunnan. *Acta Palaeontologica Sinica* **28**, 53-57.

HOU XIAN-GUANG & CHEN JUN-YUAN 1989a. Early Cambrian tentacled worm-like animal from Chengjiang, Eastern Yunnan. *Acta Palaeontologica Sinica* **28**, 32-41.

HOU XIAN-GUANG & CHEN JUN-YUAN 1989b. Early Cambrian arthropod-annelid intermediate *Luolishania longicruris* (gen. et sp. nov.) from Chengjiang, Eastern Yunnan. *Acta Palaeontologica Sinica* **28**, 207-213.

HOU XIAN-GUANG & BERGSTRÖM, J. 1991. The arthropods of the Lower Cambrian Chengjiang fauna, with relationships and evolutionary significance. In *The early evolution of Metazoa and the significance of problematic taxa* (ed. S. Conway Morris & A.M. Simonetta), pp. 179-188. Camerino: Cambridge University Press.

JELL, P.A. 1980. Two arthropods from the Lancefieldian (La1) of Central Victoria. *Alcheringa* **4**, 37-41.

KJELLESVIG-WAERING, E.N. 1986. A restudy of the fossil Scorpionida of the world. *Palaeontographica Americana* **55**, 1-287.

LAUTERBACH, K.E. 1973. Schlusselereignisse in der Evolution der Stammgruppe der Euarthropoda. *Zoologische Beiträge. Neue Folge* **19(2)**, 251-299.

LAUTERBACH, K.E. 1974. Ueber die Herkunft des Carapax der Crustaceen. *Zoologische Beiträge, Neue Folge* **20 (2)**, 273-327.

MANTON, S.M. 1977. *The Arthropoda: Habits, Functional Morphology and Evolution.* Oxford: Oxford University Press, 527 pp.

MANTON, S.M. 1979. Functional morphology and the evolution of the Exapod Classes. In *Arthopod Phylogeny* (ed. A.P. Gupta),pp. 387-465,

MCKENZIE, K.G. 1983. On the origin of Crustacea. *Memoires of the Australian Museum* **18**, 21-43.

MIKULIC, D.G., BRIGGS, D.E. & KLUESSENDORF, J. 1985. A new exceptionally preserved biota from the Lower Silurian of Wisconsin, U.S.A. *Philosophical Transactions of the Royal Society of London* **B 311**, 75-85.

MIROSHNIKOV, L.D. & KRAZOV, A.G. 1960. Rare palaeontological remains and traces of life in late Cambrian deposits of the northwestern Siberian platform (in Russian). *Trudy nauchnoissledovatel'skogo Instituta geologii Arktiki* **3**, 28-41.

MÜLLER, K.J. 1983. Crustacea with preserved soft parts from the Upper Cambrian of Sweden. *Lethaia* **16**, 93-109.

MÜLLER, K.J. & WALOSSEK 1984. Skaracaridae, a new order of Crustacea from the Upper Cambrian of Västergötland, Sweden. *Fossils and Strata (Oslo)* **17**, 1-65.

PETRUNKEVITCH, A. 1955. Arachnida. In *Treatise of Invertebrate Paleontology (ed R.C. Moore), Part P, Arthropoda* 2: 42-162.

REPINA, L.N. & OKUNEVA, O.G. 1969. Cambrian Arthropods of the maritime territory. *Paleontological Journal* 3, 95-103.

ROBISON, R.A. 1984. New occurrence of the unusual trilobite *Naraoia* from the Cambrian of Idaho and Utah. *University of Kansas Paleontological Contributions* **112**, 1-8.

ROBISON, R.A. 1985. Affinities of *Aysheaia* (Onychophora) with description of a new Cambrian species. *Journal of Paleontology* **59**, 226-235.

ROBISON, R.A. 1990. Earliest-known uniramous arthropod. *Nature* **343**, 163-164.

SCHNEIDER, J. 1983. *Euthycarcinus martensi* n. sp. - ein neuer Arthropode aus dem mitteleuropaischen Rotliegenden (Perm) mit Bemerkungen zu limnischen Arthropoden-Assoziationen. *Freiberger Forschungshefte* **C 384**, *Deutsche Verlag für Grundstoffindustrie Leipzig*, pp. 49-55.

SCHRAM, F.R. 1969. Poliphyly in the Eumalacostraca? *Crustaceana* **16**, 243-250.

SCHRAM, F.R. 1978. Arthropods: a convergent phenomenon. *Fieldiana Geology* **39**, 61-108.

SCHRAM, F.R. 1986. *Crustacea*. New York - Oxford: Oxford Universtity Press.

SCHRAM, F. & ROLFE, J. 1982. New euthycarcinoid arthropods from the Upper Pennsylvanian of France and Illinois. *Journal of Palaeontology* **56 (6)**, 1434-1450.

SCOTT-RAM, N.R. 1990. *Transformed cladistics, taxonomy and evolution.* Cambridge-New York: Cambridge University Press.

SIMONETTA, A.M. 1962. Note sugli artropodi non trilobiti della Burgess Shale, Cambriano medio della Columbia Britannica (Canada). I° Contributo: Il genere *Marrella* Walcott 1912. *Monitore Zoologico Italiano* **69**, 172-185.

SIMONETTA, A.M. 1963. Osservazioni sugli artropodi non trilobiti della "Burgess Shale" (Cambriano medio). II° Contributo: I generi *Sidneya* e *Amiella* Walcott 1911. *Monitore Zoologico Italiano* **71**, 97-108.

SIMONETTA, A.M. 1964. Osservazioni sugli artropodi non trilobiti della "Burgess Shale" (Cambriano medio). III° Contributo: I generi *Molaria, Habelia, Emeraldella, Parahabelia* (nov.), *Emeraldoides* (nov.). *Monitore Zoologico Italiano*, **72**, 216-231.

SIMONETTA, A. 1970. Studies on the non Trilobite Arthropods of the Burgess Shale (Middle Cambrian): The genera *Leanchoilia, Alalcomenaeus, Opabinia, Burgessia, Yohoia* and *Actaeus. Palaeontographia Italica* **66 (N.S. 36)**, 35-45.

SIMONETTA, A. 1975. Remarks on the origin of the Arthropoda. *Atti Società Toscana di Scienze Naturali, Memorie, Serie B, (1976)* **82**, 112-134.

SIMONETTA A.M. & DELLE CAVE L. 1977. Proposta di interpretazione della filogenesi degli Artropodi paleozoici. *Unione Zoologica Italiana, XLV convegno, Dimostrazioni: Riassunti*, p. 81.

SIMONETTA A.M. & DELLE CAVE L. 1981. I grandi Phyla nel Cambriano ed i loro riflessi sistematici. *Bollettino di Zoologia* **48 suppl.**, 103.

SIMONETTA, A. 1983. The myth of objective taxonomy and cladism. Much ado about nothing. *Atti Società Toscana di Scienze Naturali, Memorie, Serie B* **89**, 175-176.

SIMONETTA, A. 1987. Logic, Taxinomy, Taxa and Reality. *Contributi Faunistici ed Ecologici, Camerino* **N. 2**, 1-35.

SIMONETTA, A. 1988. Logica, tassinomia e realtà. In *Il problema biologico della specie* (eds.G. Ghiara & Coll.), *Collana UZI. Problemi di Biologia e storia della Natura* **Vol. 1**, pp. 59-78. Modena: Mucchi.

SIMONETTA, A. in press. Miscellaneous thoughts on ecology and evolution. *Bollettino di Zoologia*

SIMONETTA, A. & DELLE CAVE, L. 1975. The Cambrian non trilobite arthropods from the Burgess shale of British Columbia. A study of their comparative morphology, taxinomy and evolutionary significance. *Palaeontographia Italica* **69 (n.s. 39)**, 1-37.

SIMONETTA, A. & DELLE CAVE, L. 1978a. Una possibile interpretazione filogenetica degli artropodi paleozoici. *Bollettino di Zoologia* **45**, 87-90.

SIMONETTA, A. & DELLE CAVE, L. 1978b. Notes on new and strange Burgess Shale fossils (Middle Cambrian of British Columbia). *Atti Società Toscana Scienze Naturali, Memorie, Ser. A* **84**, 45-49.

SIMONETTA, A. & DELLE CAVE, L. 1980: The phylogeny of the palaeozoic Arthropods. *Bollettino di Zoologia* **47, Suppl. I**, 1-19.

SIMONETTA, A. & DELLE CAVE, L. 1981. An essay in the comparative and evolutionary morphology of Palaeozoic Arthropods. Origine dei grandi Phyla dei Metazoi. *Accademia Nazionale dei Lincei, Atti dei Convegni Lincei* **49**, 389-439.

SIMONETTA, A. & DELLE CAVE L. 1982. New fossil animals from the Middle Cambrian. *Bollettino di Zoologia*, **49**, 107-114.

STAROBOGATOV, YA.I. 1985. O Sisteme Trilobitoobraziyh organizmov. Byulleten Moskovskogo obshchestva Ispўtateleĭ prirodў, otd. geol. **60**, 88-98.

STAROBOGATOV, YA.I. 1986. Sistema Rakoobraznhh. *Zoologicheskĭ* Zhurnal, **65**, 1769-1781.

Størmer, L. 1970. Arthropods from the Lower Devonian (Lower Emsian) of Alken and der Mosel, Germany. Part 1: Arachnida. *Senckenbergiana lethaea* **51**, 335-369.

THOM, R. 1980. *Stabilità strutturale e morfogenesi. Saggio di una teoria generale dei modelli.* Torino: Einaudi.

TIEGS, O.W. & MANTON, S.M. 1958. The evolution of the Arthropoda. *Biological Reviews* **33**, 255-337.

TOWE, K.M. 1970. Oxygen-collagen priority and the early metazoan fossil record. *Proceedings of National Academy of Science* **65**, 781-788.

VALENTINE, J.W. & ERWIN, D.H. 1987. Interpreting great developmental experiments. The fossil record. In *Development as an evolutionary process* (eds. R.A. Raff & E.C. Raff), pp. 71-108, New York: Alan Liss Inc.

WALLCOTT, C.D. 1912. Cambrian geology and paleontology II, no. 6 - Middle Cambrian Branchiopoda, Malacostraca, Trilobita and Merostomata. *Smithsonian Miscellaneous Collections* **57(6)**, 145-228.

WATERSTON, C.D. 1975. Gill structures in Lower Devonian eurypterid *Tarsopterella scotica. Fossils and strata* **4**, 241-270.

WHITTINGTON, H.B. 1971. Redescription of *Marrella splendens* (Trilobitoidea) from the Burgess Shale, Middle Cambrian, British Columbia.*Geological Survey of Canada, Bulletin* **209**, 1-24.

WHITTINGTON, H.B. 1972. What is a Trilobitoid? In *Palaeontological Association Circular, Abstracts for annual meeting*, p. 8. Oxford (non vidimus).

WHITTINGTON, H.B. 1974. *Yohoia* Walcott and *Plenocaris* n. gen., arthropods from the Burgess Shale, Middle Cambrian, British Columbia. *Geological Survey of Canada, Bulletin* **231**, 1-21.

WHITTINGTON, H.B. 1975a. The enigmatic animal *Opabinia regalis*, Middle Cambrian, Burgess Shale, British Columbia. *Philosophical Transactions of the Royal Society of London* **B271**, 1-43.

WHITTINGTON, H.B. 1975b. Trilobites with appendages from the Middle Cambrian, Burgess Shale, British Columbia. *Fossils & Strata* **4**, 97-136.

WHITTINGTON, H.B. 1977a. The Middle Cambrian trilobite *Naraoia*, Burgess Shale, British Columbia. *Philosophical Transactions of the Royal Society of London* **B 280**, 409-443.

WHITTINGTON, H.B. 1977b. *Aysheaia* and arthropod relations. *Journal of Paleontology* **51, Suppl. 2, Pt. 3**, 31 (Abstr.).
WHITTINGTON, H.B. 1978. The lobopod animal *Aysheaia pedunculata.* Middle Cambrian, Burgess Shale, British Columbia. *Philosophical Transactions of the Royal Society of London* **B 284**, 165-197.
WHITTINGTON, H.B. 1980. Exoskeleton, moult stage, appendage morphology, and habits of the Middle Cambrian trilobite *Olenoides serratus. Palaeontology* **23(1)**, 171-204.
WHITTINGTON, H.B. 1981. Rare Arthropods from the Burgess Shale, Middle Cambrian, British Columbia. *Philosophical Transactions of the Royal Society of London*, **B 292, 329-357.**
WHITTINGTON, H.B. 1985. *Tegopelte gigas*, a second soft-bodied trilobite from the Burgess Shale, Middle Cambrian, British Columbia. *Journal of Paleontology* **59(5)**: 1251-1274.
WHITTINGTON, H.B. & BRIGGS, D.E.G. 1982. A new conundrum from the Middle Cambrian Burgess Shale. *Proceedings of the Third North American Paleontological Convention* **2**, 573-574.
WHITTINGTON, H.B. & BRIGGS, D.E.G. 1985. The largest Cambrian animal, *Anomalocaris*, Burgess Shale, British Columbia. *Philosophical Tansactions of the Royal Society of London* **B 309**, 569-609.

New observations on the Thylacocephala (Arthropoda, Crustacea)

Anna Alessandrello, Paolo Arduini, Giovanni Pinna & Giorgio Teruzzi[1]

Abstract

Some aspects of the anatomy of the thylacocephalan crustaceans are discussed. Their characteristic "cephalic sac", regarded as an enormous eye by some authors, is thought to be a highly sclerotized muscular structure, with an intimate relationship to the feeding behaviour of these animals. We consider Thylacocephala to have been epibenthic scavengers, living in shallow water just above soft substrates.

Introduction

Study of the remarkably preserved Osteno fauna from the Sinemurian of Lombardy, brought to light about 10 years ago representatives of an unusual group of crustaceans, represented in the Osteno deposit by the species *Ostenocaris cypriformis*. In a first note (Arduini *et al.*, 1980) these crustaceans were attributed to the cirripeds, but in a second note (Pinna *et al.*, 1982) it seemed better to establish for these forms the new class Thylacocephala, within the superclass Crustacea. In that second note, we included in the new class not only the Osteno forms, but some other fossil species previously attributed to various taxa of malacostracans (phyllocarids,stomatopods etc..).

Thylacocephala had a basic crustacean body-plan, but the body was almost completely included in a univalved carapace, roughly subtrapezoidal and laterally compressed (Fig. 1). A "sac", of notably large proportion in comparison with the animal size, protrudes from the anterior concave margin of the carapace; three pairs of appendages arise from the first third of the lower margin; posteriorly the body shows two distinct tagmata, that by analogy we have called thorax and abdomen, the first showing eight segments almost fully covered by the carapace but for the appendages, the second being smaller with an uncertain number of segments. Segments are also concealed beneath the carapace, from which only the distal extremities of appendages of uncertain form protrude.

A number of publications issued in the past few years have dealt with organisms analogous with those found in Osteno. The most comprehensive of these are those devoted to the specimens from the Gogo Formation from the Devonian of Western Australia, as well as those describing specimens from the Callovian (Middle Jurassic) of la Voulte-sur-Rhone in Southern France. Furthermore, there are records of new finds or reattributions to the Thylacocephala of isolated carapaces, which had been assigned formerly to other crustacean taxa. In particular, examples include assignment to the Phyllocarida, as in the case of *Clausocaris* Polz 1989 (= *Clausia* Oppenheim 1888) from the Kimmeridgian of Solnhofen, or of the forms from the Cenomanian of Lebanon, both identified in the past as larvae of stomatopods. On the basis of existing knowledge, the Thylacocephala represent a group that extend possibly from the Cambrian, and certainly the Devonian, to the Cretaceous.

They were particularly plentiful in some lagerstätten, where they constituted a proportion that is unusually high when compared with most other groups of invertebrates. Such is the case of the aforementioned deposits of the Gogo Formation, as well as at Osteno and La Voulte-sur-Rhone, and some other lagerstätten such as those in the Upper Triassic of the Lombardian Prealps.

There is no disagreements among the various authors of the fact that the thylacocephalans as a group show definitely peculiar characteristics in terms of arthropod organization, that have no counterpart in any of the other known groups, be they living or fossil forms. However, there is considerable disagreement on the functional interpretation of some structures and their precise anatomical arrangement, as well as on the ecology of these animals. This is not surprising because it is impossible to make a direct comparison with living forms. Our palaeontological expertise in Milan is now partly devoting itself to the detailed study of the anatomy of such animals. Above all we are concentrating on the Osteno specimens because they, in our opinion, seem to be particularly suited to the purpose owing to their exceptionally good state of preservation. At present, we cannot present a detailed and exhaustive picture of our results. Nevertheless, we believe it would be useful to add some observations to what has already been affirmed, in particular by Rolfe (1985), in order to give a better idea of the present areas of consensus and disagreement.

Accordingly this communication depends critically in its main points upon the correct interpretation of the preservational state of the Osteno material, which in turn has a direct bearing on our interpretation of the functional anatomy of Thylacocephala and their taxonomic attribution within the Arthropoda.

As far as the state of preservation is concerned, in the Osteno deposit fossils are compressed laterally. The matrix is composed of spongolitic micrite, and fossilization occurred through the diagenetic replacement of the original structures with calcium phosphate. Along with the mineralized parts of exoskeletons, soft parts of organisms such as polychaetes, nematodes, enteropneustans, teuthoid cephalopods, and crustaceans are also preserved. In the specific context of the Thylacocephala, we can observe gills, several types of muscles, some remnants of maybe parts of the nervous system, and cirri whose insertion coincides in with thoracic metameres. The cephalic sac is also preserved:

[1]Museo Civico di Storia Naturale di Milano, Corso Venezia 55 - 20121 Milano - Italy.

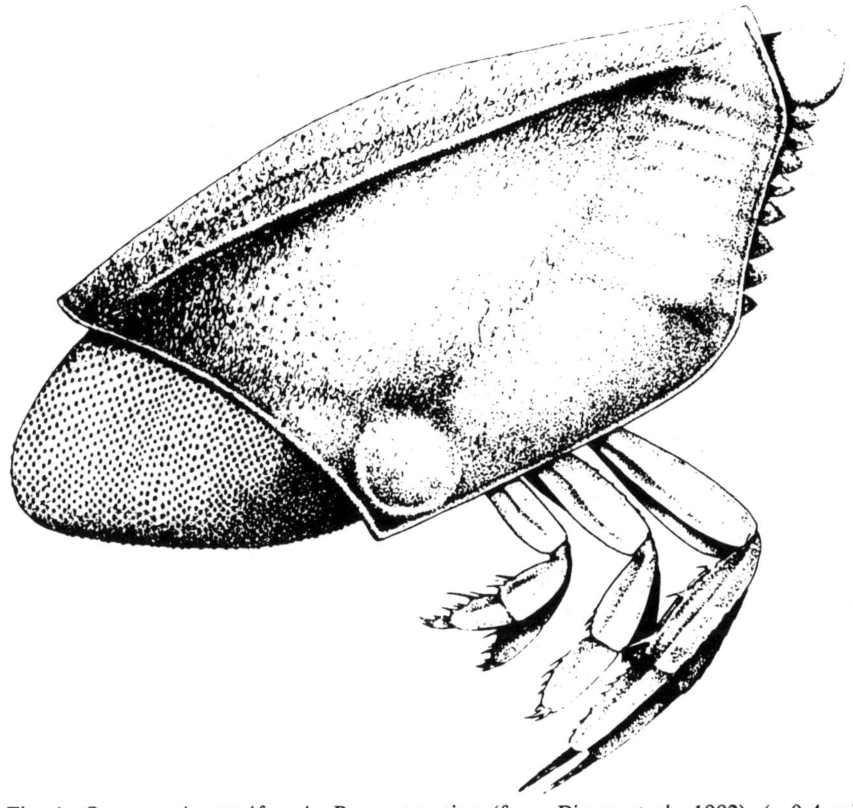

Fig. 1. *Ostenocaris cypriformis.* Reconstruction (from Pinna *et al.*, 1982). (x 0,4 ca)

it is often scarcely mineralized in small individuals but much better preserved in larger individuals. In accordance with the proposed function of the cephalic sac, we can observe that stomach residues are also often preserved.

Let us focus on the points of contention concerning the anatomy of these organisms. The body is divided into three tagmata. The cephalic tagma is the most developed and the one whose outline poses the most difficult problems for reconstruction: the cephalic sac and the three pairs of elongated appendages belong to this tagma. Thorax and abdomen, the latter being quite small, are more compact and quite distinct, even though the abdomen-thorax junction can be seen in only very few specimens. Abdominal and thoracic appendages are scarcely developed; the abdominal ones have a slightly elongated distal extremity. Gills are quite large and very similar to those of benthic decapods. The greatest controversy revolves around the interpretation of the big frontal structure, that we have termed the "cephalic sac". Part of the difficulties is that this structure is complex in form. In large specimens, where it is better preserved probably on account of a greater degree of original mineralization, the cephalic sac demonstrates a roughly hexagonal surface, under which there are one or two layers consisting of acicular sclerites of varying size and orientation. The sac has a strongly developed muscular system, especially in its median and upper part, as well as near the point of attachment to the carapace. The sac appears to be bilobate; in a specimen preserved in frontal view , it seems to be medially strengthened by an internal thin chitinous plate and gives the impression of a solution of continuity, i.e. of the possible presence of an opening. This cephalic structure has been interpreted previously by us (and we still maintain this opinion) as the homologue (even if not in a functional sense) of the peduncle of lepadomorph cirripeds, which consists of an antenna modified for attachment to the substratum. In any event, it has an evident macrostructural affinity with the peduncle of lepadomorph cirripeds. (Figs. 2 and 3). On the basis of the cephalic position and of the relationship with the other cephalic appendages, we are inclined to consider the sac as a modified antennae, even though the analogy with lepadomorphs cannot be pursued further because according to our scheme the sac does not seem to perform the same function as the peduncle in lepadomorphs.

Different interpretations have been given by other authors, in particular by Secretan, (1985) and Rolfe (1985). According to these authors, what we have defined as a cephalic sac would be interpreted better as an enormous eye. According to their analysis, which is mainly based on observations on specimens from La Voulte-sur-Rhone, the hexagons and sclerites would have to be part of ommatidia. We do not think it is possible that structures so similar in animals from different deposits can possibly correspond to two different types of structure. This point is confirmed by comparisons between the Osteno specimens and those of la Voulte-sur-Rhone . There are several reasons why we do not agree with the optical hypothesis put forward by Rolfe (1985) and Secretan (1985). In the first place, the cephalic sac has an extremely complex- and presumably multipurpose structure. In addition, apart from a few features, it does not show any affinity with a compound eye. We can observe, for instance, that there is a close connection between the stomach residues (see example Fig. 4), the sac muscular system, and the outer hexagonal layer. In addition, the sclerites that ac-

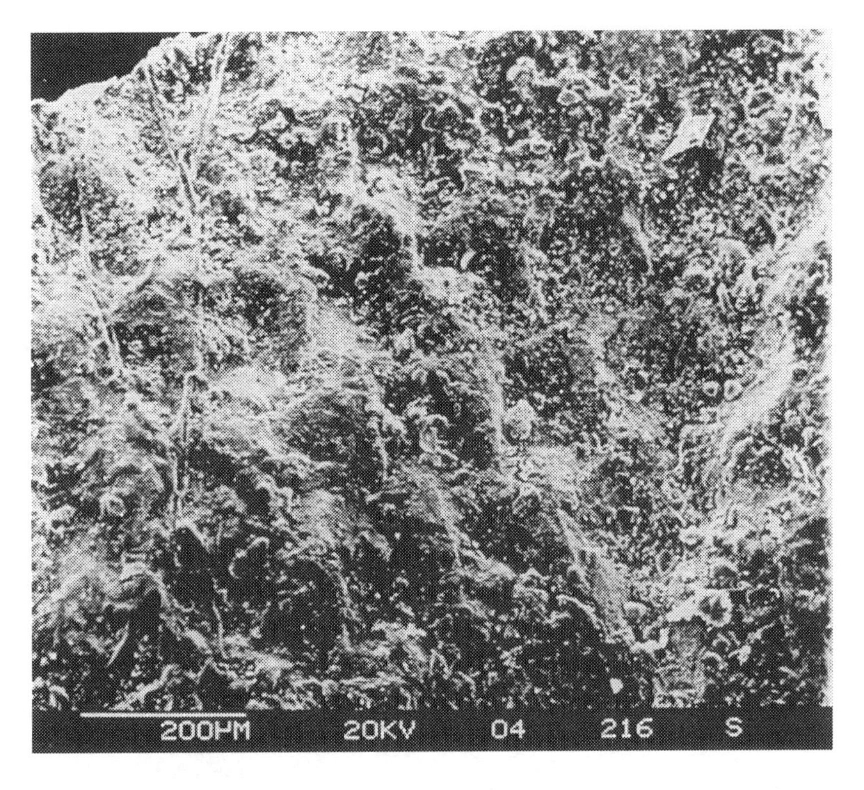

Fig. 2. *Ostenocaris cypriformis*. SEM photograph of the surface of the cephalic sac. (x 115)

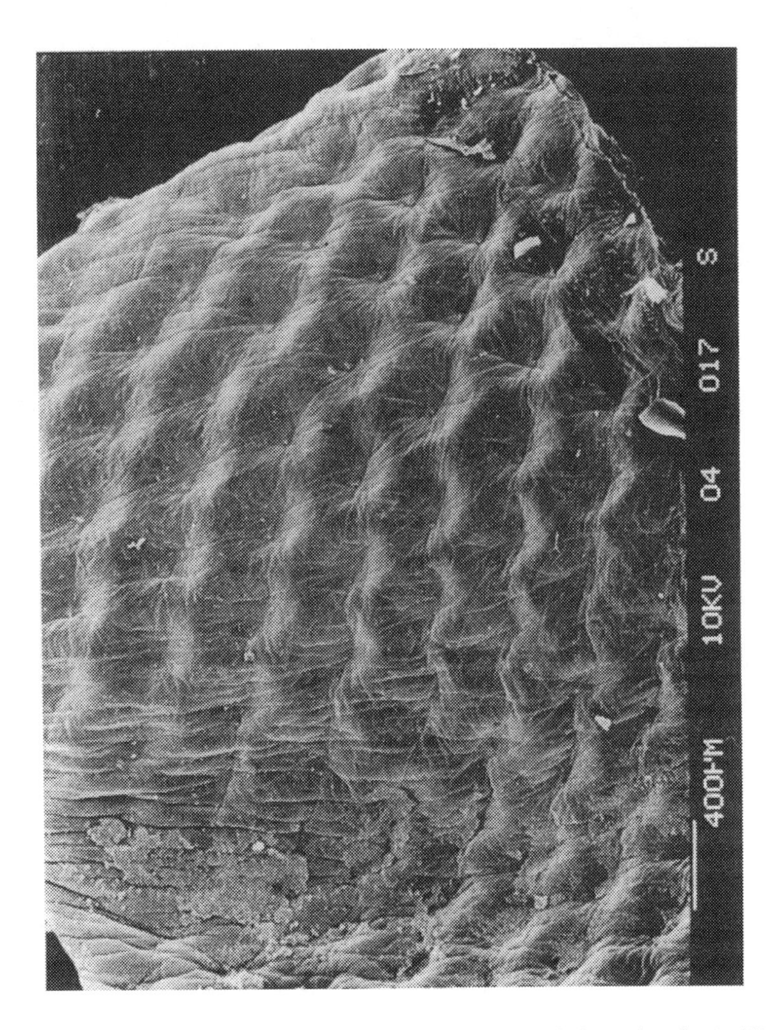

Fig. 3. *Lepas* sp.. SEM photograph of the surface of the peduncle. (x 30)

Fig. 4. *Ostenocaris cypriformis*. Alimentary residues, consisting of fragmented carapace of a small Thylacocephala, in close connection with the reticulated structure of the cephalic sac. (x 8)

Fig. 5. *Ostenocaris cypriformis*. Portion of the cephalic sac showing sclerites interstitial to the hexagon of the surface (x 50 ca).

cording to Rolfe and Secretan's hypothesis should correspond to rhabdoms are in fact interstitial to the rough hexagons themselves (Fig. 5). As a result they do not lie at the centre of the individual ommatidium as would be expected in compound eyes, but in contrast between adjacent ommatidia. In the second place, there is the above mentioned structural analogy with the cirriped peduncle.

We believe, therefore, that it is essential to regard this organ in the context of the stomach residues. These residues take the form of roughly circular structures on the median part of the front margin of the carapace. They bear a striking similarity with boluses and as already noted elsewhere they contain vertebrae of a small shark that is rather common in the Osteno deposit, hooks of teuthoids, and remains of the exoskeletons of decapod crustaceans. It appears, therefore, that the Thylacocephala swallowed coarse bits of food. Moreover, the absence in the fossils of any external structures that appear to be suitable for comminuting food into small pieces evidently is not due to poor state of preservation, but to a genuine absence of these structures in the animal itself. The conclusion is inevitable that only the big cephalic appendages somehow were used to cut food pieces into coarse chunks whose residues are still evident in the gut. If we consider the occurrence of these types of boluses, then we can imagine that there must have been a sort of storage chamber, in the form of a blind extension to the gut. We regard these boluses as being preserved in life position in the fossil (Fig. 6), and we dispute Rolfe's (1985) assertion that they have been dislocated during fossilization. We base this conclusion on the observation that other parts of the animals body do not show any deformation, apart from that caused by flattening, and they accordingly maintain the relative positions they had in life. One consequence of the alternative hypothesis is that a position of the stomach between eyes seems to be somewhat unusual. In some specimens, these food boluses occur in a position displaced from the normal one in that they are more forwardly placed towards the centre of the sac. Even in such a case the specimens do not feature any other peculiar deformation and we infer that the bolus has still been preserved in situ. At this juncture one is intitled to wonder what function these sac-antennae possessed. Even though we are still far from being reasonably sure, our hypothesis is that the position of the food bolus is not independent at the function of the sac-antenna, but has a strict correlation with it.

Thus it is possible that the marked mineralization of the sacs and their concentration of muscular material have a connection with the animal's alimentary function, at least so far as the functions of breaking the food into pieces and of rejecting indigestible portions. It is likely, therefore, that the position of the above mentioned boluses is not fortuitous, but that it corresponds to a genuine function of expulsion. This would be in keeping with the apparent impossibility of expelling the bolus through the small midgut, the extension of which can be traced into the abdomen of at least one specimen. Expulsion of indigestible residues would also be consistent with the necessity to prevent lacerations of inside tissues by hard parts like decapod

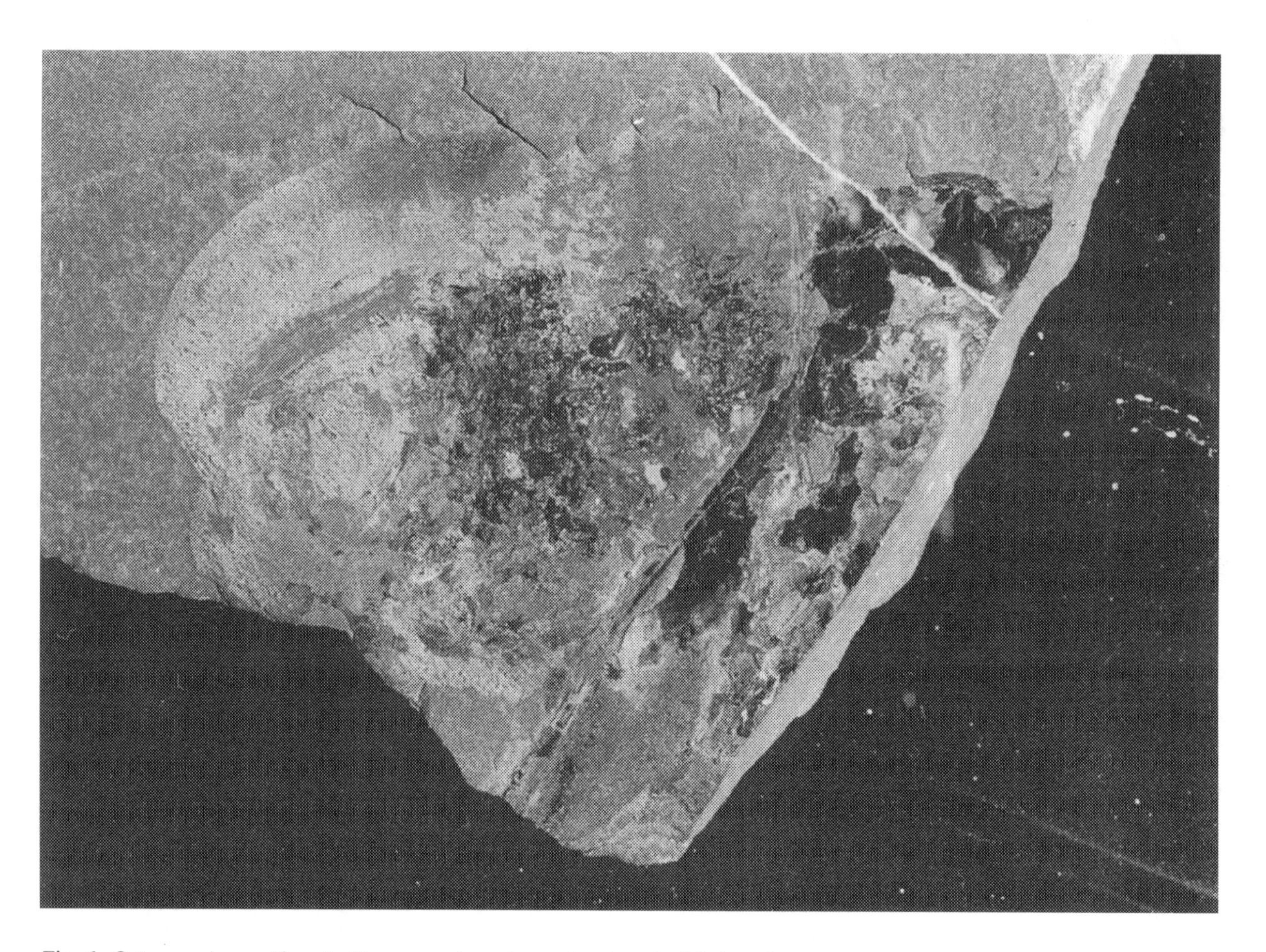

Fig. 6. *Ostenocaris cypriformis.* Fragmented specimen showing the bilobated cephalic sac, an alimentary bolus (centre), and part of the anterior margin of the carapace. (x 1,4)

exoskeletons, cephalopod hooks, and fish spines. This interpretation also is in keeping with the occurrence, in the Osteno sediments, of whole and isolated analogous forms of these boluses, as if they had been reguitated by the thylacocephalans. Such a hypothesis must be further checked. First, it is necessary to identify the correct position of the mouth, which is still fixed with any reasonable certainty. Rolfe (1985), reconciled his optical hypothesis with the supposition that these structures must have been huge eyes, enormous if compared with the body size. Drawing comparisons with certain abyssal hyperiid isopod crustaceans, he explained the gigantism of these structures by supposing that Thylacocephala lived below the photic zone. Along with the remarks already made on the structure of the sac, we would point out that all the Thylacocephala known so far come from shallow water (no more than a few dozens of metres), where the development of such large eyes is not consistent with a supposed scarcity of light. In addition, none of the associated forms show adaptative characteristics such as to suggest the occurrence of conditions comparable to those obtaining at abyssal depths.

We think there is little doubt that Thylacocephala in general, or at least those belonging to *Ostenocaris*, were necrophagous organisms. This hypothesis is supported from the presence of remains of nektonic sharks and cephalopods in the alimentary residues. Thylacocephala neither have such natant structures as would enable them to chase fast swimming animals, nor possess raptorial and offensive structures strong enough to overwhelm a shark, whatever its size. Even the three biggest pairs of cephalic appendages are not especially robust and not equipped with chelae (Secretan (1985) reconstructed *Dollocaris* with a cheliferous first appendage, but this occurrence is not, in our opinion, definitely confirmed).

Rolfe (1985) reconstructed *Dollocaris* as a semiplanktonic organism. This interpretation does not convince us for several reasons. In the first place, the general appearance, the dimensions (some specimens from Osteno are more than 30 cm in length) and the sturdiness of the animal do not appear to be consistent with those of a natant organism. The three pairs of cephalic appendages are not suitable for swimming, and the thoracic and abdominal appendages do not seem appropriate for protracted swimmming. On the other hand, it is more likely that the cephalic appendages had some type of locomotory functions. In some recently found specimens, the appendages are indeed in a position consistent with both one of our reconstructions (Pinna *et al.*, 1982, p. 479, fig. 4), and in a position that could be adopted if they had a locomotory function.

Some specimens occur with the thorax-abdomen complex protruding from the carapace. This also seems to be a natural position of the living animal, and not a dislocation that arose during fossilization. The entire cephalic unit is still in place, and the thorax-abdomen complex rotates as an integral part; it is likely that the long dorsal muscle could control this type of movement.

The last point we would like to deal with briefly here concerns the appropriate systematic assignment of the Thylacocephala.

At first we considered them close to cirripeds because of the marked similarities with the cypris larval stage of cirripeds. Consequently we examined the possibility of evolution by paedomorphosis of Thylacocephala crustaceans from larval forms similar to cypris, in a manner similar to those processes that are believed to have had an important role in crustacean diversification (Schram, 1986).

But there could well be another possibility (F.M. Scudo, pers. com.). Thylacocephala did not necessarily originate from cirripeds, but the contrary could be true. Thus, if we think of Thylacocephala in term of head-down crustaceans, semi-sessile on a soft bottom, as a first step toward head-down cirripedia that become sessile on hard substrates. The inferred stratigraphic distribution of Thylacocephala, if confirmed, fits well with this hypothesis. An even closer analogy with cirripeds has been recorded by Starobogatov (1986) who included Thylacocephala and cirripeds in his subclass Lepadioni.

But there are also differences between the Thylacocephala and cirripeds: the number of thoracic segments is five in cirripeds and eight in Thylacocephala. Other differences are the presence of the three pairs of big cephalic appendages, adaptation to a necrophagous diet, and the cephalic sac that in cirripedes is not bilobate. Such differences lead us to believe that Thylacocephala are best treated as a class of their own, to be included in the superclass Crustacea on the basis of the occurrence of antennae and antennules (granted our interpretation is correct), as well as morphological characteristics of the carapace. Thylacocephala are certainly peculiar crustaceans, with adaptations unrecorded in any other living or fossil group.

Acknowledgements

We wish to thank Simon Conway Morris for his accurate revision of the manuscript.

REFERENCES

ARDUINI, P., PINNA, G. & TERUZZI, G. 1980. A new and unusual Lower Jurassic cirriped from Osteno in Lombardy: *Ostenia cypriformis* n.g.n.sp. (Preliminary note). *Atti Società Italiana di Scienze Naturali e Museo Civico di Storia Naturale di Milano* **121**, 360-370.

ARDUINI, P., PINNA, G. & TERUZZI, G. 1984. *Ostenocaris* nom. nov. pro *Ostenia* Arduini, Pinna & Teruzzi, 1980. *Atti Società Italiana di Scienze Naturali e Museo Civico di Storia Naturale di Milano* **125**, 48.

OPPENHEIM, P. 1888. Neue Crustaceenlarven aus dem lithographischen Schiefer Bayerns. *Zeitschrift der Deutschen geologischen Gesellschaft* **40**, 709-719.

PINNA, G., ARDUINI, P., PESARINI, C. & TERUZZI, G. 1982. Thylacocephala: una nuova classe di crostacei fossili. *Atti Società Italiana di Scienze Naturali e Museo Civico di Storia Naturale di Milano* **123**, 469-482.

PINNA, G., ARDUINI, P., PESARINI, C. & TERUZZI, G. 1985. Some controversial aspects of the morphology and anatomy of *Ostenocaris cypriformis* (Crustacea, Thylacocephala). *Transactions of the Royal Society of Edinburgh: Earth Sciences* **76**, 373-379.

POLZ, H. 1989. *Clausocaris* nom. nov. pro *Clausia* Oppenheim 1888. *Archaeopteryx* **7**, 73.

ROLFE, W.D.I. 1985. Form and function in Thylacocephala, Conchyliocarida and Concavicarida (?Crustacea): a problem of interpretation. *Transactions of the Royal Society of Edinburgh: Earth Sciences* **76**, 391-399.

SECRETAN, S. 1985. Conchyliocarida, a class of fossil crustaceans: relationships to Malacostraca and postulated behaviour. *Transactions of the Royal Society of Edinburgh: Earth Sciences* **76**, 381-389.

SCHRAM, F. R. 1986. *Crustacea*. New York, Oxford: Oxford University Press.

STAROBOGATOV, I. 1986. Sistema rakoodraznoekh [The System of Crustacea, in Russian with English summary]. *Zoologicheskii Zhurnal* **65**, 1769-1781.

Fossil Acarida
Contribution of palaeontological data to acarid evolutionary history

Fabio Bernini[1]

Abstract

The acarid fossil record dates back to the Lower Devonian and consists of species belonging to two different groups, both included in the Actinotrichida lineage. These animals were probably broadly phytophagous. This can be deduced from the feeding habits of the recent, but primitive representatives. Their leg type and body structure were pre-adapted for interstitial and edaphic ways of life. These earliest actinotrichids probably followed their plant food onto dry land in edaphic environments, and then in due course exploited the opportunities available for the soil fauna. Recent discoveries in palaeosols and other edaphic remains suggest that simple soil communities appeared in the Silurian or Upper Ordovician.

In contrast to actinotrichids, the anactinotrichid lineage does not appear in the fossil record until the Cainozoic. The present-day representatives of this branch are mainly predators and carnivores, and their ancestors presumably had similar feeding habits. It may not be true that they did not exist before the Eocene-Oligocene and are of later origin than actinotrichids. Nevertheless the main adaptive radiations of these mites seem to occur in times, and for ecological reasons, quite different with respect to the other lineage.

The "modernity" of the earliest mites precludes any indication about their ancestors because there are no transitional characters linking them to earlier fossil chelicerates. The extraordinarily unvarying nature of the chelicerates, especially the terrestrial groups, as well as the scarcity and, in some cases, absence of available fossils impede the discovery of the early phases of their evolutionary history.

The evolutionary patterns occurring in both acarid lineages have been considered in order to explain the sudden appearance of their body plan and its maintenance for at least 400 Myr. It is suggested that progenesis played a role in acarid evolution.

Introduction

There are two principal hypotheses about the evolutionary history of acarids (mites and ticks). These animals are regarded by most authors as a natural monophyletic group having the rank of subclass and in which 2 (or 3) main lineages are identifiable [for reviews see Lindquist (1984) and Bernini (1986)]. The first, the Actinotrichida, includes Acaridida (= Astigmata), Actinedida (= Prostigmata) and Oribatida (= Cryptostigmata). The other, the Anactinotrichida, includes the Ixodida (= Metastigmata), Gamasida (= Mesostigmata), and Holothyrida (= Tetrastigmata). The Opilioacarida (= Notostigmata) are regarded by Grandjean (1970) as a third lineage but most recent authors maintain that they can be included in the Anactinotrichida, on the grounds of similar spermatological characters (Alberti, 1984).

Contrary to the monophyletic idea, Hammen (1977) argues very persuasively in favour of the polyphyletic origin of acarids. In the framework of a complete phylogenetic revision of the Chelicerata, he separates the two main acarid lineages, the Actinotrichida and the Anactinotrichida (Opilioacarida included), and places them in two different subclasses: the former he assigns (together with the Palpigradi) to the Epimerata; the latter related to the Ricinulei he puts in the Cryptognomae (Hammen, 1979, 1982, 1989). No cladistic (Lindquist, 1984) or classical phylogenetic analysis (Grandjean, 1970) offers conclusive indications as to which hypothesis is correct. However, the most recent studies published by Hammen (1986, 1989), based on the idea that the organisms are functioning wholes, seems convincing and tips the scale of opinion towards biphyly. Elements such as body segmentation, pattern and fate of the coxal glands, morphology of mouthparts and respiratory system, feeding characteristics and sperm transfer seem to be significant and phylogenetically weighted in favour of the independent origin of the two acarid lineages.

It is singular that, with sporadic exceptions (Bernini, 1986), the palaeontological aspect has never been introduced into the discussion of acarid evolution. All authors admit that the acarids as a group are very ancient (at least Silurian) and stable in their body plan. It is true that up to now the findings of fossil acarid have been extremely limited in number and quality. Nevertheless in the last few years some new discoveries have enriched the available material. The environmental scenario of these early animals has also gradually become clearer. All this data are probably not yet sufficient to awaken the interest of palaeontologists in this small and ostensibly unpromising animal group. Nevertheless, I am convinced that some points of their evolutionary history deserve at least preliminary attention in order to resolve some general questions and suggest new areas of research in this field.

If the mono- or biphyly of the acarids is questionable, then we can hardly speculate about the origin of these creatures from any basic chelicerate stock.

Accordingly, the aim of my paper is to summarize the available palaeontological evidence in acarids and earliest chelicerates, speculate on the events which have presumably determined their evolutionary history, and discuss the possible patterns of their evolution and the contribution of fossils to the resolution of the main problems in these topics.

Fossil evidence in Acarida

The contribution of fossil data to the solution of the problems of acarid phylogeny was recently examined (Bernini, 1986) but even more recent results have brought new evidence (Krivolutsky & Druk, 1986; Norton *et al.*, 1988). Hence I first review and summarize all the fossil acarid findings in the following scheme (Fig. 1) (Bernini, in preparation).

As is well-known, the oldest fossil mite, *Protacarus crani* Hirst, 1923, from the Scottish Rhynie Chert,

[1]Department of Evolutionary Biology, University of Siena, via P.A. Mattioli 4, Siena, Italy.

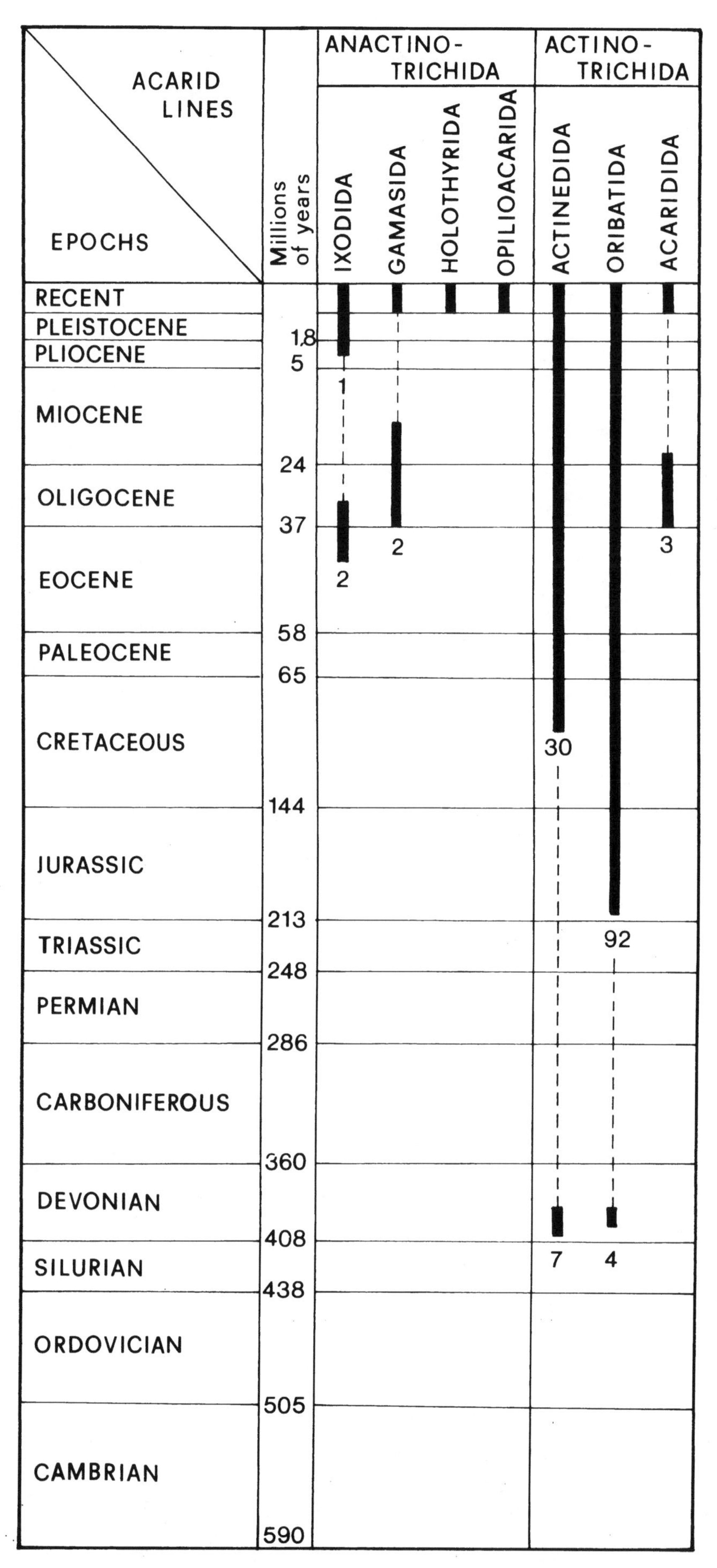

Fig. 1. Diagram showing the fossil remains divided into the several acarid groups. The numbers at the base of the lineages indicate the number of the species described up to now. The dashed lines refer to the gaps in the fossil record.

dates back to the Middle Lower Devonian (Siegenian, from 394 to 401 Myr ago). The fact that the five specimens, regarded by Hirst (1923) as belonging to a single species, were considered by Dubinin (1962) to represent five different taxa to be placed in four families all belonging to the Actinedida is less widely known. This is because no single treatise or text-book has reported Dubinin's new species and genera or attempted to introduce them into the systematic framework of this group. Indeed, a glance at the illustrations of *P. crani* is sufficient to justify Dubinin's case for splitting, even though doubts exist on the proposed systematic arrangement. For instance, in Kethley's opinion (Norton *et al.*, 1989), *Protacarus crani, Pseudoprotacarus scoticus, Paraprotacarus hirsti* and *Palaeotydeus devonicus* should all be included in the family Bimichaelidae, while *Protospeleorchestes pseudoprotacarus* remains in Nanorchestidae.

The Devonian record of acarid remains has recently increased (Norton *et al.*, 1988) with the discovery of 6 other species from the Gilboa (New York) deposit. These mites, found together with "the oldest known centipede, arthropleurids, pseudoscorpion, possible scorpions, abundant trigonotarbids, a possible insect, a probable amphibious eurypterid" and "plant fossils, primarily lycopods" (Norton *et al.*, 1988, p. 259) are, from the systematic point of view, the early derivative members of the Oribatida (4 spp.) and the Actinedida (2 spp.). All Devonian fossil mites can therefore be included in the actinotrichid lineage.

Subsequent to these findings, there is a large gap in the documentation of fossil acarids: there are no records for the rest of the Palaeozoic era, and the most ancient Mesozoic mite is a species of *Hydrozetes* (Oribatida) from the Lower Jurassic of southern Sweden (Sivhed & Wallwork, 1978). The gap is partially filled by indirect evidence of acarid life in the forest biome. The activity of microarthropod fauna, in particular oribatid mites (box mites), seems to be testified by animal products such as fossil coprolites found within the decaying stem of marattiales, the arboreous ferns of the Upper Carboniferous (Rothwell & Scott, 1983; Scott & Taylor, 1983).

Since the Lower Jurassic, the acarid record becomes more numerous and continuous. It consists essentially of oribatid, but also actinedid, mites collected from the Middle Jurassic of Siberia and the Soviet Far East, the Cretaceous of northern Siberia, Sachalin and Manitoba (Canada), the Upper Paleocene of Alberta (Canada) and Wyoming (USA), the Lower-Middle Eocene of southern Australia (Baker & Wighton, 1984; Bulanova-Zakhvatkina, 1974; Ewing, 1937; Krivolutsky & Druk, 1986; Southcott & Lange, 1971; Vercammen-Grandjean, 1973; Zacharda & Krivolutsky, 1985) to the Lower Oligocene deposits of Dominican and, above all, Baltic Amber (Larsson. 1978; Sellnick, 1931). The latter have provided a great number of the best preserved acarid fossil remains. The thousands of fossil arthropods collected give us an indication of the faunistic and ecological scenario of a particular Oligocene habitat: a multi-layered forest in the then tropical or subtropical belt. In spite of the under-representation of edaphic groups and the preponderance of the small flying insects in amber, we can still discern the faunal composition and ecological relationships of some microenvironments.

In the different classical collections examined by Sellnick (1931) there are about 850 mites: 380 of them belong to the Oribatida, more than 400 are Actinedida and the rest are distributed among the other acarid lineages. The same percentage is reported from the Danish amber collection in the Zoological Museum of Copenhagen (Larsson, 1978). 108 of the 400 specimens have been identified as oribatid mites, about 280 have been assigned to the different actinedid families, and only 10 are gamasids.

For the first time representatives of Gamasida and Ixodida, that is, of the anactinotrichid lineage are found from the Baltic (Lower Oligocene), Mexican (Oligocene) and Dominican (Upper Eocene-Oligocene) amber deposits. The earliest representative of the fifth largest acarid branch, the Acaridida, was finally discovered in Oligocenic amber deposits from Chiapas (Mexico) (Turk, 1963).

The continuity of the acarid lineages up to the present time is proved by rare Miocene and Pliocene specimens found in Disodile di Melilli in Sicily, Galicia (Poland), the Burmese and Mexican ambers and northern Siberia (Cockerell, 1917; Golosova et al., 1985; Pampaloni, 1902; Woolley, 1969). In these epochs the most frequent acarid group again seems to be the Actinotrichida and the Oribatida are the only mites identified.

Fossil and subfossil mites are also found in Pleistocene deposits, but this is beyond the scope of the present cursory review. In addition, the many acarid species are often not precisely dated or determined (Krivolutsky & Druk, 1986).

Finally, there is interesting new data which suggests the existence of soil mites, or at least suitable conditions for their life, earlier than the Lower Devonian. Palaeosols dating back to the Upper Ordovician have recently been described: they show many traces of non-vascular land plants and soil animals (Retallack, 1985; Retallack & Feakes, 1987). However, the dimensions of the burrows, presumably inhabited by arthropods, indicate that they were not related to the later (Silurian and Devonian) soil fauna. Later palaeosols from the Upper Silurian again show traces of soil life: at that time the land plants were undoubtedly vascular and the soil animals smaller than those of the Ordovician strata (Retallack, 1985). The presence of soil organisms in that period can also be inferred from descriptions of structures similar to fecal pellets of microarthropods in close association with fungal hyphae (Sherwood-Pike & Gray, 1985).

What can be inferred from this palaeontological evidence? I would like to consider four main points.

First of all, in spite of these new discoveries, the acarid fossil record remains scarce and incomplete mainly because the terrestrial environments are poorly represented (Rolfe, 1985).

The second point is the "modernity" of the fossil mites. A simple glance at the illustrations of the oldest fossils reveals their similarity to specimens of today; so much the case that some authors believed them to be a modern contamination (Crowson, 1970, 1985). This seems unlikely for many good reasons (Greenslade, 1985; Rolfe, 1980), so this "modernity" must reflect an actual state and characteristic of this animal group. The conservation of a body plan for 400 Myr is one of the most extraordinary cases of slow morphological transformation (Bernini, 1986). Unfortunately this obstructs the search for acarid ancestors because there are no transitional characters with respect to the earlier fossil chelicerates. The only exception to this seems to be *Paraprotacarus hirsti* Dubinin, 1962. This specimen is the only one visible from the ventral side. It seems to show some features that are so interesting that, if confirmed, ought to be sufficient to exclude it from the Actino- and Anactinotrichida (Bernini, 1986). This specimen could therefore represent another chelicerate lineage having some earlier derivative characters with respect to the two acarid stocks, but without clear links between them and a possible ancestor. It seems, however, to be on a parallel branch rather than at the base of the acarid lineage(s).

The third point is the different age of the fossils of the two lineages. All the earliest mite remains (Palaeozoic and Mesozoic) belong to the actinotrichid lineage. Fossils of Anactinotrichida have been found in different parts of the world but always after the Upper Eocene - Lower Oligocene. The most obvious explanation of this could be the small number of searches made or, more simply, bad luck. However, it should also be borne in mind that when anactinotrichids appear in the Baltic Oligocene deposits, they are always extremely scarce with respect to actinotrichid mites. This is also true when compared to the present-day tropical (or subtropical) edaphic biocenoses. This is understandable because ticks and all parasitic gamasids obviously are not easily trapped by amber drops on account of their peculiar life style. In addition the early derivative anactinotrichid groups Opilioacarida and Holothyrida today are so rare and localized that there is not much chance of their leaving fossils. Despite this it is possible that the Anactinotrichida are younger than the Actinotrichida and/or their main adaptive radiation was relatively more recent. In either case this means that the two phyletic lineages had a very long, separate evolutionary history.

The final point arising from palaeontological analysis is the high specific diversity which is evidently a characteristic from the first mites up to the present biocenoses. The 5 Rhynie Chert specimens belong to 5 species. The other Devonian occurrence, that of Gilboa, yields 6 species for the 14 specimens collected. The largest collection of fossil acarids from Baltic amber confirms this: 72 species for the 373 oribatids collected. This means that the acarid presence in the terrestrial edaphic biocenoses has presumably always had the same ecological and demographic weight.

Evolutionary trends in Chelicerates

Chelicerates are mainly characterized by two body tagmata called prosoma and opisthosoma, a preoral (but embryologically post-oral) usually chelate appendage, the chelicera, and five other post-oral prosomal appendages. The ancestral number of body segments seems to vary from 16 to 19 to which a spine-like tail, the telson, is added in some taxa (Hammen, 1985).

This body plan was reached very early in geological history of arthropods. Both merostome fossil representatives of the xiphosurid and eurypterid lineages have been found in sedimentary strata of the Ordovician (500 Myr ago) (Fig. 2). Caution, nevertheless, is required in the interpretation of presumed "chelicerates" from the Cambrian, such as the very incomplete *Eolimulus alatus* Bergström, the supposed eurypterid *Kodymirus* (Bergström, 1975), and especially the Aglaspida. Confirmation through further studies and findings is also needed. So, for the time being I regard this evidence as inadequate for any phylogenetic discussion. Until recently Aglaspida in particular seemed to have an important phylogenetic position suggesting affinities between Trilobita and Chelicerata (Bergström, 1979; Fortey & Whittington, 1989; Kraus, 1976; Lauterbach, 1980; Størmer, 1944; Weygoldt, 1986). These two groups were included in the Arachnomorpha (or Arachnata) on the basis of intermediates such as the aglaspidids and olenellid trilobites, of the ontogeny of *Limulus*, and the homology of the mesosomal appendages in *Limulus* and trilobites (Whittington, 1979). The only aglaspidid species in which the appendages were known (*Aglaspis spinifer* Raasch, 1939) was recently restudied (Briggs *et al.*, 1979), and removed from merostomes. This eliminates much evidence of the existence of Cambrian chelicerates and introduces doubts on their close relationship with the trilobites.

The origin of the chelicerates has very recently been put back to the Cambrian following the description of a new fossil from Mount Stephen, British Columbia (adjacent to the Burgess Shale, Middle Cambrian): *Sanctacaris uncata* Briggs & Collins, 1988. In spite of the absence of chelicera and the presence of biramous appendages the authors consider this form as a plesion, that is a primitive sister group of the chelicerates (Briggs & Collins, 1988).

Other Middle Cambrian non-trilobite arthropods from the Burgess Shale are believed to belong to basic pre-chelicerate stock. According to Briggs (pers. comm.) this stock included *Yohoia, Burgessia* and *Leanchoilia*. In contrast, Simonetta & Delle Cave (1981) considered that the pre-chelicerate representatives were to be found in the group of "emeraldellids", particularly in the genera *Strabops, Molaria, Habelia* etc. Nevertheless, none of them is more closely related to chelicerates than *Sanctacaris uncata* (Briggs & Collins, 1988). Whatever the ancestor(s) of the chelicerates it seems clear that this group also originated from one of the numerous non-trilobite arthropod lineages produced by the explosive Cambrian radiation, but there is no evidence of their early evolution.

Therefore, the first fossil remains with an unmistakably chelicerate body plan only occur in the Ordovician and belong to the Eurypterida (Whittington, 1979) (Fig. 2). This lineage lasted and radiated for more than 200 Myr, dying out in the Permian mass extinction. Its close affinity with scorpions was recently suggested by the discovery of the ancestral aquatic habits of the latter chelicerate group. It is now generally accepted that modern scorpions are an offspring of eurypterids which have survived to the present time in terrestrial refuges (Bergström, 1979; Kjellesvig-Waering, 1986; Simonetta & Delle Cave, 1981).

The other well-known merostome group, the Xiphosura, can also only be traced back with certainty to the end of the Ordovician. After a series of repeated radiations mainly in the Silurian and Carboniferous, it has survived as a single lineage (the Limulicina) to the present day (Eldredge, 1974; Fisher, 1984).

Both these groups seem to be specialized for marginal marine habitats such as shallow water (Xiphosura) or brackish and fresh water (some Xiphosura and the Eurypterida) (Fisher, 1984; Selden, 1984). Both these groups and the marine *Sanctacaris uncata* have a preda-

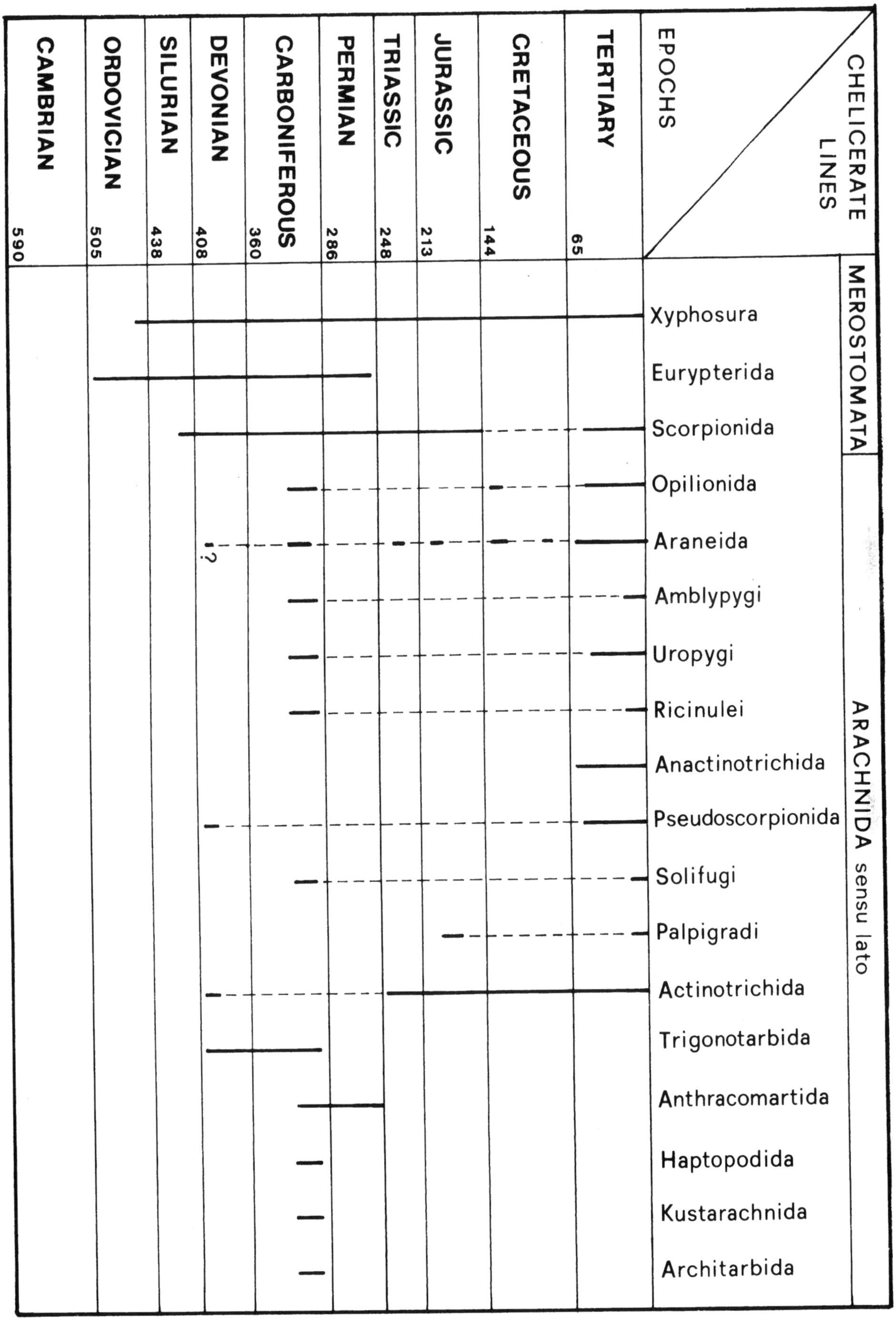

EPOCHS	Age
TERTIARY	65
CRETACEOUS	144
JURASSIC	213
TRIASSIC	248
PERMIAN	286
CARBONIFEROUS	360
DEVONIAN	408
SILURIAN	438
ORDOVICIAN	505
CAMBRIAN	590

CHELICERATE LINES

MEROSTOMATA
Xyphosura
Eurypterida
Scorpionida

ARACHNIDA sensu lato
Opilionida
Araneida ?
Amblypygi
Uropygi
Ricinulei
Anactinotrichida
Pseudoscorpionida
Solifugi
Palpigradi
Actinotrichida
Trigonotarbida
Anthracomartida
Haptopodida
Kustarachnida
Architarbida

Fig. 2. Diagram showing the fossil remains of the major groups of chelicerates. The dashed lines indicate gaps in the fossil record.

tory life-style (Briggs & Collins, 1988). This feeding habit may explain the scarcity of the fossil record because the representatives of this level of the community trophic chain are generally few in number. According to Bergström (1979), another explanation of this point may be that the evolution of chelicerates mainly occurred in the less known freshwater and brackish environments which in general do not favour the preservation of fossils.

The availability of these fossils has made possible a tentative picture of the evolutionary history of the merostomes. This is not so in the case of the land chelicerates, of which no related fossil records have been reported. Only the Middle Cambrian *Beckwithia typa* Resser, 1931, no trace of the appendages of which has been found, and the Devonian *Diploaspis casteri* Størmer, 1972, have been proposed as representatives (or late representatives) of the aquatic stock of the nonmerostome chelicerates (Bergström, 1979). Hence all phylogenetic hypotheses on the early phases of their evolutionary history are equally plausible, the only constraints being our limited knowledge of land fossils preserved since the Lower Devonian (Rolfe, 1980; Shear *et al.*, 1984) and the neontological data.

The extraordinary conservative nature of these groups suggests that the land arachnids have an extremely long evolutionary history in water. They probably branched from the first stock of the true chelicerates in the Ordovician or at the end of the Cambrian separately from the merostomes and it may be that the different lineages crossed the water-land boundary repeatedly and independently (Bergström, 1979; Kraus, 1976). If this is correct then it is quite plausible that Arachnida is not a clade but a grade (Hammen, 1986).

All the chelicerate lines of descent characterized by a predatory life style and carnivorous feeding habits, obviously excluding the phytophagous and saprophagous acarids and opilionids, have some points of their evolutionary histories in common. Spiders, pseudoscorpions and, above all, trigonotarbids appeared and radiated on land in the Lower Devonian, presumably preying on actinotrichids, myriapods and the earliest Apterygota. The remaining groups appeared in the Carboniferous, probably coinciding with the radiation of the first flying insects. Such common points are periods of radical evolutionary "experimentation" in chelicerate structural plan due to the exploitation of empty niches and new modes of life. Usually this high capacity for "experimentation" leads to wide radiation and a strong increase in species diversity. In the chelicerates this occurred only in spiders, opilionids and mites; in the others, this "experimentation" produced several taxonomic groups characterized by different structural types, each of low diversity but separated from the others by very large morphological gaps.

Following this brief review of the fossil record and evolutionary history of chelicerates, certain common patterns are evident and deserve comment.

First of all, the fossil chelicerates generally seem to be linked to the tropical environments for the entire Palaeozoic in the Laurentian and Baltic continents. To my knowledge, the only exceptions are the very scarce Silurian and Lower Devonian eurypterids of Australia and Brazil respectively (Caster & Kjellesvig-Waering, 1953), and a Lower Devonian xiphosurid from Bolivia (Eldredge, 1974).

In land chelicerates, this trend is particularly evident because no fossils except for a giant Carboniferous spider from Argentina (Hünicken, 1980), have been found in southern continents in the Palaeozoic or Mesozoic (Savory, 1977). This is important because tropical habitats were particularly well represented in the Gondwanian part of the Mesozoic world. This may be due to lack of research, available fossiliferous sites and of favourable conditions for the fossil preservation in the southern continents, rather than to a true scarcity or absence of chelicerates. Moreover during the Palaeozoic, Gondwana was generally characterized by temperate and cold climates which were presumably not those preferred by these animals. However there were other land arthropods, i.e. insects, the remains of which reveal a more generalized and widespread occupancy of the terrestrial niches of the Upper Palaeozoic and Mesozoic worlds (Jeannel, 1949; Whalley, 1988).

In second place, chelicerate remains are scarce in the whole of the Mesozoic. Except for a few spiders (Savory, 1977; Selden, 1988) and opilionids (Petrunkevitch, 1955), some mites (Krivolutsky & Druk, 1986), only one palpigrade (Savory, 1977), the aquatic xiphosurids and scorpions (the pulmonate scorpions also show a Mesozoic gap) (Kjellesvig-Waering, 1986), no other groups are represented in the Mesozoic deposits of the southern or northern continents.

The explanations given above apply equally to this case. Here we run the risk of reconstructing the biogeography and ecology of the available fossiliferous sites rather than of the once-living chelicerates! The small number of occurrences sounds a note of caution and warns against generalizations. These observations emphasize the gaps in our present knowledge of chelicerate palaeontology and indicate the necessity of future research. Once we have more data, distribution and ecology will reveal certain important aspects of chelicerate history.

Early phases of acarid evolutionary history

Although the connection with the basic chelicerate stock cannot be traced, we can briefly examine the kind of organism the earliest acarids might have been. Some indications are given by the study of the modern feeding habits of the early derivative Actinotrichida, which are the closest relatives of the Devonian mites of the Rhynie Chert and Gilboa deposit. Most Endeostigmata and Palaeosomata seem to feed on vegetable materials (including algae) or fungal mycelia but there are also carnivorous members (Bernini, 1986; Krantz & Lindquist, 1979). All the present Oribatida and many Actinedida are still in a broad sense phytophagous (saprophagous, mycetophagous, etc). Their Devonian ancestors presumably had similar feeding habits. The evolutionary history of these animals thus seems closely linked to that of land plants and fungi which constituted their food and, together with bacteria, formed the soil in which they live.

The fossil plants at Rhynie Chert offer evidence of the phytophagous habits of the first terrestrial organisms. Stems and sporangia show signs of being eaten and scarring, that could only have been inflicted by lymph-sucker phytophytes (Kevan *et al.*, 1975). This kind of evidence has also been described in successive periods such as the Carboniferous (Taylor & Scott, 1983). Among the fossil animals found at the Rhynie

Chert, only the springtail and the several species of mites were plausibly phytophagous (Kevan et al., 1975). The earliest actinotrichids therefore seem to have been phytophagous in a strict sense and saprophagous (Scott & Taylor, 1983). Nevertheless, the existence of some actinedid lines which maintained the ancestral carnivorous chelicerate feeding habits is also plausible (Krantz & Lindquist, 1979) and from which the living predator groups presumably originated.

Other aspects deducible from the Devonian mites and their modern relatives are the absence of a thick coloured cuticle and of respiratory apparatus. The actinotrichid ancestors probably lived at the bottom of the sea and/or rivers near estuaries (Rolfe, 1980) feeding on algae, fungal mycelia, bacteria and other vegetable material and/or organic detritus. They might have moved towards the terrestrial niches long before the Devonian, following their food source, making the transition from a benthic, perhaps interstitial, to an edaphic way of life. It may be that the transition from carnivorous to broad phytophagous feeding had already occurred in the aquatic environments. It is, therefore, quite probable that the ancestors of the Actinotrichida were already pre-adapted to the edaphic life style by virtue of their small size, shortened body lacking segmentation and opisthosomal appendages (Bernini, 1986), feeding flexibility due to the gnathosomization process, and leg type adapted for locomotion in interstitial and soil environments (Hammen, 1985).

Thus in the terrestrial biocenoses of the Lower Devonian a trophic chain was established: the role of primary consumer (mainly phytophagous and saprophagous) was presumably played by actinotrichid mites, Collembola, Myriapoda and possibly soft worms, while the predators were spiders, pseudoscorpions (at Gilboa), above all trigonotarbids (now extinct arachnids), possibly amphibian scorpions and eurypterids (Kevan *et al.*, 1975; Norton *et al.*, 1988).

Despite the large gap in the fossil evidence, it is possible to outline the subsequent evolutionary history of the edaphic and phytophagous actinotrichid lineages thanks to recent palaeoecological data (Retallack, 1985; Scott & Taylor, 1983). Although the ecological niches related to phytophagy were empty, adaptive radiation does not seem to have occurred at once. This is indicated by the absence of fossil remains of these animals in the numerous deposits of the Middle and Upper Palaeozoic. Palaeosols of the Upper Devonian show many clear traces of vascular plants (mainly ferns) forming streamside gallery forests, but also a singular reduction of traces attributable to a soil fauna (Retallack, 1985). According to the same author, this might be due to phenolic substances leached from the canopy of *Archaeopteris*. These were toxic for herbivores and decomposers and probably counteracted their complete exploitation of this biota. On the other hand, this was probably only one of the many morphological and physiological mechanisms developed by plants to defend themselves from the impact of the herbivorous arthropods. Of course, the animals too progressively evolved counteracting techniques. However, it was probably the change in vegetational composition of the late Palaeozoic and especially Mesozoic forests in favour of wind-pollinated plants (gymnosperms) that produced more appropriate conditions suitable for the large adaptive radiation of edaphic fauna.

All the earliest fossil Actinotrichida have been found in the northern hemisphere and precisely, if we consider the pattern of palaeogeography, on the shores of Laurentia and Baltica, namely the Old Red Sandstone continent. It is true that few searches have been made in southern regions, but the present-day distributions relative to the early derivative members of the actinotrichid lineage, e.g. Endeostigmata and Palaeosomata, are generally palearctic or cosmopolitan and there are very few genuine Gondwanian high taxonomic categories (families) in these higher taxa. The most plausible hypothesis arising from the available data is that the Actinotrichida originated in the then tropical land masses of the Old Red Sandstone continent (in Laurussia) and after the glaciations of most of Gondwanaland in the Permo-Carboniferous, spread to and speciated in the empty ecological niches of the saprophages and phytophages of the southern continents. It is probably not by chance that most genera of the Oribatida, for instance, are restricted to Gondwanaland and the present-day southern continents.

Because no Anactinotrichida fossils have been found from before the Cainozoic, indications of their different evolutionary history must be sought from zoogeography, ecology and the physiological aspects of the early-derivative members of this lineage.

Groups such as Holothyrida and Opilioacarida are now limited to Gondwana, like one of the earliest derivative families of Gamasida, the Ichthyosomatogasteridae (Bernini, 1986). Considering that many land chelicerate lineages like the Ricinulei, Solifugae etc. show similar distributions despite their apparent origin in the northern continents in the Upper Palaeozoic, these data are not strictly indicative that Anactinotrichida are of Gondwanian origin but rather suggests a relict tropical distribution. For the time being, in the absence of any kind of fossil remains, their origin may be taken to date back to the Middle-Upper Palaeozoic. The fact that the Anactinotrichida have not yet been found in the Palaeozoic or Mesozoic is no proof that they did not exist then and that they are of later origin. Too many aspects of their structure and habits seem to run against such an assumption. On the other hand, the origin of another chelicerate lineage, the Pseudoscorpionida, has recently been antedated to the Devonian (Norton *et al.*, 1988) from the previously record of Oligocene. The absence of anactinotrichids in earlier formations may be due to their small size or to their restricted ecological importance in past communities.

Some present-day features such as leg type suggest ancestral pre-adaptations realized in the aquatic life for swimming and/or walking, that is for different modes with respect to those of the Actinotrichida (Hammen, 1985). This lineage seems to have generally and primitively retained the ancestral arachnid predatory way of life although an early derivative line, the Opilioacarida, had (and still has) an omnivorous diet characterized by ingestion of solid food and internal ingestion ensured by a simple hypertrophied mouth seta, the rutellum (Grandjean, 1957). This plasticity of the gnathosoma allowed the eventual exploitation of many evolutionary opportunities by this kind of animal.

The adaptive radiation of the anactinotrichid lineage, however, does not seem to have involved the Eurasian edaphic biocenoses until the Oligocene.

Whatever the time of their origin, Anactinotrichida probably underwent their major adaptive radiation in connection with the "revolution" of terrestrial environments caused by the coevolution of insects and angiosperms and the success of mammals and birds in the Lower Cainozoic. This is compatible with the diet of these animals: anactinotrichids are prevalently primary carnivores and their predatory behaviour gave rise to a great variety of feeding habits (saprophagous, phytophagous and above all ecto- and endoparasitic). They thus began to prey on small insects (and the predominantly vegetarian actinotrichids!) and to infest mammals and birds which distributed them all over the world.

Evolutionary patterns

A further aspect arising from the above data and problems is that of the evolutionary pattern. What could have been the genetic (and/or ecological) events that determined an evolutionary history of this kind? If the origin of chelicerates can be dated back to the Cambrian and that of actinotrichids to the Lower Devonian (or perhaps to the Upper Silurian), we run out of time for gradualistic evolution, especially if we also consider the very slow velocity of morphological changes in the subsequent 400 Myr.

It is, therefore, reasonable to believe that the evolutionary pattern of the acarid lineages (and more generally the other terrestrial chelicerate branches) can be viewed as an episodic and sudden breakthrough across adaptive barriers into "open niches", such as those suitable for small animals. During such episodes, the morphological changes with respect to the ancestors could have been quite considerable (punctuated equilibria?). Then, when the niches (benthic and/or then edaphic) were occupied, stabilizing selection would have tended to prevent morphological changes, except for the tracking of new conditions as they arose (Valentine, 1985), i.e. invasion of all the countless open land niches. In this sense soil, which seems to be the primary environment of all the land chelicerate lineages, could have played the role of a "reservoir" from which the different branches evolved, transformed and specialized to their present forms.

The sudden appearance of the new body plan may have been caused by a form of paedomorphosis, such as progenesis, i.e. sexual maturity at an early stage of somatic development. For instance, the most peculiar characteristics of the creatures presumably pre-adapted to invade soil, i.e. small size, soft body wall and consequently cutaneous respiration (in actinotrichids!) might have been due to progenesis. This hypothesis is more appropriate than might seem so at first sight because in the acarids this phenomenon seems to have occurred in Acaridida starting from a branch of the primitive oribatid mites (O'Connor, 1984). The great similarity between the early derivative Acaridida and the immature stases of the Malaconothridae seems to prove this (Norton, personal communication, September, 1988).

Concluding remarks

In conclusion, palaeontological data suggest some working-hypotheses about the patterns and processes of the evolutionary history of the two acarid lineages.

They are as follows:

1) There are no indications about the ancestors of the acarid lineages nor those of the chelicerates.

2) The maintenance of the body plan for at least 400 Myr is one of the best examples of slow morphological transformation.

3) The acarid lineages have presumably always been highly speciose.

4) The evolution of the actinotrichids in the terrestrial biocenoses may have started in the Upper Silurian, before known fossil records.

5) The idea of biphyletism of Acarida is slightly reinforced by palaeontological data, and in any event the two lineages must have had at least very long and separate evolutionary histories.

6) Progenesis may have played a role in acarid evolution.

I realize that there is perhaps too little available material and that this paper might be included in the "untested-just-so-stories" category, but I agree with what Darwin wrote to A.R. Wallace on December 22, 1857 "I am a firm believer that without speculation there is no good and original observation".

Acknowledgements

I would like to thank Prof. L. Delle Cave (Dipartimento di Scienze della Terra dell'Università di Firenze) who provided some bibliographic information and Prof. R. Dallai (Dipartimento di Biologia Evolutiva dell'Università di Siena) for reading the manuscript and supplying some helpful ideas.

This work has been financed in part by the Italian CNR and in part by the MPI (40% and 60% funds).

REFERENCES

Alberti, G., 1984. The contribution of comparative spermatology to problems of Acarine systematics. In: *Acarology VI*, vol. I (ed. D.A. Griffiths and C.E. Bowman), pp. 479-490. Chichester: Ellis Horwood Ltd.

Baker, G.T. & Wighton, D.C., 1984. Fossil aquatic oribatid mites (Acari, Oribatida, Hydrozetidae, *Hydrozetes*) from the Paleocene of South Central Alberta, Canada. *Canadian Entomologist*, **116**, 773-776.

Bergström, J., 1975. Functional morphology and evolution of xiphosurids. *Fossils and Strata*, **4**, 291-305.

Bergström, J., 1979. Morphology of Fossils Arthropods as a Guide to Phylogenetic Relationships. In: *Arthropod Phylogeny* (ed. A.P. Gupta), pp. 3-56. New York: Van Nostrand Reinhold Co.

Bernini, F., 1986. Current ideas on the phylogeny and the adaptive radiations of Acarida. *Bollettino di Zoologia*, **53**, 279-313.

Briggs, D.E.G. & Collins, D., 1988. A Middle Cambrian chelicerate from Mount Stephen, British Columbia. *Palaeontology*, **31**, 779- 798.

Briggs, D.E.G., Bruton, D.L. & Whittington, H.B., 1979. Appendages of the arthropod *Aglaspis spinifera* (Upper Cambrian, Wisconsin) and their significance. *Palaeontology*, **22**, 167-180.

Bulanova-Zakhvatkina, Ye.M., 1974. Novyj rodkleščá (Acariformes, Oribatei) iz verhiego mela Tajamyra [A new genus of Oribatei (Acariformes) from the Upper Cretaceous of Taymir]. *Paleontologicheskii Zhurnal*, **2**, 141-144.

Caster, K.E. & Kjellesvig-Waering, E.N., 1953. *Melbournopterus*, a new Silurian Eurypterid from Australia. *Journal of Paleontology*, **27**, 153-156.

Cockerell, T.D.A., 1917. Arthropods in Burmese amber. *Psyche*, **24**, 40-45.

CROWSON, R.A., 1970. *Classification and biology*. London: Heinemann.

CROWSON, R.A., 1985. Comments on Insecta of the Rhynie Chert. *Entomologia Generalis*, **11**, 97-98.

DUBININ, V.B., 1962. Klass Acaromorpha. Klešči ili gatosomiye helitserovye [Class Acaromorpha: mites, or gnathosomic chelicerate arthropods]. In: Osnov'i paleontologii [*Fundamentals of Paleontology*]. (ed. B.B. Rodendorf). pp. 447-473. Moskow: Academy of Sciences of the USSR.

ELDREDGE, N., 1974. Revision of the Suborder Synziphosurina (Chelicerata, Merostomata), with Remarks on Merostome Phylogeny. *American Museum Novitates, no.* **2543**, 1-41.

EWING, H.E., 1937. Arachnida. Order Acarina. In: *Insects and Arachnids from Canadian Amber* (ed. F.M. Carpenter). *University of Toronto Studies, Geological Series*, **40**, 56-62.

FISHER, D.C., 1984. The Xiphosurida: Archetypes of Braditely? In: *Living Fossils*. (ed. N. Eldredge and S.M. Stanley), pp. 196-213. New York, Berlin, Heidelberg, Tokyo: Springer Verlag.

GOLOSOVA, L.D., DRUK, A.JA., KARPPINEN, E. & KISILJOV, S.V., 1985. Subfossil Oribatid mites (Acarina, Oribatei) of northern Siberia. *Annales Entomologici Fennici*, **51**, 3-18.

GRANDJEAN, F., 1957. L'infracapitulum et la manducation chez les Oribates et d'autres Acariens. *Annales des Sciences Naturelles, Zoologie, 11ème série*, **19**, 234-279.

GRANDJEAN, F., 1970. Stases. Actinopiline. Rappel de ma classification des Acariens en 3 groupes majeurs. Terminologie en soma. *Acarologia*, **11**, 796-827.

GREENSLADE, P.J.M., 1985. Reply to R.A. Crowson's "Comments on Insecta of the Rhynie Chert" (1985 Entomol. Gener. 11(1/2): 097-098). *Entomologia Generalis*, **13**, 115-117.

HAMMEN, VAN DER, L., 1977. A new classification of Chelicerata. *Zoologische Mededelingen (Leiden)*, **51**, 307-319.

HAMMEN, VAN DER, L., 1979. Comparative studies in Chelicerata. I. The Cryptognomae (Ricinulei, Architarbi and Anactinotrichida). *Zoologische Verhandelingen (Leiden)*, **174**, 1-62.

HAMMEN, VAN DER, L., 1982. Comparative studies in Chelicerata. II. Epimerata (Palpigradi and Actinotrichida). *Zoologische Verhandelingen (Leiden)*, **196**, 3-70.

HAMMEN, VAN DER, L., 1985. Functional morphology and affinities of extant Chelicerata in evolutionary perspective. *Transactions of the Royal Society of Edinburgh*, **76**, 137-146.

HAMMEN, VAN DER, L., 1986. Acarological and arachnological notes. *Zoologische Mededelingen (Leiden)*, **60**, 217-230.

HAMMEN, VAN DER, L., 1989. *An introduction to comparative arachnology*. The Hague: SPB Academic Publishing bv.

HIRST, S., 1923. On some arachnid remains from the Old Red Sandstone (Rhynie Chert Bed, Aberdeenshire). *Annals & Magazine of Natural History*, **12**, 455-474.

HÜNICKEN, M., 1980. A giant fossil spider (*Megarachne servinei*) from Bajo de Vliz, Upper Carboniferous, Argentina. *Boletin de la Academia Nacional de Ciencias (Cordoba)*, **53**, 317-340.

JEANNEL, R., 1949. Les Insectes. Classification et phylogénie. Les insectes fossiles. Evolution et géonémie. In: *Traité de Zoologie*, vol. 9 (ed. P.P. Grassé), pp. 1-110, Paris: Masson.

KEVAN, P.G., CHALONER, W.G. & SAVILE, D.B.O., 1975. Interrelationships of early terrestrial arthropods and plants. *Palaeontology*, **18**, 391-417.

KJELLESVIG-WAERING, E.N., 1986. A restudy of the fossil Scorpionida of the world. *Palaentographica Americana*, **55**, 5-287.

KRANTZ, G.W. & LINDQUIST, E.E., 1979. Evolution of phytophagous mites (Acari). *Annual Review of Entomology*, **24**, 121-158.

KRAUS, O., 1976. Zur phylogenetischen Stellung und Evolution der Chelicerata. *Entomologica Germanica*, **3**, 1-12.

KRIVOLUTSKY, D.A. & DRUK, A.YA., 1986. Fossil Oribatid Mites. *Annual Review of Entomology*, **31**, 533-545.

LARSSON, S.G., 1978. *Baltic Amber - a Palaeobiological Study*. Klampenborg: Scandinavian Science Press Ltd.

LAUTERBACH, K.E., 1980. Schlusselereignisse in der Evolution des Grundplans der Arachnata (Arthropoda). *Abhandlungen des Naturwissenschaftlichen Vereins in Hamburg*, **23**, 163-327.

LINDQUIST, E.E., 1984. Current theories on the evolution of major groups of Acari and on their relationships with other groups of Arachnida, with consequent implications for their classification. In: *Acarology VI*, vol.I (ed. D.A. Griffiths & C.E. Bowman), pp. 28-62. Chichester: Ellis Horwood Ltd.

NORTON, R.A., BONAMO, P.M., GRIERSON, J.D. & SHEAR, W.A., 1988. Oribatid mite fossils from a terrestrial Devonian deposit near Gilboa, New York. *Journal of Paleontology*, **62**, 259-269.

NORTON, R.A., BONAMO, P.M., GRIERSON, J.D. & SHEAR, W.A., 1989. Fossil Mites from the Devonian of New York State. In *Progress in Acarology*, vol. 1 (ed. G.P. Channabasavanna and C.A. Viraktamath), pp. 271-277. New Delhi, Bombay and Calcutta: Oxford & IBH Publisher.

O'CONNOR, B.M., 1984. Phylogenetic relationships among higher taxa in the Acariformes with particular reference to the Astigmata. In: *Acarology VI*, vol. I (ed. D.A. Griffiths & C.E. Bowman), pp. 19-27. Chichester: Ellis Horwood Ltd.

PAMPALONI, L., 1902. I resti organici nel disodile di Melilli in Sicilia. *Palaeontographia Italica*, **8**, 121-130.

PETRUNKEVITCH, A., 1955. Arachnida. In: *Treatise on Invertebrate Paleontology, Part P, Arthropoda 2* (ed. R.C. Moore), pp. 42-162. Geological Society of America and University of Kansas Press.

RAASCH, G.O., 1939. Cambrian Merostomata. *Special papers of the Geological Society of America*, **19**, 1-146.

RESSER, C.E., 1931. A new Middle Cambrian merostome crustacean. *Proceedings of the United States National Museum*, **79**, 1-4.

RETALLACK, G.J., 1985. Fossil soils as grounds for interpreting the advent of large plants and animals on land. *Philosophical Transactions of the Royal Society of London*, **B 309**, 105-142.

RETALLACK, G.J. & FEAKES, C.R., 1987. Trace fossil evidence for Late Ordovician animals on land. *Science*, **235**, 61-63.

ROLFE, W.D.I., 1980. Early invertebrate terrestrial faunas. In: *The terrestrial environment and the origin of land vertebrates*. (ed. A.L. Panchen), pp. 117-157. London and New York: Academic Press.

ROLFE, W.D.I., 1985. Early terrestrial arthropods: a fragmentary record. *Philosophical Transactions of the Royal Society of London*, **B 309**, 207-218.

ROTHWELL, G.W. & SCOTT, A.C., 1983. Coprolites within marattiaceous fern stems (*Pasaronius magnificus*) from the upper Pennsylvanian of the Appalachian basin, USA. *Palaeogeography Palaeoclimatology Palaeoecology*, **41**, 227-232.

SAVORY, T., 1977. *Arachnida*. London, New York and San Francisco: Academic Press.

SCOTT, A.C. & TAYLOR, J.N., 1983. Plant/animal interactions during the Upper Carboniferous. *Botanical Review*, **49**, 259-307.

SELDEN, P.A., 1984. Autecology of Silurian Eurypterids. *Special Papers in Palaeontology*, **32**, 39-54.

SELDEN, P.A., 1988. The Arachnid Fossil Record. *British Journal of Entomological Natural History*, **1**, 15-18.

SELLNICK, M., 1931. Milben in Bernstein. *Bernstein-Forschungen*, **2**, 148-180.

SHEAR, W.A., BONAMO, P.M., GRIERSON, J.D., ROLFE, W.D.I., SMITH, E.L. & NORTON, R.A., 1984. Early Land Animals in North America: Evidence from Devonian Age Arthropods from Gilboa, New York. *Science*, **224**, 492-494.

SHERWOOD-PIKE, M.A. & GRAY, J., 1985. Silurian fungal remains: oldest records of the Class Ascomycetes? *Lethaia*, **18**, 1-20.

SIMONETTA, A. & DELLE CAVE, L., 1981. An essay in the comparative and evolutionary morphology of Palaeozoic Arthropods. *Atti dei Convegni Lincei*, **49**, 389-439.

SIVHED, V. & WALLWORK, J.A., 1978. An early jurassic oribatid mite from southern Sweden. *Geologiska Foreningens i Stockholm Forhandingar*, **100**, 65-70.

SOUTHCOTT, R.V. & LANGE, R.T., 1971. Acarine and other micro fossils from the Maslin Eocene, South Australia. *Records of the South Australian Museum (Adelaide)*, **16**, 1-21.

STØRMER, L., 1944. On the relationships and phylogeny of fossil and recent Arachnomorpha. A comparative study on Arachnida, Xiphosura, Eurypterida, Trilobita, and other fossil Arthropoda. *Skrifter av det Norske Videnskaps-Akademi i Oslo I Matematisk-Naturvidenskapelig Klasse*, **5**, 1-158.

STÜRMER, L., 1972. Arthropods from the Lower Devonian (Lower Emsian) of Alken-an-der-Mosel, Germany. Part 2: Xiphosura. *Senckenbergiana Lethaia*, **53**, 1-29.

TAYLOR, J.N. & SCOTT, A.C., 1983. Interactions of plants and animals during the Carboniferous. *Bioscience*, **33**, 488-493.

TURK, E., 1963. A new tyroglyphid deutonymph in amber from Chiapas, Mexico. *University of California Publications in Entomology*, **31**, 49-51.

VALENTINE, J.W., 1985. Biotic diversity and clade diversity. In: *Phanerozoic diversity patterns. Profiles in macroevolution* (ed. J.W. Valentine), pp. 419-424. Princeton, New Jersey: Princeton University Press.

VERCAMMEN-GRANDJEAN, P.H., 1973. Study of the "Erythraeidae,

R.O.M. NO.8" of Ewing, 1937. In: *Proceedings of the third International Congress of Acarology held in Prague.* (ed. M. Daniel and B. Rosicky), pp. 329-335. The Hague: Dr. W. Junk Publ.

Weygoldt, P., 1986. Arthropod interrelationships-the phylogenetic-systematic approach. *Zeitschrift für Zoologische Systematik und Evolutionsforschung,* **24**, 19-35.

Whalley, P., 1988. Insect Evolution During the Extinction of the Dinosauria. *Entomologia Generalis,* **13**, 119-124.

Whittington, H.B., 1979. Early Arthropods, their appendages and relationships. In: *The origin of major Invertebrate Groups* (ed. M.R. House), pp. 253-268. London and New York: Academic Press.

Woolley, T.A., 1969. Fossil Oribatid mites in amber from Chiapas, Mexico (Acarina: Oribatei = Cryptostigmata). *University of California Publications in Entomology,* **63**, 91-99.

Zacharda, M. & Krivolutsky, D.A., 1985. Prostigmatic mites (Acarina: Prostigmata) from the Upper Cretaceous and Paleogene amber of the USSR. *Vestnik Ceskoslovenske Spolecnosti Zoologicke,* **49**, 147-152.

Are Protura really insects

R. Dallai[1]

Abstract

Protura are a very odd group of arthropods. The re-examination of many morphological features leads to the conclusion that they share only a few characters with insects.

Protura are a small group of terrestrial arthropods first recognized at the beginning of the century by Silvestri (1907). He placed *Acerentomon doderoi* among wingless insects, but erected a new order (Protura) on account of the position of the genital opening between the 11th and the 12th urosternites. This position of the genital opening, which is very unusual for an insect, together with various other features (e.g. lack of antennae, number of abdominal segments, and anamorphosis, that is an increase in the number of segments after emergence from the egg) led Berlese (1908) to regard *A. doderoi* not as an insect, but rather as a representative of a new taxon, the Myrientomata, a group close to Myriapoda.

Despite the precise morphological description given by Berlese (1909), it has long been assumed that the pseudoculi, minute structures on either side of the head of Protura, were vestigial antennae (Tuxen, 1959) (Figs. 1-3). The work of François (1959), showing that the structure is innervated by the protocerebrum and that it lacks muscles, followed by the studies of Bedini & Tongiorgi (1971) and more recently of Yin *et al.* (1986), have confirmed that although pseudoculi are sensory organs, they are not remnants of antennae nor they seem to be homologous to the Tömösvary organ of myriapods (Altner and Thies, 1976). It can be concluded, therefore, that Protura lack antennae; what is more, they seem to be arthropods that never had them. The sensory functions, proper to antennae, are shifted to the forelegs which are not used for locomotion but held aloft on the head (Sixl *et al.*, 1974; Muller, 1976; Dallai and Nosek, 1981). Protura are thus tetrapods as far as the function of walking is concerned.

This gives rise to a serious problem in the classification of Protura among the Antennata. Can a group which is primarily without antennae, belong to the Antennata? Evidently not. On the other hand, if we consider the structure of the mouth parts of Protura, there is indeed no doubt that they are clearly related to Mandibulata and this is sufficient to exclude them from the Chelicerata. According to Tuxen (1959), Protura have mandibles and maxillae (Fig. 4) that become internal (entognathy) by the development of pleural folds and their fusion with both labrum and labium. However, as Manton (1977) reminds us, entognathy is a convergent feature which has appeared in widely different taxa and can hardly be regarded as a taxonomic character indicating affinity. Moreover, in contrast with Snodgrass (1950), Manton (1964), showed that the mandibles of Crustacea are not homologous with those of Myriapoda and Insecta. In Chelicerata the lack of antennae is accompanied by lack of deutocerebrum, that is the middle part of the brain. According to François (1969) there is also a reduction in size of the deutocerebrum in Protura, but the absence of information on the embryology of Protura makes it impossible, at the moment, to determine whether this reduction is the consequence of the absence of antennae or it is a feature typical of the group.

In any case we agree that if Protura are to be placed among insects, they are unique for lack of antennae.

Anamorphosis is a characteristic that Protura shared with Myriapoda. The first stage (the praelarva) hatches with nine abdominal segments, the second stage (larva I) also has nine segments, the third (larva II) has ten segments and the fourth (maturus junior), like the following praeimago and imago, has twelve segments (Tuxen, 1949; François, 1960; Imadaté, 1961; Yin, 1981). Thus there are 5 successive post-embryonic developmental stages, at the end of which we have an adult with 12 abdominal segments. In this respect, as first recognized by Berlese (1908), the group resembles the myriapods and here too stands in contrast to all other insects. Another interesting feature is the presence of paired appendages, with eversible vesicles, on the first three abdominal segments (Fig. 5).

Among the insects, Protura are the group with the greatest number of abdominal segments; the last abdominal segment is sometimes called pigidio (Berlese, 1909) or telson (Nosek, 1973) (Fig. 5). I think it is a bit misleading to say that their "abdomen is composed of 11 segments and a well developed telson" (Imms, 1973), because this does not bring out the difference between insects and Protura. Moreover, authors have often presented Protura as having only 11 abdominal segments, the twelfth being omitted in the description.

Recent studies on Protura have showed other important discrepancies with insects. The cuticle of Protura, according to Bilinski and Klag (1978) only has pore canals in the median plate of the sternite as in Acarina. As for the respiratory system, especially the number and position of the spiracles, we know that in the different groups of Tracheata the situation is variable. Insects generally have 9-10 pairs: 2 pairs on the thorax and the others on the abdomen. Myriapods have more: diplopods, for instance, have 2 pairs on each segment. Moreover all the spiracles are placed on the pleurite or sternite of the animal. In Protura the species with a respiratory system have only 2 pairs of spiracles on the meso- and metanotum (Yin, 1984). Such a situation is known only in the Opilioacarida (Acari), which have

[1]Department of Evolutionary Biology, University of Siena, Italy.

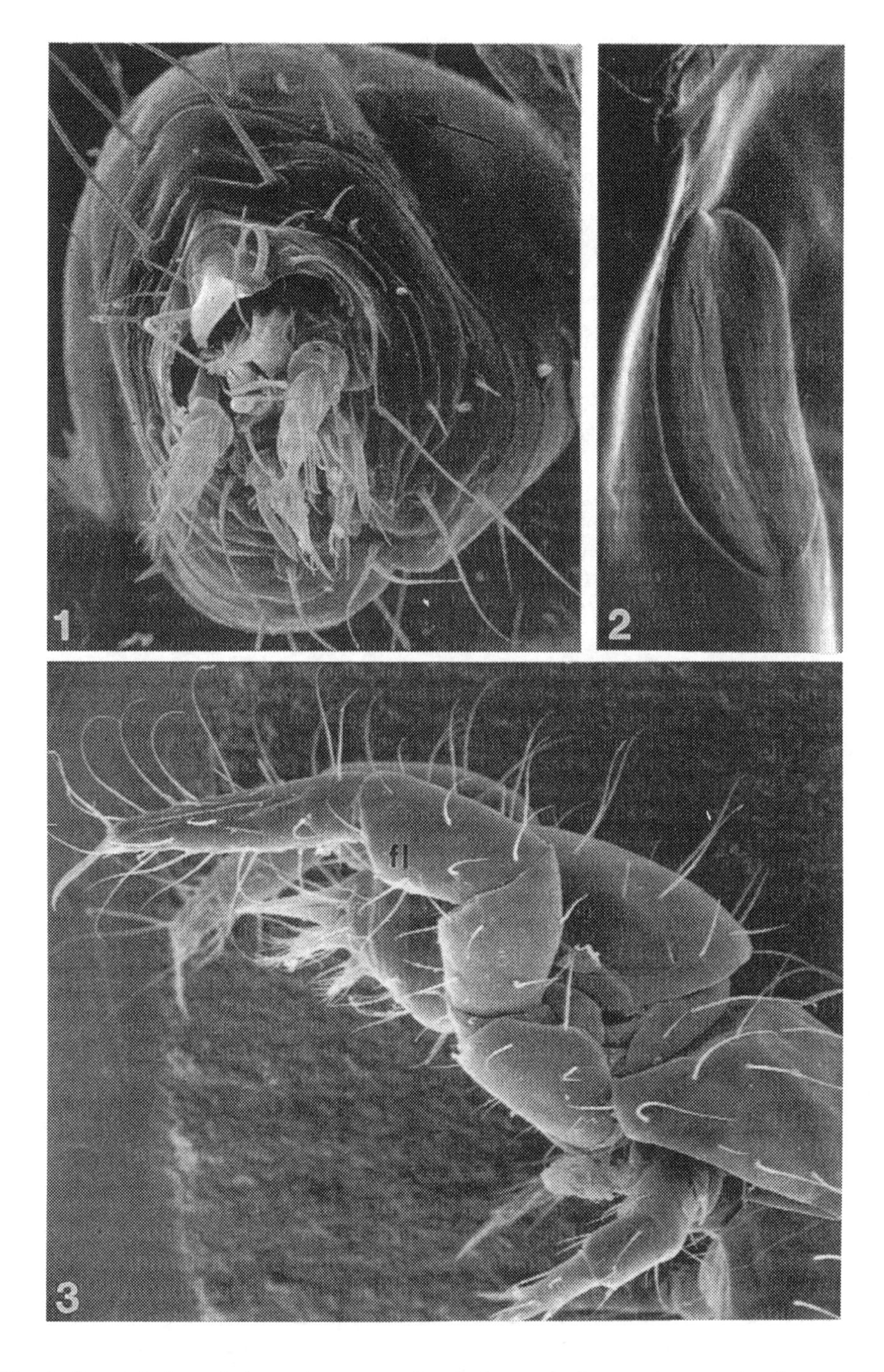

Fig. 1. Frontal view of *Acerentomon* sp. head. Arrow indicates the pseudoculus. x 1.200.

Fig. 2. Detail of the pseudoculus of *Eosentomon* sp. x 6.500.

Fig. 3. Habitus of *Acerentomon* sp. showing the forelegs (fl) held aloft on the head. x 500.

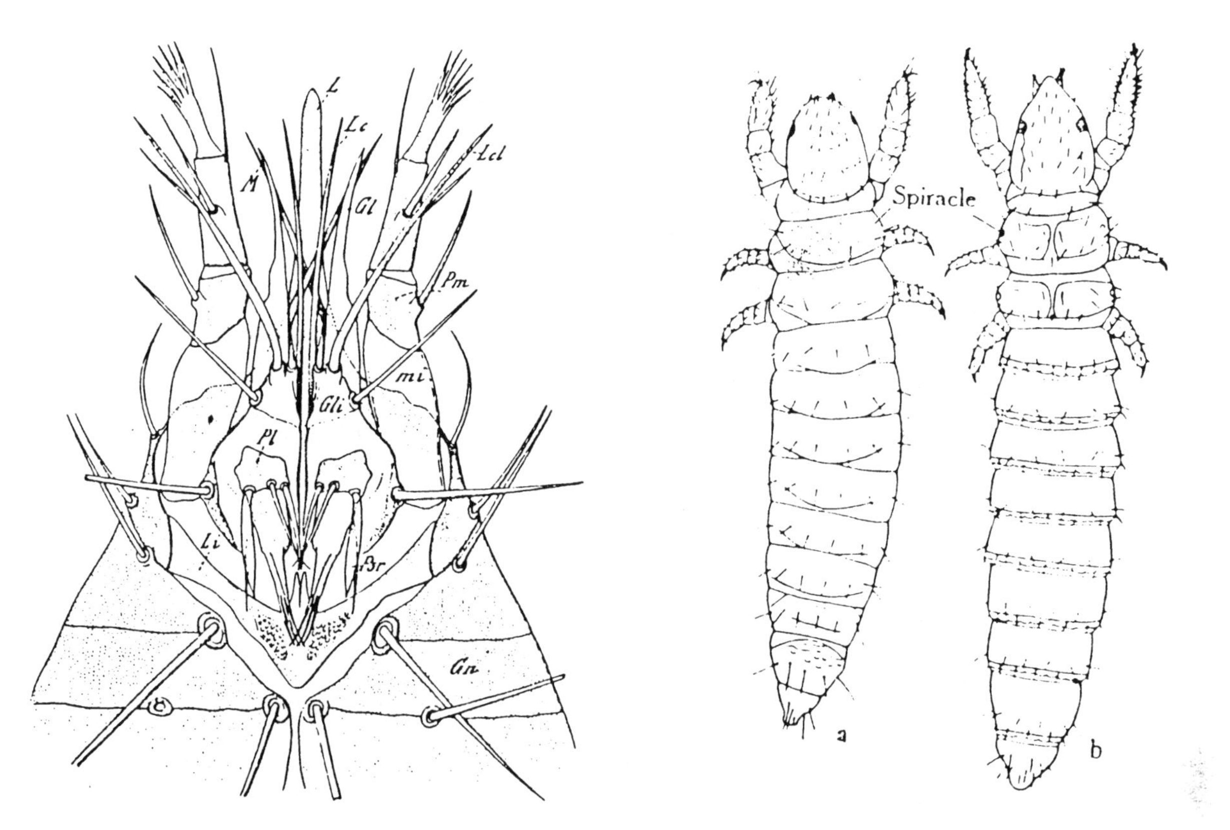

Fig. 4. Mouth parts of *Acerentomon doderoi* (from Berlese, 1909). M, mandibles; Gl, galea; Lc, lacinia; L, labrum, Li, labium.

Fig. 6. Spiracles of tracheal system in the first larval stage of *Sinentomon erythranum* (from Yin, 1984).

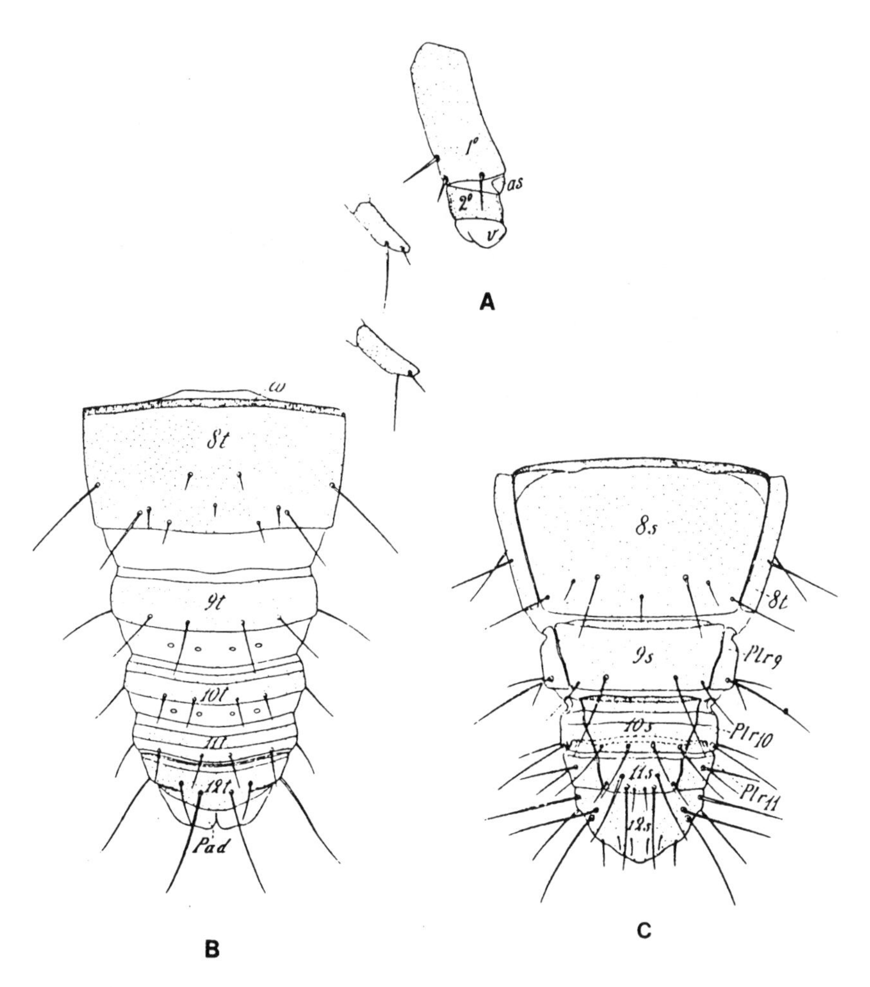

Fig. 5. A, appendages of the first three abdominal segments of *Acerentomon doderoi*; B and C, last abdominal segments in *Eosentomon transitorium* and *E. ribagoi*, respectively, from dorsal and ventral views (from Berlese, 1909).

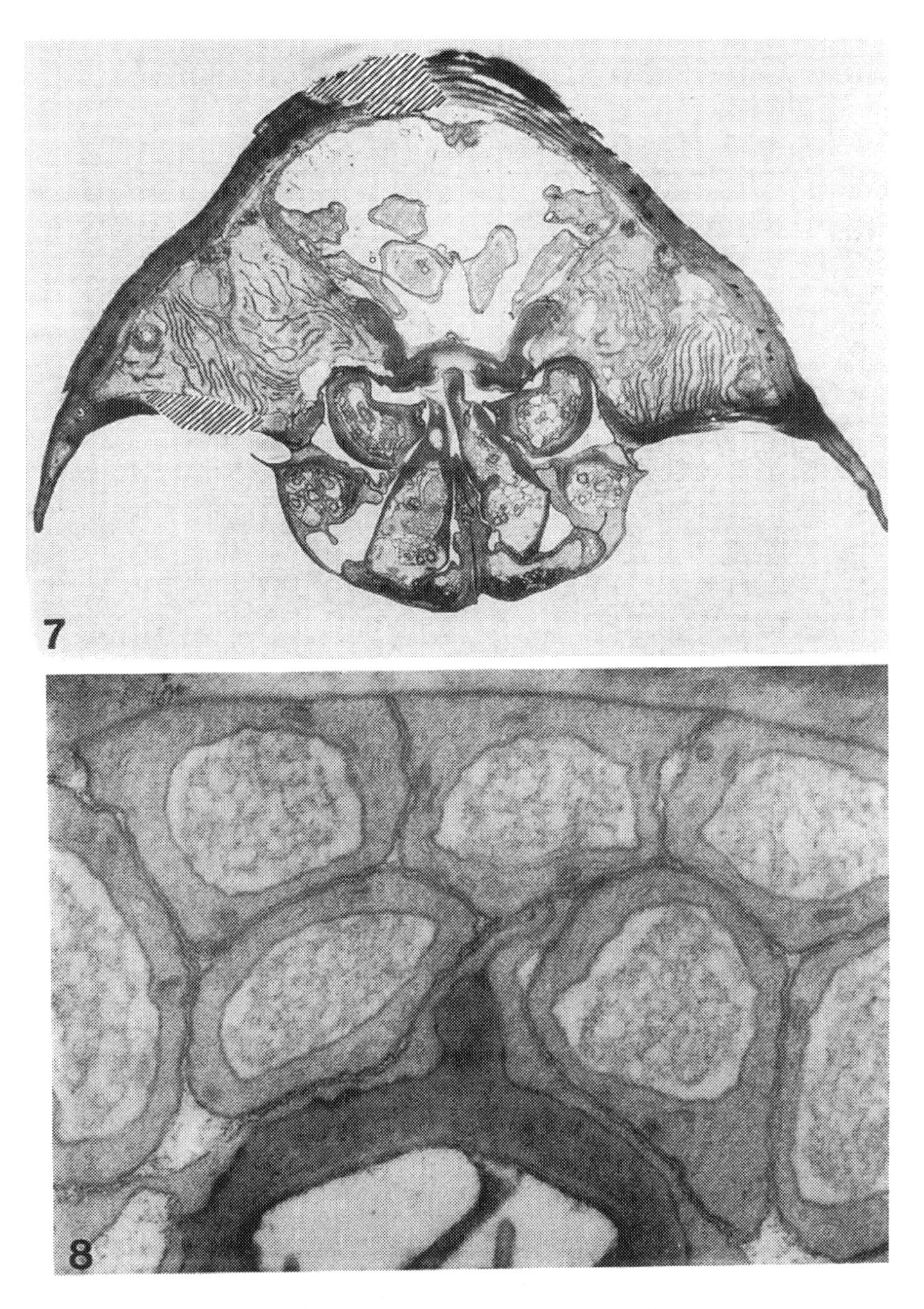

Fig. 7. Cross section of the head anterior region of *Acerentomon* sp. Seven canaliculated cells are seen in the clypeo-labrum cavity. x 6.000.

Fig. 8. Detail of canaliculated cells. x 45.000.

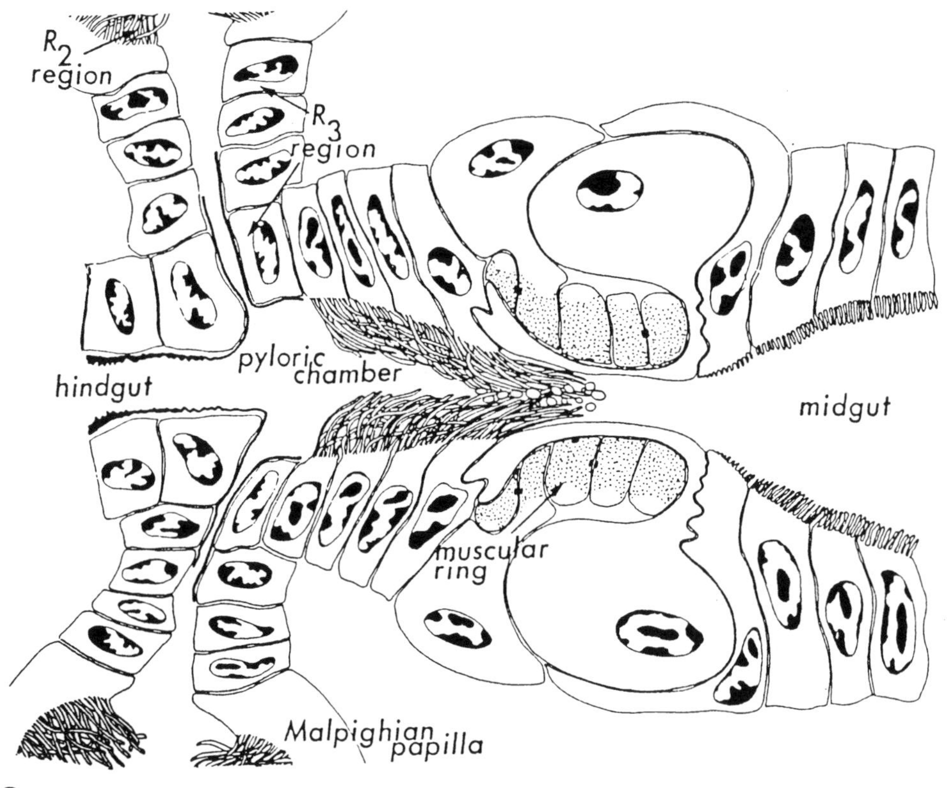

9 Fig. 9. Schematic drawing of the proturan pyloric region (from Dallai, 1976).

2 pairs of dorsal spiracles on the podosoma. It is important to remember that Eosentomidae and Sinentomidae are the only Protura with a respiratory system. According to Yin (1981; 1984), in *Sinentomon erythranum* the spiracles and tracheal system develop only after the first moult, when the first larval stage appears (Fig. 6); the same was found to be the case in *Eosentomon*. This means that the above two groups, regarded by Tuxen (1963) as the most primitive Protura because of their possessing a supposed rudiment of a respiratory organ of a hypothetical ancestor, could instead be representatives of an independent and specialized evolutionary trend. Thus the tracheal system and spiracles of Protura should not be regarded as vestigial features, but as important new acquisitions.

Additional information has recently been obtained on the ultrastructure of different organs of the Protura. The epidermal glands are organized as in other arthropods, with secretory and duct forming cells (Dallai & Burroni, 1981). Dallai & François (1985) described the peculiar structure of the clypeo-labrum and the unusual presence of hypopharyngeal sensilla (Dallai & François, 1985, 1986) (Figs. 7,8). François and Dallai (l986a,b) clarified the structure of the defensive abdominal glands which open at the 8th segment and the fine morphology of the maxillary glands. Dallai (1976, 1977a,b) and Dallai *et al.* (1987), in a series of studies on the gut structure, recognized that the pyloric region of Protura is organized in a very peculiar way. Behind a true sphincter, which can close the intestinal lumen thanks also to the adaptation of the posterior-most midgut cells, there is a special structure called the pyloric chamber (Dallai, 1976; Dallai et al., 1987) (Fig. 9). The epithelial cells of this region have long microvilli which point anteriorly towards the midgut. Into the pyloric chamber, six malpighian papillae secrete their products. This region is followed by the hindgut.

Finally, the ultrastructure of the spermatozoa of Protura reveals an aberrant situation (Dallai, 1974; Dallai & Yin, 1983; Yin *et al.*, 1986). Flagellate and motile spermatozoa are widely held to be the primitive model of the sperm cell (Baccetti & Afzelius, 1976; Dallai, 1979). Later in evolution, there is a tendency towards sperm aflagellarity and immotility in some groups of the major animal phyla. Protura are no exception to this principle. Eosentomidae and Sinentomidae have aflagellate spermatozoa; the other families have, instead, flagellate, immotile spermatozoa with a variable number of microtubular doublets in the axoneme (Figs. 10,11). In the genus *Hesperentomon* the situation is most peculiar: different numbers of microtubular doublets have been observed in the sperm axonemes of the same individual (Yin & Xué, 1987). A similar feature was described by Van Deurs (1974) in Pycnogonida, a group of atypical Chelicerata.

In my opinion the real position of Protura has been obscured by their presumed relationship with the springtail Collembola. Hennig (1969) and then Tuxen (1970, 1972), regarded Protura and Collembola as a "sister-group" of Diplura. According to Hennig (1969) the similarity was supported by a list of advanced characteristics present in Protura and Collembola. This view was revised by Kristensen (1975) who concluded that only the common features of entognathy, the modified tibiotarsi and the absence of abdominal spiracles and cerci could be regarded as true synapomorphies. In addition, François (1969) indicated the presence of a ventral cuticular groove all along a longitudinal line starting from the labium and passing through the thorax (linea ventralis). Unfortunately we

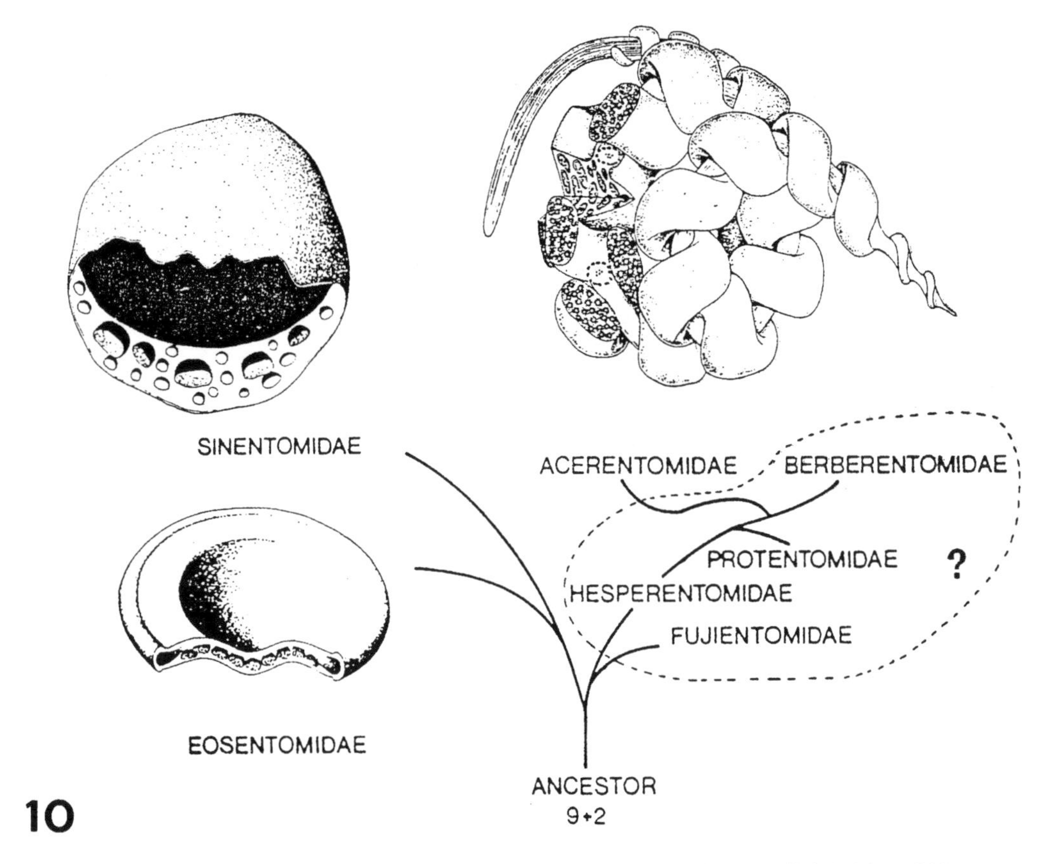

Fig. 10. Phylogenetic trends in the evolution of proturan sperm (from Dallai & Yin, 1983).

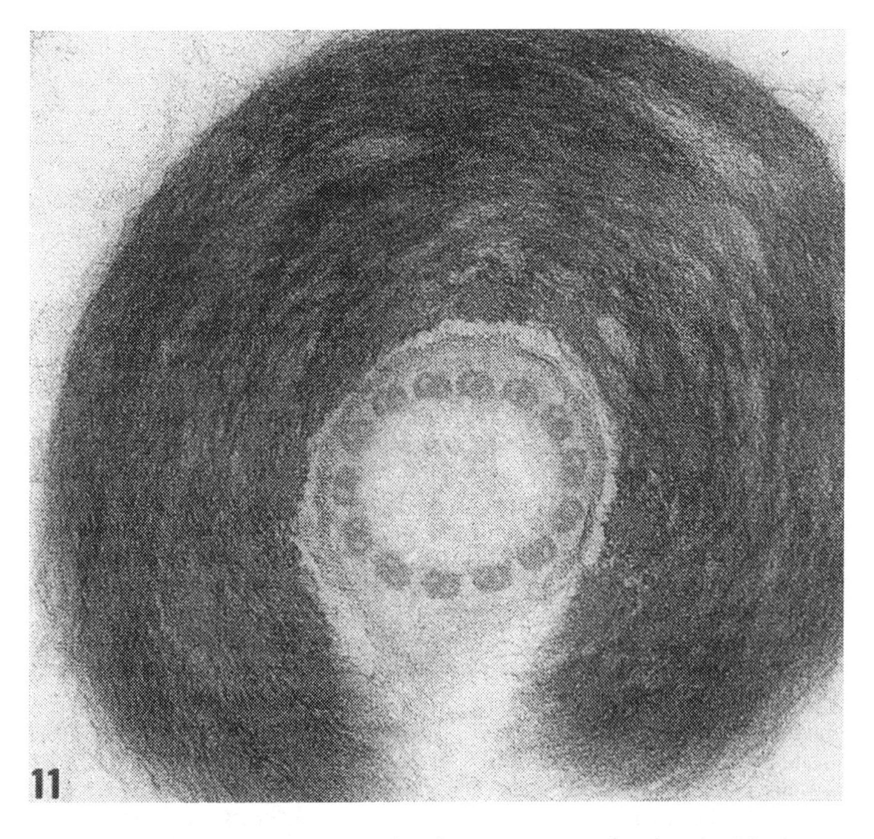

Fig. 11. Cross section of the *Acerentomon majus*. Sperm axoneme showing the 14+0 pattern. x 150.000.

have no information on the ancestors of Protura, nor have fossils of this group yet been found. They most certainly exist among the rich Palaeozoic arthropod fauna which, according to Simonetta & Delle Cave (1980) had many more branches than is commonly realized.

In conclusion we can say that the modern Protura share only a few characters with insects, while many others features are quite different. The most important reason to include Protura among the insects is their hexapody with the thorax carrying three pairs of legs. Is this really sufficient to consider them as insects? The answer to this question can only be subjective.

REFERENCES

ALTNER, H. & THIES, G. 1976. The postantennal organ: a specialized unicellular sensory input to the protocerebrum in Apterygotan Insects (Collembola). *Cell Tissue Research*, **167**, 97-110.

BACCETTI, B. & AFZELIUS, B.A. 1976. The biology of the sperm cell. In *Monographs in Developmental Biology*, Vol. 10, Basel: S. Karger AG.

BEDINI, C. & TONGIORGI, P. 1971. The fine structure of the pseudoculus of acerentomids Protura (Insecta, Apterygota). *Monitore Zoologico Italiano. (N.S.)* **5**, 25-38.

BERLESE, A. 1908. Nuovi Acerentomidi. *Redia* **5**, 16-19.

BERLESE, A. 1909. Monografia dei Myrientomata. *Redia* **6**, 1-182.

BILINSKI, S. & KLAG, J. 1978. Ultrastructural studies of *Acerentomon gallicum* Jonescu (Protura). Phylogenetic conclusions. In *Proceedings of the First International Seminar on Apterygota*, Siena (ed. R. Dallai), pp. 167- 173.

DALLAI, R., 1974. Spermatozoa and phylogenesis. A few data on Insecta Apterygota. *Pedobiologia* **14**, 148-156.

DALLAI, R., 1976. Fine structure of the pyloric region and Malpighian papillae of Protura (Insecta, Apterygota). *Journal of Morphology* **150**, 727-762.

DALLAI, R. 1977a. Comparative analysis of the hindgut in two species of Protura (Insecta, Apterygota). *Journal of Morphology* **151**, 165-186.

DALLAI, R. 1977b. Fine structure of Protura intestine. *Révue d'Ecologie et de Biologie du Sol* **14**, 139-152.

DALLAI, R. 1979. An overview of atypical spermatozoa in Insects. In *The Spermatozoon* (eds. D.W. Fawcett and J.M. Bedford), pp. 253-265, Munich: Baltimora.

DALLAI, R. & BURRRONI, D. 1981. La struttura delle ghiandole tegumentali delle zampe anteriori dei Proturi (Insecta). *Redia* **64**, 165-172.

DALLAI, R., & FRANÇOIS, J. 1985. The ultrastructure of the clypeolabrum of Protura (Insecta, Apterygota). Redia **68**, 125-133.

DALLAI, R. & FRANÇOIS, J, 1986. Fine structure of the Proturan hypopharyngeal sensilla. In *Proc. Second International Seminar on Apterygota*, Siena (ed. R. Dallai), pp. 287-293.

DALLAI, R. & NOSEK, J., 1981. Ultrastructure of sensillum t_1 on the foretarsus of *Acerentomon majus* Berlese (Protura: Acerentomontidae). *International Journal of Insect Morphology and Embryology*, **10**, 321-330.

DALLAI, R. & YIN, W.-Y. 1983. Sperm structure of *Sinentomon* (Protura) and phylogenetic considerations. *Pedobiologia*, **25**, 313-316.

DALLAI, R., YIN, W.-Y., XUÉ, L. & FANCIULLI, P.P. 1987. Fine structure of the pyloric region and of the hindgut in the Proturan *Neocondeellum dolichotarsum* (Insecta, Protura). *Journal of Morphology* **194**, 173-186.

FRANÇOIS, J. 1959. Squelette et musculature céphaliques d'*Acerentomon propinquum* (Condé) (Ins. Protoures). *Travaux du Laboratoire de Zoologie et de la Station aquicole Grimaldi de la Faculté des Science, de Dijon* **29**, 1-57.

FRANÇOIS, J. 1960. Developpement postembryonnaire d'un Protoure du genre *Acerentomon* Silv. *Travaux du Laboratoire de zoologie et de la Station aquicole de Grimaldi de la Faculté des Sciences de Dijon* **33**, 1-11.

FRANÇOIS, J. 1969. Anatomie et morphologie céphalique des Protoures (Insecta Apterygota). *Memoires du Muséum National d'Histoire Naturelle, Paris* **59**, 1-144.

FRANÇOIS, J. & DALLAI, R., 1986a. Les glandes abdominales des Protoures. in *Proceedings od the Second International Seminar on Apterygota*, Siena (ed. R. Dallai), pp. 273-280.

FRANÇOIS, J. & DALLAI, R. 1986b. Ultrastructure des glandes maxillaires d'*Acerentomon affine* Bagn. et d'*Eosentomon transitorium* Berl. (Apterygota: Protura). *International Journal of Insect Morphology and Embryology* **15**, 201-212.

HENNIG W. 1969. *Die Stammesgeschichte der Insekten*. Frankfurt am Main: Kramer.

IMADATÉ, G. 1961. Three new species of the genus *Acerentulus* Berlese (Protura) from Japan. *Kontyû, Tokyo* **29**, 226-233.

IMMS, A.D. 1973. *A General Textbook of Entomology*. London: Chapman and Hall.

KRISTENSEN, N.P. 1975. The phylogeny of hexapod orders. A critical review of recent accounts. *Zeitschrift für Zoologische Systematik und Evolutions-forschung* **13**, 1-44.

MANTON S.M. 1964. Mandibular mechanisms and the evolution of the Arthropods. *Philosophical Transaction of the Royal Society* **B247**, 1-183.

MANTON S.M. 1977. *The Arthropoda: habits, functional morphology and evolution*. Oxford: Clarendon Press.

MULLER, H. VON 1976. Feinbau der Sensillen an den Vorderbeinen der Proturen. *Zoologischer Anzeiger* **197**, 151-178.

NOSEK, J. 1973. *The European Protura. Their taxonomy, ecology and distribution with keys for determination*. Geneve: Muséum d'Histoire Naturelle.

SILVESTRI, F. 1907. Descrizione di un nuovo genere di Insetti Apterigoti, rappresentante di un nuovo ordine. *Bollettino del Laboratorio di Zoologia generale e agraria della Facoltà di Agraria di Portici*, **1**: 296-311.

SIMONETTA, A. & DELLE CAVE, L. 1980. The phylogeny of the Palaeozoic arthropods. *Bollettino di Zoologia* **47**, 1-19.

SIXL, W., NOSEK, J. & WALTINGER, H. 1974. Sensory hairs and tegument in Protura. *Pedobiologia* **14**, 109-112.

SNODGRASS, R.E. 1950. Comparative studies on the jaws of mandibulate arthropods. *Smithsonian Miscellaneous Collections* **116 (1)**, 85 pp.

TUXEN, S.L. 1949. Über den Lebenszyklus und die postembryonale Entwicklung zweier danischer Proturengattungen. *Kongelike Danske Videnskabernes Biologiske Skrifter* **6**, 1-49.

TUXEN, S.L. 1959. The phylogenetic significance of entognathy in entognathous apterygotes. *Smithsonian Miscellaneous Collections* **137**, 379-416.

TUXEN, S L. 1963. Phylogenetical trends in the Protura as shown by relationship between recent genera. *Zeitschrift für Zoologische Systematik und Evolutions-forschung* **1**, 277-310.

TUXEN, S.L. 1970. The systematic position of entognathous apterygotes. *Anales de la Escuela Nacional de Ciencias biologicas, Mexico* **17**, 65-79 (1968).

TUXEN, S.L. 1972. Filogenesi degli atterigoti. *Atti del IX Congresso Nazionale Italiano di Entomologia, Siena*, pp. 193- 205.

VAN DEURS, B. 1974. Picnogonid sperm. An example of inter- and intraspecies axonemal variation. *Cell Tissue Research* **149**, 105-111.

YIN, W.-Y. 1981. On bionomics of *Sinentomon* (Protura) and its systematic position. *Contributions of the Shanghai Institute of Entomology*, pp. 161-170.

YIN, W.-Y. 1984. A new idea on phylogeny of Protura with approach to its origin and systematic position. *Scientia Sinica* **27**, 149-160.

YIN, W.-Y., XUÉ, L. & TANG, B., 1986. A comparative study on pseudoculus of Protura. In *Proceedings of the Second International Seminar on Apterygota*, Siena (ed. R. Dallai), pp. 249-256.

YIN, W.-Y. & XUÉ, L. 1987. A variable axonemal pattern sperm and spermiogenesis in *Hesperentomon hwashanensis* (Protura: Hesperentomidae). *Contributions to the Shanghai Institute of Entomology* **7**, 111-115.

The Tully Monster and a new approach to analyzing problematica

Bret S. Beall[1]

Abstract

Reinvestigation of the Upper Carboniferous problematic Tully monster, *Tullimonstrum gregarium*, from Illinois, USA, emphasizes that 1) an unequivocal interpretation of the morphology of a fossil often may be unattainable, and multiple working hypotheses of morphology must be evaluated; 2) a combined cladistic-phenetic approach using a phylogenetic computer program permits identifying sister group relationships and testing hypotheses of affinities; and 3) this approach permits classification in a repeatable and testable manner.

Tullimonstrum had dorsoventrally, not laterally, oriented posterior fins that were slightly asymmetrical. The rigid transverse bar may be interpreted as bearing eyes or statocysts, as paired copulatory organs, as highly modified setae, or as benthic support structures. The proboscis bore simple cone-shaped stylets each with a basal cavity. When these and other alternative interpretations of 21 characters of *Tullimonstrum* were compared with character states in 13 other OTU's plus an outgroup in preliminary analyses using a phylogenetic computer program, most of the resulting cladograms indicate close affinities with the Conodonta; almost as many corroborate Foster's (1979) hypothesis of a molluscan affinity with *Tullimonstrum*. Incorporating these results into a classification requires additional ingroup analyses of the sister groups of *Tullimonstrum* as identified in this analysis.

Introduction

Studying the origin and diversification of Metazoa involves a number of different questions, including rates of morphological divergence and detailed descriptions of anagenetic events. However, these are questions derived from a particular understanding of phylogenetic relationships between organisms. Any understanding of the processes regarding the diversification of Metazoa requires a reconstruction of the relationships of the major groups of organisms on the planet, past and present. This situation requires first recognizing to what extent these organisms form discrete morphological groups. Evaluating the range of morphology present among metazoans represents the stage of analysis at which it becomes important to consider those organisms regarded as problematica. However, why are some forms regarded as problematica while others are not?

Three problems that potentially exist for any study of fossil organisms are responsible for the need to recognize problematica as such (Figure 1). First, how should the morphology of these fossilized organisms be interpreted? Secondly, given an interpretation(s), how can the affinities of the organism be recognized? Finally, how should these organisms be fit into the preexisting system of taxonomic classification? In a purely semantic sense, when an interpretation of morphology is such that affinities are obscure, the organism can be added to the roster of problematica. However, workers recently have tended to restrict application of the term "problematica" to taxa whose affinities are obscure at the level of phylum (sensu Bengtson, 1986; Hoffman & Nitecki, 1986).

These three problems also represent three levels of analysis, with the assumptions and inferences of each level influencing the next. This paper applies this hierarchical analysis to one of the most famous elements of problematica, the Tully monster. Not only is the Tully monster interesting from a strictly morphological perspective, but it is unusual in being one of the relatively few problematica to survive beyond the end of the Cambrian. Does this represent some bizarre taphonomic bias against the earlier and later history of the Tully monster, or does it indicate the presence of some macroevolutionary event during the Carboniferous, or is Tully monster not really as monstrous as some have believed?

Recognizing a hierarchy of questions and analyses permits formalizing the arguments used to interpret problematica, which will help to focus future criticisms of any particular analysis. By outlining not only the empirical evidence but also the rationale(s) involved with analyzing the morphology of fossils (including problematica), differences of opinion between workers can be understood logically rather than emotionally. In this reinvestigation of the Tully monster, *Tullimonstrum gregarium* Richardson, 1966, the first and lowest level of the hierarchy is the basic description and interpretation of the morphology preserved in the fossils. The intermediate level is the analysis of phylogenetic relationships. The final level of investigation in this hierarchy is incorporating the phylogenetic inferences into a system of taxonomic classification. This paper is organized according to these three levels of interpretation.

Reinterpreting morphology

Richardson (1966) described *Tullimonstrum* based on specimens collected by Francis Tully on waste piles produced by strip mining for coal in Pit 11 of the Mazon Creek region of northeastern Illinois. For decades, this middle Late Carboniferous (Westphalian D) deposit had produced large numbers of siderite (iron carbonate) concretions that contained a high diversity of plants with rare non-marine animals. When the Peabody Coal Company opened Pit 11 to the south of then-existing strip mines, the discovery of abundant marine animals led Richardson to recognize the need for a thorough investigation of the fauna.

Within the fauna was an abundant animal which Richardson (1966) noted had a body consisting of three primary regions (the anterior proboscis, medial trunk,

[1]Department of Geology, Field Museum of Natural History, Roosevelt Road at Lake Shore Drive, Chicago, Illinois 60605-2496, USA

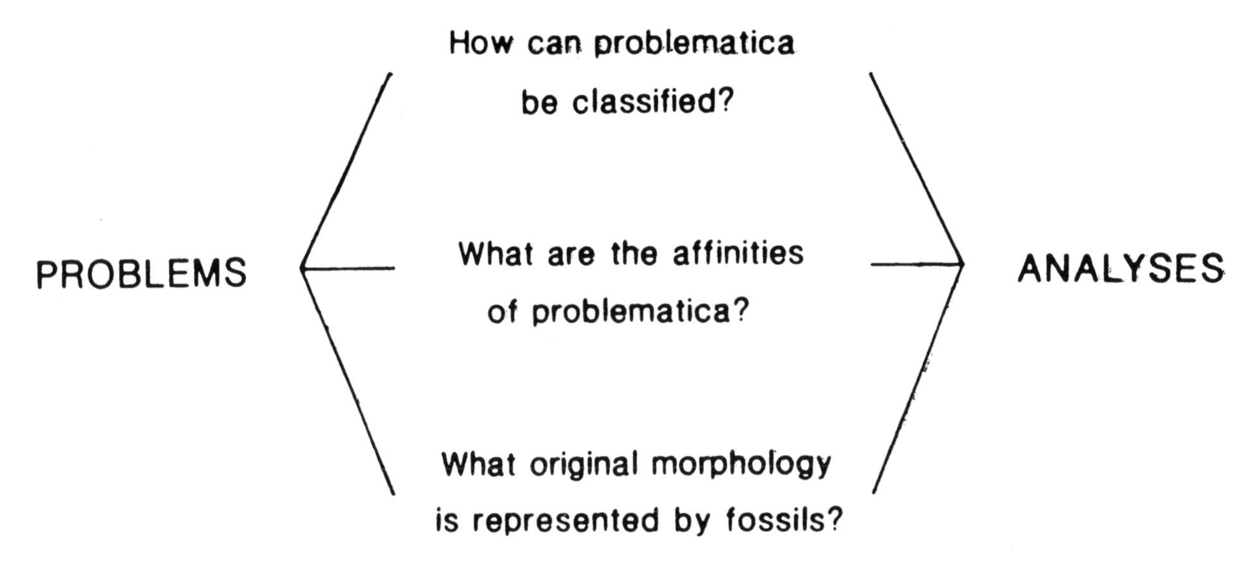

Fig. 1. Diagrammatic representation of the parallel hierarchies involved with interpreting fossils, including problematica. One hierarchy consists of a series of problems represented by the three questions in the central column; the second hierarchy consists of the analyses required to address the central column of questions.

and posterior tail). The proboscis bore an anterior "claw" or "jaw" which contained up to 14 "teeth" or stylets. The bulbous trunk was flattened dorsoventrally, with transverse segments along its length; a rigid bar bearing putative eyes crossed the ventral surface of the anterior end of the trunk. The tail bore a pair of flexible lateral fins, much like those of a squid. This unusual combination of characters led Richardson to refer cleverly to the newfound beasts as *Tullimonstrum gregarium*, "the abundant monster of Tully".

Johnson & Richardson (1969) used additional fossil material to expand the original description. Among the information that they added to the description was that an analysis of the carbonaceous material at the ends of the rigid bar was probably organic in origin, and a discussion of possible interpretations of the internal anatomy of *Tullimonstrum*. In other respects they agreed with the original description by Richardson (1966).

Foster (1979) made a significant contribution to the interpretation of the morphology of the Tully monster when he suggested that the banding that Richardson (1966) and Johnson & Richardson (1969) had used to infer that the beast was externally segmented was in fact probably the expression of strictly internal segmentation that was visible only in specimens that had decayed prior to burial. Foster also suggested that the anterior "claw" or "jaw" was the taphonomic product of compressing a cylindrical buccal mass bearing a radula. Furthermore, Foster (1979) found Tully monsters in strip mines dumps in other areas of Illinois that are stratigraphically equivalent to the Mazon Creek deposit, thus expanding the known palaeobiogeographic range and reducing the endemism of this beast.

Several aspects of the morphology of *Tullimonstrum* can be reinterpreted. The most significant reinterpretation involves the orientation of the "tail". Richardson (1966), Johnson & Richardson (1969), and Foster (1979) favoured interpreting the tail as bearing lateral or horizontal "fins," being flattened in the same plane as the dorsoventrally compressed body. At the same time, they recognized existence of an oblique "line" that ran from the anterior margin of the intersection of one "fin" toward the sagittal axis of the trunk. Foster (1979) suggested that this "line" was a crease, and thought that the posterior fins may have been oriented dorsoventrally, although he believed that "the evidence for either orientation is equivocal" (p. 273). Examination of nearly 2000 specimens of *Tullimonstrum* and comparison with modes of preservation in better understood organisms supports Foster's suggestion that this lateral "line" is a crease (Figure 2). One important factor supporting reorienting the tail is that the crease is most visible when the two tail fins are clearly visible and flattened. The crease is difficult to observe in those specimens that do not clearly exhibit the tail fins, as if the posterior portion of the creature had been mangled. This is exactly the taphonomic pattern that would be predicted if the tail fins were oriented dorsoventrally during life, rather than laterally as in previous reconstructions. The crease is easily explained as the taphonomic product of a twist in the tail induced by the weight of the sediment compressing the body; similar arguments were presented by Aldridge *et al.* (1986) in a discussion of the Carboniferous conodont animal. In the few specimens in which neither the tail fin nor the crease is visible, the tail may have been preserved in life orientation, with breakage of the concretion along the bedding plane cutting across the extremities of the fins. The value of considering the influence of taphonomic factors in the interpretation of fossils has been exemplified by Whittington (1969) and Briggs & Williams (1981) for arthropods and graptolites, and by Rex (1983) and Rex & Chaloner (1983) for various Carboniferous plants.

The shape of the "fins" of *Tullimonstrum* provides additional evidence for reorienting the posterior "fins" dorsoventrally. Richardson (1966, fig. 3), Johnson & Richardson (1969, fig. 79a), and Foster (1979, fig. 3) all illustrate well-preserved tails that show clearly that the fins are asymmetrical; Foster (1979) mentioned this asymmetry in the figure caption. Such asymmetry has several functional ramifications related to "fin" orientation. Although the asymmetry is slight (Figure 3), the difference in fin shape probably would have hindered attempts to swim in a straight line if the fins were

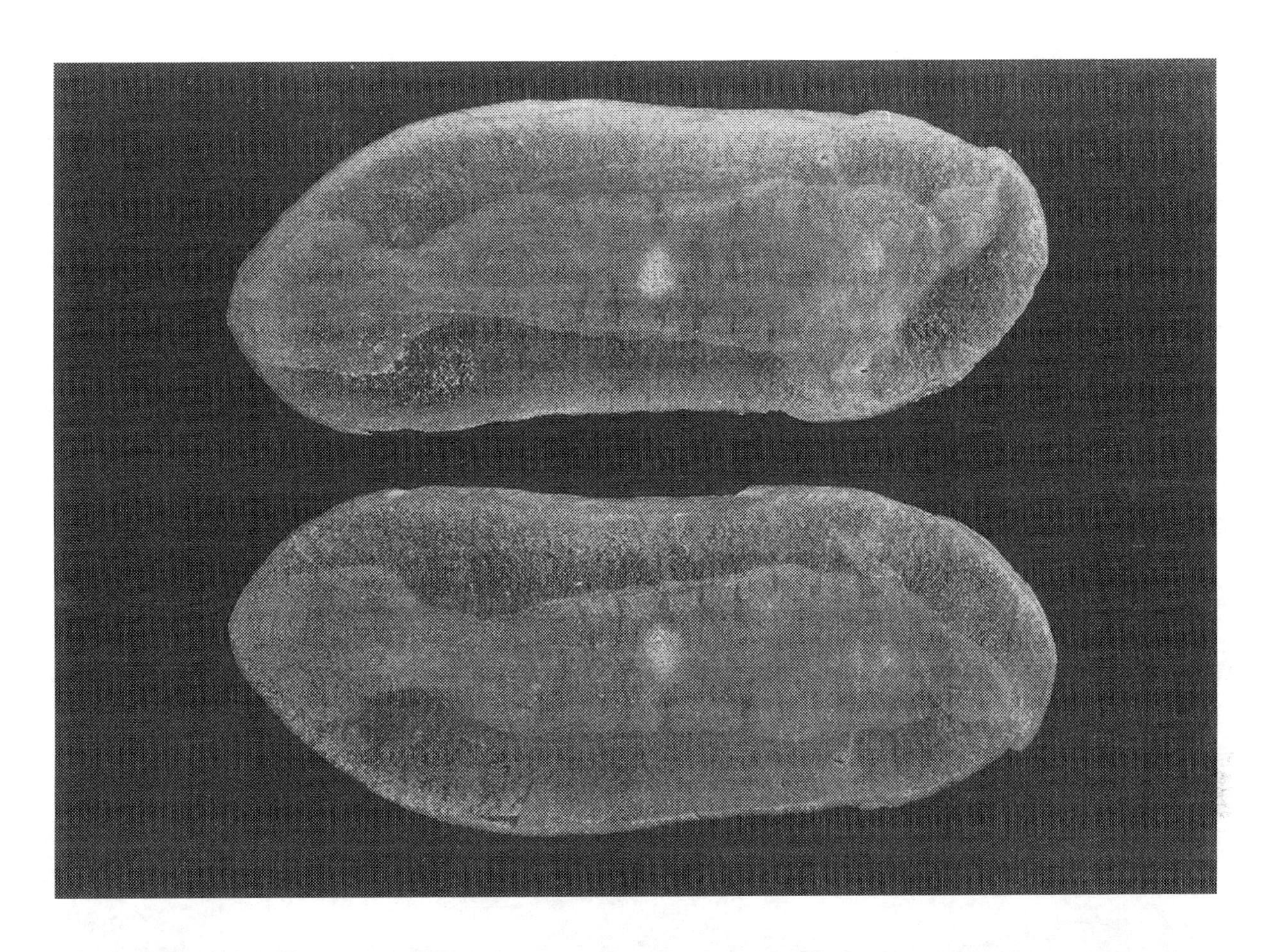

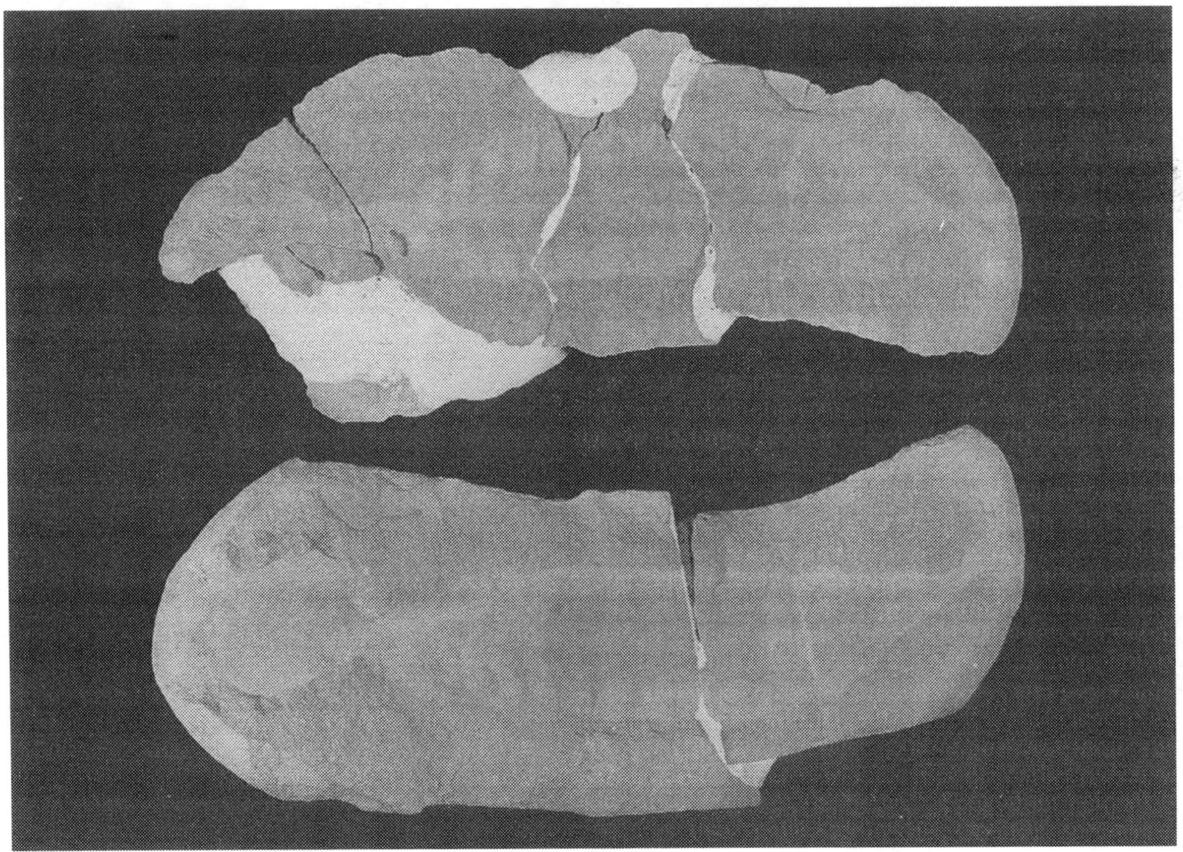

Fig. 2.a. Holotype of *Tullimonstrum gregarium* Richardson (1966), FMNH PE10504, exhibiting the oblique crease on the posterior half of the body used as evidence to reorient the tail so that the fins are oriented dorsoventrally. b. FMNH PE 10616, exhibiting a similar crease.

Fig. 3.a. A tail of *Tullimonstrum*, FMNH PE 7051, illustrating slight asymmetry. b. Another tail of *Tullimonstrum*, FMNH PE10610, illustrating the slight asymmetry, and the flexible nature of the fins which, when compressed, give the impression of rays. This asymmetry would interfere with swimming if the tail fins were oriented laterally (horizontally), but not if the fins were oriented dorsoventrally.

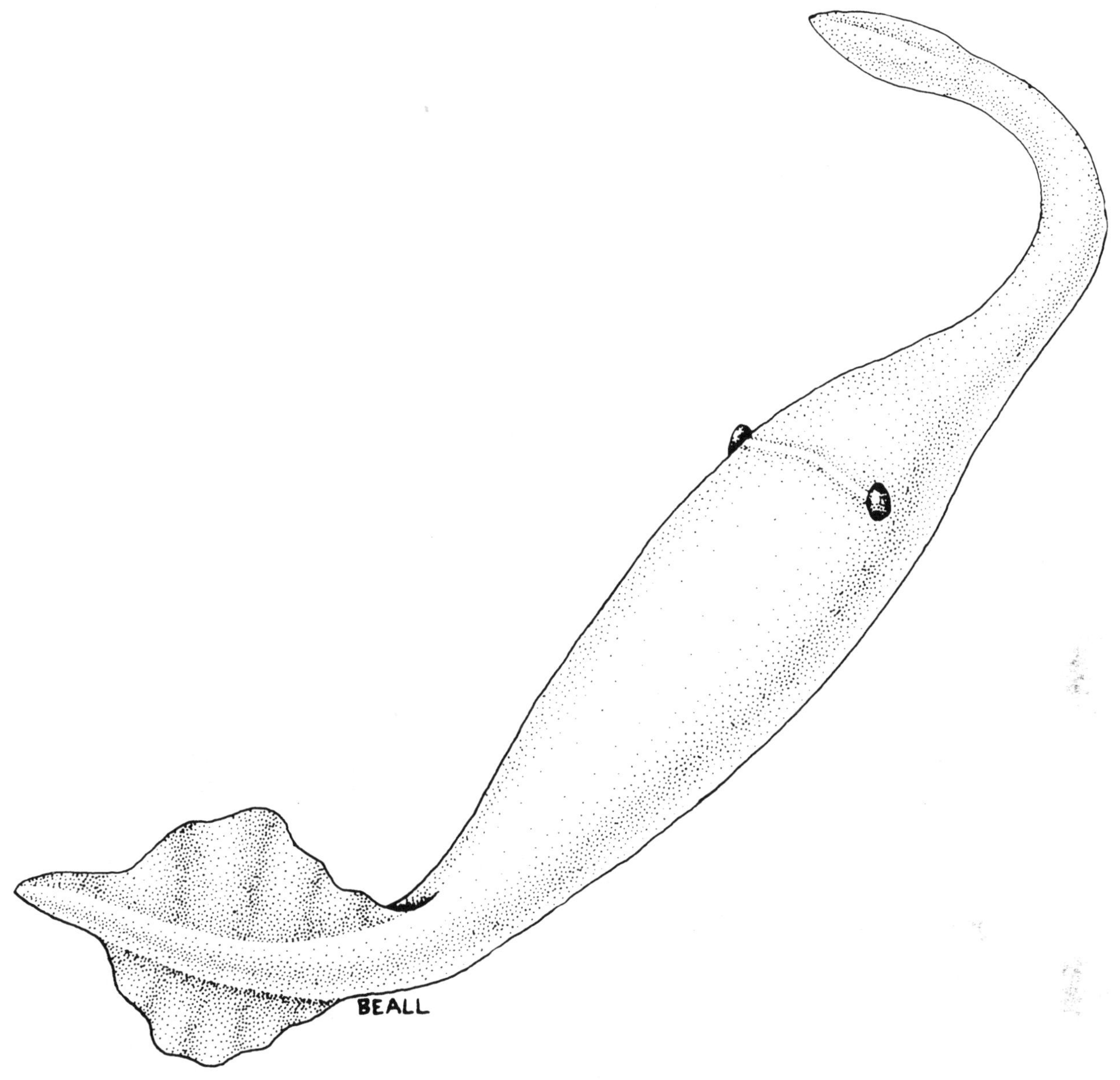

Fig. 4. A revised reconstruction of *Tullimonstrum gregarium*, with the anterior "claw" represented as a tubular buccal mass (following Foster [1979]), and with the posterior fins reoriented dorsoventrally. The bar organ has been placed dorsally, although a ventral orientation is equally likely; the terminal organs of the bar are interpreted here as either sessile eyes or as structures analogous to statocysts. Alternative interpretations of the bar organ and its terminations are very likely; see text for additional discussion.

oriented laterally. Whether the longer fin is dorsal or ventral cannot be evaluated. Nevertheless, evaluation of the oblique crease of the posterior half of the body of *Tullimonstrum* and of the tail asymmetry strongly supports reorienting the posterior fins of the Tully monster into a dorsoventral orientation. Not only does this reorientation have significant implications for the functional morphology and thus autecology of *Tullimonstrum*, but it also influences interpretations of putative relationship, as discussed below.

Several other features can be interpreted alternatively. For one thing, an assumption of bilateral symmetry permits arguing for right and left sides, and the orientation of one end bearing stylets and the opposite end presumably responsible for locomotion permits interpreting anterior and posterior directions. However, no evidence can be used to argue convincingly for which side represents dorsal and which is ventral (Johnson & Richardson, 1969). Thus, these two alternative interpretations must be considered in any discussion of relationship.

The transverse bar organ also can be reinterpreted. In fact, multiple interpretations are possible, with no single interpretation obviously being "correct." Johnson & Richardson (1969) favoured interpreting the structures at the ends of the bar as eyes. They also noted that "colleagues have suggested that the bar and bar organs might have functioned as stabilizers or otocysts" (p. 139); although Johnson & Richardson (1969) disagreed with this interpretation, their arguments reduce to a qualitative probability statement about the optimality of the bar organ functioning as a stabilizer, and so this hypothesis of function has not been falsified yet. Nitecki (private communication, 1988) observed considerable similarity between the bar organ of *Tullimonstrum* and the unique paired copulatory organs of the extant oligochaete *Alma*, suggesting similar functions. Additionally, the bar organ is similar to the pair of hardened anterior setae present in some echiurans. Finally, this unusual structure may have been used as a locomotory or feeding structure for a benthic ecology (Carman, 1989). As stated above, it is impossible to determine which surface of *Tullimonstrum* is dorsal and which is ventral, so the placement as well as the function of the bar and its terminal structures are equivocal. Furthermore, some specimens preserve the terminations of the bar organ in association with the body wall of *Tullimonstrum*, while in other specimens the terminations extend beyond the body. All of these interpretations need to be considered when evaluating the affinities of *Tullimonstrum*.

Aspects of this revised description of *Tullimonstrum* suggest a new hypothesis of its affinities. Namely, there is a great deal of similarity, and thus putative affinity, with known conodont animals. This similarity is outlined here, and this new hypothesis of affinity can be tested in the next section. The soft part morphology of only three genera of conodont animals is known. The first described specimen (Briggs *et al.*, 1983) of the conodont animal, from the Lower Carboniferous of Granton, Scotland, and attributed to the genus *Clydagnathus*, is still the best preserved of the four specimens from that locality (Aldridge *et al.*, 1986). The Granton specimens and the Tully monster both are segmented internally, bear stylets with basal cavities, have-an anterior buccal mass preserved as bifurcate anterior lobes, and have dorsoventrally-oriented asymmetrical posterior fins. At first, homology between some of these states may be questioned due to slight differences in the shapes of the stylets, the anterior bifurcate lobes, and the internal segments (also see Aldridge & Briggs (1989) for a revised reconstruction of a Granton conodont animal). Specifically, the Granton conodont animals bear bars, blades and platform elements rather than the simple cones (stylets) of *Tullimonstrum*; the anterior bifurcate lobes of the conodont animals are much shorter and more rounded than in *Tullimonstrum*. The bodies of the Granton animals also appear to have been laterally flattened in life, facilitating their preservation in a lateral perspective; in this perspective, the internal segments are chevron-shaped. However, all Tully monsters are preserved in dorsoventral aspect, suggesting that they were flattened to some degree in this orientation during life, unlike the Granton conodont animals; in this orientation, the internal segments of *Tullimonstrum* have straight margins (however, the segments in the tail, which was flattened laterally in life, are shaped somewhat like broadly open chevrons). Finally, the posterior fins of *Tullimonstrum* are proportionally larger than, and lack the parallel rays of, the Granton conodont animals. However, a qualitative survey of morphological variation in Recent orders, classes and phyla indicates that none of these differences is sufficient to reject affinities between *Tullimonstrum* and the Conodonta. Moreover, there is evidence that the range in morphology of the bodies of conodont animals is greater than indicated by the Granton specimens. Mikulic *et al.* (1985a, b) described a conodont animal from the Brandon Bridge Formation (Silurian) near Waukesha, Wisconsin. Because this organism bears an assemblage of unbroken conodonts attributable to the genus *Panderodus*, it is unquestionably a conodont animal. More importantly, the differences between this specimen and the Granton specimens make it resemble certain aspects of *Tullimonstrum* more closely. Particularly, the conodont assemblage present in the specimen of *Panderodus* consists entirely of coniform elements that are very similar to the stylets of *Tullimonstrum*, including a lateral groove (each side of which could be interpreted as a ridge which accommodates the description by Foster (1979); see cross-sections of panderodontid elements in Smith *et al.* (1987), Figure 6.4). The stylets of *Tullimonstrum* also have basal cavities like conodont elements; these are preserved as convex infillings, since only molds of the stylets remain. Also, the specimen appears to be compressed dorsoventrally, and Aldridge *et al.* (1986) even suggested that the Silurian conodont animal may have been "dorsoventrally flattened in life" (p. 287), as in *Tullimonstrum*. Additionally, the segmentation in this dorsoventrally flattened orientation appears to have straight transverse boundaries like the body segmentation of the Tully monster. Finally, while the Granton conodont animals have bodies of fairly constant width, Smith *et al.* (1987) noted that "the dimensions of the [Silurian specimen] suggest a larger and more tapered trunk" (p. 93), a condition like the anterior portion of the Tully monster. Unfortunately, the Silurian specimen is too poorly preserved to provide more comparative information, especially since the posterior half of the body is missing. The size differences between the conodont animals and most specimens of *Tullimonstrum* are insufficient to warrant placement in different higher taxa. However, the absence of any Upper Carboniferous coniform conodonts is difficult to reconcile with this interpretation of affinity; a number of possible explanations for this discrepancy will be presented elsewhere (Beall, in prep.) so that they can be treated in detail and not as *ad hoc* assertions. It is interesting to note that the stylets of *Tullimonstrum* bear some resemblance to certain elements of the Carboniferous conodont genus *Idioprioniodus* (see Sweet, 1988, p. 83), species of which have been recovered from the stratigraphically and geographically adjacent Mecca Shale of Indiana (FMNH collections). Any additional information about *Tullimonstrum*, especially regarding the composition and mode of growth of its stylets, would be particularly useful for testing this hypothesis of affinity. Nevertheless, this and other hypotheses of affinity will be tested in the following section using a new analytical method that incorporates elements of both phenetics and cladistics.

Affinities of Tullimonstrum

As stated in the introduction, the second level of the hierarchy of problems of problematica is inferring the phylogenetic position of a particular problematic organism. A great deal of success has been achieved using a basically phenetic methodology. That is, workers have looked at the overall morphology of the problematic taxon, and then argued for its being most similar to some other group. While this approach has been highly successful, there are instances where no consensus of interpretation exists among workers. Such is the case with *Tullimonstrum*. In situations like this, an alternative method of analysis may be useful in resolving differences of opinion, especially if the alternative emphasizes aspects of the morphology not emphasized by the initial method. To this end, I suggest the method described below, a combination of phenetics and cladistic methodology, applied to interpreting the affinities of the Tully monster.

Before presenting the alternative, a brief review of the results of the initial methodology is necessary. Both Richardson (1966) and Johnson & Richardson (1969) compared *Tullimonstrum* to a number of bilaterally symmetrical phyla, but in the end preferred to leave the phylum designation of the monster as "uncertain." Johnson & Richardson (1969, 119) stated that "it appears that *Tullimonstrum* represents an extinct and previously unknown phylum." Foster (1979) made a compelling case for *Tullimonstrum* being a pelagic gastropod similar to extant heteropods. He used several lines of evidence, including the presence of an elongate proboscis bearing stylets similar to the radular elements of heteropods, an elongate body with a broad, flattened posterior region, and eyes in a position near the "eyes" of *Tullimonstrum*. These interpretations represent hypotheses that can now be tested using the alternative methodology mentioned above.

As stated above, this method is a combination of phenetics and cladistic methodology. The phenetic aspect comes in the initial part of the analysis, particularly in the choice of characters and operational taxonomic units (OTU's). The essentials of the component involving cladistics (or phylogenetic systematics; Hennig, 1966, 1981) are that relationships between taxa should be inferred from the shared presence of homologous and evolutionarily modified (derived) character states (synapomorphies); if the shared character states are primitive (symplesiomorphies), they do not provide information about the most recent branching points between taxa. The upshot of recognizing this dichotomy of character states is that, by using cladistic arguments, even if two organisms share great "similarity," they would only be interpreted as being closely related (i.e., "sister groups") if the similarities are synapomorphies. Cladistic philosophy has been applied to problematic taxa before, in a rigorous, smaller analysis of graptolite affinities, by Urbanek (1986). The results of a preliminary analysis designed to investigate possible sister group relationships of *Tullimonstrum* are described below.

The first step of this analysis was to select the other OTU's to be compared with *Tullimonstrum*. The choices of OTU's were made in two ways. First, several taxa had been mentioned by previous authors (Richardson, 1966; Johnson & Richardson, 1969; Foster, 1979) as possibly having affinities with *Tullimonstrum*; I used these taxa as OTU's in order to test the hypotheses of putative relationship. Specifically, although Richardson (1966) avoided any comparisons with other taxa, Johnson & Richardson (1969) suggested that *Tullimonstrum* was at the same "grade" of morphological evolution as protostomes such as echiurans and sipunculans, but higher than the nemerteans to which it bore some "superficial resemblances" (p. 148); therefore, I added both echiurans and sipunculans to the analysis. They also rejected any affinity between Tully monster and mollusks, echinoderms, chaetognaths, hemichordates or chordates. Foster (1979) discussed the possible affinities of *Tullimonstrum* in great detail, emphasizing similarities (and evaluating differences) with bathypelagic nemerteans, polychaete annelids, the extinct and equally problematic *Opabinia* (Mutchinson, 1930; Conway Morris *et al.*, 1982), and heteropod mollusks. These four hypotheses of Foster accounted for four additional OTU's used in the analysis. Secondly, I selected other OTU's using a phenetic approach as the previous workers had. The only objective requirement was that each OTU be bilaterally symmetrical, since this condition is present in the Tully monster and is a fundamental division of animal grades. I also required all of the OTU's to be similar to *Tullimonstrum* in some subjective way; consequently, I selected as OTU's several of the taxa that had been rejected by Johnson and Richardson (1969), specifically chaetognaths and hemichordates. Other taxa added as a consequence of the criterion of "similarity" included kalyptorhynch platyhelminths, priapulids, annelids (especially the oligochaete *Alma*), the extinct *Amiskwia* from the Burgess Shale (Conway Morris *et al.*, 1982), and nematodes. Following reasons discussed in the preceding section, I added the equally controversial and famous conodont animal. These criteria of bilateral symmetry and general similarity resulted in a total of 14 OTU's, plus an outgroup (see below; Appendix I). I excluded several bilaterally symmetrical taxa from the analysis because they were either "too" dissimilar to the Tully monster (i.e., kinorhynchs), because they had a completely parasitic mode of life (i.e., gastrotrichs and aschelminths) or because their diagnoses included synapomorphies that were clearly absent from *Tullimonstrum* (i.e., vertebrates and arthropods).

Once the OTU's were selected, the next step was to "break" the morphology of *Tullimonstrum* into characters. Technically, the characters had to be selected after the OTU's in this case because use of particular OTU's often suggested alternative interpretations of some aspect of the Tully monster. For example, as discussed in the section on reinterpreting morphology, the structures at the ends of the rigid bar organ have been interpreted as stalked eyes, statocysts, external copulatory organs, and highly modified anterior setae. Because comparisons should be made between homologous character states in any phylogenetic method, the choice of characters had to reflect these different interpretations. A character such as "presence of anterior transverse organ" would have little value biologically. This procedure resulted in the 21 characters in Appendix II. The particular selection of characters was meant to be intentionally ambiguous and general in order to impart as little bias as possible on the analysis. Once the characters were selected, the states for each of the OTU's were recorded.

To this point, the analysis has not differed significantly from phenetic approaches. However, applying cladistic methodology to interpreting the affinities of *Tullimonstrum* involves the additional step of establishing the polarities of the character states of each character. In order to accomplish this, I arranged the states in morphological transformation series. The transformation series in this particular analysis are purely geometric (Figure 5; Appendix II). They are arranged in such a way that shape changes from state to state are minimized. Support for this operational approach involves a parsimony argument, in that the number of *ad hoc* hypotheses needed to explain changes from state to state is minimized. This part of the method acknowledges that evolution is a Markovian

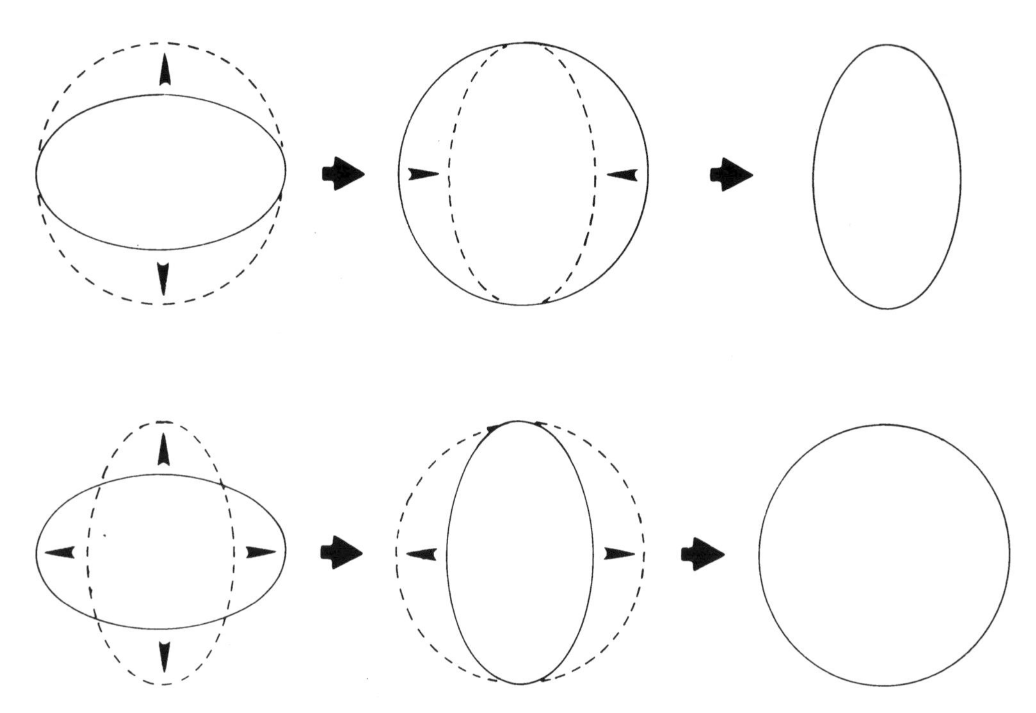

Fig. 5. Transformation series of states of character 8, Body flattening. a. Cross section of body with primitive state, body flattened dorsoventrally. Dotted lines and arrows indicate the two changes needed to transform the primitive state to the intermediate state. b. Cross section of body with intermediate state, body unflattened. Dotted lines and arrows indicate the two changes needed to transform the intermediate state to the most derived state. c. Cross section of body with most derived state, body flattened laterally. This series requires a total of four changes. The alternative series, from a to c to b would require four changes between a and c and another two from c to b. This alternative series requires a total of six changes, making it less parsimonious than the first series.

process in that the options of morphological change are limited by pre-existing morphology, which is a valid defense of the method. However, involving cladistics in this way also makes two assumptions. The first, as alluded to above, is that the different states of the characters being compared are homologous at the level of character. The second is that evolution of the character states being compared has progressed linearly. These assumptions are clarified below.

There are three reasons why assuming *a priori* that the states are homologous is inappropriate. First, aside from the characters being in similar positions in the different organisms, there is no independent evidence (such-as from embryological or other developmental studies) to argue for the homology of structures in the extinct OTU's with those in extant groups. Secondly, several of the OTU's represent highly derived members of their respective phyla; the states being compared in this analysis generally are absent in the morphology of the most primitive members of the phyla, which suggests that the character states are independently derived. Finally, detailed comparison of the states in different OTU's indicates that there is only a slight similarity in many cases; inclusion of these character comparisons is intended *only* to provide a variety of working hypotheses. However, because *application* of this methodology to inferring affinities of problematica must be very restricted (see below), arguments against this assumption do not invalidate the method.

The second major assumption of this method is that evolution of the OTU's has progressed linearly; the fact that the states being compared come from the derived members of each OTU and so are probably independently derived precludes linear evolution within the *entire* series. However, it does not preclude linear evolution among *adjacent* members of the series (see below). Again, this assumption does not invalidate the method.

The restricted application of this method referred to above is a limitation imposed by the assumptions. This method is meant to identify only the sister group(s) of the problematic taxon. It will NOT provide the phylogenetic relationships of the other OTU's to one another. If this analysis were using the most primitive members of each phylum as OTU's, then resulting cladograms would reflect the phylogenetic history of these groups. However, the fact that this approach selects specific, derived members of each phylum to be compared means that a considerable amount of evolution has occurred between the primitive and derived members of each phylum that would not be accounted for in inferences of relationship based on this analysis.

Consequently, only those taxa that are *adjacent* to *Tullimonstrum* can be assumed to be homologous validly. Because the goals of this method are sister group relationships, any single comparison between *Tullimonstrum* and some other OTU can be assumed to represent the comparison of homologous characters. The only parts of the cladograms that will be valid are those pairs of taxa of which the Tully monster is one of the sister taxa. Other pairs of taxa may, in fact, be closely related based on the comparisons of characters that can be argued to be homologous based on independent lines of data. However, these same kinds of data indicate that the characters being compared between most pairs of taxa probably are not homologous. So why is the method valid when only pairs including *Tullimonstrum* are retained? The following example from the analysis will help to explain this point.

Annelids and mollusks are related at some level

of analysis, so we might predict that they would fall out as sister groups in this analysis. However, this relationship is not especially close, and the taxa used to select character states are highly derived members of their respective clades. Consequently, there is evidence (comparative morphological and embryological) that the similarity in the states being compared between derived annelids and derived mollusks can be attributed to convergence rather than homology. However, if they *were* homologous, the fact that a phylogenetic computer program (such as PAUP [Phylogenetic Analysis Using Parsimony; Swofford, 1985]) grouped them together could be regarded as a strongly supported phylogenetic hypothesis. More relevant to this particular case is the fact that these kinds of detailed data about comparative morphology and embryology are not available for *Tullimonstrum*. There are no data to say that the conditions in *Tullimonstrum* are not homologous with the conditions in derived members of either Annelida or Mollusca. When PAUP finds the most parsimonious arrangement (cladogram) of taxa, homology is assumed, but that assumption must be ignored when the cladograms are interpreted. Because its biology is rather poorly known, *Tullimonstrum* can act as a sort of systematic "wild card". That is, because there is no evidence that the states in *Tullimonstrum* are not homologous with the states in any other OTU, the conditions in pairs of taxa of which *Tullimonstrum* is a member can be assumed to be homologous. This assumption is necessary in order to take advantage of the cladistic philosophy of grouping taxa using derived character states, and the advantage of doing this grouping using a phylogenetic computer program such as PAUP. As long as the restrictions of the method are maintained, no theoretical violations have been committed.

Once the character states of each character are polarized in transformation series for all OTU's, it is necessary to root the series. In this analysis, I have selected outgroup analysis to organize the series. There is a consensus of opinion that platyhelminths (flatworms) are the most primitive members of the Bilateria (Barnes, 1980; Meglitsch, 1967; but see Rieger *et al.* this volume, for an alternative view). Therefore, I used the purportedly most primitive members of the Platyhelminthes as the outgroup. In those cases in which the flatworms were missing anything comparable with particular characters, the most primitive state would be "absent".

After rooting the morphological transformation series using the outgroup, the character matrices could be analyzed using PAUP. As with the geometric transformation series minimizing the number of *ad hoc* changes between morphological states of a single character, PAUP minimizes the number of *ad hoc* transitions between states of all characters. An essential aspect of this approach also involves the use of multiple character matrices. Using multiple matrices permits comparison among the OTU's for each of the different analyses of the morphology of *Tullimonstrum*. For example, a single matrix cannot accommodate all of the different interpretations of the transverse bar organ and its terminal structures. Therefore, a separate matrix must be analyzed for each interpretation. Similarly, because it is impossible to evaluate which orientation of the Tully monster is dorsal and which is ventral, these options must be included, which automatically doubles the number of matrices. I analyzed a total of 48 matrices, considering different aspects of the morphology of *Tullimonstrum*; the basic configuration of each matrix, excluding those characters which are variable in *Tullimonstrum*, is shown in Appendix III. Analysis of the 48 matrices produced a total of 727 cladograms (most matrices produced more than one most parsimonious cladogram).

Since, as stated earlier, only those OTU's that lie immediately adjacent to *Tullimonstrum* should be considered as having possible affinities with it, it was possible to eliminate those branches of the cladograms that did not pertain to *Tullimonstrum*. Consequently, the original 727 cladograms could be reorganized as 49 branch topologies of which *Tullimonstrum* was a part. The number of cladograms producing any particular topology was highly variable, ranging from 2 to 151. Because of the preliminary nature of this work, only the five most frequently encountered topologies will be discussed. These are illustrated in Figures 6-10.

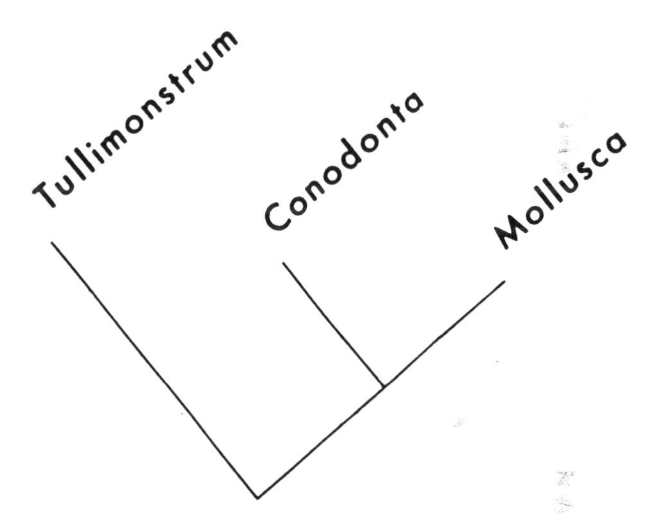

Fig. 6. The most frequently encountered tree topology involving *Tullimonstrum*, obtained in 151 out of 727 cladograms. This tree indicates that *Tullimonstrum* is equally closely related to Conodonta or Mollusca. See text for further explanations and discussion.

The most common cladogram (Figure 6), if read in a traditional manner, indicates that *Tullimonstrum* is the sister of Conodonta + Mollusca. This presents an excellent opportunity to discuss how a different approach is required to interpret the trees derived from this analysis. First, if the 21 characters used in this analysis represented everything that was known about Mollusca or Conodonta, interpreting these two taxa as sisters would be supported by this particular tree topology, corroborating the Tillier and Cuif (1986) hypothesis that conodonts and aplacophoran mollusks have affinities. However, as Briggs *et al.* (1987) have pointed out, there are additional data that indicate that both conodonts and aplacophorans have sufficient morphology to indicate that each is more closely related to other taxa than they are to each other. It is merely a consequence of leaving these other taxa out of this analysis that conodonts and mollusks fall out as sister taxa. Therefore, the trees must be read in an alternative way. Since the analysis was designed to look *only* at affinities of *Tullimonstrum*, all comparisons must be with

Tullimonstrum. Consequently, this most common tree should be interpreted as indicating that particular interpretations of the morphology of the Tully monster as evolutionarily derived makes it most similar to either conodonts or mollusks. If those aspects of morphology are homologous, then affinities between *Tullimonstrum* and either conodonts or mollusks is indicated. Thus, there are two sets of sister taxa.

The logic of this interpretation can be understood by using an "if...then" statement. Specifically, *if* the character states that are used to link *Tullimonstrum* with either conodonts or mollusks are homologous with the corresponding characters of these two taxa, *then* *Tullimonstrum* is equally closely related to either taxon. This statement is identical to the logical structure of any cladistic analysis; the only difference is that there is more evidence *a priori* that the conditions being compared *are* homologous. This does NOT imply that the conditions in the Tully monster *must* be homologous to *both* conodonts and mollusks; rather, the conditions can be homologous to *either*. If there were more information about the anatomy of *Tullimonstrum*, it might be possible to restrict the possible interpretations of homology. However, without those data, these two hypotheses of affinity must be considered equally valid.

The next most frequently encountered tree topology (Figure 7), obtained in 84 of the cladograms, places

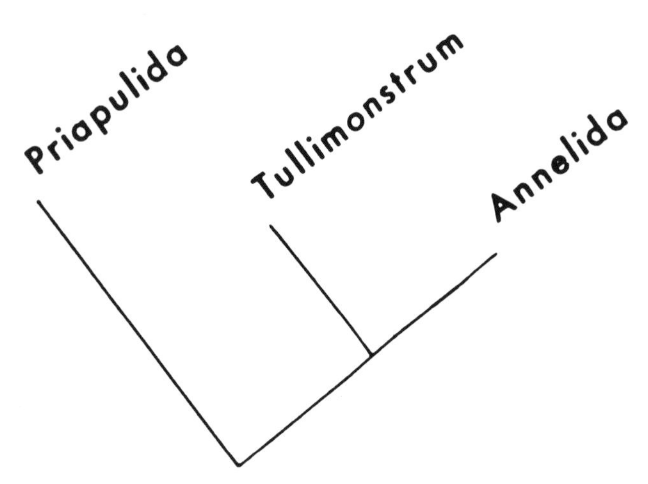

Fig. 7. The second most frequently encountered tree topology involving *Tullimonstrum*, obtained in 84 out of 727 cladograms. This tree indicates that *Tullimonstrum* is equally closely related to Annelida or Priapulida. See text for further explanations and discussion.

the Tully monster as the sister to either Priapulida or Annelida [primarily the oligochaete *Alma*, following Stephenson (1930) and Brinkhurst & Jamieson (1971)]. As for Conodonta and Mollusca, there are numerous data that suggest that the Priapulida and the Annelida are not closely related. Although Tully monster lies between Priapulida and Annelida on the cladogram, the temptation to make statements such as "*Tullimonstrum* lies phylogenetically between the Priapulida and Annelida" must be avoided. Rather, this tree topology merely suggests that given particular morphological interpretations, the Tully monster can be regarded as closely related to *either* Priapulida or Annelida.

The third and fourth trees have topologies that are identical to the first, except that each adds a different

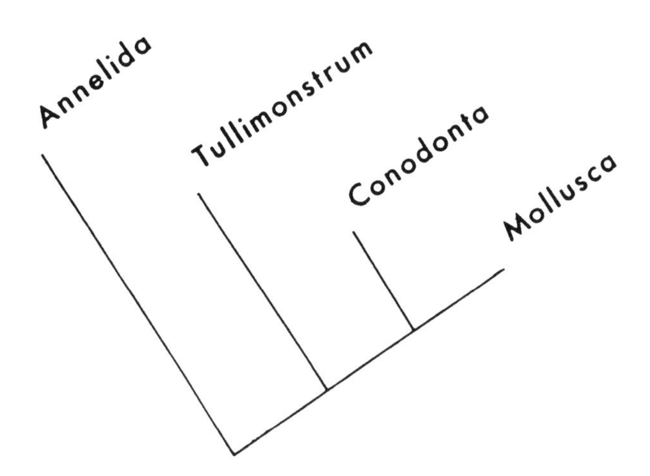

Fig. 8. The third most frequently encountered tree topology involving *Tullimonstrum*, obtained in 48 out of 727 cladograms. This tree indicates that *Tullimonstrum* is equally closely related to Conodonta *or* Mollusca *or* Annelida. See text for further explanations and discussion.

taxon below *Tullimonstrum*; consequently, these two topologies provide support for the interpretation of affinities between *Tullimonstrum* and either Conodonta or Mollusca. However, the third tree (Figure 8) also supports inferences of affinities with Annelida by placing Annelida as the sister to the group. Occurring in 48 cladograms, this tree is only slightly more frequently encountered than the fourth tree (Figure 9), which appears in 44 cladograms. The fourth tree replaces Annelida with the Burgess Shale problematicum *Opabinia*, indicating yet another possible affinity.

The last tree (Figure 10) places the Tully monster as the single sister of the Annelida (again, particularly the oligochaete *Alma*) in 44 different cladograms. Of the five topologies discussed here, only this tree permits a single interpretation of affinity, specifically Annelida + *Tullimonstrum* as two sister taxa constituting a single clade.

These five cladograms cumulatively support affinities between *Tullimonstrum* and *Opabinia*, 44 times; and Priapulida, 84 times; and Annelida, 176 times; and Mollusca, 243 times; and Conodonta, also 243 times. This suggests that the two serious hypotheses to be considered are affinities with either Mollusca and Conodonta. Tallying the distribution of taxa on all 727 cladograms from the analysis (Appendix IV) corroborates this interpretation, with affinities to Conodonta supported by 305 cladograms and to Mollusca by 295 cladograms. Affinities with *Opabinia* and Annelida rank third and fourth, with 231 and 217 trees, respectively. An affinity with the Priapulida ranks fifth, with 206 cladograms; affinities with the remaining OTU's were supported by even fewer cladograms (Appendix IV).

The above discussion has emphasized using the frequency of particular tree topologies to argue for particular interpretations of affinity; that is, the most frequently encountered topologies are to be preferred over those that are not as common. It should be remembered that these 49 tree topologies (and their 727 original cladograms) are the result of analyzing 48 data matrices that represent 48 different interpretations of the morphology of Tully monster. Even though a single data

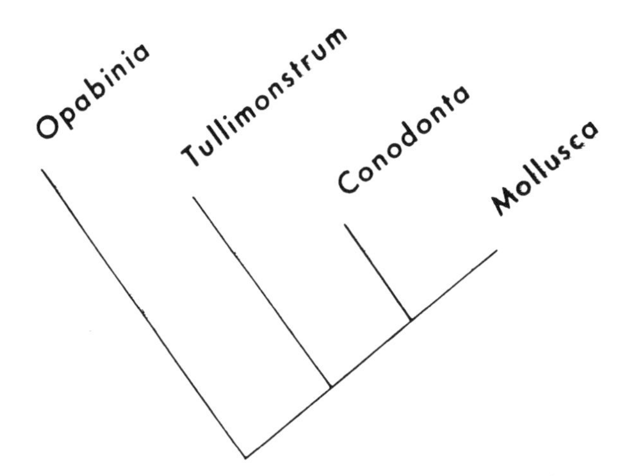

Fig. 9. The fourth most frequently encountered tree topology involving *Tullimonstrum*, obtained in 44 out of 727 cladograms. This tree indicates that *Tullimonstrum* is equally closely related to Conodonta *or* Mollusca *or Opabinia*. See text for further explanations and discussion.

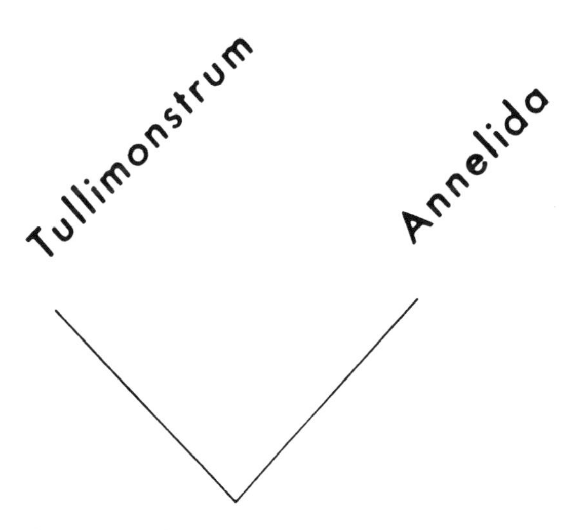

Fig. 10. The fifth most frequently encountered tree topology involving *Tullimonstrum*, obtained in 44 out of 727 cladograms. This tree indicates that *Tullimonstrum* is the sister group to the Annelida. See text for further explanations and discussion.

set usually produced more than one most parsimonious cladogram, each cladogram can still be regarded as an independent entity, so tallying the number of cladograms indicating a particular affinity with *Tullimonstrum* is a valid approach to evaluating the results. Nevertheless, an alternative way of selecting between the nine interpretations of affinity involves tallying the number of data sets that produce a particular interpretation. The results of this tally also are presented in Appendix IV. As with the previous tally, both Conodonta and Mollusca were indicated as having affinities with *Tullimonstrum* by the majority of data sets, a total of 29. However, affinity with *Opabinia* was indicated by 28 data sets. Affinity with Annelida was indicated by only 20 data sets, and the remaining five OTU's had even less support. This alternative method of data evaluation and the high number of cladograms producing congruent results indicate that *Opabinia* should also be considered seriously as possibly having affinity with *Tullimonstrum*. However, to facilitate presentation of methodological arguments in this preliminary paper, *Opabinia* will not be considered further, and only Conodonta and Mollusca will be discussed.

The upshot of this evaluation is that the analyses overwhelmingly indicate that *Tullimonstrum* has affinities with either Conodonta or Mollusca. Furthermore, the evaluation indicates that particular interpretations of *Tullimonstrum*'s morphology can be preferred over another, although this does not mean that the less preferred choices can be rejected. The ramifications of these relationships, as well as of the other cladograms, are discussed in the following section. The important point to make as a conclusion to this section is that by analyzing several data matrices that consider multiple hypotheses (interpretations) of the morphology of *Tullimonstrum*, it is possible to present alternative hypotheses of relationship in an objective manner. Admittedly, this approach may seem labour intensive and unwieldy. However, if the alternatives are evaluated explicitly using the criteria presented above, the resulting interpretations of affinity will be based on repeatable analyses and objective criteria. As this is one of the goals of scientific analysis, the extra work involved should not be objectionable.

Classification of problematica

Given a phylogenetic analysis that indicates that taxon A has affinities with taxon B or taxon C, how can this information be incorporated into a classification system? This is a fundamental problem for all organisms, not just problematica. Obviously, no single classification can be expected to accommodate two hypotheses of relationship. Therefore, the questions at hand are, "How can the relationship between taxon A and taxon B be incorporated into a classification system?" and "How can the relationship between taxon A and taxon C be incorporated into another classification system?" In the example at hand, *Tullimonstrum* is taxon A, and Conodonta and Mollusca are taxa B and C, respectively. As stated in the introduction, problematica are usually regarded as those organisms that do not fit readily into previously recognized phyla (Bengtson, 1986; Hoffman & Nitecki, 1986), so the OTU's selected to compare with *Tullimonstrum* represent phyla *sensu lato* (but see Aldridge *et al.*, 1986 for an alternative view of conodont classification). Even so, the particular character states used to represent each phylum came from particular members of the phylum. Therefore, the OTU's being compared also represent clades within their designated phyla, so that comparisons are with that intraphylar clade and not with the phylum as a whole. Also, there are problems regarding the homogeneity of character states within a phylum. For example, the data used to identify character states for conodonts come from a total of five specimens [four from the Lower Carboniferous of Scotland, representing the genus *Clydagnathus* and at least one other genus (Aldridge *et al.*, 1986), and one from the Silurian of Wisconsin, USA, representing the genus *Panderodus* (Smith *et al.*, 1987)]; the soft parts of these three taxa suggest that a substantial amount of morphological heterogeneity is present among the Conodonta. Also, the character states used to represent the phylum Mollusca were derived primarily from the gastropod order Heteropoda in order to test Foster's (1979) hypothesis of affinities. What is needed to transfer the data on affinities into a classification system?

Foster (1979) recognized that the above observations present four options of classification for the mollusks. When reformulated for a general theoretical discussion, Foster's four options can be described by the following three options:

1) Taxon A could be included within taxon B;

2) Taxon A and taxon B could be sister taxa of equivalent rank within some higher taxon that is within the phylum of taxon B;

3) Taxon A could be a representative of some higher taxon within the phylum of taxon B that is of equivalent rank to a different higher taxon containing taxon B.

Because the level of phylum is fundamental to a discussion of problematica, there is a fourth theoretical option:

4) Taxon A could represent a phylum that is a sister to the phylum of which taxon B is a representative.

What is necessary to select among these options? Do we expand the concept of an existing taxon (including phylum) to accommodate the autapomorphies of the "problematic" forms, or do we create a new higher taxon (even a phylum) for the problematica? Both options have been tried by various workers, but in general theory, neither seems to be more appropriate *a priori*. There needs to be some objective means of evaluating which option is more appropriate given a particular situation. The applications of the four classification options will be treated in the following discussion considering the problem of classifying *Tullimonstrum gregarium*.

Richardson (1966) and Johnson & Richardson (1969) took the most conservative route by classifying *Tullimonstrum* as "Phylum uncertain." Foster (1979) provided numerous lines of evidence to support his statement (p. 270) that "*Tullimonstrum gregarium* is believed to represent an earlier prosobranch gastropod invasion of an adaptive zone later exploited by heteropods." In coming to this conclusion, Foster considered alternative options in classification similar to those outlined above. The use of cladistic methodology and PAUP, which provide explicit statements about the number and nature of synapomorphies, makes the evaluation of these options more objective, as demonstrated below with the two favoured interpretations of affinity, with Conodonta and with Mollusca.

The first step is to determine which characters link *Tullimonstrum* with Conodonta and Mollusca. That is, what synapomorphies link each pair of taxa? Because the topology of the most common tree, and most of the other trees that link *Tullimonstrum* with Conodonta and Mollusca, does not place *Tullimonstrum* as the sister of one or the other taxon, identifying synapomorphies is identical for both pairs. Also, not only is it necessary to look at *which* characters are linking *Tullimonstrum* with Conodonta and Mollusca, but it is also important to evaluate *how many* characters are linking the two. This is necessary in order to select from among the options of classification presented above. If the branch bearing taxon A is not characterized by any autapomorphies, then it can be said to be indistinguishable from the most primitive condition of either taxon B or C, and so can be included within either of these taxa, which is option 1. If taxon A were united with taxa B and C in an unresolved trichotomy, and if taxa B and C are sister phyla (based on more complete phylogenetic information than is present in this analysis), then and only then is erecting a new phylum (option 4) justifiable. Without additional phylogenetic information, an unresolved trichotomy resulting from an analysis such as this one would still have to be considered with regard to options 2 and 3 before option 4 is acceptable. Options 2 and 3 essentially consider that taxon A belongs within the phyla represented by taxa B and C, but that they are grouped with taxa B and C in some way intermediate between the level of identity with B or C and the level of new phylum. The kind of taxonomic resolution that is necessary for a selection between options B and C is only possible with more detailed ingroup analyses of the phyla that include B and C. These ingroup analyses should be organized as standard phylogenetic analyses that seek relationships among all components, not like the specialized analysis suggested in this paper. Additionally, taxon A should be included as one of the OTU's; undoubtedly, data for all of the characters will not be available, and so in this secondary analysis, it would be necessary to code such data as missing (in PAUP, a "9" designates missing data) rather than to remove the character as I have suggested for the primary analysis. The results of these secondary analyses will determine, given a particular hypothesis of affinity (with a particular phylum), how taxon A can be accommodated in the existing classification system. These rationales for selecting from among options 1-4 are discussed below.

A survey of the cladograms indicating affinity between the Tully monster and Conodonta or Mollusca indicates that most of them have the three taxa united by characters 1 (position of mouth), 2 (stylets), 3 (presence of anterior bifurcate lobes), 4 (flexibility of proboscis), 5 (position of digestive tract), 6 (shape of digestive tract), 7 (relative body width), 19 (position of anus), 20 (nature of posterodorsoventral asymmetrical "fins") and 21 (flexibility of "fins"). However, depending on the topology of the tree and the distribution of other OTU's, different combinations of these characters united *Tullimonstrum*, Conodonta and Mollusca on different cladograms. Since *Tullimonstrum* has several autapomorphies in each analysis, option 1 can be rejected. Also, since unresolved trichotomies were very rare among the 727 cladograms, option 4 can be rejected; this is an example of how the erection of a new phylum may be rejected using an objective, repeatable (and valid) criterion. Thus, options 2 and 3 are favoured. Unfortunately, at this time, the ingroup analyses needed to select between options 2 and 3 are unavailable. Only after completing ingroup analyses of the Mollusca and Conodonta (or Chordata, specifically Craniata, *sensu* Aldridge *et al.* 1986, and Briggs *et al.* 1987) will it be possible to resolve the dichotomy between options 2 and 3 and finally answer the question of classification given a particular phylogenetic hypothesis.

The case for associating *Tullimonstrum* with the Mollusca was presented compellingly by Foster (1979). No new evidence about the morphology of the Tully monster is relevant to this particular hypothesis, but a brief review of the arguments is appropriate. Foster (1979) noted that both *Tullimonstrum* and heteropod gastropods were similarly sized with elongate bodies, had long probosces, and if the terminations of the bar organ were interpreted as eyes, had well developed optic organs. Additionally, Foster convincingly reinterpreted the anterior "claw" or "jaw" as a buccal mass with the buccal cavity exposed, or as a highly modified buccal mass; the associated stylets are curved with a median ridge, rather like the radula of some heteropods. Differences in character distribution include Tully monster's possession of a rigid transverse bar, rigid eye stalks, transverse musculature/internal segmentation, and a probable terminal anus, as well as its absence of a non-terminal swimming fin, an obvious visceral mass and statocysts (following Foster's interpretation). These differences are not especially important, given the range of morphology present among the Mollusca. Also, there is no reason that *Tullimonstrum* is not within the range of possible morphologies of shell-less mollusks, especially since the morphology is quite consistent with a molluscan affinity as demonstrated by the analysis presented above. However, placing *Tullimonstrum* within the Mollusca will only be possible with detailed phylogenetic analyses of other members of the phylum. Falsification of a molluscan affinity will only be possible by the chance discovery of additional information about *Tullimonstrum* that, when used in a reanalysis of the Tully monster's affinities, would not allow these two taxa to fall out in adjacent positions. Such information might include the chemical composition of the stylets, better understanding of the function of the transverse bar and its terminal organs, and data on the organization of the internal anatomy and coelomic structure of *Tullimonstrum* (but see Strathmann, this volume, and Rieger *et al.* this volume, for discussions of the importance of particular characters for determining phylogenetic level).

Regardless of the precise affinity of *Tullimonstrum*, this analysis and that of Foster (1979) indicate that, while being extremely unusual from a variety of evolutionary and ecological perspectives, the Tully monster does not require a series of contrived scenarios to explain its existence. An unusual taphonomic bias does not need to be invoked to destroy the relatives of *Tullimonstrum* selectively, nor must *Tullimonstrum* be regarded as the product of some extraordinary Late Palaeozoic macroevolutionary event. This is not to imply that either of these events could not have occurred, but in isolation, *Tullimonstrum* does not support either hypothesis given its apparent affinities with either the Mollusca or the Conodonta.

Summary

When a hierarchy of questions and analyses regarding the interpretation of a problematic fossil organism is recognized, it is possible to consider the validity of the evidence for interpretations of problematica in an explicit and repeatable way. Consideration of the morphology of *Tullimonstrum gregarium* from a taphonomic perspective permits arguing that the posterior "fins" were oriented dorsoventrally in life, not laterally as they have been illustrated traditionally; the asymmetry of the fins has functional ramifications that further supports this reinterpretation of the fin orientation. Additionally, multiple interpretations of the terminal structures of the transverse bar organ are consistent with its morphology, and so these must each be considered in analyses of affinities of *Tullimonstrum*. The inability to identify which features are dorsal or ventral must also be considered in such analyses.

The second level of the hierarchy, analyzing relationship, was accomplished using a combined cladistic-phenetic approach. The available data on the morphology of Tully monster were used to create 21 characters. The states were then compared with analogous and possibly homologous states in 14 other taxa (generally phyla), including an outgroup (primitive platyhelminths), selected on the basis of general similarity and to test previous speculation on affinity. Character matrices were organized that kept the states constant for virtually all characters of all OTU's except for some of the characters of *Tullimonstrum*. In order to accommodate the multiple interpretations of morphology, a total of 48 matrices were analyzed using PAUP (Swofford, 1985). These matrices produced a total of 727 cladograms. Since only the taxa adjacent to *Tullimonstrum* were relevant to this analysis, the topologies of the cladograms could be summarized as 49 trees. The available evidence overwhelmingly indicated affinity with either Conodonta or Mollusca, although there is some evidence of affinity with either Annelida or *Opabinia*, as well.

Incorporating this evidence of affinity into a classification, the third level of the hierarchy, requires data on the in-group relationships of both Mollusca and Conodonta (or Craniata) that are currently unavailable. However, the requirements for classifying *Tullimonstrum* with either of these taxa were outlined. Evidence from the analyses indicates that *Tullimonstrum* is not synonymous with either of the particular taxa used to represent the Conodonta and Mollusca in the analysis, nor is it justifiable to place *Tullimonstrum* in its own phylum using any sort of objective criteria. Therefore, the morphology of the Tully monster is sufficient to place it within either Conodonta (Craniata?) or Mollusca. Although this method might appear unwieldy, the assumptions of each step are either justified or explained explicitly so that each part of the analysis can be repeated objectively, thus imparting greater testability to this type of interpretation.

Future work must concentrate on in-group analyses of the Conodonta and Mollusca. Furthermore, new specimens of *Tullimonstrum* must be examined in order to determine if they bear previously unrecognized morphology or anatomy. Detailed study of the moulds of the stylets of all available specimens of *Tullimonstrum* will help to document the range of morphology present, and to make morphometric comparisons with known conodonts (especially *Idioprioniodus*) and heteropod mollusk radular elements. Additionally, given the similarities indicated by Aldridge *et al.* (1986) and Smith *et al.* (1987) between conodont animals and myxinoid craniates, the initial cladistic analyses will be run again with the Myxinoidea added as an OTU, and removing those OTU's that never appeared as a sister taxon to *Tullimonstrum* in this analysis.

Acknowledgements

Many people have contributed to this paper. M. H. Nitecki has provided encouragement and advice during all aspects of this progress. M.R. Carman made a number of suggestions regarding selection of OTU's and characters. Discussions with T. Baumiller, C. Labandeira, M. Listokin and D. Miller forced me to better focus my arguments on phylogenetic inference. C. Forster, C. Labandeira, M. Listokin and M. Nitecki read an early draft of the paper; their comments are greatly appreciated. E. Zeiger provided technical assistance. The library staff of FMNH, especially M. Calhoun, C. Callen, J. Devine, and D. Rogers, helped resolve many problems and locate many eccentric references. Lastly, I need to thank all of my friends who were patient when told that they would not get any attention "until the Tully monster paper is finished!" All errors remain my responsibility. Participation in this symposium was, further, made possible by financial aid from the University of Camerino and other sponsors of this symposium.

APPENDICES

APPENDIX I

Operational taxonomic units used in the data matrices; data on OTU's extracted from Barnes (1980) and Meglitsch (1967) except where noted.

1. primitive Platyhelminthes (outgroup)
2. Kalyptorhyncha (Platyhelminthes)
3. *Tullimonstrum gregarium*
4. Nemertea (Rhynchocoela)
5. Priapulida
6. Mollusca (Heteropoda)
7. Annelida (primarily *Alma*, an oligochaete; Stephenson, 1930; Brinkhurst and Jamieson, 1971)
8. Hemichordata
9. Chaetognatha
10. Conodonta (after Briggs *et al.* 1983; Aldridge *et al.* 1986; Smith *et al.* 1987)
11. *Opabinia regalis* (after Whittington, 1975; Briggs and Conway Morris, 1986)
12. *Amiskwia sagittiformis* (after Briggs and Conway Morris, 1986)
13. Nematoda (especially ascarids)
14. Sipuncula (after Stephen and Edmonds, 1972)
15. Echiura (after Stephen and Edmonds, 1972)

APPENDIX II

Characters and character states used in data matrices. The states for which multiple interpretations are possible in *Tullimonstrum gregarium* are indicated by an "x" following the state.

1. Position of mouth
   0. anteroventral
   1. terminal
2. Stylets
   0. absent
   1. posteriorly curved
   2. straight
   3. anteriorly curved
3. Anterior bifurcated lobes
   0. absent
   1. present
4. Flexibility of proboscis
   0. absent
   1. stiff
   2. flexible
5. Position of digestive tract
   0. dorsal (x)
   1. ventral (x)
6. Shape of digestive tract
   0. straight
   1. coiled
7. Relative body width
   0. consistent width
   1. increasing width posteriorly
8. Body flattening
   0. flattened dorsoventrally
   1. unflattened
   2. flattened laterally
9. Degree of external segmentation
   0. somites undifferentiated (x)
   1. somites differentiated by transverse grooves (x)
10. Degree and orientation of internal segmentation
   0. internal somites undifferentiated (x)
   1. internal somites separated orthogonally (x)
   2. internal somites separated by anteriorly oriented v's
11. Position of centralized ganglion
   0. anterior
   1. medial
12. Position of eyes
   0. dorsal
   1. lateral (x)
   2. absent (x)
13. Ocular attachment
   0. sessile (x)
   1. stalks (x)
   2. absent (x)
14. Hardened anterior setae
   0. absent (x)
   1. present (x)
15. Number of statocysts
   0. zero (x)
   1. one
   2. two (x)
16. Placement of statocysts
   0. absent (x)
   1. internal
   2. external (x)
17. Nature of benthic support structures
   0. absent (x)
   1. paired (x)
18. Paired copulatory structures
   0. absent (x)
   1. flexible
   2. rigid (x)
19. Position of anus
   0. absent
   1. terminal (x)
   2. lateral (x)
20. Nature of posterodorsoventral asymmetrical "fins"
   0. absent
   1. slightly asymmetrical
   2. highly asymmetrical
21. Flexibility of "fins"
   0. absent
   1. flexible
   2. stiff

APPENDIX III

Character state matrix with base states used in all analyses with PAUP (Swofford, 1985). Primitive states indicated by 0, with increasing numerical values indicating increasing derivation; missing data indicated by 9. Variable states for *Tullimonstrum gregarium* indicated by an "x"; refer to Appendix II for the range of values that could be substituted for the "x."

```
PARAM  NOTU=15  NCHAR=21  MISSING=9
OUTWIDTH=80;
DATA (A8, A1, 21I1);
platyhel 000000000000001100000
kalyptor 011100000000001100000
tullimon 1212x010xx1xxxxxxxx11
nemertea 020210000010000000100
priapuli 110100011102200000100
mollusca 111200120010102100121
annelida 101200011102200001100
hemichor 000110010002200000100
chaetogn 131210001100000000200
conodont 111190020299909909122
opabinia 110299121190009919100
amiskwia 100090100092209909100
nematoda 120010010002200000100
sipuncul 100211110112200000200
echiura 001211010102210000100
```

APPENDIX IV

Summary of evidence for particular affinities of *Tullimonstrum gregarium* based on the analysis of 48 data sets using PAUP (Swofford, 1985). Column 1 lists only those OTU's indicated to have some affinity with *Tullimonstrum*. Column 2 lists the number of trees from the analyses that indicated a particular affinity. Column 3 indicates the number of data sets that produced a tree indicating a particular affinity.

TAXON	TREES	DATA SETS
Conodonta	305	29
Mollusca	296	29
Opabinia	231	28
Annelida	217	20
Priapulida	206	16
Sipuncula	100	19
Echiura	37	14
Chaetognatha	18	9
Nemertea	14	6
Unspecified	12	4

REFERENCES

Aldridge, R.J. & Briggs, D.E.G. 1989. A soft body of evidence. *Natural History* **5/B89**, 6-11.

Aldridge, R.J., Briggs, D.E.G., Clarkson, E.N.K. & Smith, M. P. 1986. The affinities of conodonts new evidence from the Carboniferous of Edinburgh, Scotland. *Lethaia* **19**, 279-292.

Barnes, R. 1980. *Invertebrate zoology*, 4th ed. Philadelphia: Saunders College/Holt, Rinehart and Winston, 1089 pp.

Bengtson, S. 1986. Introduction: the problem of the problematica. In *Problematic Fossil Taxa* (eds. A. Hoffman and M. H. Nitecki), pp. 3-11. New York: Oxford University Press.

Briggs, D.E.G. & Conway Morris, S. 1986. Problematica from the Middle Cambrian Burgess Shale of British Columbia. In *Problematic Fossil Taxa* (eds A. Hoffman and M. H. Nitecki). pp. 167-183, New York: Oxford University Press.

Briggs, D.E.G. & Williams, S.H. 1981. The restoration of flattened fossils. *Lethaia* **14**, 157-164.

Briggs, D.E.G., Aldridge, R.J. & Smith, M.P 1987. Conodonts are not aplacophoran mollusks. *Lethaia* **20**, 381-382.

Briggs, D.E.G., Clarkson, E.N.K. & Aldridge, R.J. 1983. The conodont animal. *Lethaia* **16**, 1-14.

Brinkhurst, R.O. & Jamieson, B.J.M. 1971. *Aquatic Oligochaeta of the World*. Edinburgh: Oliver and Boyd.

Canaan, M.R. 1989. The monster of Illinois: paleontology and politics. *Rocks and Minerals* **64**, 36-41.

Conway Morris, S. 1979. The Burgess Shale (Middle Cambrian) fauna. *Annual Review of Ecology and Systematics* **10**, 327-49.

Conway Morris S., Whittington, H.B., Briggs, D.E.G., Hughes, C.P. & Bruton, D.L. 1982. *Atlas of the Burgess Shale*. London: Palaeontological Association.

Foster, M.W. 1979. A reappraisal of *Tullimonstrum gregarium*. In *Mazon Creek Fossils* (ed. M.H. Nitecki) pp. 269-301. New York: Academic Press.

Hennig, W. 1966. *Phylogenetic systematics* (translated by D.D. Davis and R. Zangerl). Urbana: University of Illinois Press.

Hennig, W. 1981. *Insect phylogeny* (translated by A.C. Pont). Chichester: John Wiley & Sons.

Hoffman, A. & Nitecki, M.H. 1986. Foreword: Why and how to do problematica. In *Problematic Fossil Taxa* (eds. A. Hoffman & M.H. Nitecki), pp. v-viii. New York: Oxford University Press.

Hutchinson, G.E. 1930. Restudy of some Burgess Shale fossils. *United States National Museum Proceedings* **78** (11), 1-24.

Johnson, R.G. & Richardson, E.S. Jr. 1969. Pennsylvanian invertebrates of the Mazon Creek area, Illinois: the morphology and affinities of *Tullimonstrum. Fieldiana (Geology)* **12**, 119-49.

Meglitsch, P.A. 1967. *Invertebrate Zoology*. London: Oxford University Press.

Mikulic, D.G., Briggs, D.E.G. & Kluessendof, J. 1985a. A Silurian soft bodied biota. *Science* **228**, 715-7.

Mikulic, D.G., Broggs, D.E.G. & Kluessendorf, J. 1985a. A new exceptionally preserved biota from the Lower Silurian of Wisconsin, USA. *Philosophical Transactions of the Royal Society of London,* **B 311**, 78-85.

Rex, G. 1983. The compression state of preservation of Carboniferous lepidodendrid leaves. *Review of Palaeobotany and Palynology* **39**, 65-85.

Rex, G. & Chaloner, W. 1983. The experimental formation of plant compression fossils. *Palaeontology* **26**, 231-252.

Richardson, E.S., Jr. Wormlike fossil from the Pennsylvanian of Illinois. *Science* 151, 75-76.

Smith, M.P., Briggs, D.E. . & Aldridge, R.J. 1986. A conodont animal from the lower Silurian of Wisconsin, USA, and the apparatus architecture of panderodontid conodonts. In *Palaeobiology of conodonts* (ed. R.J. Aldridge), pp. 91-104. Chichester: Ellis Horwood.

Stephen, A.C. & Edmonds, S.J. 1972. *The Phyla Sipuncula and Echiura*. Trustees of the British Museum (Natural History).

Stephenson, J. 1930. *The Oligochaeta*. Oxford: Clarendon Press.

Sweet, W. 1988. *The Conodonta: morphology, taxonomy, paleoecology, and evolutionary history of a long-extinct animal phylum*. New York: Oxford University Press.

Swofford, D. 1985. *PAUP Phylogenetic Analysis Using Parsimony*, Version 4.2. Champaign: Illinois Natural History Survey.

Tillier, S. & Cuif, J.P. 1986. L'animal conodonte est-il un mollusc Aplacophore? *Comptes Rendus des Séances de l'Academie des Sciences*, **série II, 7**, 627-732.

Urbanek, A. 1986. The enigma of graptolite ancestry: lesson from a phylogenetic debate. In *Problematic Fossil Taxa* (eds. A. Hoffman & M.H. Nitecki). pp. 184-226, New York: Oxford University Press.

Whittington, H.B. 1969. The Burgess Shale: history of research and preservation of fossils. *North American Paleontological Convention Proceedings I*, pp. 1170-1201.

Whittington, H.B. 1975. The enigmatic animal *Opabinia regalis*, Middle Cambrian, Burgess Shale, British Columbia. *Philosophical Transactions of the Royal Society of London* **B 271**, 1-43.

Problematica/Incertae Sedis

Ellis L. Yochelson[1]

Abstract

Identifying, naming, and classifying of organisms are intertwined, but treating these as distinct steps may help clarify their different functions in palaeontology. Some palaeontologists use "problematicum" to indicate uncertainty in biologic placement; others use "incertae sedis". In certain geologic settings, determination that an object is organic, even if it should not be formally named, is important. Almost all problematica turn out to be common fossils that are incomplete or preserved in an atypical manner and most problematica do not warrant Linnean names. In Linnean nomenclature, scientific names are applied, and these names are placed in a hierarchial classification. Incertae sedis is a formal expression of a degree of uncertainty in classification; it may be applied at a variety of taxonomic levels. Neontologists use incertae sedis rarely, mainly in connection with placement of species or genera that do not fit well into a family-level classification; palaeontologists more commonly use this term to express uncertainty at higher taxonomic levels. Many incertae sedis may have no descendants in the living biota. At what level to place enigmatic taxa - which are most abundant in the early part of the record - remains an intellectual problem with no easy answers.

Introduction

The conference organized by symposiarchs Simonetta and Conway Morris seemed to be the ideal place to present this particular essay. For almost a decade, it has been read by the technical critics of at least six journals. Invariably, they have said that part of it is obvious, and part of it is good, but none of it is appropriate for their journal. When personal circumstances prevented my attending, Prof. Simonetta kindly sent me the manuscripts prepared by some of the participants so that I would have the benefit of their views and could add those that were particularly germane to my opinions and prejudices.

It seems appropriate to use a title expressing uncertainty, as at least two different concepts — identification and classification — are being discussed; the issue of naming fossils lies in the area between them. The line between these two concepts is not always clear, and in some respects is comparable to the level of indeterminacy found in expressions such as "she/he", "in/out", or "and/or" that mar present-day writing.

After presumptive fossils are collected, they are subject to three processes that may take place nearly instantaneously or may be spaced out during a longer interval. An object is identified, a biological concept is named in accordance with rules of nomenclature, and the name is placed within the hierarchial structure of a classification. This is more of a formal division of labour than is usually made, but it may shed light to consider each process as a distinct operational level; in my view, the three operations have to be treated in the order listed. The distinction between concepts and objects may now be so obvious that only one conference paper (Simonetta and Della Cave, this volume) mentioned it and then only in a footnote, but it is fundamental. A bad object or a bad description of an object adds greatly to the amount of uncertainty that is inherent in a concept.

A recent publication (Hoffman and Nitecki, 1986; see especially the Foreward and Chapters 1 and 2) has treated several of the issues discussed below. Notwithstanding that, occasionally going over the same ground from a slightly different perspective may be instructive. The term "Problematicum" is defined in the Glossary of Geology, but incertae sedis is not, whereas incertae sedis is defined in the International Code of Zoological Nomenclature and problematicum is not. Discussion of the meanings of these terms with half a dozen colleagues resulted in six different interpretations. Some eminent authorities (Caster, 1957; Brown, 1959) use "problematica" (the plural) as a general term for what I have chosen to call "incertae sedis". Perhaps more palaeontologists use "problematica or problematic taxon" for named fossils of uncertain biologic classification than use "incertae sedis", but I remain mumpsimus.**

Problematica sensu stricto

The Latin term "problematicum" means "in the nature of a problem", but there are many kinds of problems. The minimum criterion for a fossil is that it must be of organic origin. However, terms such as "fossil soil", or "fossil fault" are acceptable profes-

[1]Research Associate, Department of Paleobiology, National Museum of Natural History, Washington, D.C. 20560, U.S.A.

**This is a perfectly good word, meaning in effect to cling to one's opinion in the face of authority, as I cling to my view that there are quite a few extinct classes within the Mollusca (Peel, this volume) and that these classes are not closely related (Yochelson, 1963). (Some definitions of mumpsimus give it as: "A bigoted adherent to exposed but customary error" but that obviously does not apply to me).

Much as I abhor footnotes, occasionally they are useful in brightening a paper. One can employ a writing style of the third person obtuse, but I remain mumpsimus that occasionally a serious subject should be treated in a light-hearted manner. Authority is a dangerous concept and no less so in science than in other fields. Palaeontologists should examine the objects (fossils) on which concepts are based and not be subservient to authority.

After all, the discovery of an unusual fossil in the Ordovician of Cincinnati, Ohio, did lead to the word "serendipity" becoming a household expression in some circles of American palaeontology. It took almost as long to find the word "mumpsimus" as it did to discover that the slash in "and/or" is referred to as a "virgule". In a sense this has little to do with the subject at hand, but in another real sense the issue is about trying to make vague notions more rigorous. This may be impossible, but as a start, I believe that using the correct word is as important in discussion as using the correct taxa in proposing a phylogeny.

sional jargon. Many geologists use "problematicum" to indicate uncertainty whether an object in question is organic or inorganic, but some palaeontologists consulted use the word only for organic items. If one does not know whether an object is a fossil, in my minority usage, it is problematic, though not a problematicum, but for many palaeontologists, a problematicum need not be organic in origin. Hofmann (1971) made a great step forward in clarifying the palaeontology of fossils in Precambrian rocks when he carefully and precisely divided forms described and/or named in the older literature among categories of "fossils", "dubiofossils", and "pseudofossils". This approach can be applied quite usefully to material from younger rocks.

One may be reasonably confident that objects collected by igneous geologists are not problematica; that is, they are inorganic in origin. However, "lava trees" do occur, and, rarely, leaf impressions may be found at the base of ash layers; even the cast of a Rhinoceros has been found in lava. Intrusive igneous rocks may contain peculiar crystallization phenomena, and hand-sized specimens containing large crystals may be mistaken for colonial corals or bryozoans. A thin section of the matrix occasionally might help in answering the fundamental question of biologic origin.

The chance of a problematicum, that is a fossil, being found in metamorphic rocks is low, but is many orders of magnitude higher than it occurring in igneous rocks. Palaeontologists should make an effort to have more contact with metamorphic geologists. It is better to look at 100 pseudofossils than to miss one fossil from a metamorphic terrane, for the scientific payoff in a correct identification may be high. Years ago I wrote to a metamorphic geologist that holes in a rock he had submitted were productoid brachiopod cross sections, and could only be dated as mid- to late Palaeozoic. His reply was three pages of effusive outpouring, for he had not even been certain the unit was originally sedimentary, let alone its general age.

Fossils have been collected from both exceedingly badly deformed and quite high-grade metamorphic rocks; they are not necessarily destroyed by metamorphic processes. The fossils are there to be found, and more effort should go into searching for them. Too few people understand such basics as looking for places where cleavage and bedding are parallel. It is up to the palaeontologist to offer his/her services and to educate the metamorphically oriented field geologist to search for fossils.

Pseudofossils in metamorphic rocks may cause problems if misidentified, but there is another embarrassing source of palaeontologic difficulty. Man has left a fossil record of his bones and footprints (Brown, 1947), but he also has left much other material, and ancient artifacts may be "fossils of the mind" (K. E. Caster, written commun., 1981). No one should ever have any difficulty in distinguishing a recent manmade object from a fossil, except that it has already happened at least once; *Calcophysoides balli* Berry upon further study turned out to be part of a "Cape Cod" firelighter (Berry, 1937). Perhaps it is poor form to mention such a mistake, but this might happen again, and if nothing else, the instance is a fine example of an author admitting error when he was wrong.

One may argue that since man (an organism) manufactures machines that drill the holes seen in the walls of many road cuts, these holes could be future problematica, but there is no need for such an exercise in tortured logic. Even shatter cones produced by blasting in quarries that simulate colonial organisms should be ignored. None of these traces of man's construction are comparable with tracks and burrows produced directly by an organism.

In addition to man, other organisms may confuse the issue of fossil or pseudofossil. Present-day plant roots are found in the cracks of outcrops. The wing of a modern insect in travertine looks like an authentic fossil but lacks the age component, though no authority will say how old an organic object must be to be judged a fossil. When insects burrow into ancient rocks (Cloud, Gustafson, and Watson, 1980) or make bee-nest impressions on rock surfaces (Sando, 1972), they leave traces that are clearly of organic origin, but are not fossils. The epitome of such complication seems to be a reported fossil ear of maize, which turned out to be an Indian-made replica of an ear of maize (Brown, 1934). As Brown wrote, "Be not deceived by external appearances".

Problematica in sedimentary rocks

Unusual objects do occur in rocks. According to legend, possibly true, James Hall heard of a farmer in northern New York who was charging visitors to see "the trail of the serpent from the Garden of Eden". Rudolf Ruedemann was sent to investigate, and after observing the mark at the end of the trail where Adam trod on the snake, he made disparaging remarks about the size of Adam's foot. Hall had to leave Albany and spend time soothing the farmer before the latter would part with these specimens of *Climactichnites*. To this day, the Late Cambrian organism which made the trace given that name remains unknown.

Not everything in a sedimentary rock is as it seems at first glance. Some shales contain concretions of many shapes; others contain compression marks and nodules. Chemical gardens that grow in geodes (Brown, 1957) are called dendrites, because they bear such a striking resemblance to plant fossils. In some limestones, secondary silicification forms rinds in the shape of brachiopods; in others, gas-bubble tracks look like small worm tubes. Boxwork weathering resembles tabulate corals; Liesegang rings in sandstone look like jellyfish. As the King of Siam said, "et cetera, et cetera".

Mud cracks break up and may be reincorporated in sediments. When thin mud films dry and roll up in ripple troughs, strange structures result (Hofmann, 1967); such features have been given a Linnean name at least three times. A depression in a hardground could be a burrow, but it might not. The literature contains vigorous discussions on whether certain marks are organic or whether they were made by the scratching of stones and other inorganic tools; in Precambrian rocks this issue is particularly critical (Milton, 1966).

Today, we are more sophisticated in our interpretations than was the New York farmer - perhaps. Seldom does an experienced palaeontologist confuse pseudofossils with fossils, especially when they are in a field setting rather than in a laboratory, but sooner or later everyone has a bad day and makes an error. Most fossils in sedimentary rocks are easy to recognize for what they are. Perhaps it is the way they break, or their colour, or the surface texture. For those who have had even a minimum experience with sedimentary rocks, fossils register on the mind almost instantaneously.

If there is a pause in the recognition process, it may well be that the brain synapses are flashing a caution light. At such times, a second look is in order; a hand lens does wonders in providing additional data. Wetting a specimen sometimes brings out a wealth of detail. The eye is good in restoring the shape if more than a third of a specimen is present, but it does less well in envisioning, for example, a shell layer that is badly worn or broken off. If one has doubts, the best thing is to look for a second or third specimen on the outcrop, although some pseudofossil objects do occur in profusion.

After one has determined on the outcrop that a problematicum is identifiable to class or phylum, usually the best thing to do is discard it. Poorly preserved fossils stored for years are still poor fossils. Computer people long ago recognized "garbage in, garbage out". The basic principle is important; leave the organic junk on the outcrop. In exceptional circumstances, poor fossils are the only geologic documentation available at a locality and in such a circumstance they ought to be preserved. Those who curate collections should recognize when to retain such documentation and when to discard junk. However, good fossils, even if they are common forms, always should be collected and should be retained by curators; fossils are a nonrenewable resource.

Very, very rarely, well-preserved fossils are found which are undeniably organic, but do not fit into an obvious systematic category; they should be cherished. Unusual fossils may occur in rocks of all ages, but are more commonly observed in older fossiliferous rocks, in part because most palaeontologists have at least a slight knowledge of the living biota but do not recognize common Palaeozoic fossils in uncommon preservation. Another obvious point is that older fossils are different. Sepkoski (1981) recognized three Phanerozoic "faunas"; there is a fourth fundamentally different assemblage in the late Precambrian, and there may be more in the Cambrian. These earlier fossils are quite different from the sort of organisms to which biologists are accustomed.

The work of organisms, that is, tracks, trails, burrows, and such, poses more problems in primary identification than do body fossils, because a variety of sedimentary structures mimic organic activity. It is important to ask the help of sedimentologists and sedimentary petrologists before going too deeply into presumed biology. Unusual ripple sets indicate one thing, but if they are interpreted incorrectly as tadpole nests, this might lead us to infer something different concerning the environment of deposition. Caster (1957) documented some of the confusion between vertebrate and invertebrate trackways. Trace fossils are difficult and the novice should not rush into print with any dramatic new "discovery".

Turbidites form a rock facies that has produced both organic and nonorganic structures. Palaeontologists who have had experience only in shallow-water environments ought to be careful about making pronouncements when examining them. Likewise, few people have had experience with freshwater and terrestrial deposits, which do not follow the normal rules for marine shelf deposits, and the biota which might be preserved is not in our normal experience. The similarity between some worm tubes and caddis fly larval tubes comes immediately to mind.

Another caution is that those who investigate Precambrian sediments work very hard and, after a great deal of looking, understandably they think that they deserve to find fossils. The thought may be the mother to the deed and non-fossils may be called fossils. A respectable record now exists of metazoan and simpler life in the Precambrian. This record has been purged of many dubious identifications and downright errors, and care should be taken not to repeat the mistakes of the past.

These two sources of difficulty can be combined. Walcott compared stromatolites to lake balls and assumed that most of the Belt Series was deposited in fresh-water. From that Walcott was led to the view that enigmatic scraps he named *Beltina* were fragments of an arthropod in a tongue of marine rock (Yochelson, 1979a). *Beltina* is not an arthropod, but at first glance, and even on closer inspection, specimens do resemble arthropod cuticle.

"How to do" problematica

In America, persons who write appropriate "how to do it" books gain financial reward and public admiration. It would be nice to write such a work concerning fossils, but how to do it escapes me. The best I am able to produce is a few platitudes, to be considered after one is sure that an authentic organic problematicum is at hand.

First, if material in the problematic class is poorly preserved, one had better be cautious in pushing a determination too far; identify in haste, and repent at leisure. Second, do not ever identify a problematicum from photographs; this is a very hard-earned piece of advice. Third, if possible, look at the problematicum itself rather than a cast; colour and texture of the object may reveal additional data. Fourth, take time for preparation of the sample; making an impression of an external mould, removing a bit of matrix, or sawing could solve a mystery. Fifth, use your imagination and your ingenuity as to the unexpected forms that might occur in a particular ecological setting; very rarely centipedes are preserved with plants, and more commonly plant fragments do float into marine deposits. Sixth, let the problematica accumulate over the years in a collection; occasionally, viewing several different specimens will lead to insight unobtainable from a single individual, no matter how exhaustively it is examined. Seventh, if a person who has collected a problematic fossil is unwilling or unable to provide details on geographic location and/or age, do not worry too much over the fossil; any geologist serious about potential fossils ought to give these clues freely. Eighth, consider the prospect that one has only a fragment of the hard part; tubercles from the claw of a lobster immediately spring to mind as one example. Ninth, likewise consider that only one hard part of several may have been preserved; many people recognize isolated teeth, but they discard single polyplacophoran plates, judging them to be indeterminate fragments. Tenth, microfossil-sized problematica pose far greater difficulties and uncertainties than do megafossil-sized objects; I would be skeptical of microfossils in strongly metamorphic rocks, though they do occur.

Ever since d'Orbigny identified foraminifers as cephalopods, palaeontology has faced the issue of scale in identifications. Too few people are cautious about staying within the size range normally associated with a particular type of animal or plant. For example, giant Early Cambrian cephalopods and minute Early Cambrian gastropods are equally suspect because they exceed the commonly known size limits of younger fossils. Microfossils and particularly ultramicrofossils require even more care than megafossils. The widespread windborne soot balls emitted from modern coal-burning powerplants are the sort of objects that may be confused with true microfossils. Spores and resting stages of minute organic objects are diverse enough to convince me that one should not stray from the size range of fossils one knows best. Even specialists may disagree among themselves, and the recent literature is full of arguments about Precambrian microfossils or microfossil-sized objects.

Ager (1970) noted that in general the best geologist is the one who has seen the most rocks, another fine basic principle. Although no palaeontologist has seen all kinds of fossils from all age rocks, one should at least pass a specimen around among onès colleagues for ideas on possible identification. One person's problematicum may be another's favourite fossil. Prof. Wyatt Durham graphically demonstrated to me that some people attend conventions just to socialize, but that a specialist can be identified at scientific meetings because he/she carries a hand lens to examine any specimens brought by colleagues. Remember, however, that specialists tend to have tunnel vision, and if one is told by a coral specialist that the object in question might be a bryozoan, it is a good idea to consult a bryozoan specialist before publishing that opinion.

The best rule of thumb is to use common sense in trying to identify a problematicum. The chances are nearly overwhelming that a fossil is the remains of a moderately common organism. Cretaceous shelf fungus can turn out to be reworked and worn Paleozoic syringoporoid corals (Brown, 1938). Dubious seed capsules can turn out to be fish-egg capsules (Brown, 1946). If misapplied, a little knowledge is a dangerous thing. A prime example of this is given in the report by Okamura (1980) on "fossil dragons", "minihumans", and other creatures observed in thin-sections.

Perhaps another caution is to be careful about venturing into new techniques to rapidly. Fauchald, Stürmer & Yochelson (1988) contains an error, because Yochelson did not understand enough of radiographs and of Devonian shale faunas. After Professor Stürmer died, I was left with some photographs of a peculiar object and felt responsible to have them published. I asked Kristian Fauchald, a zoologist, not a palaeontologist, what kind of "worm" was shown. He noted the appearance of an apparent plate with holes and suggested from his knowledge that it was a tunicate sieve-plate. It never occurred to me that this was a *Phacops* compound eye, until after publication when a German colleague wrote me. The error was mine entirely, for I did not consider alternative identifications and thus I led the zoologist down the wrong path. Once the notion of the world's first fossil tunicate was lodged in my brain, all other thoughts vanished; the only saving grace was that no generic name was applied to this dramatic discovery. Sadly for me, it is not the only time I have been mislead in interpretation of radiographs.

A moral I can draw is that when one is sure a mistake has been made, it is better to correct it than have a colleague point this out (Yochelson, 1989); (on the other hand, I study Gastropoda, and one learns from them that the first step in a movement from one place to another to to stick the neck out). Another moral is to be careful about being carried away by a new idea. There is a story about the zoologist Patten who argued over lunch with colleagues that the evidence for derivation of the vertebrates from any invertebrate group was weak. To prove his point that night he compared the features of *Limulus* with vertebrates. However, he then convinced himself that this was possible and wrote a book on the subject. I liked the idea of helping to describe the first fossil tunicate and never gave a second thought.

Another common-sense aspect of this issue is knowing at what level to cease the identification. To state that a bizarre object is a fragment of a trilobite eye rather than a bryozon fragment is one thing, but to say that it is *Phacops rana* Green is quite another matter. Identification of fossils seems to have become a bit more conservative or reasonable during the last part of this century. Likewise, in the last few decades, many of the formally named taxa have been better described and illustrated and based on more nearly complete material than in the past.

Nomina dubia/Problematica

Lately, the expression nomina dubia has appeared in the literature. It is in the ICZN glossary but is not used in the text of the Code; in my view it is an unfortunate addition, but I doubt that it will be removed from the next edition. No matter how it may sound, a nomen dubium is not a dubious name. It is a name in perfectly good standing under the International Code of Zoological Nomenclature. As I understand the term, a nomen dubium is the name of a poorly known or poorly described taxon. This may be because of inadequate descriptive work, or it may be because the taxon was based on poor material, or both. The International Code of Botanical Nomenclature does not use this term; this is clear thinking on the part of the framers of that code.

Whether a taxon is a nomen dubium is a subjective manner based on biological knowledge. Some palaeontologists of past generations thought that it was necessary to identify every fossil to the species level, and if no available name fit, to name and describe it. The result was nomenclaturally valid taxa that are not well founded biologically, that is, nomina dubia. To repeat my earlier opinion, most problematica are poor specimens and should be discarded, not named; the naming of a poor specimen results in a nomen dubium.

There is a difference between nomen dubium, which is a biological concept, and nomen nudum, a taxonomic concept. The latter, as everyone should know, and avoid, is a formal taxonomic name that does not fulfill all the requirements for nomenclatural availability. Unfortunately, interest in the formal aspects of zoological nomenclature is waning. Because there are many more new taxa yet to name, entomologists and palaeontologists are those most concerned with ICZN and its legalistic interpretations. More attention to the mechanics of naming is a strong foundation to the issue of classification.

An intermediate level of usage exists for those who believe that formal names should be applied to fossils, but who are forced to deal with less than satisfactory material. Open nomenclature (Matthews, 1973) is a useful practice. Thus, *Olenellus* species indeterminate, indicates a poor specimen probably belonging to the genus. If we were rigorous in our identifications, the most common specific name used for fossils probably would be *sp. indet.* Fossil material that can be identified to the generic level with even a modicum of certainty is out of the problematicum range.

I see a profound distinction between informal names for fossils and Linnean nomenclature. Never forget that a trilobite is an object and that *Olenellus* is a concept; those two terms are philosophically quite different. Another obvious point, often overlooked, is that nomenclature is concerned with names, but biology is concerned with organisms. One may be too much of a purist in trying to separate the two, but in my view, problematica do not have Linnean names. Once a formal name is given, a problematicum becomes a concept.

Organ/form genus and/or parataxa

From time to time, systematists who are passionately devoted to a particular group of organisms that does not fit readily into an established classification or that seems too fragmentary to assign under prevailing biologic schemes decide that they must have their own system of classification. For years, palaeobotanists recognized that leaves or pieces of bark were insufficient for a biologically meaningful classification, so they conceived of the terms "organ genus and form genus" for names based on plant species that were particularly fragmentary. Because of slowly increasing knowledge and the occasional lucky find, some problems of classification have been resolved and others are known to be resolvable. The terms "organ genus" and "form genus" will probably be removed from the Botanical Code when no longer needed.

A few palaeontologists have tried a separate system for crinoid columnals (Croneis, 1938) and a separate code for ichnotaxa (Sarjeant & Kennedy, 1973), and some palaeontologists believe that classification for fossils should be different from that for living organisms (Weller, 1949). Perhaps the most interesting example of an attempt to set up a separate classification has concerned "parataxa", a term introduced primarily to deal with conodonts. In the late 1950's, parataxa were nearly established because the conodont workers insisted the concepts of genera and species that they used were artificial. Like so many bad ideas, the concept of parataxa, once defeated, did not just disappear. It lay moribund for two decades but has recently risen again (Melville, 1981).

Now, by a turn of fate, it was the conodont workers who lead the fight against parataxa. In 20 years, there has been a change from the notion that conodonts are variously shaped "things" occurring in rocks, to the notion that conodonts are extinct organisms apparently represented by only a limited part of the anatomy of the living organism. The placement of conodonts in a phylum still provokes argument, but at lower taxonomic levels, a respectable and seemingly biologically sound classification has been constructed. I have faith that those who work on other groups of tiny objects - hystricosphaerids, titinnids, and others yet to be found - will take a scientific approach. Real palaeontologists do not have a shoe clerk approach to classification in which each shoe box must fit into a preordained slot.

The moral here is that progress is made in classification if a palaeontologist works in the proper scientific spirit. We palaeontologists know that no matter how much data is available from living organisms no neontologist ever has all the data to make an ultimate classification. Accordingly, no palaeontologist should ever feel inhibited for not having quite as much of the same kind of data as the neontologist. However, to ignore biology and classify outside its systems is not good. An artificial classification may provide some better understanding of facies or sedimentation rate or other aspects of geology. All these factors can be applied to a palaeontologic explanation, but no matter how it is rationalized, an artificial classification is not palaeontology and it can stifle the accumulation of inferences and new data needed for a better biological understanding of fossils.

Loose pieces, if one may use that phrase, are suddenly becoming of considerable significance to paleontologists concerned with the early part of the animal record. *Wiwaxia* from the Burgess Shale has now become the model for reinterpretation of a number of the "small shelly fossils"; recall that its type is a misidentified species of the hyolith taxon *Orthotheca*. An even greater challenge is presented by *Microdictyon* (Chen & Erdtmann, this volume), wherein the hard parts form a small and scattered part of the whole fossil. I believe that use of the term "shelly" has clouded our thinking, and has been subjectively equated with the term "shell". Had the concept of parataxa been used, the concept of multi-hardparted animals early in the record would have been far more difficult to accept.

Although it may not sound like much of a point to argue about today, reference to a generic name that is based ultimately on a trace as a "collective group name" rather than as an "ichnogenus" is the difference between a biologic and a non-biologic approach to its interpretation. A trace fossil is one aspect of a former living organism, and fortunately today some specialists in that group or groups are more biologically inclined than in past years.

Incertae sedis sensu stricto

It is sometimes useful to generalize before trying to define. The system of Linnaeus provided biology with two distinct concepts, first, that of binominal nomenclature and, second, that of a hierarchical arrangement in classification. Names in Linnean nomenclature are ranked within a hierarchy of decreasing levels -- kingdom, phylum, class, order, family, and genus. If we wish to speak of any of these various categories simply as categories regardless of level, we use the word "taxon". Taxon was invented by botanists and has been accepted by zoologists as a helpful word.

Combining the Latin *incertus* (uncertain, unsettled) with *sedeo* (seat, dwelling place) gives *incertae sedis*, of uncertain position. If one does not know where to place a Linnean name within the hierarchy, incertae sedis is used for the level of classification that is uncertain. Like "taxon", it is an elastic multi-level term. For example, on one end of the scale, a species of dinoflagellate may be incertae sedis at the Kingdom level, if we choose to argue the question whether these organisms are either animals or plants. At the other end of the scale, one genus of flies, *Tipula* Linnaeus, is divided into several subgenera; a few species of the genus are readily identified but cannot be placed in any of the subgenera, so they are incertae sedis at that level.

Linnean names express concepts. A query can be used to express a limited degree of uncertainty as to placement of a taxon within a hierarchy. Commonly, a query is used at a generic or specific level, but occasionally a worker may question placement of a genus within a family-level taxon. It is my understanding that the use of incertae sedis does not apply to this kind of uncertainty. Occasionally when authors deal with systematically low-level taxa that have been poorly defined, it is difficult to draw a line between a formal hierarchical use of incertae sedis and application of a question mark. However, one might generalize by suggesting that commonly incertae sedis is used mainly when one is constructing or revising a classification at or above the family level.

Classification of biological taxa is a tricky business, for a neontologist tends to look through one end of the telescope and a palaeontologist the other; accordingly, they approach incertae sedis from different viewpoints. Although there have been some changes recently, neontologists have a fairly rigid superstructure at the higher classificatory levels. New genera are still being described, but the neopilinid mollusk and similar finds in the biota cause a great deal of excitement, precisely because they are novel and do not fit well with common living forms. In contrast, the palaeontologist has a higher level classification that is anchored in quicksand, and as a matter of course, he/she should expect to recognize higher level groups unknown to or unrecognized by earlier workers (Yochelson, 1971). To use geologic analogy, classification of the living biota is nearly steadystate uniformitarianism, whereas classification of fossils tends toward more abrupt changes through time -- catastrophism, as it were. Discovery of deepsea vent and oil seep fauna is as revolutionary for neontology as discovery of the Ediacara fauna was for palaeontology.

The use and occurrence of incertae sedis

Because palaeontologists and neontologists have different backgrounds and somewhat different kinds of material to study, incertae sedis may be used in at least two slightly different ways. Some workers may apply it at a relatively low level; by this, I mean familial level or lower. For example, one might describe a new productoid brachiopod genus that does not fit into any of the currently accepted subfamilies or families. For a variety of reasons, the worker may desire not to name a new family-level taxon. He/she might then place the genus in a hierarchy of Brachiopoda, Productoidea, Productacea, Incertae Sedis. On the other hand, incertae sedis is used by others to express uncertainty above the family level. To repeat, my impression is that most palaeontologists who use it think of this term in the latter sense, for taxa that do not clearly fit into recognized high-level taxa, that is, orders, classes, or phyla.

Too many people - especially textbook writers - assume that classification is static. They confuse the Linnean hierarchy with the periodic chart of elements. For practical purposes in the world outside the high-energy physics laboratory, the periodic chart never changes. In contrast, the Linnean hierarchy is in continuous movement with each systematic paper published. This is not easy to see for in the great mass of taxa and the complexity of biology, the average paper makes only a tiny change. Systematics is different from many other sciences in that it is accumulative.

The concepts of the biologist and classification of the neontological biota do tend to preoccupy and bias the palaeobiological systematist, though not nearly so much as they did two generations ago. By using incertae sedis, the palaeobiologist performs a valuable service by waving a red/white flag to attract the attention of the neontologist. They/we do not know it all.

Because classification is a fluid matrix, occasionally strange things happen to *incertae sedis*. A fossil taxon may be well classified in the older classification, but subsequently moved to an area of uncertainty. To give only one example, the late Palaeozoic *Chaetetes* was a coral in fairly good standing, although it was not directly in the main line of coral evolution. With the rediscovery of living sclerosponges, less than two decades ago, the concept of this major group has been revised and extended by some workers to include *Chaetetes*. Sclerospongia itself may be incertae sedis, and *Chaetetes* may fall within Coelenterata or may fall within Sclerospongia, which in turn may be in the Porifera, or may not. There is even a view that the plant kingdom could be more appropriate for *Chaetetes* than the animal. None of this higher level uncertainty affects the identification of species or even genera of chaetetids, and it has no effect on stratigraphic palaeontology.

Thus, there is a distinction between identifying and classifying. It may be foolish to write such an obvious statement, but occasionally the obvious is overlooked. Not all change in classification is progress, but without change there is no progress. The computer people also recognize "nothing in, nothing out".

Not all *incertae sedis* groups have the same attributes. Some organisms such as titinnids, conodonts, or conulariids are locally abundant to moderately common. Various major groups continued through long periods of time but never reached the Holocene. Their position in the hierarchy changes irregularly, but one has the sense that with accumulation of knowledge on distribution and the rare lucky morphologic find, these groups will eventually be placed in a hierarchy based exclusively on fossils.

Other *incertae sedis* are bunched at those rare windows to the past, which, because of unusual preservation, give us an insight into past life that the normal rocks do not. The Burgess Shale, Rhynie Chert, and Hunsrück Slate come to mind, but such windows are not confined to the early or middle Palaeozoic, as Mazon Creek and Solenhofen attest. Lagerstätten is a new word in palaeontology (Robison; Chen and Ertmann, this volume), but a most important one. In these occurrences, commonly, though not always, specimens are rare but well preserved. Because they are often deposited in unusual environments, the fossils present do not match our normal marine hard-part fossils. Whittington (1981) has nicely contrasted these two different modes of *incertae sedis* occurrences.

One of the most interesting issues in classification of fossils is the occurrence of a fossil in geologic time (Strathmann, this volume). This point has colored our thinking on classification. *Tullimonstrum* occurs at an instant in time and seemingly in our present state of knowledge at only one place. It is difficult to place and could be so distinct as to warrant placement in its own phylum. In general, we are now willing to accept the notion of novel experiments early in the record and accord them high systematic rank, but we are more reluctant to accord the same consideration to younger fossils.

How to deal with incertae sedis

Just as I could not give advice on how to deal with *Problematica*, I find it virtually impossible to deal with the question of how to remove taxa from *incertae sedis*. About all I can contribute is the old supposedly humerous story about the tourist asking the native of New Haven, Connecticut, how to get to Yale University and receiving the reply, "Study". One should also add patience and persistence to study as the way to approach the uncertain. If a palaeontologist can convince his/her peers as to the merits of a particular classification, one can move some taxa out of *incertae sedis*. As an example to consider of whether it is better to move a genus into a phylum and class, regardless of its quite atypical morphology, or leave it as *incertae sedis*, I give you *Janospira* (Runnegar,1977; Yochelson, 1977a).

As a matter of fact, many taxa classified as incertae sedis are to a large extent based on fossils that are out of the mainstream of interest of most paleontologists. Many palaeontologists study trilobites, but few study *Caryocaris*. This fossil is a particularly good example, for Rolfe (1969) sorted the various taxa attributed to phyllocarid crustaceans into those that were well understood, those that were *incertae sedis* on the subordinal level, those that might or might not be crustaceans, and, finally, *incertae sedis* that Rolfe was satisfied did not fall within the Arthropoda, but that were not otherwise assigned by him.

Occasionally, a remarkable specimen will clarify classification dramatically. One example is *Conchopeltis*, a redescription of which (Knight, 1937) suggested that the genus was allied to the conulariids rather than to the Mollusca. Discovery of a specimen with marginal tentacles definitely removed *Conchopeltis* from the Mollusca (Harrington & Moore, 1956), though its relationship to conulariids was considered suspect. Oliver (1984) removed the genus from the conulariids to the Cnidaria, where its position within that phylum is still not firmly established.

As the tentaculate specimen discussed above shows, we should always keep collecting, for who knows what the next outcrop will yield. It is for this process that the word "serendipity" was resurrected from oblivion by Kenneth Caster; remember, however, not to look blindly, for fortune favours the prepared mind, and in some regards that is the essence of serendipity in dealing with fossils. Although Caster (1942, p. 61) remarked, "long scrutiny of problematic objects has been known to engender hallucination...", he did not really mean it. As keen a palaeontologist as he is, I think he meant to observe carefully and to avoid mind set.

The process of identification seems to work in two different ways. In the first, one sees that an object is a calcareous prism from a thick *Inoceramus* shell, a piece of a *Receptaculites*, or a similar unexpected organic object. In the other, one decides that because this object is not a brachiopod spine or an elongate foraminifer, it could be a worm tube. The positive and negative approaches to identification are not at all the same. They should not be considered to have the same degree of significance, but both processes help arrive at an identification.

Occasionally, a flash of insight will reveal a new biologic arrangement of taxa, or a new approach to interpretation (Bengtson & Missarzhevsky, 1981), but more often than not, slow painful study eventually resolves *incertae sedis*. The isolated genera eventually come together to form families, and these link to make higher groups. Consider the Archaeocyatha. One genus was named more than a century ago. For three-quarters of a century, only one or two workers per generation devoted any attention to the archaeocyathids. Although a phylum name was proposed for this group by Okulitch, it was largely ignored. Then suddenly in the late 1940s and 1950s many persons began to work on Lower Cambrian rocks and recognized the need to study Archaeocyatha in detail. Major uncertainties still exist in some of the lower levels of classification, but archaeocyathids are not *incertae sedis*, for many workers recognize them as an extinct phylum. Because classification is fluid, now there has been a swing back toward placing them within the sponges; currently there is an argument as to their position at the Kingdom level.

I hope I may be forgiven a personal example. To understand the Mollusca, it is necessary to remove non-Mollusca from consideration. One Early Cambrian fossil had been a problem for decades. After sufficient data had accumulated, I took the extreme step of placing *Salterella* in a phylum containing only this single genus (Yochelson, 1977b). Thus, *Salterella* is no longer *incertae sedis* in my view but belongs in the phylum Agmata. Let me add immediately that I would certainly not suggest that each *incertae sedis* taxon necessarily deserves placement in a phylum all by itself.

Of course, classification is subjective, and therefore one does not deal with truth/error, but rather with the quite different issue of acceptance/ rejection. I cannot tell you whether my proposal of phylum Agmata is accepted; ask the question again in a decade. The many factors that lead to acceptance or rejection of a particular classification of fossils proposed by a specialist are murky waters indeed. It is a curious point that any particular classification is the work of a specialist, but for it to be generally accepted depends ultimately on whether writers of textbooks, who of necessity are generalists, read, understand, and accept the conclusions presented in a particular paper among the piles of literature turned out each year.

Discussion

To return to the conulariids, removal of *Conchopeltis* has simplified this group to some extent, though to be honest that genus was never very close morphologically to Conularia. In this volume, two specialists (Babcock; Van Iten) present two different views as to the placement of conulariids. In these views, either it is a phylum which has become extinct or it falls somewhere within the scyphozoan Cnidaria. Both may be wrong, but clearly both cannot be right. If a political compromise were appropriate, one might suggest that a new class within an established phylum is a solution. Classification is not so much a matter of compromise, as it is of evidence. Both writers present new data and new interpretations. In such a situation, the wisest course is to remain silent, but one does not always choose this course. In my judgement, Babcock has made the stronger case.

Another interesting problem is presented by *Tullimonstrum* (Beall; Schram, this volume). They have both used cladistic analysis and have arrived at different views. If I would draw a moral, it is that while cladistics is a useful tool, it is simply one more tool to use in reaching a subjective decision. I began as a palaeontologist just before statistics were to be the answer to problems in classification. Later, I was advised that numerical taxonomy was the answer to difficult problems. On the basis of experience with these two methods of resolution, I doubt that cladistics will provide the the way to an ultimate solution, yet I admit it is a helpful approach.

Likewise, I do not understand much of molecular biology and therefore fear it, but it too can provide some interesting data (Christen *et al.*, this volume). It just seems to me that if 500 MY of evolution has modified the outsides of animals, it may have also affected their insides. Further, the structure of DNA in *Tullimonstrum* is not known.

Conulariids and *Tullimonstrum* are significant because they do not occur in the early part of the animal record, but are in younger rocks from which the fossils we have used to develop many of our ideas on classification occur. My choice would be to place *Tullimonstrum* in its own phylum. My reason for this is the same as for placing *Salterella* in its own phylum, and the conulariids in their own phylum. The preserved morphology of the organism appears so fundamentally different from any other known morphology that it goes nowhere else in the animal kingdom. All are different at the highest level of classification within a kingdom. The concept of *Tullimonstrum* is also well enough known, that one need no longer have to use *incertae sedis at the phylum level.*

I have stated before that I judge there to be a basic difference between distinction and diversity of fossil taxa. Distinctiveness is not based on either abundance or distribution in time. Diversity is a measure of success, if that is the appropriate word, as indicated by proliferation of lower level taxa within a high-level taxon. To assume that a form must have a long evolutionary history or must be diverse before it can be considered an order, class, or phylum seems to beg the issue of placement. To repeat my prejudice, if the fossil is fundamentally different from other forms, it deserves a high rank, be it rare or abundant, and regardless of its occurrence or geologic range.

This brings up another issue, that of the complexity of what is preserved. Some organisms have more preservable parts than others. The chitinous-like multiparted exoskeleton and jointed appendages provide a wealth of morphologic details. It is precisely because there is a wealth of detail that so many of the symposium papers are concerned with the issue of Arthropoda and fossils which look like arthropods. The emerging discovery that the unique Burgess Shale is not unique in time or space should encourage the establishment of high-level systematic categories for fossils which look like Arthropoda but may not be.

It is futile for me to evaluate or comment on the many contributions, herein, to the classification of these organisms. Besides, I have no backround in studying that form, or bauplan, of animal, and any opinion I might venture therefore should carry little weight. It is at least a little comfort to the palaeontologist to know that even within a group as much studied as the living Insecta, there are problems of classification at a high level (Dallai, this volume).

Fortunately for me, I chose to study a group which does not have as many parts preserved as fossils as do the Arthropoda *sensu lato*. Nevertheless, the issue of complexity does remain. If one deals only with the exterior, my impression is that fossil Gastropoda are more complex than fossil Cephalopoda. When information on septation is added, the situation is changed. The basis for recognition of genera is different in the two classes and comparision of rates of evolution is difficult. Notwithstanding that, there is at least one example in the Palaeozoic of species of gastropods seemingly evolving more rapidly than cephalopods (Gordon & Yochelson, 1987).

If Mollusca have few parts and therefore seem to be superficially simple in their evolution compared to the Arthropoda, the early "jellyfish" pose an even greater problem. What can be preserved and how it is preserved (Bruton, this volume) return in part to my starting point on trying to distinguish organic from inorganic. At the same time, terminology in describing enigmatic foms has to approached carefully. Use of the term "valves" for *Velumbrella* (Dzik, this volume) may be as misleading as "shelly".

It is difficult to say whether model building, which necessarily contains a large amount of intuition, is more popular than cladistic analysis, but certainly it has had an impact on our opinion in classification (Bergström; Valentine, this volume). A model is helpful, but it is not a fact. So far as it is known, cross-lamellar shell structure is unique to living Mollusca and unique to the fossils traditionally placed within the Mollusca. Hyolitha have been shown to have cross-lamellar structure. On the other hand, according to the model of some specialists, the early mollusk had a dorsal integument. It is obvious that the tube-like shell of a hyolith cannot be dorsal. Because the fossil does not fit the model, the Hyolitha have been raised to phylum level. There may be other reasons for according phylum status, but this approach does not appeal to me. A model is supposed to assist our thinking, not control it.

By their very nature, major changes in classification ought to come about gradually. After many years, the phylum Molluscoidea was eventually discarded in favour of the phyla Brachiopoda and Bryozoa, and no one has ever gone back to the earlier view for decades. The concensus view was that it was a step forward. On the other hand, the proposed molluscan Class Eopteropoda Termier & Termier appeared briefly and disappeared rapidly, perhaps because other palaeontologists could readily distribute its constituents through several different phyla.

As a point of view of general philosophy, I best understand systematic groups that have reasonably compact limits, and, within my specialty, I can comprehend family-level taxa more or less readily. It is less easy to group families into orders, but even they can be comprehended if one plays close attention. The overall classification makes better sense if the lower level building blocks are nice and firm. Irregular blocks or peculiar taxa that do not show obvious similarities should be left to one side rather than put into the foundation. This is why I welcome many classes of mollusks containing only extinct forms and I particular welcome the removal from the Monoplacophora of genera to form a new class (Peel, this volume).

Some palaeontologists argue that every taxon has to go somewhere. That may be so, but it should not always be done immediately, as witness *Janospira*. If concepts of morphology are too broadly interpreted, the result is large, unwieldy, biologically artificial categories. My example is the Class Pteropoda of Miller (1889). Though Miller clearly stated that it was artificial, the generations who followed him used the class heading uncritically for genera that would have been better off as *incertae sedis* (Yochelson, 1979).

On rare occasions it is appropriate to repeat the obvious. Beware of homeomorphy. To give an example, if we say that small coiled shells are gastropods without a more critical inquiry into what the concept of Gastropoda implies, perhaps we might confuse coiled worm tubes with them (Yochelson, 1975). The farther back in time, the less the biota resembles the present-day biota; that is evolution. I think we are making a mistake by placing some early fossils into generally accepted higher level taxa.

If we study hard, keep an open mind, find the right specimens, and make some inspired guesses, eventually each *incertae sedis* may find a satisfactory resting place in the biological hierarchy, even if it turns out to be a systematic niche unknown in the present-day biota. Until then, it is a wise idea to use this term to attract attention where attention is needed.

Evolution and revolution are a function of the time scale one uses. George Gaylord Simpson long ago pointed out that the rates of evolution vary among groups and vary through time. True enough. He also said there were no extinct phyla, and that seems to be false. A new philosophy is sweeping through the classification of fossils. Macroevolution is a serious area of investigation. Palaeontologists having such disparate backgrounds as Olson (1981) and Jaanusson (1981) support the view that gaps between high-level taxa are real, not artifacts of the fossil record.

The times have changed and there is now more of a willingness to accept major changes. For me, it has resulted in a strange turn of affairs. When I proposed an extinct phylum which contained one genus, I was viewed as a wildeyed radical. Now when I am unwilling to immediately accept the hyoliths as representing an extinct phylum, I am viewed as a moss-covered reactionary. My position has not changed because I based my defintion of the position of these fossils on the evidence of the preserved hard parts. To date, there has not been a definition of hyoliths which differentiates them from other forms, living and fossil, placed in the Mollusca. If, as, and when, such a definition is provided, I would like to think I am broad-minded enough to consider it and perhaps revise my stand. There is considerably more liberty today to propose high-level taxa. Liberty is not the same as license to generate wild ideas. Classification is a game and any game has rules, even if the rules are as obtuse as some of the features of the fossils.

Curiously enough, other fossil groups equally difficult to place, such as tentaculitids (Yochelson, 1979b) or hyolithellids, seem to have played no part in the game of how many phyla might there be in the fossil record. They too should be considered; so long as they are removed from the Mollusca, I do not care where specialists place them.

A logical development that is already occurring is the recognition that appending some fossil taxa to living taxa is confusing our study of evolution. Perhaps conulariids are not coelenterates, nor are graptolites, pterobranchs. I predict that a generation from now, class-order taxa based on extinct forms will be the rule rather than the exception in the biological classification of many phyla, but even I dare not hazard a guess on how many extinct phyla will be recognized.

One last point. Gould (1989) has written a masterly account of the Burgess Shale trying to convey the importance of this faunule to the general public. Along the way, he downplays the importance of C. D. Walcott, the discover, the collector, and first to describe most of the taxa. He gives full credit for this, but denigrates his thinking as attempting to shoehorn taxa in the high-level groups where they did not fit. In recent years about 50 man-years of work have gone in study of this fauna and that cannot be compared to the limited time Walcott had. Further, Walcott is branded as a conservative in taxonomy, yet it was Walcott who moved the Trilobita from out of the Crustacea and recognized them as a class based on extinct organisms, fifteen years before the Burgess fossils were found. Surely in the context of its time, that was as much of a revolution in classification as is the present ferment. It is important to have a bit of humility and recognize that no matter how good our classification is today, those who follow will be surprised at our bizarre notions.

Addendum

Richard G. Bromley (1990, Trace Fossils, biology and taphonomy, Unwin Hyman, London) has written a delighfull and insightfull book on trace fossils. In my casual remarks about ichnogenera, I may have been too extreme. Dr. Bromley gives a detailed discussion of how several different kinds of animals might form the same trace fossils or how the collective activities of several different animals might form a single trace.

Notwithstanding all the problems faced in biological analysis, Dr. Bromley is an optimist and uses the word "biology" in this subtitle. I remain mumpsimus that for the paleontologist the concept behind collective group will prevail over ichnogenus.

REFERENCES

AGER, D.V. 1970. On seeing the most rocks. *Proceedings of the Geologists' Association* **81**, 421-427.

Bengtson, S. & Missarzhevinsky, V.V. 1981. Coeloscleritophora - a major group of enigmatic Cambrian metazoans. *United States Geological Survey Open-file Report* **81-793**, 19-21.
Berry, E.W. 1937. A correction. *Torreya* **37**, 108.
Brown, R.W. 1934. The supposed fossil ear of maize from Cuzco, Peru. *Journal of the Washington Academy of Sciences* **24**, 293-296.
Brown, R.W. 1938. Two fossils misidentified as shelf-fungi. *Journal of the Washington Academy of Sciences* **28**, 130-131.
Brown, R.W. 1946. Baffling fossils. *Scientific Monthly* **63**, 149-151.
Brown, R.W. 1947. Fossil plants and human footprints in Nicaragua. *Journal of Paleontology* **21**, 38-50.
Brown, R.W. 1957. Plantlike features in thundereggs and geodes. *Annual Report of the Smithsonian Institution for Fiscal Year 1956*, 329-339.
Brown, R.W. 1959. Some paleobotanical Problematica. *Journal of Paleontology* **33**, 120-124.
Caster, K.E. 1942. Two siphonophores from the Paleozoic. *Paleontographica Americana* **11 (14)**, 56-90.
Caster, K.E. 1957. Problematica. In *Treatise on marine ecology and paleoecology*. vol. 2 (ed. H.W. Ladd). *Geological Society of America Memoir* **67**, 1021-1033.
Cloud, P.E., Gustafson, L.B. & Watson J.A.L. 1980. The works of living social insects as pseudofossils and the age of the oldest known metozoa. *Science* **210**, 1013-1015.
Croneis, C. 1938. Utilitarian classification for fragmentary fossils. *Journal of Geology* **46**, 975-984.
Fauchald, K., Stürmer, W. & Yochelson, E.L. 1988, Two worm-like organisms from the Hunshrück Slate (Lower Devonian), southern Germany. *Paläontologisches Zeitschrift* **62**, 201-215.
Gordon, M., jr. & Yochelson, E.L. 1987. Late Mississippian Gastropoda of the Chainman Shale, West Central Utah. *United States Geological Survey Professional Paper* **1368**, 1-112.
Gould, S.J. 1989. *Wonderful life: The Burgess Shale and the Nature of History*. New York: W.H. Norton, 346 pp.
Harrington, H.J. & Moore R.C. 1956. Scyphozoa. In *Treatise of Invertebrate Paleontology* (ed. R.C. Moore) Part F, *Coelenterata*, pp.F27-F38. Lawrence, Kansas: Geological Society of America and University of Kansas Press.
Hoffmann, A. & Nitecki M.H. (eds.), 1986. Problematic fossil taxa. New York: Oxford University Press, 267 p.
Hoffmann, H.J. 1967. Precambrian fossils near Elliot Lake, Ontario. *Science* **156**, 500-504.
Hoffmann, H.J. 1971. Precambrian fossils, pseudofossils, and problematica in Canada. *Geological Survey of Canada Bulletin* **189**, 1-146.
Jaanusson, V. 1981. Functional thresholds in evolutionary progress. *Lethaia*: **14**, 251-260.
Knight, J.B. 1937. *Conchopeltis* Walcott, an Ordovician genus of the Conulariida. *Journal of Paleontology* **11**, 186-188.
Matthews, S.C. 1973. Notes on open nomenclature and on synonymy lists. *Palaeontology* **16**, 713-719.
Melville, R.V. 1981. The International Code of Zoological Nomenclature - Result of vote on proposals for substantive amendments (fifth Installment) Z.N.(S) 1973. *Bulletin of Zoological Nomenclature* **38(1)**, 24-48.
Miller, S.A. 1889. *North American geology and palaeontology for the use of amateurs, students, and scientists*. Cincinnati, Ohio: Western Methodist Book Concern.
Milton, D.L. 1966. Drifting organisms in the Precambrian sea. *Science* **153**, 293-294.
Okamura, C. 1980. Period of the far eastern minicreatures. *Original report of the Okamura Fossil Laboratory* **14**, 165-346.
Oliver, W.A. jr. 1984. *Conchopeltis*: its affinities and significance. *Palaeontographica Americana* **54**,141-147.
Olson, E.C. 1981. The problem of missing links: today and yesterday. *Quarterly Review of Biologyy* **56**, 405-442.
Rolfe, W.D.I. 1969. Phyllocarida. In *Treatise of invertebrate paleontology* (ed. R.C. Moore), part R. *Arthropoda 4*, pp. R296-R331. Lawrence, Kansas: Geological Society of America and University of Kansas Press.
Sando, W.J. 1972. Bee-nest pseudofossils from Montana, Wyoming, and south west Africa. *Journal of Paleontology* **52**, 421-425.
Sarjeant, W.A.S. & Kennedy W.J. 1973. Proposal of a code for the nomenclature of trace-fossils. *Canadian Journal of Earth Sciences* **10**, 460-475.
Sepkoski, J.J. jr. 1981. The uniqueness of the Cambrian fauna. *United States Geological Survey Open-file Report* **81-743**, 203-207.
Weller, J.M. 1949. Paleontologic classification. *Journal of Paleontology* **23**: 680-690.
Whittington, H.B. 1981. Cambrian animals: their ancestors and descendents. *Proceedings of the Linnaean Society of New South Wales* **105(2)**, 79-87.
Yochelson, E.L. 1963. Problems of the early history of the Mollusca. *Proceedings of the XVI International Congress of Zoology, Washington, D.C., Aug. 20-27, 1963*, **vol. 2**, 187.
Yochelson, E.L. 1971. Phylum and class nomenclature in systematics. *Systematic Zoology* **20**, 245-249.
Yochelson, E.L. Discussion of early Cambrian "Mollusca". *Journal of the Geological Society of London* **131**, 662-663.
Yochelson, E.L. 1977a. Comments on *Janospira*. *Lethaia* **10**, 204.
Yochelson, E.L. 1977b. Agmata, a proposed extinct phylum of early Cambrian age. *Journal of Paleontology* **51**, 437-454.
Yochelson, E.L. 1979a. Charles D. Walcott - America's pioneer in Precambrian paleontology and stratigraphy. In *History of Concepts in Precambrian geology* (eds. W.O. Kupsch & W.A.S. Sarjeant). *Geological Association of Canada Special Paper* **19**, 261-292.
Yochelson, E.L. 1979. Early radiation of Mollusca and mollusc-like groups. In *The Origin of Major Invertebrate Groups* (ed. M.E. House). *Systematics Association Special Volume* **12**, 323-352.
Yochelson, E.L. 1989. Reconsideration of possible soft parts in dacryoconarids (incertae sedis) from Hunsrück-Schiefer in Western Germany. *Senckenbergiana lethea* **69**, 381-390.